STUDENT SOLUTIONS MANUAL

C TRIMBLE & ASSOCIATES

INTRODUCTORY ALGEBRA

FIFTH EDITION

Elayn Martin-Gay

University of New Orleans

PEARSON

Boston Columbus Indianapolis New York San Francisco

Amsterdam Cape Town Dubai London Madrid Milan Munich Paris Montreal Toronto

Delhi Mexico City São Paulo Sydney Hong Kong Seoul Singapore Taipei Tokyo

Reproduced by Pearson from electronic files supplied by the author.

Publishing as Pearson, 501 Boylston Street, Boston, MA 02116.

ISBN-13: 978-0-13-386517-2
ISBN-10: 0-13-386517-7

1 2 3 4 5 6 OPM 20 19 18 17 16

www.pearsonhighered.com

PEARSON

Contents

Chapter R

Section R.1 Practice Exercises

1. $10 = 1 \cdot 10$, $10 = 2 \cdot 5$
   The factors of 10 are 1, 2, 5, and 10.

2. $18 = 1 \cdot 18$, $18 = 2 \cdot 9$, $18 = 3 \cdot 6$
   The factors of 18 are 1, 2, 3, 6, 9, and 18.

3. 5 is a prime number. Its factors are 1 and 5 only.
   16 is a composite number. Its factors are 1, 2, 4, 8, and 16.
   23 is a prime number. Its factors are 1 and 23 only.
   42 is a composite number. Its factors are 1, 2, 3, 6, 7, 14, 21, and 42.

4. $44 = 4 \cdot 11 = 2 \cdot 2 \cdot 11$
   The prime factorization of 44 is $2 \cdot 2 \cdot 11$.

5. $60 = 4 \cdot 15 = 2 \cdot 2 \cdot 3 \cdot 5$
   The prime factorization of 60 is $2 \cdot 2 \cdot 3 \cdot 5$.

6. $$\begin{array}{r} 11 \\ 3\overline{)\ 33} \\ 3\overline{)\ 99} \\ 3\overline{)297} \end{array}$$
   The prime factorization of 297 is $3 \cdot 3 \cdot 3 \cdot 11$.

7. $14 = 2 \cdot 7$
   $35 = 5 \cdot 7$
   $\text{LCM} = 2 \cdot 5 \cdot 7 = 70$

8. $5 = 5$
   $9 = 3 \cdot 3$
   $\text{LCM} = 3 \cdot 3 \cdot 5 = 45$

9. $4 = 2 \cdot 2$
   $15 = 3 \cdot 5$
   $10 = 2 \cdot 5$
   $\text{LCM} = 2 \cdot 2 \cdot 3 \cdot 5 = 60$

Vocabulary, Readiness & Video Check R.1

1. The number 40 equals $2 \cdot 2 \cdot 2 \cdot 5$. Since each factor is prime, we call $2 \cdot 2 \cdot 2 \cdot 5$ the prime factorization of 40.

2. A natural number, other than 1, that is not prime is called a composite number.

3. A natural number that has exactly two different factors, 1 and itself, is called a prime number.

4. The least common multiple of a list of numbers is the smallest number that is a multiple of all the numbers in the list.

5. To factor means to write as a product.

6. A multiple of a number is the product of that number and any natural number.

7. No, the natural number 1 is neither prime nor composite.

8. We may write factors in different order, but every natural number has only one prime factorization.

9. The least common multiple, LCM, of a list of numbers is the smallest number that is a multiple of each number in the list.

Exercise Set R.1

1. $9 = 1 \cdot 9$, $9 = 3 \cdot 3$
   The factors of 9 are 1, 3, and 9.

3. $24 = 1 \cdot 24$, $24 = 2 \cdot 12$, $24 = 3 \cdot 8$, $24 = 4 \cdot 6$
   The factors of 24 are 1, 2, 3, 4, 6, 8, 12, and 24.

5. $42 = 1 \cdot 42$, $42 = 2 \cdot 21$, $42 = 3 \cdot 14$, $42 = 6 \cdot 7$
   The factors of 42 are 1, 2, 3, 6, 7, 14, 21, and 42.

7. $80 = 1 \cdot 80$, $80 = 2 \cdot 40$, $80 = 4 \cdot 20$, $80 = 5 \cdot 16$, $80 = 8 \cdot 10$
   The factors of 80 are 1, 2, 4, 5, 8, 10, 16, 20, 40, and 80.

9. $19 = 1 \cdot 19$
   The factors of 19 are 1 and 19.

11. 13 is a prime number. Its factors are only 1 and 13.

13. 39 is a composite number. Its factors are 1, 3, 13, and 39.

15. 41 is a prime number. Its factors are only 1 and 41.

17. 201 is composite. Its factors are 1, 3, 67, and 201.

19. 2065 is a composite number. Its factors are 1, 5, 7, 35, 59, 295, 413, and 2065.

21. $18 = 2 \cdot 9 = 2 \cdot 3 \cdot 3$
The prime factorization of 18 is $2 \cdot 3 \cdot 3$.

23. $20 = 4 \cdot 5 = 2 \cdot 2 \cdot 5$
The prime factorization of 20 is $2 \cdot 2 \cdot 5$.

25. $56 = 2 \cdot 28 = 2 \cdot 2 \cdot 14 = 2 \cdot 2 \cdot 2 \cdot 7$
The prime factorization of 56 is $2 \cdot 2 \cdot 2 \cdot 7$.

27. $81 = 3 \cdot 27 = 3 \cdot 3 \cdot 9 = 3 \cdot 3 \cdot 3 \cdot 3$
The prime factorization of 81 is $3 \cdot 3 \cdot 3 \cdot 3$.

29.
$$\begin{array}{r} 5 \\ 5\overline{)\ 25} \\ 3\overline{)\ 75} \\ 2\overline{)150} \\ 2\overline{)300} \end{array}$$
The prime factorization of 300 is $2 \cdot 2 \cdot 3 \cdot 5 \cdot 5$.

31.
$$\begin{array}{r} 7 \\ 7\overline{)\ 49} \\ 3\overline{)147} \\ 2\overline{)294} \\ 2\overline{)588} \end{array}$$
The prime factorization of 588 is $2 \cdot 2 \cdot 3 \cdot 7 \cdot 7$.

33. $48 = 1 \cdot 48$, $48 = 2 \cdot 24$, $48 = 3 \cdot 16$, $48 = 4 \cdot 12$, $48 = 6 \cdot 8$
The factors of 48 are 1, 2, 3, 4, 6, 8, 12, 16, and 48, which is choice d.

35. $3 = 3$
$4 = 2 \cdot 2$
$\text{LCM} = 2 \cdot 2 \cdot 3 = 12$

37. $6 = 2 \cdot 3$
$14 = 2 \cdot 7$
$\text{LCM} = 2 \cdot 3 \cdot 7 = 42$

39. $20 = 2 \cdot 2 \cdot 5$
$30 = 2 \cdot 3 \cdot 5$
$\text{LCM} = 2 \cdot 2 \cdot 3 \cdot 5 = 60$

41. $5 = 5$
$7 = 7$
$\text{LCM} = 5 \cdot 7 = 35$

43. $9 = 3 \cdot 3$
$12 = 2 \cdot 2 \cdot 3$
$\text{LCM} = 2 \cdot 2 \cdot 3 \cdot 3 = 36$

45. $16 = 2 \cdot 2 \cdot 2 \cdot 2$
$20 = 2 \cdot 2 \cdot 5$
$\text{LCM} = 2 \cdot 2 \cdot 2 \cdot 2 \cdot 5 = 80$

47. $40 = 2 \cdot 2 \cdot 2 \cdot 5$
$90 = 2 \cdot 3 \cdot 3 \cdot 5$
$\text{LCM} = 2 \cdot 2 \cdot 2 \cdot 3 \cdot 3 \cdot 5 = 360$

49. $24 = 2 \cdot 2 \cdot 2 \cdot 3$
$36 = 2 \cdot 2 \cdot 3 \cdot 3$
$\text{LCM} = 2 \cdot 2 \cdot 2 \cdot 3 \cdot 3 = 72$

51. $2 = 2$
$8 = 2 \cdot 2 \cdot 2$
$15 = 3 \cdot 5$
$\text{LCM} = 2 \cdot 2 \cdot 2 \cdot 3 \cdot 5 = 120$

53. $2 = 2$
$3 = 3$
$7 = 7$
$\text{LCM} = 2 \cdot 3 \cdot 7 = 42$

55. $8 = 2 \cdot 2 \cdot 2$
$24 = 2 \cdot 2 \cdot 2 \cdot 3$
$48 = 2 \cdot 2 \cdot 2 \cdot 2 \cdot 3$
$\text{LCM} = 2 \cdot 2 \cdot 2 \cdot 2 \cdot 3 = 48$

57. $8 = 2 \cdot 2 \cdot 2$
$18 = 2 \cdot 3 \cdot 3$
$30 = 2 \cdot 3 \cdot 5$
$\text{LCM} = 2 \cdot 2 \cdot 2 \cdot 3 \cdot 3 \cdot 5 = 360$

59. a. $40 = 2 \cdot 20 = 2 \cdot 2 \cdot 10 = 2 \cdot 2 \cdot 2 \cdot 5$
The prime factorization of 40 is $2 \cdot 2 \cdot 2 \cdot 5$.

b. $40 = 4 \cdot 10 = 2 \cdot 2 \cdot 10 = 2 \cdot 2 \cdot 2 \cdot 5$
The prime factorization of 40 is $2 \cdot 2 \cdot 2 \cdot 5$.

c. answers may vary

61. $5 = 5$
$7 = 7$
$\text{LCM} = 5 \cdot 7 = 35$
They have the same night off every 35 days.

63. $315 = 3 \cdot 3 \cdot 5 \cdot 7$
$504 = 2 \cdot 2 \cdot 2 \cdot 3 \cdot 3 \cdot 7$
$\text{LCM} = 2 \cdot 2 \cdot 2 \cdot 3 \cdot 3 \cdot 5 \cdot 7 = 2520$

Section R.2 Practice Exercises

1. $\frac{4}{4}=1$ since $4 \div 4 = 1$.

2. $\frac{9}{3}=3$ since $9 \div 3 = 3$.

3. $\frac{10}{10}=1$ since $10 \div 10 = 1$.

4. $\frac{5}{1}=5$ since $5 \div 1 = 5$.

5. $\frac{0}{11}=0$ since $0 \cdot 11 = 0$.

6. $\frac{11}{0}$ is undefined because there is no number that when multiplied by 0 gives 11.

7. $\frac{1}{4}=\frac{1}{4}\cdot\frac{5}{5}=\frac{1\cdot 5}{4\cdot 5}=\frac{5}{20}$

8. $\frac{20}{35}=\frac{2\cdot 2\cdot 5}{5\cdot 7}=\frac{4}{7}$

9. $\frac{7}{20}$ is already simplified.

10. $\frac{12}{40}=\frac{4\cdot 3}{4\cdot 10}=\frac{3}{10}$

11. $\frac{3}{4}\cdot\frac{8}{9}=\frac{3\cdot 8}{4\cdot 9}=\frac{3\cdot 4\cdot 2}{4\cdot 3\cdot 3}=\frac{2}{3}$

12. $\frac{2}{9}\div\frac{3}{4}=\frac{2}{9}\cdot\frac{4}{3}=\frac{2\cdot 4}{9\cdot 3}=\frac{8}{27}$

13. $\frac{8}{11}\div 24=\frac{8}{11}\div\frac{24}{1}=\frac{8}{11}\cdot\frac{1}{24}=\frac{8\cdot 1}{11\cdot 8\cdot 3}=\frac{1}{33}$

14. $\frac{5}{4}\div\frac{15}{8}=\frac{5}{4}\cdot\frac{8}{15}=\frac{5\cdot 4\cdot 2}{4\cdot 5\cdot 3}=\frac{2}{3}$

15. $\frac{2}{11}+\frac{5}{11}=\frac{2+5}{11}=\frac{7}{11}$

16. $\frac{1}{8}+\frac{3}{8}=\frac{1+3}{8}=\frac{4}{8}=\frac{4}{2\cdot 4}=\frac{1}{2}$

17. $\frac{7}{6}-\frac{2}{6}=\frac{7-2}{6}=\frac{5}{6}$

18. $\frac{13}{10}-\frac{3}{10}=\frac{13-3}{10}=\frac{10}{10}=1$

19. $\frac{3}{8}+\frac{1}{20}=\frac{3}{8}\cdot\frac{5}{5}+\frac{1}{20}\cdot\frac{2}{2}=\frac{15}{40}+\frac{2}{40}=\frac{17}{40}$

20. $\frac{8}{15}-\frac{1}{3}=\frac{8}{15}-\frac{1}{3}\cdot\frac{5}{5}=\frac{8}{15}-\frac{5}{15}=\frac{3}{15}=\frac{3}{3\cdot 5}=\frac{1}{5}$

21. $5\frac{1}{6}=\frac{6\cdot 5+1}{6}=\frac{31}{6}$; $4\frac{2}{5}=\frac{5\cdot 4+2}{5}=\frac{22}{5}$

$$5\frac{1}{6}\cdot 4\frac{2}{5}=\frac{31}{6}\cdot\frac{22}{5}=\frac{31\cdot 2\cdot 11}{2\cdot 3\cdot 5}=\frac{341}{15}=22\frac{11}{15}$$

22. $7\frac{3}{8}+6\frac{3}{4}=\frac{59}{8}+\frac{27}{4}=\frac{59}{8}+\frac{54}{8}=\frac{113}{8}=14\frac{1}{8}$

23. $$\begin{array}{r} 76\frac{1}{12} \\ -\ 35\frac{1}{4} \\ \hline \end{array} \quad \begin{array}{r} 76\frac{1}{12} \\ -\ 35\frac{3}{12} \\ \hline \end{array} \quad \begin{array}{r} 75\frac{13}{12} \\ -\ 35\frac{3}{12} \\ \hline 40\frac{10}{12} \end{array} = 40\frac{5}{6}$$

Vocabulary, Readiness & Video Check R.2

1. The number $\frac{17}{31}$ is called a fraction. The number 31 is called its denominator and 17 is called its numerator.

2. The fraction $\frac{8}{3}$ is called an improper fraction, the fraction $\frac{3}{8}$ is called a proper fraction, and $10\frac{3}{8}$ is called a mixed number.

3. In $\frac{11}{48}$, since 11 and 48 have no common factors other than 1, $\frac{11}{48}$ is in simplest form.

4. Fractions that represent the same portion of a whole are called equivalent fractions.

5. To multiply two fractions, we write $\frac{a}{b}\cdot\frac{c}{d}=\underline{\frac{a\cdot c}{b\cdot d}}$.

6. Two numbers are <u>reciprocals</u> of each other if their product is 1.

7. To divide two fractions, we write $\frac{a}{b}\div\frac{c}{d}=\underline{\frac{a\cdot d}{b\cdot c}}$.

8. $\frac{a}{b}+\frac{c}{b}=\underline{\frac{a+c}{b}}$ and $\frac{a}{b}-\frac{c}{b}=\underline{\frac{a-c}{b}}$.

9. The smallest positive number divisible by all the denominators of a list of fractions is called the <u>least common denominator (LCD)</u>.

10. The LCD for $\frac{1}{6}$ and $\frac{5}{8}$ is <u>24</u>.

11. The fraction is equal to 1.

12. Equivalent fractions represent the same <u>quantity</u>.

13. wrote both the numerator and denominator as products of prime numbers

14. $\frac{20}{1}$ or 20

15. When adding or subtraction fractions, we must have common denominators. When multiplying or dividing fractions, we do not.

16. Our original sum, $4\frac{7}{6}$, is not in proper form because the fraction part, $\frac{7}{6}$, is an improper fraction.

Exercise Set R.2

1. $\frac{14}{14}=1$ since $14\cdot 1=14$.

3. $\frac{20}{2}=10$ since $2\cdot 10=20$.

5. $\frac{13}{1}=13$ since $1\cdot 13=13$.

7. $\frac{0}{9}=0$ since $9\cdot 0=0$.

9. $\frac{9}{0}$ is undefined.

11. $\frac{7}{10}=\frac{7\cdot 3}{10\cdot 3}=\frac{21}{30}$

13. $\frac{2}{9}=\frac{2\cdot 2}{9\cdot 2}=\frac{4}{18}$

15. $\frac{4}{5}=\frac{4\cdot 4}{5\cdot 4}=\frac{16}{20}$

17. $\frac{2}{4}=\frac{2\cdot 1}{2\cdot 2}=\frac{1}{2}$

19. $\frac{10}{15}=\frac{2\cdot 5}{3\cdot 5}=\frac{2}{3}$

21. $\frac{3}{7}$ cannot be simplified further.

23. $\frac{18}{30}=\frac{3\cdot 6}{5\cdot 6}=\frac{3}{5}$

25. $\frac{16}{20}=\frac{4\cdot 4}{4\cdot 5}=\frac{4}{5}$

27. $\frac{66}{48}=\frac{6\cdot 11}{6\cdot 8}=\frac{11}{8}$

29. $\frac{120}{244}=\frac{4\cdot 30}{4\cdot 61}=\frac{30}{61}$

31. $\frac{192}{264}=\frac{8\cdot 24}{11\cdot 24}=\frac{8}{11}$

33. $\frac{1}{2}\cdot\frac{3}{4}=\frac{1\cdot 3}{2\cdot 4}=\frac{3}{8}$

35. $\frac{2}{3}\cdot\frac{3}{4}=\frac{2\cdot 3}{3\cdot 4}=\frac{2\cdot 3\cdot 1}{2\cdot 3\cdot 2}=\frac{1}{2}$

37. $\frac{1}{2}\div\frac{7}{12}=\frac{1}{2}\cdot\frac{12}{7}=\frac{1\cdot 12}{2\cdot 7}=\frac{2\cdot 6}{2\cdot 7}=\frac{6}{7}$

39. $\frac{3}{4}\div\frac{1}{20}=\frac{3}{4}\cdot\frac{20}{1}=\frac{3\cdot 20}{4\cdot 1}=\frac{3\cdot 4\cdot 5}{4\cdot 1}=\frac{15}{1}=15$

41. $5\frac{1}{9}\cdot 3\frac{2}{3}=\frac{46}{9}\cdot\frac{11}{3}=\frac{506}{27}=18\frac{20}{27}$

43. $8\frac{3}{5}\div 2\frac{9}{10}=\frac{43}{5}\div\frac{29}{10}$
$=\frac{43}{5}\cdot\frac{10}{29}$
$=\frac{43\cdot 5\cdot 2}{5\cdot 29}$
$=\frac{86}{29}$
$=2\frac{28}{29}$

45. $\frac{4}{5}+\frac{1}{5}=\frac{4+1}{5}=\frac{5}{5}=1$

47. $\frac{4}{15}-\frac{1}{12}=\frac{4}{15}\cdot\frac{4}{4}-\frac{1}{12}\cdot\frac{5}{5}$
$=\frac{16}{60}-\frac{5}{60}$
$=\frac{16-5}{60}$
$=\frac{11}{60}$

49. $\frac{2}{3}+\frac{3}{7}=\frac{2\cdot 7}{3\cdot 7}+\frac{3\cdot 3}{7\cdot 3}=\frac{14}{21}+\frac{9}{21}=\frac{14+9}{21}=\frac{23}{21}$

51. $\frac{10}{3}-\frac{5}{21}=\frac{10\cdot 7}{3\cdot 7}-\frac{5}{21}$
$=\frac{70}{21}-\frac{5}{21}$
$=\frac{70-5}{21}$
$=\frac{65}{21}$

53.
$$\begin{array}{r} 8\frac{1}{8} \\ -\,6\frac{3}{8} \\ \hline \end{array}\qquad \begin{array}{r} 7\frac{9}{8} \\ -\,6\frac{3}{8} \\ \hline 1\frac{6}{8}=1\frac{3}{4} \end{array}$$

55.
$$\begin{array}{r} 1\frac{1}{2} \\ +\,3\frac{2}{3} \\ \hline \end{array}\qquad \begin{array}{r} 1\frac{3}{6} \\ +\,3\frac{4}{6} \\ \hline 4\frac{7}{6}=4+1\frac{1}{6}=5\frac{1}{6} \end{array}$$

57. $\frac{23}{105}+\frac{4}{105}=\frac{23+4}{105}=\frac{27}{105}=\frac{3\cdot 9}{3\cdot 35}=\frac{9}{35}$

59. $\frac{17}{21}-\frac{10}{21}=\frac{17-10}{21}=\frac{7}{21}=\frac{1\cdot 7}{3\cdot 7}=\frac{1}{3}$

61. $\frac{7}{10}\cdot\frac{5}{21}=\frac{7\cdot 5}{10\cdot 21}=\frac{7\cdot 5}{5\cdot 2\cdot 7\cdot 3}=\frac{1}{2\cdot 3}=\frac{1}{6}$

63. $\frac{9}{20}\div 12=\frac{9}{20}\div\frac{12}{1}=\frac{9}{20}\cdot\frac{1}{12}=\frac{3\cdot 3\cdot 1}{20\cdot 3\cdot 4}=\frac{3}{80}$

65. $\frac{5}{22}-\frac{5}{33}=\frac{5\cdot 3}{22\cdot 3}-\frac{5\cdot 2}{33\cdot 2}$
$=\frac{15}{66}-\frac{10}{66}$
$=\frac{15-10}{66}$
$=\frac{5}{66}$

67. $17\frac{2}{5}+30\frac{2}{3}=\frac{87}{5}+\frac{92}{3}$
$=\frac{87\cdot 3}{5\cdot 3}+\frac{92\cdot 5}{3\cdot 5}$
$=\frac{261}{15}+\frac{460}{15}$
$=\frac{721}{15}$
$=48\frac{1}{15}$

69. $7\frac{2}{5}\div\frac{1}{5}=\frac{37}{5}\div\frac{1}{5}=\frac{37}{5}\cdot\frac{5}{1}=\frac{37\cdot 5}{5\cdot 1}=\frac{37}{1}=37$

71. $4\frac{2}{11}\cdot 2\frac{1}{2}=\frac{46}{11}\cdot\frac{5}{2}=\frac{2\cdot 23\cdot 5}{11\cdot 2}=\frac{115}{11}=10\frac{5}{11}$

73. $\frac{12}{5}-1=\frac{12}{5}-\frac{5}{5}=\frac{12-5}{5}=\frac{7}{5}$

75. $\begin{array}{r} 8\frac{11}{12} \\ -1\frac{5}{6} \\ \hline \end{array}$ $\quad \begin{array}{r} 8\frac{11}{12} \\ -1\frac{10}{12} \\ \hline 7\frac{1}{12} \end{array}$

77. $\frac{2}{3}-\frac{5}{9}+\frac{5}{6}=\frac{2\cdot 6}{3\cdot 6}-\frac{5\cdot 2}{9\cdot 2}+\frac{5\cdot 3}{6\cdot 3}$
$=\frac{12}{18}-\frac{10}{18}+\frac{15}{18}$
$=\frac{12-10+15}{18}$
$=\frac{17}{18}$

79. The work is incorrect.
$\frac{12}{24}=\frac{2\cdot 2\cdot 3}{2\cdot 2\cdot 2\cdot 3}=\frac{1}{2}$

81. The work is incorrect.
$\frac{2}{7}+\frac{9}{7}=\frac{2+9}{7}=\frac{11}{7}$

83. answers may vary

85. $1-\frac{1}{3}-\frac{5}{12}-\frac{1}{6}=\frac{12}{12}-\frac{4}{12}-\frac{5}{12}-\frac{2}{12}$
$=\frac{12-4-5-2}{12}$
$=\frac{1}{12}$

The unknown part is $\frac{1}{12}$.

87. $1-\frac{3}{11}-\frac{2}{11}=\frac{11}{11}-\frac{3}{11}-\frac{2}{11}=\frac{11-3-2}{11}=\frac{6}{11}$

The unknown part is $\frac{6}{11}$.

89. $\begin{array}{r} 11\frac{1}{4} \\ -3\frac{3}{5} \\ \hline \end{array}$ $\quad \begin{array}{r} 11\frac{5}{20} \\ -3\frac{12}{20} \\ \hline \end{array}$ $\quad \begin{array}{r} 10\frac{25}{20} \\ -3\frac{12}{20} \\ \hline 7\frac{13}{20} \end{array}$

On average, Tucson gets $7\frac{13}{20}$ inches more rain than Yuma.

91. The piece representing education is labeled $\frac{6}{100}$.
$\frac{6}{100}=\frac{2\cdot 3}{2\cdot 50}=\frac{3}{50}$
$\frac{3}{50}$ of entering college freshmen plan to major in education.

93. answers may vary

95. The piece representing National Memorials is labeled $\frac{8}{100}$.
$\frac{8}{100}=\frac{4\cdot 2}{4\cdot 25}=\frac{2}{25}$
$\frac{2}{25}$ of National Park Service areas are National Memorials.

97. answers may vary

99. $A=lw=\frac{2}{5}\cdot\frac{3}{11}=\frac{2\cdot 3}{5\cdot 11}=\frac{6}{55}$

The area is $\frac{6}{55}$ square meter.

Section R.3 Practice Exercises

1. $0.27=\frac{27}{100}$

2. $5.1=\frac{51}{10}$

3. $7.685=\frac{7685}{1000}$

4. a. $\begin{array}{r} 7.19 \\ 19.782 \\ +\ 1.006 \\ \hline 27.978 \end{array}$

b. $\begin{array}{r} 12. \\ 0.79 \\ +\ 0.03 \\ \hline 12.82 \end{array}$

5. a. $\begin{array}{r} 84.230 \\ -\ 26.982 \\ \hline 57.248 \end{array}$

b.
$$\begin{array}{r} 90.00 \\ -\ 0.19 \\ \hline 89.81 \end{array}$$

6. a.
$$\begin{array}{r} 0.31 \\ \times\ 4.6 \\ \hline 186 \\ 1\ 24 \\ \hline 1.426 \end{array}$$

b.
$$\begin{array}{r} 1.26 \\ \times\ 0.03 \\ \hline 0.0378 \end{array}$$

7. a.
$$\begin{array}{r} 43.5 \\ 0.5\overline{)\ 21.75} \\ -20 \\ \hline 1\ 7 \\ -1\ 5 \\ \hline 25 \\ -25 \\ \hline 0 \end{array}$$

b.
$$\begin{array}{r} 2600 \\ 0.006\overline{)\ 15.600} \\ -12 \\ \hline 3\ 6 \\ -3\ 6 \\ \hline 0 \end{array}$$

8. 12.9187 rounded to the nearest hundredth is 12.92.

9. 245.348 rounded to the nearest tenth is 245.3.

10.
$$\begin{array}{r} 0.4 \\ 5\overline{)\ 2.0} \\ -2\ 0 \\ \hline 0 \end{array}$$

$\frac{2}{5} = 0.4$

11.
$$\begin{array}{r} 0.833 \\ 6\overline{)\ 5.000} \\ -4\ 8 \\ \hline 20 \\ -18 \\ \hline 20 \\ -18 \\ \hline 2 \end{array}$$

$\frac{5}{6} = 0.833... = 0.8\overline{3}$

12.
$$\begin{array}{r} 0.1111 \\ 9\overline{)1.0000} \\ -9 \\ \hline 10 \\ -9 \\ \hline 10 \\ -9 \\ \hline 10 \\ -9 \\ \hline 1 \end{array}$$

$\frac{1}{9} = 0.1111... \approx 0.111$

13. a. $20\% = 20.\% = 0.20$

b. $1.2\% = 01.2\% = 0.012$

c. $465\% = 465.\% = 4.65$

14. a. $0.42 = 0.42 = 42\%$

b. $0.003 = 0.003 = 0.3\%$

c. $2.36 = 2.36 = 236\%$

d. $0.7 = 0.70 = 70\%$

Vocabulary, Readiness & Video Check R.3

1. Like fractional notation, <u>decimal</u> notation is used to denote a part of a whole.

2. To write fractions as decimals, divide the <u>numerator</u> by the <u>denominator</u>.

3. To add or subtract decimals, write the decimals so that the decimal points line up <u>vertically</u>.

4. When multiplying decimals, the decimal point in the product is placed so that the number of decimal places in the product is equal to the sum of the number of decimal places in the factors.

5. Percent means "per hundred."

6. 100% = 1

7. The % symbol is read as percent.

8. To write a percent as a *decimal*, drop the % symbol and move the decimal point two places to the left.

9. To write a decimal as a *percent*, move the decimal point two places to the right and attach the % symbol.

10. Reading a decimal correctly gives us the correct place value, which tells us the denominator of our equivalent fraction.

11. When adding or subtracting decimal numbers, we do line up decimal points. When multiplying decimal numbers, we do not need to line up decimal points.

12. when rounding whole numbers, digits to the right of the rounding place are replaced by zeros; when rounding decimal numbers to the right of the decimal point, digits to the right of the rounding place are not replaced by zeros

13. To write a fraction as a decimal, we divide the numerator by the denominator.

14. 1

Exercise Set R.3

1. $0.6 = \frac{6}{10}$

3. $1.86 = \frac{186}{100}$

5. $0.114 = \frac{114}{1000}$

7. $123.1 = \frac{1231}{10}$

9. $$\begin{array}{r} 5.7 \\ +\ 1.13 \\ \hline 6.83 \end{array}$$

11. $$\begin{array}{r} 24.6 \\ 2.39 \\ +\ 0.0678 \\ \hline 27.0578 \end{array}$$

13. $$\begin{array}{r} 8.8 \\ -\ 2.3 \\ \hline 6.5 \end{array}$$

15. $$\begin{array}{r} 18.00 \\ -\ 2.78 \\ \hline 15.22 \end{array}$$

17. $$\begin{array}{r} 0.2 \\ \times\ 0.6 \\ \hline 0.12 \end{array}$$

19. $$\begin{array}{r} 0.063 \\ \times\ \ 4.2 \\ \hline 126 \\ 252 \\ \hline 0.2646 \end{array}$$

21. $$\begin{array}{r} 1.68 \\ 5\overline{)\ 8.40} \\ -5 \\ \hline 3\ 4 \\ -3\ 0 \\ \hline 40 \\ -40 \\ \hline 0 \end{array}$$

23. $$\begin{array}{r} 5.8 \\ 0.82\overline{)\ 4.756} \\ -4\ 10 \\ \hline 656 \\ -656 \\ \hline 0 \end{array}$$

25. $$\begin{array}{r} 45.02 \\ 3.006 \\ +\ 8.405 \\ \hline 56.431 \end{array}$$

27.
$$\begin{array}{r} 6.75 \\ \times\, 10 \\ \hline 67.5 \end{array}$$

29.
$$\begin{array}{r} 7\,0. \\ 0.6\overline{)\ 42.0} \\ -42 \\ \hline 00 \end{array}$$

31.
$$\begin{array}{r} 654.90 \\ -\ 56.67 \\ \hline 598.23 \end{array}$$

33.
$$\begin{array}{r} 5.62 \\ \times\ \ 7.7 \\ \hline 3\,934 \\ 39\,34\ \ \\ \hline 43.274 \end{array}$$

35.
$$\begin{array}{r} 840. \\ 0.063\overline{)\ 52.920} \\ -50\,4\ \ \\ \hline 252\ \\ -252\ \\ \hline 00 \end{array}$$

37.
$$\begin{array}{r} 16.003 \\ \times\ \ \ 5.31 \\ \hline 16003 \\ 4\,8009\ \ \\ 80\,015\ \ \ \ \\ \hline 84.97593 \end{array}$$

39. 0.57 rounded to the nearest tenth is 0.6.

41. 0.234 rounded to the nearest hundredth is 0.23.

43. 0.5945 rounded to the nearest thousandth is 0.595.

45. 98,207.23 rounded to the nearest tenth is 98,207.2.

47. 12.347 rounded to the nearest hundredth is 12.35.

49. $\frac{3}{4} = \frac{3 \cdot 25}{4 \cdot 25} = \frac{75}{100} = 0.75$

51.
$$\begin{array}{r} 0.333... \\ 3\overline{)1.000} \\ -9\ \ \ \\ \hline 10\ \ \\ -9\ \ \\ \hline 10\ \\ -9\ \\ \hline 1\ \end{array}$$

$\frac{1}{3} = 0.\overline{3} \approx 0.33$

53. $\frac{7}{16} = \frac{7 \cdot 625}{16 \cdot 625} = \frac{4375}{10{,}000} = 0.4375$

55.
$$\begin{array}{r} 0.5454... \\ 11\overline{)\ 6.0000} \\ -5\,5\ \ \ \ \ \\ \hline 50\ \ \ \\ -44\ \ \ \\ \hline 60\ \ \\ -55\ \ \\ \hline 50\ \\ -44\ \\ \hline 6\ \end{array}$$

$\frac{6}{11} = 0.\overline{54} \approx 0.55$

57.
$$\begin{array}{r} 4.833... \\ 6\overline{)\ 29.000} \\ -24\ \ \ \ \ \ \\ \hline 5\,0\ \ \ \\ -4\,8\ \ \ \\ \hline 20\ \ \\ -18\ \ \\ \hline 20\ \\ -18\ \\ \hline 2\ \end{array}$$

$\frac{29}{6} = 4.8\overline{3} \approx 4.83$

59. 28% = 0.28

61. 3.1% = 0.031

63. 135% = 1.35

65. 200% = 2.00 or 2

67. 96.55% = 0.9655

69. $0.1\% = 0.001$

71. $15.8\% = 0.158$

73. $0.68 = 68\%$

75. $0.876 = 87.6\%$

77. $1 = 1.00 = 100\%$

79. $0.5 = 50\%$

81. $1.92 = 192\%$

83. $0.004 = 0.4\%$

85. $0.781 = 78.1\%$

87. $0.5\% = 0.005$
$$0.5\% = \frac{0.5}{100} = \frac{5}{1000} = \frac{1}{200}$$

89. $14.2\% = 0.142$
$$14.2\% = \frac{14.2}{100} = \frac{142}{1000} = \frac{71}{500}$$

91. **a.** In the number 3.659, the place value of the 6 is tenths.

b. In the number 3.659, the place value of the 9 is thousandths.

c. In the number 3.659, the place value of the 3 is ones.

93. answers may vary

95.
$$\begin{array}{r} 213.4 \\ -\ 30.8 \\ \hline 182.6 \end{array}$$
The consumption of fluid milk products is 182.6 pounds more than cheese products.

97. **a.** 52.8647% rounded to the nearest tenth percent is 52.9%.

b. 52.8647% rounded to the nearest hundredth percent is 52.86%.

99. **a.** $6.5\% = 0.065 \neq 0.65$

b. $7.8\% = 0.078$

c. $120\% = 1.20 \neq 0.12$

d. $0.35\% = 0.0035$

b and **d** are correct.

101. $45\% + 40\% + 11\% = 96\%$
$100\% - 96\% = 4\%$
4% of the U.S. population has AB blood type.

103. The longest bar corresponds to network systems and data communications analysts, so that is predicted to be the fastest growing occupation.

105. $35\% = 0.35$

107. answers may vary

Chapter R Vocabulary Check

1. To factor means to write as a product.
2. A multiple of a number is the product of that number and any natural number.
3. A composite number is a natural number greater than 1 that is not prime.
4. The word percent means per 100.
5. Fractions that represent the same portion of a whole are called equivalent fractions.
6. An improper fraction is a fraction whose numerator is greater than or equal to its denominator.
7. A prime number is a natural number greater than 1 whose only factors are 1 and itself.
8. A fraction is simplified when the numerator and the denominator have no factors in common other than 1.
9. A proper fraction is one whose numerator is less than its denominator.
10. A mixed number contains a whole number part and a fraction part.

Chapter R Review

1. $42 = 2 \cdot 21 = 2 \cdot 3 \cdot 7$
2. $$\begin{aligned} 800 &= 2 \cdot 400 \\ &= 2 \cdot 2 \cdot 200 \\ &= 2 \cdot 2 \cdot 2 \cdot 100 \\ &= 2 \cdot 2 \cdot 2 \cdot 2 \cdot 50 \\ &= 2 \cdot 2 \cdot 2 \cdot 2 \cdot 2 \cdot 25 \\ &= 2 \cdot 2 \cdot 2 \cdot 2 \cdot 2 \cdot 5 \cdot 5 \end{aligned}$$

3. $12 = 2 \cdot 2 \cdot 3$
$30 = 2 \cdot 3 \cdot 5$
$\text{LCM} = 2 \cdot 2 \cdot 3 \cdot 5 = 60$

4. $7 = 7$
$42 = 2 \cdot 3 \cdot 7$
$\text{LCM} = 2 \cdot 3 \cdot 7 = 42$

5. $4 = 2 \cdot 2$
$6 = 2 \cdot 3$
$10 = 2 \cdot 5$
$\text{LCM} = 2 \cdot 2 \cdot 3 \cdot 5 = 60$

6. $2 = 2$
$5 = 5$
$7 = 7$
$\text{LCM} = 2 \cdot 5 \cdot 7 = 70$

7. $\frac{5}{8} = \frac{5 \cdot 3}{8 \cdot 3} = \frac{15}{24}$

8. $\frac{2}{3} = \frac{2 \cdot 20}{3 \cdot 20} = \frac{40}{60}$

9. $\frac{8}{20} = \frac{2 \cdot 2 \cdot 2}{2 \cdot 2 \cdot 5} = \frac{2}{5}$

10. $\frac{15}{100} = \frac{3 \cdot 5}{2 \cdot 2 \cdot 5 \cdot 5} = \frac{3}{20}$

11. $\frac{12}{6} = \frac{2 \cdot 6}{6} = 2$

12. $\frac{8}{8} = 1$

13. $\frac{1}{7} \cdot \frac{8}{11} = \frac{1 \cdot 8}{7 \cdot 11} = \frac{8}{77}$

14.
$$\begin{aligned}\frac{5}{12} + \frac{2}{15} &= \frac{5 \cdot 5}{12 \cdot 5} + \frac{2 \cdot 4}{15 \cdot 4}\\ &= \frac{25}{60} + \frac{8}{60}\\ &= \frac{25+8}{60}\\ &= \frac{33}{60}\\ &= \frac{11 \cdot 3}{20 \cdot 3}\\ &= \frac{11}{20}\end{aligned}$$

15. $\frac{3}{10} \div 6 = \frac{3}{10} \div \frac{6}{1} = \frac{3}{10} \cdot \frac{1}{6} = \frac{3 \cdot 1}{10 \cdot 6} = \frac{3 \cdot 1}{10 \cdot 2 \cdot 3} = \frac{1}{20}$

16. $\frac{7}{9} - \frac{1}{6} = \frac{7 \cdot 2}{9 \cdot 2} - \frac{1 \cdot 3}{6 \cdot 3} = \frac{14}{18} - \frac{3}{18} = \frac{14-3}{18} = \frac{11}{18}$

17. $3\frac{3}{8} \cdot 4\frac{1}{4} = \frac{27}{8} \cdot \frac{17}{4} = \frac{27 \cdot 17}{8 \cdot 4} = \frac{459}{32} = 14\frac{11}{32}$

18. $2\frac{1}{3} - 1\frac{5}{6} = \frac{7}{3} - \frac{11}{6} = \frac{14}{6} - \frac{11}{6} = \frac{14-11}{6} = \frac{3}{6} = \frac{1}{2}$

19.
$$\begin{array}{rr} 16\frac{9}{10} & 16\frac{27}{30} \\ +\,3\frac{2}{3} & +\,3\frac{20}{30} \\ \hline & 19\frac{47}{30} \end{array}$$
$19\frac{47}{30} = 19 + 1\frac{17}{30} = 20\frac{17}{30}$

20.
$$\begin{aligned}6\frac{2}{7} \div 2\frac{1}{5} &= \frac{44}{7} \div \frac{11}{5}\\ &= \frac{44}{7} \cdot \frac{5}{11}\\ &= \frac{44 \cdot 5}{7 \cdot 11}\\ &= \frac{4 \cdot 11 \cdot 5}{7 \cdot 11}\\ &= \frac{20}{7}\\ &= 2\frac{6}{7}\end{aligned}$$

21. $A = lw = \frac{11}{12} \cdot \frac{3}{5} = \frac{11 \cdot 3}{12 \cdot 5} = \frac{11 \cdot 3}{3 \cdot 4 \cdot 5} = \frac{11}{20}$

The area is $\frac{11}{20}$ square mile.

22. $A = \frac{1}{2}bh = \frac{1}{2} \cdot \frac{5}{4} \cdot \frac{1}{2} = \frac{1 \cdot 5 \cdot 1}{2 \cdot 4 \cdot 2} = \frac{5}{16}$

The area is $\frac{5}{16}$ square meter.

23. $1.81 = \frac{181}{100}$

24. $0.035 = \frac{35}{1000}$

25.
$$\begin{array}{r} 76.358 \\ +\ 18.76 \\ \hline 95.118 \end{array}$$

26.
$$\begin{array}{r} 35 \\ 0.02 \\ +\ 1.765 \\ \hline 36.785 \end{array}$$

27.
$$\begin{array}{r} 18.00 \\ -\ 4.62 \\ \hline 13.38 \end{array}$$

28.
$$\begin{array}{r} 804.062 \\ -\ 112.489 \\ \hline 691.573 \end{array}$$

29.
$$\begin{array}{r} 7.6 \\ \times\ 12 \\ \hline 15\ 2 \\ 76 \\ \hline 91.2 \end{array}$$

30.
$$\begin{array}{r} 14.63 \\ \times\ \ 3.2 \\ \hline 2\ 926 \\ 43\ 89 \\ \hline 46.816 \end{array}$$

31.
$$\begin{array}{r} 28.6 \\ 27\overline{)\ 772.2} \\ -54 \\ \hline 232 \\ -216 \\ \hline 16\ 2 \\ -16\ 2 \\ \hline 0 \end{array}$$

32.
$$\begin{array}{r} 230. \\ 0.06\overline{)\ 13.80} \\ -12 \\ \hline 1\ 8 \\ -1\ 8 \\ \hline 00 \end{array}$$

33. 0.7652 rounded to the nearest hundredth is 0.77.

34. 25.6293 rounded to the nearest tenth is 25.6.

35. $\frac{1}{2} = \frac{1 \cdot 5}{2 \cdot 5} = \frac{5}{10} = 0.5$

36. $\frac{3}{8} = \frac{3 \cdot 125}{8 \cdot 125} = \frac{375}{1000} = 0.375$

37.
$$\begin{array}{r} 0.3636... \\ 11\overline{)\ 4.0000} \\ -3\ 3 \\ \hline 70 \\ -66 \\ \hline 40 \\ -33 \\ \hline 70 \\ -66 \\ \hline 4 \end{array}$$

$\frac{4}{11} = 0.\overline{36} \approx 0.364$

38.
$$\begin{array}{r} 0.833... \\ 6\overline{)\ 5.000} \\ -4\ 8 \\ \hline 20 \\ -18 \\ \hline 20 \\ -18 \\ \hline 2 \end{array}$$

$\frac{5}{6} = 0.8\overline{3} \approx 0.833$

39. 29% = 0.29

40. 1.4% = 0.014

41. 0.39 = 39%

42. 1.2 = 120%

43. 68.3% = 0.683
The decimal is 0.683.

44. 2.3% = 0.023
5 = 500%
40% = 0.4
The true statement is b.

Chapter R Test

1. $\begin{array}{r} 3 \\ 3\overline{)9} \\ 2\overline{)18} \\ 2\overline{)36} \\ 2\overline{)72} \end{array}$

   The prime factorization of 72 is $2 \cdot 2 \cdot 2 \cdot 3 \cdot 3$.

2. $5 = 5$
   $18 = 2 \cdot 3 \cdot 3$
   $20 = 2 \cdot 2 \cdot 5$
   $\text{LCM} = 2 \cdot 2 \cdot 3 \cdot 3 \cdot 5 = 180$

3. $\frac{5}{12} = \frac{5 \cdot 5}{12 \cdot 5} = \frac{25}{60}$

4. $\frac{15}{20} = \frac{3 \cdot 5}{4 \cdot 5} = \frac{3}{4}$

5. $\frac{48}{100} = \frac{4 \cdot 12}{4 \cdot 25} = \frac{12}{25}$

6. $1.3 = 1\frac{3}{10} = \frac{13}{10}$

7. $\frac{5}{8} + \frac{7}{10} = \frac{5 \cdot 5}{8 \cdot 5} + \frac{7 \cdot 4}{10 \cdot 4} = \frac{25}{40} + \frac{28}{40} = \frac{25+28}{40} = \frac{53}{40}$

8. $\frac{2}{3} \cdot \frac{27}{49} = \frac{2 \cdot 27}{3 \cdot 49} = \frac{2 \cdot 3 \cdot 9}{3 \cdot 49} = \frac{18}{49}$

9. $$\begin{aligned} \frac{9}{10} \div 18 &= \frac{9}{10} \div \frac{18}{1} \\ &= \frac{9}{10} \cdot \frac{1}{18} \\ &= \frac{9 \cdot 1}{10 \cdot 18} \\ &= \frac{9 \cdot 1}{10 \cdot 9 \cdot 2} \\ &= \frac{1}{20} \end{aligned}$$

10. $\frac{8}{9} - \frac{1}{12} = \frac{8 \cdot 4}{9 \cdot 4} - \frac{1 \cdot 3}{12 \cdot 3} = \frac{32}{36} - \frac{3}{36} = \frac{32-3}{36} = \frac{29}{36}$

11. $1\frac{2}{9} + 3\frac{2}{3} = \frac{11}{9} + \frac{11}{3} = \frac{11}{9} + \frac{33}{9} = \frac{11+33}{9} = \frac{44}{9} = 4\frac{8}{9}$

12. $\begin{array}{r} 5\frac{6}{11} \\ -\,3\frac{7}{22} \\ \hline \end{array} \qquad \begin{array}{r} 5\frac{12}{22} \\ -\,3\frac{7}{22} \\ \hline 2\frac{5}{22} \end{array}$

13. $6\frac{7}{8} \div \frac{1}{8} = \frac{55}{8} \div \frac{1}{8} = \frac{55}{8} \cdot \frac{8}{1} = \frac{55 \cdot 8}{8 \cdot 1} = \frac{55}{1} = 55$

14. $2\frac{1}{10} \cdot 6\frac{1}{2} = \frac{21}{10} \cdot \frac{13}{2} = \frac{21 \cdot 13}{10 \cdot 2} = \frac{273}{20} = 13\frac{13}{20}$

15. $\begin{array}{r} 43 \\ 0.21 \\ +\,1.9 \\ \hline 45.11 \end{array}$

16. $\begin{array}{r} 123.60 \\ -\,57.72 \\ \hline 65.88 \end{array}$

17. $\begin{array}{r} 7.93 \\ \times\ 1.6 \\ \hline 4758 \\ 793 \\ \hline 12.688 \end{array}$

18. $\begin{array}{r} 320. \\ 0.25\overline{)\,80.00} \\ -75 \\ \hline 5\,0 \\ -5\,0 \\ \hline 00 \end{array}$

19. 23.7272 rounded to the nearest hundredth is 23.73.

20. $\frac{7}{8} = \frac{7 \cdot 125}{8 \cdot 125} = \frac{875}{1000} = 0.875$

21. $6\overline{)1.000}$ quotient $0.166...$

$$\begin{array}{r} -6 \\ \hline 40 \\ -36 \\ \hline 40 \\ -36 \\ \hline 4 \end{array}$$

$\frac{1}{6} = 0.1\overline{6} \approx 0.167$

22. $63.2\% = 0.632$

23. $0.09 = 9\%$

24. $\frac{3}{4} = \frac{3 \cdot 25}{4 \cdot 25} = \frac{75}{100} = 0.75 = 75\%$

25. $\frac{3}{4}$ of the fresh water is icecaps and glaciers.

26. $\frac{1}{200}$ of the fresh water is active water.

27. $1 - \frac{3}{4} - \frac{1}{200} = \frac{200}{200} - \frac{150}{200} - \frac{1}{200}$

$= \frac{200 - 150 - 1}{200}$

$= \frac{49}{200}$

$\frac{49}{200}$ of the fresh water is groundwater.

28. $1 - \frac{1}{200} = \frac{200}{200} - \frac{1}{200} = \frac{200 - 1}{200} = \frac{199}{200}$

$\frac{199}{200}$ of the fresh water is groundwater or icecaps and glaciers.

29. Area $= \frac{1}{2}$(base)(height)

$= \frac{1}{2} \cdot \frac{3}{4} \cdot \frac{1}{3}$

$= \frac{1 \cdot 3 \cdot 1}{2 \cdot 4 \cdot 3}$

$= \frac{1}{8}$

The area is $\frac{1}{8}$ square foot.

30. $A = lw = \frac{9}{8} \cdot \frac{7}{8} = \frac{9 \cdot 7}{8 \cdot 8} = \frac{63}{64}$

The area is $\frac{63}{64}$ square centimeter.

Chapter 1

Section 1.2 Practice Exercises

1. Since 8 is to the right of 6 on a number line, a statement $8 < 6$ is false.

2. Since 100 is to the right of 10 on a number line, a statement $100 > 10$ is true.

3. Since $21 = 21$, the statement $21 \le 21$ is true.

4. Since $21 = 21$, the statement $21 \ge 21$ is true.

5. Since neither $0 > 5$ nor $0 = 5$ is true, the statement $0 \ge 5$ is false.

6. Since $25 > 22$, the statement $25 \ge 22$ is true.

7. **a.** Fourteen is greater than or equal to fourteen is written as $14 \ge 14$.

   b. Zero is less than five is written as $0 < 5$.

   c. Nine is not equal to 10 is written as $9 \ne 10$.

8. The integer -8 represents 8 feet below sea level.

9. $\frac{5}{4} = 1\frac{1}{4}$

   $-2\frac{1}{2}$ $-\frac{2}{3}$ $\frac{1}{5}$ $\frac{5}{4}$ 2.25

   -5 -4 -3 -2 -1 0 1 2 3 4 5

10. **a.** $-11 < -9$ since -11 is to the left of -9 on a number line.

    b. By comparing digits in the same places, we find that $4.511 > 4.151$, since $0.5 > 0.1$.

    c. By dividing, we find that $\frac{7}{8} = 0.875$ and $\frac{2}{3} = 0.66....$ Since $0.875 > 0.66...$, then $\frac{7}{8} > \frac{2}{3}$.

11. **a.** The natural numbers are 6 and 913.

    b. The whole numbers are 0, 6, and 913.

    c. The integers are -100, 0, 6, and 913.

    d. The rational numbers are -100, $-\frac{2}{5}$, 0, 6, and 913.

    e. The irrational number is π.

    f. All numbers in the given set are real numbers.

12. **a.** $|7| = 7$ since 7 is 7 units from 0 on a number line.

    b. $|-8| = 8$ since -8 is 8 units from 0 on a number line.

    c. $\left|\frac{2}{3}\right| = \frac{2}{3}$

    d. $|0| = 0$ since 0 is 0 units from 0 on a number line.

    e. $|-3.06| = 3.06$

13. **a.** $|-4| = 4$

    b. $-3 < |0|$ since $-3 < 0$.

    c. $|-2.7| > |-2|$ since $2.7 > 2$.

    d. $|-6| \le |-16|$ since $6 < 16$.

    e. $|10| < \left|-10\frac{1}{3}\right|$ since $10 < 10\frac{1}{3}$.

Vocabulary, Readiness & Video Check 1.2

1. The <u>whole</u> numbers are $\{0, 1, 2, 3, 4, ...\}$.

2. The <u>natural</u> numbers are $\{1, 2, 3, 4, 5, ...\}$.

3. The symbols $\ne$, $\le$, and $>$ are called <u>inequality</u> symbols.

4. The <u>integers</u> are $\{..., -3, -2, -1, 0, 1, 2, 3, ...\}$.

5. The <u>real</u> numbers are {all numbers that correspond to points on a number line}.

6. The <u>rational</u> numbers are $\left\{\frac{a}{b} \middle| a \text{ and } b \text{ are integers}, b \ne 0\right\}$.

7. The integer <u>0</u> is neither positive nor negative.

8. The point on a number line halfway between 0 and $\frac{1}{2}$ can be represented by $\underline{\frac{1}{4}}$.

9. The distance between a real number a and 0 is called the <u>absolute value</u> of a.

10. The absolute value of a is written in symbols as $\underline{|a|}$.

11. To form a true statement: $0 < 7$.

12. Five is greater than or equal to 4; $5 \geq 4$

13. 0 belongs to the whole numbers, the integers, the rational numbers, and the real numbers; since 0 is a rational number, it cannot also be an irrational number.

14. The <u>absolute value</u> of a real number a, denoted by $|a|$, is the distance between a and 0 on a number line.

Exercise Set 1.2

1. Since 4 is to the left of 10 on a number line, $4 < 10$.

3. Since 7 is to the right of 3 on a number line, $7 > 3$.

5. $6.26 = 6.26$

7. Since 0 is to the left of 7 on a number line, $0 < 7$.

9. Since 32 is to the left of 212 on a number line, $32 < 212$.

11. Since 30 is to the left of 45 on a number line, $30 \leq 45$.

13. Since $11 = 11$, the statement $11 \leq 11$ is true.

15. Since -11 is to the left of -10 on a number line, $-11 > -10$ is false.

17. Comparing digits with the same place value, we have $0.0 < 0.9$. Thus the statement $5.092 < 5.902$ is true.

19. Rewrite the fractions with a common denominator and compare numerators.
$\frac{9}{10} = \frac{81}{90}$; $\frac{8}{9} = \frac{80}{90}$
Since $81 > 80$, then $\frac{9}{10} \leq \frac{8}{9}$ is false.

21. $25 \geq 20$ has the same meaning as $20 \leq 25$.

23. $0 < 6$ has the same meaning as $6 > 0$.

25. $-10 > -12$ has the same meaning as $-12 < -10$.

27. Seven is less than eleven is written as $7 < 11$.

29. Five is greater than or equal to four is written as $5 \geq 4$.

31. Fifteen is not equal to negative two is written as $15 \neq -2$.

33. The integer 14,494 represents 14,494 feet above sea level. The integer -282 represents 282 feet below sea level.

35. The integer $-27,724$ represents 27,724 fewer students.

37. The integer 475 represents a \$475 deposit. The integer -195 represents a \$195 withdrawal.

39. 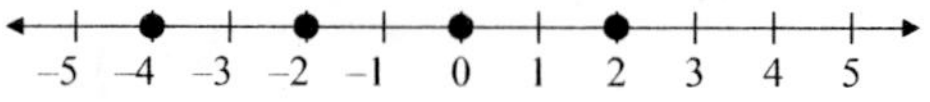
-5 -4 -3 -2 -1 0 1 2 3 4 5

41. 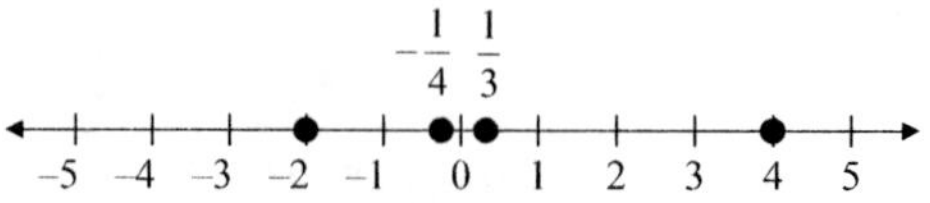
$-\frac{1}{4}$ $\frac{1}{3}$
-5 -4 -3 -2 -1 0 1 2 3 4 5

43. 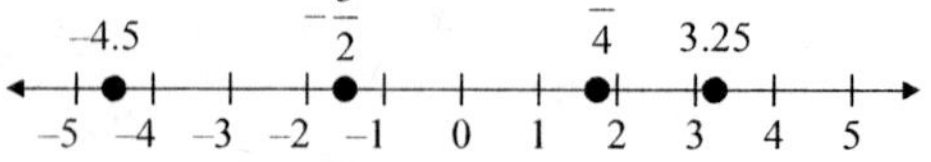
-4.5 $-\frac{3}{2}$ $\frac{7}{4}$ 3.25
-5 -4 -3 -2 -1 0 1 2 3 4 5

45. 0 is a whole number, an integer, a rational number, and a real number.

47. -7 is an integer, a rational number, and a real number.

49. 265 is a natural number, a whole number, an integer, a rational number, and a real number.

51. $\frac{2}{3}$ is a rational number and a real number.

53. False; the rational number $\frac{2}{3}$ is not an integer.

55. True; 0 is a real number.

57. False; the negative number $-\sqrt{2}$ is not a rational number.

59. False; the real number $\sqrt{7}$ is not a rational number.

61. $|8.9| = 8.9$ since 8.9 is 8.9 units from 0 on a number line.

63. $|-20| = 20$ since -20 is 20 units from 0 on a number line.

65. $\left|\frac{9}{2}\right| = \frac{9}{2}$ since $\frac{9}{2}$ is $\frac{9}{2}$ units from 0 on a number line.

67. $\left|-\frac{12}{13}\right| = \frac{12}{13}$ since $-\frac{12}{13}$ is $\frac{12}{13}$ unit from 0 on a number line.

69. $|-5| = 5$
$-4 = -4$
Since 5 is to the right of -4 on a number line, $|-5| > -4$.

71. $\left|-\frac{5}{8}\right| = \frac{5}{8}$
$\left|\frac{5}{8}\right| = \frac{5}{8}$
Since $\frac{5}{8} = \frac{5}{8}$, then $\left|-\frac{5}{8}\right| = \left|\frac{5}{8}\right|$.

73. $|-2| = 2$
$|-2.7| = 2.7$
Since 2 is to the left of 2.7 on a number line, $|-2| < |-2.7|$.

75. $|0| = 0$
$|-8| = 8$
Since 0 is to the left of 8 on a number line, $|0| < |-8|$.

77. The 2012 cranberry production in Oregon was 40 million pounds, while the 2012 cranberry production in Washington was 14 million pounds.
40 million > 14 million or
40,000,000 > 14,000,000

79. The 2012 cranberry production in Washington was 14 million pounds, while the 2012 cranberry production in New Jersey was 54 million pounds.
$54 - 14 = 40$
The production in Washington was 40 million pounds less or -40 million.

81. Since -0.04 is to the right of -26.7 on a number line, $-0.04 > -26.7$.

83. Sun: -26.7
Arcturus: -0.04
Since $-26.7 < -0.04$, the sun is brighter than Arcturus.

85. Since the brightest star corresponds to the smallest apparent magnitude, which is -26.7, the brightest star is the sun.

87. answers may vary

Section 1.3 Practice Exercises

1. **a.** $4^2 = 4 \cdot 4 = 16$

b. $2^2 = 2 \cdot 2 = 4$

c. $3^4 = 3 \cdot 3 \cdot 3 \cdot 3 = 81$

d. $9^1 = 9$

e. $\left(\frac{2}{5}\right)^3 = \left(\frac{2}{5}\right)\left(\frac{2}{5}\right)\left(\frac{2}{5}\right) = \frac{2 \cdot 2 \cdot 2}{5 \cdot 5 \cdot 5} = \frac{8}{125}$

f. $(0.8)^2 = (0.8)(0.8) = 0.64$

2. $3 \cdot 2 + 4^2 = 3 \cdot 2 + 16 = 6 + 16 = 22$

3. $28 \div 7 \cdot 2 = 4 \cdot 2 = 8$

4. $\frac{9}{5} \cdot \frac{1}{3} - \frac{1}{3} = \frac{9}{15} - \frac{1}{3} = \frac{9}{15} - \frac{5}{15} = \frac{4}{15}$

5. $5+3[2(3\cdot 4+1)-20]=5+3[2(12+1)-20]$
$=5+3[2(13)-20]$
$=5+3[26-20]$
$=5+3[6]$
$=5+18$
$=23$

6. $\dfrac{1+|7-4|+3^2}{8-5}=\dfrac{1+|3|+3^2}{8-5}$
$=\dfrac{1+3+3^2}{3}$
$=\dfrac{1+3+9}{3}$
$=\dfrac{13}{3}$

7. a. Replace y with 4.
$3y^2=3\cdot(4)^2=3\cdot 16=48$

b. Replace x with 1 and y with 4.
$2y-x=2(4)-1=8-1=7$

c. Replace x with 1 and y with 4.
$\dfrac{11x}{3y}=\dfrac{11\cdot 1}{3\cdot 4}=\dfrac{11}{12}$

d. Replace x with 1 and y with 4.
$\dfrac{x}{y}+\dfrac{6}{y}=\dfrac{1}{4}+\dfrac{6}{4}=\dfrac{7}{4}$

e. Replace x with 1 and y with 4.
$y^2-x^2=4^2-1^2=16-1=15$

8. $5x-10=x+2$
$5(3)-10\stackrel{?}{=}3+2$
$15-10\stackrel{?}{=}5$
$5=5$ True
3 is a solution.

9. a. $5\cdot x$ and $5x$ are both ways to denote the product of 5 and x.

b. A number added to 7 is denoted by $7+x$.

c. A number divided by 11.2 is denoted by $x\div 11.2$ or $\dfrac{x}{11.2}$.

d. A number subtracted from 8 is denoted by $8-x$.

e. Twice a number, plus 1 is denoted by $2x+1$.

10. a. The ratio of a number and 6 is 24 is written as $\dfrac{x}{6}=24$.

b. The difference of 10 and a number is 18 is written as $10-x=18$.

c. One less than twice a number is 99 is written as $2x-1=99$.

Calculator Explorations

1. $5^3=125$

2. $7^4=2401$

3. $9^5=59{,}049$

4. $8^6=262{,}144$

5. $2(20-5)=30$

6. $3(14-7)+21=42$

7. $24(862-455)+89=9857$

8. $99+(401+962)=1462$

9. $\dfrac{4623+129}{36-34}=2376$

10. $\dfrac{956-452}{89-86}=168$

Vocabulary, Readiness & Video Check 1.3

1. In 2^5, the 2 is called the base and the 5 is called the exponent.

2. True or false: 2^5 means 2.5. false

3. To simplify $8+2\cdot 6$, which operation should be performed first? multiplication

4. To simplify $(8+2)\cdot 6$, which operation should be performed first? addition

5. To simplify $9(3-2)\div 3+6$, which operation should be performed first? subtraction

6. To simplify $8 \div 2 \cdot 6$, which operation should be performed first? division

7. A combination of operations on letters (variables) and numbers is an expression.

8. A letter that represents a number is a variable.

9. $3x - 2y$ is called an expression and the letters x and y are variables.

10. Replacing a variable in an expression by a number and then finding the value of the expression is called evaluating the expression.

11. A statement of the form "expression = expression" is called an equation.

12. A value for the variable that makes the equation a true statement is called a solution.

13. The order in which we perform operation does matter! We came up with an order of operations to avoid getting more than one answer when evaluating an expression.

14. The replacement value for z is not used because it's not needed—there is no variable z in the given algebraic expression.

15. No; the variable was replaced with 0 in the equation to see if a true statement occurred, and it did not.

16. We translate phrases to mathematical expressions and sentences to mathematical equations.

Exercise Set 1.3

1. $3^5 = 3 \cdot 3 \cdot 3 \cdot 3 \cdot 3 = 243$

3. $3^3 = 3 \cdot 3 \cdot 3 = 27$

5. $1^5 = 1 \cdot 1 \cdot 1 \cdot 1 \cdot 1 = 1$

7. $5^1 = 5$

9. $7^2 = 7 \cdot 7 = 49$

11. $\left(\frac{2}{3}\right)^4 = \left(\frac{2}{3}\right)\left(\frac{2}{3}\right)\left(\frac{2}{3}\right)\left(\frac{2}{3}\right) = \frac{2 \cdot 2 \cdot 2 \cdot 2}{3 \cdot 3 \cdot 3 \cdot 3} = \frac{16}{81}$

13. $\left(\frac{1}{5}\right)^3 = \left(\frac{1}{5}\right)\left(\frac{1}{5}\right)\left(\frac{1}{5}\right) = \frac{1 \cdot 1 \cdot 1}{5 \cdot 5 \cdot 5} = \frac{1}{125}$

15. $(1.2)^2 = 1.2 \cdot 1.2 = 1.44$

17. $(0.7)^3 = 0.7 \cdot 0.7 \cdot 0.7 = 0.343$

19. $5 \cdot 5 = 5^2$ square meters

21. $5 + 6 \cdot 2 = 5 + 12 = 17$

23. $4 \cdot 8 - 6 \cdot 2 = 32 - 12 = 20$

25. $18 \div 3 \cdot 2 = 6 \cdot 2 = 12$

27. $2 + (5 - 2) + 4^2 = 2 + 3 + 4^2 = 2 + 3 + 16 = 21$

29. $5 \cdot 3^2 = 5 \cdot 9 = 45$

31. $\frac{1}{4} \cdot \frac{2}{3} - \frac{1}{6} = \frac{2}{12} - \frac{1}{6} = \frac{1}{6} - \frac{1}{6} = 0$

33. $\frac{6-4}{9-2} = \frac{2}{7}$

35.
$$\begin{aligned} 2[5 + 2(8 - 3)] &= 2[5 + 2(5)] \\ &= 2[5 + 10] \\ &= 2[15] \\ &= 30 \end{aligned}$$

37. $\frac{19 - 3 \cdot 5}{6 - 4} = \frac{19 - 15}{2} = \frac{4}{2} = 2$

39. $\frac{|6-2| + 3}{8 + 2 \cdot 5} = \frac{4+3}{8+10} = \frac{7}{18}$

41. $\frac{3 + 3(5+3)}{3^2 + 1} = \frac{3 + 3(8)}{9+1} = \frac{3 + 24}{10} = \frac{27}{10}$

43.
$$\begin{aligned} \frac{6 + |8-2| + 3^2}{18 - 3} &= \frac{6 + |6| + 9}{15} \\ &= \frac{6 + 6 + 9}{15} \\ &= \frac{21}{15} \\ &= \frac{7}{5} \end{aligned}$$

45. $2+3[10(4\cdot 5-16)-30]$
$=2+3[10(20-16)-30]$
$=2+3[10(4)-30]$
$=2+3[40-30]$
$=2+3[10]$
$=2+30$
$=32$

47. $\left(\frac{2}{3}\right)^3+\frac{1}{9}+\frac{1}{3}\cdot\frac{4}{3}=\frac{8}{27}+\frac{1}{9}+\frac{1}{3}\cdot\frac{4}{3}$
$=\frac{8}{27}+\frac{1}{9}+\frac{4}{9}$
$=\frac{8}{27}+\frac{3}{27}+\frac{12}{27}$
$=\frac{23}{27}$

49. Replace y with 3.
$3y=3(3)=9$

51. Replace x with 1 and z with 5.
$\frac{z}{5x}=\frac{5}{5(1)}=\frac{5}{5}=1$

53. Replace x with 1.
$3x-2=3(1)-2=3-2=1$

55. Replace x with 1 and y with 3.
$|2x+3y|=|2(1)+3(3)|=|2+9|=|11|=11$

57. Replace x with 1, y with 3, and z with 5.
$xy+z=1(3)+5=3+5=8$

59. Replace y with 3.
$5y^2=5(3^2)=5(9)=45$

61. Replace x with 12, y with 8, and z with 4.
$\frac{x}{z}+3y=\frac{12}{4}+3(8)=3+3(8)=3+24=27$

63. Replace x with 12 and y with 8.
$x^2-3y+x=(12)^2-3(8)+12$
$=144-3(8)+12$
$=144-24+12$
$=120+12$
$=132$

65. Replace x with 12, y with 8, and z with 4.
$\frac{x^2+z}{y^2+2z}=\frac{(12)^2+4}{(8)^2+2(4)}$
$=\frac{144+4}{64+2(4)}$
$=\frac{148}{64+8}$
$=\frac{148}{72}$
$=\frac{4\cdot 37}{4\cdot 18}$
$=\frac{37}{18}$

67. $3x-6=9$
$3(5)-6 \stackrel{?}{=} 9$
$15-6 \stackrel{?}{=} 9$
$9=9$ True
Since the result is true, 5 is a solution of the given equation.

69. $2x+6=5x-1$
$2(0)+6 \stackrel{?}{=} 5(0)-1$
$0+6 \stackrel{?}{=} 0-1$
$6=-1$ False
Since the result is false, 0 is not a solution of the given equation.

71. $2x-5=5$
$2(8)-5 \stackrel{?}{=} 5$
$16-5 \stackrel{?}{=} 5$
$11=5$ False
Since the result is false, 8 is not a solution of the given equation.

73. $x+6=x+6$
$2+6 \stackrel{?}{=} 2+6$
$8=8$ True
Since the result is true, 2 is a solution of the given equation.

75. $x=5x+15$
$0 \stackrel{?}{=} 5(0)+15$
$0 \stackrel{?}{=} 0+15$
$0=15$ False
Since the result is false, 0 is not a solution of the given equation.

77. $\frac{1}{3}x = 9$

$\frac{1}{3}(27) \stackrel{?}{=} 9$

$9 = 9$ True

Since the result is true, 27 is a solution of the given equation.

79. Fifteen more than a number is written as $x + 15$.

81. Five subtracted from a number is written as $x - 5$.

83. The ratio of a number and 4 is written as $\frac{x}{4}$.

85. Three times a number, increased by 22 is written as $3x + 22$.

87. One increased by two equals the quotient of nine and three is written as $1 + 2 = 9 \div 3$.

89. Three is not equal to four divided by two is written as $3 \neq 4 \div 2$.

91. The sum of 5 and a number is 20 is written as $5 + x = 20$.

93. The product 7.6 and a number is 17 is written as $7.6x = 17$.

95. Thirteen minus three times a number is 13 is written as $13 - 3x = 13$.

97. no; answers may vary

99. **a.** $(6 + 2) \cdot (5 + 3) = 8 \cdot 8 = 64$

b. $(6 + 2) \cdot 5 + 3 = 8 \cdot 5 + 3 = 40 + 3 = 43$

c. $6 + 2 \cdot 5 + 3 = 6 + 10 + 3 = 19$

d. $6 + 2 \cdot (5 + 3) = 6 + 2 \cdot 8 = 6 + 16 = 22$

	Length, l	Width, w	Perimeter of Rectangle: $2l + 2w$	Area of Rectangle: lw
101.	4 in.	3 in.	$2l + 2w$ = 2(4 in.) + 2(3 in.) = 8 in. + 6 in. = 14 in.	lw = (4 in.)(3 in.) = 12 sq in.
103.	5.3 in.	1.7 in.	$2l + 2w$ = 2(5.3 in.) + 2(1.7 in.) = 10.6 in. + 3.4 in. = 14 in.	lw = (5.3 in.)(1.7 in.) = 9.01 sq in.

105. Rectangles with the same perimeter can have different areas.

107. $(20 - 4) \cdot 4 \div 2 = 16 \cdot 4 \div 2 = 64 \div 2 = 32$

109. **a.** $5x + 6$ is an expression since it does not contain the equal symbol, "=."

b. $2a = 7$ is an equation since it contains the equal symbol.

c. $3a + 2 = 9$ is an equation since it contains the equal symbol.

d. $4x + 3y - 8z$ is an expression since it does not contain the equal symbol.

e. $5^2 - 2(6-2)$ is an expression since it does not contain the equal symbol.

111. answers may vary

113. answers may vary; for example, $-2(5) - 1$: $-2(5) - 1 = -10 - 1 = -11$

Section 1.4 Practice Exercises

1. $-2 + (-4) = -6$

2.

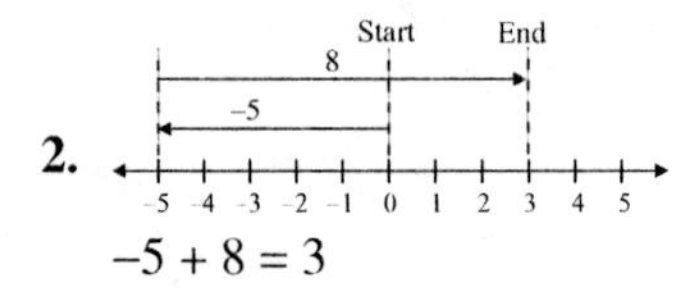

$-5 + 8 = 3$

3.

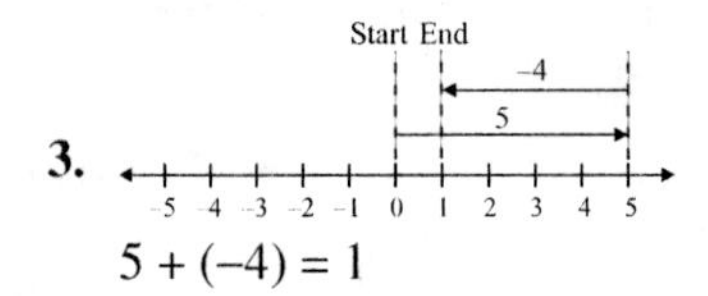

$5 + (-4) = 1$

4. $-8 + (-5) = -13$

5. $(-14) + 6 = -8$

6. $(-17) + (-10) = -27$

7. $(-4) + 12 = 8$

8. $1.5 + (-3.2) = -1.7$

9. $-\frac{5}{12} + \left(-\frac{1}{12}\right) = -\frac{6}{12} = -\frac{6 \cdot 1}{6 \cdot 2} = -\frac{1}{2}$

10. $12.1 + (-3.6) = 8.5$

11. $-\frac{4}{5} + \frac{2}{3} = -\frac{12}{15} + \frac{10}{15} = -\frac{2}{15}$

12. a. $16 + (-9) + (-9) = 7 + (-9) = -2$

b. $[3+(-13)]+[-4+(-7\} = [-10]+[-11]$
$= -21$

13. The opposite of -35 is 35.

14. The opposite of 12 is -12.

15. The opposite of $-\frac{3}{11}$ is $\frac{3}{11}$.

16. The opposite of 1.9 is -1.9.

17. a. $-(-22) = 22$

b. $-\left(-\frac{2}{7}\right) = \frac{2}{7}$

c. $-(-x) = x$

d. $-|-14| = -14$

e. $-|2.3| = -2.3$

18. $30 + (-30) = 0$

19. $-81 + 81 = 0$

20. $x + 3y = -6 + 3(2) = -6 + 6 = 0$

21. $x + y = -13 + (-9) = -22$

22. Temperature at 8 a.m. $= -7 + (+4) + (+7)$
$= -3 + (+7)$
$= 4$

The temperature was 4°F at 8 a.m.

Vocabulary, Readiness & Video Check 1.4

1. If n is a number, then $-n + n = \underline{0}$.

2. Since $x + n = n + x$, we say that addition is <u>commutative</u>.

3. If a is a number, then $-(-a) = \underline{a}$.

4. Since $n + (x + a) = (n + x) + a$, we say that addition is <u>associative</u>.

5. To add two numbers with the same sign, add their <u>absolute values</u> and use their common sign as the sign of the sum.

6. Negative; when you add two numbers with different signs, the sign of the sum is the same as the sign of the number with the larger absolute value and -8.4 has a larger absolute value than 6.3.

7. Example 12 is an example of the opposite of the *absolute value* of $-a$, not the opposite of $-a$. The absolute value of $-a$ is positive, so its opposite is negative. Therefore the answers to Examples 11 and 12 have different signs.

8. The algebraic expressions are $x + y$ and $3x + y$. For $3x + y$, the variable x is multiplied by 3.

9. Depths below the surface; the diver's position is 231 feet below the surface.

Exercise Set 1.4

1. $6 + (-3) = 3$

3. $-6 + (-8) = -14$

5. $8 + (-7) = 1$

7. $-14 + 2 = -12$

9. $-2 + (-3) = -5$

11. $-9 + (-3) = -12$

13. $-7 + 3 = -4$

15. $10 + (-3) = 7$

17. $5 + (-7) = -2$

19. $-16 + 16 = 0$

21. $27 + (-46) = -19$

23. $-18 + 49 = 31$

25. $-33 + (-14) = -47$

27. $6.3 + (-8.4) = -2.1$

29. $117 + (-79) = 38$

31. $-9.6 + (-3.5) = -13.1$

33. $-\frac{3}{8}+\frac{5}{8}=\frac{2}{8}=\frac{2\cdot 1}{2\cdot 4}=\frac{1}{4}$

35. $-\frac{7}{16}+\frac{1}{4}=-\frac{7}{16}+\frac{4}{16}=-\frac{3}{16}$

37. $-\frac{7}{10}+\left(-\frac{3}{5}\right)=-\frac{7}{10}+\left(-\frac{6}{10}\right)=-\frac{13}{10}$

39. $|-8| + (-16) = 8 + (-16) = -8$

41. $-15 + 9 + (-2) = -6 + (-2) = -8$

43. $-21 + (-16) + (-22) = -37 + (-22) = -59$

45. $-23 + 16 + (-2) = -7 + (-2) = -9$

47. $|5 + (-10)| = |-5| = 5$

49. $6 + (-4) + 9 = 2 + 9 = 11$

51. $[-17+(-4)]+[-12+15]=[-21]+[3]=-18$

53.
$$\begin{aligned}|9+(-12)|+|-16| &= |-3|+|-16|\\ &= 3+16\\ &= 19\end{aligned}$$

55.
$$\begin{aligned}-13+[5+(-3)+4] &= -13+[2+4]\\ &= -13+[6]\\ &= -7\end{aligned}$$

57. $-38 + 12 = -26$
The sum of -38 and 12 is -26.

59. The additive inverse of 6 is -6.

61. The additive inverse of -2 is 2.

63. The additive inverse of 0 is 0.

65. Since $|-6| = 6$, the additive inverse of $|-6|$ is -6.

67. $-|-2| = -2$

69. $-(-7) = 7$

71. $-(-7.9) = 7.9$

73. $-(-5z) = 5z$

75. $\left|-\frac{2}{3}\right|=\frac{2}{3}$

77. $x + y = -20 + (-50) = -70$

79. $3x + y = 3(2) + (-3) = 6 + (-3) = 3$

81. The sum of -6 and 25 is $-6 + 25 = 19$.

83. The sum of -31, -9, and 30 is
$-31 + (-9) + 30 = -40 + 30 = -10$.

85. $0 + (-215) + (-16) = -215 + (-16) = -231$
The diver's final depth is 231 feet below the surface.

87. $-35 + 142 = 107$
The highest recorded temperature in Massachusetts was 107°F.

89. $-411 + 316 = -95$
Your elevation is −95 meters.

91. $$\begin{aligned} -3+2+(-4)+(-3) &= -1+(-4)+(-3) \\ &= -5+(-3) \\ &= -8 \end{aligned}$$
His total score was −8.

93. $8.9 + 13.7 + (-3.5) + (-2.2)$
$= 22.6 + (-3.5) + (-2.2)$
$= 19.1 + (-2.2)$
$= 16.9$
The total net income for fiscal year 2013 was $16.9 million.

95. The highest bar corresponds to July, so the highest low temperature was in July.

97. The shortest bar above 0 corresponds to October.

99. $$\frac{-9.1+14.4+8.8}{3} = \frac{5.3+8.8}{3} = \frac{14.1}{3} = 4.7$$
The average for the months of April, May, and October is 4.7°F.

101. answers may vary

103. $7 + (-10) = -3$

105. $-10 + (-12) = -22$

107. True

109. False; for example, $4 + (-2) = 2 > 0$.

111. answers may vary

Section 1.5 Practice Exercises

1. a. $-20 - 6 = -20 + (-6) = -26$

b. $3 - (-5) = 3 + 5 = 8$

c. $7 - 17 = 7 + (-17) = -10$

d. $-4 - (-9) = -4 + 9 = 5$

2. $9.6 - (-5.7) = 15.3$

3. $$-\frac{4}{9}-\frac{2}{9} = -\frac{4}{9}+\left(-\frac{2}{9}\right) = -\frac{6}{9} = -\frac{2}{3}$$

4. $$-\frac{1}{4}-\left(-\frac{2}{5}\right) = -\frac{1}{4}+\frac{2}{5} = -\frac{5}{20}+\frac{8}{20} = \frac{3}{20}$$

5. a. $-11 - 7 = -11 + (-7) = -18$

b. $35 - (-25) = 35 + 25 = 60$

6. a. $$\begin{aligned} -20-5+12-(-3) &= -20+(-5)+12+3 \\ &= -10 \end{aligned}$$

b. $$\begin{aligned} 5.2-(-4.4)+(-8.8) &= 5.2+4.4+(-8.8) \\ &= 0.8 \end{aligned}$$

7. a. $$\begin{aligned} -9+[(-4-1)-10] &= -9+[(-4+(-1))-10] \\ &= -9+[(-5)-10] \\ &= -9+[-5+(-10)] \\ &= -9+[-15] \\ &= -24 \end{aligned}$$

b. $$\begin{aligned} 5^2-20+[-11-(-3)] &= 5^2-20+[-11+3] \\ &= 5^2-20+[-8] \\ &= 25-20+(-8) \\ &= 25+(-20)+(-8) \\ &= 5+(-8) \\ &= -3 \end{aligned}$$

8. a. Replace x with 1 and y with −4.
$$\frac{x-y}{14+x} = \frac{1-(-4)}{14+1} = \frac{1+4}{15} = \frac{5}{15} = \frac{1}{3}$$

b. Replace x with 1 and y with −4.
$$x^2 - y = (1)^2 - (-4) = 1-(-4) = 1+4 = 5$$

9. $$\begin{aligned} -1+x &= 1 \\ -1+(-2) &\stackrel{?}{=} 1 \\ -3 &= 1 \text{ False} \end{aligned}$$
−2 is not a solution.

10. $29{,}028 - (-1312) = 29{,}028 + 1312 = 30{,}340$
Mount Everest is 30,340 feet higher than the Dead Sea.

11. a. These angles are supplementary, so their sum is 180°. This means that $m\angle x$ is 180° − 78°.
$m\angle x = 180° - 78° = 102°$

b. These angles are complementary, so their sum is 90°. This means that $m\angle y$ is $90° - 81°$.
$m\angle y = 90° - 81° = 9°$

Vocabulary, Readiness & Video Check 1.5

1. It is true that $a - b = \underline{a + (-b)}$. b

2. The opposite of n is $\underline{-n}$. a

3. To evaluate $x - y$ for $x = -10$ and $y = -14$, we replace x with -10 and y with -14 and evaluate $\underline{-10 - (-14)}$. d

4. The expression $-5 - 10$ equals $\underline{-5 + (-10)}$. c

5. To subtract two real numbers, change the operation to <u>addition</u> and take the <u>opposite</u> of the second number.

6. $-10 + (8) + (-4) + (-20)$; it's rewritten to change the subtraction operations to addition and turn the expression into an addition of real numbers problem.

7. There's a minus sign in the numerator and the replacement value is negative (notice parentheses are used around the replacement value), and it's always good to be careful when working with negative signs.

8. -4 is NOT a solution of $x - 9 = 5$

9. This means that the overall vertical altitude change of the jet is actually a decrease in altitude from when the Example started.

10. In Example 10, you have two supplementary angles and know the measure of one of them. From the definition, you know that the two supplementary angles must sum to 180°. Therefore we can subtract the known angle measure from 180° to get the measure of the other angle.

Exercise Set 1.5

1. $-6 - 4 = -6 + (-4) = -10$

3. $4 - 9 = 4 + (-9) = -5$

5. $16 - (-3) = 16 + 3 = 19$

7. $7 - (-4) = 7 + 4 = 11$

9. $-26 - (-18) = -26 + 18 = -8$

11. $-6 - 5 = -6 + (-5) = -11$

13. $16 - (-21) = 16 + 21 = 37$

15. $-6 - (-11) = -6 + 11 = 5$

17. $-44 - 27 = -44 + (-27) = -71$

19. $-21 - (-21) = -21 + 21 = 0$

21. $-\frac{3}{11} - \left(-\frac{5}{11}\right) = -\frac{3}{11} + \frac{5}{11} = \frac{2}{11}$

23. $9.7 - 16.1 = 9.7 + (-16.1) = -6.4$

25. $-2.6 - (-6.7) = -2.6 + 6.7 = 4.1$

27. $\frac{1}{2} - \frac{2}{3} = \frac{1}{2} + \left(-\frac{2}{3}\right) = \frac{3}{6} + \left(-\frac{4}{6}\right) = -\frac{1}{6}$

29. $$\begin{aligned} -\frac{1}{6} - \frac{3}{4} &= -\frac{1}{6} + \left(-\frac{3}{4}\right) \\ &= -\frac{2}{12} + \left(-\frac{9}{12}\right) \\ &= -\frac{11}{12} \end{aligned}$$

31. $8.3 - (-0.62) = 8.3 + 0.62 = 8.92$

33. $0 - 8.92 = 0 + (-8.92) = -8.92$

35. $8 - (-5) = 8 + 5 = 13$
-5 subtracted from 8 is 13.

37. $-6 - (-1) = -6 + 1 = -5$
The difference between -6 and -1 is -5.

39. $7 - 8 = 7 + (-8) = -1$
8 subtracted from 7 is -1.

41. $-8 - 15 = -8 + (-15) = -23$
-8 decreased by 15 is -23.

43. $$\begin{aligned} &-10 - (-8) + (-4) - 20 \\ &= -10 + 8 + (-4) + (-20) \\ &= -2 + (-4) + (-20) \\ &= -6 + (-20) \\ &= -26 \end{aligned}$$

45. $5-9+(-4)-8-8$
$=5+(-9)+(-4)+(-8)+(-8)$
$=-4+(-4)+(-8)+(-8)$
$=-8+(-8)+(-8)$
$=-16+(-8)$
$=-24$

47. $-6-(2-11)=-6-(-9)=-6+9=3$

49. $3^3-8\cdot 9=27-72=27+(-72)=-45$

51. $2-3(8-6)=2-3[8+(-6)]$
$=2-3(2)$
$=2-6$
$=2+(-6)$
$=-4$

53. $(3-6)+4^2=[3+(-6)]+16=[-3]+16=13$

55. $-2+[(8-11)-(-2-9)]$
$=-2+[(8+(-11))-(-2+(-9))]$
$=-2+[(-3)-(-11)]$
$=-2+[-3+11]$
$=-2+8$
$=6$

57. $|-3|+2^2+[-4-(-6)]=|-3|+4+[-4+6]$
$=3+4+2$
$=7+2$
$=9$

59. Replace x with -5 and y with 4.
$x-y=-5-4=-5+(-4)=-9$

61. Replace x with -5 and y with 4.
$\frac{9-x}{y+6}=\frac{9-(-5)}{4+6}=\frac{9+5}{10}=\frac{14}{10}=\frac{2\cdot 7}{2\cdot 5}=\frac{7}{5}$

63. Replace x with -5, y with 4, and t with 10.
$|x|+2t-8y=|-5|+2(10)-8(4)$
$=5+20-32$
$=25-32$
$=25+(-32)$
$=-7$

65. Replace x with -5 and y with 4.
$y^2-x=4^2-(-5)=16+5=21$

67. Replace x with -5 and t with 10.
$\frac{|x-(-10)|}{2t}=\frac{|-5-(-10)|}{2(10)}$
$=\frac{|-5+10|}{2(10)}$
$=\frac{|5|}{2(10)}$
$=\frac{5}{20}$
$=\frac{1\cdot 5}{4\cdot 5}$
$=\frac{1}{4}$

69. $x-9=5$
$-4-9 \stackrel{?}{=} 5$
$-4+(-9) \stackrel{?}{=} 5$
$-13=5$ False

Since the result is false, -4 is not a solution of the given equation.

71. $-x+6=-x-1$
$-(-2)+6 \stackrel{?}{=} -(-2)-1$
$2+6 \stackrel{?}{=} 2-1$
$8 \stackrel{?}{=} 2+(-1)$
$8=1$ False

Since the result is false, -2 is not a solution of the given equation.

73. $-x-13=-15$
$-2-13 \stackrel{?}{=} -15$
$-2+(-13) \stackrel{?}{=} -15$
$-15=-15$ True

Since the result is true, 2 is a solution of the given equation.

75. $134-(-129)=134+129=263$
Therefore, 134°F is 263°F warmer than −129°F.

77. $13{,}796-(-21{,}857)=13{,}796+21{,}857=35{,}653$
The difference in elevation is 35,653 feet.

79. Complementary angles sum to 90°.
$90°-60°=x$
$90°+(-60°)=x$
$30°=x$

81. $-250+120-178=-250+120+(-178)$
$=-130+(-178)$
$=-308$

The overall vertical change is −308 feet.

83. $19{,}340 - (-512) = 19{,}340 + 512 = 19{,}852$
Mt. Kilimanjaro is 19,852 feet higher than Lake Assal.

85. $y = 180° - 50°$
$y = 180° + (-50°)$
$y = 130°$

87. The sum of -5 and a number is $-5 + x$.

89. Subtract a number from -20 is $-20 - x$.

91.

Month	Monthly Increase or Decrease
February	$-23.7 - (-19.3) = -23.7 + 19.3 = -4.4°$
March	$-21.1 - (-23.7) = -21.1 + 23.7 = 2.6°$
April	$-9.1 - (-21.1) = -9.1 + 21.1 = 12°$
May	$14.4 - (-9.1) = 14.4 + 9.1 = 23.5°$
June	$29.7 - 14.4 = 29.7 + (-14.4) = 15.3°$

93. The largest positive number corresponds to May.

95. answers may vary.

97. $9 - (-7) = 9 + 7 = 16$

99. $10 - 30 = 10 + (-30) = -20$

101. true; answers may vary

103. false; answers may vary.

105. Since 56,875 is less than 87,262, the answer is negative.
$56{,}875 - 87{,}262 = -30{,}387$

Integrated Review

1. The opposite of a positive number is a <u>negative</u> number.

2. The sum of two negative numbers is a <u>negative</u> number.

3. The absolute value of a negative number is a <u>positive</u> number.

4. The absolute value of zero is <u>0</u>.

5. The sum of two positive numbers is a <u>positive</u> number.

6. The sum of a number and its opposite is <u>0</u>.

7. The absolute value of a positive number is a <u>positive</u> number.

8. The reciprocal of a negative number is a <u>negative</u> number.

9. The opposite of $\frac{1}{7}$ is $-\frac{1}{7}$.

The absolute value of $\frac{1}{7}$ is $\frac{1}{7}$.

10. The opposite of $-\frac{12}{5}$ is $\frac{12}{5}$.

The absolute value of $-\frac{12}{5}$ is $\frac{12}{5}$.

11. The number whose opposite is −3 is 3.
The absolute value of 3 is 3.

12. The number whose opposite is $\frac{9}{11}$ is $-\frac{9}{11}$.

The absolute value of $-\frac{9}{11}$ is $\frac{9}{11}$.

13. $-19 + (-23) = -42$

14. $7 - (-3) = 7 + 3 = 10$

15. $-15 + 17 = 2$

16. $-8 - 10 = -8 + (-10) = -18$

17. $18 + (-25) = -7$

18. $-2 + (-37) = -39$

19. $-14 - (-12) = -14 + 12 = -2$

20. $5 - 14 = 5 + (-14) = -9$

21. $4.5 - 7.9 = 4.5 + (-7.9) = -3.4$

22. $-8.6 - 1.2 = -8.6 + (-1.2) = -9.8$

23. $-\frac{3}{4}-\frac{1}{7}=-\frac{3}{4}+\left(-\frac{1}{7}\right)=-\frac{21}{28}+\left(-\frac{4}{28}\right)=-\frac{25}{28}$

24. $\frac{2}{3}-\frac{7}{8}=\frac{2}{3}+\left(-\frac{7}{8}\right)=\frac{16}{24}+\left(-\frac{21}{24}\right)=-\frac{5}{24}$

25.
$$\begin{aligned}-9-(-7)+4-6&=-9+7+4+(-6)\\&=-2+4+(-6)\\&=2+(-6)\\&=-4\end{aligned}$$

26.
$$\begin{aligned}11-20+(-3)-12&=11+(-20)+(-3)+(-12)\\&=-9+(-3)+(-12)\\&=-12+(-12)\\&=-24\end{aligned}$$

27.
$$\begin{aligned}24-6(14-11)&=24-6[14+(-11)]\\&=24-6(3)\\&=24+(-6)(3)\\&=24+(-18)\\&=6\end{aligned}$$

28.
$$\begin{aligned}30-5(10-8)&=30-5[10+(-8)]\\&=30-5(2)\\&=30+(-5)(2)\\&=30+(-10)\\&=20\end{aligned}$$

29.
$$\begin{aligned}(7-17)+4^2&=[7+(-17)]+4^2\\&=-10+4^2\\&=-10+16\\&=6\end{aligned}$$

30.
$$\begin{aligned}9^2+(10-30)&=9^2+[10+(-30)]\\&=9^2+(-20)\\&=81+(-20)\\&=61\end{aligned}$$

31.
$$\begin{aligned}|-9|+3^2+(-4-20)&=|-9|+3^2+[-4+(-20)]\\&=9+9+(-24)\\&=18+(-24)\\&=-6\end{aligned}$$

32.
$$\begin{aligned}|-4-5|+5^2+(-50)&=|-4+(-5)|+5^2+(-50)\\&=|-9|+5^2+(-50)\\&=9+25+(-50)\\&=34+(-50)\\&=-16\end{aligned}$$

33.
$$\begin{aligned}&-7+[(1-2)+(-2-9)]\\&=-7+[(1+(-2))+(-2+(-9))]\\&=-7+[(-1)+(-11)]\\&=-7+(-12)\\&=-19\end{aligned}$$

34.
$$\begin{aligned}&-6+[(-3+7)+(4-15)]\\&=-6+[(-3+7)+(4+(-15))]\\&=-6+[4+(-11)]\\&=-6+(-7)\\&=-13\end{aligned}$$

35. $1 - 5 = 1 + (-5) = -4$

36. $-3 - (-2) = -3 + 2 = -1$

37. $\frac{1}{4} - \left(-\frac{2}{5}\right) = \frac{1}{4} + \frac{2}{5} = \frac{5}{20} + \frac{8}{20} = \frac{13}{20}$

38. $-\frac{5}{8} - \frac{1}{10} = -\frac{5}{8} + \left(-\frac{1}{10}\right) = -\frac{25}{40} + \left(-\frac{4}{40}\right) = -\frac{29}{40}$

39.
$$\begin{aligned} &2(19-17)^3 - 3(-7+9)^2 \\ &= 2[19+(-17)]^3 + (-3)(-7+9)^2 \\ &= 2(2)^3 + (-3)(2)^2 \\ &= 2(8) + (-3)(4) \\ &= 16 + (-12) \\ &= 4 \end{aligned}$$

40.
$$\begin{aligned} &3(10-9)^2 + 6(20-19)^3 \\ &= 3[10+(-9)]^2 + 6[20+(-19)]^3 \\ &= 3(1)^2 + 6(1)^3 \\ &= 3(1) + 6(1) \\ &= 3 + 6 \\ &= 9 \end{aligned}$$

41. Replace x with -2 and y with -1.
$x - y = -2 - (-1) = -2 + 1 = -1$

42. Replace x with -2 and y with -1.
$x + y = -2 + (-1) = -3$

43. Replace y with -1 and z with 9.
$y + z = -1 + 9 = 8$

44. Replace y with -1 and z with 9.
$z - y = 9 - (-1) = 9 + 1 = 10$

45. Replace x with -2, y with -1, and z with 9.
$$\begin{aligned} \frac{|5z - x|}{y - x} &= \frac{|5(9) - (-2)|}{-1 - (-2)} \\ &= \frac{|45 + 2|}{-1 + 2} \\ &= \frac{|47|}{1} \\ &= \frac{47}{1} \\ &= 47 \end{aligned}$$

46. Replace x with -2, y with -1, and z with 9.
$$\begin{aligned} \frac{|-x - y + z|}{2z} &= \frac{|-(-2) - (-1) + 9|}{2(9)} \\ &= \frac{|2 + 1 + 9|}{18} \\ &= \frac{|12|}{18} \\ &= \frac{12}{18} \\ &= \frac{2 \cdot 6}{3 \cdot 6} \\ &= \frac{2}{3} \end{aligned}$$

Section 1.6 Practice Exercises

1. $-8(3) = -24$

2. $5(-30) = -150$

3. $-4(-12) = 48$

4. $-\frac{5}{6} \cdot \frac{1}{4} = -\frac{5 \cdot 1}{6 \cdot 4} = -\frac{5}{24}$

5. $6(-2.3) = -13.8$

6. $-15(-2) = 30$

7. a. $5(0)(-3) = 0(-3) = 0$

b. $(-1)(-6)(-7) = (6)(-7) = -42$

c.
$$\begin{aligned} (-2)(4)(-8)(-1) &= (-8)(-8)(-1) \\ &= 64(-1) \\ &= -64 \end{aligned}$$

8. a. $(-2)^4 = (-2)(-2)(-2)(-2) = 16$

b. $-2^4 = -(2 \cdot 2 \cdot 2 \cdot 2) = -16$

c. $(-1)^5 = (-1)(-1)(-1)(-1)(-1) = -1$

d. $-1^5 = -(1 \cdot 1 \cdot 1 \cdot 1 \cdot 1) = -1$

e. $\left(-\frac{7}{9}\right)^2 = \left(-\frac{7}{9}\right)\left(-\frac{7}{9}\right) = \frac{49}{81}$

9. a. The reciprocal of 13 is $\frac{1}{13}$ since $13 \cdot \frac{1}{13} = 1$.

b. The reciprocal of $\frac{7}{15}$ is $\frac{15}{7}$ since $\frac{7}{15}\cdot\frac{15}{7}=1.$

c. The reciprocal of -5 is $-\frac{1}{5}$ since $-5\cdot -\frac{1}{5}=1.$

d. The reciprocal of $-\frac{8}{11}$ is $-\frac{11}{8}$ since $-\frac{8}{11}\cdot -\frac{11}{8}=1.$

e. The reciprocal of 7.9 is $\frac{1}{7.9}$ since $7.9\cdot\frac{1}{7.9}=1.$

10. a. $-12\div 4=-12\cdot\frac{1}{4}=-3$

b. $\frac{-20}{-10}=-20\cdot -\frac{1}{10}=2$

c. $\frac{36}{-4}=36\cdot -\frac{1}{4}=-9$

11. a. $\frac{-25}{5}=-5$

b. $\frac{-48}{-6}=8$

c. $\frac{50}{-2}=-25$

d. $\frac{-72}{0.2}=-360$

12. $-\frac{5}{9}\div\frac{2}{3}=-\frac{5}{9}\cdot\frac{3}{2}=-\frac{15}{18}=-\frac{3\cdot 5}{3\cdot 6}=-\frac{5}{6}$

13. $-\frac{2}{7}\div\left(-\frac{1}{5}\right)=-\frac{2}{7}\cdot\left(-\frac{5}{1}\right)=\frac{10}{7}$

14. a. $\frac{-7}{0}$ is undefined.

b. $\frac{0}{-2}=0$

15. a. $\frac{0(-5)}{3}=\frac{0}{3}=0$

b. $-3(-9)-4(-4)=27-(-16)=27+16=43$

c.
$$\begin{aligned}&(-3)^2+2\big[(5-15)-|-4-1|\big]\\&=(-3)^2+2\big[(-10)-|-5|\big]\\&=(-3)^2+2[-10-5]\\&=(-3)^2+2(-15)\\&=9+(-30)\\&=-21\end{aligned}$$

d. $\frac{-7(-4)+2}{-10-(-5)}=\frac{28+2}{-10+5}=\frac{30}{-5}=-6$

e.
$$\begin{aligned}\frac{5(-2)^3+52}{-4+1}&=\frac{5(-8)+52}{-3}\\&=\frac{-40+52}{-3}\\&=\frac{12}{-3}\\&=-4\end{aligned}$$

16. a. Replace x with -1 and y with -5.
$\frac{3y}{45x}=\frac{3(-5)}{45(-1)}=\frac{-15}{-45}=\frac{15\cdot 1}{15\cdot 3}=\frac{1}{3}$

b. Replace x with -1 and y with -5.
$$\begin{aligned}x^2-y^3&=(-1)^2-(-5)^3\\&=1-(-125)\\&=1+125\\&=126\end{aligned}$$

c. Replace x with -1 and y with -5.
$\frac{x+y}{3x}=\frac{-1+(-5)}{3(-1)}=\frac{-6}{-3}=2$

17. $\frac{x}{4}-3=x+3$
$\frac{-8}{4}-3 \stackrel{?}{=} -8+3$
$-2-3 \stackrel{?}{=} -5$
$-5=-5$ True
-8 is a solution.

18. total score $= 4 \cdot (-13) = -52$
The card player's total score was -52.

Calculator Explorations

1. $-38(26-27)=38$

2. $-59(-8)+1726=2198$

3. $134+25(68-91)=-441$

4. $45(32)-8(218)=-304$

5. $\frac{-50(294)}{175-205}=490$

6. $\frac{-444-444.8}{-181-(-181)}$ is undefined.

7. $9^5-4550=54{,}499$

8. $5^8-6259=384{,}366$

9. $(-125)^2=15{,}625$

10. $-125^2=-15{,}625$

Vocabulary, Readiness & Video Check 1.6

1. The product of a negative number and a positive number is a <u>negative</u> number.

2. The product of two negative numbers is a <u>positive</u> number.

3. The quotient of two negative numbers is a <u>positive</u> number.

4. The quotient of a negative number and a positive number is a <u>negative</u> number.

5. The product of a negative number and zero is <u>0</u>.

6. The reciprocal of a negative number is a <u>negative</u> number.

7. The quotient of 0 and a negative number is <u>0</u>.

8. The quotient of a negative number and 0 is <u>undefined</u>.

9. The parentheses, or lack of them, determine the base of the expression. In Example 6, $(-2)^4$, the base is -2 and all of -2 is raised to the 4th power. In Example 7, -2^4, the base is 2 and only 2 is raised to the 4th power.

10. Remember, the product of a number and its reciprocal is 1, *not* -1. $\frac{2}{3}\cdot\frac{3}{2}=1$, as needed.

11. Yes; because division of real numbers is defined in terms of multiplication.

12. The replacement values are negative and one of them will be squared. If a negative number is squared, it must be placed in parentheses so the entire value, including the negative, is squared.

13. Yes; a true statement results when x is replaced with 5.

14. The football team lost 4 yards on each play and a loss of yardage is represented by a negative number.

Exercise Set 1.6

1. $-6(4)=-24$

3. $2(-1)=-2$

5. $-5(-10)=50$

7. $-3\cdot 15=-45$

9. $-\frac{1}{2}\left(-\frac{3}{5}\right)=\frac{3}{10}$

11. $5(-1.4)=-7$

13. $(-1)(-3)(-5)=3(-5)=-15$

15. $(2)(-1)(-3)(0)=-2(-3)(0)=6(0)=0$

17. $(-4)^2=(-4)(-4)=16$

19. $-4^2=-4\cdot 4=-16$

21. $\left(-\frac{3}{4}\right)^2 = \left(-\frac{3}{4}\right)\left(-\frac{3}{4}\right) = \frac{9}{16}$

23. $-0.7^2 = -0.7 \cdot 0.7 = -0.49$

25. The reciprocal of $\frac{2}{3}$ is $\frac{3}{2}$ since $\frac{2}{3} \cdot \frac{3}{2} = 1$.

27. The reciprocal of -14 is $-\frac{1}{14}$ since $-14 \cdot \left(-\frac{1}{14}\right) = 1$.

29. The reciprocal of $-\frac{3}{11}$ is $-\frac{11}{3}$ since $-\frac{3}{11} \cdot \left(-\frac{11}{3}\right) = 1$.

31. The reciprocal of 0.2 is $\frac{1}{0.2}$ since $0.2\left(\frac{1}{0.2}\right) = 1$.

33. $\frac{18}{-2} = -9$

35. $-48 \div 12 = -48 \cdot \frac{1}{12} = -4$

37. $\frac{0}{-4} = 0$

39. $\frac{5}{0}$ is undefined.

41. $\frac{6}{7} \div \left(-\frac{1}{3}\right) = \frac{6}{7} \cdot \left(-\frac{3}{1}\right) = -\frac{18}{7}$

43. $-3.2 \div -0.02 = \frac{-3.2}{-0.02} = 160$

45. $(-8)(-8) = 64$

47. $\frac{2}{3}\left(-\frac{4}{9}\right) = -\frac{8}{27}$

49. $\frac{-12}{-4} = 3$

51. $\frac{30}{-2} = -15$

53. $(-5)^3 = (-5)(-5)(-5) = 25(-5) = -125$

55. $(-0.2)^3 = (-0.2)(-0.2)(-0.2)$
$= 0.04(-0.2)$
$= -0.008$

57. $\left(-\frac{3}{4}\right)\left(-\frac{8}{9}\right) = \frac{24}{36} = \frac{2 \cdot 12}{3 \cdot 12} = \frac{2}{3}$

59. $-\frac{5}{9} \div \left(-\frac{3}{4}\right) = -\frac{5}{9} \cdot \left(-\frac{4}{3}\right) = \frac{20}{27}$

61. $-2.1(-0.4) = 0.84$

63. $\frac{-48}{1.2} = -40$

65. $(-3)^4 = (-3)(-3)(-3)(-3)$
$= 9(-3)(-3)$
$= -27(-3)$
$= 81$

67. $-1^7 = -1 \cdot 1 \cdot 1 \cdot 1 \cdot 1 \cdot 1 \cdot 1 = -1$

69. $-11 \cdot 11 = -121$

71. $-\frac{4}{9} \div \frac{4}{9} = -\frac{4}{9} \cdot \frac{9}{4} = -\frac{36}{36} = -1$

73. $-9 - 10 = -9 + (-10) = -19$

75. $-9(-10) = 90$

77. $7(-12) = -84$

79. $7 + (-12) = -5$

81. $\frac{-9(-3)}{-6} = \frac{27}{-6} = -\frac{9 \cdot 3}{2 \cdot 3} = -\frac{9}{2}$

83. $-3(2 - 8) = -3[2 + (-8)] = -3(-6) = 18$

85. $-7(-2) - 3(-1) = 14 - (-3) = 14 + 3 = 17$

87. $2^2 - 3[(2-8)-(-6-8)]$
$= 2^2 - 3[(-6)-(-14)]$
$= 2^2 - 3[-6+14]$
$= 4 - 3(8)$
$= 4 - 24$
$= 4 + (-24)$
$= -20$

89. $\frac{-6^2+4}{-2} = \frac{-36+4}{-2} = \frac{-32}{-2} = 16$

91. $\frac{-3-5^2}{2(-7)} = \frac{-3-25}{-14}$
$= \frac{-3+(-25)}{-14}$
$= \frac{-28}{-14}$
$= \frac{-14 \cdot 2}{-14 \cdot 1}$
$= 2$

93. $\frac{22+(3)(-2)^2}{-5-2} = \frac{22+3(4)}{-5+(-2)}$
$= \frac{22+12}{-7}$
$= \frac{34}{-7}$
$= -\frac{34}{7}$

95. $\frac{(-4)^2-16}{4-12} = \frac{16-16}{4-12} = \frac{16+(-16)}{4+(-12)} = \frac{0}{-8} = 0$

97. $\frac{6-2(-3)}{4-3(-2)} = \frac{6-(-6)}{4-(-6)} = \frac{6+6}{4+6} = \frac{12}{10} = \frac{2 \cdot 6}{2 \cdot 5} = \frac{6}{5}$

99. $\frac{|5-9|+|10-15|}{|2(-3)|} = \frac{|5+(-9)|+|10+(-15)|}{|-6|}$
$= \frac{|-4|+|-5|}{|-6|}$
$= \frac{4+5}{6}$
$= \frac{9}{6}$
$= \frac{3 \cdot 3}{3 \cdot 2}$
$= \frac{3}{2}$

101. $\frac{-7(-1)+(-3)4}{(-2)(5)+(-6)(-8)} = \frac{7+(-12)}{-10+48} = \frac{-5}{38} = -\frac{5}{38}$

103. Replace x with -5 and y with -3.
$\frac{2x-5}{y-2} = \frac{2(-5)-5}{-3-2}$
$= \frac{-10-5}{-3-2}$
$= \frac{-10+(-5)}{-3+(-2)}$
$= \frac{-15}{-5}$
$= 3$

105. Replace x with -5 and y with -3.
$\frac{6-y}{x-4} = \frac{6-(-3)}{-5-4} = \frac{6+3}{-5+(-4)} = \frac{9}{-9} = -1$

107. Replace x with -5 and y with -3.
$\frac{4-2x}{y+3} = \frac{4-2(-5)}{-3+3}$
$= \frac{4-(-10)}{0}$
$= \frac{4+10}{0}$
$= \frac{14}{0}$ is undefined.

109. Replace x with -5 and y with -3.
$\frac{x^2+y}{3y} = \frac{(-5)^2+(-3)}{3(-3)}$
$= \frac{25+(-3)}{-9}$
$= \frac{22}{-9}$
$= -\frac{22}{9}$

111. $-3x-5=-20$
$-3(5)-5 \stackrel{?}{=} -20$
$-15-5 \stackrel{?}{=} -20$
$-15+(-5) \stackrel{?}{=} -20$
$-20=-20$ True
Since the result is true, 5 is a solution of the given equation.

113. $\frac{x}{5}+2=-1$

$\frac{15}{5}+2 \stackrel{?}{=} -1$

$3+2 \stackrel{?}{=} -1$

$5=-1$ False

Since the result is false, 15 is not a solution of the given equation.

115. $\frac{x-3}{7}=-2$

$\frac{-11-3}{7} \stackrel{?}{=} -2$

$\frac{-11+(-3)}{7} \stackrel{?}{=} -2$

$\frac{-14}{7} \stackrel{?}{=} -2$

$-2=-2$ True

Since the result is true, −11 is a solution of the given equation.

117. The product of −71 and a number is $-71 \cdot x$ or $-71x$.

119. Subtract a number from −16 is $-16 - x$.

121. −29 increased by a number is $-29 + x$.

123. Divide a number by −33 is $\frac{x}{-33}$ or $x \div (-33)$.

125. A loss of 4 yards is represented by −4.
$3 \cdot (-4) = -12$
The team had a total loss of 12 yards.

127. Each move of 20 feet down is represented by −20.
$5 \cdot (-20) = -100$
The diver is at a depth of 100 feet.

129. True since the product of an odd number of negative numbers is negative.

131. False since the product of an even number of negative integers is positive.

133. $2(-81) = -162$
The surface temperature of Jupiter is −162°F.

135. A loss of $33 million can be represented by −$33 million.

$\frac{-\$33 \text{ million}}{3} = -\$11 \text{ million}$

The loss would have been −$11 million per month.

137. answers may vary

139. 1 and −1; answers may vary

141. $\frac{0}{5}-7=0-7=0+(-7)=-7$

143. $-8(-5) + (-1) = 40 + (-1) = 39$

Section 1.7 Practice Exercises

1. a. $7 \cdot y = y \cdot 7$

b. $4 + x = x + 4$

2. a. $5 \cdot (-3 \cdot 6) = (5 \cdot -3) \cdot 6$

b. $(-2 + 7) + 3 = -2 + (7 + 3)$

c. $(q + r) + 17 = q + (r + 17)$

d. $(ab) \cdot 21 = a \cdot (b \cdot 21)$

3. Since the order of two numbers was changed but their grouping was not, the statement is true by the commutative property of multiplication.

4. Since the grouping of the numbers was changed and their order was not, the statement is true by the associative property of addition.

5. $$\begin{aligned}(-3+x)+17 &= -3+(x+17)\\ &= -3+(17+x)\\ &= (-3+17)+x\\ &= 14+x\end{aligned}$$

6. $4(5x) = (4 \cdot 5) \cdot x = 20x$

7. $5(x + y) = 5(x) + 5(y) = 5x + 5y$

8. $-3(2 + 7x) = -3(2) + (-3)(7x) = -6 - 21x$

9. $$\begin{aligned}4(x+6y-2z) &= 4(x)+4(6y)-4(2z)\\ &= 4x+24y-8z\end{aligned}$$

10. $-1(3 - a) = (-1)(3) - (-1)(a) = -3 + a$

11. $$\begin{aligned}-(8+a-b) &= -1(8+a-b)\\ &= (-1)(8)+(-1)(a)-(-1)(b)\\ &= -8-a+b\end{aligned}$$

12. $$\begin{aligned}\frac{1}{2}(2x+4)+9 &= \frac{1}{2}(2x)+\frac{1}{2}(4)+9\\ &= 1x+2+9\\ &= x+11\end{aligned}$$

13. $9 \cdot 3 + 9 \cdot y = 9(3 + y)$

14. $4x + 4y = 4(x + y)$

15. $7(a + b) = 7 \cdot a + 7 \cdot b$ illustrates the distributive property.

16. $12 + y = y + 12$ illustrates the commutative property of addition.

17. $-4 \cdot (6 \cdot x) = (-4 \cdot 6) \cdot x$ illustrates the associative property of multiplication.

18. $6 + (z + 2) = 6 + (2 + z)$ illustrates the commutative property of addition.

19. $3\left(\frac{1}{3}\right) = 1$ illustrates the multiplicative inverse property.

20. $(x + 0) + 23 = x + 23$ illustrates the identity element for addition.

21. $(7 \cdot y) \cdot 10 = y \cdot (7 \cdot 10)$ illustrates the commutative and associative properties of multiplication.

Vocabulary, Readiness & Video Check 1.7

1. $x + 5 = 5 + x$ is a true statement by the commutative property of addition.

2. $x \cdot 5 = 5 \cdot x$ is a true statement by the commutative property of multiplication.

3. $3(y + 6) = 3 \cdot y + 3 \cdot 6$ is true by the distributive property.

4. $2 \cdot (x \cdot y) = (2 \cdot x) \cdot y$ is a true statement by the associative property of multiplication.

5. $x + (7 + y) = (x + 7) + y$ is a true statement by the associative property of addition.

6. The numbers $-\frac{2}{3}$ and $-\frac{3}{2}$ are called reciprocals or multiplicative inverses.

7. The numbers $-\frac{2}{3}$ and $\frac{2}{3}$ are called opposites or additive inverses.

8. order; grouping

9. 2 is outside the parentheses, so the point is made that you should only distribute the −9 to the terms within the parentheses and not also to the 2.

10. The identity element for addition is 0 because if we add 0 to any real number, the result is that real number.
The identity element for multiplication is 1 because any real number times 1 gives a result of that original real number.

Exercise Set 1.7

1. $x + 16 = 16 + x$ by the commutative property of addition.

3. $-4 \cdot y = y \cdot (-4)$ by the commutative property of multiplication.

5. $xy = yx$ by the commutative property of multiplication.

7. $2x + 13 = 13 + 2x$ by the commutative property of addition.

9. $(xy) \cdot z = x \cdot (yz)$ by the associative property of multiplication.

11. $2 + (a + b) = (2 + a) + b$ by the associative property of addition.

13. $4 \cdot (ab) = 4a \cdot (b)$ by the associative property of multiplication.

15. $(a + b) + c = a + (b + c)$ by the associative property of addition.

17. $8 + (9 + b) = (8 + 9) + b = 17 + b$

19. $4(6y) = (4 \cdot 6)y = 24y$

21. $\frac{1}{5}(5y) = \left(\frac{1}{5} \cdot 5\right)y = 1y = y$

23.
$$\begin{aligned}(13 + a) + 13 &= (a + 13) + 13 \\ &= a + (13 + 13) \\ &= a + 26\end{aligned}$$

25. $-9(8x) = (-9 \cdot 8)x = -72x$

27. $\frac{3}{4}\left(\frac{4}{3}s\right) = \left(\frac{3}{4} \cdot \frac{4}{3}\right)s = 1s = s$

29. $-\frac{1}{2}(5x) = \left(-\frac{1}{2}\cdot 5\right)x = -\frac{5}{2}x$

31. $4(x + y) = 4(x) + 4(y) = 4x + 4y$

33. $9(x - 6) = 9(x) - 9(6) = 9x - 54$

35. $2(3x + 5) = 2(3x) + 2(5) = 6x + 10$

37. $7(4x - 3) = 7(4x) - 7(3) = 28x - 21$

39. $3(6 + x) = 3(6) + 3(x) = 18 + 3x$

41. $-2(y - z) = -2(y) - (-2)z = -2y + 2z$

43. $-\frac{1}{3}(3y+5) = -\frac{1}{3}(3y) - \frac{1}{3}(5) = -y - \frac{5}{3}$

45. $5(x+4m+2) = 5(x)+5(4m)+5(2)$
$= 5x+20m+10$

47. $-4(1-2m+n)+4 = -4(1)-4(-2m)-4(n)+4$
$= -4+8m-4n+4$
$= (-4+4)+8m-4n$
$= 0+8m-4n$
$= 8m-4n$

49. $-(5x+2) = -1(5x+2)$
$= -1(5x)+(-1)(2)$
$= -5x-2$

51. $-(r-3-7p) = -1(r-3-7p)$
$= -1(r)-1(-3)-1(-7p)$
$= -r+3+7p$

53. $\frac{1}{2}(6x+7)+\frac{1}{2} = \frac{1}{2}(6x)+\frac{1}{2}(7)+\frac{1}{2}$
$= \left(\frac{1}{2}\cdot 6\right)x+\frac{7}{2}+\frac{1}{2}$
$= 3x+\frac{8}{2}$
$= 3x+4$

55. $-\frac{1}{3}(3x-9y) = -\frac{1}{3}(3x)-\frac{1}{3}(-9y) = -x+3y$

57. $3(2r+5)-7 = 3(2r)+3(5)-7$
$= 6r+15-7$
$= 6r+8$

59. $-9(4x+8)+2 = -9(4x)-9(8)+2$
$= -36x-72+2$
$= -36x-70$

61. $-0.4(4x+5)-0.5 = -0.4(4x)+(-0.4)(5)-0.5$
$= -1.6x-2-0.5$
$= -1.6x-2.5$

63. $4 \cdot 1 + 4 \cdot y = 4(1 + y)$

65. $11x + 11y = 11 \cdot x + 11 \cdot y = 11(x + y)$

67. $(-1) \cdot 5 + (-1) \cdot x = -1(5 + x) = -(5 + x)$

69. $30a + 30b = 30 \cdot a + 30 \cdot b = 30(a + b)$

71. $3 \cdot 5 = 5 \cdot 3$ illustrates the commutative property of multiplication.

73. $2 + (x + 5) = (2 + x) + 5$ illustrates the associative property of addition.

75. $(x + 9) + 3 = (9 + x) + 3$ illustrates the commutative property of addition.

77. $(4 \cdot y) \cdot 9 = 4 \cdot (y \cdot 9)$ illustrates the associative property of multiplication.

79. $0 + 6 = 6$ illustrates the identity property of addition.

81. $-4(y + 7) = -4 \cdot y + (-4) \cdot 7$ illustrates the distributive property.

83. $6\cdot\frac{1}{6} = 1$ illustrates the multiplicative inverse property.

85. $-6 \cdot 1 = -6$ illustrates the identity element for multiplication.

87. The opposite of 8 is -8.

The reciprocal of 8 is $\frac{1}{8}$.

89. The opposite of x is $-x$.

The reciprocal of x is $\frac{1}{x}$.

91. The expression is the reciprocal of $\frac{1}{2x}$ or $2x$.

The opposite of $2x$ is $-2x$.

93. False; the opposite of $-\frac{a}{2}$ is $\frac{a}{2}$. $-\frac{2}{a}$ is the reciprocal of $-\frac{a}{2}$.

95. "Taking a test" and "studying for the test" are not commutative, since the order in which they are performed affects the outcome.

97. "Putting on your left shoe" and "putting on your right shoe" are commutative, since the order in which they are performed does not affect the outcome.

99. "Mowing the lawn" and "trimming the hedges" are commutative, since the order in which they are performed does not affect the outcome.

101. "Feeding the dog" and "feeding the cat" are commutative, since the order in which they are performed does not affect the outcome.

103. **a.** The property illustrated is the commutative property of addition since the order in which they are added changed.

b. The property illustrated is the commutative property of addition since the order in which they are added changed.

c. The property illustrated is the associative property of addition since the grouping of addition changed.

105. answers may vary

107. answers may vary

Section 1.8 Practice Exercises

1. **a.** The numerical coefficient of $-4x$ is -4.

b. The numerical coefficient of $15y^3$ is 15.

c. The numerical coefficient of x is 1, since $x = 1x$.

d. The numerical coefficient of $-y$ is -1, since $-y = -1y$.

e. The numerical coefficient of $\frac{z}{4}$ is $\frac{1}{4}$, since $\frac{z}{4}$ is $\frac{1}{4}\cdot z$.

2. **a.** $7x^2$ and $-6x^3$ are unlike terms, since the exponents on x are not the same.

b. $3x^2y^2$, $-x^2y^2$, and $4x^2y^2$ are like terms, since each variable and its exponent match.

c. $-5ab$ and $3ba$ are like terms, since $ab = ba$ by the commutative property.

d. $2x^3$ and $4y^3$ are unlike terms, since the variables are not the same.

e. $-7m^4$ and $7m^4$ are like terms, since the variable and its exponent match.

3. **a.** $9y - 4y = (9 - 4)y = 5y$

b. $11x^2 + x^2 = (11+1)x^2 = 12x^2$

c. $$\begin{aligned}5y-3x+4x &= 5y+(-3+4)x\\ &= 5y+1x\\ &= 5y+x\end{aligned}$$

d. $14m^2 - m^2 + 3m^2 = (14-1+3)m^2 = 16m^2$

4. $7y + 2y + 6 + 10 = (7 + 2)y + (6 + 10) = 9y + 16$

5. $$\begin{aligned}-2x+4+x-11 &= -2x+x+4-11\\ &= (-2+1)x+(4-11)\\ &= -x-7\end{aligned}$$

6. The terms $3z$ and $-3z^2$ cannot be combined because they are unlike terms.

7. $8.9y + 4.2y - 3 = (8.9 + 4.2)y - 3 = 13.1y - 3$

8. $3(11y + 6) = 3(11y) + 3(6) = 33y + 18$

9. $$\begin{aligned}-4(x+0.2y-3) &= -4(x)+(-4)(0.2y)-(-4)(3)\\ &= -4x-0.8y+12\end{aligned}$$

10. $$\begin{aligned}&-(3x+2y+z-1)\\ &= -1(3x+2y+z-1)\\ &= -1(3x)+(-1)(2y)+(-1)(z)-(-1)(1)\\ &= -3x-2y-z+1\end{aligned}$$

11. $4(4x - 6) + 20 = 16x - 24 + 20 = 16x - 4$

12. $$\begin{aligned}5-(3x+9)+6x &= 5-3x-9+6x\\ &= -3x+6x+5-9\\ &= 3x-4\end{aligned}$$

13. $-3(7x+1)-(4x-2)=-21x-3-4x+2$
$=-25x-1$

14. $8+11(2y-9)=8+22y-99=-91+22y$

15. $(4x-3)-(9x-10)=4x-3-9x+10=-5x+7$

16. Three times a number subtracted from 10 is written as $10 - 3x$. This expression cannot be simplified.

17. The sum of a number and 2, divided by 5, is written as $(x + 2) \div 5$ or $\frac{x+2}{5}$. This expression cannot be simplified.

18. Three times the sum of a number and 6, is written as $3(x + 6)$.
$3(x + 6) = 3 \cdot x + 3 \cdot 6 = 3x + 18$

19. Seven times the difference of a number and 4 is written as $7(x - 4)$.
$7(x - 4) = 7x - 28$

Vocabulary, Readiness & Video Check 1.8

1. $14y^2+2x-23$ is called an expression while $14y^2$, $2x$, and -23 are each called a term.

2. To multiply $3(-7x + 1)$, we use the distributive property.

3. To simplify an expression like $y + 7y$, we combine like terms.

4. The term z has an understood numerical coefficient of 1.

5. The terms $-x$ and $5x$ are like terms and the terms $5x$ and $5y$ are unlike terms.

6. For the term $-3x^2y$, -3 is called the numerical coefficient.

7. Although these terms have exactly the same variables, the exponents on each are not exactly the same—the exponents on x differ in each term.

8. distributive property

9. -1

10. The sum of 5 times a number and -2, added to 7 times the number; $5x + (-2) + 7x$; because there are like terms

Exercise Set 1.8

1. The numerical coefficient of $-7y$ is -7.

3. The numerical coefficient of x is 1, since $x = 1x$.

5. The numerical coefficient of $17x^2y$ is 17.

7. $5y$ and $-y$ are like terms, since the variable and its exponent match.

9. $2z$ and $3z^2$ are unlike terms, since the exponents on z are not the same.

11. $8wz$ and $\frac{1}{7}zw$ are like terms, since $wz = zw$ by the commutative property.

13. $7y + 8y = (7 + 8)y = 15y$

15. $8w-w+6w=8w-1w+6w$
$=(8-1+6)w$
$=13w$

17. $3b-5-10b-4=3b-10b-5-4$
$=(3-10)b+(-5-4)$
$=-7b-9$

19. $m-4m+2m-6=(1-4+2)m-6$
$=-1m-6$
$=-m-6$

21. $5g-3-5-5g=5g-5g-3-5$
$=(5-5)g+(-3-5)$
$=0g-8$
$=-8$

23. $6.2x-4+x-1.2=6.2x+1x-4-1.2$
$=(6.2+1)x+(-4-1.2)$
$=7.2x-5.2$

25. $2k-k-6=2k-1k-6$
$=(2-1)k-6$
$=1k-6$
$=k-6$

27. $-9x+4x+18-10x=-9x+4x-10x+18$
$=(-9+4-10)x+18$
$=-15x+18$

29. $6x-5x+x-3+2x=6x-5x+x+2x-3$
$=(6-5+1+2)x-3$
$=4x-3$

31. $7x^2+8x^2-10x^2=(7+8-10)x^2=5x^2$

33. $3.4m-4-3.4m-7=3.4m-3.4m-4-7$
$=(3.4-3.4)m+(-4-7)$
$=0m-11$
$=-11$

35. $6x+0.5-4.3x-0.4x+3$
$=6x-4.3x-0.4x+0.5+3$
$=(6-4.3-0.4)x+(0.5+3)$
$=1.3x+3.5$

37. $5(y+4)=5(y)+5(4)=5y+20$

39. $-2(x+2)=-2(x)-2(2)=-2x-4$

41. $-5(2x-3y+6)$
$=-5(2x)-(-5)(3y)+(-5)(6)$
$=-10x+15y-30$

43. $-(3x-2y+1)=-1(3x-2y+1)$
$=-1(3x)-1(-2y)-1(1)$
$=-3x+2y-1$

45. $7(d-3)+10=7d-21+10=7d-11$

47. $-4(3y-4)+12y=-12y+16+12y$
$=0y+16$
$=16$

49. $3(2x-5)-5(x-4)=6x-15-5x+20$
$=6x-5x-15+20$
$=x+5$

51. $-2(3x-4)+7x-6=-6x+8+7x-6$
$=-6x+7x+8-6$
$=1x+2$
$=x+2$

53. $5k-(3k-10)=5k-3k+10=2k+10$

55. $(3x+4)-(6x-1)=3x+4-6x+1$
$=3x-6x+4+1$
$=-3x+5$

57. $5(x+2)-(3x-4)=5x+10-3x+4$
$=5x-3x+10+4$
$=2x+14$

59. $\frac{1}{3}(7y-1)+\frac{1}{6}(4y+7)=\frac{7}{3}y-\frac{1}{3}+\frac{4}{6}y+\frac{7}{6}$
$=\frac{7}{3}y+\frac{2}{3}y-\frac{2}{6}+\frac{7}{6}$
$=\frac{9}{3}y+\frac{5}{6}$
$=3y+\frac{5}{6}$

61. $2+4(6x-6)=2+24x-24$
$=2-24+24x$
$=-22+24x$

63. $0.5(m+2)+0.4m=0.5m+1+0.4m$
$=0.5m+0.4m+1$
$=0.9m+1$

65. $10-3(2x+3y)=10-6x-9y$

67. $6(3x-6)-2(x+1)-17x$
$=18x-36-2x-2-17x$
$=18x-2x-17x-36-2$
$=-1x-38$
$=-x-38$

69. $\frac{1}{2}(12x-4)-(x+5)=6x-2-x-5$
$=6x-x-2-5$
$=5x-7$

71. $(4x-10)+(6x+7)=4x-10+6x+7$
$=4x+6x-10+7$
$=10x-3$

73. $(3x-8)-(7x+1)=3x-8-7x-1$
$=3x-7x-8-1$
$=-4x-9$

75. $(m-9)-(5m-6)=m-9-5m+6$
$=m-5m-9+6$
$=-4m-3$

77. Twice a number, decreased by four is written as $2x-4$.

79. Three-fourths of a number, increased by twelve is written as $\frac{3}{4}x+12$.

81. The sum of 5 times a number and −2, added to 7 times the number is written as $5x+(-2)+7x=5x+7x-2=12x-2$.

83. Eight times the sum of a number and six is written as $8(x + 6) = 8x + 48$.

85. Double a number minus the sum of the number and ten is written as
$2x - (x + 10) = 2x - x - 10 = x - 10$.

87. Since 1 cone balances 1 cube and 1 cylinder balances 2 cubes, 1 cone and 1 cylinder balances $1 + 2 = 3$ cubes. The scale shown is balanced.

89. Since 1 cylinder balances 2 cubes, the left side is equivalent to 5 cubes. Since 1 cone balances 1 cube, the right side is equivalent to 5 cubes. The scale is balanced.

91. answers may vary

93.
$$\begin{aligned} 2[(4x-1)+5x] &= 2(4x-1+5x) \\ &= 2(9x-1) \\ &= 18x-2 \end{aligned}$$
The perimeter is $(18x - 2)$ feet.

95. The length of the first board in inches is $12(x + 2)$.
$$\begin{aligned} 12(x+2)+(3x-1) &= 12x+24+3x-1 \\ &= 12x+3x+24-1 \\ &= 15x+23 \end{aligned}$$
The total length is $(15x + 23)$ inches.

97. answers may vary

Chapter 1 Vocabulary Check

1. The symbols ≠, <, and > are called inequality symbols.

2. A mathematical statement that two expressions are equal is called and equation.

3. The absolute value of a number is the distance between that number and 0 on a number line.

4. A symbol used to represent a number is called a variable.

5. Two numbers that are the same distance from 0 but lie on opposite sides of 0 are called opposites.

6. The number in a fraction above the fraction bar is called the numerator.

7. A solution of an equation is a value for the variable that makes the equation a true statement.

8. Two numbers whose product is 1 are called reciprocals.

9. In 2^3, the 2 is called the base and the 3 is called the exponent.

10. The numerical coefficient of a term is its numerical factor.

11. The number in a fraction below the fraction bar is called the denominator.

12. Parentheses and brackets are examples of grouping symbols.

13. A term is a number or the product of a number and variables raised to powers.

14. Terms with the same variables raised to the same powers are called like terms.

15. If terms are not like terms, then they are unlike terms.

Chapter 1 Review

1. Since 8 is to the left of 10 on a number line, $8 < 10$.

2. Since 7 is to the right of 2 on a number line, $7 > 2$.

3. Since −4 is to the right of −5 on a number line, $-4 > -5$.

4. Since $\frac{12}{2} = 6$ is to the right of −8 on a number line, $\frac{12}{2} > -8$.

5. Since $|-7| = 7$ is to the left of $|-8| = 8$ on a number line, $|-7| < |-8|$.

6. Since $|-9| = 9$ is to the right of −9 on a number line, $|-9| > -9$.

7. $-|-1| = -1$

8. Since $|-14| = 14$ and $-(-14) = 14$, $|-14| = -(-14)$.

9. Since 1.2 is to the right of 1.02 on a number line, $1.2 > 1.02$.

10. Since $-\frac{3}{2} = -\frac{6}{4}$ and $-\frac{6}{4}$ is to the left of $-\frac{3}{4}$ on a number line, $-\frac{3}{2} < -\frac{3}{4}$.

11. Four is greater than or equal to negative three is written as $4 \geq -3$.

12. Six is not equal to five is written as $6 \neq 5$.

13. 0.03 is less than 0.3 is written as $0.03 < 0.3$.

14. Since 1729 is to the left of 2870 on a number line, $1729 < 2870$.

15. **a.** The natural numbers are 1, 3.

b. The whole numbers are 0, 1, 3.

c. The integers are −6, 0, 1, 3.

d. The rational numbers are −6, 0, 1, $1\frac{1}{2}$, 3, 9.62.

e. The irrational number is π.

f. The real numbers are all numbers in the set.

16. **a.** The natural numbers are 2, 5.

b. The whole numbers are 2, 5.

c. The integers are −3, 2, 5.

d. The rational numbers are −3, −1.6, 2, 5, $\frac{11}{2}$, 15.1.

e. The irrational numbers are $\sqrt{5}$, 2π.

f. The real numbers are all numbers in the set.

17. Since $-4 < -2$, the most negative number is −4 which corresponds to Friday. Thus, Friday showed the greatest loss.

18. The greatest positive number is +5 which corresponds to Wednesday. Thus, Wednesday showed the greatest gain.

19. $6 \cdot 3^2 + 2 \cdot 8 = 6 \cdot 9 + 2 \cdot 8 = 54 + 16 = 70$
The answer is c.

20. $68 - 5 \cdot 2^3 = 68 - 5 \cdot 8 = 68 - 40 = 68 + (-40) = 28$
The answer is b.

21.
$$\begin{aligned} 3(1+2\cdot 5)+4 &= 3(1+10)+4 \\ &= 3(11)+4 \\ &= 33+4 \\ &= 37 \end{aligned}$$

22.
$$\begin{aligned} 8+3(2\cdot 6-1) &= 8+3(12-1) \\ &= 8+3[12+(-1)] \\ &= 8+3(11) \\ &= 8+33 \\ &= 41 \end{aligned}$$

23.
$$\begin{aligned} \frac{4+|6-2|+8^2}{4+6\cdot 4} &= \frac{4+|6+(-2)|+8^2}{4+6\cdot 4} \\ &= \frac{4+|4|+8^2}{4+6\cdot 4} \\ &= \frac{4+4+64}{4+24} \\ &= \frac{72}{28} \\ &= \frac{18\cdot 4}{7\cdot 4} \\ &= \frac{18}{7} \end{aligned}$$

24.
$$\begin{aligned} 5[3(2+5)-5] &= 5[3(7)-5] \\ &= 5[21-5] \\ &= 5[21+(-5)] \\ &= 5(16) \\ &= 80 \end{aligned}$$

25. The difference of twenty and twelve is equal to the product of two and four is written as $20 - 12 = 2 \cdot 4$.

26. The quotient of nine and two is greater than negative five is written as $\frac{9}{2} > -5$.

27. Replace x with 6 and y with 2.
$2x + 3y = 2(6) + 3(2) = 12 + 6 = 18$

28. Replace x with 6, y with 2, and z with 8.
$x(y + 2z) = 6[2 + 2(8)] = 6(2 + 16) = 6(18) = 108$

29. Replace x with 6, y with 2, and z with 8.
$\frac{x}{y} + \frac{z}{2y} = \frac{6}{2} + \frac{8}{2(2)} = \frac{6}{2} + \frac{8}{4} = 3 + 2 = 5$

30. Replace x with 6 and y with 2.
$$\begin{aligned} x^2 - 3y^2 &= 6^2 - 3(2)^2 \\ &= 36 - 3(4) \\ &= 36 - 12 \\ &= 36 + (-12) \\ &= 24 \end{aligned}$$

31. Replace a with 37 and b with 80.
$$\begin{aligned} 180 - a - b &= 180 - 37 - 80 \\ &= 180 + (-37) + (-80) \\ &= 143 + (-80) \\ &= 63 \end{aligned}$$
The measure of the unknown angle is 63°.

32. Replace a with 93, b with 80, and c with 82.
$$\begin{aligned} 360 - a - b - c &= 360 - 93 - 80 - 82 \\ &= 360 + (-93) + (-80) + (-82) \\ &= 267 + (-80) + (-82) \\ &= 187 + (-82) \\ &= 105 \end{aligned}$$
The measure of the unknown angle is 105°.

33.
$$\begin{aligned} 7x - 3 &= 18 \\ 7(3) - 3 &\stackrel{?}{=} 18 \\ 21 - 3 &\stackrel{?}{=} 18 \\ 21 + (-3) &\stackrel{?}{=} 18 \\ 18 &= 18 \end{aligned}$$
Since the results is true, 3 is a solution of the given equation.

34.
$$\begin{aligned} 3x^2 + 4 &= x - 1 \\ 3(1)^2 + 4 &\stackrel{?}{=} 1 - 1 \\ 3(1) + 4 &\stackrel{?}{=} 0 \\ 3 + 4 &\stackrel{?}{=} 0 \\ 7 &= 0 \end{aligned}$$
Since the result is false, 1 is not a solution of the given equation.

35. The opposite of −9 is 9.

36. The opposite of $\frac{2}{3}$ is $-\frac{2}{3}$.

37. The opposite of $|-2| = 2$ is −2.

38. The opposite of $-|-7| = -7$ is 7.

39. $-15 + 4 = -11$

40. $-6 + (-11) = -17$

41. $\frac{1}{16} + \left(-\frac{1}{4}\right) = \frac{1}{16} + \left(-\frac{4}{16}\right) = -\frac{3}{16}$

42. $-8 + |-3| = -8 + 3 = -5$

43. $-4.6 + (-9.3) = -13.9$

44. $-2.8 + 6.7 = 3.9$

45. $6 - 20 = 6 + (-20) = -14$

46. $-3.1 - 8.4 = -3.1 + (-8.4) = -11.5$

47. $-6 - (-11) = -6 + 11 = 5$

48. $4 - 15 = 4 + (-15) = -11$

49.
$$\begin{aligned} -21 - 16 + 3(8 - 2) &= -21 + (-16) + 3[8 + (-2)] \\ &= -21 + (-16) + 3(6) \\ &= -21 + (-16) + 18 \\ &= -37 + 18 \\ &= -19 \end{aligned}$$

50.
$$\begin{aligned} \frac{11 - (-9) + 6(8 - 2)}{2 + 3 \cdot 4} &= \frac{11 + 9 + 6[8 + (-2)]}{2 + 3 \cdot 4} \\ &= \frac{11 + 9 + 6(6)}{2 + 3 \cdot 4} \\ &= \frac{11 + 9 + 36}{2 + 12} \\ &= \frac{56}{14} \\ &= 4 \end{aligned}$$

51. Replace x with 3, y with −6, and z with −9.
$$\begin{aligned} 2x^2 - y + z &= 2(3)^2 - (-6) + (-9) \\ &= 2(9) + 6 + (-9) \\ &= 18 + 6 + (-9) \\ &= 24 + (-9) \\ &= 15 \end{aligned}$$
The answer is a.

52. Replace x with 3 and y with −6.

$$\frac{|y-4x|}{2x}=\frac{|-6-4(3)|}{2(3)}=\frac{|-6-12|}{6}=\frac{|-6+(-12)|}{6}=\frac{|-18|}{6}=\frac{18}{6}=3$$

The answer is a.

53. $50+1+(-2)+5+1+(-4)$
$=51+(-2)+5+1+(-4)$
$=49+5+1+(-4)$
$=54+1+(-4)$
$=55+(-4)$
$=51$
The price at the end of the week is \$51.

54. $50+1+(-2)+5=51+(-2)+5=49+5=54$
The price at the end of the day on Wednesday is \$54.

55. The reciprocal of −6 is $-\frac{1}{6}$.

56. The reciprocal of $\frac{3}{5}$ is $\frac{5}{3}$.

57. $6(-8)=-48$

58. $(-2)(-14)=28$

59. $\frac{-18}{-6}=3$

60. $\frac{42}{-3}=-14$

61. $-3(-6)(-2)=18(-2)=-36$

62. $(-4)(-3)(0)(-6)=12(0)(-6)=0(-6)=0$

63. $\frac{4(-3)+(-8)}{2+(-2)}=\frac{-12+(-8)}{2+(-2)}=\frac{-20}{0}$ is undefined.

64. $$\frac{3(-2)^2-5}{-14}=\frac{3(4)-5}{-14}=\frac{12-5}{-14}=\frac{12+(-5)}{-14}=\frac{7}{-14}=-\frac{1}{2}$$

65. $-6+5=5+(-6)$ illustrates the commutative property of addition.

66. $6\cdot 1=6$ illustrates the multiplicative identity property.

67. $3(8-5)=3\cdot 8-3\cdot 5$ illustrates the distributive property.

68. $4+(-4)=0$ illustrates the additive inverse property.

69. $2+(3+9)=(2+3)+9$ illustrates the associative property of addition.

70. $2\cdot 8=8\cdot 2$ illustrates the commutative property of multiplication.

71. $6(8+5)=6\cdot 8+6\cdot 5$ illustrates the distributive property.

72. $(3\cdot 8)\cdot 4=3\cdot(8\cdot 4)$ illustrates the associative property of multiplication.

73. $4\cdot\frac{1}{4}=1$ illustrates the multiplicative inverse property.

74. $8+0=8$ illustrates the additive identity property.

75. $4(8+3)=4(3+8)$ illustrates the commutative property of addition.

76. $5(2+1)=5\cdot 2+5\cdot 1$ illustrates the distributive property.

77. $5x-x+2x=5x-1x+2x=(5-1+2)x=6x$

78. $0.2z-4.6z-7.4z=(0.2-4.6-7.4)z=-11.8z$

79. $\frac{1}{2}x+3+\frac{7}{2}x-5=\frac{1}{2}x+\frac{7}{2}x+3-5$
$=\left(\frac{1}{2}+\frac{7}{2}\right)x+3+(-5)$
$=\frac{8}{2}x+(-2)$
$=4x-2$

80. $\frac{4}{5}y+1+\frac{6}{5}y+2=\frac{4}{5}y+\frac{6}{5}y+1+2$
$=\left(\frac{4}{5}+\frac{6}{5}\right)y+3$
$=\frac{10}{5}y+3$
$=2y+3$

81. $2(n-4)+n-10=2n-8+n-10$
$=2n+n-8-10$
$=3n-18$

82. $3(w+2)-(12-w)=3w+6-12+w$
$=3w+w+6-12$
$=4w-6$

83. $(x+5)-(7x-2)=x+5-7x+2$
$=x-7x+5+2$
$=-6x+7$

84. $(y-0.7)-(1.4y-3)=y-0.7-1.4y+3$
$=y-1.4y-0.7+3$
$=-0.4y+2.3$

85. Three times a number decreased by 7 is written as $3x-7$.

86. Twice the sum of a number and 2.8, added to 3 times the number is written as
$2(x+2.8)+3x=2x+5.6+3x$
$=2x+3x+5.6$
$=5x+5.6.$

87. $-|-11|=-11$
$|11.4|=11.4$
So, $-|-11|<|11.4|$.

88. Since $-1\frac{1}{2}$ is to the right of $-2\frac{1}{2}$ on the number line, $-1\frac{1}{2}>-2\frac{1}{2}$.

89. $-7.2+(-8.1)=-15.3$

90. $14-20=14+(-20)=-6$

91. $4(-20)=-80$

92. $\frac{-20}{4}=-5$

93. $-\frac{4}{5}\left(\frac{5}{16}\right)=-\frac{20}{80}=-\frac{1\cdot 20}{4\cdot 20}=-\frac{1}{4}$

94. $-0.5(-0.3)=0.15$

95. $8\div 2\cdot 4=4\cdot 4=16$

96. $(-2)^4=(-2)(-2)(-2)(-2)$
$=4(-2)(-2)$
$=-8(-2)$
$=16$

97. $\frac{-3-2(-9)}{-15-3(-4)}=\frac{-3+18}{-15+12}=\frac{15}{-3}=-5$

98. $5+2[(7-5)^2+(1-3)]$
$=5+2[(7+(-5))^2+(1+(-3))]$
$=5+2[(2)^2+(-2)]$
$=5+2[4+(-2)]$
$=5+2(2)$
$=5+4$
$=9$

99. $-\frac{5}{8}\div\frac{3}{4}=-\frac{5}{8}\cdot\frac{4}{3}=-\frac{20}{24}=-\frac{5\cdot 4}{6\cdot 4}=-\frac{5}{6}$

100. $\frac{-15+(-4)^2+|-9|}{10-2\cdot 5}=\frac{-15+16+9}{10-10}=\frac{10}{0}$ is undefined.

101. $7(3x-3)-5(x+4)=21x-21-5x-20$
$=21x-5x-21-20$
$=16x-41$

102. $8+2(9x-10)=8+18x-20$
$=18x+8-20$
$=18x-12$

Chapter 1 Test

1. The absolute value of negative seven is greater than five is written as $|-7|>5$.

2. The sum of nine and five is greater than or equal to four is written as $9 + 5 \geq 4$.

3. $-13 + 8 = -5$

4. $-13 - (-2) = -13 + 2 = -11$

5. $6 \cdot 3 - 8 \cdot 4 = 18 - 32 = 18 + (-32) = -14$

6. $(13)(-3) = -39$

7. $(-6)(-2) = 12$

8. $\dfrac{|-16|}{-8} = \dfrac{16}{-8} = -2$

9. $\dfrac{-8}{0}$ is undefined.

10. $\dfrac{|-6|+2}{5-6} = \dfrac{6+2}{5-6} = \dfrac{8}{-1} = -8$

11. $\dfrac{1}{2} - \dfrac{5}{6} = \dfrac{3}{6} - \dfrac{5}{6} = -\dfrac{2}{6} = -\dfrac{1}{3}$

12. $-1\dfrac{1}{8} + 5\dfrac{3}{4} = -1\dfrac{1}{8} + 5\dfrac{6}{8} = 4\dfrac{5}{8}$

13. $-\dfrac{3}{5} + \dfrac{15}{8} = -\dfrac{24}{40} + \dfrac{75}{40} = \dfrac{51}{40}$ or $1\dfrac{11}{40}$

14. $3(-4)^2 - 80 = 3(16) - 80 = 48 - 80 = -32$

15.
$$\begin{aligned} 6[5+2(3-8)-3] &= 6[5+2(3+(-8))-3] \\ &= 6[5+2(-5)-3] \\ &= 6[5+(-10)+(-3)] \\ &= 6[-5+(-3)] \\ &= 6[-8] \\ &= -48 \end{aligned}$$

16. $\dfrac{-12+3\cdot 8}{4} = \dfrac{-12+24}{4} = \dfrac{12}{4} = 3$

17. $\dfrac{(-2)(0)(-3)}{-6} = \dfrac{0(-3)}{-6} = \dfrac{0}{-6} = 0$

18. Since -3 is to the right of -7 on a number line, $-3 > -7$.

19. Since 4 is to the right of -8 on a number line, $4 > -8$.

20. Since $|-3| = 3$ is to the right of 2 on a number line, $|-3| > 2$.

21. $|-2| = 2$
$-1 - (-3) = -1 + 3 = 2$
Since $2 = 2$, $|-2| = -1 - (-3)$.

22. **a.** The natural numbers are 1, 7.

b. The whole numbers are 0, 1, 7.

c. The integers are $-5, -1, 0, 1, 7$.

d. The rational numbers are $-5, -1, \dfrac{1}{4}, 0, 1, 7, 11.6$.

e. The irrational numbers are $\sqrt{7}, 3\pi$.

f. The real numbers are $-5, -1, \dfrac{1}{4}, 0, 1, 7, 11.6, \sqrt{7}, 3\pi$.

23. Replace x with 6 and y with -2.
$x^2 + y^2 = 6^2 + (-2)^2 = 36 + 4 = 40$

24. Replace x with 6, y with -2, and z with -3.
$x + yz = 6 + (-2)(-3) = 6 + 6 = 12$

25. Replace x with 6 and y with -2.
$$\begin{aligned} 2+3x-y &= 2+3(6)-(-2) \\ &= 2+18+2 \\ &= 20+2 \\ &= 22 \end{aligned}$$

26. Replace x with 6, y with -2, and z with -3.
$$\begin{aligned} \frac{y+z-1}{x} &= \frac{-2+(-3)-1}{6} \\ &= \frac{-5-1}{6} \\ &= \frac{-5+(-1)}{6} \\ &= \frac{-6}{6} \\ &= -1 \end{aligned}$$

27. $8 + (9 + 3) = (8 + 9) + 3$ illustrates the associative property of addition.

28. $6 \cdot 8 = 8 \cdot 6$ illustrates the commutative property of multiplication.

29. $-6(2 + 4) = -6 \cdot 2 + (-6) \cdot 4$ illustrates the distributive property.

30. $\frac{1}{6}(6) = 1$ illustrates the multiplicative inverse property.

31. The opposite of -9 is 9.

32. The reciprocal of $-\frac{1}{3}$ of -3.

33. Losses of yardage occurred on the second and third downs. -10 indicates a loss of 10 yards while -2 indicates a loss of 2 yards, so the greatest loss of yardage occurred on the second down.

34. $$\begin{aligned} 5+(-10)+(-2)+29 &= -5+(-2)+29 \\ &= -7+29 \\ &= 22 \end{aligned}$$
Since the team was 22 yards from the goal, a touchdown was scored.

35. $-14 + 31 = 17$
The temperature was 17° at noon.

36. $-1.5(280) = -420$
She lost $420.

37. $$\begin{aligned} 2y-6-y-4 &= 2y-y-6-4 \\ &= 1y-10 \\ &= y-10 \end{aligned}$$

38. $$\begin{aligned} 2.7x+6.1+3.2x-4.9 &= 2.7x+3.2x+6.1-4.9 \\ &= 5.9x+1.2 \end{aligned}$$

39. $$\begin{aligned} 4(x-2)-3(2x-6) &= 4x-8-6x+18 \\ &= 4x-6x-8+18 \\ &= -2x+10 \end{aligned}$$

40. $$\begin{aligned} -5(y+1)+2(3-5y) &= -5y-5+6-10y \\ &= -5y-10y-5+6 \\ &= -15y+1 \end{aligned}$$

Chapter 2

Section 2.1 Practice Exercises

1.
$$x-5=8$$
$$x-5+5=8+5$$
$$x=13$$
Check: $x-5=8$
$$13-5 \stackrel{?}{=} 8$$
$$8=8 \quad \text{True}$$
The solution is 13.

2.
$$y+1.7=0.3$$
$$y+1.7-1.7=0.3-1.7$$
$$y=-1.4$$
Check: $y+1.7=0.3$
$$-1.4+1.7 \stackrel{?}{=} 0.3$$
$$0.3=0.3 \quad \text{True}$$
The solution is -1.4.

3.
$$\frac{7}{8}=y-\frac{1}{3}$$
$$\frac{7}{8}+\frac{1}{3}=y-\frac{1}{3}+\frac{1}{3}$$
$$\frac{7}{8}\cdot\frac{3}{3}+\frac{1}{3}\cdot\frac{8}{8}=y$$
$$\frac{21}{24}+\frac{8}{24}=y$$
$$\frac{29}{24}=y$$
Check: $\frac{7}{8}=y-\frac{1}{3}$
$$\frac{7}{8} \stackrel{?}{=} \frac{29}{24}-\frac{1}{3}$$
$$\frac{7}{8} \stackrel{?}{=} \frac{29}{24}-\frac{8}{24}$$
$$\frac{7}{8} \stackrel{?}{=} \frac{21}{24}$$
$$\frac{7}{8}=\frac{7}{8} \quad \text{True}$$
The solution is $\frac{29}{24}$.

4.
$$3x+10=4x$$
$$3x+10-3x=4x-3x$$
$$10=x$$
Check: $3x+10=4x$
$$3(10)+10 \stackrel{?}{=} 4(10)$$
$$30+10 \stackrel{?}{=} 40$$
$$40=40 \quad \text{True}$$
The solution is 10.

5.
$$10w+3-4w+4=-2w+3+7w$$
$$6w+7=5w+3$$
$$-5w+6w+7=-5w+5w+3$$
$$w+7=3$$
$$w+7-7=3-7$$
$$w=-4$$
Check:
$$10w+3-4w+4=-2w+3+7w$$
$$10(-4)+3-4(-4)+4 \stackrel{?}{=} -2(-4)+3+7(-4)$$
$$-40+3+16+4 \stackrel{?}{=} 8+3-28$$
$$-17=-17 \quad \text{True}$$
The solution is -4.

6.
$$3(2w-5)-(5w+1)=-3$$
$$3(2w)-3(5)-1(5w)-1(1)=-3$$
$$6w-15-5w-1=-3$$
$$w-16=-3$$
$$w-16+16=-3+16$$
$$w=13$$
Check: $3(2w-5)-(5w+1)=-3$
$$3(2\cdot 13-5)-(5\cdot 13+1) \stackrel{?}{=} -3$$
$$3(26-5)-(65+1) \stackrel{?}{=} -3$$
$$3(21)-66 \stackrel{?}{=} -3$$
$$63-66 \stackrel{?}{=} -3$$
$$-3=-3 \quad \text{True}$$
The solution is 13.

7.
$$12-y=9$$
$$12-y-12=9-12$$
$$-y=-3$$
$$y=3$$
Check: $12-y=9$
$$12-3 \stackrel{?}{=} 9$$
$$9=9 \quad \text{True}$$
The solution is 3.

8. a. If the sum of two numbers is 11 and one number is 4, find the other number by subtracting 4 from 11. The other number is 11 − 4, or 7.

b. If the sum of two numbers is 11 and one number is x, find the other number by subtracting x from 11. The other number is $11 - x$.

c. If the sum of two numbers is 56 and one number is a, find the other number by subtracting a from 56. The other number is $56 - a$.

9. Mike received 100,445 more votes than Zane, who received n votes. So, Mike received $(n + 100{,}445)$ votes.

Vocabulary, Readiness & Video Check 2.1

1. A combination of operations on variables and numbers is called an expression.

2. A statement of the form "expression = expression" is called an equation.

3. An equation contains an equal sign (=).

4. An expression does not contain an equal sign (=).

5. An expression may be simplified and evaluated while an equation may be solved.

6. A solution of an equation is a number that when substituted for a variable makes the equation a true statement.

7. Equivalent equations have the same solution.

8. By the addition property of equality, the same number may be added to or subtracted from both sides of an equation without changing the solution of the equation.

9. $x+4=6$
$x=2$

10. $x+7=17$
$x=10$

11. $n+18=30$
$n=12$

12. $z+22=40$
$z=18$

13. $b-11=6$
$b=17$

14. $d-16=5$
$d=21$

15. The addition property of equality means that if we have an equation, we can add the same real number to both sides of an equation and have an equivalent equation.

16. $15x - 14 = 14x - 1$

17. $\frac{1}{7}x$

Exercise Set 2.1

1. $x+7=10$
$x+7-7=10-7$
$x=3$
Check: $x+7=10$
$3+7 \stackrel{?}{=} 10$
$10=10$ True
The solution is 3.

3. $x-2=-4$
$x-2+2=-4+2$
$x=-2$
Check: $x-2=-4$
$-2-2 \stackrel{?}{=} -4$
$-4=-4$ True
The solution is -2.

5. $-11=3+x$
$-11-3=3+x-3$
$-14=x$
Check: $-11=3+x$
$-11 \stackrel{?}{=} 3+(-14)$
$-11=-11$ True
The solution is -14.

7. $r-8.6=-8.1$
$r-8.6+8.6=-8.1+8.6$
$r=0.5$
Check: $r-8.6=-8.1$
$0.5-8.6 \stackrel{?}{=} -8.1$
$-8.1=-8.1$ True
The solution is 0.5.

9.
$$x - \frac{2}{5} = -\frac{3}{20}$$
$$x - \frac{2}{5} + \frac{2}{5} = -\frac{3}{20} + \frac{2}{5}$$
$$x = -\frac{3}{20} + \frac{8}{20}$$
$$x = \frac{5}{20}$$
$$x = \frac{1}{4}$$

Check:
$$x - \frac{2}{5} = -\frac{3}{20}$$
$$\frac{1}{4} - \frac{2}{5} \stackrel{?}{=} -\frac{3}{20}$$
$$\frac{5}{20} - \frac{8}{20} \stackrel{?}{=} -\frac{3}{20}$$
$$-\frac{3}{20} = -\frac{3}{20} \quad \text{True}$$

The solution is $\frac{1}{4}$.

11.
$$\frac{1}{3} + f = \frac{3}{4}$$
$$-\frac{1}{3} + \frac{1}{3} + f = -\frac{1}{3} + \frac{3}{4}$$
$$f = -\frac{4}{12} + \frac{9}{12}$$
$$f = \frac{5}{12}$$

Check:
$$\frac{1}{3} + f = \frac{3}{4}$$
$$\frac{1}{3} + \frac{5}{12} \stackrel{?}{=} \frac{3}{4}$$
$$\frac{4}{12} + \frac{5}{12} \stackrel{?}{=} \frac{3}{4}$$
$$\frac{9}{12} \stackrel{?}{=} \frac{3}{4}$$
$$\frac{3}{4} = \frac{3}{4} \quad \text{True}$$

The solution is $\frac{5}{12}$.

13.
$$7x + 2x = 8x - 3$$
$$9x = 8x - 3$$
$$9x - 8x = 8x - 3 - 8x$$
$$x = -3$$

Check:
$$7x + 2x = 8x - 3$$
$$7(-3) + 2(-3) \stackrel{?}{=} 8(-3) - 3$$
$$-21 - 6 \stackrel{?}{=} -24 - 3$$
$$-27 = -27 \quad \text{True}$$

The solution is -3.

15.
$$\frac{5}{6}x + \frac{1}{6}x = -9$$
$$\frac{6}{6}x = -9$$
$$x = -9$$

Check:
$$\frac{5}{6}x + \frac{1}{6}x = -9$$
$$\frac{5}{6}(-9) + \frac{1}{6}(-9) \stackrel{?}{=} -9$$
$$-\frac{45}{6} - \frac{9}{6} \stackrel{?}{=} -9$$
$$-\frac{54}{6} \stackrel{?}{=} -9$$
$$-9 = -9 \quad \text{True}$$

The solution is -9.

17.
$$2y + 10 = 5y - 4y$$
$$2y + 10 = y$$
$$2y + 10 - 2y = y - 2y$$
$$10 = -y$$
$$-10 = y$$

Check:
$$2y + 10 = 5y - 4y$$
$$2(-10) + 10 \stackrel{?}{=} 5(-10) - 4(-10)$$
$$-20 + 10 \stackrel{?}{=} -50 + 40$$
$$-10 = -10 \quad \text{True}$$

The solution is -10.

19.
$$-5(n - 2) = 8 - 4n$$
$$-5n + 10 = 8 - 4n$$
$$5n - 5n + 10 = 5n + 8 - 4n$$
$$10 = n + 8$$
$$10 - 8 = n + 8 - 8$$
$$2 = n$$

Check:
$$-5(n - 2) = 8 - 4n$$
$$-5(2 - 2) \stackrel{?}{=} 8 - 4(2)$$
$$-5(0) \stackrel{?}{=} 8 - 8$$
$$0 = 0 \quad \text{True}$$

The solution is 2.

21.
$$\begin{aligned}\frac{3}{7}x+2&=-\frac{4}{7}x-5\\ \frac{3}{7}x+2+\frac{4}{7}x&=-\frac{4}{7}x-5+\frac{4}{7}x\\ x+2&=-5\\ x+2-2&=-5-2\\ x&=-7\end{aligned}$$

Check:
$$\begin{aligned}\frac{3}{7}x+2&=-\frac{4}{7}x-5\\ \frac{3}{7}(-7)+2&\stackrel{?}{=}-\frac{4}{7}(-7)-5\\ -3+2&\stackrel{?}{=}4-5\\ -1&=-1 \quad \text{True}\end{aligned}$$

The solution is −7.

23.
$$\begin{aligned}5x-6&=6x-5\\ -5x+5x-6&=-5x+6x-5\\ -6&=x-5\\ -6+5&=x-5+5\\ -1&=x\end{aligned}$$

Check:
$$\begin{aligned}5x-6&=6x-5\\ 5(-1)-6&\stackrel{?}{=}6(-1)-5\\ -5-6&\stackrel{?}{=}-6-5\\ -11&=-11 \quad \text{True}\end{aligned}$$

The solution is −1.

25.
$$\begin{aligned}8y+2-6y&=3+y-10\\ 2y+2&=y-7\\ 2y+2-y&=y-7-y\\ y+2&=-7\\ y+2-2&=-7-2\\ y&=-9\end{aligned}$$

Check:
$$\begin{aligned}8y+2-6y&=3+y-10\\ 8(-9)+2-6(-9)&\stackrel{?}{=}3+(-9)-10\\ -72+2+54&\stackrel{?}{=}3-9-10\\ -16&=-16 \quad \text{True}\end{aligned}$$

The solution is −9.

27.
$$\begin{aligned}-3(x-4)&=-4x\\ -3x+12&=-4x\\ 3x-3x+12&=3x-4x\\ 12&=-x\\ -12&=x\end{aligned}$$

Check:
$$\begin{aligned}-3(x-4)&=-4x\\ -3(-12-4)&\stackrel{?}{=}-4(-12)\\ -3(-16)&\stackrel{?}{=}48\\ 48&=48 \quad \text{True}\end{aligned}$$

The solution is −12.

29.
$$\begin{aligned}\frac{3}{8}x-\frac{1}{6}&=-\frac{5}{8}x-\frac{2}{3}\\ \frac{3}{8}x-\frac{1}{6}+\frac{5}{8}x&=-\frac{5}{8}x-\frac{2}{3}+\frac{5}{8}x\\ x-\frac{1}{6}&=-\frac{2}{3}\\ x-\frac{1}{6}+\frac{1}{6}&=-\frac{2}{3}+\frac{1}{6}\\ x&=-\frac{4}{6}+\frac{1}{6}\\ x&=-\frac{3}{6}\\ x&=-\frac{1}{2}\end{aligned}$$

Check:
$$\begin{aligned}\frac{3}{8}x-\frac{1}{6}&=-\frac{5}{8}x-\frac{2}{3}\\ \frac{3}{8}\left(-\frac{1}{2}\right)-\frac{1}{6}&\stackrel{?}{=}-\frac{5}{8}\left(-\frac{1}{2}\right)-\frac{2}{3}\\ -\frac{3}{16}-\frac{1}{6}&\stackrel{?}{=}\frac{5}{16}-\frac{2}{3}\\ -\frac{9}{48}-\frac{8}{48}&\stackrel{?}{=}\frac{15}{48}-\frac{32}{48}\\ -\frac{17}{48}&=-\frac{17}{48} \quad \text{True}\end{aligned}$$

The solution is $-\frac{1}{2}$.

31.
$$\begin{aligned}2(x-4)&=x+3\\ 2x-8&=x+3\\ -x+2x-8&=-x+x+3\\ x-8&=3\\ x-8+8&=3+8\\ x&=11\end{aligned}$$

Check:
$$\begin{aligned}2(x-4)&=x+3\\ 2(11-4)&\stackrel{?}{=}11+3\\ 2(7)&\stackrel{?}{=}14\\ 14&=14 \quad \text{True}\end{aligned}$$

The solution is 7.

33.
$$\begin{aligned}3(n-5)-(6-2n)&=4n\\ 3n-15-6+2n&=4n\\ 5n-21&=4n\\ 5n-21-5n&=4n-5n\\ -21&=-n\\ 21&=n\end{aligned}$$

Check:
$$\begin{aligned} 3(n-5)-(6-2n) &= 4n \\ 3(21-5)-(6-2\cdot 21) &\stackrel{?}{=} 4(21) \\ 3(21-5)-(6-42) &\stackrel{?}{=} 84 \\ 3(16)-(-36) &\stackrel{?}{=} 84 \\ 48+36 &\stackrel{?}{=} 84 \\ 84 &= 84 \quad \text{True} \end{aligned}$$

The solution is 21.

35.
$$\begin{aligned} -2(x+6)+3(2x-5) &= 3(x-4)+10 \\ -2x-12+6x-15 &= 3x-12+10 \\ 4x-27 &= 3x-2 \\ -3x+4x-27 &= -3x+3x-2 \\ x-27 &= -2 \\ x-27+27 &= -2+27 \\ x &= 25 \end{aligned}$$

Check:
$$\begin{aligned} -2(x+6)+3(2x-5) &= 3(x-4)+10 \\ -2(25+6)+3(2\cdot 25-5) &\stackrel{?}{=} 3(25-4)+10 \\ -2(31)+3(50-5) &\stackrel{?}{=} 3(21)+10 \\ -62+3(45) &\stackrel{?}{=} 63+10 \\ -62+135 &\stackrel{?}{=} 73 \\ 73 &= 73 \quad \text{True} \end{aligned}$$

The solution is 25.

37.
$$\begin{aligned} 13x-3 &= 14x \\ 13x-3-13x &= 14x-13x \\ -3 &= x \end{aligned}$$

39.
$$\begin{aligned} 5b-0.7 &= 6b \\ 5b-0.7-5b &= 6b-5b \\ -0.7 &= b \end{aligned}$$

41.
$$\begin{aligned} 3x-6 &= 2x+5 \\ 3x-6+6 &= 2x+5+6 \\ 3x &= 2x+11 \\ 3x-2x &= 2x+11-2x \\ x &= 11 \end{aligned}$$

43.
$$\begin{aligned} 13x-9+2x-5 &= 12x-1+2x \\ 15x-14 &= 14x-1 \\ 15x-14-14x &= 14x-1-14x \\ x-14 &= -1 \\ x-14+14 &= -1+14 \\ x &= 13 \end{aligned}$$

45.
$$\begin{aligned} 7(6+w) &= 6(2+w) \\ 42+7w &= 12+6w \\ 42+7w-6w &= 12+6w-6w \\ 42+w &= 12 \\ 42+w-42 &= 12-42 \\ w &= -30 \end{aligned}$$

47.
$$\begin{aligned} n+4 &= 3.6 \\ n+4-4 &= 3.6-4 \\ n &= -0.4 \end{aligned}$$

49.
$$\begin{aligned} 10-(2x-4) &= 7-3x \\ 10-2x+4 &= 7-3x \\ 14-2x &= 7-3x \\ 14-2x+3x &= 7-3x+3x \\ 14+x &= 7 \\ 14+x-14 &= 7-14 \\ x &= -7 \end{aligned}$$

51.
$$\begin{aligned} \frac{1}{3} &= x+\frac{2}{3} \\ \frac{1}{3}-\frac{2}{3} &= x+\frac{2}{3}-\frac{2}{3} \\ -\frac{1}{3} &= x \end{aligned}$$

53.
$$\begin{aligned} -6.5-4x-1.6-3x &= -6x+9.8 \\ -8.1-7x &= -6x+9.8 \\ -8.1-7x+7x &= -6x+9.8+7x \\ -8.1 &= x+9.8 \\ -8.1-9.8 &= x+9.8-9.8 \\ -17.9 &= x \end{aligned}$$

55. If the sum of the two numbers is 20 and one number is p, then the other number is $20 - p$.

57. If the sum of the lengths of the two pieces is 10 feet and one piece is x feet, then the other piece has a length of $(10 - x)$ feet.

59. If the sum of the measures of two angles is 180° and one angle measures $x°$, then the other angle measures $(180 - x)°$.

61. If the number of undergraduate students is n, and the number of graduate students is 29,000 fewer than n, then the number of graduate students is $n - 29{,}000$.

63. If the area of the Gobi Desert is x square miles and the area of the Sahara Desert is 7 times the area of the Gobi Desert, then the area of the Sahara Desert is $7x$ square miles.

65. The multiplicative inverse of $\frac{5}{8}$ is $\frac{8}{5}$, since $\frac{5}{8}\cdot\frac{8}{5}=1$.

67. The multiplicative inverse of 2 is $\frac{1}{2}$, since

$2\cdot\frac{1}{2}=1.$

69. The multiplicative inverse of $-\frac{1}{9}$ is -9, since

$-\frac{1}{9}\cdot(-9)=1.$

71. $\frac{3x}{3}=\frac{3\cdot x}{3\cdot 1}=\frac{x}{1}=x$

73. $-5\left(-\frac{1}{5}y\right)=\left[-5\cdot\left(-\frac{1}{5}\right)\right]y=1y=y$

75. $\frac{3}{5}\left(\frac{5}{3}x\right)=\left(\frac{3}{5}\cdot\frac{5}{3}\right)x=1x=x$

77. answers may vary

79.
$$\begin{aligned} x-4&=-9\\ x-4+4&=-9+4\\ x&=-5 \end{aligned}$$

81. answers may vary

83.
$$\begin{aligned} 180-x-(2x+7)&=180-x-2x-7\\ &=173-3x \end{aligned}$$
The measure of the third angle is $(173 - 3x)°$.

85. answers may vary

87.
$$\begin{aligned} 36.766+x&=-108.712\\ 36.766+x-36.766&=-108.712-36.766\\ x&=-145.478 \end{aligned}$$

Section 2.2 Practice Exercises

1.
$$\begin{aligned} \frac{3}{7}x&=9\\ \frac{7}{3}\cdot\left(\frac{3}{7}x\right)&=\frac{7}{3}\cdot 9\\ \left(\frac{7}{3}\cdot\frac{3}{7}\right)x&=\frac{7}{3}\cdot 9\\ 1x&=21\\ x&=21 \end{aligned}$$

Check:
$$\begin{aligned} \frac{3}{7}x&=9\\ \frac{3}{7}(21)&\stackrel{?}{=}9\\ 9&=9 \quad \text{True} \end{aligned}$$
The solution is 21.

2.
$$\begin{aligned} 7x&=42\\ \frac{7x}{7}&=\frac{42}{7}\\ 1\cdot x&=6\\ x&=6 \end{aligned}$$

Check:
$$\begin{aligned} 7x&=42\\ 7\cdot 6&\stackrel{?}{=}42\\ 42&=42 \quad \text{True} \end{aligned}$$
The solution is 6.

3.
$$\begin{aligned} -4x&=52\\ \frac{-4x}{-4}&=\frac{52}{-4}\\ 1x&=-13\\ x&=-13 \end{aligned}$$

Check:
$$\begin{aligned} -4x&=52\\ -4(-13)&\stackrel{?}{=}52\\ 52&=52 \quad \text{True} \end{aligned}$$
The solution is -13.

4.
$$\begin{aligned} \frac{y}{5}&=13\\ \frac{1}{5}y&=13\\ 5\cdot\frac{1}{5}y&=5\cdot 13\\ 1y&=65\\ y&=65 \end{aligned}$$

Check:
$$\begin{aligned} \frac{y}{5}&=13\\ \frac{65}{5}&\stackrel{?}{=}13\\ 13&=13 \quad \text{True} \end{aligned}$$
The solution is 65.

5.
$$\begin{aligned} 2.6x&=13.52\\ \frac{2.6x}{2.6}&=\frac{13.52}{2.6}\\ x&=5.2 \end{aligned}$$

Check:
$$\begin{aligned} 2.6x&=13.52\\ 2.6(5.2)&\stackrel{?}{=}13.52\\ 13.52&=13.52 \quad \text{True} \end{aligned}$$
The solution is 5.2.

6.
$$-\frac{5}{6}y = -\frac{3}{5}$$
$$-\frac{6}{5}\cdot -\frac{5}{6}y = -\frac{6}{5}\cdot -\frac{3}{5}$$
$$y = \frac{18}{25}$$

Check:
$$-\frac{5}{6}y = -\frac{3}{5}$$
$$-\frac{5}{6}\left(\frac{18}{25}\right) \stackrel{?}{=} -\frac{3}{5}$$
$$-\frac{3}{5} = -\frac{3}{5} \quad \text{True}$$

The solution is $\frac{18}{25}$.

7.
$$-x+7 = -12$$
$$-x+7-7 = -12-7$$
$$-x = -19$$
$$\frac{-x}{-1} = \frac{-19}{-1}$$
$$1x = 19$$
$$x = 19$$

Check:
$$-x+7 = -12$$
$$-19+7 \stackrel{?}{=} -12$$
$$-12 = -12 \quad \text{True}$$

The solution is 19.

8.
$$-7x+2x+3-20 = -2$$
$$-5x-17 = -2$$
$$-5x-17+17 = -2+17$$
$$-5x = 15$$
$$\frac{-5x}{-5} = \frac{15}{-5}$$
$$x = -3$$

Check:
$$-7x+2x+3-20 = -2$$
$$-7(-3)+2(-3)+3-20 \stackrel{?}{=} -2$$
$$21-6+3-20 \stackrel{?}{=} -2$$
$$-2 = -2 \quad \text{True}$$

The solution is -3.

9.
$$10x-4 = 7x+14$$
$$10x-4-7x = 7x+14-7x$$
$$3x-4 = 14$$
$$3x-4+4 = 14+4$$
$$3x = 18$$
$$\frac{3x}{3} = \frac{18}{3}$$
$$x = 6$$

Check:
$$10x-4 = 7x+14$$
$$10(6)-4 \stackrel{?}{=} 7(6)+14$$
$$60-4 \stackrel{?}{=} 42+14$$
$$56 = 56 \quad \text{True}$$

The solution is 6.

10.
$$4(3x-2) = -1+4$$
$$4(3x)-4(2) = -1+4$$
$$12x-8 = 3$$
$$12x-8+8 = 3+8$$
$$12x = 11$$
$$\frac{12x}{12} = \frac{11}{12}$$
$$x = \frac{11}{12}$$

Check:
$$4(3x-2) = -1+4$$
$$4\left(3\cdot\frac{11}{12}-2\right) \stackrel{?}{=} -1+4$$
$$4\left(\frac{11}{4}-2\right) \stackrel{?}{=} -1+4$$
$$11-8 \stackrel{?}{=} 3$$
$$3 = 3 \quad \text{True}$$

The solution is $\frac{11}{12}$.

11. a. If x is the first integer, then $x + 1$ is the second integer.
Their sum is $x + (x + 1) = x + x + 1 = 2x + 1$.

b. If x is the first odd integer, then $x + 2$ is the second consecutive odd integer.
Their sum is $x + (x + 2) = x + x + 2 = 2x + 2$.

Vocabulary, Readiness & Video Check 2.2

1. By the multiplication property of equality, both sides of an equation may be multiplied or divided by the same nonzero number without changing the solution of the equation.

2. By the addition property of equality, the same number may be added to or subtracted from both sides of an equation without changing the solution of the equation.

3. An equation may be solved while an expression may be simplified and evaluated.

4. An equation contains an equal sign (=) while an expression does not.

5. Equivalent equations have the same solution.

6. A <u>solution</u> of an equation is a number that when substituted for a variable makes the equation a true statement.

7. $3a = 27$
$a = 9$

8. $9c = 54$
$c = 6$

9. $5b = 10$
$b = 2$

10. $7t = 14$
$t = 2$

11. $6x = -30$
$x = -5$

12. $8r = -64$
$r = -8$

13. We can multiply both sides of an equation by the <u>same</u> nonzero number and have an equivalent equation.

14. addition property; multiplication property; answers may vary

15. $(x+1)+(x+3) = 2x+4$

Exercise Set 2.2

1. $-5x = -20$
$$\frac{-5x}{-5} = \frac{-20}{-5}$$
$x = 4$
Check: $-5x = -20$
$-5(4) \stackrel{?}{=} -20$
$-20 = -20$ True
The solution is 4.

3. $3x = 0$
$$\frac{3x}{3} = \frac{0}{3}$$
$x = 0$
Check: $3x = 0$
$3 \cdot 0 \stackrel{?}{=} 0$
$0 = 0$ True
The solution is 0.

5. $-x = -12$
$$\frac{-x}{-1} = \frac{-12}{-1}$$
$x = 12$
Check: $-x = -12$
$-12 = -12$ True
The solution is 12.

7. $\frac{2}{3}x = -8$
$$\frac{3}{2} \cdot \frac{2}{3}x = \frac{3}{2} \cdot (-8)$$
$x = -12$
Check: $\frac{2}{3}x = -8$
$\frac{2}{3}(-12) \stackrel{?}{=} -8$
$-8 = -8$ True
The solution is -12.

9. $\frac{1}{6}d = \frac{1}{2}$
$$6 \cdot \frac{1}{6}d = 6 \cdot \frac{1}{2}$$
$d = 3$
Check: $\frac{1}{6}d = \frac{1}{2}$
$\frac{1}{6}(3) \stackrel{?}{=} \frac{1}{2}$
$\frac{1}{2} = \frac{1}{2}$ True
The solution is 3.

11. $\frac{a}{2} = 1$
$$2 \cdot \frac{a}{2} = 2 \cdot 1$$
$a = 2$
Check: $\frac{a}{2} = 1$
$\frac{2}{2} \stackrel{?}{=} 1$
$1 = 1$ True
The solution is 2.

13. $\frac{k}{-7} = 0$
$$-7\left(\frac{k}{-7}\right) = -7(0)$$
$k = 0$

Check: $\frac{k}{-7} = 0$

$\frac{0}{-7} \stackrel{?}{=} 0$

$0 = 0$ True

The solution is 0.

15. $1.7x = 10.71$

$\frac{1.7x}{1.7} = \frac{10.71}{1.7}$

$x = 6.3$

Check: $1.7x = 10.71$

$1.7(6.3) \stackrel{?}{=} 10.71$

$10.71 = 10.71$ True

The solution is 6.3.

17. $2x - 4 = 16$

$2x - 4 + 4 = 16 + 4$

$2x = 20$

$\frac{2x}{2} = \frac{20}{2}$

$x = 10$

Check: $2x - 4 = 16$

$2(10) - 4 \stackrel{?}{=} 16$

$20 - 4 \stackrel{?}{=} 16$

$16 = 16$ True

The solution is 10.

19. $-x + 2 = 22$

$-x + 2 - 2 = 22 - 2$

$-x = 20$

$\frac{-x}{-1} = \frac{20}{-1}$

$x = -20$

Check: $-x + 2 = 22$

$-(-20) + 2 \stackrel{?}{=} 22$

$20 + 2 \stackrel{?}{=} 22$

$22 = 22$ True

The solution is −20.

21. $6a + 3 = 3$

$6a + 3 - 3 = 3 - 3$

$6a = 0$

$\frac{6a}{6} = \frac{0}{6}$

$a = 0$

Check: $6a + 3 = 3$

$6(0) + 3 \stackrel{?}{=} 3$

$0 + 3 \stackrel{?}{=} 3$

$3 = 3$ True

The solution is 0.

23. $\frac{x}{3} - 2 = -5$

$\frac{x}{3} - 2 + 2 = -5 + 2$

$\frac{x}{3} = -3$

$3 \cdot \frac{x}{3} = 3 \cdot (-3)$

$x = -9$

Check: $\frac{x}{3} - 2 = -5$

$\frac{-9}{3} - 2 \stackrel{?}{=} -5$

$-3 - 2 \stackrel{?}{=} -5$

$-5 = -5$ True

The solution is −9.

25. $6z - 8 - z + 3 = 0$

$5z - 5 = 0$

$5z - 5 + 5 = 0 + 5$

$5z = 5$

$\frac{5z}{5} = \frac{5}{5}$

$z = 1$

Check: $6z - 8 - z + 3 = 0$

$6(1) - 8 - 1 + 3 \stackrel{?}{=} 0$

$6 - 8 - 1 + 3 \stackrel{?}{=} 0$

$0 = 0$ True

The solution is 1.

27. $1 = 0.4x - 0.6x - 5$

$1 = -0.2x - 5$

$1 + 5 = -0.2x - 5 + 5$

$6 = -0.2x$

$\frac{6}{-0.2} = \frac{-0.2x}{-0.2}$

$-30 = x$

Check: $1 = 0.4x - 0.6x - 5$

$1 \stackrel{?}{=} 0.4(-30) - 0.6(-30) - 5$

$1 \stackrel{?}{=} -12 + 18 - 5$

$1 = 1$ True

The solution is −30.

29.
$$\frac{2}{3}y-11=-9$$
$$\frac{2}{3}y-11+11=-9+11$$
$$\frac{2}{3}y=2$$
$$\frac{3}{2}\left(\frac{2}{3}y\right)=\frac{3}{2}(2)$$
$$y=3$$
Check: $\frac{2}{3}y-11=-9$
$$\frac{2}{3}(3)-11\stackrel{?}{=}-9$$
$$2-11\stackrel{?}{=}-9$$
$$-9=-9 \quad \text{True}$$
The solution is 3.

31.
$$\frac{3}{4}t-\frac{1}{2}=\frac{1}{3}$$
$$\frac{3}{4}t-\frac{1}{2}+\frac{1}{2}=\frac{1}{3}+\frac{1}{2}$$
$$\frac{3}{4}t=\frac{2}{6}+\frac{3}{6}$$
$$\frac{3}{4}t=\frac{5}{6}$$
$$\frac{4}{3}\cdot\frac{3}{4}t=\frac{4}{3}\cdot\frac{5}{6}$$
$$t=\frac{10}{9}$$
Check: $\frac{3}{4}t-\frac{1}{2}=\frac{1}{3}$
$$\frac{3}{4}\cdot\frac{10}{9}-\frac{1}{2}\stackrel{?}{=}\frac{1}{3}$$
$$\frac{5}{6}-\frac{3}{6}\stackrel{?}{=}\frac{1}{3}$$
$$\frac{2}{6}\stackrel{?}{=}\frac{1}{3}$$
$$\frac{1}{3}=\frac{1}{3} \quad \text{True}$$
The solution is $\frac{10}{9}$.

33.
$$8x+20=6x+18$$
$$8x+20-6x=6x+18-6x$$
$$2x+20=18$$
$$2x+20-20=18-20$$
$$2x=-2$$
$$\frac{2x}{2}=\frac{-2}{2}$$
$$x=-1$$

35.
$$3(2x+5)=-18+9$$
$$6x+15=-18+9$$
$$6x+15=-9$$
$$6x+15-15=-9-15$$
$$6x=-24$$
$$\frac{6x}{6}=\frac{-24}{6}$$
$$x=-4$$

37.
$$2x-5=20x+4$$
$$2x-5-20x=20x+4-20x$$
$$-18x-5=4$$
$$-18x-5+5=4+5$$
$$-18x=9$$
$$\frac{-18x}{-18}=\frac{9}{-18}$$
$$x=-\frac{1}{2}$$

39.
$$2+14=-4(3x-4)$$
$$2+14=-12x+16$$
$$16=-12x+16$$
$$16-16=-12x+16-16$$
$$0=-12x$$
$$\frac{0}{-12}=\frac{-12x}{-12}$$
$$0=x$$

41.
$$-6y-3=-5y-7$$
$$-6y-3+6y=-5y-7+6y$$
$$-3=y-7$$
$$-3+7=y-7+7$$
$$4=y$$

43.
$$\frac{1}{2}(2x-1)=-\frac{1}{7}-\frac{3}{7}$$
$$x-\frac{1}{2}=-\frac{4}{7}$$
$$x-\frac{1}{2}+\frac{1}{2}=-\frac{4}{7}+\frac{1}{2}$$
$$x=-\frac{8}{14}+\frac{7}{14}$$
$$x=-\frac{1}{14}$$

45.
$$\begin{aligned} -10z-0.5&=-20z+1.6\\ -10z-0.5+20z&=-20z+1.6+20z\\ 10z-0.5&=1.6\\ 10z-0.5+0.5&=1.6+0.5\\ 10z&=2.1\\ \frac{10z}{10}&=\frac{2.1}{10}\\ z&=0.21 \end{aligned}$$

47.
$$\begin{aligned} -4x+20&=4x-20\\ -4x-4x+20&=-4x+4x-20\\ -8x+20&=-20\\ -8x+20-20&=-20-20\\ -8x&=-40\\ \frac{-8x}{-8}&=\frac{-40}{-8}\\ x&=5 \end{aligned}$$

49.
$$\begin{aligned} 42&=7x\\ \frac{42}{7}&=\frac{7x}{7}\\ 6&=x \end{aligned}$$

51.
$$\begin{aligned} 4.4&=-0.8x\\ \frac{4.4}{-0.8}&=\frac{-0.8x}{-0.8}\\ -5.5&=x \end{aligned}$$

53.
$$\begin{aligned} 6x+10&=-20\\ 6x+10-10&=-20-10\\ 6x&=-30\\ \frac{6x}{6}&=\frac{-30}{6}\\ x&=-5 \end{aligned}$$

55.
$$\begin{aligned} 5-0.3k&=5\\ -5+5-0.3k&=-5+5\\ -0.3k&=0\\ \frac{-0.3k}{-0.3}&=\frac{0}{-0.3}\\ k&=0 \end{aligned}$$

57.
$$\begin{aligned} 13x-5&=11x-11\\ 13x-5+5&=11x-11+5\\ 13x&=11x-6\\ 13x-11x&=11x-11x-6\\ 2x&=-6\\ \frac{2x}{2}&=\frac{-6}{2}\\ x&=-3 \end{aligned}$$

59.
$$\begin{aligned} 9(3x+1)&=4x-5x\\ 27x+9&=-x\\ -27x+27x+9&=-27x-x\\ 9&=-28x\\ \frac{9}{-28}&=\frac{-28x}{-28}\\ -\frac{9}{28}&=x \end{aligned}$$

61.
$$\begin{aligned} -\frac{3}{7}p&=-2\\ -\frac{7}{3}\left(-\frac{3}{7}p\right)&=-\frac{7}{3}(-2)\\ p&=\frac{14}{3} \end{aligned}$$

63.
$$\begin{aligned} -\frac{4}{3}x&=12\\ -\frac{3}{4}\cdot\left(-\frac{4}{3}x\right)&=-\frac{3}{4}\cdot 12\\ x&=-9 \end{aligned}$$

65.
$$\begin{aligned} -2x-\frac{1}{2}&=\frac{7}{2}\\ -2x-\frac{1}{2}+\frac{1}{2}&=\frac{7}{2}+\frac{1}{2}\\ -2x&=\frac{8}{2}\\ -2x&=4\\ \frac{-2x}{-2}&=\frac{4}{-2}\\ x&=-2 \end{aligned}$$

67.
$$\begin{aligned} 10&=2x-1\\ 10+1&=2x-1+1\\ 11&=2x\\ \frac{11}{2}&=\frac{2x}{2}\\ \frac{11}{2}&=x \end{aligned}$$

69.
$$\begin{aligned} 10-3x-6-9x&=7\\ 4-12x&=7\\ 4-12x-4&=7-4\\ -12x&=3\\ \frac{-12x}{-12}&=\frac{3}{-12}\\ x&=-\frac{1}{4} \end{aligned}$$

71.
$$\begin{aligned} z-5z &= 7z-9-z \\ -4z &= 6z-9 \\ -4z-6z &= 6z-9-6z \\ -10z &= -9 \\ \frac{-10z}{-10} &= \frac{-9}{-10} \\ z &= \frac{9}{10} \end{aligned}$$

73.
$$\begin{aligned} -x-\frac{4}{5} &= x+\frac{1}{2}+\frac{2}{5} \\ -x-\frac{4}{5} &= x+\frac{5}{10}+\frac{4}{10} \\ -x-\frac{4}{5} &= x+\frac{9}{10} \\ -x-\frac{4}{5}+x &= x+\frac{9}{10}+x \\ -\frac{4}{5} &= 2x+\frac{9}{10} \\ -\frac{4}{5}-\frac{9}{10} &= 2x \\ -\frac{8}{10}-\frac{9}{10} &= 2x \\ -\frac{17}{10} &= 2x \\ \frac{1}{2}\left(-\frac{17}{10}\right) &= \frac{1}{2}(2x) \\ -\frac{17}{20} &= x \end{aligned}$$

75.
$$\begin{aligned} -15+37 &= -2(x+5) \\ 22 &= -2x-10 \\ 22+10 &= -2x-10+10 \\ 32 &= -2x \\ \frac{32}{-2} &= \frac{-2x}{-2} \\ -16 &= x \end{aligned}$$

77. If x represents the first of two consecutive odd integers, then $x + 2$ represents the second. Thus, the sum is represented by $x + x + 2 = 2x + 2$.

79. If x represents the first integer, then $x + 1$, $x + 2$, and $x + 3$ represent the second, third, and fourth integers, respectively. The sum of the first and third integers is represented by $x + x + 2 = 2x + 2$.

81. If x represents the number on the first door, then the next four door numbers are represented by $x + 2$, $x + 4$, $x + 6$, and $x + 8$.
The sum of the numbers is
$x + x + 2 + x + 4 + x + 6 + x + 8 = 5x + 20$.

83.
$$\begin{aligned} 5x+2(x-6) &= 5x+2\cdot x+2\cdot(-6) \\ &= 5x+2x-12 \\ &= 7x-12 \end{aligned}$$

85.
$$\begin{aligned} 6(2z+4)+20 &= 6\cdot 2z+6\cdot 4+20 \\ &= 12z+24+20 \\ &= 12z+44 \end{aligned}$$

87.
$$\begin{aligned} -(x-1)+x &= -x+1+x \\ &= -x+x+1 \\ &= 0+1 \\ &= 1 \end{aligned}$$

89. If the solution is -8, then replacing x by -8 results in a true statement.
$6x = 6(-8) = -48$
The missing number is -48.

91. answers may vary

93. answers may vary

95.
$$\begin{aligned} 0.07x-5.06 &= -4.92 \\ 0.07x-5.06+5.06 &= -4.92+5.06 \\ 0.07x &= 0.14 \\ \frac{0.07x}{0.07} &= \frac{0.14}{0.07} \\ x &= 2 \end{aligned}$$

Section 2.3 Practice Exercises

1.
$$\begin{aligned} 5(3x-1)+2 &= 12x+6 \\ 15x-5+2 &= 12x+6 \\ 15x-3 &= 12x+6 \\ 15x-3-12x &= 12x+6-12x \\ 3x-3 &= 6 \\ 3x-3+3 &= 6+3 \\ 3x &= 9 \\ \frac{3x}{3} &= \frac{9}{3} \\ x &= 3 \end{aligned}$$

Check:
$$\begin{aligned} 5(3x-1)+2 &= 12x+6 \\ 5[3(3)-1]+2 &\stackrel{?}{=} 12(3)+6 \\ 5(9-1)+2 &\stackrel{?}{=} 36+6 \\ 5(8)+2 &\stackrel{?}{=} 42 \\ 40+2 &\stackrel{?}{=} 42 \\ 42 &= 42 \quad \text{True} \end{aligned}$$

The solution is 3.

2.
$$9(5-x)=-3x$$
$$45-9x=-3x$$
$$45-9x+9x=-3x+9x$$
$$45=6x$$
$$\frac{45}{6}=\frac{6x}{6}$$
$$\frac{15}{2}=x$$

Check:
$$9(5-x)=-3x$$
$$9\left(5-\frac{15}{2}\right)\stackrel{?}{=}-3\left(\frac{15}{2}\right)$$
$$9\left(\frac{10}{2}-\frac{15}{2}\right)\stackrel{?}{=}-\frac{45}{2}$$
$$9\left(-\frac{5}{2}\right)\stackrel{?}{=}-\frac{45}{2}$$
$$-\frac{45}{2}=-\frac{45}{2}\quad\text{True}$$

The solution is $\frac{15}{2}$.

3.
$$\frac{5}{2}x-1=\frac{3}{2}x-4$$
$$2\left(\frac{5}{2}x-1\right)=2\left(\frac{3}{2}x-4\right)$$
$$5x-2=3x-8$$
$$5x-2-3x=3x-8-3x$$
$$2x-2=-8$$
$$2x-2+2=-8+2$$
$$2x=-6$$
$$\frac{2x}{2}=\frac{-6}{2}$$
$$x=-3$$

Check:
$$\frac{5}{2}x-1=\frac{3}{2}x-4$$
$$\frac{5}{2}(-3)-1\stackrel{?}{=}\frac{3}{2}(-3)-4$$
$$-\frac{15}{2}-1\stackrel{?}{=}-\frac{9}{2}-4$$
$$-\frac{15}{2}-\frac{2}{2}\stackrel{?}{=}-\frac{9}{2}-\frac{8}{2}$$
$$-\frac{17}{2}=-\frac{17}{2}\quad\text{True}$$

The solution is −3.

4.
$$\frac{3(x-2)}{5}=3x+6$$
$$5\cdot\frac{3(x-2)}{5}=5(3x+6)$$
$$3(x-2)=5(3x+6)$$
$$3x-6=15x+30$$
$$3x-6-3x=15x+30-3x$$
$$-6=12x+30$$
$$-6-30=12x+30-30$$
$$-36=12x$$
$$\frac{-36}{12}=\frac{12x}{12}$$
$$-3=x$$

Check:
$$\frac{3(x-2)}{5}=3x+6$$
$$\frac{3(-3-2)}{5}\stackrel{?}{=}3(-3)+6$$
$$\frac{3(-5)}{5}\stackrel{?}{=}-9+6$$
$$\frac{-15}{5}\stackrel{?}{=}-3$$
$$-3=-3$$

The solution is −3.

5.
$$0.06x-0.10(x-2)=-0.16$$
$$100[0.06x-0.10(x-2)]=100[-0.16]$$
$$6x-10(x-2)=-16$$
$$6x-10x+20=-16$$
$$-4x+20=-16$$
$$-4x+20-20=-16-20$$
$$-4x=-36$$
$$\frac{-4x}{-4}=\frac{-36}{-4}$$
$$x=9$$

To check, replace x with 9 in the original equation. The solution is 9.

6.
$$5(2-x)+8x=3(x-6)$$
$$10-5x+8x=3x-18$$
$$10+3x=3x-18$$
$$10+3x-3x=3x-18-3x$$
$$10=-18$$

Since the statement $10=-18$ is false, the equation has no solution.

7.
$$\begin{aligned}-6(2x+1)-14&=-10(x+2)-2x\\-12x-6-14&=-10x-20-2x\\-12x-20&=-12x-20\\12x-12x-20&=12x-12x-20\\-20&=-20\end{aligned}$$
Since $-20 = -20$ is a true statement, every real number is a solution.

Calculator Explorations

1. $2x = 48 + 6x$

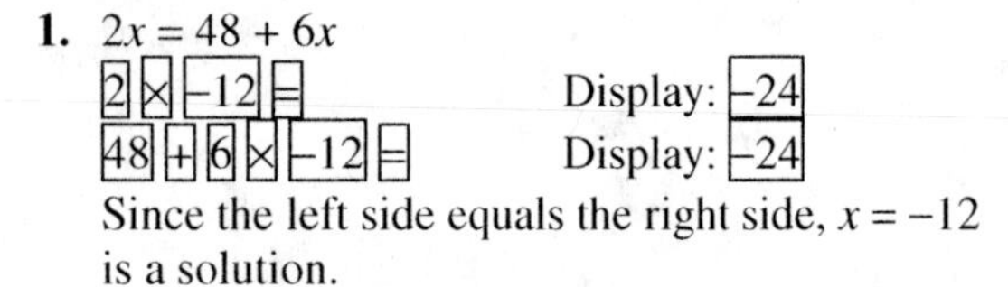

Since the left side equals the right side, $x = -12$ is a solution.

2. $-3x - 7 = 3x - 1$

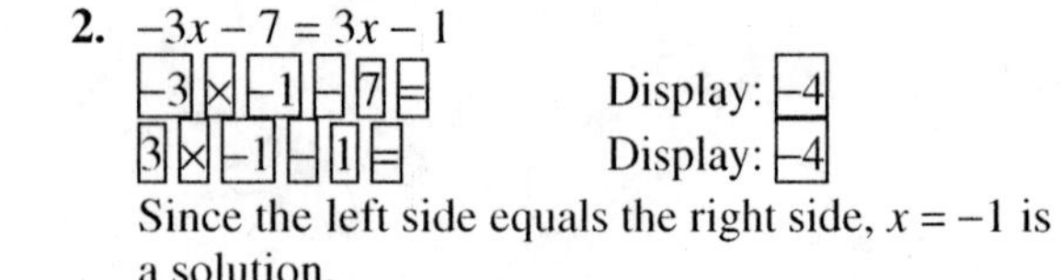

Since the left side equals the right side, $x = -1$ is a solution.

3. $5x - 2.6 = 2(x + 0.8)$

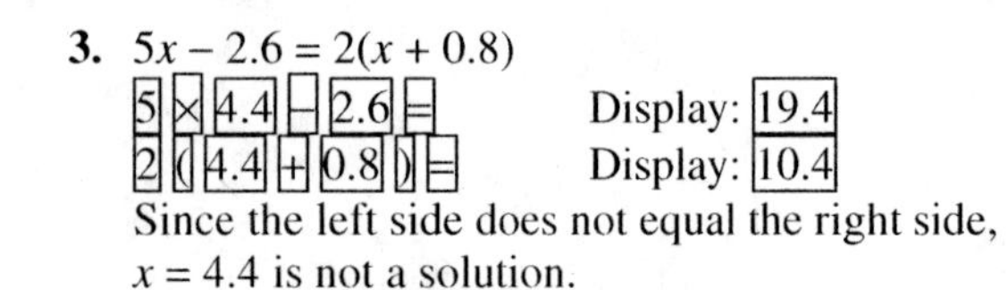

Since the left side does not equal the right side, $x = 4.4$ is not a solution.

4. $-1.6x - 3.9 = -6.9x - 25.6$

−1.6 × 5 − 3.9 = Display: −11.9
−6.9 × 5 − 25.6 = Display: −60.1

Since the left side does not equal the right side, $x = 5$ is not a solution.

5. $\frac{564x}{4} = 200x - 11(649)$

(564 × 121) ÷ 4 = Display: 17061
200 × 121 − 11 × 649 = Display: 17061

Since the left side equals the right side, $x = 121$ is a solution.

6. $20(x - 39) = 5x - 432$

20 (23.2 − 39) = Display: −316
5 × 23.2 − 432 = Display: −316

Since the left side equals the right side, $x = 23.2$ is a solution.

Vocabulary, Readiness & Video Check 2.3

1. $x = -7$ is an <u>equation</u>.

2. $x - 7$ is an <u>expression</u>.

3. $4y - 6 + 9y + 1$ is an <u>expression</u>.

4. $4y - 6 = 9y + 1$ is an <u>equation</u>.

5. $\frac{1}{x} - \frac{x-1}{8}$ is an <u>expression</u>.

6. $\frac{1}{x} - \frac{x-1}{8} = 6$ is an <u>equation</u>.

7. $0.1x + 9 = 0.2x$ is an <u>equation</u>.

8. $0.1x^2 + 9y - 0.2x^2$ is an <u>expression</u>.

9. 3; distributive property, addition property of equality, multiplication property of equality

10. Since both sides have more than one term, you need to apply the distributive property to make sure you multiply every single term in the equation by the LCD.

11. The number of decimal places in each number helps you determine what power of 10 you can multiply through by so you are no longer dealing with decimals

12. a. If you have a true statement, then the equation has <u>all real numbers as a</u> solution.

b. If you have a false statement, then the equation has <u>no</u> solutions.

Exercise Set 2.3

1.
$$\begin{aligned}-4y+10&=-2(3y+1)\\-4y+10&=-6y-2\\-4y+10-10&=-6y-2-10\\-4y&=-6y-12\\-4y+6y&=-6y-12+6y\\2y&=-12\\\frac{2y}{2}&=\frac{-12}{2}\\y&=-6\end{aligned}$$

3.
$$\begin{aligned}15x-8&=10+9x\\15x-8-9x&=10+9x-9x\\6x-8&=10\\6x-8+8&=10+8\\6x&=18\\\frac{6x}{6}&=\frac{18}{6}\\x&=3\end{aligned}$$

5.
$$\begin{aligned}-2(3x-4)&=2x\\-6x+8&=2x\\-6x+8+6x&=2x+6x\\8&=8x\\\frac{8}{8}&=\frac{8x}{8}\\1&=x\end{aligned}$$

7.
$$\begin{aligned}5(2x-1)-2(3x)&=1\\10x-5-6x&=1\\-5+4x&=1\\5-5+4x&=5+1\\4x&=6\\\frac{4x}{4}&=\frac{6}{4}\\x&=\frac{3}{2}\end{aligned}$$

9.
$$\begin{aligned}-6(x-3)-26&=-8\\-6x+18-26&=-8\\-6x-8&=-8\\-6x-8+8&=-8+8\\-6x&=0\\\frac{-6x}{-6}&=\frac{0}{-6}\\x&=0\end{aligned}$$

11.
$$\begin{aligned}8-2(a+1)&=9+a\\8-2a-2&=9+a\\-2a+6&=9+a\\-2a+6-a&=9+a-a\\-3a+6&=9\\-3a+6-6&=9-6\\-3a&=3\\\frac{-3a}{-3}&=\frac{3}{-3}\\a&=-1\end{aligned}$$

13.
$$\begin{aligned}4x+3&=-3+2x+14\\4x+3&=11+2x\\4x+3-2x&=11+2x-2x\\2x+3&=11\\2x+3-3&=11-3\\2x&=8\\\frac{2x}{2}&=\frac{8}{2}\\x&=4\end{aligned}$$

15.
$$\begin{aligned}-2y-10&=5y+18\\-2y-10+10&=5y+18+10\\-2y&=5y+28\\-2y-5y&=5y+28-5y\\-7y&=28\\\frac{-7y}{-7}&=\frac{28}{-7}\\y&=-4\end{aligned}$$

17.
$$\begin{aligned}\frac{2}{3}x+\frac{4}{3}&=-\frac{2}{3}\\3\left(\frac{2}{3}x+\frac{4}{3}\right)&=3\left(-\frac{2}{3}\right)\\2x+4&=-2\\2x+4-4&=-2-4\\2x&=-6\\\frac{2x}{2}&=\frac{-6}{2}\\x&=-3\end{aligned}$$

19.
$$\begin{aligned}\frac{3}{4}x-\frac{1}{2}&=1\\4\left(\frac{3}{4}x-\frac{1}{2}\right)&=4(1)\\3x-2&=4\\3x-2+2&=4+2\\3x&=6\\\frac{3x}{3}&=\frac{6}{3}\\x&=2\end{aligned}$$

21.
$$\begin{aligned}0.50x+0.15(70)&=35.5\\50x+15(70)&=3550\\50x+1050&=3550\\50x+1050-1050&=3550-1050\\50x&=2500\\\frac{50x}{50}&=\frac{2500}{50}\\x&=50\end{aligned}$$

23.
$$\frac{2(x+1)}{4} = 3x-2$$
$$4\left[\frac{2(x+1)}{4}\right] = 4(3x-2)$$
$$2(x+1) = 4(3x-2)$$
$$2x+2 = 12x-8$$
$$2x+2+8 = 12x-8+8$$
$$2x+10 = 12x$$
$$2x+10-2x = 12x-2x$$
$$10 = 10x$$
$$\frac{10}{10} = \frac{10x}{10}$$
$$1 = x$$

25.
$$x+\frac{7}{6} = 2x-\frac{7}{6}$$
$$6\left(x+\frac{7}{6}\right) = 6\left(2x-\frac{7}{6}\right)$$
$$6x+7 = 12x-7$$
$$6x+7+7 = 12x-7+7$$
$$6x+14 = 12x$$
$$6x+14-6x = 12x-6x$$
$$14 = 6x$$
$$\frac{14}{6} = \frac{6x}{6}$$
$$\frac{7}{3} = x$$

27.
$$0.12(y-6)+0.06y = 0.08y-0.7$$
$$12(y-6)+6y = 8y-70$$
$$12y-72+6y = 8y-70$$
$$18y-72 = 8y-70$$
$$18y-72-8y = 8y-70-8y$$
$$10y-72 = -70$$
$$10y-72+72 = -70+72$$
$$10y = 2$$
$$\frac{10y}{10} = \frac{2}{10}$$
$$y = 0.2$$

29.
$$4(3x+2) = 12x+8$$
$$12x+8 = 12x+8$$
Since both sides of the equation are identical, the equation is an identity and every real number is a solution.

31.
$$\frac{x}{4}+1 = \frac{x}{4}$$
$$\frac{x}{4}+1-\frac{x}{4} = \frac{x}{4}-\frac{x}{4}$$
$$1 = 0$$
Since the statement 1 = 0 is false, the equation has no solution.

33.
$$3x-7 = 3(x+1)$$
$$3x-7 = 3x+3$$
$$3x-7-3x = 3x+3-3x$$
$$-7 = 3$$
Since the statement −7 = 3 is false, the equation has no solution.

35.
$$-2(6x-5)+4 = -12x+14$$
$$-12x+10+4 = -12x+14$$
$$-12x+14 = -12x+14$$
Since both sides of the equation are identical, the equation is an identity and every real number is a solution.

37.
$$\frac{6(3-z)}{5} = -z$$
$$5\cdot\frac{6(3-z)}{5} = 5(-z)$$
$$6(3-z) = -5z$$
$$18-6z = -5z$$
$$18-6z+6z = -5z+6z$$
$$18 = z$$

39.
$$-3(2t-5)+2t = 5t-4$$
$$-6t+15+2t = 5t-4$$
$$-4t+15 = 5t-4$$
$$-4t+15+4t = 5t-4+4t$$
$$15 = 9t-4$$
$$15+4 = 9t-4+4$$
$$19 = 9t$$
$$\frac{19}{9} = \frac{9t}{9}$$
$$\frac{19}{9} = t$$

41.
$$\begin{aligned}5y+2(y-6)&=4(y+1)-2\\5y+2y-12&=4y+4-2\\7y-12&=4y+2\\7y-12+12&=4y+2+12\\7y&=4y+14\\7y-4y&=4y+14-4y\\3y&=14\\\frac{3y}{3}&=\frac{14}{3}\\y&=\frac{14}{3}\end{aligned}$$

43.
$$\begin{aligned}\frac{3(x-5)}{2}&=\frac{2(x+5)}{3}\\6\left[\frac{3(x-5)}{2}\right]&=6\left[\frac{2(x+5)}{3}\right]\\9(x-5)&=4(x+5)\\9x-45&=4x+20\\9x-45+45&=4x+20+45\\9x&=4x+65\\9x-4x&=4x+65-4x\\5x&=65\\\frac{5x}{5}&=\frac{65}{5}\\x&=13\end{aligned}$$

45.
$$\begin{aligned}0.7x-2.3&=0.5\\7x-23&=5\\7x-23+23&=5+23\\7x&=28\\\frac{7x}{7}&=\frac{28}{7}\\x&=4\end{aligned}$$

47.
$$\begin{aligned}5x-5&=2(x+1)+3x-7\\5x-5&=2x+2+3x-7\\5x-5&=5x-5\end{aligned}$$

Since both sides of the equation are identical, the equation is an identity and every real number is a solution.

49.
$$\begin{aligned}4(2n+1)&=3(6n+3)+1\\8n+4&=18n+9+1\\8n+4&=18n+10\\8n+4-10&=18n+10-10\\8n-6&=18n\\8n-6-8n&=18n-8n\\-6&=10n\\\frac{-6}{10}&=\frac{10n}{10}\\-\frac{3}{5}&=n\end{aligned}$$

51.
$$\begin{aligned}x+\frac{5}{4}&=\frac{3}{4}x\\4\left(x+\frac{5}{4}\right)&=4\left(\frac{3}{4}x\right)\\4x+5&=3x\\4x+5-4x&=3x-4x\\5&=-x\\\frac{5}{-1}&=\frac{-x}{-1}\\-5&=x\end{aligned}$$

53.
$$\begin{aligned}\frac{x}{2}-1&=\frac{x}{5}+2\\10\left(\frac{x}{2}-1\right)&=10\left(\frac{x}{5}+2\right)\\5x-10&=2x+20\\5x-10+10&=2x+20+10\\5x&=2x+30\\5x-2x&=2x+30-2x\\3x&=30\\\frac{3x}{3}&=\frac{30}{3}\\x&=10\end{aligned}$$

55.
$$\begin{aligned}2(x+3)-5&=5x-3(1+x)\\2x+6-5&=5x-3-3x\\2x+1&=2x-3\\2x+1-2x&=2x-3-2x\\1&=-3\end{aligned}$$

Since the statement $1=-3$ is false, the equation has no solution.

57.
$$\begin{aligned} 0.06-0.01(x+1) &= -0.02(2-x) \\ 6-1(x+1) &= -2(2-x) \\ 6-x-1 &= -4+2x \\ 5-x &= -4+2x \\ 5-x+x &= -4+2x+x \\ 5 &= -4+3x \\ 5+4 &= -4+3x+4 \\ 9 &= 3x \\ \frac{9}{3} &= \frac{3x}{3} \\ 3 &= x \end{aligned}$$

59.
$$\begin{aligned} \frac{9}{2}+\frac{5}{2}y &= 2y-4 \\ 2\left(\frac{9}{2}+\frac{5}{2}y\right) &= 2(2y-4) \\ 9+5y &= 4y-8 \\ 9+5y-4y &= 4y-8-4y \\ 9+y &= -8 \\ 9+y-9 &= -8-9 \\ y &= -17 \end{aligned}$$

61.
$$\begin{aligned} \frac{3}{4}x-1+\frac{1}{2}x &= \frac{5}{12}x+\frac{1}{6} \\ 12\left(\frac{3}{4}x-1+\frac{1}{2}x\right) &= 12\left(\frac{5}{12}x+\frac{1}{6}\right) \\ 9x-12+6x &= 5x+2 \\ 15x-12 &= 5x+2 \\ 15x-12-5x &= 5x+2-5x \\ 10x-12 &= 2 \\ 10x-12+12 &= 2+12 \\ 10x &= 14 \\ \frac{10x}{10} &= \frac{14}{10} \\ x &= \frac{7}{5} \end{aligned}$$

63.
$$\begin{aligned} 3x+\frac{5}{16} &= \frac{3}{4}-\frac{1}{8}x-\frac{1}{2} \\ 16\left(3x+\frac{5}{16}\right) &= 16\left(\frac{3}{4}-\frac{1}{8}x-\frac{1}{2}\right) \\ 48x+5 &= 12-2x-8 \\ 48x+5 &= 4-2x \\ 48x+5+2x &= 4-2x+2x \\ 50x+5 &= 4 \\ 50x+5-5 &= 4-5 \\ 50x &= -1 \\ \frac{50x}{50} &= \frac{-1}{50} \\ x &= -\frac{1}{50} \end{aligned}$$

65. The perimeter is the sum of the lengths of the sides.
$$\begin{aligned} x+(2x-3)+(3x-5) &= x+2x-3+3x-5 \\ &= 6x-8 \end{aligned}$$
The perimeter is $(6x - 8)$ meters.

67. A number subtracted from -8 is $-8 - x$.

69. The sum of -3 and twice a number is $-3 + 2x$.

71. The product of 9 and the sum of a number and 20 is $9(x + 20)$.

73. **a.** Since both sides of the equation are identical, the equation is an identity and every real number is a solution.

b. answers may vary

c. answers may vary

75. $5x + 1 = 5x + 1$
Since both sides of the equation are identical, the equation is an identity and every real number is a solution. The choice is a.

77.
$$\begin{aligned} 2x-6x-10 &= -4x+3-10 \\ -4x-10 &= -4x-7 \\ -4x-10+4x &= -4x-7+4x \\ -10 &= -7 \end{aligned}$$
Since the statement $-10 = -7$ is false, the equation has no solution. The choice is b.

79.
$$\begin{aligned}9x-20&=8x-20\\9x-20-8x&=8x-20-8x\\x-20&=-20\\x-20+20&=-20+20\\x&=0\end{aligned}$$
The choice is c.

81. answers may vary

83. a. The perimeter is the sum of the lengths of the sides.
$x+x+x+2x+2x=28$

b.
$$\begin{aligned}x+x+x+2x+2x&=28\\7x&=28\\\frac{7x}{7}&=\frac{28}{7}\\x&=4\end{aligned}$$

c. The sides of length x cm are 4 cm and the sides of length $2x$ cm are 2(4 cm) = 8 cm.

85. answers may vary

87.
$$\begin{aligned}1000(7x-10)&=50(412+100x)\\7000x-10,000&=20,600+5000x\\7000x-10,000-5000x&=20,600+5000x-5000x\\2000x-10,000&=20,600\\2000x-10,000+10,000&=20,600+10,000\\2000x&=30,600\\\frac{2000x}{2000}&=\frac{30,600}{2000}\\x&=15.3\end{aligned}$$

89.
$$\begin{aligned}0.035x+5.112&=0.010x+5.107\\35x+5112&=10x+5107\\35x+5112-10x&=10x+5107-10x\\25x+5112&=5107\\25x+5112-5112&=5107-5112\\25x&=-5\\\frac{25x}{25}&=\frac{-5}{25}\\x&=-\frac{1}{5}\\x&=-0.2\end{aligned}$$

Integrated Review

1.
$$\begin{aligned}x-10&=-4\\x-10+10&=-4+10\\x&=6\end{aligned}$$

2.
$$\begin{aligned}y+14&=-3\\y+14-14&=-3-14\\y&=-17\end{aligned}$$

3.
$$\begin{aligned} 9y &= 108 \\ \frac{9y}{9} &= \frac{108}{9} \\ y &= 12 \end{aligned}$$

4.
$$\begin{aligned} -3x &= 78 \\ \frac{-3x}{-3} &= \frac{78}{-3} \\ x &= -26 \end{aligned}$$

5.
$$\begin{aligned} -6x+7 &= 25 \\ -6x+7-7 &= 25-7 \\ -6x &= 18 \\ \frac{-6x}{-6} &= \frac{18}{-6} \\ x &= -3 \end{aligned}$$

6.
$$\begin{aligned} 5y-42 &= -47 \\ 5y-42+42 &= -47+42 \\ 5y &= -5 \\ \frac{5y}{5} &= \frac{-5}{5} \\ y &= -1 \end{aligned}$$

7.
$$\begin{aligned} \frac{2}{3}x &= 9 \\ \frac{3}{2}\cdot\frac{2}{3}x &= \frac{3}{2}\cdot 9 \\ x &= \frac{27}{2} \end{aligned}$$

8.
$$\begin{aligned} \frac{4}{5}z &= 10 \\ \frac{5}{4}\cdot\frac{4}{5}z &= \frac{5}{4}\cdot 10 \\ z &= \frac{50}{4} \\ z &= \frac{25}{2} \end{aligned}$$

9.
$$\begin{aligned} \frac{r}{-4} &= -2 \\ -4\cdot\frac{r}{-4} &= -4\cdot(-2) \\ r &= 8 \end{aligned}$$

10.
$$\begin{aligned} \frac{y}{-8} &= 8 \\ -8\cdot\frac{y}{-8} &= -8\cdot 8 \\ y &= -64 \end{aligned}$$

11.
$$\begin{aligned} 6-2x+8 &= 10 \\ -2x+14 &= 10 \\ -2x+14-14 &= 10-14 \\ -2x &= -4 \\ \frac{-2x}{-2} &= \frac{-4}{-2} \\ x &= 2 \end{aligned}$$

12.
$$\begin{aligned} -5-6y+6 &= 19 \\ -6y+1 &= 19 \\ -6y+1-1 &= 19-1 \\ -6y &= 18 \\ \frac{-6y}{-6} &= \frac{18}{-6} \\ y &= -3 \end{aligned}$$

13.
$$\begin{aligned} 2x-7 &= 6x-27 \\ 2x-7+7 &= 6x-27+7 \\ 2x &= 6x-20 \\ 2x-6x &= 6x-20-6x \\ -4x &= -20 \\ \frac{-4x}{-4} &= \frac{-20}{-4} \\ x &= 5 \end{aligned}$$

14.
$$\begin{aligned} 3+8y &= 3y-2 \\ 3+8y-3y &= 3y-2-3y \\ 3+5y &= -2 \\ -3+3+5y &= -3-2 \\ 5y &= -5 \\ \frac{5y}{5} &= \frac{-5}{5} \\ y &= -1 \end{aligned}$$

15.
$$\begin{aligned} 9(3x-1) &= -4+49 \\ 27x-9 &= 45 \\ 27x-9+9 &= 45+9 \\ 27x &= 54 \\ \frac{27x}{27} &= \frac{54}{27} \\ x &= 2 \end{aligned}$$

16.
$$\begin{aligned} 12(2x+1) &= -6+66 \\ 24x+12 &= 60 \\ 24x+12-12 &= 60-12 \\ 24x &= 48 \\ \frac{24x}{24} &= \frac{48}{24} \\ x &= 2 \end{aligned}$$

17.
$$\begin{aligned} -3a+6+5a &= 7a-8a \\ 6+2a &= -a \\ 6+2a-2a &= -a-2a \\ 6 &= -3a \\ \frac{6}{-3} &= \frac{-3a}{-3} \\ -2 &= a \end{aligned}$$

18.
$$\begin{aligned} 4b-8-b &= 10b-3b \\ 3b-8 &= 7b \\ -3b+3b-8 &= -3b+7b \\ -8 &= 4b \\ \frac{-8}{4} &= \frac{4b}{4} \\ -2 &= b \end{aligned}$$

19.
$$\begin{aligned} -\frac{2}{3}x &= \frac{5}{9} \\ -\frac{3}{2}\cdot\left(-\frac{2}{3}x\right) &= -\frac{3}{2}\cdot\frac{5}{9} \\ x &= -\frac{15}{18} \\ x &= -\frac{5}{6} \end{aligned}$$

20.
$$\begin{aligned} -\frac{3}{8}y &= -\frac{1}{16} \\ -\frac{8}{3}\cdot\left(-\frac{3}{8}y\right) &= -\frac{8}{3}\cdot\left(-\frac{1}{16}\right) \\ y &= \frac{1}{6} \end{aligned}$$

21.
$$\begin{aligned} 10 &= -6n+16 \\ 10-16 &= -6n+16-16 \\ -6 &= -6n \\ \frac{-6}{-6} &= \frac{-6n}{-6} \\ 1 &= n \end{aligned}$$

22.
$$\begin{aligned} -5 &= -2m+7 \\ -5-7 &= -2m+7-7 \\ -12 &= -2m \\ \frac{-12}{-2} &= \frac{-2m}{-2} \\ 6 &= m \end{aligned}$$

23.
$$\begin{aligned} 3(5c-1)-2 &= 13c+3 \\ 15c-3-2 &= 13c+3 \\ 15c-5 &= 13c+3 \\ 15c-5+5 &= 13c+3+5 \\ 15c &= 13c+8 \\ 15c-13c &= 13c+8-13c \\ 2c &= 8 \\ \frac{2c}{2} &= \frac{8}{2} \\ c &= 4 \end{aligned}$$

24.
$$\begin{aligned} 4(3t+4)-20 &= 3+5t \\ 12t+16-20 &= 3+5t \\ 12t-4 &= 3+5t \\ 12t-4-5t &= 3+5t-5t \\ 7t-4 &= 3 \\ 7t-4+4 &= 3+4 \\ 7t &= 7 \\ \frac{7t}{7} &= \frac{7}{7} \\ t &= 1 \end{aligned}$$

25.
$$\begin{aligned} \frac{2(z+3)}{3} &= 5-z \\ 3\left[\frac{2(z+3)}{3}\right] &= 3(5-z) \\ 2(z+3) &= 3(5-z) \\ 2z+6 &= 15-3z \\ 2z+6+3z &= 15-3z+3z \\ 6+5z &= 15 \\ 6+5z-6 &= 15-6 \\ 5z &= 9 \\ \frac{5z}{5} &= \frac{9}{5} \\ z &= \frac{9}{5} \end{aligned}$$

26.
$$\begin{aligned}\frac{3(w+2)}{4}&=2w+3\\ 4\left[\frac{3(w+2)}{4}\right]&=4(2w+3)\\ 3(w+2)&=4(2w+3)\\ 3w+6&=8w+12\\ 3w+6-6&=8w+12-6\\ 3w&=8w+6\\ 3w-8w&=8w+6-8w\\ -5w&=6\\ \frac{-5w}{-5}&=\frac{6}{-5}\\ w&=-\frac{6}{5}\end{aligned}$$

27.
$$\begin{aligned}-2(2x-5)&=-3x+7-x+3\\ -4x+10&=-4x+10\end{aligned}$$
Since both sides of the equation are identical, the equation is an identity and every real number is a solution.

28.
$$\begin{aligned}-4(5x-2)&=-12x+4-8x+4\\ -20x+8&=-20x+8\end{aligned}$$
Since both sides of the equation are identical, the equation is an identity and every real number is a solution.

29.
$$\begin{aligned}0.02(6t-3)&=0.04(t-2)+0.02\\ 2(6t-3)&=4(t-2)+2\\ 12t-6&=4t-8+2\\ 12t-6&=4t-6\\ 12t-6-4t&=4t-6-4t\\ 8t-6&=-6\\ 8t-6+6&=-6+6\\ 8t&=0\\ \frac{8t}{8}&=\frac{0}{8}\\ t&=0\end{aligned}$$

30.
$$\begin{aligned}0.03(m+7)&=0.02(5-m)+0.03\\ 3(m+7)&=2(5-m)+3\\ 3m+21&=10-2m+3\\ 3m+21&=13-2m\\ 3m+21+2m&=13-2m+2m\\ 5m+21&=13\\ 5m+21-21&=13-21\\ 5m&=-8\\ \frac{5m}{5}&=\frac{-8}{5}\\ m&=-1.6\end{aligned}$$

31.
$$\begin{aligned}-3y&=\frac{4(y-1)}{5}\\ 5(-3y)&=5\left[\frac{4(y-1)}{5}\right]\\ -15y&=4(y-1)\\ -15y&=4y-4\\ -15y-4y&=4y-4-4y\\ -19y&=-4\\ \frac{-19y}{-19}&=\frac{-4}{-19}\\ y&=\frac{4}{19}\end{aligned}$$

32.
$$\begin{aligned}-4x&=\frac{5(1-x)}{6}\\ 6(-4x)&=6\cdot\frac{5(1-x)}{6}\\ -24x&=5(1-x)\\ -24x&=5-5x\\ -24x+5x&=5-5x+5x\\ -19x&=5\\ \frac{-19x}{-19}&=\frac{5}{-19}\\ x&=-\frac{5}{19}\end{aligned}$$

33.
$$\begin{aligned}\frac{5}{3}x-\frac{7}{3}&=x\\ 3\left(\frac{5}{3}x-\frac{7}{3}\right)&=3x\\ 5x-7&=3x\\ -5x+5x-7&=-5x+3x\\ -7&=-2x\\ \frac{-7}{-2}&=\frac{-2x}{-2}\\ \frac{7}{2}&=x\end{aligned}$$

34.
$$\begin{aligned}\frac{7}{5}n+\frac{3}{5}&=-n\\ 5\left(\frac{7}{5}n+\frac{3}{5}\right)&=5(-n)\\ 7n+3&=-5n\\ -7n+7n+3&=-7n-5n\\ 3&=-12n\\ \frac{3}{-12}&=\frac{-12n}{-12}\\ -\frac{1}{4}&=n\end{aligned}$$

35.
$$\frac{1}{10}(3x-7)=\frac{3}{10}x+5$$
$$\frac{3}{10}x-\frac{7}{10}=\frac{3}{10}x+5$$
$$-\frac{3}{10}x+\frac{3}{10}x-\frac{7}{10}=-\frac{3}{10}x+\frac{3}{10}x+5$$
$$-\frac{7}{10}=5$$

Since the statement $-\frac{7}{10}=5$ is false, the equation has no solution.

36.
$$\frac{1}{7}(2x-5)=\frac{2}{7}x+1$$
$$7\cdot\frac{1}{7}(2x-5)=7\left(\frac{2}{7}x+1\right)$$
$$2x-5=2x+7$$
$$2x-5-2x=2x+7-2x$$
$$-5=7$$

Since the statement $-5 = 7$ is false, the equation has no solution.

37.
$$5+2(3x-6)=-4(6x-7)$$
$$5+6x-12=-24x+28$$
$$6x-7=-24x+28$$
$$24x+6x-7=24x-24x+28$$
$$30x-7=28$$
$$30x-7+7=28+7$$
$$30x=35$$
$$\frac{30x}{30}=\frac{35}{30}$$
$$x=\frac{7}{6}$$

38.
$$3+5(2x-4)=-7(5x+2)$$
$$3+10x-20=-35x-14$$
$$10x-17=-35x-14$$
$$10x-17+35x=-35x-14+35x$$
$$45x-17=-14$$
$$45x-17+17=-14+17$$
$$45x=3$$
$$\frac{45x}{45}=\frac{3}{45}$$
$$x=\frac{1}{15}$$

Section 2.4 Practice Exercises

1. Let x represent the number.
$$3x-6=2x+3$$
$$3x-6-2x=2x+3-2x$$
$$x-6=3$$
$$x-6+6=3+6$$
$$x=9$$
The number is 9.

2. Let x represent the number.
$$3(x-5)=2x-3$$
$$3x-15=2x-3$$
$$3x-15-2x=2x-3-2x$$
$$x-15=-3$$
$$x-15+15=-3+15$$
$$x=12$$
The number is 12.

3. Let x represent the length of the shorter piece. Then $5x$ represents the length of the longer piece. Their sum is 18 feet.
$$x+5x=18$$
$$6x=18$$
$$\frac{6x}{6}=\frac{18}{6}$$
$$x=3$$
The shorter piece is 3 feet and the longer piece is 5(3) = 15 feet.

4. Let x represent the number of votes for Texas. Then $x + 17$ represents the number of votes for California. Their sum is 93.
$$x+x+17=93$$
$$2x+17=93$$
$$2x+17-17=93-17$$
$$2x=76$$
$$\frac{2x}{2}=\frac{76}{2}$$
$$x=38$$
Texas has 38 electoral votes and California has 38 + 17 = 55 electoral votes.

5. Let x represent the number of miles driven. The cost for x miles is $0.15x$. The daily cost is $28.
$$0.15x+28=52$$
$$0.15x+28-28=52-28$$
$$0.15x=24$$
$$\frac{0.15x}{0.15}=\frac{24}{0.15}$$
$$x=160$$
You drove 160 miles.

6. Let x represent the measure of the smallest angle. Then $2x$ represents the measure of the second angle and $3x$ represents the measure of the third angle. The sum of the measures of the angles of a triangle equals 180.

$$x+2x+3x=180$$
$$6x=180$$
$$\frac{6x}{6}=\frac{180}{6}$$
$$x=30$$

If $x = 30$, then $2x = 2(30) = 60$ and $3x = 3(30) = 90$.

The smallest is 30°, second is 60°, and third is 90°.

7. If x is the first even integer, then $x + 2$ and $x + 4$ are the next two even integers.

$$x+x+2+x+4=144$$
$$3x+6=144$$
$$3x+6-6=144-6$$
$$3x=138$$
$$\frac{3x}{3}=\frac{138}{3}$$
$$x=46$$

If $x = 46$, then $x + 2 = 48$ and $x + 4 = 50$. The integers are 46, 48, 50.

Vocabulary, Readiness & Video Check 2.4

1. If x is the number, then "double the number" is $2x$, and "double the number, decreased by 31" is $2x - 31$.

2. If x is the number, then "three times the number" is $3x$, and "three times the number, increased by 17" is $3x + 17$.

3. If x is the number, then "the sum of the number and 5" is $x + 5$, and "twice the sum of the number and 5" is $2(x + 5)$.

4. If x is the number, then "the difference of the number and 11" is $x - 11$, and "seven times the difference of the number and 11" is $7(x - 11)$.

5. If y is the number, then "the difference of 20 and the number" is $20 - y$, and "the difference of 20 and the number, divided by 3" is $\frac{20-y}{3}$ or $(20 - y) \div 3$.

6. If y is the number, then "the sum of -10 and the number" is $-10 + y$, and "the sum of -10 and the number, divided by 9" is $\frac{(-10+y)}{9}$ or $(-10 + y) \div 9$.

7. in the statement of the application

8. The original application asks for the measure of two supplementary angles. The solution of $x = 43$ only gives us the measure of one of the angles.

9. That the 3 angle measures are consecutive even integers and that they sum to 180°.

Exercise Set 2.4

1.
$$2x+7=x+6$$
$$2x+7-x=x+6-x$$
$$x+7=6$$
$$x+7-7=6-7$$
$$x=-1$$

The number is -1.

3.
$$3x-6=2x+8$$
$$3x-6-2x=2x+8-2x$$
$$x-6=8$$
$$x-6+6=8+6$$
$$x=14$$

The number is 14.

5.
$$2(x-8)=3(x+3)$$
$$2x-16=3x+9$$
$$2x-16-2x=3x+9-2x$$
$$-16=x+9$$
$$-16-9=x+9-9$$
$$-25=x$$

The number is -25.

7.
$$2x(3)=5x-\frac{3}{4}$$
$$6x=5x-\frac{3}{4}$$
$$6x-5x=5x-\frac{3}{4}-5x$$
$$x=-\frac{3}{4}$$

The number is $-\frac{3}{4}$.

9. The sum of the three lengths is 25 inches.

$$x+2x+1+5x=25$$
$$1+8x=25$$
$$1+8x-1=25-1$$
$$8x=24$$
$$\frac{8x}{8}=\frac{24}{8}$$
$$x=3$$

$2x = 2(3) = 6$
$1 + 5x = 1 + 5(3) = 1 + 15 = 16$
The lengths are 3 inches, 6 inches, and 16 inches.

11. Let x be the length of the first piece. Then the second piece is $2x$ and the third piece is $5x$. The sum of the lengths is 40 inches.

$$x+2x+5x=40$$
$$8x=40$$
$$\frac{8x}{8}=\frac{40}{8}$$
$$x=5$$

$2x = 2(5) = 10$
$5x = 5(5) = 25$
The 1st piece is 5 inches, 2nd piece is 10 inches, and 3rd piece is 25 inches.

13. Let x represent the amount, in millions of pounds, of apples produced in Pennsylvania. then $x + 226$ represents the amount of apples produced in New York.

$$x+x+226=1214$$
$$2x+226=1214$$
$$2x+226-226=1214-226$$
$$2x=988$$
$$\frac{2x}{2}=\frac{988}{2}$$
$$x=494$$

$x + 226 = 494 + 226 = 720$
Pennsylvania produced 494 million pounds of apples and New York produced 720 pounds.

15. Let x be the number of miles. Then the cost for x miles is $0.29x$. Each day costs \$24.95.

$$0.29x+2(24.95)=100$$
$$0.29x+49.9=100$$
$$0.29x+49.9-49.9=100-49.9$$
$$0.29x=50.1$$
$$\frac{0.29x}{0.29}=\frac{50.1}{0.29}$$
$$x\approx 172.8$$

You can drive 172 whole miles on a \$100 budget.

17. Let x be the number of miles. Then the total fare is $3 + 0.8x + 4.5$.

$$\begin{aligned} 3+0.8x+4.5 &= 27.5 \\ 30+8x+45 &= 275 \\ 8x+75 &= 275 \\ 8x+75-75 &= 275-75 \\ 8x &= 200 \\ \frac{8x}{8} &= \frac{200}{8} \\ x &= 25 \end{aligned}$$

You can travel 25 miles from the airport by taxi for \$27.50.

19. Let x be the measure of each of the two equal angles. Then $2x + 30$ is the measure of the third angle. Their sum is 180°.

$$\begin{aligned} x+x+2x+30 &= 180 \\ 4x+30 &= 180 \\ 4x+30-30 &= 180-30 \\ 4x &= 150 \\ \frac{4x}{4} &= \frac{150}{4} \\ x &= 37.5 \end{aligned}$$

$2x + 30 = 2(37.5) + 30 = 75 + 30 = 105$

The 1st angle measures 37.5°, the 2nd angle measures 37.5°, and the 3rd angle measures 105°.

21. Angles A and D both measure $x°$, while angles C and B both measure $(2x)°$. The sum of the angle measures is 360°.

$$\begin{aligned} x+2x+x+2x &= 360 \\ 6x &= 360 \\ \frac{6x}{6} &= \frac{360}{6} \\ x &= 60 \end{aligned}$$

$2x = 2(60) = 120$

Angles A and D measure 60°; angles B and C measure 120°.

	First Integer	Next Integers			Indicated Sum
23.	x	$x + 1$	$x + 2$		$x + (x + 1) + (x + 2) = 3x + 3$
25.	x	$x + 2$	$x + 4$		$x + (x + 4) = 2x + 4$
27.	x	$x + 1$	$x + 2$	$x + 3$	$x + (x + 1) + (x + 2) + (x + 3) = 4x + 6$
29.	x	$x + 2$	$x + 4$		$(x + 2) + (x + 4) = 2x + 6$

31. If x is the first integer, the next consecutive integer is $x + 1$.

$$\begin{aligned} x+x+1 &= 469 \\ 2x+1 &= 469 \\ 2x+1-1 &= 469-1 \\ 2x &= 468 \\ \frac{2x}{2} &= \frac{468}{2} \\ x &= 234 \end{aligned}$$

The page numbers are 234 and 234 + 1 = 235.

33. If x is the first integer, the next two consecutive integers are $x + 1$ and $x + 2$.

$$x+x+1+x+2=99$$
$$3x+3=99$$
$$3x+3-3=99-3$$
$$3x=96$$
$$\frac{3x}{3}=\frac{96}{3}$$
$$x=32$$

The code for Belgium is 32, France is $32 + 1 = 33$, and Spain is $32 + 2 = 34$.

35. Let x be the length of the shorter piece. Then $2x + 2$ is the length of the longer piece. The measures sum to 17 feet.

$$x+2x+2=17$$
$$3x+2=17$$
$$3x+2-2=17-2$$
$$3x=5$$
$$\frac{3x}{3}=\frac{15}{3}$$
$$x=5$$

$2x + 2 = 2(5) + 2 = 10 + 2 = 12$

The pieces measure 5 feet and 12 feet.

37. Let x represent the speed of the TGV. Then the speed of the Maglev is $x + 3.8$.

$$x+x+3.8=718.2$$
$$2x+3.8=718.2$$
$$2x+3.8-3.8=718.2-3.8$$
$$2x=714.4$$
$$\frac{2x}{2}=\frac{714.4}{2}$$
$$x=357.2$$

$x + 3.8 = 357.2 + 3.8 = 361$

The speed of the TGV is 357.2 miles per hour and the speed of the Maglev is 361 miles per hour.

39. Let x be the measure of the smaller angle. Then the larger angle measures $3x + 8$. Their sum is 180°.

$$x+3x+8=180$$
$$4x+8=180$$
$$4x+8-8=180-8$$
$$4x=172$$
$$\frac{4x}{4}=\frac{172}{4}$$
$$x=43$$

$3x + 8 = 3(43) + 8 = 129 + 8 = 137$

The angles measure 43° and 137°.

41. Let x be the first even integer. Then the next two consecutive even integers are $x + 2$ and $x + 4$. The sum of the measures of the angles of a triangle is 180°.

$$x+x+2+x+4=180$$
$$3x+6=180$$
$$3x+6-6=180-6$$
$$3x=174$$
$$\frac{3x}{3}=\frac{174}{3}$$
$$x=58$$

$x + 2 = 58 + 2 = 60$

$x + 4 = 58 + 4 = 62$

The angles measure 58°, 60°, and 62°.

43.

$$\frac{1}{5}+2x=3x-\frac{4}{5}$$
$$\frac{1}{5}+2x-2x=3x-\frac{4}{5}-2x$$
$$\frac{1}{5}=x-\frac{4}{5}$$
$$\frac{1}{5}+\frac{4}{5}=x-\frac{4}{5}+\frac{4}{5}$$
$$\frac{5}{5}=x$$
$$1=x$$

The number is 1.

45. Let x be the number of miles. Then the charge for driving x miles in one day is $39 + 0.2x$.

$$39+0.2x=95$$
$$390+2x=950$$
$$390+2x-390=950-390$$
$$2x=560$$
$$\frac{2x}{2}=\frac{560}{2}$$
$$x=280$$

You drove 280 miles.

47. Let x represent the number of points scored by Wisconsin. Then Stanford scored $x + 6$ points. Their combined scores totaled 34 points.

$$x+x+6=34$$
$$2x+6=34$$
$$2x+6-6=34-6$$
$$2x=28$$
$$\frac{2x}{2}=\frac{28}{2}$$
$$x=14$$

$x + 6 = 14 + 6 = 20$

Wisconsin scored 14 points and Stanford scored 20 points.

49. Let x represent the number of counties in Montana. Then $x + 2$ represents the number of counties in California.

$$x+x+2=114$$
$$2x+2=114$$
$$2x+2-2=114-2$$
$$2x=112$$
$$\frac{2x}{2}=\frac{112}{2}$$
$$x=56$$

$x + 2 = 56 + 2 = 58$

Montana has 56 counties and California has 58 counties.

51. Let x represent the number of satellites for Neptune. Then $x + 13$ represents the number of satellites for Uranus and $4x + 6$ represents the number of satellites for Saturn. The total number of satellites is 103.

$$x+x+13+4x+6=103$$
$$6x+19=103$$
$$6x+19-19=103-19$$
$$6x=84$$
$$\frac{6x}{6}=\frac{84}{6}$$
$$x=14$$

$x + 13 = 14 + 13 = 27$

$4x + 6 = 4(14) + 6 = 56 + 6 = 62$

Neptune has 14 satellites, Uranus has 27 satellites, and Saturn has 62 satellites.

53.

$$3(x+5)=2x-1$$
$$3x+15=2x-1$$
$$3x+15-2x=2x-1-2x$$
$$x+15=-1$$
$$x+15-15=-1-15$$
$$x=-16$$

The number is -16.

55. Let x represent the area of the Gobi Desert, in square miles. Then $7x$ represents the area of the Sahara Desert.

$$x+7x=4,000,000$$
$$8x=4,000,000$$
$$\frac{8x}{8}=\frac{4,000,000}{8}$$
$$x=500,000$$

$7x = 7(500,000) = 3,500,000$

The Gobi Desert's area is 500,000 square miles and the Sahara Desert's area is 3,500,000 square miles.

57. Let x represent the number of gold medals won by Jamaica. Then Cuba won $x + 1$ medals and New Zealand won $x + 2$ medals.

$$x+x+1+x+2=15$$
$$3x+3=15$$
$$3x+3-3=15-3$$
$$3x=12$$
$$\frac{3x}{3}=\frac{12}{3}$$
$$x=4$$

$x + 1 = 4 + 1 = 5$

$x + 2 = 4 + 2 = 6$

Jamaica won 4 gold medals, Cuba won 5 gold medals, and New Zealand won 6 gold medals.

59. Let x represent the number of male students. Then $x + 1580$ represents the number of female students enrolled.

$$x+x+1580=58,788$$
$$2x+1580=58,788$$
$$2x+1580-1580=58,788-1580$$
$$2x=57,208$$
$$\frac{2x}{2}=\frac{57,208}{2}$$
$$x=28,604$$

$x + 1580 = 28,604 + 1580 = 30,184$

There were 28,604 male students and 30,184 female students enrolled.

61. Let x be the measure of the two equal angles. Then $x + 76.5$ is the measure of the third angle.

$$x+x+x+76.5=180$$
$$3x+76.5=180$$
$$3x+76.5-76.5=180-76.5$$
$$3x=103.5$$
$$\frac{3x}{3}=\frac{103.5}{3}$$
$$x=34.5$$

$x + 76.5 = 34.5 + 76.5 = 111$

The three angles measure 34.5°, 34.5°, and 111°.

63. The longest bar represents Hawaii, so Hawaii spent the most money on tourism.

65. Let x represent the amount, in millions of dollars, spent on tourism by California. Then Florida spent $x + 6$.

$$x+x+6=106$$
$$2x+6=106$$
$$2x+6-6=106-6$$
$$2x=100$$
$$\frac{2x}{2}=\frac{100}{2}$$
$$x=50$$

$x + 6 = 50 + 6 = 56$

California spent \$50 million on tourism and Florida spent \$56 million.

67. answers may vary

69. Replace W by 7 and L by 10.

$2W + 2L = 2(7) + 2(10) = 14 + 20 = 34$

71. Replace r by 15.

$\pi r^2 = \pi(15)^2 = \pi(225) = 225\pi$

73. Let x represent the width. Then $1.6x$ represents the length. The perimeter is

$2 \cdot \text{length} + 2 \cdot \text{width}$.

$$2(1.6x)+2x=78$$
$$3.2x+2x=78$$
$$5.2x=78$$
$$\frac{5.2x}{5.2}=\frac{78}{5.2}$$
$$x=15$$

$1.6x = 1.6(15) = 24$

The dimensions of the garden are 15 feet by 24 feet.

75. 90 chirps every minute is $\frac{90 \text{ chirps}}{1 \text{ min}}$. There are 60 minutes in one hour.

$$\frac{90 \text{ chirps}}{1 \text{ min}} \cdot 60 \text{ min} = 5400 \text{ chirps}$$

At this rate, there are 5400 chirps each hour.

$24 \cdot 5400 = 129,600$

There are 129,600 chirps in one 24-hour day.

$365 \cdot 129,600 = 47,304,000$

There are 47,304,000 chirps in one year.

77. answers may vary

79. answers may vary

81. Measurements may vary. Rectangle (c) best approximates the shape of the golden rectangle.

Section 2.5 Practice Exercises

1. Use $d = rt$ when $d = 1180$ and $r = 50$.

$$d=rt$$
$$1180=50t$$
$$\frac{1180}{50}=\frac{50t}{50}$$
$$23.6=t$$

They will spend 23.6 hours driving.

2. Use $A = lw$ when $w = 18$.

$$A=lw$$
$$450=l\cdot 18$$
$$\frac{450}{18}=\frac{18l}{18}$$
$$25=l$$

The length of the deck is 25 feet.

3. Use $F=\frac{9}{5}C+32$ with $C = 5$.

$$F=\frac{9}{5}C+32$$
$$F=\frac{9}{5}\cdot 5+32$$
$$F=9+32$$
$$F=41$$

Thus, 5°C is equivalent to 41°F.

4. Let x be the width. Then $4x + 1$ is the length. The perimeter is 52 meters.

$$P=2l+2w$$
$$52=2(4x+1)+2x$$
$$52=8x+2+2x$$
$$52=10x+2$$
$$52-2=10x+2-2$$
$$50=10x$$
$$\frac{50}{10}=\frac{10x}{10}$$
$$5=x$$

$4x + 1 = 4(5) + 1 = 20 + 1 = 21$

The width is 5 meters and the length is 21 meters.

5.

$$C=2\pi r$$
$$\frac{C}{2\pi}=\frac{2\pi r}{2\pi}$$
$$\frac{C}{2\pi}=r \text{ or } r=\frac{C}{2\pi}$$

6.
$$\begin{aligned} P &= 2l + 2w \\ P - 2w &= 2l + 2w - 2w \\ P - 2w &= 2l \\ \frac{P-2w}{2} &= \frac{2l}{2} \\ \frac{P-2w}{2} &= l \text{ or } l = \frac{P-2w}{2} \end{aligned}$$

7.
$$\begin{aligned} P &= 2a + b - c \\ P + c &= 2a + b - c + c \\ P + c &= 2a + b \\ P + c - b &= 2a + b - b \\ P + c - b &= 2a \\ \frac{P+c-b}{2} &= a \text{ or } a = \frac{P-b-c}{2} \end{aligned}$$

8.
$$\begin{aligned} A &= \frac{a+b}{2} \\ 2A &= 2 \cdot \frac{a+b}{2} \\ 2A &= a + b \\ 2A - a &= a + b - a \\ 2A - a &= b \text{ or } b = 2A - a \end{aligned}$$

Vocabulary, Readiness & Video Check 2.5

1. A formula is an equation that describes known <u>relationships</u> among quantities.

2. This is a distance, rate, and time problem. The rate is given in miles per hour (mph) and the time is given in hours, so the distance that we are finding must be in miles.

3. To show that the process of solving this equation for x—dividing both sides by 5, the coefficient of x—is the same process used to solve a formula for a specific variable. Treat whatever is multiplied by that specific variable as the coefficient—the coefficient is all the factors except that specific variable.

Exercise Set 2.5

1. Use $A = bh$ when $A = 45$ and $b = 15$.
$$\begin{aligned} A &= bh \\ 45 &= 15 \cdot h \\ \frac{45}{15} &= \frac{15h}{15} \\ 3 &= h \end{aligned}$$

3. Use $S = 4lw + 2wh$ when $S = 102$, $l = 7$, and $w = 3$.
$$\begin{aligned} S &= 4lw + 2wh \\ 102 &= 4 \cdot 7 \cdot 3 + 2 \cdot 3 \cdot h \\ 102 &= 84 + 6h \\ 102 - 84 &= 84 + 6h - 84 \\ 18 &= 6h \\ \frac{18}{6} &= \frac{6h}{6} \\ 3 &= h \end{aligned}$$

5. Use $A = \frac{1}{2}h(B+b)$ when $A = 180$, $B = 11$, and $b = 7$.
$$\begin{aligned} A &= \frac{1}{2}h(B+b) \\ 180 &= \frac{1}{2}h(11+7) \\ 180 &= \frac{1}{2}h(18) \\ 180 &= 9h \\ \frac{180}{9} &= \frac{9h}{9} \\ 20 &= h \end{aligned}$$

7. Use $P = a + b + c$ when $P = 30$, $a = 8$, and $b = 10$.
$$\begin{aligned} P &= a + b + c \\ 30 &= 8 + 10 + c \\ 30 &= 18 + c \\ 30 - 18 &= 18 + c - 18 \\ 12 &= c \end{aligned}$$

9. Use $C = 2\pi r$ when $C = 15.7$ and 3.14 is used as an approximation for π.
$$\begin{aligned} C &= 2\pi r \\ 15.7 &= 2(3.14)r \\ 15.7 &= 6.28r \\ \frac{15.7}{6.28} &= \frac{6.28r}{6.28} \\ 2.5 &= r \end{aligned}$$

11.
$$\begin{aligned} f &= 5gh \\ \frac{f}{5g} &= \frac{5gh}{5g} \\ \frac{f}{5g} &= h \end{aligned}$$

13. $V = lwh$
$\frac{V}{lh} = \frac{lwh}{lh}$
$\frac{V}{lh} = w$

15. $3x + y = 7$
$3x + y - 3x = 7 - 3x$
$y = 7 - 3x$

17. $A = P + PRT$
$A - P = P + PRT - P$
$A - P = PRT$
$\frac{A-P}{PT} = \frac{PRT}{PT}$
$\frac{A-P}{PT} = R$

19. $V = \frac{1}{3}Ah$
$3V = 3 \cdot \frac{1}{3}Ah$
$3V = Ah$
$\frac{3V}{h} = \frac{Ah}{h}$
$\frac{3V}{h} = A$

21. $P = a + b + c$
$P - b - c = a + b + c - b - c$
$P - b - c = a$

23. $S = 2\pi rh + 2\pi r^2$
$S - 2\pi r^2 = 2\pi rh + 2\pi r^2 - 2\pi r^2$
$S - 2\pi r^2 = 2\pi rh$
$\frac{S - 2\pi r^2}{2\pi r} = \frac{2\pi rh}{2\pi r}$
$\frac{S - 2\pi r^2}{2\pi r} = h$

25. Use $A = lw$ when $A = 10{,}080$ and $w = 84$.
$A = lw$
$10{,}080 = l(84)$
$\frac{10{,}080}{84} = \frac{84l}{84}$
$120 = l$
The length (height) of the sign is 120 feet.

27. a. $\text{Area} = \frac{1}{2}h(B + b)$
$= \frac{1}{2} \cdot 12(56 + 24)$
$= 6(80)$
$= 480$
Perimeter = 24 + 20 + 56 + 20 = 120
The area is 480 square inches and the perimeter is 120 inches.

b. The frame goes around the edges of the picture, so it involves perimeter. The glass covers the picture, so it involves area.

29. a. Area $= l \cdot w = (11.5)(9) = 103.5$
$\text{Perimeter} = 2l + 2w$
$= 2(11.5) + 2(9)$
$= 23 + 18$
$= 41$
The area is 103.5 square feet and the perimeter is 41 feet.

b. The baseboard goes around the edges of the room, so it involves the perimeter. The carpet covers the floor of the room, so it involves area.

31. Use $F = \frac{9}{5}C + 32$ when $F = 14$.
$F = \frac{9}{5}C + 32$
$14 = \frac{9}{5}C + 32$
$14 - 32 = \frac{9}{5}C + 32 - 32$
$-18 = \frac{9}{5}C$
$\frac{5}{9} \cdot (-18) = \frac{5}{9} \cdot \frac{9}{5}C$
$-10 = C$
Thus, 14°F is equivalent to −10°C.

33. Use $d = rt$ when $d = 25{,}000$ and $r = 4000$.
$d = rt$
$25{,}000 = 4000t$
$\frac{25{,}000}{4000} = \frac{4000t}{4000}$
$6.25 = t$
It will take the X-30 6.25 hours to travel around the Earth.

35. Let x be the length. Then $\frac{2}{3}x$ is the width. Use $P = 2 \cdot \text{length} + 2 \cdot \text{width}$ when $P = 260$.

$$P = 2 \cdot \text{length} + 2 \cdot \text{width}$$
$$260 = 2x + 2 \cdot \frac{2}{3}x$$
$$260 = 2x + \frac{4}{3}x$$
$$260 = \frac{6}{3}x + \frac{4}{3}x$$
$$260 = \frac{10}{3}x$$
$$\frac{3}{10} \cdot 260 = \frac{3}{10} \cdot \frac{10}{3}x$$
$$78 = x$$

The length is 78 feet and the width is $\frac{2}{3} \cdot 78 = 52$ feet.

37. Let x represent the length of the shortest side. Then the second side has length $2x$ and the third side has length $30 + x$. The perimeter is the sum of the lengths of the sides.

$$x + 2x + 30 + x = 102$$
$$4x + 30 = 102$$
$$4x + 30 - 30 = 102 - 30$$
$$4x = 72$$
$$\frac{4x}{4} = \frac{72}{4}$$
$$x = 18$$

$2x = 2(18) = 36$
$30 + x = 30 + 18 = 48$
The flower bed has sides of length 18 feet, 36 feet, and 48 feet.

39. Use $d = rt$ when $r = 55$ and $t = 2\frac{1}{2}$.

$$d = rt$$
$$d = 55 \cdot 2\frac{1}{2}$$
$$d = 55 \cdot 2.5$$
$$d = 137.5$$

The distance between Bar Harbor and Yarmouth is 137.5 miles.

41. Use $N = 86$.

$$T = 50 + \frac{N-40}{4}$$
$$T = 50 + \frac{86-40}{4}$$
$$T = 50 + \frac{46}{4}$$
$$T = 50 + 11.5$$
$$T = 61.5$$

The temperature is 61.5° Fahrenheit.

43. Use $T = 55$.

$$T = 50 + \frac{N-40}{4}$$
$$55 = 50 + \frac{N-40}{4}$$
$$55 - 50 = 50 + \frac{N-40}{4} - 50$$
$$5 = \frac{N-40}{4}$$
$$4 \cdot 5 = 4 \cdot \frac{N-40}{4}$$
$$20 = N - 40$$
$$20 + 40 = N - 40 + 40$$
$$60 = N$$

There are 60 chirps per minute.

45. As the number of cricket chirps per minute increases, the air temperature of their environment increases.

47. To find the amount of water in the tank, use $V = lwh$ with $l = 8$, $w = 3$, and $h = 6$.
$V = lwh = 8 \cdot 3 \cdot 6 = 144$
The tank holds 144 cubic feet of water. Let x represent the number of piranhas the tank could hold. Then $1.5x = 144$.

$$1.5x = 144$$
$$\frac{1.5x}{1.5} = \frac{144}{1.5}$$
$$x = 96$$

The tank could hold 96 piranhas.

49. Use $A = \frac{1}{2}h(B+b)$ to find the area of the lawn.

$$A = \frac{1}{2}h(B+b)$$
$$A = \frac{1}{2}(60)(130+70) = 30(200) = 6000$$

Let x be the number of bags of fertilizer.

$$4000x = 6000$$
$$\frac{4000x}{4000} = \frac{6000}{4000}$$
$$x = 1.5$$

Since $\frac{1}{2}$ bag cannot be purchased, 2 bags must be purchased to cover the lawn.

51. Use $A = \pi r^2$ to find the area of a pizza.

For the 16-inch pizza, $r = \frac{16}{2} = 8.$

$A = \pi r^2 = \pi(8)^2 = 64\pi$

For a 10-inch pizza, $r = \frac{10}{2} = 5.$

$A = \pi r^2 = \pi(5)^2 = 25\pi$

Two 10-inch pizzas have an area of $2 \cdot 25\pi = 50\pi$ square inches. Since $50\pi < 64\pi$, you get more pizza by buying the 16-inch pizza.

53. Use $d = rt$ when $r = 552$ and $d = 42.8$.

$$d = rt$$
$$42.8 = 552t$$
$$\frac{42.8}{552} = \frac{552t}{552}$$
$$0.0775 = t$$

It would last 0.0775 hour or $0.0775(60) \approx 4.65$ minutes.

55. Let s represent the length of one side of the square. Then the perimeter of the square is $4s$. A side of the triangle is $s + 5$ and the triangle's perimeter is $3(s + 5)$.

$$3(s+5) = 4s+7$$
$$3s+15 = 4s+7$$
$$3s+15-3s = 4s+7-3s$$
$$15 = s+7$$
$$15-7 = s+7-7$$
$$8 = s$$

$s + 5 = 8 + 5 = 13$

Each side of the triangle has length 13 inches.

57. Use $d = rt$ when $d = 135$ an $r = 60$.

$$d = rt$$
$$135 = 60t$$
$$\frac{135}{60} = \frac{60t}{60}$$
$$2.25 = t$$

It will take 2.25 hours.

59. Use $A = lw$ when $A = 1{,}813{,}500$ and $w = 150$.

$$A = lw$$
$$1{,}813{,}500 = l(150)$$
$$\frac{1{,}813{,}500}{150} = \frac{150l}{150}$$
$$12{,}090 = l$$

The length of the runway is 12,090 feet (more than 2 miles!).

61. Use $F = \frac{9}{5}C + 32$ when $F = 122$.

$$122 = \frac{9}{5}C + 32$$
$$122 - 32 = \frac{9}{5}C + 32 - 32$$
$$90 = \frac{9}{5}C$$
$$\frac{5}{9} \cdot 90 = \frac{5}{9} \cdot \frac{9}{5}C$$
$$50 = C$$

Thus, 122°F is equivalent to 50°C.

63. Use $V = lwh$ when $l = 199$, $w = 78.5$, and $h = 33$.

$V = lwh = 199(78.5)(33) = 515{,}509.5$

The smallest possible shipping crate has a volume of 515,509.5 cubic inches.

65. Use $V = \frac{4}{3}\pi r^3$ when $r = \frac{9.5}{2} = 4.75$ and $\pi = 3.14$.

$V = \frac{4}{3}\pi r^3 = \frac{4}{3}(3.14)(4.75)^3 \approx 449$

The volume of the sphere is 449 cubic inches.

67. Use $F = \frac{9}{5}C + 32$ when $C = 167$.

$$F = \frac{9}{5}C + 32$$
$$= \frac{9}{5}(167) + 32$$
$$= 300.6 + 32$$
$$= 332.6$$
$$\approx 333$$

The average temperature on the planet Mercury is 333°F.

69. $32\% = 0.32$

71. $200\% = 2.00$ or 2

73. $0.17 = 0.17(100\%) = 17\%$

75. $7.2 = 7.2(100\%) = 720\%$

77.
$$\begin{aligned} N &= R + \frac{V}{G} \\ N - R &= R + \frac{V}{G} - R \\ N - R &= \frac{V}{G} \\ G(N - R) &= G \cdot \frac{V}{G} \\ G(N - R) &= V \end{aligned}$$

79. Use $V = lwh$. If the length is doubled, the new length is $2l$. If the width and height are doubled, the new width and height are $2w$ and $2h$, respectively.
$V = (2l)(2w)(2h) = 2 \cdot 2 \cdot 2lwh = 8lwh$
The volume of the box is multiplied by 8.

81. Replace T with N and solve for N.
$$\begin{aligned} T &= 50 + \frac{N-40}{4} \\ N &= 50 + \frac{N-40}{4} \\ N - 50 &= 50 + \frac{N-40}{4} - 50 \\ N - 50 &= \frac{N-40}{4} \\ 4(N-50) &= 4 \cdot \frac{N-40}{4} \\ 4N - 200 &= N - 40 \\ 4N - 200 - N &= N - 40 - N \\ 3N - 200 &= -40 \\ 3N - 200 + 200 &= -40 + 200 \\ 3N &= 160 \\ \frac{3N}{3} &= \frac{160}{3} \\ N &= 53\frac{1}{3} \end{aligned}$$
They are the same when the number of cricket chirps per minute is $53\frac{1}{3}$.

83. ▲ − ● · ▮ = ■
− ● · ▮ = ■ − ▲
● = (▲ − ■) / ▮

85. $\dfrac{20 \text{ miles}}{1 \text{ hour}} \cdot \dfrac{5280 \text{ feet}}{1 \text{ mile}} \cdot \dfrac{1 \text{ hour}}{60 \text{ minutes}} \cdot \dfrac{1 \text{ minute}}{60 \text{ seconds}}$
$= \dfrac{20 \cdot 5280 \text{ feet}}{60 \cdot 60 \text{ seconds}}$
≈ 29.3 feet/second
Use $d = rt$ when $d = 1300$ and $r = 29.3$.
$$\begin{aligned} d &= rt \\ 1300 &= 29.3t \\ \frac{1300}{29.3} &= \frac{29.3t}{29.3} \\ 44.3 &\approx t \end{aligned}$$
It took 44.3 seconds to travel that distance.

87. Use $I = PRT$ when $I = 1{,}056{,}000$, $R = 0.055$, and $T = 6$.
$$\begin{aligned} I &= PRT \\ 1{,}056{,}000 &= P(0.055)(6) \\ 1{,}056{,}000 &= 0.33P \\ \frac{1{,}056{,}000}{0.33} &= \frac{0.33P}{0.33} \\ 3{,}200{,}000 &= P \end{aligned}$$

89. Use $V = \frac{4}{3}\pi r^3$ when $r = 3$.
$$\begin{aligned} V &= \frac{4}{3}\pi \cdot 3^3 \\ V &\approx 113.1 \end{aligned}$$

Section 2.6 Practice Exercises

1. Let x be the unknown percent.
$$\begin{aligned} 22 &= x \cdot 40 \\ 22 &= 40x \\ \frac{22}{40} &= \frac{40x}{40} \\ 0.55 &= x \\ 55\% &= x \end{aligned}$$
The number 22 is 55% of 40.

2. Let x be the unknown number.
$$\begin{aligned} 150 &= 40\% \cdot x \\ 150 &= 0.4x \\ \frac{150}{0.4} &= \frac{0.4x}{0.4} \\ 375 &= x \end{aligned}$$
The number 150 is 40% of 375.

3. a. From the graph, we see 66% are for solely pleasure.

b. From the graph, 66% are for pleasure and 4% are for combined business/pleasure.
The sum is 66% + 4% = 70%.

c. Find 66% of 250.
$0.66(250) = 165$
We expect 165 people to be traveling solely for pleasure.

4.
$$\begin{aligned} \text{discount} &= \text{percent} \cdot \text{original price} \\ &= 40\% \cdot \$400 \\ &= 0.40 \cdot \$400 \\ &= \$160 \end{aligned}$$
$$\begin{aligned} \text{new price} &= \text{original price} - \text{discount} \\ &= \$400 - \$160 \\ &= \$240 \end{aligned}$$
The discount in price is \$160 and the new price is \$240.

5. increase = new − old = 200 − 120 = 80
Let x be the percent of increase.
$$\begin{aligned} 80 &= x \cdot 120 \\ \frac{80}{120} &= \frac{120x}{120} \\ 0.667 &\approx x \\ 66.7\% &\approx x \end{aligned}$$
The percent of increase is 66.7%.

6. Let x be the original price.
$$\begin{aligned} x - 0.20x &= 46 \\ 0.8x &= 46 \\ \frac{0.8x}{0.8} &= \frac{46}{0.8} \\ x &= 57.5 \end{aligned}$$
The original price is \$57.50.

7. Let x represent the liters of 20% solution.

	Number of Liters	Dye Strength	Amount
20% solution	x	20%	$0.2x$
50% solution	$6 - x$	50%	$0.5(6 - x)$
40% solution	6	40%	$0.4(6)$

$$\begin{aligned} 0.2x + 0.5(6 - x) &= 0.4(6) \\ 0.2x + 3 - 0.5x &= 2.4 \\ -0.3x + 3 &= 2.4 \\ -0.3x + 3 - 3 &= 2.4 - 3 \\ -0.3x &= -0.6 \\ \frac{-0.3x}{-0.3} &= \frac{-0.6}{-0.3} \\ x &= 2 \end{aligned}$$
$6 - x = 6 - 2 = 4$
If 2 liters of 20% solution are mixed with 4 liters of 50% solution, the result is 6 liters of 40% solution.

Vocabulary, Readiness & Video Check 2.6

1. no; $25\% + 25\% + 40\% \neq 100\%$

2. no; $30\% + 30\% + 30\% \neq 100\%$

3. yes; $25\% + 25\% + 25\% + 25\% = 100\%$

4. yes; $40\% + 50\% + 10\% = 100\%$

5. **a.** equals; =

   b. multiplication; ·

   c. Drop the percent symbol and move the decimal point two places to the left.

6. **a.** You also find a discount amount by multiplying the (discount) percent by the original price.

   b. For discount, the new price is the original price minus the discount amount, so you *subtract* from the original price rather than *add* as with mark-up.

7. You must first find the actual amount of increase in price by subtracting the original price from the new price.

8.

Alloy	Ounces	Copper Strength	Amount of Copper
10%	x	0.10	$0.10x$
30%	400	0.30	0.30(400)
20%	$x + 400$	0.20	$0.20(x + 400$

$0.10x + 0.30(400) = 0.20(x + 400)$

Exercise Set 2.6

1. Let x be the unknown number.
$x = 16\% \cdot 70$
$x = 0.16 \cdot 70$
$x = 11.2$
11.2 is 16% of 70.

3. Let x be the unknown percent.
$28.6 = x \cdot 52$
$\frac{28.6}{52} = \frac{52x}{52}$
$0.55 = x$
$55\% = x$
The number 28.6 is 55% of 52.

5. Let x be the unknown number.
$45 = 25\% \cdot x$
$45 = 0.25 \cdot x$
$\frac{45}{0.25} = \frac{0.25x}{0.25}$
$180 = x$
45 is 25% of 180.

7. From the graph, 15% of overnight stays were made in RVs.

9. 23% of overnight stays were made in lodges.
$23\% \cdot 1{,}350{,}000 = 0.23 \cdot 1{,}350{,}000 = 310{,}500$
You would expect 310,500 of the overnight stays in Yellowstone National Park to have been made in lodges.

11.
$$\begin{aligned}\text{discount} &= \text{percent} \cdot \text{original price}\\ &= 8\% \cdot \$18{,}500\\ &= 0.08 \cdot \$18{,}500\\ &= \$1480\end{aligned}$$
$$\begin{aligned}\text{new price} &= \text{original price} - \text{discount}\\ &= \$18{,}500 - \$1480\\ &= \$17{,}020\end{aligned}$$
The discount is \$1480 and the new price is \$17,020.

13. $15\% \cdot 40.50 = 0.15 \cdot 40.5 = 6.075$
The tip is \$6.08.
$40.5 + 6.08 = 46.58$
The total cost of the meal is \$46.58.

15.
$$\begin{aligned}\text{percent of decrease} &= \frac{\text{amount of decrease}}{\text{original amount}}\\ &= \frac{290{,}000 - 263{,}000}{290{,}000}\\ &= \frac{27{,}000}{290{,}000}\\ &\approx 0.093\end{aligned}$$
The number of complaints decreased by 9.3%.

17.
$$\begin{aligned}\text{percent of increase} &= \frac{\text{amount of increase}}{\text{original amount}}\\ &= \frac{40-28}{28}\\ &= \frac{12}{28}\\ &\approx 0.429\end{aligned}$$
The area increased by 42.9%.

19. Let x represent the original price.
$$\begin{aligned}x - 25\% \cdot x &= 78\\ x - 0.25x &= 78\\ 0.75x &= 78\\ \frac{0.75x}{0.75} &= \frac{78}{0.75}\\ x &= 104\end{aligned}$$
The original price of the shoes was \$104.

21. Let x represent last year's salary.

$$x+4\%\cdot x=44,200$$
$$x+0.04x=44,200$$
$$1.04x=44,200$$
$$\frac{1.04x}{1.04}=\frac{44,200}{1.04}$$
$$x=42,500$$

Last year's salary was $42,500.

23. Let x represent the number of gallons of pure acid.

	Number of Gallons ·	Acid Strength =	Amount of Acid
Pure Acid	x	100%	$1x$
40% Acid Solution	2	40%	$0.4x$
70% Acid Solution Needed	$x+2$	70%	$0.7(x+2)$

The amount of acid being combined must be the same as that in the mixture.

$$x+0.4x=0.7(x+2)$$
$$1.4x=0.7x+1.4$$
$$1.4x-0.7x=0.7x+1.4-0.7x$$
$$0.7x=1.4$$
$$\frac{0.7x}{0.7}=\frac{1.4}{0.7}$$
$$x=2$$

Thus, 2 gallons of pure acid should be used.

25. Let x represent the number of pounds of coffee worth $7 a pound.

	Number of pounds ·	Cost per pound =	Value
$7/lb coffee	x	7	$7x$
$4/lb coffee	14	4	$4\cdot 14=56$
$5/lb coffee wanted	$x+14$	5	$5(x+14)$

The value of the coffee being combined must be the same as the value of the mixture.

$$7x+56=5(x+14)$$
$$7x+56=5x+70$$
$$7x+56-5x=5x+70-5x$$
$$2x+56=70$$
$$2x+56-56=70-56$$
$$2x=14$$
$$\frac{2x}{2}=\frac{14}{2}$$
$$x=7$$

7 pounds of the $4 a pound coffee should be used.

27. $23\%\cdot 20=0.23\cdot 20=4.6$

29. Let x represent the unknown number.

$$40 = 80\% \cdot x$$
$$40 = 0.80 \cdot x$$
$$\frac{40}{0.8} = \frac{0.8x}{0.8}$$
$$50 = x$$

40 is 80% of 50.

31. Let x represent the unknown percent.

$$144 = x \cdot 480$$
$$\frac{144}{480} = \frac{480x}{480}$$
$$0.3 = x$$
$$30\% = x$$

144 is 30% of 480.

33. From the graph, it appears that 71% of the population of Fairbanks, Alaska shops by catalog.

35. 65% of 291,800 = 0.65 · 291,800 = 189,670
We predict 189,670 catalog shoppers live in Anchorage.

37.

Top Cranberry-Producing States in 2012 (in millions of pounds)		
	Millions of Pounds	Percent of Total (rounded to nearest percent)
Wisconsin	450	$\frac{450}{768} \approx 0.586 \approx 59\%$
Oregon	40	$\frac{40}{768} \approx 0.052 \approx 5\%$
Massachusetts	210	$\frac{210}{768} \approx 0.273 \approx 27\%$
Washington	14	$\frac{14}{768} \approx 0.018 \approx 2\%$
New Jersey	54	$\frac{54}{768} \approx 7\%$
Total	768	

39.

$$\text{percent of increase} = \frac{\text{amount of increase}}{\text{original amount}}$$
$$= \frac{70-40}{40}$$
$$= \frac{30}{40}$$
$$= 0.75$$

The price increased by 75%.

41. Let x represent the amount Charles paid for the car.

$$x + 20\% \cdot x = 4680$$
$$x + 0.20x = 4680$$
$$1.2x = 4680$$
$$\frac{1.2x}{1.2} = \frac{4680}{1.2}$$
$$x = 3900$$

Charles paid $3900 for the car.

43. $$\text{percent of increase} = \frac{\text{amount of increase}}{\text{original amount}} = \frac{144-36}{36} = \frac{108}{36} = 3$$

The area increased by 300%.

45. Markup = 5% · 2.20 = 0.05 · 2.2 = 0.11
New price = 2.20 + 0.11 = 2.31
The markup is $0.11 and the new price is $2.31.

47. Let x be the ounces of alloy that is 20% copper.

	ounces	concentration	amount
20% copper	x	20%	$0.2x$
50% copper	200	50%	0.5(200)
30% copper	$200 + x$	30%	$0.3(200 + x)$

The amount of copper being combined must be the same as that in the mixture.

$$0.2x + 0.5(200) = 0.3(200 + x)$$
$$0.2x + 100 = 60 + 0.3x$$
$$0.2x + 100 - 0.2x = 60 + 0.3x - 0.2x$$
$$100 = 60 + 0.1x$$
$$100 - 60 = 60 + 0.1x - 60$$
$$40 = 0.1x$$
$$\frac{40}{0.1} = \frac{0.1x}{0.1}$$
$$400 = x$$

Thus 400 ounces should be used.

49. $$\text{percent of decrease} = \frac{\text{amount of decrease}}{\text{original amount}} = \frac{67{,}000 - 58{,}000}{67{,}000} = \frac{9000}{67{,}000} \approx 0.134$$

The number of milk cow operations decreased by 13.4%.

51. Let x be the prior number of employees.

$$x - 0.35x = 78$$
$$0.65x = 78$$
$$\frac{0.65x}{0.65} = \frac{78}{0.65}$$
$$x = 120$$

There were 120 employees prior to the layoffs.

53. decrease = 25% · 256 = 0.25 · 256 = 64
256 − 64 = 192
The price of the coat decreased by \$64. The sale price was \$192.

55. increase = 48% · 577 = 0.48 · 577 = 276.96
577 + 276.96 = 853.96
The Naga Jolokia pepper measures 854 thousand Scoville units.

57. 55% ·240 million = 0.55 · 240 million = 132 million
You would expect that 132 million U.S. adults owned a smartphone in 2013.

59. Let x be the ounces of self-tanning lotion.

	ounces	cost (\$)	value
self-tanning	x	3	$3x$
everyday	800	0.30	0.3(800)
experimental	$800 + x$	1.20	$1.2(800 + x)$

The value of those being combined must be the same as the value of the mixture.

$$3x + 0.3(800) = 1.2(800 + x)$$
$$3x + 240 = 960 + 1.2x$$
$$3x + 240 - 1.2x = 960 + 1.2x - 1.2x$$
$$1.8x + 240 = 960$$
$$1.8x + 240 - 240 = 960 - 240$$
$$1.8x = 720$$
$$\frac{1.8x}{1.8} = \frac{720}{1.8}$$
$$x = 400$$

Therefore, 400 ounces of the self-tanning lotion should be used.

61. $-5 > -7$ since -5 is to the right of -7 on a number line.

63. $|-5| = 5$
$-(-5) = 5$
$|-5| = -(-5)$ since $5 = 5$.

65. $(-3)^2 = (-3)(-3) = 9$

$-3^2 = -(3 \cdot 3) = -9$

Since $9 > -9$, $(-3)^2 > -3^2$.

67. no; answers may vary

69. 230 mg is what percent of 2400 mg?
Let x represent the unknown percent.

$$x \cdot 2400 = 230$$
$$\frac{2400x}{2400} = \frac{230}{2400}$$
$$x = 0.0958\overline{3}$$

This food contains 9.6% of the daily value of sodium in one serving.

71. 35 is what percent of 130? Let x be the unknown percent.

$$35 = x \cdot 130$$
$$\frac{35}{130} = \frac{130x}{130}$$
$$0.269 \approx x$$

The percent calories from fat is 26.9%. Yes, this food satisfies the recommendation since $26.9\% \le 30\%$.

73. 12 g · 4 calories/gram = 48 calories
48 of the 280 calories come from protein.

$$\frac{48}{280} \approx 0.171$$

17.1% of the calories in this food come from protein.

Section 2.7 Practice Exercises

1. $x \ge -2$

−5 −4 −3 −2 −1 0 1 2 3 4 5

2. $5 > x$ or $x < 5$

−5 −4 −3 −2 −1 0 1 2 3 4 5

3. $-3 \le x < 1$

−5 −4 −3 −2 −1 0 1 2 3 4 5

4.
$$x - 6 \ge -11$$
$$x - 6 + 6 \ge -11 + 6$$
$$x \ge -5$$

−5 −4 −3 −2 −1 0 1 2 3 4 5

5.
$$-3x \le 12$$
$$\frac{-3x}{-3} \ge \frac{12}{-3}$$
$$x \ge -4$$

−5 −4 −3 −2 −1 0 1 2 3 4 5

6.
$$5x > -20$$
$$\frac{5x}{5} > \frac{-20}{5}$$
$$x > -4$$

−5 −4 −3 −2 −1 0 1 2 3 4 5

7.
$$-3x + 11 \le -13$$
$$-3x + 11 - 11 \le -13 - 11$$
$$-3x \le -24$$
$$\frac{-3x}{-3} \ge \frac{-24}{-3}$$
$$x \ge 8$$

$\{x | x \ge 8\}$

−10 −8 −6 −4 −2 0 2 4 6 8 10

8.
$$2x - 3 > 4(x - 1)$$
$$2x - 3 > 4x - 4$$
$$2x - 3 - 4x > 4x - 4 - 4x$$
$$-2x - 3 > -4$$
$$-2x - 3 + 3 > -4 + 3$$
$$-2x > -1$$
$$\frac{-2x}{-2} < \frac{-1}{-6}$$
$$x < \frac{1}{2}$$

$\left\{x \middle| x < \frac{1}{2}\right\}$

$\frac{1}{2}$

−5 −4 −3 −2 −1 0 1 2 3 4 5

9.
$$3(x + 5) - 1 \ge 5(x - 1) + 7$$
$$3x + 15 - 1 \ge 5x - 5 + 7$$
$$3x + 14 \ge 5x + 2$$
$$3x + 14 - 5x \ge 5x + 2 - 5x$$
$$-2x + 14 \ge 2$$
$$-2x + 14 - 14 \ge 2 - 14$$
$$-2x \ge -12$$
$$\frac{-2x}{-2} \le \frac{-12}{-2}$$
$$x \le 6$$

$\{x | x \le 6\}$

10. Let x be the unknown number.

$$35 - 2x > 15$$
$$35 - 2x - 35 > 15 - 35$$
$$-2x > -20$$
$$\frac{-2x}{-2} < \frac{-20}{-2}$$
$$x < 10$$

All numbers less than 10 make the statement true.

11. Let x represent the minimum sales.

$$600 + 0.04x \geq 3000$$
$$0.04x \geq 2400$$
$$x \geq 60{,}000$$

Alex must have minimum sales of \$60,000.

Vocabulary, Readiness & Video Check 2.7

1. $6x - 7(x + 9)$ is an <u>expression</u>.

2. $6x = 7(x + 9)$ is an <u>equation</u>.

3. $6x < 7(x + 9)$ is an <u>inequality</u>.

4. $5y - 2 \geq -38$ is an <u>inequality</u>.

5. $\frac{9}{7} = \frac{x+2}{14}$ is an <u>equation</u>.

6. $\frac{9}{7} - \frac{x+2}{14}$ is an <u>expression</u>.

7. $x \geq -3$
   -5 is not a solution.

8. $x < 6$
   $|-6| = 6$ is not a solution.

9. $x < 4.01$
   4.1 is not a solution.

10. $x \geq -3$
    -4 is not a solution.

11. An open circle indicates > or <; a closed circle indicates ≥ or ≤.

12. addition property of equality

13. $\{x | x \geq -2\}$

14. The multiplication property of inequality is applied at this step when we divide by the coefficient of x. The coefficient is positive and so the inequality symbol remains the same, but if the coefficient had been negative, the direction of the inequality symbol would have been reversed.

15. is greater than; >

Exercise Set 2.7

1. 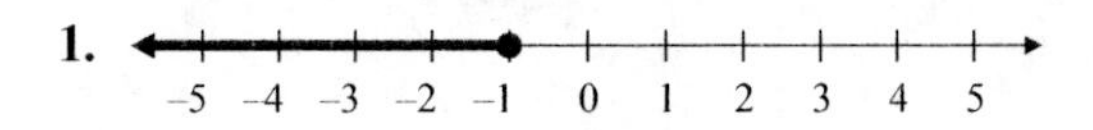

3. $\frac{1}{2}$

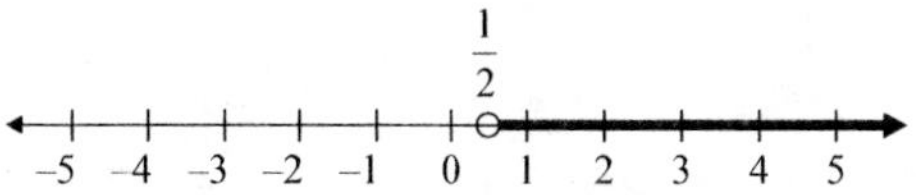

5.

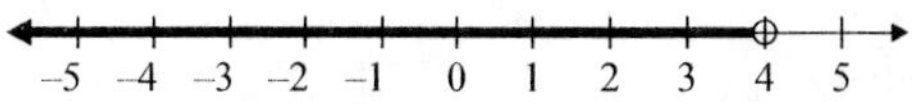

7.

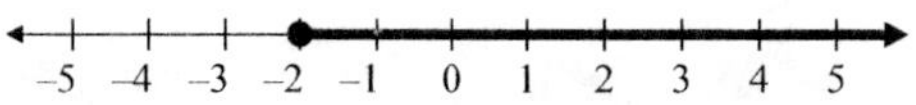

9.

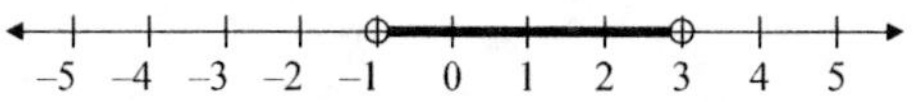

11.

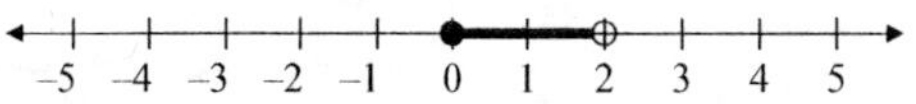

13.

$$x - 2 \geq -7$$
$$x - 2 + 2 \geq -7 + 2$$
$$x \geq -5$$

$\{x | x \geq -5\}$

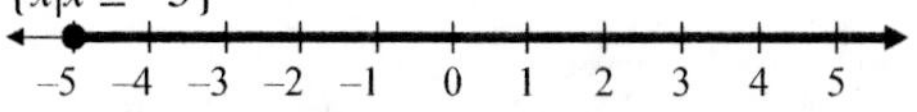

15.

$$-9 + y < 0$$
$$9 - 9 + y < 9 + 0$$
$$y < 9$$

$\{y | y < 9\}$

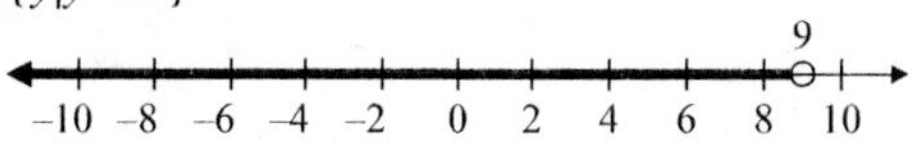

17.

$$3x - 5 > 2x - 8$$
$$3x - 5 - 2x > 2x - 8 - 2x$$
$$x - 5 > -8$$
$$x - 5 + 5 > -8 + 5$$
$$x > -3$$

$\{x | x > -3\}$

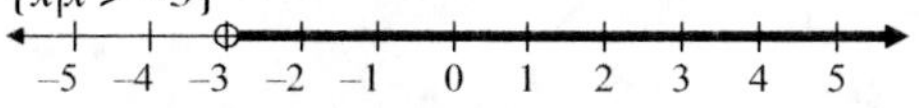

19.

$$4x - 1 \leq 5x - 2x$$
$$4x - 1 \leq 3x$$
$$4x - 1 - 4x \leq 3x - 4x$$
$$-1 \leq -x$$
$$\frac{-1}{-1} \geq \frac{-x}{-1}$$
$$1 \geq x \text{ or } x \leq 1$$

$\{x | x \leq 1\}$

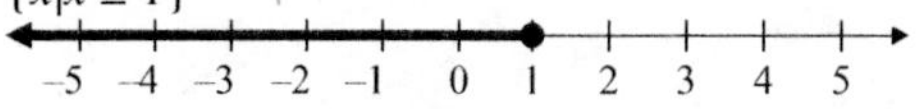

21. $2x < -6$

$\frac{2x}{2} < \frac{-6}{2}$

$x < -3$

$\{x|x < -3\}$

(number line from −5 to 5: open circle at −3, shaded to the left)

23. $-8x \le 16$

$\frac{-8x}{-8} \ge \frac{16}{-8}$

$x \ge -2$

$\{x|x \ge -2\}$

(number line from −5 to 5: closed circle at −2, shaded to the right)

25. $-x > 0$

$(-1)(-x) < (-1)(0)$

$x < 0$

$\{x|x < 0\}$

(number line from −5 to 5: open circle at 0, shaded to the left)

27. $\frac{3}{4}y \ge -2$

$\frac{4}{3} \cdot \frac{3}{4}y \ge \frac{4}{3} \cdot (-2)$

$y \ge -\frac{8}{3}$

$\left\{y \middle| y \ge -\frac{8}{3}\right\}$

(number line from −5 to 5: closed circle at $-\frac{8}{3}$, shaded to the right)

29. $-0.6y < -1.8$

$\frac{-0.6y}{-0.6} > \frac{-1.8}{-0.6}$

$y > 3$

$\{y|y > 3\}$

(number line from −5 to 5: open circle at 3, shaded to the right)

31. $-8 < x + 7$

$-8 - 7 < x + 7 - 7$

$-15 < x$

$\{x|x > -15\}$

33. $7(x+1) - 6x \ge -4$

$7x + 7 - 6x \ge -4$

$x + 7 \ge -4$

$x + 7 - 7 \ge -4 - 7$

$x \ge -11$

$\{x|x \ge -11\}$

35. $4x > 1$

$\frac{4x}{4} > \frac{1}{4}$

$x > \frac{1}{4}$

$\left\{x \middle| x > \frac{1}{4}\right\}$

37. $-\frac{2}{3}y \le 8$

$-\frac{3}{2}\left(-\frac{2}{3}y\right) \ge -\frac{3}{2}(8)$

$y \ge -12$

$\{y|y \ge -12\}$

39. $4(2z+1) < 4$

$8z + 4 < 4$

$8z + 4 - 4 < 4 - 4$

$8z < 0$

$\frac{8z}{8} < \frac{0}{8}$

$z < 0$

$\{z|z < 0\}$

41. $3x - 7 < 6x + 2$

$3x - 7 - 3x < 6x + 2 - 3x$

$-7 < 3x + 2$

$-7 - 2 < 3x + 2 - 2$

$-9 < 3x$

$\frac{-9}{3} < \frac{3x}{3}$

$-3 < x$

$\{x|x > -3\}$

43.
$$\begin{aligned} 5x-7x &\le x+2 \\ -2x &\le x+2 \\ -2x-x &\le x+2-x \\ -3x &\le 2 \\ \frac{-3x}{-3} &\ge \frac{2}{-3} \\ x &\ge -\frac{2}{3} \end{aligned}$$
$\left\{x \middle| x \ge -\frac{2}{3}\right\}$

45.
$$\begin{aligned} -6x+2 &\ge 2(5-x) \\ -6x+2 &\ge 10-2x \\ -6x+2+6x &\ge 10-2x+6x \\ 2 &\ge 10+4x \\ 2-10 &\ge 10+4x-10 \\ -8 &\ge 4x \\ \frac{-8}{4} &\ge \frac{4x}{4} \\ -2 &\ge x \end{aligned}$$
$\{x|x \le -2\}$

47.
$$\begin{aligned} 3(x-5) &< 2(2x-1) \\ 3x-15 &< 4x-2 \\ 3x-15-3x &< 4x-2-3x \\ -15 &< x-2 \\ -15+2 &< x-2+2 \\ -13 &< x \end{aligned}$$
$\{x|x > -13\}$

49.
$$\begin{aligned} 4(3x-1) &\le 5(2x-4) \\ 12x-4 &\le 10x-20 \\ 12x-4-10x &\le 10x-20-10x \\ 2x-4 &\le -20 \\ 2x-4+4 &\le -20+4 \\ 2x &\le -16 \\ \frac{2x}{2} &\le \frac{-16}{2} \\ x &\le -8 \end{aligned}$$
$\{x|x \le -8\}$

51.
$$\begin{aligned} 3(x+2)-6 &> -2(x-3)+14 \\ 3x+6-6 &> -2x+6+14 \\ 3x &> -2x+20 \\ 3x+2x &> -2x+20+2x \\ 5x &> 20 \\ \frac{5x}{5} &> \frac{20}{5} \\ x &> 4 \end{aligned}$$
$\{x|x > 4\}$

53.
$$\begin{aligned} -5(1-x)+x &\le -(6-2x)+6 \\ -5+5x+x &\le -6+2x+6 \\ -5+6x &\le 2x \\ -5+6x-6x &\le 2x-6x \\ -5 &\le -4x \\ \frac{-5}{-4} &\ge \frac{-4x}{-4} \\ \frac{5}{4} &\ge x \end{aligned}$$
$\left\{x \middle| x \le \frac{5}{4}\right\}$

55.
$$\begin{aligned} \frac{1}{4}(x+4) &< \frac{1}{5}(2x+3) \\ 20\cdot\frac{1}{4}(x+4) &< 20\cdot\frac{1}{5}(2x+3) \\ 5(x+4) &< 4(2x+3) \\ 5x+20 &< 8x+12 \\ 5x+20-5x &< 8x+12-5x \\ 20 &< 3x+12 \\ 20-12 &< 3x+12-12 \\ 8 &< 3x \\ \frac{8}{3} &< \frac{3x}{3} \\ \frac{8}{3} &< x \end{aligned}$$
$\left\{x \middle| x > \frac{8}{3}\right\}$

57.
$$\begin{aligned} -5x+4 &\le -4(x-1) \\ -5x+4 &\le -4x+4 \\ -5x+4+4x &\le -4x+4+4x \\ -x+4 &\le 4 \\ -x+4-4 &\le 4-4 \\ -x &\le 0 \\ -1(-x) &\ge -1(0) \\ x &\ge 0 \end{aligned}$$
$\{x|x \ge 0\}$

59. Let x be the number.
$$\begin{aligned} 2x+6 &> -14 \\ 2x+6-6 &> -14-6 \\ 2x &> -20 \\ \frac{2x}{2} &> \frac{-20}{2} \\ x &> -10 \end{aligned}$$
All numbers greater than −10 make this statement true.

61. Use $P = 2l + 2w$ when $w = 15$ and $P \le 100$.

$$2l + 2(15) \le 100$$
$$2l + 30 \le 100$$
$$2l + 30 - 30 \le 100 - 30$$
$$2l \le 70$$
$$\frac{2l}{2} \le \frac{70}{2}$$
$$l \le 35$$

The maximum length of the rectangle is 35 cm.

63. Let x be the score in his third game.

$$\frac{146 + 201 + x}{3} \ge 180$$
$$\frac{347 + x}{3} \ge 180$$
$$3 \cdot \frac{347 + x}{3} \ge 3 \cdot 180$$
$$347 + x \ge 540$$
$$347 + x - 347 \ge 540 - 347$$
$$x \ge 193$$

He must bowl at least 193 on the third game.

65. Let x represent the number of people. Then the cost is $50 + 34x$.

$$50 + 34x \le 3000$$
$$50 + 34x - 50 \le 3000 - 50$$
$$34x \le 2950$$
$$\frac{34x}{34} \le \frac{2950}{34}$$
$$x \le \frac{2950}{34} \approx 86.76$$

They can invite at most 86 people.

67. Let x represent the number of minutes.

$$5.8x \ge 200$$
$$\frac{5.8x}{5.8} \ge \frac{200}{5.8}$$
$$x \ge \frac{200}{5.8} \approx 35$$

The person must walk at least 35 minutes.

69. $3^4 = 3 \cdot 3 \cdot 3 \cdot 3 = 81$

71. $1^8 = 1 \cdot 1 \cdot 1 \cdot 1 \cdot 1 \cdot 1 \cdot 1 \cdot 1 = 1$

73. $\left(\frac{7}{8}\right)^2 = \left(\frac{7}{8}\right)\left(\frac{7}{8}\right) = \frac{49}{64}$

75. The lowest point on the graph corresponds to 2006, so there were the fewest Starbucks locations in 2006.

77. The greatest increase occurred between 2012 and 2013.

79. The number of Starbucks locations first rose above 11,500 in 2008.

81. Since $3 < 5$, $3(-4) > 5(-4)$.

83. If $m \le n$, then $-2m \ge -2n$.

85. Reverse the direction of the inequality symbol when multiplying or dividing by a negative number.

87. Let x be the score on his final exam. Since the final counts as two tests, his final course average is $\frac{75 + 83 + 85 + 2x}{5}$.

$$\frac{75 + 83 + 85 + 2x}{5} \ge 80$$
$$\frac{243 + 2x}{5} \ge 80$$
$$5\left(\frac{243 + 2x}{5}\right) \ge 5(80)$$
$$243 + 2x \ge 400$$
$$243 + 2x - 243 \ge 400 - 243$$
$$2x \ge 157$$
$$\frac{2x}{2} \ge \frac{157}{2}$$
$$x \ge 78.5$$

His final exam score must be at least 78.5 for him to get a B.

Chapter 2 Vocabulary Check

1. A <u>linear equation in one variable</u> can be written in the form $Ax + B = C$.

2. Equations that have the same solution are called <u>equivalent equations</u>.

3. An equation that describes a known relationship among quantities is called a <u>formula</u>.

4. A <u>linear inequality in one variable</u> can be written in the form $ax + b < c$, (or $>$, $\le$, $\ge$).

5. The solutions to the equation $x + 5 = x + 5$ are <u>all real numbers</u>.

6. The solution to the equation $x + 5 = x + 4$ is <u>no solution</u>.

7. If both sides of an inequality are multiplied or divided by the same positive number, the direction of the inequality symbol is the same.

8. If both sides of an inequality are multiplied or divided by the same negative number, the direction of the inequality symbol is reversed.

Chapter 2 Review

1. $$\begin{aligned} 8x+4&=9x \\ 8x+4-8x&=9x-8x \\ 4&=x \end{aligned}$$

2. $$\begin{aligned} 5y-3&=6y \\ 5y-3-5y&=6y-5y \\ -3&=y \end{aligned}$$

3. $$\begin{aligned} \frac{2}{7}x+\frac{5}{7}x&=6 \\ \frac{7}{7}x&=6 \\ 1x&=6 \\ x&=6 \end{aligned}$$

4. $$\begin{aligned} 3x-5&=4x+1 \\ 3x-5-3x&=4x+1-3x \\ -5&=x+1 \\ -5-1&=x+1-1 \\ -6&=x \end{aligned}$$

5. $$\begin{aligned} 2x-6&=x-6 \\ 2x-6-x&=x-6-x \\ x-6&=-6 \\ x-6+6&=-6+6 \\ x&=0 \end{aligned}$$

6. $$\begin{aligned} 4(x+3)&=3(1+x) \\ 4x+12&=3+3x \\ 4x+12-3x&=3+3x-3x \\ 12+x&=3 \\ -12+12+x&=-12+3 \\ x&=-9 \end{aligned}$$

7. $$\begin{aligned} 6(3+n)&=5(n-1) \\ 18+6n&=5n-5 \\ 18+6n-5n&=5n-5-5n \\ 18+n&=-5 \\ -18+18+n&=-18-5 \\ n&=-23 \end{aligned}$$

8. $$\begin{aligned} 5(2+x)-3(3x+2)&=-5(x-6)+2 \\ 10+5x-9x-6&=-5x+30+2 \\ -4x+4&=-5x+32 \\ 5x-4x+4&=5x-5x+32 \\ x+4&=32 \\ x+4-4&=32-4 \\ x&=28 \end{aligned}$$

9. If the sum is 10 and one number is x, then the other number is $10 - x$. The choice is b.

10. Since Mandy is 5 inches taller than Melissa, and x represents Mandy's height, then $x - 5$ represents Melissa's height. The choice is a.

11. Complementary angles sum to 90°. The complement of angle x is $90 - x$. The choice is b.

12. Supplementary angles sum to 180°. The supplement to $(x + 5)°$ is $180 - (x + 5) = 180 - x - 5 = 175 - x$. The choice is c.

13. $$\begin{aligned} \frac{3}{4}x&=-9 \\ \frac{4}{3}\cdot\frac{3}{4}x&=\frac{4}{3}\cdot(-9) \\ x&=-12 \end{aligned}$$

14. $$\begin{aligned} \frac{x}{6}&=\frac{2}{3} \\ 6\cdot\frac{x}{6}&=6\cdot\frac{2}{3} \\ x&=4 \end{aligned}$$

15. $$\begin{aligned} -5x&=0 \\ \frac{-5x}{-5}&=\frac{0}{-5} \\ x&=0 \end{aligned}$$

16. $$\begin{aligned} -y&=7 \\ \frac{-y}{-1}&=\frac{7}{-1} \\ y&=-7 \end{aligned}$$

17. $$\begin{aligned} 0.2x&=0.15 \\ 20x&=15 \\ \frac{20x}{20}&=\frac{15}{20} \\ x&=0.75 \end{aligned}$$

18.
$$\begin{aligned}\frac{-x}{3} &= 1\\ -3\left(\frac{-x}{3}\right) &= -3(1)\\ x &= -3\end{aligned}$$

19.
$$\begin{aligned}-3x+1 &= 19\\ -3x+1-1 &= 19-1\\ -3x &= 18\\ \frac{-3x}{-3} &= \frac{18}{-3}\\ x &= -6\end{aligned}$$

20.
$$\begin{aligned}5x+25 &= 20\\ 5x+25-25 &= 20-25\\ 5x &= -5\\ \frac{5x}{5} &= \frac{-5}{5}\\ x &= -1\end{aligned}$$

21.
$$\begin{aligned}7(x-1)+9 &= 5x\\ 7x-7+9 &= 5x\\ 7x+2 &= 5x\\ -7x+7x+2 &= -7x+5x\\ 2 &= -2x\\ \frac{2}{-2} &= \frac{-2x}{-2}\\ -1 &= x\end{aligned}$$

22.
$$\begin{aligned}7x-6 &= 5x-3\\ 7x-6-5x &= 5x-3-5x\\ 2x-6 &= -3\\ 2x-6+6 &= -3+6\\ 2x &= 3\\ \frac{2x}{2} &= \frac{3}{2}\\ x &= \frac{3}{2} \text{ or } 1\frac{1}{2}\end{aligned}$$

23.
$$\begin{aligned}-5x+\frac{3}{7} &= \frac{10}{7}\\ 7\left(-5x+\frac{3}{7}\right) &= 7\cdot\frac{10}{7}\\ -35x+3 &= 10\\ -35x+3-3 &= 10-3\\ -35x &= 7\\ \frac{-35x}{-35} &= \frac{7}{-35}\\ x &= -\frac{1}{5}\end{aligned}$$

24.
$$\begin{aligned}5x+x &= 9+4x-1+6\\ 6x &= 4x+14\\ 6x-4x &= 4x+14-4x\\ 2x &= 14\\ \frac{2x}{2} &= \frac{14}{2}\\ x &= 7\end{aligned}$$

25. Let x be the first integer. Then $x + 1$ and $x + 2$ are the next two consecutive integers. Their sum is $x + x + 1 + x + 2 = 3x + 3$.

26. Let x be the first even integer. Then $x + 2$, $x + 4$, and $x + 6$ are the 2nd, 3rd, and 4th consecutive even integers. The sum of the first and fourth is $x + x + 6 = 2x + 6$.

27.
$$\begin{aligned}\frac{5}{3}x+4 &= \frac{2}{3}x\\ 3\left(\frac{5}{3}x+4\right) &= 3\left(\frac{2}{3}x\right)\\ 5x+12 &= 2x\\ 5x+12-5x &= 2x-5x\\ 12 &= -3x\\ \frac{12}{-3} &= \frac{-3x}{-3}\\ -4 &= x\end{aligned}$$

28.
$$\begin{aligned}\frac{7}{8}x+1 &= \frac{5}{8}x\\ 8\left(\frac{7}{8}x+1\right) &= 8\left(\frac{5}{8}x\right)\\ 7x+8 &= 5x\\ 7x+8-7x &= 5x-7x\\ 8 &= -2x\\ \frac{8}{-2} &= \frac{-2x}{-2}\\ -4 &= x\end{aligned}$$

29.
$$\begin{aligned}-(5x+1) &= -7x+3\\ -5x-1 &= -7x+3\\ -5x-1+7x &= -7x+3+7x\\ 2x-1 &= 3\\ 2x-1+1 &= 3+1\\ 2x &= 4\\ \frac{2x}{2} &= \frac{4}{2}\\ x &= 2\end{aligned}$$

30.
$$-4(2x+1)=-5x+5$$
$$-8x-4=-5x+5$$
$$-8x-4+8x=-5x+5+8x$$
$$-4=3x+5$$
$$-4-5=3x+5-5$$
$$-9=3x$$
$$\frac{-9}{3}=\frac{3x}{3}$$
$$-3=x$$

31.
$$-6(2x-5)=-3(9+4x)$$
$$-12x+30=-27-12x$$
$$12x-12x+30=12x-27-12x$$
$$30=-27$$
Since the statement $30=-27$ is false, the equation has no solution.

32.
$$3(8y-1)=6(5+4y)$$
$$24y-3=30+24y$$
$$24y-3-24y=30+24y-24y$$
$$-3=30$$
Since the statement $-3=30$ is false, the equation has no solution.

33.
$$\frac{3(2-z)}{5}=z$$
$$5\left[\frac{3(2-z)}{5}\right]=5\cdot z$$
$$3(2-z)=5z$$
$$6-3z=5z$$
$$6-3z+3z=5z+3z$$
$$6=8z$$
$$\frac{6}{8}=\frac{8z}{8}$$
$$\frac{3}{4}=z$$

34.
$$\frac{4(n+2)}{5}=-n$$
$$5\left[\frac{4(n+2)}{5}\right]=5(-n)$$
$$4(n+2)=-5n$$
$$4n+8=-5n$$
$$4n+8-4n=-5n-4n$$
$$8=-9n$$
$$\frac{8}{-9}=\frac{-9n}{-9}$$
$$-\frac{8}{9}=n$$

35.
$$0.5(2n-3)-0.1=0.4(6+2n)$$
$$5(2n-3)-1=4(6+2n)$$
$$10n-15-1=24+8n$$
$$10n-16=24+8n$$
$$10n-16-8n=24+8n-8n$$
$$2n-16=24$$
$$2n-16+16=24+16$$
$$2n=40$$
$$\frac{2n}{2}=\frac{40}{2}$$
$$n=20$$

36.
$$-9-5a=3(6a-1)$$
$$-9-5a=18a-3$$
$$9-5a+5a=18a-3+5a$$
$$-9=23a-3$$
$$-9+3=23a-3+3$$
$$-6=23a$$
$$\frac{-6}{23}=\frac{23a}{23}$$
$$-\frac{6}{23}=a$$

37.
$$\frac{5(c+1)}{6}=2c-3$$
$$6\left[\frac{5(c+1)}{6}\right]=6(2c-3)$$
$$5(c+1)=6(2c-3)$$
$$5c+5=12c-18$$
$$5c+5-5c=12c-18-5c$$
$$5=7c-18$$
$$5+18=7c-18+18$$
$$23=7c$$
$$\frac{23}{7}=\frac{7c}{7}$$
$$\frac{23}{7}=c$$

38.
$$\frac{2(8-a)}{3}=4-4a$$
$$3\left[\frac{2(8-a)}{3}\right]=3(4-4a)$$
$$2(8-a)=3(4-4a)$$
$$16-2a=12-12a$$
$$16-2a+12a=12-12a+12a$$
$$16+10a=12$$
$$16+10a-16=12-16$$
$$10a=-4$$
$$\frac{10a}{10}=\frac{-4}{10}$$
$$a=-\frac{2}{5}$$

39.
$$200(70x-3560)=-179(150x-19,300)$$
$$14,000x-712,000=-26,850x+3,454,700$$
$$14,000x-712,000+26,850x=-26,850x+3,454,700+26,850x$$
$$40,850x-712,000=3,454,700$$
$$40,850x-712,000+712,000=3,454,700+712,000$$
$$40,850x=4,166,700$$
$$\frac{40,850x}{40,850}=\frac{4,166,700}{40,850}$$
$$x=102$$

40.
$$1.72y-0.04y=0.42$$
$$172y-4y=42$$
$$168y=42$$
$$\frac{168y}{168}=\frac{42}{168}$$
$$y=0.25$$

41. Let x be the length of the side of the square base. Then the height is $10x + 50.5$. The sum is 7327.
$$x+10x+50.5=7327$$
$$11x+50.5=7327$$
$$11x+50.5-50.5=7327-50.5$$
$$11x=7276.5$$
$$\frac{11x}{11}=\frac{7276.5}{11}$$
$$x=661.5$$
$$10x+50.5=10(661.5)+50.5$$
$$=6615+50.5$$
$$=6665.5$$
The height is 6665.5 inches.

42. Let x be the length of the short piece. Then $2x$ is the length of the long piece. The lengths sum to 12.

$$x+2x=12$$
$$3x=12$$
$$\frac{3x}{3}=\frac{12}{3}$$
$$x=4$$
$$2x=2(4)=8$$

The short piece is 4 feet and the long piece is 8 feet.

43. Let x represent the number of national battlefields. Then there were $3x - 4$ national memorials. The total is 40.

$$x+3x-4=40$$
$$4x-4=40$$
$$4x-4+4=40+4$$
$$4x=44$$
$$\frac{4x}{4}=\frac{44}{4}$$
$$x=11$$
$$3x-4=3(11)-4=33-4=29$$

There were 11 national battlefields and 29 national memorials in 2013.

44. Let x be the first integer. Then $x + 1$ and $x + 2$ are the next two consecutive integers. Their sum is -114.

$$x+x+1+x+2=-114$$
$$3x+3=-114$$
$$3x+3-3=-114-3$$
$$3x=-117$$
$$\frac{3x}{3}=\frac{-117}{3}$$
$$x=-39$$
$$x+1=-39+1=-38$$
$$x+2=-39+2=-37$$

The integers are -39, -38, and -37.

45.
$$\frac{x}{3}=x-2$$
$$3\cdot\frac{x}{3}=3(x-2)$$
$$x=3x-6$$
$$x-3x=3x-6-3x$$
$$-2x=-6$$
$$\frac{-2x}{-2}=\frac{-6}{-2}$$
$$x=3$$

The number is 3.

46.
$$2(x+6)=-x$$
$$2x+12=-x$$
$$-2x+2x+12=-2x-x$$
$$12=-3x$$
$$\frac{12}{-3}=\frac{-3x}{-3}$$
$$-4=x$$

The number is -4.

47. Use $P = 2l + 2w$ when $P = 46$ and $l = 14$.

$$P=2l+2w$$
$$46=2(14)+2w$$
$$46=28+2w$$
$$46-28=28+2w-28$$
$$18=2w$$
$$\frac{18}{2}=\frac{2w}{2}$$
$$9=w$$

48. Use $V = lwh$ when $V = 192$, $l = 8$, and $w = 6$.

$$V=lwh$$
$$192=8\cdot6\cdot h$$
$$192=48h$$
$$\frac{192}{48}=\frac{48h}{48}$$
$$4=h$$

49.
$$y=mx+b$$
$$y-b=mx+b-b$$
$$y-b=mx$$
$$\frac{y-b}{x}=\frac{mx}{x}$$
$$\frac{y-b}{x}=m$$

50.
$$r=vst-5$$
$$r+5=vst-5+5$$
$$r+5=vst$$
$$\frac{r+5}{vt}=\frac{vst}{vt}$$
$$\frac{r+5}{vt}=s$$

51.
$$2y-5x=7$$
$$-2y+2y-5x=-2y+7$$
$$-5x=-2y+7$$
$$\frac{-5x}{-5}=\frac{-2y+7}{-5}$$
$$x=\frac{2y-7}{5}$$

52.
$$3x-6y=-2$$
$$-3x+3x-6y=-3x-2$$
$$-6y=-3x-2$$
$$\frac{-6y}{-6}=\frac{-3x-2}{-6}$$
$$y=\frac{3x+2}{6}$$

53.
$$C=\pi D$$
$$\frac{C}{D}=\frac{\pi D}{D}$$
$$\frac{C}{D}=\pi$$

54.
$$C=2\pi r$$
$$\frac{C}{2r}=\frac{2\pi r}{2r}$$
$$\frac{C}{2r}=\pi$$

55. Use $V = lwh$ when $V = 900$, $l = 20$ and $h = 3$.
$$V=lwh$$
$$900=20\cdot w\cdot 3$$
$$900=60w$$
$$\frac{900}{60}=\frac{60w}{60}$$
$$15=w$$
The width is 15 meters.

56. Let x be the width. Then the length is $x + 6$. Use $P = 2 \cdot \text{length} + 2 \cdot \text{width}$ when $P = 60$.
$$P=2\cdot \text{length}+2\cdot \text{width}$$
$$60=2(x+6)+2x$$
$$60=2x+12+2x$$
$$60=4x+12$$
$$60-12=4x+12-12$$
$$48=4x$$
$$\frac{48}{4}=\frac{4x}{4}$$
$$12=x$$
$x + 6 = 12 + 6 = 18$

The dimensions of the billboard are 12 feet by 18 feet.

57. Use $d = rt$ when d = 10K or 10,000 m and $r = 125$.
$$d=rt$$
$$10{,}000=125t$$
$$\frac{10{,}000}{125}=\frac{125t}{125}$$
$$80=t$$
The time is 80 minutes or $\frac{80}{60}=1\frac{1}{3}$ hours or 1 hour and 20 minutes.

58. Use $F=\frac{9}{5}C+32$ when $F = 113$.
$$F=\frac{9}{5}C+32$$
$$113=\frac{9}{5}C+32$$
$$113-32=\frac{9}{5}C+32-32$$
$$81=\frac{9}{5}C$$
$$\frac{5}{9}\cdot 81=\frac{5}{9}\cdot\frac{9}{5}C$$
$$45=C$$
Thus, 113°F is equivalent to 45°C.

59. Let x be the unknown percent.
$$9=x\cdot 45$$
$$\frac{9}{45}=\frac{45x}{45}$$
$$0.2=x$$
$$20\%=x$$
9 is 20% of 45.

60. Let x be the unknown percent.
$$59.5=x\cdot 85$$
$$\frac{59.5}{85}=\frac{85x}{85}$$
$$0.7=x$$
$$70\%=x$$
59.5 is 70% of 85.

61. Let x be the unknown number.
$$137.5=125\%\cdot x$$
$$137.5=1.25x$$
$$\frac{137.5}{1.25}=\frac{1.25x}{1.25}$$
$$110=x$$
137.5 is 125% of 110.

62. Let x be the unknown number.
$768 = 60\% \cdot x$
$768 = 0.6x$
$\frac{768}{0.6} = \frac{0.6x}{0.6}$
$1280 = x$
768 is 60% of 1280.

63. increase = $11\% \cdot 1900 = 0.11 \cdot 1900 = 209$
new price = $1900 + 209 = 2109$
The mark-up is $209 and the new price is $2109.

64. Find 85% of 108,000.
$85\% \cdot 108{,}000 = 0.85 \cdot 108{,}000 = 91{,}800$
You would expect 91,800 motion picture and television industry businesses to have fewer than 10 employees.

65. Let x be the number of gallons of 40% solution. Then $30 - x$ is the number of gallons of 10% solution.

	gallons	concentration	amount
40% solution	x	40%	$0.4x$
10% solution	$30 - x$	10%	$0.1(30 - x)$
20% solution	30	20%	$0.2(30)$

The amount of acid in the combined solutions must be the same as in the mixture.
$0.4x + 0.1(30 - x) = 0.2(30)$
$0.4x + 3 - 0.1x = 6$
$3 + 0.3x = 6$
$3 + 0.3x - 3 = 6 - 3$
$0.3x = 3$
$\frac{0.3x}{0.3} = \frac{3}{0.3}$
$x = 10$
$30 - x = 30 - 10 = 20$
Mix 10 gallons of 40% solution with 20 gallons of 10% solution.

66. $\text{percent of increase} = \frac{\text{amount of increase}}{\text{original amount}}$
$= \frac{7.96 - 6.03}{6.03}$
$= \frac{1.93}{6.03}$
≈ 0.320
The percent of increase was 32%.

67. From the graph, 18% of motorists who use a cell phone while driving have almost hit another car.

68. The tallest bar represents the most common effect. Therefore, swerving is the most common effect of cell phone use on driving.

69. 21% of drivers cut off someone. Find 21% of 4600.
$21\% \cdot 4600 = 0.21 \cdot 4600 = 966$
You expect 966 customers to cut someone off while driving and talking on their cell phones.

70. $46\% + 41\% + 21\% + 18\% = 126\%$
No, the percents do not sum to 100%.
Answers may vary.

71. 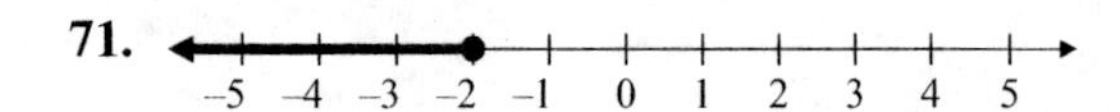

72. 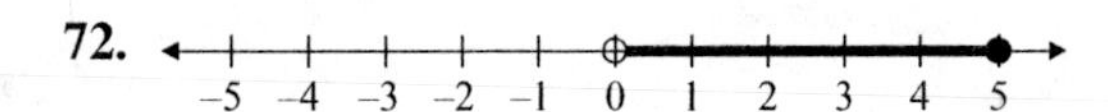

73.
$$\begin{aligned} x-5 &\le -4 \\ x-5+5 &\le -4+5 \\ x &\le 1 \end{aligned}$$
$\{x|x \le 1\}$

74.
$$\begin{aligned} x+7 &> 2 \\ x+7-7 &> 2-7 \\ x &> -5 \end{aligned}$$
$\{x|x > -5\}$

75.
$$\begin{aligned} -2x &\ge -20 \\ \frac{-2x}{-2} &\le \frac{-20}{-2} \\ x &\le 10 \end{aligned}$$
$\{x|x \le 10\}$

76.
$$\begin{aligned} -3x &> 12 \\ \frac{-3x}{-3} &< \frac{12}{-3} \\ x &< -4 \end{aligned}$$
$\{x|x < -4\}$

77.
$$\begin{aligned} 5x-7 &> 8x+5 \\ 5x-7-8x &> 8x+5-8x \\ -3x-7 &> 5 \\ -3x-7+7 &> 5+7 \\ -3x &> 12 \\ \frac{-3x}{-3} &< \frac{12}{-3} \\ x &< -4 \end{aligned}$$
$\{x|x < -4\}$

78.
$$\begin{aligned} x+4 &\ge 6x-16 \\ x+4-6x &\ge 6x-16-6x \\ -5x+4 &\ge -16 \\ -5x+4-4 &\ge -16-4 \\ -5x &\ge -20 \\ \frac{-5x}{-5} &\le \frac{-20}{-5} \\ x &\le 4 \end{aligned}$$
$\{x|x \le 4\}$

79.
$$\begin{aligned} \frac{2}{3}y &> 6 \\ \frac{3}{2}\cdot\frac{2}{3}y &> \frac{3}{2}\cdot 6 \\ y &> 9 \end{aligned}$$
$\{y|y > 9\}$

80.
$$\begin{aligned} -0.5y &\le 7.5 \\ \frac{-0.5y}{-0.5} &\ge \frac{7.5}{-0.5} \\ y &\ge -15 \end{aligned}$$
$\{y|y \ge -15\}$

81.
$$\begin{aligned} -2(x-5) &> 2(3x-2) \\ -2x+10 &> 6x-4 \\ -2x+10-6x &> 6x-4-6x \\ -8x+10 &> -4 \\ -8x+10-10 &> -4-10 \\ -8x &> -14 \\ \frac{-8x}{-8} &< \frac{-14}{-8} \\ x &< \frac{7}{4} \end{aligned}$$
$\left\{x \middle| x < \frac{7}{4}\right\}$

82.
$$\begin{aligned} 4(2x-5) &\le 5x-1 \\ 8x-20 &\le 5x-1 \\ 8x-20-5x &\le 5x-1-5x \\ 3x-20 &\le -1 \\ 3x-20+20 &\le -1+20 \\ 3x &\le 19 \\ \frac{3x}{3} &\le \frac{19}{3} \\ x &\le \frac{19}{3} \end{aligned}$$
$\left\{x \middle| x \le \frac{19}{3}\right\}$

83. Let x be the sales. Her weekly earnings are $175 + 0.05x$.
$$\begin{aligned} 175+0.05x &\ge 300 \\ 175+0.05x-175 &\ge 300-175 \\ 0.05x &\ge 125 \\ \frac{0.05x}{0.05} &\ge \frac{125}{0.05} \\ x &\ge 2500 \end{aligned}$$
She must have weekly sales of at least \$2500.

84. Let x be his score on the fourth round.
$$\begin{aligned} \frac{76+82+79+x}{4} &< 80 \\ \frac{237+x}{4} &< 80 \\ 4\cdot\frac{237+x}{4} &< 4\cdot 80 \\ 237+x &< 320 \\ 237+x-237 &< 320-237 \\ x &< 83 \end{aligned}$$
His score must be less than 83.

85.
$$\begin{aligned} 6x+2x-1 &= 5x+11 \\ 8x-1 &= 5x+11 \\ 8x-1-5x &= 5x+11-5x \\ 3x-1 &= 11 \\ 3x-1+1 &= 11+1 \\ 3x &= 12 \\ \frac{3x}{3} &= \frac{12}{3} \\ x &= 4 \end{aligned}$$

86.
$$\begin{aligned} 2(3y-4) &= 6+7y \\ 6y-8 &= 6+7y \\ 6y-8-6y &= 6+7y-6y \\ -8 &= 6+y \\ -8-6 &= 6+y-6 \\ -14 &= y \end{aligned}$$

87.
$$\begin{aligned} 4(3-a)-(6a+9) &= -12a \\ 12-4a-6a-9 &= -12a \\ 3-10a &= -12a \\ 3-10a+10a &= -12a+10a \\ 3 &= -2a \\ \frac{3}{-2} &= \frac{-2a}{-2} \\ -\frac{3}{2} &= a \end{aligned}$$

88.
$$\frac{x}{3} - 2 = 5$$
$$\frac{x}{3} - 2 + 2 = 5 + 2$$
$$\frac{x}{3} = 7$$
$$3 \cdot \frac{x}{3} = 3 \cdot 7$$
$$x = 21$$

89. $2(y+5) = 2y+10$
$$2y+10 = 2y+10$$
Since both sides of the equation are identical, the equation is an identity and every real number is a solution.

90. $7x - 3x + 2 = 2(2x-1)$
$$4x+2 = 4x-2$$
$$4x+2-4x = 4x-2-4x$$
$$2 = -2$$
Since the statement $2 = -2$ is false, there is no solution.

91. Let x be the number.
$$6+2x = x-7$$
$$6+2x-x = x-7-x$$
$$6+x = -7$$
$$6+x-6 = -7-6$$
$$x = -13$$
The number is -13.

92. Let x be the length of the shorter piece. Then $4x + 3$ is the length of the longer piece. The lengths sum to 23.
$$x+4x+3 = 23$$
$$5x+3 = 23$$
$$5x+3-3 = 23-3$$
$$5x = 20$$
$$\frac{5x}{5} = \frac{20}{5}$$
$$x = 4$$
$4x + 3 = 4(4) + 3 = 16 + 3 = 19$
The shorter piece is 4 inches and the longer piece is 19 inches.

93.
$$V = \frac{1}{3}Ah$$
$$3V = 3 \cdot \frac{1}{3}Ah$$
$$3V = Ah$$
$$\frac{3V}{A} = \frac{Ah}{A}$$
$$\frac{3V}{A} = h$$

94. Let x be the number.
$x = 26\% \cdot 85$
$x = 0.26 \cdot 85$
$x = 22.1$
22.1 is 26% of 85.

95. Let x be the unknown number.
$$72 = 45\% \cdot x$$
$$72 = 0.45x$$
$$\frac{72}{0.45} = \frac{0.45x}{0.45}$$
$$160 = x$$
72 is 45% of 160.

96.
$$\text{percent of increase} = \frac{\text{amount of increase}}{\text{original amount}} = \frac{282-235}{235} = \frac{47}{235} = 0.2$$
The percent of increase is 20%.

97.
$$4x-7 > 3x+2$$
$$4x-7-3x > 3x+2-3x$$
$$x-7 > 2$$
$$x-7+7 > 2+7$$
$$x > 9$$
$\{x|x > 9\}$

98. $-5x < 20$
$$\frac{-5x}{5} > \frac{20}{-5}$$
$$x > -4$$
$\{x|x > -4\}$

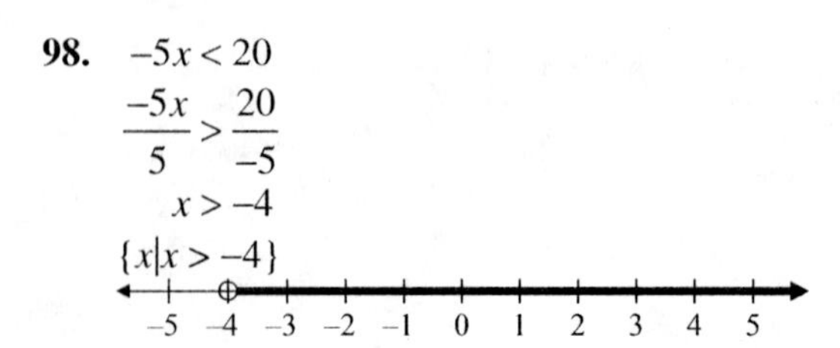

99. $-3(1+2x)+x \geq -(3-x)$

$$-3-6x+x \geq -3+x$$
$$-3-5x \geq -3+x$$
$$-3-5x-x \geq -3+x-x$$
$$-3-6x \geq -3$$
$$-3-6x+3 \geq -3+3$$
$$-6x \geq 0$$
$$\frac{-6x}{-6} \leq \frac{0}{-6}$$
$$x \leq 0$$

$\{x|x \leq 0\}$

−5 −4 −3 −2 −1 0 1 2 3 4 5

Chapter 2 Test

1.
$$-\frac{4}{5}x = 4$$
$$-\frac{5}{4}\left(-\frac{4}{5}x\right) = -\frac{5}{4}(4)$$
$$x = -5$$

2.
$$4(n-5) = -(4-2n)$$
$$4n-20 = -4+2n$$
$$4n-20-2n = -4+2n-2n$$
$$2n-20 = -4$$
$$2n-20+20 = -4+20$$
$$2n = 16$$
$$\frac{2n}{2} = \frac{16}{2}$$
$$n = 8$$

3.
$$5y-7+y = -(y+3y)$$
$$6y-7 = -y-3y$$
$$6y-7 = -4y$$
$$6y-7-6y = -4y-6y$$
$$-7 = -10y$$
$$\frac{-7}{-10} = \frac{-10y}{-10}$$
$$\frac{7}{10} = y$$

4.
$$4z+1-z = 1+z$$
$$3z+1 = 1+z$$
$$3z+1-z = 1+z-z$$
$$2z+1 = 1$$
$$2z+1-1 = 1-1$$
$$2z = 0$$
$$\frac{2z}{2} = \frac{0}{2}$$
$$z = 0$$

5.
$$\frac{2(x+6)}{3} = x-5$$
$$3\left(\frac{2(x+6)}{3}\right) = 3(x-5)$$
$$2(x+6) = 3(x-5)$$
$$2x+12 = 3x-15$$
$$2x+12-2x = 3x-15-2x$$
$$12 = x-15$$
$$12+15 = x-15+15$$
$$27 = x$$

6.
$$\frac{4(y-1)}{5} = 2y+3$$
$$5\left[\frac{4(y-1)}{5}\right] = 5(2y+3)$$
$$4(y-1) = 5(2y+3)$$
$$4y-4 = 10y+15$$
$$4y-4-10y = 10y+15-10y$$
$$-6y-4 = 15$$
$$-6y-4+4 = 15+4$$
$$-6y = 19$$
$$\frac{-6y}{-6} = \frac{19}{-6}$$
$$y = -\frac{19}{6}$$

7.
$$\frac{1}{2}-x+\frac{3}{2} = x-4$$
$$-x+\frac{4}{2} = x-4$$
$$-x+2 = x-4$$
$$-x+2+x = x-4+x$$
$$2 = 2x-4$$
$$2+4 = 2x-4+4$$
$$6 = 2x$$
$$\frac{6}{2} = \frac{2x}{2}$$
$$3 = x$$

8.
$$\frac{1}{3}(y+3) = 4y$$
$$3 \cdot \frac{1}{3}(y+3) = 3 \cdot 4y$$
$$y+3 = 12y$$
$$y+3-y = 12y-y$$
$$3 = 11y$$
$$\frac{3}{11} = \frac{11y}{11}$$
$$\frac{3}{11} = y$$

9.
$$\begin{aligned}
-0.3(x-4)+x &= 0.5(3-x)\\
-0.3(x-4)+1.0x &= 0.5(3-x)\\
-3(x-4)+10x &= 5(3-x)\\
-3x+12+10x &= 15-5x\\
7x+12 &= 15-5x\\
7x+12+5x &= 15-5x+5x\\
12x+12 &= 15\\
12x+12-12 &= 15-12\\
12x &= 3\\
\frac{12x}{12} &= \frac{3}{12}\\
x &= \frac{1}{4} = 0.25
\end{aligned}$$

10.
$$\begin{aligned}
-4(a+1)-3a &= -7(2a-3)\\
-4a-4-3a &= -14a+21\\
-4-7a &= -14a+21\\
-4-7a+14a &= -14a+21+14a\\
-4+7a &= 21\\
-4+7a+4 &= 21+4\\
7a &= 25\\
\frac{7a}{7} &= \frac{25}{7}\\
a &= \frac{25}{7}
\end{aligned}$$

11.
$$\begin{aligned}
-2(x-3) &= x+5-3x\\
-2x+6 &= -2x+5\\
2x-2x+6 &= 2x-2x+5\\
6 &= 5
\end{aligned}$$
Since the statement $6 = 5$ is false, there is no solution.

12. Let x be the number.
$$\begin{aligned}
x+\frac{2}{3}x &= 35\\
\frac{3}{3}x+\frac{2}{3}x &= 35\\
\frac{5}{3}x &= 35\\
\frac{3}{5}\cdot\frac{5}{3}x &= \frac{3}{5}\cdot 35\\
x &= 21
\end{aligned}$$
The number is 21.

13. $A = lw = (35)(20) = 700$
The area of the deck is 700 square feet. To paint two coats of water seal means covering $2 \cdot 700 = 1400$ square feet.
$$1400 \text{ sq ft} \cdot \frac{1 \text{ gal}}{200 \text{ sq ft}} = 7 \text{ gal}$$
7 gallons of water seal are needed.

14. Use $y = mx + b$ when $y = -14$, $m = -2$, and $b = -2$.
$$\begin{aligned}
y &= mx+b\\
-14 &= -2x+(-2)\\
-14+2 &= -2x+(-2)+2\\
-12 &= -2x\\
\frac{-12}{-2} &= \frac{-2x}{-2}\\
6 &= x
\end{aligned}$$

15.
$$\begin{aligned}
V &= \pi r^2 h\\
\frac{V}{\pi r^2} &= \frac{\pi r^2 h}{\pi r^2}\\
\frac{V}{\pi r^2} &= h
\end{aligned}$$

16.
$$\begin{aligned}
3x-4y &= 10\\
3x-4y-3x &= 10-3x\\
-4y &= 10-3x\\
\frac{-4y}{-4} &= \frac{10-3x}{-4}\\
y &= \frac{3x-10}{4}
\end{aligned}$$

17.
$$\begin{aligned}
3x-5 &\ge 7x+3\\
3x-5-3x &\ge 7x+3-3x\\
-5 &\ge 4x+3\\
-5-3 &\ge 4x+3-3\\
-8 &\ge 4x\\
\frac{-8}{4} &\ge \frac{4x}{4}\\
-2 &\ge x
\end{aligned}$$
$\{x \mid x \le -2\}$

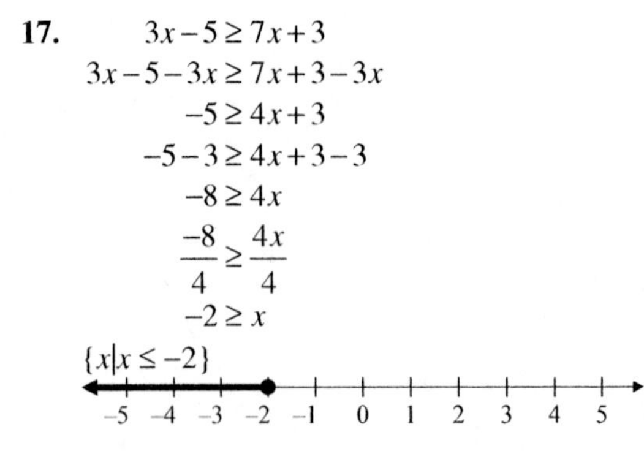

18.
$$\begin{aligned}
x+6 &> 4x-6\\
x+6-4x &> 4x-6-4x\\
-3x+6 &> -6\\
-3x+6-6 &> -6-6\\
-3x &> -12\\
\frac{-3x}{-3} &< \frac{-12}{-3}\\
x &< 4
\end{aligned}$$
$\{x \mid x < 4\}$

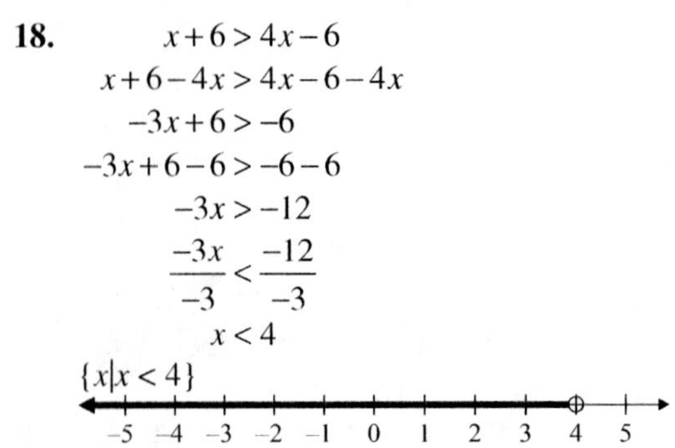

19. $-0.3x \ge 2.4$

$$\frac{-0.3x}{-0.3} \le \frac{2.4}{-0.3}$$
$$x \le -8$$
$$\{x|x \le -8\}$$

20. $-5(x-1)+6 \le -3(x+4)+1$

$$-5x+5+6 \le -3x-12+1$$
$$-5x+11 \le -3x-11$$
$$-5x+11+3x \le -3x-11+3x$$
$$-2x+11 \le -11$$
$$-2x+11-11 \le -11-11$$
$$-2x \le -22$$
$$\frac{-2x}{-2} \ge \frac{-22}{-2}$$
$$x \ge 11$$
$$\{x|x \ge 11\}$$

21. $\frac{2(5x+1)}{3} > 2$

$$3 \cdot \frac{2(5x+1)}{3} > 3(2)$$
$$2(5x+1) > 6$$
$$10x+2 > 6$$
$$10x+2-2 > 6-2$$
$$10x > 4$$
$$\frac{10x}{10} > \frac{4}{10}$$
$$x > \frac{2}{5}$$
$$\left\{x \middle| x > \frac{2}{5}\right\}$$

22. From the graph, 69% are classified as weak.
Find 69% of 800.
$69\% \cdot 800 = 0.69 \cdot 800 = 552$
You would expect 552 of the 800 to be classified as weak.

23. Let x be the unknown percent.

$$72 = x \cdot 180$$
$$\frac{72}{180} = \frac{180x}{180}$$
$$0.4 = x$$

72 is 40% of 180.

24. Let x = area code 1, then $2x$ = area code 2.

$$x+2x = 1203$$
$$3x = 1203$$
$$\frac{3x}{3} = \frac{1203}{3}$$
$$x = 401$$

$2x = 2(401) = 802$
The area codes are 401 and 802.

25. Let x represent the number of public libraries in Ohio. Then there are $x + 387$ public libraries in California.

$$x+x+387 = 1827$$
$$2x+387 = 1827$$
$$2x+387-387 = 1827-387$$
$$2x = 1440$$
$$\frac{2x}{2} = \frac{1440}{2}$$
$$x = 720$$

$x + 387 = 720 + 387 = 1107$
Ohio has 720 public libraries and California has 1107.

Cumulative Review Chapters 1–2

1. Since $8 = 8$, the statement $8 \ge 8$ is true.

2. Since -4 is to the right of -6 on a number line, the statement $-4 < -6$ is false.

3. Since $8 = 8$, the statement $8 \le 8$ is true.

4. Since 3 is to the right of -3 on a number line, the statement $3 > -3$ is true.

5. Since neither $23 < 0$ nor $23 = 0$ is true, the statement $23 \le 0$ is false.

6. Since $-8 = -8$, the statement $-8 \ge -8$ is true.

7. Since $0 < 23$ is true, the statement $0 \le 23$ is true.

8. Since $-8 = -8$, the statement $-8 \le -8$ is true.

9. a. $|0| < 2$ since $|0| = 0$ and $0 < 2$.

b. $|-5| = 5$

c. $|-3| > |-2|$ since $3 > 2$.

d. $|-9| < |-9.7|$ since $9 < 9.7$.

e. $\left|-7\frac{1}{6}\right| > |7|$ since $7\frac{1}{6} > 7$.

10. a. $|5| = 5$ since 5 is 5 units from 0 on a number line.

b. $|-8| = 8$ since -8 is 8 units from 0 on a number line.

c. $\left|-\frac{2}{3}\right| = \frac{2}{3}$ since $-\frac{2}{3}$ is $\frac{2}{3}$ unit from 0 on a number line.

11. $\frac{3+|4-3|+2^2}{6-3} = \frac{3+|1|+2^2}{6-3}$
$= \frac{3+1+2^2}{3}$
$= \frac{3+1+4}{3}$
$= \frac{8}{3}$

12. $1+2(9-7)^3+4^2 = 1+2(2)^3+4^2$
$= 1+2(8)+16$
$= 1+16+16$
$= 33$

13. $(-8) + (-11) = -19$

14. $-2 + (-8) = -10$

15. $(-2) + 10 = 8$

16. $-10 + 20 = 10$

17. $0.2 + (-0.5) = -0.3$

18. $1.2 + (-1.2) = 0$

19. a. $-3+[(-2-5)-2] = -3+[(-2+(-5))-2]$
$= -3+[(-7)-2]$
$= -3+[-7+(-2)]$
$= -3+[-9]$
$= -12$

b. $2^3-10+[-6-(-5)] = 2^3-10+[-6+5]$
$= 2^3-10+[-1]$
$= 8-10+(-1)$
$= 8+(-10)+(-1)$
$= -2+(-1)$
$= -3$

20. a. $-(-5) = 5$

b. $-\left(-\frac{2}{3}\right) = \frac{2}{3}$

c. $-(-a) = a$

d. $-|-3| = -3$

21. a. $7(0)(-6) = 0(-6) = 0$

b. $(-2)(-3)(-4) = (6)(-4) = -24$

c. $(-1)(-5)(-9)(-2) = 5(-9)(-2)$
$= (-45)(-2)$
$= 90$

22. a. $-2.7 - 8.4 = -2.7 + (-8.4) = -11.1$

b. $-\frac{4}{5}-\left(-\frac{3}{5}\right) = -\frac{4}{5}+\frac{3}{5} = -\frac{1}{5}$

c. $\frac{1}{4}-\left(-\frac{1}{2}\right) = \frac{1}{4}+\frac{1}{2} = \frac{1}{4}+\frac{2}{4} = \frac{3}{4}$

23. a. $-18 \div 3 = -18 \cdot \frac{1}{3} = -6$

b. $\frac{-14}{-2} = -14 \cdot -\frac{1}{2} = 7$

c. $\frac{20}{-4} = 20 \cdot -\frac{1}{4} = -5$

24. a. $(4.5)(-0.08) = -0.36$

b. $-\frac{3}{4} \cdot -\frac{8}{17} = \frac{3 \cdot 8}{4 \cdot 17} = \frac{6}{17}$

25. $-5(-3+2z) = -5(-3)+(-5)(2z) = 15-10z$

26. $2(y-3x+4) = 2(y)-2(3x)+2(4)$
$= 2y-6x+8$

27. $\frac{1}{2}(6x+14)+10 = \frac{1}{2}(6x)+\frac{1}{2}(14)+10$
$= 3x+7+10$
$= 3x+17$

28. $-(x+4)+3(x+4) = -1(x+4)+3(x+4)$
$= -1 \cdot x+(-1)(4)+3 \cdot x+3 \cdot 4$
$= -x-4+3x+12$
$= -x+3x-4+12$
$= 2x+8$

29. a. $2x$ and $3x^2$ are unlike terms, since the exponents on x are not the same.

b. $4x^2y$, xy^2, and $-2x^2y$ are like terms, since each variable and its exponent match.

c. $-2yz$ and $-3zy$ are like terms, since $zy = yz$ by the commutative property.

d. $-x^4$ and x^4 are like terms. The variable and its exponent match.

e. $-8a^5$ and $8a^5$ are like terms. The variable and its exponent match.

30. a. $\frac{-32}{8} = -4$

b. $\frac{-108}{-12} = 9$

c. $\frac{-5}{7} \div \left(\frac{-9}{2}\right) = \frac{-5}{7} \cdot \left(\frac{2}{-9}\right) = \frac{5 \cdot 2}{7 \cdot 9} = \frac{10}{63}$

31. $(2x - 3) - (4x - 2) = 2x - 3 - 4x + 2 = -2x - 1$

32.
$$\begin{aligned}(-5x+1)-(10x+3) &= -5x+1-10x-3\\ &= -15x-2\end{aligned}$$

33.
$$\begin{aligned}x-7 &= 10\\ x-7+7 &= 10+7\\ x &= 17\end{aligned}$$

34.
$$\begin{aligned}\frac{5}{6}+x &= \frac{2}{3}\\ \frac{5}{6}+x-\frac{5}{6} &= \frac{2}{3}-\frac{5}{6}\\ x &= \frac{4}{6}-\frac{5}{6}\\ x &= -\frac{1}{6}\end{aligned}$$

35.
$$\begin{aligned}-z-4 &= 6\\ -z-4+4 &= 6+4\\ -z &= 10\\ \frac{-z}{-1} &= \frac{10}{-1}\\ z &= -10\end{aligned}$$

36.
$$\begin{aligned}-3x+1-(-4x-6) &= 10\\ -3x+1+4x+6 &= 10\\ x+7 &= 10\\ x+7-7 &= 10-7\\ x &= 3\end{aligned}$$

37.
$$\begin{aligned}\frac{2(a+3)}{3} &= 6a+2\\ 3\cdot\frac{2(a+3)}{3} &= 3(6a+2)\\ 2(a+3) &= 3(6a+2)\\ 2a+6 &= 18a+6\\ 2a+6-18a &= 18a+6-18a\\ -16a+6 &= 6\\ -16a+6-6 &= 6-6\\ -16a &= 0\\ \frac{-16a}{-16} &= \frac{0}{-16}\\ a &= 0\end{aligned}$$

38.
$$\begin{aligned}\frac{x}{4} &= 18\\ 4\cdot\frac{x}{4} &= 4\cdot 18\\ x &= 72\end{aligned}$$

39. Let x be the number of Democrats. Then $x + 34$ is the number of Republicans. The total number is 432.
$$\begin{aligned}x+x+34 &= 432\\ 2x+34 &= 432\\ 2x+34-34 &= 432-34\\ 2x &= 398\\ \frac{2x}{2} &= \frac{398}{2}\\ x &= 199\end{aligned}$$
$x + 34 = 199 + 34 = 233$
There were 199 Democrats and 233 Republicans.

40.
$$\begin{aligned}6x+5 &= 4(x+4)-1\\ 6x+5 &= 4x+16-1\\ 6x+5 &= 4x+15\\ 6x+5-4x &= 4x+15-4x\\ 2x+5 &= 15\\ 2x+5-5 &= 15-5\\ 2x &= 10\\ \frac{2x}{2} &= \frac{10}{2}\\ x &= 5\end{aligned}$$

41. Use $d = rt$ when $d = 31,680$ and $r = 400$.
$$\begin{aligned}d &= rt\\ 31,680 &= 400t\\ \frac{31,680}{400} &= \frac{400t}{400}\\ 79.2 &= t\end{aligned}$$
It will take the ice 79.2 years to reach the lake.

42.
$$x+4=3x-8$$
$$x+4-3x=3x-8-3x$$
$$-2x+4=-8$$
$$-2x+4-4=-8-4$$
$$-2x=-12$$
$$\frac{-2x}{-2}=\frac{-12}{-2}$$
$$x=6$$
The number is 6.

43. Let x be the unknown percent.
$$63=x\cdot 72$$
$$\frac{63}{72}=\frac{72x}{72}$$
$$0.875=x$$
$$87.5\%=x$$
63 is 87.5% of 72.

44.
$$C=2\pi r$$
$$\frac{C}{2\pi}=\frac{2\pi r}{2\pi}$$
$$\frac{C}{2\pi}=r \text{ or } r=\frac{C}{2\pi}$$

45.
$$5(2x+3)=-1+7$$
$$5(2x)+5(3)=-1+7$$
$$10x+15=6$$
$$10x+15-15=6-15$$
$$10x=-9$$
$$\frac{10x}{10}=\frac{-9}{10}$$
$$x=-\frac{9}{10}$$

46.
$$x-3>2$$
$$x-3+3>2+3$$
$$x>5$$
$$\{x|x>5\}$$

47. $-1>x$ or $x<-1$

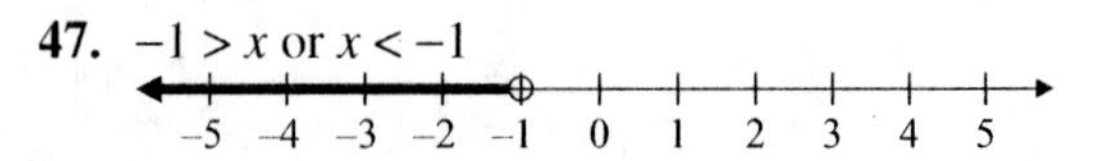

48.
$$3x-4\le 2x-14$$
$$3x-4-2x\le 2x-14-2x$$
$$x-4\le -14$$
$$x-4+4\le -14+4$$
$$x\le -10$$
$$\{x|x\le -10\}$$

49.
$$2(x-3)-5\le 3(x+2)-18$$
$$2x-6-5\le 3x+6-18$$
$$2x-11\le 3x-12$$
$$-x-11\le -12$$
$$-x\le -1$$
$$\frac{-x}{-1}\ge\frac{-1}{-1}$$
$$x\ge 1$$
$$\{x|x\ge 1\}$$

50.
$$-3x\ge 9$$
$$\frac{-3x}{-3}\le\frac{9}{-3}$$
$$x\le -3$$
$$\{x|x\le -3\}$$

Chapter 3

Section 3.1 Practice Exercises

1. $3^4 = 3 \cdot 3 \cdot 3 \cdot 3 = 81$

2. $7^1 = 7$

3. $(-2)^3 = (-2)(-2)(-2) = -8$

4. $-2^3 = -(2 \cdot 2 \cdot 2) = -8$

5. $\left(\frac{2}{3}\right)^2 = \frac{2}{3} \cdot \frac{2}{3} = \frac{4}{9}$

6. $5 \cdot 6^2 = 5 \cdot 36 = 180$

7. **a.** When x is 4, $\begin{aligned} 3x^2 &= 3 \cdot 4^2 \\ &= 3 \cdot (4 \cdot 4) \\ &= 3 \cdot 16 \\ &= 48. \end{aligned}$

   b. When x is -2, $\begin{aligned} \frac{x^4}{-8} &= \frac{(-2)^4}{-8} \\ &= \frac{(-2)(-2)(-2)(-2)}{-8} \\ &= \frac{16}{-8} \\ &= -2. \end{aligned}$

8. $7^3 \cdot 7^2 = 7^{3+2} = 7^5$

9. $x^4 \cdot x^9 = x^{4+9} = x^{13}$

10. $r^5 \cdot r = r^{5+1} = r^6$

11. $s^6 \cdot s^2 \cdot s^3 = s^{6+2+3} = s^{11}$

12. $(-3)^9 \cdot (-3) = (-3)^{9+1} = (-3)^{10}$

13. $\begin{aligned} (6x^3)(-2x^9) &= (6 \cdot x^3) \cdot (-2 \cdot x^9) \\ &= (6 \cdot -2) \cdot (x^3 \cdot x^9) \\ &= -12x^{12} \end{aligned}$

14. $\begin{aligned} (m^5n^{10})(mn^8) &= (m^5 \cdot m) \cdot (n^{10} \cdot n^8) \\ &= m^6 \cdot n^{18} \\ &= m^6n^{18} \end{aligned}$

15. $\begin{aligned} (-x^9y)(4x^2y^{11}) &= (-1 \cdot 4) \cdot (x^9 \cdot x^2) \cdot (y \cdot y^{11}) \\ &= -4x^{11}y^{12} \end{aligned}$

16. $(9^4)^{10} = 9^{4 \cdot 10} = 9^{40}$

17. $(z^6)^3 = z^{6 \cdot 3} = z^{18}$

18. $(xy)^7 = x^7 \cdot y^7 = x^7y^7$

19. $(3y)^4 = 3^4 \cdot y^4 = 81y^4$

20. $\begin{aligned} (-2p^4q^2r)^3 &= (-2)^3 \cdot (p^4)^3 \cdot (q^2)^3 \cdot (r^1)^3 \\ &= -8p^{12}q^6r^3 \end{aligned}$

21. $\begin{aligned} (-a^4b)^7 &= (-1a^4b)^7 \\ &= (-1)^7 \cdot (a^4)^7 \cdot (b^1)^7 \\ &= -1a^{28}b^7 \\ &= -a^{28}b^7 \end{aligned}$

22. $\left(\frac{r}{s}\right)^6 = \frac{r^6}{s^6},\ s \neq 0$

23. $\left(\frac{5x^6}{9y^3}\right)^2 = \frac{5^2 \cdot (x^6)^2}{9^2 \cdot (y^3)^2} = \frac{25x^{12}}{81y^6},\ y \neq 0$

24. $\frac{y^7}{y^3} = y^{7-3} = y^4$

25. $\frac{5^9}{5^6} = 5^{9-6} = 5^3 = 125$

26. $\frac{(-2)^{14}}{(-2)^{10}} = (-2)^{14-10} = (-2)^4 = 16$

27. $\begin{aligned} \frac{7a^4b^{11}}{ab} &= 7 \cdot \frac{a^4}{a^1} \cdot \frac{b^{11}}{b^1} \\ &= 7 \cdot (a^{4-1}) \cdot (b^{11-1}) \\ &= 7a^3b^{10} \end{aligned}$

28. $8^0 = 1$

29. $(2r^2s)^0 = 1$

30. $(-7)^0 = 1$

31. $-7^0 = -1\cdot 7^0 = -1\cdot 1 = -1$

32. $7y^0 = 7\cdot y^0 = 7\cdot 1 = 7$

33. a. $\frac{x^7}{x^4} = x^{7-4} = x^3$

b. $(3y^4)^4 = 3^4\cdot (y^4)^4 = 81y^{16}$

c. $\left(\frac{x}{4}\right)^3 = \frac{x^3}{4^3} = \frac{x^3}{64}$

Vocabulary, Readiness & Video Check 3.1

1. Repeated multiplication of the same factor can be written using an exponent.

2. In 5^2, the 2 is called the exponent and the 5 is called the base.

3. To simplify $x^2\cdot x^7$, keep the base and add the exponents.

4. To simplify $(x^3)^6$, keep the base and multiply the exponents.

5. The understood exponent on the term y is 1.

6. If $x^{\square} = 1$, the exponent is 0.

7. In 3^2, the base is 3 and the exponent is 2.

8. In $(-3)^6$, the base is -3 and the exponent is 6.

9. In -4^2, the base is 4 and the exponent is 2.

10. In $5\cdot 3^4 = 5^1\cdot 3^4$, the base 5 has exponent 1 and the base 3 has exponent 4.

11. In $5x^2 = 5^1x^2$, the base 5 has exponent 1 and the base x has exponent 2.

12. In $(5x)^2$, the base is $5x$ and the exponent is 2.

13. Example 4 can be written as $-4^2 = -1\cdot 4^2$, which is similar to Example 7, $4\cdot 3^2$, and shows why the negative sign should not be considered part of the base when there are no parentheses.

14. The properties allow us to reorder and regroup factors and put factors with common bases together, making it easier to apply the product rule.

15. Be careful not to confuse the power rule with the product rule. The power rule involves a power raised to a power (exponents are multiplied), and the product rule involves a product (exponents are added).

16. Remember to raise the -2 (or any number) to the power along with the variables.

17. the quotient rule

18. No, Example 30 is a fraction cubed and the quotient rule for exponents is not needed to simplify.

Exercise Set 3.1

1. $7^2 = 7\cdot 7 = 49$

3. $(-5)^1 = -5$

5. $-2^4 = -(2\cdot 2\cdot 2\cdot 2) = -16$

7. $(-2)^4 = (-2)(-2)(-2)(-2) = 16$

9. $\left(\frac{1}{3}\right)^3 = \frac{1}{3}\cdot\frac{1}{3}\cdot\frac{1}{3} = \frac{1}{27}$

11. $7\cdot 2^4 = 7(2\cdot 2\cdot 2\cdot 2) = 7\cdot 16 = 112$

13. When x is -2, $x^2 = (-2)^2 = (-2)(-2) = 4$.

15. When x is 3,
$5x^3 = 5(3)^3 = 5(3)(3)(3) = 5(27) = 135.$

17. When $x = 3$ and $y = -5$,

$$\begin{aligned} 2xy^2 &= 2(3)(-5)^2 \\ &= 2(3)(-5)(-5) \\ &= 2(3)(25) \\ &= 150. \end{aligned}$$

19. When z is -2,

$$\frac{2z^4}{5} = \frac{2(-2)^4}{5} = \frac{2(-2)(-2)(-2)(-2)}{5} = \frac{2(16)}{5} = \frac{32}{5}.$$

21. $x^2 \cdot x^5 = x^{2+5} = x^7$

23. $(-3)^3 \cdot (-3)^9 = (-3)^{3+9} = (-3)^{12}$

25. $(5y^4)(3y) = (5 \cdot 3)(y^4 \cdot y) = 15y^{4+1} = 15y^5$

27. $(x^9y)(x^{10}y^5) = (x^9 \cdot x^{10})(y \cdot y^5) = (x^{9+10})(y^{1+5}) = x^{19}y^6$

29. $(-8mn^6)(9m^2n^2) = (-8 \cdot 9)(m \cdot m^2)(n^6 \cdot n^2) = -72m^{1+2}n^{6+2} = -72m^3n^8$

31. $(4z^{10})(-6z^7)(z^3) = (4 \cdot -6 \cdot 1)(z^{10} \cdot z^7 \cdot z^3) = -24z^{10+7+3} = -24z^{20}$

33. Area = (length)(width)
= ($5x^3$ feet)($4x^2$ feet)
= $(5 \cdot 4)(x^3 \cdot x^2)$ square feet
= $20x^{3+2}$ square feet
= $20x^5$ square feet

35. $(x^9)^4 = x^{9 \cdot 4} = x^{36}$

37. $(pq)^8 = p^8 \cdot q^8 = p^8q^8$

39. $(2a^5)^3 = (2)^3(a^5)^3 = 2^3a^{15} = 8a^{15}$

41. $(x^2y^3)^5 = (x^2)^5(y^3)^5 = x^{10}y^{15}$

43. $(-7a^2b^5c)^2 = (-7)^2(a^2)^2(b^5)^2(c)^2 = 49a^4b^{10}c^2$

45. $\left(\frac{r}{s}\right)^9 = \frac{r^9}{s^9}$

47. $\left(\frac{mp}{n}\right)^9 = \frac{m^9p^9}{n^9}$

49. $$\left(\frac{-2xz}{y^5}\right)^2 = \frac{(-2)^2(x)^2(z)^2}{(y^5)^2} = \frac{4x^2z^2}{y^{5 \cdot 2}} = \frac{4x^2z^2}{y^{10}}$$

51. Area = (length)(length)
= ($8z^5$ decimeters)($8z^5$ decimeters)
= $(8 \cdot 8)(z^5 \cdot z^5)$ square decimeters
= $64z^{5+5}$ square decimeters
= $64z^{10}$ square decimeters

53. Volume = (length)(width)(height)
= ($3y^4$ feet)($3y^4$ feet)($3y^4$ feet)
= $(3)^3(y^4)^3$ cubic feet
= $27y^{4 \cdot 3}$ cubic feet
= $27y^{12}$ cubic feet

55. $\frac{x^3}{x} = x^{3-1} = x^2$

57. $\frac{(-4)^6}{(-4)^3} = (-4)^{6-3} = (-4)^3 = -64$

59. $\frac{p^7q^{20}}{pq^{15}} = \frac{p^7}{p} \cdot \frac{q^{20}}{q^{15}} = p^{7-1} \cdot q^{20-15} = p^6q^5$

61. $$\frac{7x^2y^6}{14x^2y^3} = \frac{7}{14} \cdot \frac{x^2}{x^2} \cdot \frac{y^6}{y^3} = \frac{1}{2} \cdot x^{2-2} \cdot y^{6-3} = \frac{1}{2}x^0y^3 = \frac{y^3}{2}$$

63. $7^0 = 1$

65. $(2x)^0 = 1$

67. $-7x^0 = -7 \cdot x^0 = -7 \cdot 1 = -7$

69. $5^0 + y^0 = 1 + 1 = 2$

71. $-9^2 = -(9)^2 = -(9 \cdot 9) = -81$

73. $\left(\frac{1}{4}\right)^3 = \frac{1}{4} \cdot \frac{1}{4} \cdot \frac{1}{4} = \frac{1}{64}$

75. $b^4 b^2 = b^{4+2} = b^6$

77. $a^2 a^3 a^4 = a^{2+3+4} = a^9$

79. $(2x^3)(-8x^4) = (2 \cdot -8)(x^3 \cdot x^4)$
$= -16x^{3+4}$
$= -16x^7$

81. $(a^7 b^{12})(a^4 b^8) = a^7 a^4 \cdot b^{12} b^8$
$= a^{7+4} b^{12+8}$
$= a^{11} b^{20}$

83. $(-2mn^6)(-13m^8 n) = (-2)(-13)(m \cdot m^8)(n^6 \cdot n)$
$= 26m^{1+8} n^{6+1}$
$= 26m^9 n^7$

85. $(z^4)^{10} = z^{4 \cdot 10} = z^{40}$

87. $(4ab)^3 = (4)^3 a^3 b^3 = 64a^3 b^3$

89. $(-6xyz^3)^2 = (-6)^2 x^2 y^2 (z^3)^2$
$= 36x^2 y^2 z^{3 \cdot 2}$
$= 36x^2 y^2 z^6$

91. $\frac{z^{12}}{z^4} = z^{12-4} = z^8$

93. $\frac{3x^5}{x} = 3 \cdot \frac{x^5}{x} = 3x^{5-1} = 3x^4$

95. $(6b)^0 = 1$

97. $(9xy)^2 = 9^2 \cdot x^2 y^2 = 81x^2 y^2$

99. $2^3 + 2^5 = (2 \cdot 2 \cdot 2) + (2 \cdot 2 \cdot 2 \cdot 2 \cdot 2)$
$= 8 + 32$
$= 40$

101. $\left(\frac{3y^5}{6x^4}\right)^3 = \left(\frac{y^5}{2x^4}\right)^3 = \frac{(y^5)^3}{2^3 (x^4)^3} = \frac{y^{5 \cdot 3}}{8x^{4 \cdot 3}} = \frac{y^{15}}{8x^{12}}$

103. $\frac{2x^3 y^2 z}{xyz} = 2 \cdot \frac{x^3}{x} \cdot \frac{y^2}{y} \cdot \frac{z}{z}$
$= 2x^{3-1} y^{2-1} z^{1-1}$
$= 2x^2 y^1 z^0$
$= 2x^2 y$

105. $5 - 7 = 5 + (-7) = -2$

107. $3 - (-2) = 3 + 2 = 5$

109. $-11 - (-4) = -11 + 4 = -7$

111. The expression $(x^{14})^{23}$ can be simplified by multiplying the exponents; c.

113. The expression $x^{14} + x^{23}$ cannot be simplified by adding subtracting, multiplying, or dividing the exponents; e.

115. answers may vary

117. answers may vary

119. $V = x^3$
$= (7 \text{ meters})^3$
$= 7^3$ cubic meters
$= 343$ cubic meters

121. The volume of a cube measures the amount of material that the cube can hold, so to find the amount of water that a swimming pool can hold, the formula for volume should be used.

123. answers may vary

125. answers may vary

127. $x^{5a} x^{4a} = x^{5a+4a} = x^{9a}$

129. $(a^b)^5 = a^{b \cdot 5} = a^{5b}$

131. $\dfrac{x^{9a}}{x^{4a}} = x^{9a-4a} = x^{5a}$

Section 3.2 Practice Exercises

1. $5^{-3} = \dfrac{1}{5^3} = \dfrac{1}{125}$

2. $7x^{-4} = 7^1 \cdot \dfrac{1}{x^4} = \dfrac{7^1}{x^4}$ or $\dfrac{7}{x^4}$

3. $5^{-1} + 3^{-1} = \dfrac{1}{5} + \dfrac{1}{3} = \dfrac{3}{15} + \dfrac{5}{15} = \dfrac{8}{15}$

4. $(-3)^{-4} = \dfrac{1}{(-3)^4} = \dfrac{1}{(-3)(-3)(-3)(-3)} = \dfrac{1}{81}$

5. $\left(\dfrac{6}{7}\right)^{-2} = \dfrac{6^{-2}}{7^{-2}} = \dfrac{6^{-2}}{1} \cdot \dfrac{1}{7^{-2}} = \dfrac{1}{6^2} \cdot \dfrac{7^2}{1} = \dfrac{7^2}{6^2} = \dfrac{49}{36}$

6. $\dfrac{x}{x^{-4}} = \dfrac{x^1}{x^{-4}} = x^{1-(-4)} = x^5$

7. $\dfrac{y^{-9}}{z^{-5}} = y^{-9} \cdot \dfrac{1}{z^{-5}} = \dfrac{1}{y^9} \cdot z^5 = \dfrac{z^5}{y^9}$

8. $\dfrac{y^{-4}}{y^6} = y^{-4-6} = y^{-10} = \dfrac{1}{y^{10}}$

9.
$$\begin{aligned} \frac{(x^5)^3 x}{x^4} &= \frac{x^{15} \cdot x}{x^4} \\ &= \frac{x^{15+1}}{x^4} \\ &= \frac{x^{16}}{x^4} \\ &= x^{16-4} \\ &= x^{12} \end{aligned}$$

10.
$$\begin{aligned} \left(\frac{9x^3}{y}\right)^{-2} &= \frac{9^{-2}(x^3)^{-2}}{y^{-2}} \\ &= \frac{9^{-2}x^{-6}}{y^{-2}} \\ &= \frac{y^2}{9^2 x^6} \\ &= \frac{y^2}{81x^6} \end{aligned}$$

11. $(a^{-4}b^7)^{-5} = (a^{-4})^{-5}(b^7)^{-5} = a^{20}b^{-35} = \dfrac{a^{20}}{b^{35}}$

12. $\dfrac{(2x)^4}{x^8} = \dfrac{2^4 x^4}{x^8} = 2^4 x^{4-8} = 2^4 x^{-4} = \dfrac{16}{x^4}$

13. $\dfrac{y^{-10}}{(y^5)^4} = \dfrac{y^{-10}}{y^{20}} = y^{-10-20} = y^{-30} = \dfrac{1}{y^{30}}$

14. $(4a^2)^{-3} = 4^{-3}(a^2)^{-3} = 4^{-3}a^{-6} = \dfrac{1}{4^3 a^6} = \dfrac{1}{64a^6}$

15.
$$\begin{aligned} -\frac{32x^{-3}y^{-6}}{8x^{-5}y^{-2}} &= -\frac{32}{8} \cdot x^{-3-(-5)} y^{-6-(-2)} \\ &= -4x^2 y^{-4} \\ &= -\frac{4x^2}{y^4} \end{aligned}$$

16.
$$\begin{aligned} \frac{(3x^{-2}y)^{-2}}{(2x^7 y)^3} &= \frac{3^{-2}(x^{-2})^{-2}y^{-2}}{2^3(x^7)^3 y^3} \\ &= \frac{3^{-2}x^4 y^{-2}}{2^3 x^{21} y^3} \\ &= \frac{3^{-2}}{2^3} \cdot x^{4-21} y^{-2-3} \\ &= \frac{3^{-2}}{2^3} x^{-17} y^{-5} \\ &= \frac{1}{2^3 3^2 x^{17} y^5} \\ &= \frac{1}{72x^{17}y^5} \end{aligned}$$

17. a. $420{,}000 = 4.2 \times 10^5$

b. $0.00017 = 1.7 \times 10^{-4}$

c. $9,060,000,000 = 9.06 \times 10^9$

d. $0.000007 = 7.0 \times 10^{-6}$

18. a. $3.062 \times 10^{-4} = 0.0003062$

b. $5.21 \times 10^4 = 52,100$

c. $9.6 \times 10^{-5} = 0.000096$

d. $6.002 \times 10^6 = 6,002,000$

19. a.
$$\begin{aligned}(9 \times 10^7)(4 \times 10^{-9}) &= 9 \cdot 4 \cdot 10^7 \cdot 10^{-9}\\ &= 36 \times 10^{-2}\\ &= 0.36\end{aligned}$$

b.
$$\begin{aligned}\frac{8 \times 10^4}{2 \times 10^{-3}} &= \frac{8}{2} \times 10^{4-(-3)}\\ &= 4 \times 10^7\\ &= 40,000,000\end{aligned}$$

Calculator Explorations

1. 5.31×10^3 5.31 EE 3

2. -4.8×10^{14} −4.8 EE 14

3. 6.6×10^{-9} 6.6 EE −9

4. -9.9811×10^{-2} −9.9811 EE −2

5. $3,000,000 \times 5,000,000 = 1.5 \times 10^{13}$

6. $230,000 \times 1,000 = 2.3 \times 10^8$

7. $(3.26 \times 10^6)(2.5 \times 10^{13}) = 8.15 \times 10^{19}$

8. $(8.76 \times 10^{-4})(1.237 \times 10^9) = 1.083612 \times 10^6$

Vocabulary, Readiness & Video Check 3.2

1. The expression x^{-3} equals $\underline{\frac{1}{x^3}}$.

2. The expression 5^{-4} equals $\underline{\frac{1}{625}}$.

3. The number 3.021×10^{-3} is written in scientific notation.

4. The number 0.0261 is written in standard form.

5. $5x^{-2} = 5 \cdot \frac{1}{x^2} = \frac{5}{x^2}$

6. $3x^{-3} = 3 \cdot \frac{1}{x^3} = \frac{3}{x^3}$

7. $\frac{1}{y^{-6}} = y^6$

8. $\frac{1}{x^{-3}} = x^3$

9. $\frac{4}{y^{-3}} = 4 \cdot \frac{1}{y^{-3}} = 4y^3$

10. $\frac{16}{y^{-7}} = 16 \cdot \frac{1}{y^{-7}} = 16y^7$

11. A negative exponent has nothing to do with the sign of the simplified result.

12. power of a product rule, power rule for exponents, negative exponent definition, quotient rule for exponents

13. When you move the decimal point to the left, the sign of the exponent will be positive; when you move the decimal point to the right, the sign of the exponent will be negative.

14. the exponent on 10

15. the quotient rule

Exercise Set 3.2

1. $4^{-3} = \frac{1}{4^3} = \frac{1}{64}$

3. $7x^{-3} = 7 \cdot \frac{1}{x^3} = \frac{7}{x^3}$

5. $\left(-\frac{1}{4}\right)^{-3} = \frac{(-1)^{-3}}{4^{-3}}$
$= (-1)^{-3} \cdot \frac{1}{4^{-3}}$
$= \frac{1}{(-1)^3} \cdot 4^3$
$= \frac{1}{-1} \cdot 64$
$= -64$

7. $3^{-1} + 2^{-1} = \frac{1}{3} + \frac{1}{2} = \frac{2}{6} + \frac{3}{6} = \frac{5}{6}$

9. $\frac{1}{p^{-3}} = p^3$

11. $\frac{p^{-5}}{q^{-4}} = \frac{1}{p^5} \cdot \frac{q^4}{1} = \frac{q^4}{p^5}$

13. $\frac{x^{-2}}{x} = \frac{x^{-2}}{x^1} = x^{-2-1} = x^{-3} = \frac{1}{x^3}$

15. $\frac{z^{-4}}{z^{-7}} = z^{-4-(-7)} = z^{-4+7} = z^3$

17. $3^{-2} + 3^{-1} = \frac{1}{3^2} + \frac{1}{3} = \frac{1}{9} + \frac{1}{3} = \frac{4}{9}$

19. $(-3)^{-2} = \frac{1}{(-3)^2} = \frac{1}{9}$

21. $\frac{-1}{p^{-4}} = -1 \cdot \frac{1}{p^{-4}} = -1 \cdot p^4 = -p^4$

23. $-2^0 - 3^0 = -(2^0) - (3^0) = -1 - 1 = -2$

25. $\frac{x^2 x^5}{x^3} = \frac{x^{2+5}}{x^3} = \frac{x^7}{x^3} = x^{7-3} = x^4$

27. $\frac{p^2 p}{p^{-1}} = \frac{p^{2+1}}{p^{-1}} = \frac{p^3}{p^{-1}} = p^{3-(-1)} = p^{3+1} = p^4$

29. $\frac{(m^5)^4 m}{m^{10}} = \frac{m^{5\cdot 4} m^1}{m^{10}}$
$= \frac{m^{20} m^1}{m^{10}}$
$= \frac{m^{20+1}}{m^{10}}$
$= \frac{m^{21}}{m^{10}}$
$= m^{21-10}$
$= m^{11}$

31. $\frac{r}{r^{-3} r^{-2}} = \frac{r}{r^{(-3)+(-2)}} = \frac{r}{r^{-5}} = r^{1-(-5)} = r^{1+5} = r^6$

33. $(x^5 y^3)^{-3} = (x^5)^{-3} (y^3)^{-3} = x^{-15} y^{-9} = \frac{1}{x^{15} y^9}$

35. $\frac{(x^2)^3}{x^{10}} = \frac{x^6}{x^{10}} = x^{6-10} = x^{-4} = \frac{1}{x^4}$

37. $\frac{(a^5)^2}{(a^3)^4} = \frac{a^{10}}{a^{12}} = a^{10-12} = a^{-2} = \frac{1}{a^2}$

39. $\frac{8k^4}{2k} = \frac{8}{2} \cdot \frac{k^4}{k} = 4 \cdot k^{4-1} = 4k^3$

41. $\frac{-6m^4}{-2m^3} = \frac{-6}{-2} \cdot \frac{m^4}{m^3} = 3 \cdot m^{4-3} = 3m^1 = 3m$

43. $\frac{-24a^6 b}{6ab^2} = \frac{-24}{6} \cdot \frac{a^6}{a} \cdot \frac{b}{b^2}$
$= -4a^{6-1} b^{1-2}$
$= -4a^5 b^{-1}$
$= -\frac{4a^5}{b}$

45. $\dfrac{6x^2y^3}{-7x^2y^5} = \dfrac{6}{-7}\cdot\dfrac{x^2}{x^2}\cdot\dfrac{y^3}{y^5}$
$= -\dfrac{6}{7}\cdot x^{2-2}y^{3-5}$
$= -\dfrac{6}{7}x^0y^{-2}$
$= -\dfrac{6}{7}\cdot 1\cdot\dfrac{1}{y^2}$
$= -\dfrac{6}{7y^2}$

47. $(3a^2b^{-4})^3 = 3^3a^{2\cdot 3}b^{-4\cdot 3} = 27a^6b^{-12} = \dfrac{27a^6}{b^{12}}$

49. $(a^{-5}b^2)^{-6} = (a^{-5})^{-6}(b^2)^{-6} = a^{30}b^{-12} = \dfrac{a^{30}}{b^{12}}$

51. $\left(\dfrac{x^{-2}y^4}{x^3y^7}\right)^2 = \dfrac{x^{-2\cdot 2}}{x^{3\cdot 2}}\cdot\dfrac{y^{4\cdot 2}}{y^{7\cdot 2}}$
$= \dfrac{x^{-4}}{x^6}\cdot\dfrac{y^8}{y^{14}}$
$= x^{-4-6}y^{8-14}$
$= x^{-10}y^{-6}$
$= \dfrac{1}{x^{10}y^6}$

53. $\dfrac{4^2z^{-3}}{4^3z^{-5}} = \dfrac{4^2}{4^3}\cdot\dfrac{z^{-3}}{z^{-5}}$
$= 4^{2-3}z^{-3-(-5)}$
$= 4^{-1}z^{-3+5}$
$= \dfrac{1}{4}\cdot z^2$
$= \dfrac{z^2}{4}$

55. $\dfrac{3^{-1}x^4}{3^3x^{-7}} = \dfrac{3^{-1}}{3^3}\cdot\dfrac{x^4}{x^{-7}}$
$= 3^{-1-3}x^{4-(-7)}$
$= 3^{-4}x^{4+7}$
$= 3^{-4}x^{11}$
$= \dfrac{x^{11}}{3^4}$
$= \dfrac{x^{11}}{81}$

57. $\dfrac{7ab^{-4}}{7^{-1}a^{-3}b^2} = \dfrac{7}{7^{-1}}\cdot\dfrac{a^1}{a^{-3}}\cdot\dfrac{b^{-4}}{b^2}$
$= 7^{1-(-1)}a^{1-(-3)}b^{-4-2}$
$= 7^{1+1}a^{1+3}b^{-6}$
$= 7^2a^4b^{-6}$
$= \dfrac{49a^4}{b^6}$

59. $\dfrac{-12m^5n^{-7}}{4m^{-2}n^{-3}} = \dfrac{-12}{4}\cdot\dfrac{m^5}{m^{-2}}\cdot\dfrac{n^{-7}}{n^{-3}}$
$= -3m^{5-(-2)}n^{-7-(-3)}$
$= -3m^{5+2}n^{-7+3}$
$= -3m^7n^{-4}$
$= -\dfrac{3m^7}{n^4}$

61. $\left(\dfrac{a^{-5}b}{ab^3}\right)^{-4} = \dfrac{(a^{-5})^{-4}b^{-4}}{a^{-4}(b^3)^{-4}}$
$= \dfrac{a^{20}b^{-4}}{a^{-4}b^{-12}}$
$= a^{20-(-4)}b^{-4-(-12)}$
$= a^{20+4}b^{-4+12}$
$= a^{24}b^8$

63. $(5^2)(8)(2^0) = (25)(8)(1) = 200$

65. $\dfrac{(xy^3)^5}{(xy)^{-4}} = \dfrac{x^5(y^3)^5}{x^{-4}y^{-4}}$
$= \dfrac{x^5y^{15}}{x^{-4}y^{-4}}$
$= x^{5-(-4)}y^{15-(-4)}$
$= x^{5+4}y^{15+4}$
$= x^9y^{19}$

67. $\frac{(-2xy^{-3})^{-3}}{(xy^{-1})^{-1}} = \frac{-2^{-3}x^{-3}(y^{-3})^{-3}}{x^{-1}(y^{-1})^{-1}}$
$= \frac{-2^{-3}x^{-3}y^9}{x^{-1}y}$
$= -2^{-3}x^{-3-(-1)}y^{9-1}$
$= -2^{-3}x^{-3+1}y^8$
$= -2^{-3}x^{-2}y^8$
$= -\frac{y^8}{2^3x^2}$
$= -\frac{y^8}{8x^2}$

69. $\frac{(a^4b^{-7})^{-5}}{(5a^2b^{-1})^{-2}} = \frac{(a^4)^{-5}(b^{-7})^{-5}}{5^{-2}(a^2)^{-2}(b^{-1})^{-2}}$
$= \frac{a^{-20}b^{35}}{5^{-2}a^{-4}b^2}$
$= 5^2a^{-20-(-4)}b^{35-2}$
$= 25a^{-20+4}b^{33}$
$= 25a^{-16}b^{33}$
$= \frac{25b^{33}}{a^{16}}$

71. $V = s^3$
$= \left(\frac{3x^{-2}}{z} \text{ inches}\right)^3$
$= \left(\frac{3^3(x^{-2})^3}{z^3}\right)$ cubic inches
$= \frac{27x^{-6}}{z^3}$ cubic inches
$= \frac{27}{z^3x^6}$ cubic inches

73. $78,000 = 7.8\times10^4$

75. $0.00000167 = 1.67\times10^{-6}$

77. $0.00635 = 6.35\times10^{-3}$

79. $1,160,000 = 1.16\times10^6$

81. $4200 = 4.2\times10^3$

83. $8.673\times10^{-10} = 0.0000000008673$

85. $3.3\times10^{-2} = 0.033$

87. $2.032\times10^4 = 20,320$

89. $7.0\times10^8 = 700,000,000$

91. $5,700,000,000,000 = 5.7\times10^{12}$

93. $1.01\times10^{13} = 10,100,000,000,000$

95. $3,000,000,000,000 = 3\times10^{12}$

97. $(1.2\times10^{-3})(3\times10^{-2}) = 1.2\cdot3\cdot10^{-3}\cdot10^{-2}$
$= 3.6\times10^{-5}$
$= 0.000036$

99. $(4\times10^{-10})(7\times10^{-9}) = (4\cdot7)(10^{-10}\cdot10^{-9})$
$= 28\cdot10^{-19}$
$= 0.0000000000000000028$

101. $\frac{8\times10^{-1}}{16\times10^5} = \frac{8}{16}\times10^{-1-5}$
$= 0.5\times10^{-6}$
$= 5.0\times10^{-7}$
$= 0.0000005$

103. $\frac{1.4\times10^{-2}}{7\times10^{-8}} = \frac{1.4}{7}\cdot\frac{10^{-2}}{10^{-8}}$
$= 0.2\cdot10^{-2-(-8)}$
$= 0.2\cdot10^{-2+8}$
$= 0.2\cdot10^6$
$= 200,000$

105. $7.5\times10^5\cdot3600 = 7.5\times10^5\cdot3.6\times10^3$
$= 7.5\cdot3.6\cdot10^5\cdot10^3$
$= 27\times10^8$
$= 2.7\times10^9$

107. $3x - 5x + 2 = -2x + 7$

109. $y - 10 + y = y + y - 10 = 2y - 10$

111. $7x + 2 - 8x - 6 = 7x - 8x + 2 - 6 = -x - 4$

113. 900 million $= 900,000,000 = 9\times10^8$

115. 14.056 million $= 14,056,000 = 1.4056\times10^7$

117. 2.5 nanometers $= 2.5\times10^{-9}$ m

119. 310 nanometers $= 310\times10^{-9}$ m
$$= 3.1\times10^2\times10^{-9} \text{ m}$$
$$= 3.1\times10^{-7} \text{ m}$$
$$= 0.00000031$$

121. $(2a^3)^3a^4 + a^5a^8 = 2^3(a^3)^3a^4 + a^5a^8$
$$= 8a^9a^4 + a^5a^8$$
$$= 8a^{13} + a^{13}$$
$$= 9a^{13}$$

123. If $x^{\square} = \frac{1}{x^5}$, the exponent is −5.

125. answers may vary

127. a. 9.7×10^{-2} or $1.3\times10^1 \Rightarrow 1.3\times10^1$

b. 8.6×10^5 or $4.4\times10^7 \Rightarrow 4.4\times10^7$

c. 6.1×10^{-2} or $5.6\times10^{-4} \Rightarrow 6.1\times10^{-2}$

129. answers may vary

131. $(x^{-3s})^3 = x^{-9s} = \frac{1}{x^{9s}}$

133. $a^{4m+1}\cdot a^4 = a^{4m+1+4} = a^{4m+5}$

Section 3.3 Practice Exercises

1. $-6x^6 + 4x^5 + 7x^3 - 9x^2 - 1$

Term	Coefficient
$7x^3$	7
$-9x^2$	−9
$-6x^6$	−6
$4x^5$	4
−1	−1

2. $-15x^3 + 2x^2 - 5$

The term $-15x^3$ has degree 3.

The term $2x^2$ has degree 2.

The term −5 has degree 0 since −5 is $-5x^0$.

3. a. The degree of the binomial $-6x + 14$ is 1.

b. The degree of the polynomial $9x - 3x^6 + 5x^4 + 2$ is 6. The polynomial is neither a monomial, binomial, or trinomial.

c. The degree of the trinomial $10x^2 - 6x - 6$ is 2.

4. a. When $x = -1$,
$-2x + 10 = -2(-1) + 10 = 2 + 10 = 12.$

b. When $x = -1$,
$$6x^2 + 11x - 20 = 6(-1)^2 + 11(-1) - 20$$
$$= 6 - 11 - 20$$
$$= -25.$$

5. When $t = 2$ seconds,
$$-16t^2 + 592.1 = -16(2)^2 + 592.1$$
$$= -16(4) + 592.1$$
$$= -64 + 592.1$$
$$= 528.1 \text{ feet.}$$

When $t = 4$ seconds,
$$-16t^2 + 592.1 = -16(4)^2 + 592.1$$
$$= -16(16) + 592.1$$
$$= -256 + 592.1$$
$$= 336.1 \text{ feet.}$$

6. $-6y + 8y = (-6 + 8)y = 2y$

7. $14y^2 + 3 - 10y^2 - 9 = 14y^2 - 10y^2 + 3 - 9$
$$= 4y^2 - 6$$

8. $7x^3 + x^3 = 7x^3 + 1x^3 = 8x^3$

9. $23x^2 - 6x - x - 15 = 23x^2 - 7x - 15$

10. $\frac{2}{7}x^3 - \frac{1}{4}x + 2 - \frac{1}{2}x^3 + \frac{3}{8}x$
$= \frac{2}{7}x^3 - \frac{1}{2}x^3 - \frac{1}{4}x + \frac{3}{8}x + 2$
$= \frac{4}{14}x^3 - \frac{7}{14}x^3 - \frac{2}{8}x + \frac{3}{8}x + 2$
$= -\frac{3}{14}x^3 + \frac{1}{8}x + 2$

11. $\text{Area} = 5 \cdot x + x \cdot x + 4 \cdot 5 + x \cdot x + 8 \cdot x$
$= 5x + x^2 + 20 + x^2 + 8x$
$= x^2 + x^2 + 5x + 8x + 20$
$= 2x^2 + 13x + 20$

12.

Terms of Polynomial	Degree of Term
$-2x^3y^2$	3 + 2 or 5
4	0
$-8xy$	1 + 1 or 2
$3x^3y$	3 + 1 or 4
$5xy^2$	1 + 2 or 3

The degree of the polynomial is 5.

13. $11ab - 6a^2 - ba + 8b^2 = (11-1)ab - 6a^2 + 8b^2$
$= 10ab - 6a^2 + 8b^2$

14. $7x^2y^2 + 2y^2 - 4y^2x^2 + x^2 - y^2 + 5x^2$
$= 7x^2y^2 - 4x^2y^2 + 2y^2 - y^2 + x^2 + 5x^2$
$= 3x^2y^2 + y^2 + 6x^2$

15. a. $x^2 + 9 = x^2 + 0x^1 + 9$ or $x^2 + 0x + 9$

b. $9m^3 + m^2 - 5 = 9m^3 + m^2 + 0m^1 - 5$
$= 9m^3 + m^2 + 0m - 5$

c. $-3a^3 + a^4 = a^4 - 3a^3 + 0a^2 + 0a^1 + 0a^0$
$= a^4 - 3a^3 + 0a^2 + 0a + 0a^0$

Vocabulary, Readiness & Video Check 3.3

1. A binomial is a polynomial with exactly two terms.

2. A monomial is a polynomial with exactly one term.

3. A trinomial is a polynomial with exactly three terms.

4. The numerical factor of a term is called the coefficient.

5. A number term is also called a constant.

6. The degree of a polynomial is the greatest degree of any term of the polynomial.

7. 3; x^2, $-3x$, 5

8. The degree of the polynomial is the greatest degree of any of its terms, so we need to find the degree of each term first.

9. the replacement value for the variable

10. simplifying it

11. 2; $9ab$

12. 2; $0y^2$

Exercise Set 3.3

1.

Term	Coefficient
x^2	1
$-3x$	-3
5	5

3.

Term	Coefficient
$-5x^4$	-5
$3.2x^2$	3.2
x	1
-5	-5

5. $x + 2 = x^1 + 2$
This is a binomial of degree 1.

7. $9m^3 - 5m^2 + 4m - 8$
This is a polynomial of degree 3. None of these.

9. $12x^4 - x^6 - 12x^2 = -x^6 + 12x^4 - 12x^2$
This is a trinomial of degree 6.

11. $3z-5z^4=-5z^4+3z$
This is a binomial of degree 4.

13. a. $5x-6=5(0)-6=0-6=-6$

b. $5x-6=5(-1)-6=-5-6=-11$

15. a. $x^2-5x-2=(0)^2-5(0)-2=0-0-2=-2$

b. $x^2-5x-2=(-1)^2-5(-1)-2=1+5-2=4$

17. a.
$$\begin{aligned}-x^3+4x^2-15&=-(0)^3+4(0)^2-15\\&=0+0-15\\&=-15\end{aligned}$$

b.
$$\begin{aligned}-x^3+4x^2-15&=-(-1)^3+4(-1)^2-15\\&=-(-1)+4(1)-15\\&=1+4-15\\&=-10\end{aligned}$$

19.
$$\begin{aligned}-16t^2+200t&=-16(1)^2+200(1)\\&=-16+200\\&=184\end{aligned}$$
After 1 second, the height of the rocket is 184 feet.

21.
$$\begin{aligned}-16t^2+200t&=-16(7.6)^2+200(7.6)\\&=-16(57.76)+1520\\&=-924.16+1520\\&=595.84\end{aligned}$$
After 7.6 seconds, the height of the rocket is 595.84 feet.

23.
$$\begin{aligned}-15x^2+77x+499&=-15(2)^2+77(2)+499\\&=-15(4)+77(2)+499\\&=-60+154+499\\&=593\end{aligned}$$
There were 593 thousand visitors in 2012.

25.
$$\begin{aligned}3.7x^2+0.7x+241.6&=3.7(7)^2+0.7(7)+241.6\\&=3.7(49)+0.7(7)+241.6\\&=181.3+4.9+241.6\\&=427.8\end{aligned}$$
There will be 427.8 thousand cell sites in 2015.

27. $9x-20x=(9-20)\,x=-11x$

29. $14x^3+9x^3=(14+9)x^3=23x^3$

31.
$$\begin{aligned}7x^2+3+9x^2-10&=7x^2+9x^2+3-10\\&=(7+9)x^2+3-10\\&=16x^2-7\end{aligned}$$

33. $15x^2-3x^2-13=(15-3)x^2-13=12x^2-13$

35. $8s-5s+4s=(8-5+4)s=7s$

37.
$$\begin{aligned}&0.1y^2-1.2y^2+6.7-1.9\\&=(0.1-1.2)y^2+6.7-1.9\\&=-1.1y^2+4.8\end{aligned}$$

39.
$$\begin{aligned}&\frac{2}{3}x^4+12x^3+\frac{1}{6}x^4-19x^3-19\\&=\frac{2}{3}x^4+\frac{1}{6}x^4+12x^3-19x^3-19\\&=\left(\frac{4}{6}+\frac{1}{6}\right)x^4+(12-19)x^3-19\\&=\frac{5}{6}x^4-7x^3-19\end{aligned}$$

41.
$$\begin{aligned}&\frac{3}{20}x^3+\frac{1}{10}-\frac{3}{10}x-\frac{1}{5}-\frac{7}{20}x+6x^2\\&=\frac{3}{20}x^3+6x^2-\frac{3}{10}x-\frac{7}{20}x+\frac{1}{10}-\frac{1}{5}\\&=\frac{3}{20}x^3+6x^2+\left(-\frac{6}{20}-\frac{7}{20}\right)x+\frac{1}{10}-\frac{2}{10}\\&=\frac{3}{20}x^3+6x^2-\frac{13}{20}x-\frac{1}{10}\end{aligned}$$

43.
$$\begin{aligned}4x^2+7x+x^2+5x&=4x^2+x^2+7x+5x\\&=(4+1)x^2+(7+5)x\\&=5x^2+12x\end{aligned}$$

45.
$$\begin{aligned}&5x+3+4x+3+2x+6+3x+7x\\&=5x+4x+2x+3x+7x+3+3+6\\&=(5+4+2+3+7)x+12\\&=21x+12\end{aligned}$$

47. $9ab=9a^1b^1$ has degree $1+1=2$.
$-6a=-6a^1$ has degree 1.
$5b=5b^1$ has degree 1.
$-3=-3a^0b^0$ has degree 0.
$9ab-6a+5b-3$ is a polynomial of degree 2.

49. $x^3y = x^3y^1$ has degree 3 + 1 = 4.

$-6 = -6x^0y^0$ has degree 0.

$2x^2y^2$ has degree 2 + 2 = 4.

$5y^3$ has degree 3.

$x^3y - 6 + 2x^2y^2 + 5y^3$ is a polynomial of degree 4.

51. $3ab - 4a + 6ab - 7a = 3ab + 6ab - 4a - 7a$
$= (3+6)ab - (4+7)a$
$= 9ab - 11a$

53. $4x^2 - 6xy + 3y^2 - xy = 4x^2 - 6xy - xy + 3y^2$
$= 4x^2 + (-6-1)xy + 3y^2$
$= 4x^2 - 7xy + 3y^2$

55. $5x^2y + 6xy^2 - 5yx^2 + 4 - 9y^2x$
$= 5x^2y - 5x^2y + 6xy^2 - 9xy^2 + 4$
$= (5-5)x^2y + (6-9)xy^2 + 4$
$= 0x^2y - 3xy^2 + 4$
$= -3xy^2 + 4$

57. $14y^3 - 9 + 3a^2b^2 - 10 - 19b^2a^2$
$= 14y^3 - 9 - 10 + 3a^2b^2 - 19a^2b^2$
$= 14y^3 + (-9-10) + (3-19)a^2b^2$
$= 14y^3 - 19 - 16a^2b^2$

59. $7x^2 + 3 = 7x^2 + 0x + 3$

61. $x^3 - 64 = x^3 + 0x^2 + 0x - 64$

63. $5y^3 + 2y - 10 = 5y^3 + 0y^2 + 2y - 10$

65. $8y + 2y^4 = 2y^4 + 0y^3 + 0y^2 + 8y + 0$

67. $6x^5 + x^3 - 3x + 15$
$= 6x^5 + 0x^4 + x^3 + 0x^2 - 3x + 15$

69. $4 + 5(2x + 3) = 4 + 10x + 15 = 10x + 19$

71. $2(x-5) + 3(5-x) = 2x - 10 + 15 - 3x$
$= 2x - 3x - 10 + 15$
$= (2-3)x + 5$
$= -x + 5$

73. answers may vary

75. answers may vary

77. $x^4 \cdot x^9 = x^{4+9} = x^{13}$

79. $a \cdot b^3 \cdot a^2 \cdot b^7 = a^1 \cdot a^2 \cdot b^3 \cdot b^7$
$= a^{1+2}b^{3+7}$
$= a^3b^{10}$

81. $(y^5)^4 + (y^2)^{10} = y^{20} + y^{20} = 2y^{20}$

83. answers may vary

85. answers may vary

87. $1.85x^2 - 3.76x + 9.25x^2 + 10.76 - 4.21x$
$= 1.85x^2 + 9.25x^2 - 3.76x - 4.21x + 10.76$
$= (1.85 + 9.25)x^2 - (3.76 + 4.21)x + 10.76$
$= 11.1x^2 - 7.97x + 10.76$

Section 3.4 Practice Exercises

1. $(3x^5 - 7x^3 + 2x - 1) + (3x^3 - 2x)$
$= 3x^5 - 7x^3 + 2x - 1 + 3x^3 - 2x$
$= 3x^5 + (-7x^3 + 3x^3) + (2x - 2x) - 1$
$= 3x^5 - 4x^3 - 1$

2. $(5x^2 - 2x + 1) + (-6x^2 + x - 1)$
$= 5x^2 - 2x + 1 - 6x^2 + x - 1$
$= (5x^2 - 6x^2) + (-2x + x) + (1 - 1)$
$= -x^2 - x$

3.
$$\begin{array}{r} 9y^2 - 6y + 5 \\ 4y + 3 \\ \hline 9y^2 - 2y + 8 \end{array}$$

4. $(9x+5) - (4x-3) = (9x+5) + [-(4x-3)]$
$= (9x+5) + (-4x+3)$
$= 9x + 5 - 4x + 3$
$= 5x + 8$

5. $(4x^3 - 10x^2 + 1) - (-4x^3 + x^2 - 11)$
$= (4x^3 - 10x^2 + 1) + (4x^3 - x^2 + 11)$
$= 4x^3 - 10x^2 + 1 + 4x^3 - x^2 + 11$
$= 4x^3 + 4x^3 - 10x^2 - x^2 + 1 + 11$
$= 8x^3 - 11x^2 + 12$

6. $\begin{array}{r} 2y^2-2y+7 \\ \underline{-(6y^2-3y+2)} \end{array} \Rightarrow \begin{array}{r} 2y^2-2y+7 \\ \underline{-6y^2+3y-2} \\ -4y^2+\ y+5 \end{array}$

7. $[(4x-3)+(12x-5)]-(3x+1) = 4x-3+12x-5-3x-1$
$= 4x+12x-3x-3-5-1$
$= 13x-9$

8. $(2a^2-ab+6b^2)+(-3a^2+ab-7b^2) = 2a^2-ab+6b^2-3a^2+ab-7b^2$
$= -a^2-b^2$

9. $(5x^2y^2+3-9x^2y+y^2)-(-x^2y^2+7-8xy^2+2y^2) = 5x^2y^2+3-9x^2y+y^2+x^2y^2-7+8xy^2-2y^2$
$= 6x^2y^2-4-9x^2y+8xy^2-y^2$

Vocabulary, Readiness & Video Check 3.4

1. $-9y-5y=-14y$

2. $6m^5+7m^5=13m^5$

3. $x+6x=7x$

4. $7z-z=6z$

5. $5m^2+2m=5m^2+2m$

6. $8p^3+3p^2=8p^3+3p^2$

7. $-3y^2$ and $2y^2$; $-4y$ and y

8. Example 2 is a subtraction example, so the signs of the polynomial being subtracted must change when parentheses are removed.

9. We're translating a subtraction problem. Order matters when subtracting, so we need to be careful that the order of the expressions is correct.

10. Because the operation is addition.

Exercise Set 3.4

1. $(3x+7)+(9x+5) = 3x+7+9x+5$
$= 3x+9x+7+5$
$= 12x+12$

3. $(-7x+5)+(-3x^2+7x+5) = -7x+5-3x^2+7x+5$
$= -3x^2-7x+7x+5+5$
$= -3x^2+10$

5. $(-5x^2+3)+(2x^2+1) = -5x^2+3+2x^2+1$
$= -5x^2+2x^2+3+1$
$= -3x^2+4$

7. $(-3y^2-4y)+(2y^2+y-1)$
$= -3y^2-4y+2y^2+y-1$
$= -3y^2+2y^2-4y+y-1$
$= -y^2-3y-1$

9. $(1.2x^3-3.4x+7.9)+(6.7x^3+4.4x^2-10.9)$
$= 1.2x^3-3.4x+7.9+6.7x^3+4.4x^2-10.9$
$= 1.2x^3+6.7x^3+4.4x^2-3.4x+7.9-10.9$
$= 7.9x^3+4.4x^2-3.4x-3$

11. $\left(\frac{3}{4}m^2-\frac{2}{5}m+\frac{1}{8}\right)+\left(-\frac{1}{4}m^2-\frac{3}{10}m+\frac{11}{16}\right)$
$= \frac{3}{4}m^2-\frac{2}{5}m+\frac{1}{8}-\frac{1}{4}m^2-\frac{3}{10}m+\frac{11}{16}$
$= \frac{3}{4}m^2-\frac{1}{4}m^2-\frac{2}{5}m-\frac{3}{10}m+\frac{1}{8}+\frac{11}{16}$
$= \frac{3}{4}m^2-\frac{1}{4}m^2-\frac{4}{10}m-\frac{3}{10}m+\frac{2}{16}+\frac{11}{16}$
$= \frac{2}{4}m^2-\frac{7}{10}m+\frac{13}{16}$
$= \frac{1}{2}m^2-\frac{7}{10}m+\frac{13}{16}$

13.
$$\begin{array}{r} 3t^2+4 \\ \underline{5t^2-8} \\ 8t^2-4 \end{array}$$

15.
$$\begin{array}{r} 10a^3-8a^2+4a+9 \\ \underline{5a^3+9a^2-7a+7} \\ 15a^3+a^2-3a+16 \end{array}$$

17. $(2x+5)-(3x-9) = (2x+5)+(-3x+9)$
$= 2x+5-3x+9$
$= 2x-3x+5+9$
$= -x+14$

19. $(5x^2+4)-(-2x^2+4) = (5x^2+4)+(2x^2-4)$
$= 5x^2+4+2x^2-4$
$= 5x^2+2x^2+4-4$
$= 7x^2$

21. $3x-(5x-9) = 3x+(-5x+9)$
$= 3x-5x+9$
$= -2x+9$

23. $(2x^2+3x-9)-(-4x+7)$
$= (2x^2+3x-9)+(4x-7)$
$= 2x^2+3x-9+4x-7$
$= 2x^2+3x+4x-9-7$
$= 2x^2+7x-16$

25. $(5x+8)-(-2x^2-6x+8)$
$= (5x+8)+(2x^2+6x-8)$
$= 5x+8+2x^2+6x-8$
$= 2x^2+5x+6x+8-8$
$= 2x^2+11x$

27. $(0.7x^2+0.2x-0.8)-(0.9x^2+1.4)$
$= (0.7x^2+0.2x-0.8)+(-0.9x^2-1.4)$
$= 0.7x^2+0.2x-0.8-0.9x^2-1.4$
$= 0.7x^2-0.9x^2+0.2x-0.8-1.4$
$= -0.2x^2+0.2x-2.2$

29. $\left(\frac{1}{4}z^2-\frac{1}{5}z\right)-\left(-\frac{3}{20}z^2+\frac{1}{10}z-\frac{7}{20}\right)$
$= \left(\frac{1}{4}z^2-\frac{1}{5}z\right)+\left(\frac{3}{20}z^2-\frac{1}{10}z+\frac{7}{20}\right)$
$= \frac{1}{4}z^2-\frac{1}{5}z+\frac{3}{20}z^2-\frac{1}{10}z+\frac{7}{20}$
$= \frac{5}{20}z^2+\frac{3}{20}z^2-\frac{2}{10}z-\frac{1}{10}z+\frac{7}{20}$
$= \frac{8}{20}z^2-\frac{3}{10}z+\frac{7}{20}$
$= \frac{2}{5}z^2-\frac{3}{10}z+\frac{7}{20}$

31.
$$\begin{array}{r} 4z^2-8z+3 \\ \underline{-(6z^2+8z-3)} \end{array} \Rightarrow \begin{array}{r} 4z^2-8z+3 \\ \underline{-6z^2-8z+3} \\ -2z^2-16z+6 \end{array}$$

33.
$$\begin{array}{r} 5u^5-4u^2+3u-7 \\ \underline{-(3u^5+6u^2-8u+2)} \end{array} \quad \begin{array}{r} 5u^5-4u^2+3u-7 \\ \underline{-3u^5-6u^2+8u-2} \\ 2u^5-10u^2+11u-9 \end{array}$$

35. $(3x+5)+(2x-14) = 3x+5+2x-14$
$= 3x+2x+5-14$
$= 5x-9$

37. $(9x-1)-(5x+2) = (9x-1)+(-5x-2)$
$= 9x-1-5x-2$
$= 4x-3$

39. $(14y+12)+(-3y-5) = 14y+12-3y-5$
$= 11y+7$

41. $(x^2+2x+1)-(3x^2-6x+2) = (x^2+2x+1)+(-3x^2+6x-2)$
$= x^2+2x+1-3x^2+6x-2$
$= -2x^2+8x-1$

43. $(3x^2+5x-8)+(5x^2+9x+12)-(8x^2-14) = (3x^2+5x-8)+(5x^2+9x+12)+(-8x^2+14)$
$= 3x^2+5x-8+5x^2+9x+12-8x^2+14$
$= 14x+18$

45. $(-a^2+1)-(a^2-3)+(5a^2-6a+7) = (-a^2+1)+(-a^2+3)+(5a^2-6a+7)$
$= -a^2+1-a^2+3+5a^2-6a+7$
$= 3a^2-6a+11$

47. $(7x-3)-4x = 7x-3-4x = 3x-3$

49. $(4x^2-6x+1)+(3x^2+2x+1) = 4x^2-6x+1+3x^2+2x+1$
$= 7x^2-4x+2$

51. $(7x^2+3x+9)-(5x+7) = (7x^2+3x+9)+(-5x-7)$
$= 7x^2+3x+9-5x-7$
$= 7x^2-2x+2$

53. $(8y^2+7)+(6y+9)-(4y^2-6y-3) = (8y^2+7)+(6y+9)+(-4y^2+6y+3)$
$= 8y^2+7+6y+9-4y^2+6y+3$
$= 4y^2+12y+19$

55. $(x^2-9x+2)+(2x^2-6x+1)-(3x^2-4) = (x^2-9x+2)+(2x^2-6x+1)+(-3x^2+4)$
$= x^2-9x+2+2x^2-6x+1-3x^2+4$
$= -15x+7$

57. $(9a+6b-5)+(-11a-7b+6) = 9a+6b-5-11a-7b+6$
$= -2a-b+1$

59. $(4x^2+y^2+3)-(x^2+y^2-2) = 4x^2+y^2+3-x^2-y^2+2$
$= 3x^2+5$

61. $(x^2+2xy-y^2)+(5x^2-4xy+20y^2) = x^2+2xy-y^2+5x^2-4xy+20y^2$
$= 6x^2-2xy+19y^2$

63. $(11r^2s+16rs-3-2r^2s^2)-(3sr^2+5-9r^2s^2)=11r^2s+16rs-3-2r^2s^2-3r^2s-5+9r^2s^2$
$=8r^2s+16rs-8+7r^2s^2$

65. $(2x^2+5)+(4x-1)+(-x^2+3x)=2x^2+5+4x-1-x^2+3x$
$=x^2+7x+4$

The perimeter is (x^2+7x+4) feet.

67. $(2x-3)+\left(\frac{4}{5}x\right)+\left(\frac{7}{10}x-1\right)+(2x-2)+(x+4)+(3x+5)=2x-3+\frac{4}{5}x+\frac{7}{10}x-1+2x-2+x+4+3x+5$
$=\frac{19}{2}x+3$

The perimeter is $\left(\frac{19}{2}x+3\right)$ units.

69. $(4y^2+4y+1)-(y^2-10)=(4y^2+4y+1)+(-y^2+10)$
$=4y^2+4y+1-y^2+10$
$=3y^2+4y+11$

The remaining piece is $(3y^2+4y+11)$ meters long.

71. $[(1.2x^2-3x+9.1)-(7.8x^2-3.1+8)]+(1.2x-6)=(1.2x^2-3x+9.1)+(-7.8x^2+3.1-8)+(1.2x-6)$
$=1.2x^2-3x+9.1-7.8x^2+3.1-8+1.2x-6$
$=-6.6x^2-1.8x-1.8$

73. $3x(2x)=(3\cdot 2)(x\cdot x)=6x^2$

75. $(12x^3)(-x^5)=(12\cdot -1)(x^3\cdot x^5)=-12x^8$

77. $10x^2(20xy^2)=10\cdot 20\cdot(x^2\cdot x)(y^2)=200x^3y^2$

79. Since 3 + 4 = 7, $3x^2+4x^2=7x^2$ is a true statement.

81. Since 2 + 4 = 6 and 3 – 5 = –2, $2x^4+3x^3-5x^3+4x^4=6x^4-2x^3$ is a true statement.

83. $10y-6y^2-y=9y-6y^2$; b

85. $(5x-3)+(5x-3)=5x-3+5x-3=10x-6$; e

87. **a.** $z+3z=1z+3z=4z$

b. $z\cdot 3z=z^1\cdot 3z^1=3z^{1+1}=3z^2$

c. $-z-3z=-1z-3z=-4z$

d. $(-z)(-3z)=(-z^1)(-3z^1)=3z^{1+1}=3z^2$; answers may vary

89. **a.** $m\cdot m\cdot m=m^1\cdot m^1\cdot m^1=m^{1+1+1}=m^3$

b. $m+m+m=1m+1m+1m=(1+1+1)m=3m$

c. $(-m)(-m)(-m)=(-1\cdot m^1)(-1\cdot m^1)(-1\cdot m^1)$
$=(-1)(-1)(-1)(m\cdot m\cdot m)$
$=-1m^3$
$=-m^3$

d. $-m-m-m=-1m-1m-1m$
$=(-1-1-1)m$
$=-3m$; answers may vary

91. $(377x^2-720x+1003)+(538x^2+19,421x+54,762)$
$=377x^2-720x+1003+538x^2+19,421x+54,762$
$=915x^2+18,701x+55,765$

Section 3.5 Practice Exercises

1. $10x\cdot 9x=(10\cdot 9)(x\cdot x)=90x^2$

2. $8x^3(-11x^7)=(8\cdot -11)(x^3\cdot x^7)=-88x^{10}$

3. $(-5x^4)(-x)=(-5\cdot -1)(x^4\cdot x)=5x^5$

4. $4x(x^2+4x+3)=4x(x^2)+4x(4x)+4x(3)$
$=4x^3+16x^2+12x$

5. $8x(7x^4+1)=8x(7x^4)+8x(1)=56x^5+8x$

6. $-2x^3(3x^2-x+2)=-2x^3(3x^2)-2x^3(-x)-2x^3(2)$
$=-6x^5+2x^4-4x^3$

7. a. $(x+5)(x+10)=x(x+10)+5(x+10)$
$=x\cdot x+x\cdot 10+5\cdot x+5\cdot 10$
$=x^2+10x+5x+50$
$=x^2+15x+50$

b. $(4x+5)(3x-4)=4x(3x-4)+5(3x-4)$
$=4x(3x)+4x(-4)+5(3x)+5(-4)$
$=12x^2-16x+15x-20$
$=12x^2-x-20$

8. $(3x-2y)^2=(3x-2y)(3x-2y)$
$=3x(3x)+3x(-2y)+(-2y)(3x)+(-2y)(-2y)$
$=9x^2-6xy-6xy+4y^2$
$=9x^2-12xy+4y^2$

9. $(x+3)(2x^2-5x+4)=x(2x^2)+x(-5x)+x(4)+3(2x^2)+3(-5x)+3(4)$
$=2x^3-5x^2+4x+6x^2-15x+12$
$=2x^3+x^2-11x+12$

10.
$$\begin{array}{r} y^2-4y+5 \\ 3y^2+1 \\ \hline y^2-4y+5 \\ 3y^4-12y^3+15y^2 \\ \hline 3y^4-12y^3+16y^2-4y+5 \end{array}$$

11.
$$\begin{array}{r} 4x^2\ \ -x-1 \\ 3x^2+6x-2 \\ \hline -8x^2+2x+2 \\ 24x^3-\ 6x^2-6x \\ 12x^4-\ 3x^3-\ 3x^2 \\ \hline 12x^4+21x^3-17x^2-4x+2 \end{array}$$

Vocabulary, Readiness & Video Check 3.5

1. The expression $5x(3x+2)$ equals $5x \cdot 3x + 5x \cdot 2$ by the distributive property.
2. The expression $(x+4)(7x-1)$ equals $x(7x-1)+4(7x-1)$ by the distributive property.
3. The expression $(5y-1)^2$ equals $(5y-1)(5y-1)$.
4. The expression $9x \cdot 3x$ equals $27x^2$.
5. $x^3 \cdot x^5 = x^{3+5} = x^8$
6. $x^2 \cdot x^6 = x^{2+6} = x^8$
7. $x^3 + x^5$ cannot be simplified.
8. $x^2 + x^6$ cannot be simplified.
9. $x^7 \cdot x^7 = x^{7+7} = x^{14}$
10. $x^{11} \cdot x^{11} = x^{11+11} = x^{22}$
11. $x^7 + x^7 = 1x^7 + 1x^7 = (1+1)x^7 = 2x^7$
12. $x^{11} + x^{11} = 1x^{11} + 1x^{11} = (1+1)x^{11} = 2x^{11}$
13. No. The monomials are unlike terms.
14. distributive property, product rule
15. Three times: First $(a-2)$ is distributed to a and 7, and then a is distributed to $(a-2)$ and 7 is distributed to $(a-2)$.
16. Yes. The parentheses have been removed for the vertical format, but every term in the first polynomial is still distributed to every term in the second polynomial.

Exercise Set 3.5

1. $8x^2 \cdot 3x = (8\cdot 3)(x^2 \cdot x) = 24x^3$

3. $(-x^3)(-x) = (-1\cdot -1)(x^3 \cdot x) = x^4$

5. $-4n^3 \cdot 7n^7 = (-4\cdot 7)(n^3 \cdot n^7) = -28n^{10}$

7. $(-3.1x^3)(4x^9) = (-3.1\cdot 4)(x^3 \cdot x^9) = -12.4x^{12}$

9. $\left(-\frac{1}{3}y^2\right)\left(\frac{2}{5}y\right) = \left(-\frac{1}{3}\right)\left(\frac{2}{5}\right)(y^2 \cdot y) = -\frac{2}{15}y^3$

11. $(2x)(-3x^2)(4x^5) = (2\cdot -3\cdot 4)(x\cdot x^2 \cdot x^5)$
$= -24x^8$

13. $3x(2x+5) = 3x(2x)+3x(5) = 6x^2+15x$

15. $7x(x^2+2x-1) = 7x(x^2)+7x(2x)+7x(-1)$
$= 7x^3+14x^2-7x$

17. $-2a(a+4) = -2a(a)+(-2a)(4) = -2a^2-8a$

19. $3x(2x^2-3x+4) = 3x(2x^2)+3x(-3x)+3x(4)$
$= 6x^3-9x^2+12x$

21. $3a^2(4a^3+15) = 3a^2(4a^3)+3a^2(15)$
$= 12a^5+45a^2$

23. $-2a^2(3a^2-2a+3)$
$= -2a^2(3a^2)-2a^2(-2a)-2a^2(3)$
$= -6a^4+4a^3-6a^2$

25. $3x^2y(2x^3-x^2y^2+8y^3)$
$= 3x^2y(2x^3)+3x^2y(-x^2y^2)+3x^2y(8y^3)$
$= 6x^5y-3x^4y^3+24x^2y^4$

27. $-y(4x^3-7x^2y+xy^2+3y^3)=-y(4x^3)-y(-7x^2y)-y(xy^2)-y(3y^3)$
$=-4x^3y+7x^2y^2-xy^3-3y^4$

29. $\frac{1}{2}x^2(8x^2-6x+1)=\frac{1}{2}x^2(8x^2)+\frac{1}{2}x^2(-6x)+\frac{1}{2}x^2(1)$
$=4x^4-3x^3+\frac{1}{2}x^2$

31. $(x+4)(x+3)=x(x+3)+4(x+3)$
$=x(x)+x(3)+4(x)+4(3)$
$=x^2+3x+4x+12$
$=x^2+7x+12$

33. $(a+7)(a-2)=a(a-2)+7(a-2)$
$=a(a)+a(-2)+7(a)+7(-2)$
$=a^2-2a+7a-14$
$=a^2+5a-14$

35. $\left(x+\frac{2}{3}\right)\left(x-\frac{1}{3}\right)=x\left(x-\frac{1}{3}\right)+\frac{2}{3}\left(x-\frac{1}{3}\right)$
$=x(x)+x\left(-\frac{1}{3}\right)+\frac{2}{3}(x)+\frac{2}{3}\left(-\frac{1}{3}\right)$
$=x^2-\frac{1}{3}x+\frac{2}{3}x-\frac{2}{9}$
$=x^2+\frac{1}{3}x-\frac{2}{9}$

37. $(3x^2+1)(4x^2+7)=3x^2(4x^2+7)+1(4x^2+7)$
$=3x^2(4x^2)+3x^2(7)+1(4x^2)+1(7)$
$=12x^4+21x^2+4x^2+7$
$=12x^4+25x^2+7$

39. $(4x-3)(3x-5)=4x(3x-5)+(-3)(3x-5)$
$=4x(3x)+4x(-5)+(-3)(3x)+(-3)(-5)$
$=12x^2-20x-9x+15$
$=12x^2-29x+15$

41. $(1-3a)(1-4a)=1(1-4a)+(-3a)(1-4a)$
$=1(1)+1(-4a)+(-3a)(1)+(-3a)(-4a)$
$=1-4a-3a+12a^2$
$=1-7a+12a^2$

43. $(2y-4)^2=(2y-4)(2y-4)$
$=2y(2y-4)+(-4)(2y-4)$
$=2y(2y)+2y(-4)+(-4)(2y)+(-4)(-4)$
$=4y^2-8y-8y+16$
$=4y^2-16y+16$

45. $$\begin{aligned}(x-2)(x^2-3x+7) &= x(x^2-3x+7)+(-2)(x^2-3x+7)\\ &= x(x^2)+x(-3x)+x(7)+(-2)(x^2)+(-2)(-3x)+(-2)(7)\\ &= x^3-3x^2+7x-2x^2+6x-14\\ &= x^3-5x^2+13x-14\end{aligned}$$

47. $$\begin{aligned}(x+5)(x^3-3x+4) &= x(x^3-3x+4)+5(x^3-3x+4)\\ &= x(x^3)+x(-3x)+x(4)+5(x^3)+5(-3x)+5(4)\\ &= x^4-3x^2+4x+5x^3-15x+20\\ &= x^4+5x^3-3x^2-11x+20\end{aligned}$$

49. $$\begin{aligned}(2a-3)(5a^2-6a+4) &= 2a(5a^2-6a+4)+(-3)(5a^2-6a+4)\\ &= 2a(5a^2)+2a(-6a)+2a(4)+(-3)(5a^2)+(-3)(-6a)+(-3)(4)\\ &= 10a^3-12a^2+8a-15a^2+18a-12\\ &= 10a^3-27a^2+26a-12\end{aligned}$$

51. $$\begin{aligned}(7xy-y)^2 &= (7xy-y)(7xy-y)\\ &= 7xy(7xy-y)+(-y)(7xy-y)\\ &= 7xy(7xy)+7xy(-y)+(-y)(7xy)+(-y)(-y)\\ &= 49x^2y^2-7xy^2-7xy^2+y^2\\ &= 49x^2y^2-14xy^2+y^2\end{aligned}$$

53. $$\begin{array}{r} 2x-11\\ 6x+\ 1\\ \hline 2x-11\\ 12x^2-66x\\ \hline 12x^2-64x-11\end{array}$$

55. $$\begin{array}{r} 2x^2+4x-1\\ x+3\\ \hline 6x^2+12x-3\\ 2x^3+\ 4x^2-\ \ x\\ \hline 2x^3+10x^2+11x-3\end{array}$$

57. $$\begin{array}{r} x^2+5x-7\\ 2x^2-7x-9\\ \hline -9x^2-45x+63\\ -7x^3-35x^2+49x\\ 2x^4+10x^3-14x^2\\ \hline 2x^4+3x^3-58x^2+4x+63\end{array}$$

59. $-1.2y(-7y^6)=(-1.2\cdot -7)(y\cdot y^6)=8.4y^7$

61. $$\begin{aligned}-3x(x^2+2x-8) &= -3x(x^2)+(-3x)(2x)+(-3x)(-8)\\ &= -3x^3-6x^2+24x\end{aligned}$$

63. $(x+19)(2x+1) = x(2x+1)+19(2x+1)$
$= x(2x)+x(1)+19(2x)+19(1)$
$= 2x^2+x+38x+19$
$= 2x^2+39x+19$

65. $\left(x+\frac{1}{7}\right)\left(x-\frac{3}{7}\right)$
$= x\left(x-\frac{3}{7}\right)+\frac{1}{7}\left(x-\frac{3}{7}\right)$
$= x(x)+x\left(-\frac{3}{7}\right)+\frac{1}{7}(x)+\frac{1}{7}\left(-\frac{3}{7}\right)$
$= x^2-\frac{3}{7}x+\frac{1}{7}x-\frac{3}{49}$
$= x^2-\frac{2}{7}x-\frac{3}{49}$

67. $(3y+5)^2 = (3y+5)(3y+5)$
$= 3y(3y+5)+5(3y+5)$
$= 3y(3y)+3y(5)+5(3y)+5(5)$
$= 9y^2+15y+15y+25$
$= 9y^2+30y+25$

69. $(a+4)(a^2-6a+6)$
$= a(a^2-6a+6)+4(a^2-6a+6)$
$= a(a^2)+a(-6a)+a(6)+4(a^2)+4(-6a)+4(6)$
$= a^3-6a^2+6a+4a^2-24a+24$
$= a^3-2a^2-18a+24$

71. $(2x+5)(2x-5)$
$= 2x(2x-5)+5(2x-5)$
$= 2x(2x)+2x(-5)+5(2x)+5(-5)$
$= 4x^2-10x+10x-25$
$= 4x^2-25$
The area is $(4x^2-25)$ square yards.

73. Area $= \frac{1}{2}$(base)(height)
$= \frac{1}{2}(3x-2)(4x)$
$= 2x(3x-2)$
$= 2x(3x)+2x(-2)$
$= 6x^2-4x$
The area is $(6x^2-4x)$ square inches.

75. Add: $5a + 15a = (5 + 15)a = 20a$
Subtract: $5a - 15a = (5 - 15)a = -10a$
Multiply: $5a \cdot 15a = 5 \cdot 15 \cdot a^1 \cdot a^1 = 75a^{1+1} = 75a^2$
Divide: $\frac{5a}{15a} = \frac{1}{3}a^{1-1} = \frac{1}{3}a^0 = \frac{1}{3} \cdot 1 = \frac{1}{3}$

77. Add: $-3y^5+9y^4$ cannot be simplified.
Subtract: $-3y^5-9y^4$ cannot be simplified.
Multiply: $-3y^5 \cdot 9y^4 = -3 \cdot 9 \cdot y^5 \cdot y^4$
$= -27y^{5+4}$
$= -27y^9$
Divide: $\frac{-3y^5}{9y^4} = -\frac{3}{9} \cdot y^{5-4} = -\frac{1}{3}y^1 = -\frac{y}{3}$

79. a. $(3x+5)+(3x+7) = 3x+5+3x+7$
$= 6x+12$

b. $(3x+5)(3x+7)$
$= 3x(3x+7)+5(3x+7)$
$= 3x(3x)+3x(7)+5(3x)+5(7)$
$= 9x^2+21x+15x+35$
$= 9x^2+36x+35$
answers may vary

81. $(3x - 1) + (10x - 6) = 3x - 1 + 10x - 6 = 13x - 7$

83. $(3x-1)(10x-6)$
$= 3x(10x-6)+(-1)(10x-6)$
$= 3x(10x)+3x(-6)+(-1)(10x)+(-1)(-6)$
$= 30x^2-18x-10x+6$
$= 30x^2-28x+6$

85. $(3x-1)-(10x-6) = (3x-1)+(-10x+6)$
$= 3x-1-10x+6$
$= -7x+5$

87. The areas of the smaller rectangles are:
$x \cdot x = x^2$
$x \cdot 3 = 3x$
The area of the figure is x^2+3x.

89. The area of the figure is $x(1 + 2x)$.
The areas of the smaller rectangles are
$x \cdot 1 = x$
$x \cdot 2x = 2x^2$
The area of the figure is $x+2x^2$.

91. $5a + 6a = (5 + 6)a = 11a$

93. $(5x)^2 + (2y)^2 = 5^2x^2 + 2^2y^2 = 25x^2 + 4y^2$

95. a. $(a+b)(a-b) = a(a-b)+b(a-b)$
$= a(a)+a(-b)+b(a)+b(-b)$
$= a^2 - ab + ab - b^2$
$= a^2 - b^2$

b. $(2x+3y)(2x-3y)$
$= 2x(2x-3y)+3y(2x-3y)$
$= 2x(2x)+2x(-3y)+3y(2x)+3y(-3y)$
$= 4x^2 - 6xy + 6xy - 9y^2$
$= 4x^2 - 9y^2$

c. $(4x+7)(4x-7)$
$= 4x(4x-7)+7(4x-7)$
$= 4x(4x)+4x(-7)+7(4x)+7(-7)$
$= 16x^2 - 28x + 28x - 49$
$= 16x^2 - 49$

d. answers may vary

Section 3.6 Practice Exercises

1. $(x+7)(x-5)$
$= (x)(x)+(x)(-5)+(7)(x)+(7)(-5)$
$= x^2 - 5x + 7x - 35$
$= x^2 + 2x - 35$

2. $(6x-1)(x-4)$
$= 6x(x)+6x(-4)+(-1)(x)+(-1)(-4)$
$= 6x^2 - 24x - x + 4$
$= 6x^2 - 25x + 4$

3. $(2y^2+3)(y-4) = 2y^3 - 8y^2 + 3y - 12$

4. $(2x+9)^2 = (2x+9)(2x+9)$
$= (2x)(2x)+(2x)(9)+9(2x)+9(9)$
$= 4x^2 + 18x + 18x + 81$
$= 4x^2 + 36x + 81$

5. $(y+3)^2 = y^2 + 2(y)(3) + 3^2 = y^2 + 6y + 9$

6. $(r-s)^2 = r^2 - 2(r)(s) + s^2 = r^2 - 2rs + s^2$

7. $(6x+5)^2 = (6x)^2 + 2(6x)(5) + 5^2$
$= 36x^2 + 60x + 25$

8. $(x^2-3y)^2 = (x^2)^2 - 2(x^2)(3y) + (3y)^2$
$= x^4 - 6x^2y + 9y^2$

9. $(x+9)(x-9) = x^2 - 9^2 = x^2 - 81$

10. $(5+4y)(5-4y) = 5^2 - (4y)^2 = 25 - 16y^2$

11. $\left(x-\frac{1}{3}\right)\left(x+\frac{1}{3}\right) = x^2 - \left(\frac{1}{3}\right)^2 = x^2 - \frac{1}{9}$

12. $(3a-b)(3a+b) = (3a)^2 - b^2 = 9a^2 - b^2$

13. $(2x^2-6y)(2x^2+6y) = (2x^2)^2 - (6y)^2$
$= 4x^4 - 36y^2$

14. $(7x-1)^2 = (7x)^2 - 2(7x)(1) + 1^2$
$= 49x^2 - 14x + 1$

15. $(5y+3)(2y-5)$
$= (5y)(2y)+(5y)(-5)+(3)(2y)+(3)(-5)$
$= 10y^2 - 25y + 6y - 15$
$= 10y^2 - 19y - 15$

16. $(2a-1)(2a+1) = (2a)^2 - (1)^2 = 4a^2 - 1$

17. $\left(5y-\frac{1}{9}\right)^2 = (5y)^2 - 2(5y)\left(\frac{1}{9}\right) + \left(\frac{1}{9}\right)^2$
$= 25y^2 - \frac{10}{9}y + \frac{1}{81}$

Vocabulary, Readiness & Video Check 3.6

1. $(x+4)^2 = x^2 + 2(x)(4) + 4^2$
$= x^2 + 8x + 16$
$\neq x^2 + 16$
The statement is false.

2. The statement is true.

3. $(x+4)(x-4) = x^2 - 4^2 = x^2 - 16 \neq x^2 + 16$
The statement is false.

4. $(x-1)(x^3+3x-1) = x(x^3+3x-1)-1(x^3+3x-1)$
$= x^4+3x^2-x-x^3-3x+1$
$= x^4-x^3+3x^2-4x+1$
The product is a polynomial of degree 4, not 5. The statement is false.

5. a binomial times a binomial

6. FOIL order for multiplication, distributive property

7. Multiplying gives you four terms, and the two like terms will always subtract out.

8. Multiplying the sum and difference of the same two terms, squaring a binomial, and the FOIL order for multiplication when multiplying a binomial and a binomial.

Exercise Set 3.6

1. $(x+3)(x+4) = x^2+4x+3x+12 = x^2+7x+12$

3. $(x-5)(x+10) = x^2+10x-5x-50$
$= x^2+5x-50$

5. $(5x-6)(x+2) = 5x^2+10x-6x-12$
$= 5x^2+4x-12$

7. $(y-6)(4y-1) = 4y^2-y-24y+6$
$= 4y^2-25y+6$

9. $(2x+5)(3x-1) = 6x^2-2x+15x-5$
$= 6x^2+13x-5$

11. $(y^2+7)(6y+4) = 6y^3+4y^2+42y+28$

13. $\left(x-\frac{1}{3}\right)\left(x+\frac{2}{3}\right) = x^2+\frac{2}{3}x-\frac{1}{3}x-\frac{2}{9}$
$= x^2+\frac{1}{3}x-\frac{2}{9}$

15. $(0.4-3a)(0.2-5a) = 0.08-2.0a-0.6a+15a^2$
$= 0.08-2.6a+15a^2$

17. $(x+5y)(2x-y) = 2x^2-xy+10xy-5y^2$
$= 2x^2+9xy-5y^2$

19. $(x+2)^2 = x^2+2(2)(x)+2^2 = x^2+4x+4$

21. $(2a-3)^2 = (2a)^2-2(2a)(3)+(3)^2$
$= 4a^2-12a+9$

23. $(3a-5)^2 = (3a)^2-2(3a)(5)+5^2$
$= 9a^2-30a+25$

25. $(x^2+0.5)^2 = (x^2)^2+2(x^2)(0.5)+(0.5)^2$
$= x^4+x^2+0.25$

27. $\left(y-\frac{2}{7}\right)^2 = y^2-2(y)\left(\frac{2}{7}\right)+\left(\frac{2}{7}\right)^2$
$= y^2-\frac{4}{7}y+\frac{4}{49}$

29. $(2x-1)^2 = (2x)^2-2(2x)(1)+(1)^2$
$= 4x^2-4x+1$

31. $(5x+9)^2 = (5x)^2+2(5x)(9)+9^2$
$= 25x^2+90x+81$

33. $(3x-7y)^2 = (3x)^2-2(3x)(7y)+(7y)^2$
$= 9x^2-42xy+49y^2$

35. $(4m+5n)^2 = (4m)^2+2(4m)(5n)+(5n)^2$
$= 16m^2+40mn+25n^2$

37. $(5x^4-3)^2 = (5x^4)^2-2(5x^4)(3)+(3)^2$
$= 25x^8-30x^4+9$

39. $(a-7)(a+7) = a^2-7^2 = a^2-49$

41. $(x+6)(x-6) = (x)^2-(6)^2 = x^2-36$

43. $(3x-1)(3x+1) = (3x)^2-1^2 = 9x^2-1$

45. $(x^2+5)(x^2-5) = (x^2)^2-(5)^2 = x^4-25$

47. $(2y^2-1)(2y^2+1) = (2y^2)^2-1^2 = 4y^4-1$

49. $(4-7x)(4+7x) = (4)^2-(7x)^2 = 16-49x^2$

51. $\left(3x-\frac{1}{2}\right)\left(3x+\frac{1}{2}\right) = (3x)^2-\left(\frac{1}{2}\right)^2 = 9x^2-\frac{1}{4}$

53. $(9x+y)(9x-y) = (9x)^2-(y)^2 = 81x^2-y^2$

55. $(2m+5n)(2m-5n) = (2m)^2 - (5n)^2$
$= 4m^2 - 25n^2$

57. $(a+5)(a+4) = a^2 + 4a + 5a + 20 = a^2 + 9a + 20$

59. $(a-7)^2 = a^2 - 2(a)(7) + 7^2 = a^2 - 14a + 49$

61. $(4a+1)(3a-1) = 12a^2 - 4a + 3a - 1$
$= 12a^2 - a - 1$

63. $(x+2)(x-2) = x^2 - 2^2 = x^2 - 4$

65. $(3a+1)^2 = (3a)^2 + 2(3a)(1) + (1)^2 = 9a^2 + 6a + 1$

67. $(x+y)(4x-y) = 4x^2 - xy + 4xy - y^2$
$= 4x^2 + 3xy - y^2$

69. $\left(\frac{1}{3}a^2 - 7\right)\left(\frac{1}{3}a^2 + 7\right) = \left(\frac{1}{3}a^2\right)^2 - 7^2$
$= \frac{1}{9}a^4 - 49$

71. $(3b+7)(2b-5) = 6b^2 - 15b + 14b - 35$
$= 6b^2 - b - 35$

73. $(x^2+10)(x^2-10) = (x^2)^2 - (10)^2 = x^4 - 100$

75. $(4x+5)(4x-5) = (4x)^2 - 5^2 = 16x^2 - 25$

77. $(5x-6y)^2 = (5x)^2 - 2(5x)(6y) + (6y)^2$
$= 25x^2 - 60xy + 36y^2$

79. $(2r-3s)(2r+3s) = (2r)^2 - (3s)^2 = 4r^2 - 9s^2$

81. $(2x+1)^2 = (2x)^2 + 2(2x)(1) + (1)^2 = 4x^2 + 4x + 1$
The area of the rug is $(4x^2 + 4x + 1)$ square feet.

83. $\frac{50b^{10}}{70b^5} = \frac{50}{70} \cdot \frac{b^{10}}{b^5} = \frac{5}{7}b^{10-5} = \frac{5}{7}b^5 = \frac{5b^5}{7}$

85. $\frac{8a^{17}b^5}{-4a^7b^{10}} = \frac{8}{-4} \cdot \frac{a^{17}}{a^7} \cdot \frac{b^5}{b^{10}}$
$= -2a^{17-7}b^{5-10}$
$= -2a^{10}b^{-5}$
$= -\frac{2a^{10}}{b^5}$

87. $\frac{2x^4y^{12}}{3x^4y^4} = \frac{2}{3} \cdot \frac{x^4}{x^4} \cdot \frac{y^{12}}{y^4}$
$= \frac{2}{3}x^{4-4}y^{12-4}$
$= \frac{2}{3}x^0y^8$
$= \frac{2y^8}{3}$

89. $(a-b)^2 = (a)^2 - 2(a)(b) + (b)^2 = a^2 - 2ab + b^2$
which is choice c.

91. $(a+b)^2 = (a)^2 + 2(a)(b) + (b)^2 = a^2 + 2ab + b^2$
which is choice d.

93. Since $x^2 \cdot x^2 = x^4$ and $3x^2 + 7x^2 = 10x^2$, $(x^2+7)(x^2+3) = x^4 + 10x^2 + 21$ is a true statement.

95. $(x^2-1)^2 - x^2 = ((x^2)^2 - 2(x^2)(1) + 1^2) - x^2$
$= (x^4 - 2x^2 + 1) - x^2$
$= x^4 - 2x^2 + 1 - x^2$
$= x^4 - 3x^2 + 1$
The area is $(x^4 - 3x^2 + 1)$ square meters.

97. $(5x-3)^2 - (x+1)^2$
$= ((5x)^2 - 2(5x)(3) + 3^2) - (x^2 + 2x + 1^2)$
$= (25x^2 - 30x + 9) + (-x^2 - 2x - 1)$
$= 25x^2 - 30x + 9 - x^2 - 2x - 1$
$= 24x^2 - 32x + 8$
The area is $(24x^2 - 32x + 8)$ square meters.

99. answers may vary

101. answers may vary

Integrated Review

1. $(5x^2)(7x^3) = (5 \cdot 7)(x^2 \cdot x^3) = 35x^5$

2. $(4y^2)(-8y^7) = (4 \cdot -8)(y^2 \cdot y^7) = -32y^9$

3. $-4^2 = -(4^2) = -16$

4. $(-4)^2 = (-4)(-4) = 16$

5. $(x-5)(2x+1) = 2x^2 + 1x - 10x - 5 = 2x^2 - 9x - 5$

6. $(3x-2)(x+5) = 3x^2 + 15x - 2x - 10$
   $= 3x^2 + 13x - 10$

7. $(x - 5) + (2x + 1) = x - 5 + 2x + 1 = 3x - 4$

8. $(3x - 2) + (x + 5) = 3x - 2 + x + 5 = 4x + 3$

9. $\frac{7x^9y^{12}}{x^3y^{10}} = 7 \cdot x^{9-3} \cdot y^{12-10} = 7x^6y^2$

10. $\frac{20a^2b^8}{14a^2b^2} = \frac{20}{14}a^{2-2}b^{8-2} = \frac{10a^0b^6}{7} = \frac{10b^6}{7}$

11. $(12m^7n^6)^2 = 12^2 \cdot m^{7\cdot2}n^{6\cdot2} = 144m^{14}n^{12}$

12. $(4y^9z^{10})^3 = 4^3 \cdot y^{9\cdot3}z^{10\cdot3} = 64y^{27}z^{30}$

13. $(4y-3)(4y+3) = (4y)^2 - 3^2 = 16y^2 - 9$

14. $(7x-1)(7x+1) = (7x)^2 - 1^2 = 49x^2 - 1$

15. $(x^{-7}y^5)^9 = x^{-7\cdot9}y^{5\cdot9} = x^{-63}y^{45} = \frac{y^{45}}{x^{63}}$

16. $8^{-2} = \frac{1}{8^2} = \frac{1}{64}$

17. $(3^{-1}x^9)^3 = 3^{-1\cdot3}x^{9\cdot3} = 3^{-3}x^{27} = \frac{x^{27}}{3^3} = \frac{x^{27}}{27}$

18. $\frac{(r^7s^{-5})^6}{(2r^{-4}s^{-4})^4} = \frac{r^{7\cdot6}s^{-5\cdot6}}{2^4r^{-4\cdot4}s^{-4\cdot4}}$
    $= \frac{r^{42}s^{-30}}{16r^{-16}s^{-16}}$
    $= \frac{r^{42-(-16)}s^{-30-(-16)}}{16}$
    $= \frac{r^{58}s^{-14}}{16}$
    $= \frac{r^{58}}{16s^{14}}$

19. $(7x^2 - 2x + 3) - (5x^2 + 9)$
    $= (7x^2 - 2x + 3) + (-5x^2 - 9)$
    $= 7x^2 - 2x + 3 - 5x^2 - 9$
    $= 2x^2 - 2x - 6$

20. $(10x^2 + 7x - 9) - (4x^2 - 6x + 2)$
    $= (10x^2 + 7x - 9) + (-4x^2 + 6x - 2)$
    $= 10x^2 + 7x - 9 - 4x^2 + 6x - 2$
    $= 6x^2 + 13x - 11$

21. $0.7y^2 - 1.2 + 1.8y^2 - 6y + 1 = 2.5y^2 - 6y - 0.2$

22. $7.8x^2 - 6.8x - 3.3 + 0.6x^2 - 0.9$
    $= 8.4x^2 - 6.8x - 4.2$

23. $(3y^2 - 6y + 1) - (y^2 + 2)$
    $= (3y^2 - 6y + 1) + (-y^2 - 2)$
    $= 3y^2 - 6y + 1 - y^2 - 2$
    $= 2y^2 - 6y - 1$

24. $(z^2 + 5) - (3z^2 - 1) + \left(8z^2 + 2z - \frac{1}{2}\right)$
    $= (z^2 + 5) + (-3z^2 + 1) + \left(8z^2 + 2z - \frac{1}{2}\right)$
    $= z^2 + 5 - 3z^2 + 1 + 8z^2 + 2z - \frac{1}{2}$
    $= 6z^2 + 2z + \frac{11}{2}$

25. $(x+4)^2 = x^2 + 2(x)(4) + 4^2 = x^2 + 8x + 16$

26. $(y-9)^2 = y^2 - 2(y)(9) + 9^2 = y^2 - 18x + 81$

27. $(x + 4) + (x + 4) = x + 4 + x + 4 = 2x + 8$

28. $(y-9)+(y-9)=y-9+y-9=2y-18$

29. $$\begin{aligned}7x^2-6xy+4(y^2-xy)&=7x^2-6xy+4y^2-4xy\\&=7x^2-10xy+4y^2\end{aligned}$$

30. $$\begin{aligned}5a^2-3ab+6(b^2-a^2)&=5a^2-3ab+6b^2-6a^2\\&=-a^2-3ab+6b^2\end{aligned}$$

31. $$\begin{aligned}(x-3)(x^2+5x-1)&=x(x^2+5x-1)+(-3)(x^2+5x-1)\\&=x(x^2)+x(5x)+x(-1)+(-3)(x^2)+(-3)(5x)+(-3)(-1)\\&=x^3+5x^2-x-3x^2-15x+3\\&=x^3+2x^2-16x+3\end{aligned}$$

32. $$\begin{aligned}(x+1)(x^2-3x-2)&=x(x^2-3x-2)+1(x^2-3x-2)\\&=x(x^2)+x(-3x)+x(-2)+1(x^2)+1(-3x)+1(-2)\\&=x^3-3x^2-2x+x^2-3x-2\\&=x^3-2x^2-5x-2\end{aligned}$$

33. $$\begin{aligned}(2x-7)(3x+10)&=6x^2+20x-21x-70\\&=6x^2-x-70\end{aligned}$$

34. $$\begin{aligned}(5x-1)(4x+5)&=20x^2+25x-4x-5\\&=20x^2+21x-5\end{aligned}$$

35. $$\begin{aligned}(2x-7)(x^2-6x+1)&=2x(x^2-6x+1)+(-7)(x^2-6x+1)\\&=2x(x^2)+2x(-6x)+2x(1)+(-7)(x^2)+(-7)(-6x)+(-7)(1)\\&=2x^3-12x^2+2x-7x^2+42x-7\\&=2x^3-19x^2+44x-7\end{aligned}$$

36. $$\begin{aligned}(5x-1)(x^2+2x-3)&=5x(x^2+2x-3)+(-1)(x^2+2x-3)\\&=5x(x^2)+5x(2x)+5x(-3)+(-1)(x^2)+(-1)(2x)+(-1)(-3)\\&=5x^3+10x^2-15x-x^2-2x+3\\&=5x^3+9x^2-17x+3\end{aligned}$$

37. $\left(2x+\frac{5}{9}\right)\left(2x-\frac{5}{9}\right)=(2x)^2-\left(\frac{5}{9}\right)^2=4x^2-\frac{25}{81}$

38. $$\begin{aligned}\left(12y+\frac{3}{7}\right)\left(12y-\frac{3}{7}\right)&=(12y)^2-\left(\frac{3}{7}\right)^2\\&=144y^2-\frac{9}{49}\end{aligned}$$

Section 3.7 Practice Exercises

1. $\frac{25x^3+5x^2}{5x^2}=\frac{25x^3}{5x^2}+\frac{5x^2}{5x^2}=5x+1$

2. $\frac{24x^7+12x^2-4x}{4x^2}=\frac{24x^7}{4x^2}+\frac{12x^2}{4x^2}-\frac{4x}{4x^2}$
$=6x^5+3-\frac{1}{x}$

3. $\frac{12x^3y^3-18xy+6y}{3xy}=\frac{12x^3y^3}{3xy}-\frac{18xy}{3xy}+\frac{6y}{3xy}$
$=4x^2y^2-6+\frac{2}{x}$

4.
$$\begin{array}{r} x+7 \\ x+5\overline{)x^2+12x+35} \\ \underline{x^2+\ 5x} \\ 7x+35 \\ \underline{7x+35} \\ 0 \end{array}$$

Thus, $\frac{x^2+12x+35}{x+5}=x+7.$

5.
$$\begin{array}{r} 4x+3 \\ 2x-1\overline{)8x^2+2x-7} \\ \underline{8x^2-4x} \\ 6x-7 \\ \underline{6x-3} \\ -4 \end{array}$$

Thus, $\frac{8x^2+2x-7}{2x-1}=4x+3+\frac{-4}{2x-1}$ or $4x+3-\frac{4}{2x-1}.$

6. $15-2x^2=-2x^2+0x+15$
$$\begin{array}{r} -2x-6 \\ x-3\overline{)-2x^2+0x+15} \\ \underline{-2x^2+6x} \\ -6x+15 \\ \underline{-6x+18} \\ -3 \end{array}$$

Thus, $\frac{15-2x^2}{x-3}=-2x-6+\frac{-3}{x-3}$ or $-2x-6-\frac{3}{x-3}.$

7. $\frac{5-x+9x^3}{3x+2}=\frac{9x^3+0x^2-x+5}{3x+2}$
$$\begin{array}{r} 3x^2-2x+1 \\ 3x+2\overline{)9x^3+0x^2-\ x+5} \\ \underline{9x^3+6x^2} \\ -6x^2-\ x \\ \underline{-6x^2-4x} \\ 3x+5 \\ \underline{3x+2} \\ 3 \end{array}$$

Thus, $\frac{5-x+9x^3}{3x+2}=3x^2-2x+1+\frac{3}{3x+2}.$

8.
$$\begin{array}{r} x^2+x+1 \\ x-1\overline{)x^3+0x^2+0x-1} \\ \underline{x^3-\ x^2} \\ x^2+0x \\ \underline{x^2-\ x} \\ x-1 \\ \underline{x-1} \\ 0 \end{array}$$

Thus, $\frac{x^3-1}{x-1}=x^2+x+1.$

Vocabulary, Readiness & Video Check 3.7

1. In $6\overline{)18}$ (quotient 3), the 18 is the <u>dividend</u>, the 3 is the <u>quotient</u>, and the 6 is the <u>divisor</u>.

2. In $x+1\overline{)x^2+3x+2}$ (quotient $x+2$), the $x+1$ is the <u>divisor</u>, the x^2+3x+2 is the <u>dividend</u>, and the $x+2$ is the <u>quotient</u>.

3. $\frac{a^6}{a^4}=a^{6-4}=a^2$

4. $\frac{p^8}{p^3}=p^{8-3}=p^5$

5. $\frac{y^2}{y}=\frac{y^2}{y^1}=y^{2-1}=y^1=y$

6. $\dfrac{a^3}{a} = \dfrac{a^3}{a^1} = a^{3-1} = a^2$

7. the common denominator

8. Filling in missing powers helps us keep like terms lined up and our work clear and neat.

Exercise Set 3.7

1. $\dfrac{12x^4 + 3x^2}{x} = \dfrac{12x^4}{x} + \dfrac{3x^2}{x} = 12x^3 + 3x$

3. $\dfrac{20x^3 - 30x^2 + 5x + 5}{5} = \dfrac{20x^3}{5} - \dfrac{30x^2}{5} + \dfrac{5x}{5} + \dfrac{5}{5}$
$= 4x^3 - 6x^2 + x + 1$

5. $\dfrac{15p^3 + 18p^2}{3p} = \dfrac{15p^3}{3p} + \dfrac{18p^2}{3p} = 5p^2 + 6p$

7. $\dfrac{-9x^4 + 18x^5}{6x^5} = \dfrac{-9x^4}{6x^5} + \dfrac{18x^5}{6x^5} = \dfrac{-3}{2x} + 3$

9. $\dfrac{-9x^5 + 3x^4 - 12}{3x^3} = \dfrac{-9x^5}{3x^3} + \dfrac{3x^4}{3x^3} - \dfrac{12}{3x^3}$
$= -3x^2 + x - \dfrac{4}{x^3}$

11. $\dfrac{4x^4 - 6x^3 + 7}{-4x^4} = \dfrac{4x^4}{-4x^4} - \dfrac{6x^3}{-4x^4} + \dfrac{7}{-4x^4}$
$= -1 + \dfrac{3}{2x} - \dfrac{7}{4x^4}$

13.
$$\begin{array}{r|l} & x+1 \\ x+3 & x^2+4x+3 \\ & \underline{x^2+3x} \\ & \quad x+3 \\ & \quad \underline{x+3} \\ & \qquad 0 \end{array}$$

$\dfrac{x^2 + 4x + 3}{x + 3} = x + 1$

15.
$$\begin{array}{r|l} & 2x+3 \\ x+5 & 2x^2+13x+15 \\ & \underline{2x^2+10x} \\ & \qquad 3x+15 \\ & \qquad \underline{3x+15} \\ & \qquad\qquad 0 \end{array}$$

$\dfrac{2x^2 + 13x + 15}{x + 5} = 2x + 3$

17.
$$\begin{array}{r|l} & 2x+1 \\ x-4 & 2x^2-7x+3 \\ & \underline{2x^2-8x} \\ & \qquad x+3 \\ & \qquad \underline{x-4} \\ & \qquad\quad 7 \end{array}$$

$\dfrac{2x^2 - 7x + 3}{x - 4} = 2x + 1 + \dfrac{7}{x - 4}$

19.
$$\begin{array}{r|l} & 3a^2-3a+1 \\ 3a+2 & 9a^3-3a^2-3a+4 \\ & \underline{9a^3+6a^2} \\ & \quad -9a^2-3a \\ & \quad \underline{-9a^2-6a} \\ & \qquad\qquad 3a+4 \\ & \qquad\qquad \underline{3a+2} \\ & \qquad\qquad\quad 2 \end{array}$$

$\dfrac{9a^3 - 3a^2 - 3a + 4}{3a + 2} = 3a^2 - 3a + 1 + \dfrac{2}{3a + 2}$

21.
$$\begin{array}{r|l} & 4x+3 \\ 2x+1 & 8x^2+10x+1 \\ & \underline{8x^2+4x} \\ & \qquad 6x+1 \\ & \qquad \underline{6x+3} \\ & \qquad\quad -2 \end{array}$$

$\dfrac{8x^2 + 10x + 1}{2x + 1} = 4x + 3 - \dfrac{2}{2x + 1}$

23.
$$\begin{array}{r} 2x^2+6x-5 \\ x-2\overline{\big)\,2x^3+2x^2-17x+8} \\ \underline{2x^3-4x^2} \\ 6x^2-17x \\ \underline{6x^2-12x} \\ -5x+8 \\ \underline{-5x+10} \\ -2 \end{array}$$

$$\frac{2x^3+2x^2-17x+8}{x-2}=2x^2+6x-5-\frac{2}{x-2}$$

25.
$$\begin{array}{r} x+6 \\ x-6\overline{\big)\,x^2+0x-36} \\ \underline{x^2-6x} \\ 6x-36 \\ \underline{6x-36} \\ 0 \end{array}$$

$$\frac{x^2-36}{x-6}=x+6$$

27.
$$\begin{array}{r} x^2+3x+9 \\ x-3\overline{\big)\,x^3+0x^2+0x-27} \\ \underline{x^3-3x^2} \\ 3x^2+0x \\ \underline{3x^2-9x} \\ 9x-27 \\ \underline{9x-27} \\ 0 \end{array}$$

$$\frac{x^3-27}{x-3}=x^2+3x+9$$

29. $1-3x^2=-3x^2+0x+1$
$$\begin{array}{r} -3x+6 \\ x+2\overline{\big)\,-3x^2+0x+1} \\ \underline{-3x^2-6x} \\ 6x+1 \\ \underline{6x+12} \\ -11 \end{array}$$

$$\frac{1-3x^2}{x+2}=-3x+6-\frac{11}{x+2}$$

31.
$$\begin{array}{r} 2b-1 \\ 2b-1\overline{\big)\,4b^2-4b-5} \\ \underline{4b^2-2b} \\ -2b-5 \\ \underline{-2b+1} \\ -6 \end{array}$$

$$\frac{-4b+4b^2-5}{2b-1}=2b-1-\frac{6}{2b-1}$$

33. $\frac{a^2b^2-ab^3}{ab}=\frac{a^2b^2}{ab}-\frac{ab^3}{ab}=ab-b^2$

35.
$$\begin{array}{r} 4x+9 \\ 2x-3\overline{\big)\,8x^2+6x-27} \\ \underline{8x^2-12x} \\ 18x-27 \\ \underline{18x-27} \\ 0 \end{array}$$

$$\frac{8x^2+6x-27}{2x-3}=4x+9$$

37.
$$\begin{aligned}\frac{2x^2y+8x^2y^2-xy^2}{2xy}&=\frac{2x^2y}{2xy}+\frac{8x^2y^2}{2xy}-\frac{xy^2}{2xy}\\&=x+4xy-\frac{y}{2}\end{aligned}$$

39.
$$\begin{array}{r} 2b^2+b+2 \\ b+4\overline{\big)\,2b^3+9b^2+6b-4} \\ \underline{2b^3+8b^2} \\ b^2+6b \\ \underline{b^2+4b} \\ 2b-4 \\ \underline{2b+8} \\ -12 \end{array}$$

$$\frac{2b^3+9b^2+6b-4}{b+4}=2b^2+b+2-\frac{12}{b+4}$$

41.
$$\begin{array}{r} y^2+5y+10 \\ y-2\overline{)\,y^3+3y^2+\;0y+\;4} \\ \underline{y^3-2y^2\qquad\qquad} \\ 5y^2+\;0y\qquad \\ \underline{5y^2-10y\qquad} \\ 10y+\;4 \\ \underline{10y-20} \\ 24 \end{array}$$

$$\frac{y^3+3y^2+4}{y-2}=y^2+5y+10+\frac{24}{y-2}$$

43.
$$\begin{array}{r} -6x-12 \\ x-2\overline{)\,-6x^2+\;0x+\;5} \\ \underline{-6x^2+12x\qquad} \\ -12x+\;5 \\ \underline{-12x+24} \\ -19 \end{array}$$

$$\frac{5-6x^2}{x-2}=-6x-12-\frac{19}{x-2}$$

45.
$$\begin{array}{r} x^3-x^2+x \\ x^2+x\overline{)\,x^5+0x^4+0x^3+x^2} \\ \underline{x^5+\;x^4\qquad\qquad\qquad} \\ -x^4+0x^3\qquad \\ \underline{-x^4-\;x^3\qquad} \\ x^3+x^2 \\ \underline{x^3+x^2} \\ 0 \end{array}$$

$$\frac{x^5+x^2}{x^2+x}=x^3-x^2+x$$

47. $\frac{12}{4}=3$, so $12=4\cdot 3$.

49. $\frac{20}{-5}=-4$, so $20=-5\cdot -4$.

51. $\frac{9x^2}{3x}=3x$, so $9x^2=3x\cdot 3x$.

53. $\frac{36x^2}{4x}=9x$, so $36x^2=4x\cdot 9x$.

55.
$$\frac{12x^3+4x-16}{4}=\frac{12x^3}{4}+\frac{4x}{4}-\frac{16}{4}$$
$$=3x^3+x-4$$

The length of each side of the square is $(3x^3+x-4)$ feet.

57.
$$\begin{array}{r} 2x+5 \\ 5x+3\overline{)\,10x^2+31x+15} \\ \underline{10x^2+\;6x\qquad} \\ 25x+15 \\ \underline{25x+15} \\ 0 \end{array}$$

The height of the parallelogram is $(2x + 5)$ meters.

59. answers may vary

61. $\frac{a+7}{7}=\frac{a}{7}+\frac{7}{7}=\frac{a}{7}+1$ which is choice c.

Chapter 3 Vocabulary Check

1. A term is a number or the product of numbers and variables raised to powers.
2. The FOIL method may be used when multiplying two binomials.
3. A polynomial with exactly 3 terms is called a trinomial.
4. The degree of polynomial is the greatest degree of any term of the polynomial.
5. A polynomial with exactly 2 terms is called a binomial.
6. The coefficient of a term is its numerical factor.
7. The degree of a term is the sum of the exponents on the variables in the term.
8. A polynomial with exactly 1 term is called a monomial.
9. Monomials, binomials, and trinomials are all examples of polynomials.
10. The distributive property is used to multiply $2x(x-4)$.

Chapter 3 Review

1. In 3^2, the base is 3 and the exponent is 2.

2. In $(-5)^4$, the base is -5 and the exponent is 4.

3. In -5^4, the base is 5 and the exponent is 4.

4. In x^6, the base is x and the exponent is 6.

5. $8^3 = 8 \cdot 8 \cdot 8 = 512$

6. $(-6)^2 = (-6)(-6) = 36$

7. $-6^2 = -(6 \cdot 6) = -36$

8. $-4^3 - 4^0 = -64 - 1 = -65$

9. $(3b)^0 = 1$

10. $\dfrac{8b}{8b} = 1$

11. $y^2 \cdot y^7 = y^{2+7} = y^9$

12. $x^9 \cdot x^5 = x^{9+5} = x^{14}$

13. $(2x^5)(-3x^6) = -6x^{5+6} = -6x^{11}$

14. $(-5y^3)(4y^4) = -20y^{3+4} = -20y^7$

15. $(x^4)^2 = x^{4 \cdot 2} = x^8$

16. $(y^3)^5 = y^{3 \cdot 5} = y^{15}$

17. $(3y^6)^4 = 3^4 \cdot y^{6 \cdot 4} = 81y^{24}$

18. $(2x^3)^3 = 2^3 \cdot x^{3 \cdot 3} = 8x^9$

19. $\dfrac{x^9}{x^4} = x^{9-4} = x^5$

20. $\dfrac{z^{12}}{z^5} = z^{12-5} = z^7$

21. $\dfrac{a^5b^4}{ab} = a^{5-1}b^{4-1} = a^4b^3$

22. $\dfrac{x^4y^6}{xy} = x^{4-1}y^{6-1} = x^3y^5$

23. $\dfrac{3x^4y^{10}}{12xy^6} = \dfrac{3}{12}x^{4-1}y^{10-6} = \dfrac{1}{4}x^3y^4 = \dfrac{x^3y^4}{4}$

24. $\dfrac{2x^7y^8}{8xy^2} = \dfrac{1}{4}x^{7-1}y^{8-2} = \dfrac{x^6y^6}{4}$

25.
$$\begin{aligned} 5a^7(2a^4)^3 &= 5a^7(2^3 \cdot a^{4 \cdot 3}) \\ &= 5a^7(8a^{12}) \\ &= 40a^{7+12} \\ &= 40a^{19} \end{aligned}$$

26.
$$\begin{aligned} (2x)^2(9x) &= (2^2x^2)(9x) \\ &= 4x^2(9x) \\ &= 36x^{2+1} \\ &= 36x^3 \end{aligned}$$

27. $(-5a)^0 + 7^0 + 8^0 = 1 + 1 + 1 = 3$

28. $8x^0 + 9^0 = 8 \cdot 1 + 1 = 8 + 1 = 9$

29. $\left(\dfrac{3x^4}{4y}\right)^3 = \dfrac{3^3x^{4 \cdot 3}}{4^3y^{1 \cdot 3}} = \dfrac{27x^{12}}{64y^3}$; b

30. $\left(\dfrac{5a^6}{b^3}\right)^2 = \dfrac{5^2a^{6 \cdot 2}}{b^{3 \cdot 2}} = \dfrac{25a^{12}}{b^6}$; c

31. $7^{-2} = \dfrac{1}{7^2} = \dfrac{1}{49}$

32. $-7^{-2} = -\dfrac{1}{7^2} = -\dfrac{1}{49}$

33. $2x^{-4} = \dfrac{2}{x^4}$

34. $(2x)^{-4} = \dfrac{1}{(2x)^4} = \dfrac{1}{2^4x^4} = \dfrac{1}{16x^4}$

35. $\left(\dfrac{1}{5}\right)^{-3} = (5)^3 = 125$

36. $\left(\frac{-2}{3}\right)^{-2} = \left(\frac{3}{-2}\right)^2 = \frac{9}{4}$

37. $2^0 + 2^{-4} = 1 + \frac{1}{2^4} = 1 + \frac{1}{16} = \frac{17}{16}$

38. $6^{-1} - 7^{-1} = \frac{1}{6} - \frac{1}{7} = \frac{7}{42} - \frac{6}{42} = \frac{1}{42}$

39. $\frac{x^5}{x^{-3}} = x^{5-(-3)} = x^8$

40. $\frac{z^4}{z^{-4}} = z^{4-(-4)} = z^8$

41. $\frac{r^{-3}}{r^{-4}} = r^{-3-(-4)} = r^1 = r$

42. $\frac{y^{-2}}{y^{-5}} = y^{-2-(-5)} = y^3$

43. $\left(\frac{bc^{-2}}{bc^{-3}}\right)^4 = (b^{1-1}c^{-2-(-3)})^4 = (b^0c^1)^4 = c^4$

44.
$$\begin{aligned}\left(\frac{x^{-3}y^{-4}}{x^{-2}y^{-5}}\right)^{-3} &= (x^{-3-(-2)}y^{-4-(-5)})^{-3}\\ &= (x^{-1}y^1)^{-3}\\ &= x^{-1\cdot -3}y^{1\cdot -3}\\ &= x^3y^{-3}\\ &= \frac{x^3}{y^3}\end{aligned}$$

45. $\frac{x^{-4}y^{-6}}{x^2y^7} = x^{-4-2}y^{-6-7} = x^{-6}y^{-13} = \frac{1}{x^6y^{13}}$

46. $\frac{a^5b^{-5}}{a^{-5}b^5} = a^{5-(-5)}b^{(-5)-5} = a^{10}b^{-10} = \frac{a^{10}}{b^{10}}$

47. $0.00027 = 2.7\times10^{-4}$

48. $0.8868 = 8.868\times10^{-1}$

49. $80{,}800{,}000 = 8.08\times10^7$

50. $868{,}000 = 8.68\times10^5$

51. $130{,}300{,}000 = 1.303\times10^8$

52. $150{,}000 = 1.5\times10^5$

53. $8.67\times10^5 = 867{,}000$

54. $3.86\times10^{-3} = 0.00386$

55. $8.6\times10^{-4} = 0.00086$

56. $8.936\times10^5 = 893{,}600$

57. $1.43128\times10^{15} = 1{,}431{,}280{,}000{,}000{,}000$

58. $1\times10^{-10} = 0.0000000001$

59. $(8\times10^4)(2\times10^{-7}) = 16\times10^{-3} = 0.016$

60. $\frac{8\times10^4}{2\times10^{-7}} = 4\times10^{11} = 400{,}000{,}000{,}000$

61. The degree of $(y^5 + 7x - 8x^4)$ is 5.

62. The degree of $(9y^2 + 30y + 25)$ is 2.

63. The degree of $(-14x^2y - 28x^2y^3 - 42x^2y^2)$ is 2 + 3 or 5.

64. The degree of $(6x^2y^2z^2 + 5x^2y^3 - 12xyz)$ is 2 + 2 + 2 or 6.

65.
$$\begin{aligned}-16t^2 + 4000 &= -16(0)^2 + 4000\\ &= 0 + 4000\\ &= 4000\end{aligned}$$
$$\begin{aligned}-16t^2 + 4000 &= -16(1)^2 + 4000\\ &= -16 + 4000\\ &= 3984\end{aligned}$$
$$\begin{aligned}-16t^2 + 4000 &= -16(3)^2 + 4000\\ &= -144 + 4000\\ &= 3856\end{aligned}$$
$$\begin{aligned}-16t^2 + 4000 &= -16(5)^2 + 4000\\ &= -400 + 4000\\ &= 3600\end{aligned}$$

t	0 seconds	1 second	3 seconds	5 seconds
$-16t^2+4000$	4000 feet	3984 feet	3856 feet	3600 feet

66. $2x^2+20x=2(1)^2+20(1)=2+20=22$
$2x^2+20x=2(3)^2+20(3)=18+60=78$
$2x^2+20x=2(5.1)^2+20(5.1)=52.02+102$
$=154.02$
$2x^2+20x=2(10)^2+20(10)=200+200=400$

x	1	3	5.1	10
$2x^2+20x$	22	78	154.02	400

67. $7a^2-4a^2-a^2=(7-4-1)a^2=2a^2$

68. $9y+y-14y=(9+1-14)y=-4y$

69. $6a^2+4a+9a^2=6a^2+9a^2+4a=15a^2+4a$

70. $21x^2+3x+x^2+6=21x^2+x^2+3x+6$
$=22x^2+3x+6$

71. $4a^2b-3b^2-8q^2-10a^2b+7q^2$
$=4a^2b-10a^2b-3b^2-8q^2+7q^2$
$=-6a^2b-3b^2-q^2$

72. $2s^{14}+3s^{13}+12s^{12}-s^{10}$ cannot be combined.

73. $(3x^2+2x+6)+(5x^2+x)=3x^2+2x+6+5x^2+x$
$=3x^2+5x^2+2x+x+6$
$=8x^2+3x+6$

74. $(2x^5+3x^4+4x^3+5x^2)+(4x^2+7x+6)=2x^5+3x^4+4x^3+5x^2+4x^2+7x+6$
$=2x^5+3x^4+4x^3+9x^2+7x+6$

75. $(-5y^2+3)-(2y^2+4)=(-5y^2+3)+(-2y^2-4)$
$=-5y^2+3-2y^2-4$
$=-7y^2-1$

76. $(2m^7+3x^4+7m^6)-(8m^7+4m^2+6x^4)=(2m^7+3x^4+7m^6)+(-8m^7-4m^2-6x^4)$
$=2m^7+3x^4+7m^6-8m^7-4m^2-6x^4$
$=-6m^7-3x^4+7m^6-4m^2$

77. $(3x^2 - 7xy + 7y^2) - (4x^2 - xy + 9y^2)$
$= (3x^2 - 7xy + 7y^2) + (-4x^2 + xy - 9y^2)$
$= 3x^2 - 7xy + 7y^2 - 4x^2 + xy - 9y^2$
$= -x^2 - 6xy - 2y^2$

78. $(8x^6 - 5xy - 10y^2) - (7x^6 - 9xy - 12y^2)$
$= (8x^6 - 5xy - 10y^2) + (-7x^6 + 9xy + 12y^2)$
$= 8x^6 - 5xy - 10y^2 - 7x^6 + 9xy + 12y^2$
$= x^6 + 4xy + 2y^2$

79. $(-9x^2 + 6x + 2) + (4x^2 - x - 1)$
$= -9x^2 + 6x + 2 + 4x^2 - x - 1$
$= -5x^2 + 5x + 1$

80. $[(x^2 + 7x + 9) + (x^2 + 4)] - (4x^2 + 8x - 7)$
$= (x^2 + 7x + 9) + (x^2 + 4) + (-4x^2 - 8x + 7)$
$= x^2 + 7x + 9 + x^2 + 4 - 4x^2 - 8x + 7$
$= -2x^2 - x + 20$

81. $6(x + 5) = 6(x) + 6(5) = 6x + 30$

82. $9(x - 7) = 9(x) + 9(-7) = 9x - 63$

83. $4(2a + 7) = 4(2a) + 4(7) = 8a + 28$

84. $9(6a - 3) = 9(6a) + 9(-3) = 54a - 27$

85. $-7x(x^2 + 5) = (-7x)(x^2) + (-7x)(5)$
$= -7x^3 - 35x$

86. $-8y(4y^2 - 6) = (-8y)(4y^2) + (-8y)(-6)$
$= -32y^3 + 48y$

87. $-2(x^3 - 9x^2 + x)$
$= (-2)(x^3) + (-2)(-9x^2) + (-2)(x)$
$= -2x^3 + 18x^2 - 2x$

88. $-3a(a^2b + ab + b^2)$
$= (-3a)(a^2b) + (-3a)(ab) + (-3a)(b^2)$
$= -3a^3b - 3a^2b - 3ab^2$

89. $(3a^3 - 4a + 1)(-2a)$
$= (3a^3)(-2a) + (-4a)(-2a) + (1)(-2a)$
$= -6a^4 + 8a^2 - 2a$

90. $(6b^3 - 4b + 2)(7b)$
$= (6b^3)(7b) + (-4b)(7b) + (2)(7b)$
$= 42b^4 - 28b^2 + 14b$

91. $(2x + 2)(x - 7) = 2x^2 - 14x + 2x - 14$
$= 2x^2 - 12x - 14$

92. $(2x - 5)(3x + 2) = 6x^2 + 4x - 15x - 10$
$= 6x^2 - 11x - 10$

93. $(4a - 1)(a + 7) = 4a^2 + 28a - a - 7$
$= 4a^2 + 27a - 7$

94. $(6a - 1)(7a + 3) = 42a^2 + 18a - 7a - 3$
$= 42a^2 + 11a - 3$

95. $(x + 7)(x^3 + 4x - 5)$
$= x(x^3 + 4x - 5) + 7(x^3 + 4x - 5)$
$= x(x^3) + x(4x) + x(-5) + 7(x^3) + 7(4x) + 7(-5)$
$= x^4 + 4x^2 - 5x + 7x^3 + 28x - 35$
$= x^4 + 7x^3 + 4x^2 + 23x - 35$

96. $(x + 2)(x^5 + x + 1)$
$= x(x^5 + x + 1) + 2(x^5 + x + 1)$
$= x(x^5) + x(x) + x(1) + 2(x^5) + 2(x) + 2(1)$
$= x^6 + x^2 + x + 2x^5 + 2x + 2$
$= x^6 + 2x^5 + x^2 + 3x + 2$

97. $(x^2+2x+4)(x^2+2x-4) = x^2(x^2+2x-4)+2x(x^2+2x-4)+4(x^2+2x-4)$
$= x^4+2x^3-4x^2+2x^3+4x^2-8x+4x^2+8x-16$
$= x^4+4x^3+4x^2-16$

98. $(x^3+4x+4)(x^3+4x-4) = x^3(x^3+4x-4)+4x(x^3+4x-4)+4(x^3+4x-4)$
$= x^6+4x^4-4x^3+4x^4+16x^2-16x+4x^3+16x-16$
$= x^6+8x^4+16x^2-16$

99. $(x+7)^3 = (x+7)(x+7)^2$
$= (x+7)(x^2+2(x)(7)+7^2)$
$= (x+7)(x^2+14x+49)$
$= x(x^2+14x+49)+7(x^2+14x+49)$
$= x^3+14x^2+49x+7x^2+98x+343$
$= x^3+21x^2+147x+343$

100. $(2x-5)^3 = (2x-5)(2x-5)^2$
$= (2x-5)((2x)^2-(2)(2x)(5)+5^2)$
$= (2x-5)(4x^2-20x+25)$
$= 2x(4x^2-20x+25)+(-5)(4x^2-20x+25)$
$= 8x^3-40x^2+50x-20x^2+100x-125$
$= 8x^3-60x^2+150x-125$

101. $(x+7)^2 = x^2+2(x)(7)+7^2 = x^2+14x+49$

102. $(x-5)^2 = x^2-2(x)(5)+5^2 = x^2-10x+25$

103. $(3x-7)^2 = (3x)^2-2(3x)(7)+7^2$
$= 9x^2-42x+49$

104. $(4x+2)^2 = (4x)^2+2(4x)(2)+2^2$
$= 16x^2+16x+4$

105. $(5x-9)^2 = (5x)^2-2(5x)(9)+9^2$
$= 25x^2-90x+81$

106. $(5x+1)(5x-1) = (5x)^2-1^2 = 25x^2-1$

107. $(7x+4)(7x-4) = (7x)^2-4^2 = 49x^2-16$

108. $(a+2b)(a-2b) = a^2-(2b)^2 = a^2-4b^2$

109. $(2x-6)(2x+6) = (2x)^2-6^2 = 4x^2-36$

110. $$\begin{aligned}(4a^2-2b)(4a^2+2b) &= (4a^2)^2-(2b)^2\\ &= 16a^4-4b^2\end{aligned}$$

111. $$\begin{aligned}(3x-1)^2 &= (3x)^2-2(3x)(1)+1^2\\ &= 9x^2-6x+1\end{aligned}$$

The area is $(9x^2-6x+1)$ square meters.

112. $$\begin{aligned}(5x+2)(x-1) &= 5x^2-5x+2x-2\\ &= 5x^2-3x-2\end{aligned}$$

The area is $(5x^2-3x-2)$ square miles.

113. $$\begin{aligned}\frac{x^2+21x+49}{7x^2} &= \frac{x^2}{7x^2}+\frac{21x}{7x^2}+\frac{49}{7x^2}\\ &= \frac{1}{7}+\frac{3}{x}+\frac{7}{x^2}\end{aligned}$$

114. $$\begin{aligned}\frac{5a^3b-15ab^2+20ab}{-5ab} &= \frac{5a^3b}{-5ab}+\frac{-15ab^2}{-5ab}+\frac{20ab}{-5ab}\\ &= -a^2+3b-4\end{aligned}$$

115. $$\begin{array}{r} a+1 \\ a-2\overline{)a^2-\ \ a+4} \\ \underline{a^2-2a} \\ a+4 \\ \underline{a-2} \\ 6 \end{array}$$

$$\frac{a^2-a+4}{a-2} = a+1+\frac{6}{a-2}$$

116. $$\begin{array}{r} 4x \\ x+5\overline{)4x^2+20x+7} \\ \underline{4x^2+20x} \\ 7 \end{array}$$

$$\frac{4x^2+20x+7}{x+5} = 4x+\frac{7}{x+5}$$

117. $$\begin{array}{r} a^2+3a+8 \\ a-2\overline{)a^3+\ \ a^2+2a+\ \ 6} \\ \underline{a^3-2a^2} \\ 3a^2+2a \\ \underline{3a^2-6a} \\ 8a+\ \ 6 \\ \underline{8a-16} \\ 22 \end{array}$$

$$\frac{a^3+a^2+2a+6}{a-2} = a^2+3a+8+\frac{22}{a-2}$$

118. $$\begin{array}{r} 3b^2-4b \\ 3b-2\overline{)9b^3-18b^2+8b-1} \\ \underline{9b^3-\ \ 6b^2} \\ -12b^2+8b \\ \underline{-12b^2+8b} \\ -1 \end{array}$$

$$\frac{9b^3-18b^2+8b-1}{3b-2} = 3b^2-4b-\frac{1}{3b-2}$$

119. $$\begin{array}{r} 2x^3-x^2+0x+2 \\ 2x-1\overline{)4x^4-4x^3+x^2+4x-3} \\ \underline{4x^4-2x^3} \\ -2x^3+x^2 \\ \underline{-2x^3+x^2} \\ 4x-3 \\ \underline{4x-2} \\ -1 \end{array}$$

$$\frac{4x^4-4x^3+x^2+4x-3}{2x-1} = 2x^3-x^2+2-\frac{1}{2x-1}$$

120. $$\begin{array}{r} x^2-16x-117 \\ x-6\overline{)-x^3-10x^2-21x+\ \ 18} \\ \underline{-x^3+\ \ 6x^2} \\ -16x^2-21x \\ \underline{-16x^2+96x} \\ -117x+\ \ 18 \\ \underline{-117x+702} \\ -684 \end{array}$$

$$\frac{-x^3-10x^2-21x+18}{x-6} = -x^2-16x-117-\frac{684}{x-6}$$

121. $\dfrac{15x^3-3x^2+60}{3x^2}=\dfrac{15x^3}{3x^2}-\dfrac{3x^2}{3x^2}+\dfrac{60}{3x^2}$
$=5x-1+\dfrac{20}{x^2}$

The width is $\left(5x-1+\dfrac{20}{x^2}\right)$ feet.

122. $\dfrac{21a^3b^6+3a-3}{3}=\dfrac{21a^3b^6}{3}+\dfrac{3a}{3}-\dfrac{3}{3}$
$=7a^3b^6+a-1$

The length of a side is $(7a^3b^6+a-1)$ units.

123. $3^3=(3)(3)(3)=27$

124. $\left(-\dfrac{1}{2}\right)^3=\left(-\dfrac{1}{2}\right)\left(-\dfrac{1}{2}\right)\left(-\dfrac{1}{2}\right)=-\dfrac{1}{8}$

125. $(4xy^2)(x^3y^5)=4x^{1+3}y^{2+5}=4x^4y^7$

126. $\dfrac{18x^9}{27x^3}=\dfrac{2}{3}x^{9-3}=\dfrac{2x^6}{3}$

127. $\left(\dfrac{3a^4}{b^2}\right)^3=\dfrac{3^3a^{4\cdot3}}{b^{2\cdot3}}=\dfrac{27a^{12}}{b^6}$

128. $(2x^{-4}y^3)^{-4}=2^{-4}x^{-4\cdot-4}y^{3\cdot-4}=\dfrac{x^{16}}{16y^{12}}$

129. $\dfrac{a^{-3}b^6}{9^{-1}a^{-5}b^{-2}}=9a^{-3-(-5)}b^{6-(-2)}=9a^2b^8$

130. $(-y^2-4)+(3y^2-6)=-y^2-4+3y^2-6$
$=2x^2-10$

131. $(6x+2)+(5x-7)=6x+2+5x-7=11x-5$

132. $(5x^2+2x-6)-(-x-4)$
$=(5x^2+2x-6)+(x+4)$
$=5x^2+2x-6+x+4$
$=5x^2+3x-2$

133. $(8y^2-3y+1)-(3y^2+2)$
$=(8y^2-3y+1)+(-3y^2-2)$
$=8y^2-3y+1-3y^2-2$
$=5y^2-3y-1$

134. $(2x+5)(3x-2)=6x^2-4x+15x-10$
$=6x^2+11x-10$

135. $4x(7x^2+3)=4x(7x^2)+4x(3)=28x^3+12x$

136. $(7x-2)(4x-9)=28x^2-63x-8x+18$
$=28x^2-71x+18$

137. $(x-3)(x^2+4x-6)$
$=x(x^2+4x-6)+(-3)(x^2+4x-6)$
$=x^3+4x^2-6x-3x^2-12x+18$
$=x^3+x^2-18x+18$

138. $(5x+4)^2=(5x)^2+2(5x)(4)+4^2$
$=25x^2+40x+16$

139. $(6x+3)(6x-3)=(6x)^2-3^2=36x^2-9$

140. $\dfrac{8a^4-2a^3+4a-5}{2a^3}=\dfrac{8a^4}{2a^3}-\dfrac{2a^3}{2a^3}+\dfrac{4a}{2a^3}-\dfrac{5}{2a^3}$
$=4a-1+\dfrac{2}{a^2}-\dfrac{5}{2a^3}$

141.
$$\begin{array}{r}
x-3 \\
x+5\overline{)x^2+2x+10} \\
\underline{x^2+5x} \\
-3x+10 \\
\underline{-3x-15} \\
25
\end{array}$$

$\dfrac{x^2+2x+10}{x+5}=x-3+\dfrac{25}{x+5}$

142.
$$\begin{array}{r} 2x^2+7x+5 \\ 2x-3\overline{)4x^3+8x^2-11x+\ \ 4} \\ \underline{4x^3-6x^2\qquad\qquad\quad} \\ 14x^2-11x\qquad \\ \underline{14x^2-21x\qquad} \\ 10x+\ \ 4 \\ \underline{10x-15} \\ 19 \end{array}$$

$$\frac{4x^3+8x^2-11x+4}{2x-3}=2x^2+7x+5+\frac{19}{2x-3}$$

Chapter 3 Test

1. $2^5=2\cdot 2\cdot 2\cdot 2\cdot 2=32$

2. $(-3)^4=(-3)(-3)(-3)(-3)=81$

3. $-3^4=-(3\cdot 3\cdot 3\cdot 3)=-81$

4. $4^{-3}=\frac{1}{4^3}=\frac{1}{64}$

5.
$$\begin{aligned}(3x^2)(-5x^9)&=3(-5)(x^2\cdot x^9)\\&=-15x^{2+9}\\&=-15x^{11}\end{aligned}$$

6. $\frac{y^7}{y^2}=y^{7-2}=y^5$

7. $\frac{r^{-8}}{r^{-3}}=r^{-8-(-3)}=r^{-5}=\frac{1}{r^5}$

8.
$$\begin{aligned}\left(\frac{4x^2y^3}{x^3y^{-4}}\right)^2&=\frac{4^2x^{2\cdot 2}y^{3\cdot 2}}{x^{3\cdot 2}y^{-4\cdot 2}}\\&=\frac{16x^4y^6}{x^6y^{-8}}\\&=16x^{4-6}y^{6-(-8)}\\&=16x^{-2}y^{14}\\&=\frac{16y^{14}}{x^2}\end{aligned}$$

9.
$$\begin{aligned}\frac{6^2x^{-4}y^{-1}}{6^3x^{-3}y^7}&=\frac{6^2}{6^3}\cdot\frac{x^{-4}}{x^{-3}}\cdot\frac{y^{-1}}{y^7}\\&=6^{2-3}x^{-4-(-3)}y^{-1-7}\\&=6^{-1}x^{-1}y^{-8}\\&=\frac{1}{6}\cdot\frac{1}{x}\cdot\frac{1}{y^8}\\&=\frac{1}{6xy^8}\end{aligned}$$

10. $563{,}000=5.63\times 10^5$

11. $0.0000863=8.63\times 10^{-5}$

12. $1.5\times 10^{-3}=0.0015$

13. $6.23\times 10^4=62{,}300$

14. $(1.2\times 10^5)(3\times 10^{-7})=3.6\times 10^{-2}=0.036$

15. a. $4xy^2+7xyz+x^3y-2$

Term	Numerical Coefficient	Degree of Terms
$4xy^2$	4	3
$7xyz$	7	3
x^3y	1	4
-2	-2	0

b. The degree of $4xy^2+7xyz+x^3y$ is 3 + 1 or 4.

16.
$$\begin{aligned}&5x^2+4x-7x^2+11+8x\\&=(5-7)x^2+(4+8)x+11\\&=-2x^2+12x+11\end{aligned}$$

17.
$$\begin{aligned}&(8x^3+7x^2+4x-7)+(8x^3-7x-6)\\&=8x^3+7x^2+4x-7+8x^3-7x-6\\&=8x^3+8x^3+7x^2+4x-7x-7-6\\&=16x^3+7x^2-3x-13\end{aligned}$$

18.
$$\begin{array}{r} 5x^3+\ \ x^2+5x-2 \\ \underline{-\ (8x^3-4x^2+\ \ x-7)} \end{array}\qquad \begin{array}{r} 5x^3+\ \ x^2+5x-2 \\ \underline{-8x^3+4x^2-\ \ x+7} \\ -3x^3+5x^2+4x+5 \end{array}$$

19. $[(8x^2+7x+5)+(x^3-8)]-(4x+2)$
$=(8x^2+7x+5)+(x^3-8)+(-4x-2)$
$=8x^2+7x+5+x^3-8-4x-2$
$=x^3+8x^2+3x-5$

20. $(3x+7)(x^2+5x+2)$
$=3x(x^2+5x+2)+7(x^2+5x+2)$
$=3x(x^2)+3x(5x)+3x(2)+7(x^2)+7(5x)+7(2)$
$=3x^3+15x^2+6x+7x^2+35x+14$
$=3x^3+22x^2+41x+14$

21. $3x^2(2x^2-3x+7)$
$=3x^2(2x^2)+3x^2(-3x)+3x^2(7)$
$=6x^4-9x^3+21x^2$

22. $(x+7)(3x-5)=3x^2-5x+21x-35$
$=3x^2+16x-35$

23. $\left(3x-\frac{1}{5}\right)\left(3x+\frac{1}{5}\right)=(3x)^2-\left(\frac{1}{5}\right)^2=9x^2-\frac{1}{25}$

24. $(4x-2)^2=(4x)^2-2(4x)(2)+2^2$
$=16x^2-16x+4$

25. $(8x+3)^2=(8x)^2+2(8x)(3)+(3)^2$
$=64x^2+48x+9$

26. $(x^2-9b)(x^2+9b)=(x^2)^2-(9b)^2=x^4-81b^2$

27. $-16t^2+1001=-16(0)^2+1001=1001$ ft
$=-16(1)^2+1001=-16+1001$
$=985$ ft
$=-16(3)^2+1001=-144+1001$
$=857$ ft
$=-16(5)^2+1001=-400+1001$
$=601$ ft

28. $(2x-3)(2x+3)=(2x)^2-(3)^2$
$=4x^2-9$

The area is $(2x-3)(2x+3)$ or $(4x^2-9)$ square inches.

29. $\frac{4x^2+2xy-7x}{8xy}=\frac{4x^2}{8xy}+\frac{2xy}{8xy}-\frac{7x}{8xy}$
$=\frac{x}{2y}+\frac{1}{4}-\frac{7}{8y}$

30.
$$\begin{array}{r} x+2 \\ x+5\overline{)x^2+7x+10} \\ \underline{x^2+5x} \\ 2x+10 \\ \underline{2x+10} \\ 0 \end{array}$$

$\frac{x^2+7x+10}{x+5}=x+2$

31.
$$\begin{array}{r} 9x^2-6x+4 \\ 3x+2\overline{)27x^3+0x^2+0x-8} \\ \underline{27x^3+18x^2} \\ -18x^2+0x \\ \underline{-18x^2-12x} \\ 12x-8 \\ \underline{12x+8} \\ -16 \end{array}$$

$\frac{27x^3-8}{3x+2}=9x^2-6x+4-\frac{16}{3x+2}$

Cumulative Review Chapters 1–3

1. a. The natural numbers in the set are 11 and 112.

b. The whole numbers in the set are 0, 11, and 112.

c. The integers in the set are –3, –2, 0, 11, and 112.

d. The rational numbers in the set are –3, –2, 0, $\frac{1}{4}$, 11, and 112.

e. The only irrational number in the set is $\sqrt{2}$.

f. The real numbers in the set are –2, 0, $\frac{1}{4}$, 112, –3, 11, and $\sqrt{2}$.

2. a. $|-7.2|=7.2$

b. $|0|=0$

c. $\left|-\frac{1}{2}\right|=\frac{1}{2}$

3. a. $3^2=3\cdot 3=9$

b. $5^3 = 5 \cdot 5 \cdot 5 = 125$

c. $2^4 = 2 \cdot 2 \cdot 2 \cdot 2 = 16$

d. $7^1 = 7$

e. $\left(\frac{3}{7}\right)^2 = \frac{3}{7} \cdot \frac{3}{7} = \frac{9}{49}$

f. $(0.6)^2 = 0.6 \cdot 0.6 = 0.36$

4. a. $\frac{3}{4} \cdot \frac{7}{21} = \frac{1}{4} \cdot \frac{21}{21} = \frac{1}{4}$

b. $\frac{1}{2} \cdot 4\frac{5}{6} = \frac{1}{2} \cdot \frac{29}{6} = \frac{29}{12} = 2\frac{5}{12}$

5. $\frac{3}{2} \cdot \frac{1}{2} - \frac{1}{2} = \frac{3}{4} - \frac{1}{2} = \frac{3}{4} - \frac{2}{4} = \frac{1}{4}$

6. $\frac{2x-7y}{x^2} = \frac{2(5)-7(1)}{5^2} = \frac{10-7}{25} = \frac{3}{25}$

7. a. "The sum of a number and 3" is $x + 3$.

b. "The product of 3 and a number" is $3x$.

c. "The quotient of 7.3 and a number" is $7.3 \div x$ or $\frac{7.3}{x}$.

d. "10 decreased by a number" is $10 - x$.

e. "5 times a number, increased by 7" is $5x + 7$.

8.
$$\begin{aligned} 8+3(2\cdot 6-1) &= 8+3(12-1) \\ &= 8+3(11) \\ &= 8+33 \\ &= 41 \end{aligned}$$

9. $11.4 + (-4.7) = 6.7$

10.
$$\begin{aligned} 5(1)+2 &\stackrel{?}{=} 1-8 \\ 5+2 &\stackrel{?}{=} 1-8 \\ 7 &= -7 \quad \text{False} \end{aligned}$$
No, $x = 1$ is not a solution.

11. a. $\frac{x-y}{12+x} = \frac{2-(-5)}{12+2} = \frac{2+5}{12+2} = \frac{7}{14} = \frac{1}{2}$

b. $x^2 - y = 2^2 - (-5) = 4 + 5 = 9$

12. a. $7 - 40 = 7 + (-40) = -33$

b. $-5 - (-10) = -5 + 10 = 5$

13. $\frac{-30}{-10} = 3$

14. $\frac{-48}{6} = -8$

15. $\frac{42}{-0.6} = -70$

16. $\frac{-30}{-0.2} = 150$

17. $5(3x + 2) = 5(3x) + 5(2) = 15x + 10$

18. $-3(2x - 3) = (-3)(2x) + (-3)(-3) = -6x + 9$

19.
$$\begin{aligned} &-2(y+0.3z-1) \\ &= (-2)(y)+(-2)(0.3z)+(-2)(-1) \\ &= -2y-0.6z+2 \end{aligned}$$

20.
$$\begin{aligned} 4(-x^2+6x-1) &= 4(-x^2)+4(6x)+4(-1) \\ &= -4x^2+24x-4 \end{aligned}$$

21. $-(9x + y - 2z + 6) = -9x - y + 2z - 6$

22. $-(-4xy + 6y - 2) = 4xy - 6y + 2$

23.
$$\begin{aligned} 6(2a-1)-(11a+6) &= 7 \\ 12a-6-11a-6 &= 7 \\ a-12 &= 7 \\ a &= 19 \end{aligned}$$

24.
$$\begin{aligned} 2x+\frac{1}{8} &= x-\frac{3}{8} \\ 8\left(2x+\frac{1}{8}\right) &= 8\left(x-\frac{3}{8}\right) \\ 16x+1 &= 8x-3 \\ 8x+1 &= -3 \\ 8x &= -4 \\ x &= \frac{-4}{8} \\ x &= -\frac{1}{2} \end{aligned}$$

25. $\frac{y}{7} = 20$
$7 \cdot \frac{y}{7} = 20 \cdot 7$
$y = 140$

26. $10 = 5j - 2$
$12 = 5j$
$\frac{12}{5} = j$

27. $0.25x + 0.10(x-3) = 1.1$
$0.25x + 0.10x - 0.30 = 1.1$
$0.35x - 0.30 = 1.1$
$0.35x = 1.4$
$x = 4$

28. $\frac{7x+5}{3} = x+3$
$7x+5 = 3(x+3)$
$7x+5 = 3x+9$
$4x+5 = 9$
$4x = 4$
$x = 1$

29. $2(x+4) = 4x - 12$
$2x + 8 = 4x - 12$
$8 = 2x - 12$
$20 = 2x$
$10 = x$
The number is 10.

30. $(x + 7) - 2x = -x + 7$

31. $30 \cdot 2 + 2x = 140$
$60 + 2x = 140$
$2x = 80$
$x = 40$
The length is 40 feet.

32. $\frac{4(-3)+(-8)}{5+(-5)} = \frac{-12-8}{5-5} = \frac{-20}{0} = \text{undefined}$

33. $\frac{120}{x} = \frac{15}{100}$
$15x = 12{,}000$
$x = 800$

34. [Number line from −5 to 5, shaded to the left of an open circle at 5]

35. $-4x + 7 \geq -9$
$-4x \geq -16$
$x \leq 4$
$\{x | x \leq 4\}$
[Number line from −5 to 5, shaded to the left of a closed circle at 4]

36. a. $(-5)^2 = (-5)(-5) = 25$

b. $-5^2 = -(5 \cdot 5) = -25$

c. $2 \cdot 5^2 = 2 \cdot 25 = 50$

37. a. $x^7 \cdot x^4 = x^{7+4} = x^{11}$

b. $\left(\frac{t}{2}\right)^4 = \frac{t^4}{2^4} = \frac{t^4}{16}$

c. $(9y^5)^2 = 9^2 y^{5 \cdot 2} = 81y^{10}$

38. $\frac{(z^2)^3 \cdot z^7}{z^9} = \frac{z^6 \cdot z^7}{z^9} = \frac{z^{6+7}}{z^9} = \frac{z^{13}}{z^9} = z^{13-9} = z^4$

39. $\left(\frac{3a^2}{b}\right)^{-3} = \frac{3^{-3}a^{2(-3)}}{b^{-3}} = \frac{3^{-3}a^{-6}}{b^{-3}} = \frac{b^3}{27a^6}$

40. $(5x^7)(-3x^9) = -15x^{7+9} = -15x^{16}$

41. $(5y^3)^{-2} = 5^{-2}y^{3 \cdot -2} = 5^{-2}y^{-6} = \frac{1}{25y^6}$

42. $(-3)^{-2} = \frac{1}{(-3)^2} = \frac{1}{9}$

43. $9x^3 + x^3 = (9+1)x^3 = 10x^3$

44. $(5y^2 - 6) - (y^2 + 2) = (5y^2 - 6) + (-y^2 - 2)$
$= 5y^2 - 6 - y^2 - 2$
$= 4y^2 - 8$

45. $5x^2 + 6x - 9x - 3 = 5x^2 - 3x - 3$

46. $(10x^2 - 3)(10x^2 + 3) = (10x^2)^2 - (3)^2$
$= 100x^4 - 9$

47. $7x(x^2+2x+5) = 7x(x^2)+7x(2x)+7x(5)$
$= 7x^3+14x^2+35x$

48. $(10x^2+3)^2 = (10x^2)^2+2(10x^2)(3)+3^2$
$= 100x^4+60x^2+9$

49. $\dfrac{9x^5-12x^2+3x}{3x^2} = \dfrac{9x^5}{3x^2}-\dfrac{12x^2}{3x^2}+\dfrac{3x}{3x^2}$
$= 3x^3-4+\dfrac{1}{x}$

Chapter 4

Section 4.1 Practice Exercises

1. a. $45 = 3 \cdot 3 \cdot 5$
$75 = 3 \cdot 5 \cdot 5$
$\text{GCF} = 3 \cdot 5 = 15$

b. $32 = 2 \cdot 2 \cdot 2 \cdot 2 \cdot 2$
$33 = 3 \cdot 11$
There are no common prime factors; thus, the GCF is 1.

c. $14 = 2 \cdot 7$
$24 = 2 \cdot 2 \cdot 2 \cdot 3$
$60 = 2 \cdot 2 \cdot 3 \cdot 5$
$\text{GCF} = 2$

2. a. The GCF is y^4, since 4 is the smallest exponent to which y is raised.

b. The GCF is x^1 or x, since 1 is the smallest exponent on x.

3. a. $6x^2 = 2 \cdot 3 \cdot x^2$
$9x^4 = 3 \cdot 3 \cdot x^4$
$-12x^5 = -1 \cdot 2 \cdot 2 \cdot 3 \cdot x^5$
$\text{GCF} = 3 \cdot x^2 = 3x^2$

b. $-16y = -1 \cdot 2 \cdot 2 \cdot 2 \cdot 2 \cdot y$
$-20y^6 = -1 \cdot 2 \cdot 2 \cdot 5 \cdot y^6$
$40y^4 = 2 \cdot 2 \cdot 2 \cdot 5 \cdot y^4$
$\text{GCF} = 2 \cdot 2 \cdot y = 4y$

c. The GCF of a^5, a, and a^3 is a.
The GCF of b^4, b^3, and b^2 is b^2.
Thus, the GCF of a^5b^4, ab^3, and a^3b^2 is ab^2.

4. a. The GCF of terms $10y$ and 25 is 5.
$10y + 25 = 5 \cdot 2y + 5 \cdot 5 = 5(2y + 5)$

b. The GCF of x^4 and x^9 is x^4.
$x^4 - x^9 = x^4(1) - x^4(x^5) = x^4(1 - x^5)$

5. $-10x^3 + 8x^2 - 2x = 2x(-5x^2) + 2x(4x) + 2x(-1)$
$= 2x(-5x^2 + 4x - 1)$

6. $4x^3 + 12x = 4x(x^2 + 3)$

7. $\frac{2}{5}a^5 - \frac{4}{5}a^3 + \frac{1}{5}a^2 = \frac{1}{5}a^2(2a^3 - 4a + 1)$

8. $6a^3b + 3a^3b^2 + 9a^2b^4 = 3a^2b(2a + ab + 3b^3)$

9. $7(p + 2) + q(p + 2) = (p + 2)(7 + q)$

10. $7xy^3(p+q) - (p+q) = 7xy^3(p+q) - 1(p+q)$
$= (p+q)(7xy^3 - 1)$

11. $ab + 7a + 2b + 14 = (ab + 7a) + (2b + 14)$
$= a(b+7) + 2(b+7)$
$= (b+7)(a+2)$

12. $28x^3 - 7x^2 + 12x - 3 = (28x^3 - 7x^2) + (12x - 3)$
$= 7x^2(4x-1) + 3(4x-1)$
$= (4x-1)(7x^2+3)$

13. $2xy + 5y^2 - 4x - 10y$
$= (2xy + 5y^2) + (-4x - 10y)$
$= y(2x + 5y) - 2(2x + 5y)$
$= (2x + 5y)(y - 2)$

14. $3x^2 + 4xy + 3x + 4y = (3x^2 + 4xy) + (3x + 4y)$
$= x(3x + 4y) + 1(3x + 4y)$
$= (3x + 4y)(x + 1)$

15. $4x^3 + x - 20x^2 - 5 = x(4x^2 + 1) - 5(4x^2 + 1)$
$= (4x^2 + 1)(x - 5)$

16. $3xy - 4 + x - 12y = (3xy + x) + (-12y - 4)$
$= x(3y + 1) - 4(3y + 1)$
$= (3y + 1)(x - 4)$

17. $2x - 2 + x^3 - 3x^2 = 2(x - 1) + x^2(x - 3)$
There is no common binomial factor that can now be factored out. This polynomial is not factorable by grouping.

Vocabulary, Readiness & Video Check 4.1

1. Since $5 \cdot 4 = 20$, the numbers 5 and 4 are called factors of 20.

2. The greatest common factor of a list of integers is the largest integer that is a factor of all the integers in the list.

3. The greatest common factor of a list of common variables raised to powers is the variable raised to the least exponent in the list.

4. The process of writing a polynomial as a product is called factoring.

5. A factored form of $7x + 21 + xy + 3y$ is $7(x + 3) + y(x + 3)$. false

6. A factored form of $3x^3 + 6x + x^2 + 2$ is $3x(x^2 + 2)$. false

7. $14 = 2 \cdot 7$

8. $15 = 3 \cdot 5$

9. The GCF of 18 and 3 is 3.

10. The GCF of 7 and 35 is 7.

11. The GCF of 20 and 15 is 5.

12. The GCF of 6 and 15 is 3.

13. The GCF of a list of numbers is the largest number that is a factor of all numbers in the list.

14. There is no need to factor the variable parts. The GCF of common variable factors is the variable raised to the smallest exponent.

15. When factoring out a GCF, the number of terms in the other factor should be the same as the number of terms in your original polynomial.

16. Look for a GCF other than 1 or −1; four terms

Exercise Set 4.1

1. $32 = 2 \cdot 2 \cdot 2 \cdot 2 \cdot 2$
$36 = 2 \cdot 2 \cdot 3 \cdot 3$
$\text{GCF} = 2 \cdot 2 = 4$

3. $18 = 2 \cdot 3 \cdot 3$
$42 = 2 \cdot 3 \cdot 7$
$84 = 2 \cdot 2 \cdot 3 \cdot 7$
$\text{GCF} = 2 \cdot 3 = 6$

5. $24 = 2 \cdot 2 \cdot 2 \cdot 3$
$14 = 2 \cdot 7$
$21 = 3 \cdot 7$
GCF = 1 since there are no common prime factors.

7. The GCF of y^2, y^4, and y^7 is y^2.

9. The GCF of z^7, z^9, and z^{11} is z^7.

11. The GCF of x^{10}, x, and x^3 is x.
The GCF of y^2, y^2, and y^3 is y^2.
Thus, the GCF of $x^{10}y^2$, xy^2, and x^3y^3 is xy^2.

13. $14x = 2 \cdot 7 \cdot x$
$21 = 3 \cdot 7$
$\text{GCF} = 7$

15. $12y^4 = 2 \cdot 2 \cdot 3 \cdot y^4$
$20y^3 = 2 \cdot 2 \cdot 5 \cdot y^3$
$\text{GCF} = 2 \cdot 2 \cdot y^3 = 4y^3$

17. $-10x^2 = -1 \cdot 2 \cdot 5 \cdot x^2$
$15x^3 = 3 \cdot 5 \cdot x^3$
$\text{GCF} = 5 \cdot x^2 = 5x^2$

19. $12x^3 = 2 \cdot 2 \cdot 3 \cdot x^3$
$-6x^4 = -1 \cdot 2 \cdot 3 \cdot x^4$
$3x^5 = 3 \cdot x^5$
$\text{GCF} = 3 \cdot x^3 = 3x^3$

21. $-18x^2y = -1 \cdot 2 \cdot 3 \cdot 3 \cdot x^2 \cdot y$
$9x^3y^3 = 3 \cdot 3 \cdot x^3 \cdot y^3$
$36x^3y = 2 \cdot 2 \cdot 3 \cdot 3 \cdot x^3 \cdot y$
$\text{GCF} = 3 \cdot 3 \cdot x^2 \cdot y = 9x^2y$

23. $20a^6b^2c^8 = 2 \cdot 2 \cdot 5 \cdot a^6 \cdot b^2 \cdot c^8$
$50a^7b = 2 \cdot 5 \cdot 5 \cdot a^7 \cdot b$
$\text{GCF} = 2 \cdot 5 \cdot a^6 \cdot b = 10a^6b$

25. $3a + 6 = 3(a + 2)$

27. $30x - 15 = 15(2x - 1)$

29. $x^3 + 5x^2 = x^2(x + 5)$

31. $6y^4 + 2y^3 = 2y^3(3y + 1)$

33. $32xy - 18x^2 = 2x(16y - 9x)$

35. $4x - 8y + 4 = 4(x - 2y + 1)$

37. $6x^3 - 9x^2 + 12x = 3x(2x^2 - 3x + 4)$

39. $a^7b^6 - a^3b^2 + a^2b^5 - a^2b^2$
$= a^2b^2(a^5b^4 - a + b^3 - 1)$

41. $5x^3y - 15x^2y + 10xy = 5xy(x^2 - 3x + 2)$

43. $8x^5 + 16x^4 - 20x^3 + 12 = 4(2x^5 + 4x^4 - 5x^3 + 3)$

45. $\frac{1}{3}x^4 + \frac{2}{3}x^3 - \frac{4}{3}x^5 + \frac{1}{3}x = \frac{1}{3}x(x^3 + 2x^2 - 4x^4 + 1)$

47. $y(x^2 + 2) + 3(x^2 + 2) = (x^2 + 2)(y + 3)$

49. $z(y + 4) + 3(y + 4) = (y + 4)(z + 3)$

51. $r(z^2 - 6) + (z^2 - 6) = r(z^2 - 6) + 1(z^2 - 6)$
$= (z^2 - 6)(r + 1)$

53. $-2x - 14 = (-2)x + (-2)(7) = -2(x + 7)$

55. $-2x^5 + x^7 = (-x^5)(2) + (-x^5)(-x^2)$
$= -x^5[2 + (-x^2)]$
$= -x^5(2 - x^2)$

57. $-6a^4 + 9a^3 - 3a^2$
$= (-3a^2)(2a^2) + (-3a^2)(-3a) + (-3a^2)(1)$
$= -3a^2[2a^2 + (-3a) + 1]$
$= -3a^2(2a^2 - 3a + 1)$

59. $x^3 + 2x^2 + 5x + 10 = x^2(x + 2) + 5(x + 2)$
$= (x + 2)(x^2 + 5)$

61. $5x + 15 + xy + 3y = 5(x + 3) + y(x + 3)$
$= (x + 3)(5 + y)$

63. $6x^3 - 4x^2 + 15x - 10 = 2x^2(3x - 2) + 5(3x - 2)$
$= (3x - 2)(2x^2 + 5)$

65. $5m^3 + 6mn + 5m^2 + 6n$
$= m(5m^2 + 6n) + 1(5m^2 + 6n)$
$= (5m^2 + 6n)(m + 1)$

67. $2y - 8 + xy - 4x = 2(y - 4) + x(y - 4)$
$= (y - 4)(2 + x)$

69. $2x^3 + x^2 + 8x + 4 = x^2(2x + 1) + 4(2x + 1)$
$= (2x + 1)(x^2 + 4)$

71. $3x - 3 + x^3 - 4x^2 = 3(x - 1) + x^2(x - 4)$
The polynomial is not factorable by grouping.

73. $4x^2 - 8xy - 3x + 6y = 4x(x - 2y) - 3(x - 2y)$
$= (x - 2y)(4x - 3)$

75. $5q^2 - 4pq - 5q + 4p = q(5q - 4p) - 1(5q - 4p)$
$= (5q - 4p)(q - 1)$

77. $12x^2y - 42x^2 - 4y + 14$
$= 2(6x^2y - 21x^2 - 2y + 7)$
$= 2[3x^2(2y - 7) - 1(2y - 7)]$
$= 2(2y - 7)(3x^2 - 1)$

79. $6a^2 + 9ab^2 + 6ab + 9b^3$
$= 3(2a^2 + 3ab^2 + 2ab + 3b^3)$
$= 3[a(2a + 3b^2) + b(2a + 3b^2)]$
$= 3(2a + 3b^2)(a + b)$

81. $(x + 2)(x + 5) = x^2 + 5x + 2x + 10 = x^2 + 7x + 10$

83. $(b + 1)(b - 4) = b^2 - 4b + b - 4 = b^2 - 3b - 4$

85. $2 \cdot 6 = 12$
$2 + 6 = 8$
2 and 6 have a product of 12 and a sum of 8.

87. $-1 \cdot (-8) = 8$
$-1 + (-8) = -9$
-1 and -8 have a product of 8 and a sum of -9.

89. $-2 \cdot 5 = -10$
$-2 + 5 = 3$
-2 and 5 have a product of -10 and a sum of 3.

91. $-8 \cdot 3 = -24$
$-8 + 3 = -5$
-8 and 3 have a product of -24 and a sum of -5.

93. $-2x + 14 = -2(x - 7)$, which is choice d.

95. $(a + 6)(a + 2)$ is factored.

97. Since $5(2y + z) - b(2y + z) = (2y + z)(5 - b)$, the given expression is not factored.

99. a. $15x^2 - 250x + 2200$
$= 15(11)^2 - 250(11) + 2200$
$= 15(121) - 250(11) + 2200$
$= 1815 - 2750 + 2200$
$= 1265$
There were approximately 1265 million digital music tracks sold in the U.S. in 2011.

b. Let $x = 12$ for 2012.
$15x^2 - 250x + 2200$
$= 15(12)^2 - 250(12) + 2200$
$= 15(144) - 250(12) + 2200$
$= 2160 - 3000 + 2200$
$= 1360$
There were approximately 1360 million digital music tracks sold in the U.S. in 2012.

c. Let $x = 18$ for 2018.
$15x^2 - 250x + 2200$
$= 15(18)^2 - 250(18) + 2200$
$= 15(324) - 250(18) + 2200$
$= 4860 - 4500 + 2200$
$= 2560$
If the model continues to hold, there would be 2560 million digital music tracks sold in the U.S. in 2018.

d. $15x^2 - 250x + 2200 = 5(3x^2 - 50x + 440)$

101. a. $-322x^2 + 966x + 8372$
$= -322(1)^2 + 966(1) + 8372$
$= -322(1) + 966(1) + 8372$
$= -322 + 966 + 8372$
$= 9016$
The U.S. orange production in 2011 was approximately 9016 thousand tons.

b. Let $x = 3$ for 2013.
$-322x^2 + 966x + 8372$
$= -322(3)^2 + 966(3) + 8372$
$= -322(9) + 966(3) + 8372$
$= -2898 + 2898 + 8372$
$= 8372$
The U.S. orange production in 2013 was approximately 8372 thousand tons.

c. $-322x^2 + 966x + 8372 = -322(x^2 - 3x - 26)$
or $322(-x^2 + 3x + 26)$

103. The area of the circle is πx^2. Since the sides of the square have length $2x$, the area of the square is $(2x)^2 = 4x^2$. The shaded region is the region inside the square but outside the circle. The area of the shaded region is $4x^2 - \pi x^2 = x^2(4 - \pi)$.

105. Area = width · length
$5x^5 - 5x^2 = 5x^2(x^3 - 1)$
The length is $(x^3 - 1)$ units.

107. answers may vary

109. answers may vary

Section 4.2 Practice Exercises

1.

Factors of 20	Sum of Factors
1, 20	21
2, 10	12
4, 5	9

Thus, $x^2 + 12x + 20 = (x + 10)(x + 2)$.

2. a.

Factors of 22	Sum of Factors
−1, −22	−23
−2, −11	−13

$x^2 - 23x + 22 = (x - 1)(x - 22)$

b.

Factors of 50	Sum of Factors
−1, −50	−51
−2, −25	−27
−5, −10	−15

$x^2 - 27x + 50 = (x - 2)(x - 25)$

3.

Factors of −36	Sum of Factors
−1, 36	35
1, −36	−35
−2, 18	16
2, −18	−16
−3, 12	9
3, −12	−9
−4, 9	5
4, −9	−5
−6, 6	0

$x^2+5x-36=(x+9)(x-4)$

4. a. Two factors of −40 whose sum is −3 are −8 and 5.
$q^2-3q-40=(q-8)(q+5)$

b. Two factors of −48 whose sum is 2 are 8 and −6.
$y^2+2y-48=(y+8)(y-6)$

5. Since there are no two numbers whose product is 15 and whose sum is 6, the polynomial is prime.

6. a. Two factors of $14y^2$ whose sum is 9*y* are 2*y* and 7*y*.
$x^2+9xy+14y^2=(x+2y)(x+7y)$

b. Two factors of $30b^2$ whose sum is $-13b$ are $-3b$ and $-10b$.
$a^2-13ab+30b^2=(a-3b)(a-10b)$

7. Two factors of 12 whose sum is 8 are 6 and 2.
$x^4+8x^2+12=(x^2+6)(x^2+2)$

8. $48-14x+x^2=x^2-14x+48$
Two factors of 48 whose sum is −14 are −6 and −8.
$x^2-14x+48=(x-6)(x-8)$

9. a. $4x^2-24x+36=4(x^2-6x+9)$
Two factors of 9 whose sum is −6 are −3 and −3.
$4(x^2-6x+9)=4(x-3)(x-3)$ or $4(x-3)^2$

b. $x^3+3x^2-4x=x(x^2+3x-4)$
Two factors of −4 whose sum is 3 are 4 and −1.
$x(x^2+3x-4)=x(x+4)(x-1)$

10. $5x^5-25x^4-30x^3=5x^3(x^2-5x-6)$
$=5x^3(x+1)(x-6)$

Vocabulary, Readiness & Video Check 4.2

1. To factor x^2+7x+6, we look for two numbers whose product is 6 and whose sum is 7. true

2. We can write the factorization (*y* + 2)(*y* + 4) also as (*y* + 4)(*y* + 2). true

3. The factorization (4*x* − 12)(*x* − 5) is completely factored. false

4. The factorization (*x* + 2*y*)(*x* + *y*) may also be written as $(x+2y)^2$. false

5. $x^2+9x+20=(x+4)(x+5)$

6. $x^2+12x+35=(x+5)(x+7)$

7. $x^2-7x+12=(x-4)(x-3)$

8. $x^2-13x+22=(x-2)(x-11)$

9. $x^2+4x+4=(x+2)(x+2)$

10. $x^2+10x+24=(x+6)(x+4)$

11. 15 is positive, so its factors would have to be either both positive or both negative. Since the factors need to sum to −8, both factors must be negative.

12. Since the sum of the factors is 3, the factors are −2 and 5 (−2 + 5 = 3). (In other words, the factor with the smaller absolute value is negative so that the sum is positive.) If we accidentally choose factors whose sum is −3, simply "switch" the signs of the factors.

Exercise Set 4.2

1. Two factors of 6 whose sum is 7 are 6 and 1.
$x^2+7x+6=(x+6)(x+1)$

3. Two factors of 9 whose sum is −10 are −9 and −1.
$y^2 - 10y + 9 = (y-9)(y-1)$

5. Two factors of 9 whose sum is −6 are −3 and −3.
$x^2 - 6x + 9 = (x-3)(x-3)$ or $(x-3)^2$

7. Two factors of −18 whose sum is −3 are −6 and 3.
$x^2 - 3x - 18 = (x-6)(x+3)$

9. Two factors of −70 whose sum is 3 are 10 and −7.
$x^2 + 3x - 70 = (x+10)(x-7)$

11. Since there are no two numbers whose product is 2 and whose sum is 5, the polynomial is prime.

13. Two factors of $15y^2$ whose sum is $8y$ are $5y$ and $3y$.
$x^2 + 8xy + 15y^2 = (x+5y)(x+3y)$

15. Two factors of −15 whose sum is −2 are −5 and 3.
$a^4 - 2a^2 - 15 = (a^2-5)(a^2+3)$

17. $13 + 14m + m^2 = m^2 + 14m + 13$
Two factors of 13 whose sum is 14 are 13 and 1.
$13 + 14m + m^2 = m^2 + 14m + 13 = (m+13)(m+1)$

19. $10t - 24 + t^2 = t^2 + 10t - 24$
Two factors of −24 whose sum is 10 are −2 and 12.
$10t - 24 + t^2 = t^2 + 10t - 24 = (t-2)(t+12)$

21. Two factors of $16b^2$ whose sum is $-10b$ are $-2b$ and $-8b$.
$a^2 - 10ab + 16b^2 = (a-2b)(a-8b)$

23. $2z^2 + 20z + 32 = 2(z^2 + 10z + 16)$
$= 2(z+8)(z+2)$

25. $2x^3 - 18x^2 + 40x = 2x(x^2 - 9x + 20)$
$= 2x(x-5)(x-4)$

27. $x^2 - 3xy - 4y^2 = (x-4y)(x+y)$

29. $x^2 + 15x + 36 = (x+12)(x+3)$

31. $x^4 - x^2 - 2 = (x^2-2)(x^2+1)$

33. $r^2 - 16r + 48 = (r-12)(r-4)$

35. $x^2 + xy - 2y^2 = (x+2y)(x-y)$

37. $3x^2 + 9x - 30 = 3(x^2 + 3x - 10) = 3(x+5)(x-2)$

39. $3x^4 - 60x^2 + 108 = 3(x^4 - 20x^2 + 36)$
$= 3(x^2-18)(x^2-2)$

41. $x^2 - 18x - 144 = (x-24)(x+6)$

43. $r^2 - 3r + 6$ is prime.

45. $x^2 - 8x + 15 = (x-5)(x-3)$

47. $6x^3 + 54x^2 + 120x = 6x(x^2 + 9x + 20)$
$= 6x(x+4)(x+5)$

49. $4x^2y + 4xy - 12y = 4y(x^2 + x - 3)$

51. $x^2 - 4x - 21 = (x-7)(x+3)$

53. $x^2 + 7xy + 10y^2 = (x+5y)(x+2y)$

55. $64 + 24t + 2t^2 = 2t^2 + 24t + 64$
$= 2(t^2 + 12t + 32)$
$= 2(t+8)(t+4)$

57. $x^3 - 2x^2 - 24x = x(x^2 - 2x - 24)$
$= x(x-6)(x+4)$

59. $2t^5 - 14t^4 + 24t^3 = 2t^3(t^2 - 7t + 12)$
$= 2t^3(t-4)(t-3)$

61. $5x^3y - 25x^2y^2 - 120xy^3$
$= 5xy(x^2 - 5xy - 24y^2)$
$= 5xy(x-8y)(x+3y)$

63. $162 - 45m + 3m^2 = 3m^2 - 45m + 162$
$= 3(m^2 - 15m + 54)$
$= 3(m-9)(m-6)$

65. $-x^2+12x-11=-1(x^2-12x+11)$
$=-1(x-11)(x-1)$

67. $\frac{1}{2}y^2-\frac{9}{2}y-11=\frac{1}{2}(y^2-9y-22)$
$=\frac{1}{2}(y-11)(y+2)$

69. $x^3y^2+x^2y-20x=x(x^2y^2+xy-20)$
$=x(xy-4)(xy+5)$

71. $(2x+1)(x+5)=2x^2+10x+x+5$
$=2x^2+11x+5$

73. $(5y-4)(3y-1)=15y^2-5y-12y+4$
$=15y^2-17y+4$

75. $(a+3b)(9a-4b)=9a^2-4ab+27ab-12b^2$
$=9a^2+23ab-12b^2$

77. $(x-3)(x+8)=x^2+8x-3x-24=x^2+5x-24$

79. answers may vary

81. $P=2l+2w$
$=2(x^2+10x)+2(4x+33)$
$=2x^2+20x+8x+66$
$=2x^2+28x+66$
$=2(x^2+14x+33)$
$=2(x+3)(x+11)$

83. $-16t^2+64t+80=-16(t^2-4t-5)$
$=-16(t-5)(t+1)$

85. $x^2+\frac{1}{2}x+\frac{1}{16}=\left(x+\frac{1}{4}\right)\left(x+\frac{1}{4}\right)$ or $\left(x+\frac{1}{4}\right)^2$

87. $z^2(x+1)-3z(x+1)-70(x+1)$
$=(x+1)(z^2-3z-70)$
$=(x+1)(z-10)(z+7)$

89. The factors of c must sum to −16. Since c is positive, both factors must have the same sign.
$-1+(-15)=-16$; $(-1)(-15)=15$
$-2+(-14)=-16$; $(-2)(-14)=28$
$-3+(-13)=-16$; $(-3)(-13)=39$
$-4+(-12)=-16$; $(-4)(-12)=48$
$-5+(-11)=-16$; $(-5)(-11)=55$
$-6+(-10)=-16$; $(-6)(-10)=60$
$-7+(-9)=-16$; $(-7)(-9)=63$
$-8+(-8)=-16$; $(-8)(-8)=64$
The possible values of c are 15, 28, 39, 48, 55, 60, 63, and 64.

91. The factors of 20 must sum to b. Since 20 is positive and b is positive, both factors of 20 must be positive.
$1+20=21$; $(1)(20)=20$
$2+10=12$; $(2)(10)=20$
$4+5=9$; $(4)(5)=20$
The possible values of b are 9, 12, and 21.

93. $x^{2n}+8x^n-20=(x^n+10)(x^n-2)$

Section 4.3 Practice Exercises

1. a. Factors of $5x^2$: $5x^2=5x\cdot x$
Factors of 10: $10=1\cdot 10$, $10=2\cdot 5$
$5x^2+27x+10=(5x+2)(x+5)$

b. Factors of $4x^2$: $4x^2=4x\cdot x$, $4x^2=2x\cdot 2x$
Factors of 5: $5=1\cdot 5$
$4x^2+12x+5=(2x+5)(2x+1)$

2. a. Factors of $2x^2$: $2x^2=2x\cdot x$
Factors of 12: $12=-1\cdot -12$, $12=-2\cdot -6$, $12=-3\cdot -4$
$2x^2-11x+12=(2x-3)(x-4)$

b. Factors of $6x^2$: $6x^2=6x\cdot x$, $6x^2=3x\cdot 2x$
Factors of 1: $1=-1\cdot -1$
$6x^2-5x+1=(3x-1)(2x-1)$

3. a. Factors of $3x^2$: $3x^2=3x\cdot x$
Factors of −5: $-5=-5\cdot 1$, $-5=1\cdot -5$
$3x^2+14x-5=(3x-1)(x+5)$

b. Factors of $35x^2$: $35x^2=35x\cdot x$,
$35x^2=5x\cdot 7x$
Factors of −4: $-4=-1\cdot 4$, $-4=1\cdot -4$, $-4=-2\cdot 2$
$35x^2+4x-4=(5x+2)(7x-2)$

4. a. Factors of $14x^2$: $14x^2 = 14x \cdot x$,
$14x^2 = 7x \cdot 2x$
Factors of $-2y^2$: $-2y^2 = -2y \cdot y$,
$-2y^2 = 2y \cdot -y$
$14x^2 - 3xy - 2y^2 = (7x + 2y)(2x - y)$

b. Factors of $12a^2$: $12a^2 = 12a \cdot a$,
$12a^2 = 6a \cdot 2a$, $12a^2 = 3a \cdot 4a$
Factors of $-3b^2$: $-3b^2 = -3b \cdot b$,
$-3b^2 = 3b \cdot -b$
$12a^2 - 16ab - 3b^2 = (6a + b)(2a - 3b)$

5. Factors of $2x^4$: $2x^4 = 2x^2 \cdot x^2$
Factors of -7: $-7 = -7 \cdot 1$, $-7 = 7 \cdot -1$
$2x^4 - 5x^2 - 7 = (2x^2 - 7)(x^2 + 1)$

6. a. $3x^3 + 17x^2 + 10x = x(3x^2 + 17x + 10)$
Factors of $3x^2$: $3x^2 = 3x \cdot x$
Factors of 10: $10 = 1 \cdot 10$, $10 = 2 \cdot 5$
$x(3x^2 + 17x + 10) = x(3x + 2)(x + 5)$

b. $6xy^2 + 33xy - 18x = 3x(2y^2 + 11y - 6)$
Factors of $2y^2$: $2y^2 = 2y \cdot y$
Factors of -6: $-6 = 1 \cdot -6$, $-6 = -1 \cdot 6$,
$-6 = 2 \cdot -3$, $-6 = -2 \cdot 3$
$3x(2y^2 + 11y - 6) = 3x(2y - 1)(y + 6)$

7. $-5x^2 - 19x + 4 = -1(5x^2 + 19x - 4)$
Factors of $5x^2$: $5x^2 = 5x \cdot x$
Factors of -4: $-4 = -4 \cdot 1$, $-4 = 4 \cdot -1$,
$-4 = 2 \cdot -2$
$-1(5x^2 + 19x - 4) = -1(x + 4)(5x - 1)$

Vocabulary, Readiness & Video Check 4.3

1. $2x^2 + 5x + 3$ factors as $(2x + 3)(x + 1)$, which is choice **d**.

2. $7x^2 + 9x + 2$ factors as $(7x + 2)(x + 1)$, which is choice **b**.

3. $3x^2 + 31x + 10$ factors as $(3x + 1)(x + 10)$, which is choice **c**.

4. $5x^2 + 61x + 12$ factors as $(5x + 1)(x + 12)$, which is choice **a**.

5. Consider the factors of the first and last terms and the signs of the trinomial. Continue to check possible factors by multiplying until we get the middle term of the trinomial.

6. If the GCF has been factored out, then neither binomial can contain a common factor other than 1 or −1. This helps limit our choice of factors for one or both binomials since we cannot choose factors that would give the terms in either binomial a common factor.

Exercise Set 4.3

1. $5x^2 = 5x \cdot x$
$8 = 2 \cdot 4$
$5x^2 + 22x + 8 = (5x + 2)(x + 4)$

3. $50x^2 = 5x \cdot 10x$
$-2 = 2 \cdot -1$
$50x^2 + 15x - 2 = (5x + 2)(10x - 1)$

5. $20x^2 = 5x \cdot 4x$
$-6 = 2 \cdot -3$
$20x^2 - 7x - 6 = (5x + 2)(4x - 3)$

7. Factors of $2x^2$: $2x^2 = 2x \cdot x$
Factors of 15: $15 = 1 \cdot 15$, $15 = 3 \cdot 5$
$2x^2 + 13x + 15 = (2x + 3)(x + 5)$

9. Factors of $8y^2$: $8y^2 = 8y \cdot y$, $8y^2 = 4y \cdot 2y$.
Factors of 9: $9 = -1 \cdot -9$, $-3 \cdot -3$.
$8y^2 - 17y + 9 = (y - 1)(8y - 9)$

11. Factors of $2x^2$: $2x^2 = 2x \cdot x$
Factors of -5: $-5 = 1 \cdot -5$, $-5 = -1 \cdot 5$
$2x^2 - 9x - 5 = (2x + 1)(x - 5)$

13. Factors of $20r^2$: $20r^2 = 20r \cdot r$, $20r^2 = 10r \cdot 2r$,
$20r^2 = 5r \cdot 4r$.
Factors of -8: $-8 = -1 \cdot 8$, $-8 = -2 \cdot 4$,
$-8 = -4 \cdot 2$, $-8 = -8 \cdot 1$
$20r^2 + 27r - 8 = (4r - 1)(5r + 8)$

15. Factors of $10x^2$: $10x^2 = 10x \cdot x$, $10x^2 = 5x \cdot 2x$

Factors of 3: $3 = 1 \cdot 3$

$10x^2 + 31x + 3 = (10x+1)(x+3)$

17. $x + 3x^2 - 2 = 3x^2 + x - 2$

Factors of $3x^2$: $3x^2 = 3x \cdot x$

Factors of -2: $-2 = -1 \cdot 2$, $-2 = 2 \cdot -1$

$3x^2 + x - 2 = (3x-2)(x+1)$

19. Factors of $6x^2$: $6x^2 = 6x \cdot x$, $6x^2 = 3x \cdot 2x$

Factors of $5y^2$: $5y^2 = -5y \cdot -y$

$6x^2 - 13xy + 5y^2 = (3x-5y)(2x-y)$

21. Factors of $15m^2$: $15m^2 = 15m \cdot m$,

$15m^2 = 5m \cdot 3m$.

Factors of -15: $-15 = -1 \cdot 15$, $-15 = -3 \cdot 5$,

$-15 = -5 \cdot 3$, $-15 = -15 \cdot 1$

$15m^2 - 16m - 15 = (3m-5)(5m+3)$

23. $-9x + 20 + x^2 = x^2 - 9x + 20$

Factors of x^2: $x^2 = x \cdot x$

Factors of 20: $20 = -1 \cdot -20$, $20 = -2 \cdot -10$,

$20 = -4 \cdot -5$

$x^2 - 9x + 20 = (x-4)(x-5)$

25. Factors of $2x^2$: $2x^2 = 2x \cdot x$

Factors of -99: $-99 = -1 \cdot 99$, $-99 = -3 \cdot 33$,

$-99 = -9 \cdot 11$, $-99 = -11 \cdot 9$, $-99 = -33 \cdot 3$,

$-99 = -99 \cdot 1$

$2x^2 - 7x - 99 = (2x+11)(x-9)$

27. $-27t + 7t^2 - 4 = 7t^2 - 27t - 4$

Factors of $7t^2$: $7t^2 = 7t \cdot t$

Factors of -4: $-4 = -1 \cdot 4$, $-4 = 1 \cdot -4$,

$-4 = 2 \cdot -2$

$7t^2 - 27t - 4 = (7t+1)(t-4)$

29. Factors of $3a^2$: $3a^2 = 3a \cdot a$

Factors of $3b^2$: $3b^2 = b \cdot 3b$

$3a^2 + 10ab + 3b^2 = (3a+b)(a+3b)$

31. Factors of $49p^2$: $49p^2 = 49p \cdot p$,

$49p^2 = 7p \cdot 7p$

Factors of -2: $-2 = -1 \cdot 2$, $-2 = 1 \cdot -2$

$49p^2 - 7p - 2 = (7p+1)(7p-2)$

33. Factors of $18x^2$: $18x^2 = 18x \cdot x$,

$18x^2 = 9x \cdot 2x$, $18x^2 = 6x \cdot 3x$

Factors of -14: $-14 = -1 \cdot 14$, $-14 = -2 \cdot 7$,

$-14 = -7 \cdot 2$, $-14 = -14 \cdot 1$

$18x^2 - 9x - 14 = (6x-7)(3x+2)$

35. Factors of $2m^2$: $2m^2 = 2m \cdot m$

Factors of 10: $10 = 1 \cdot 10$, $10 = 2 \cdot 5$

$2m^2 + 17m + 10$ is prime.

37. Factors of $24x^2$: $24x^2 = 24x \cdot x$,

$24x^2 = 12x \cdot 2x$, $24x^2 = 8x \cdot 3x$, $24x^2 = 6x \cdot 4x$

Factors of 12: $12 = 1 \cdot 12$, $12 = 2 \cdot 6$, $12 = 3 \cdot 4$

$24x^2 + 41x + 12 = (3x+4)(8x+3)$

39. $12x^3 + 11x^2 + 2x = x(12x^2 + 11x + 2)$

Factor of $12x^2$: $12x^2 = 12x \cdot x$, $12x^2 = 6x \cdot 2x$,

$12x^2 = 4x \cdot 3x$

Factors of 2: $2 = 1 \cdot 2$

$12x^3 + 11x^2 + 2x = x(3x+2)(4x+1)$

41. $21b^2 - 48b - 45 = 3(7b^2 - 16b - 15)$

Factors of $7b^2$: $7b^2 = 7b \cdot b$

Factors of -15: $-15 = -1 \cdot 15$, $-15 = -3 \cdot 5$,

$-15 = -5 \cdot 3$, $-15 = -15 \cdot 1$

$21b^2 - 48b - 45 = 3(7b+5)(b-3)$

43. $7z + 12z^2 - 12 = 12z^2 + 7z - 12$

Factors of $12z^2$: $12z^2 = 12z \cdot z$, $12z^2 = 6z \cdot 2z$,

$12z^2 = 4z \cdot 3z$

Factors of -12: $-12 = -12 \cdot 1$, $-12 = 12 \cdot -1$,

$-12 = -6 \cdot 2$, $-12 = 6 \cdot -2$, $-12 = -4 \cdot 3$,

$-12 = 4 \cdot -3$

$12z^2 + 7z - 12 = (3z+4)(4z-3)$

45. $6x^2y^2 - 2xy^2 - 60y^2 = 2y^2(3x^2 - x - 30)$

Factors of $3x^2$: $3x^2 = 3x \cdot x$
Factors of -30: $-30 = -1 \cdot 30$, $-30 = -2 \cdot 15$, $-30 = -3 \cdot 10$, $-30 = -5 \cdot 6$, $-30 = -6 \cdot 5$, $-30 = -10 \cdot 3$, $-30 = -15 \cdot 2$, $-30 = -30 \cdot 1$
$6x^2y^2 - 2xy^2 - 60y^2 = 2y^2(3x - 10)(x + 3)$

47. Factors of $4x^2$: $4x^2 = 4x \cdot x$, $4x^2 = 2x \cdot 2x$
Factors of -21: $-21 = -1 \cdot 21$, $-21 = 1 \cdot -21$, $-21 = -7 \cdot 3$, $-21 = 7 \cdot -3$
$4x^2 - 8x - 21 = (2x - 7)(2x + 3)$

49. $3x^2 - 42x + 63 = 3(x^2 - 14x + 21)$

$x^2 - 14x + 21$ is prime, so $3(x^2 - 14x + 21)$ is a factored form of $3x^2 - 42x + 63$.

51. Factors of $8x^2$: $8x^2 = 8x \cdot x$, $8x^2 = 4x \cdot 2x$

Factors of $-27y^2$: $-27y^2 = -27y \cdot y$, $-27y^2 = 27y \cdot -y$, $-27y^2 = -9y \cdot 3y$, $-27y^2 = 9y \cdot -3y$
$8x^2 + 6xy - 27y^2 = (4x + 9y)(2x - 3y)$

53. $-x^2 + 2x + 24 = -1(x^2 - 2x - 24)$
$= -1(x - 6)(x + 4)$

55. $4x^3 - 9x^2 - 9x = x(4x^2 - 9x - 9)$

Factors of $4x^2$: $4x^2 = 4x \cdot x$, $4x^2 = 2x \cdot 2x$
Factors of -9: $-9 = -1 \cdot 9$, $-9 = 1 \cdot -9$, $-9 = 3 \cdot -3$
$4x^3 - 9x^2 - 9x = x(4x + 3)(x - 3)$

57. Factors of $24x^2$: $24x^2 = 24x \cdot x$, $24x^2 = 12x \cdot 2x$, $24x^2 = 8x \cdot 3x$, $24x^2 = 6x \cdot 4x$
Factors of 9: $9 = -1 \cdot -9$, $9 = -3 \cdot -3$
$24x^2 - 58x + 9 = (4x - 9)(6x - 1)$

59. $40a^2b + 9ab - 9b = b(40a^2 + 9a - 9)$

Factors of $40a^2$: $40a^2 = 40a \cdot a$, $40a^2 = 20a \cdot 2a$, $40a^2 = 10a \cdot 4a$, $40a^2 = 8a \cdot 5a$
Factors of -9: $-9 = -1 \cdot 9$, $-9 = 1 \cdot -9$, $-9 = 3 \cdot -3$
$40a^2b + 9ab - 9b = b(8a - 3)(5a + 3)$

61. $30x^3 + 38x^2 + 12x = 2x(15x^2 + 19x + 6)$

Factors of $15x^2$: $15x^2 = 15x \cdot x$, $15x^2 = 5x \cdot 3x$
Factors of 6: $6 = 1 \cdot 6$, $6 = 2 \cdot 3$
$30x^3 + 38x^2 + 12x = 2x(3x + 2)(5x + 3)$

63. $6y^3 - 8y^2 - 30y = 2y(3y^2 - 4y - 15)$

Factors of $3y^2$: $3y^2 = 3y \cdot y$
Factors of -15: $-15 = -1 \cdot 15$, $-15 = 1 \cdot -15$, $-15 = -3 \cdot 5$, $-15 = 3 \cdot -5$
$6y^3 - 8y^2 - 30y = 2y(3y + 5)(y - 3)$

65. $10x^4 + 25x^3y - 15x^2y^2 = 5x^2(2x^2 + 5xy - 3y^2)$

Factors of $2x^2$: $2x^2 = 2x \cdot x$
Factors of $-3y^2$: $-3y^2 = -y \cdot 3y$, $-3y^2 = -3y \cdot y$
$10x^4 + 25x^3y - 15x^2y^2 = 5x^2(2x - y)(x + 3y)$

67. $-14x^2 + 39x - 10 = -1(14x^2 - 39x + 10)$

Factors of $14x^2$: $14x^2 = 14x \cdot x$, $14x^2 = 7x \cdot 2x$
Factors of 10: $10 = -1 \cdot -10$, $10 = -2 \cdot -5$
$-14x^2 + 39x - 10 = -1(2x - 5)(7x - 2)$

69. $16p^4 - 40p^3 + 25p^2 = p^2(16p^2 - 40p + 25)$

Factors of $16p^2$: $16p^2 = 16p \cdot p$, $16p^2 = 8p \cdot 2p$, $16p^2 = 4p \cdot 4p$
Factors of 25: $25 = -1 \cdot -25$, $25 = -5 \cdot -5$
$16p^4 - 40p^3 + 25p^2 = p^2(4p - 5)(4p - 5)$ or $p^2(4p - 5)^2$

71. $-2x^2 + 9x + 5 = -1(2x^2 - 9x - 5)$

Factors of $2x^2$: $2x^2 = 2x \cdot x$
Factors of -5: $-5 = -1 \cdot 5$, $-5 = 1 \cdot -5$
$-2x^2 + 9x + 5 = -1(2x + 1)(x - 5)$

73. $-4 + 52x - 48x^2 = -48x^2 + 52x - 4$
$= -4(12x^2 - 13x + 1)$

Factors of $12x^2$: $12x^2 = 12x \cdot x$, $12x^2 = 6x \cdot 2x$, $12x^2 = 4x \cdot 3x$
Factors of 1: $1 = -1 \cdot -1$
$-4 + 52x - 48x^2 = -4(12x - 1)(x - 1)$

75. Factors of $2t^4$: $2t^4 = 2t^2 \cdot t^2$
Factors of −27: −27 = −1 · 27, −27 = 1 · −27,
−27 = −3 · 9, −27 = 3 · −9
$2t^4 + 3t^2 - 27 = (2t^2 + 9)(t^2 - 3)$

77. Factors of $5x^2y^2$: $5x^2y^2 = 5xy \cdot xy$
Factors of 1: 1 = 1 · 1
There is no combination that gives the correct middle term, so $5x^2y^2 + 20xy + 1$ is prime.

79. $6a^5 + 37a^3b^2 + 6ab^4 = a(6a^4 + 37a^2b^2 + 6b^4)$
Factors of $6a^4$: $6a^4 = 6a^2 \cdot a^2$, $6a^4 = 3a^2 \cdot 2a^2$
Factors of $6b^4$: $6b^4 = 6b^2 \cdot b^2$, $6b^4 = 3b^2 \cdot 2b^2$
$6a^5 + 37a^3b^2 + 6ab^4 = a(6a^2 + b^2)(a^2 + 6b^2)$

81. $(x-4)(x+4) = (x)^2 - (4)^2 = x^2 - 16$

83. $(x+2)^2 = x^2 + 2x(2) + 2^2 = x^2 + 4x + 4$

85. $(2x-1)^2 = (2x)^2 - 2(2x)(1) + (1)^2 = 4x^2 - 4x + 1$

87. The tallest bar corresponds to 25–34, so the 25–34 age range has the highest percentage of text message users.

89. answers may vary

91. No.
$4x^2 = 2 \cdot 2 \cdot x \cdot x$
$19x = 19 \cdot x$
$12 = 2 \cdot 2 \cdot 3$
There is no common factor (other than 1).

93. $(3x^2 + 1) + (6x + 4) + (x^2 + 15x) = 4x^2 + 21x + 5$
$= (4x+1)(x+5)$

95. $4x^2 + 2x + \frac{1}{4} = \left(2x + \frac{1}{2}\right)\left(2x + \frac{1}{2}\right)$ or $\left(2x + \frac{1}{2}\right)^2$

97. $4x^2(y-1)^2 + 25x(y-1)^2 + 25(y-1)^2$
$= (y-1)^2(4x^2 + 25x + 25)$
$= (y-1)^2(4x+5)(x+5)$

99. Factors of $3x^2$: $3x^2 = 3x \cdot x$
Factors of −5: −5 = −1 · 5, −5 = 1 · −5
$(3x-1)(x+5) = 3x^2 + 14x - 5$
$(3x+1)(x-5) = 3x^2 - 14x - 5$
$(3x-5)(x+1) = 3x^2 - 2x - 5$
$(3x+5)(x-1) = 3x^2 + 2x - 5$
Since b is positive, the possible values are 2 and 14.

101. Note that 5 + 2 = 7.
$(5x+2)(x+1) = 5x(x+1) + 2(x+1)$
$= 5x^2 + 5x + 2x + 2$
$= 5x^2 + 7x + 2$
If $c = 2$, then $5x^2 + 7x + c$ is factorable.

103. answers may vary

Section 4.4 Practice Exercises

1. a. 3 · 8 = 24
12 · 2 = 24
12 + 2 = 14
$3x^2 + 14x + 8 = 3x^2 + 12x + 2x + 8$
$= 3x(x+4) + 2(x+4)$
$= (x+4)(3x+2)$

b. 12 · 5 = 60
15 · 4 = 60
15 + 4 = 19
$12x^2 + 19x + 5 = 12x^2 + 15x + 4x + 5$
$= 3x(4x+5) + 1(4x+5)$
$= (4x+5)(3x+1)$

2. a. $30x^2 - 26x + 4 = 2(15x^2 - 13x + 2)$
15 · 2 = 30
−3 · −10 = 30
−3 + (−10) = −13
$30x^2 - 26x + 4 = 2(15x^2 - 13x + 2)$
$= 2(15x^2 - 3x - 10x + 2)$
$= 2[3x(5x-1) - 2(5x-1)]$
$= 2(5x-1)(3x-2)$

b. $6x^2y - 7xy - 5y = y(6x^2 - 7x - 5)$

$6 \cdot -5 = -30$
$3 \cdot -10 = -30$
$3 + (-10) = -7$

$$\begin{aligned} 6x^2y - 7xy - 5y &= y(6x^2 - 7x - 5) \\ &= y(6x^2 + 3x - 10x - 5) \\ &= y[3x(2x+1) - 5(2x+1)] \\ &= y(2x+1)(3x-5) \end{aligned}$$

3. $12y^5 + 10y^4 - 42y^3 = 2y^3(6y^2 + 5y - 21)$

$6 \cdot -21 = -126$
$14 \cdot -9 = -126$
$14 + (-9) = 5$

$$\begin{aligned} &12y^5 + 10y^4 - 42y^3 \\ &= 2y^3(6y^2 + 5y - 21) \\ &= 2y^3(6y^2 + 14y - 9y - 21) \\ &= 2y^3[2y(3y+7) - 3(3y+7)] \\ &= 2y^3(3y+7)(2y-3) \end{aligned}$$

Vocabulary, Readiness & Video Check 4.4

1. $a = 1, b = 6, c = 8$
$a \cdot c = 1 \cdot 8 = 8$
$4 \cdot 2 = 8$ and $4 + 2 = 6$; choice **a**.

2. $a = 1, b = 11, c = 24$
$a \cdot c = 1 \cdot 24 = 24$
$8 \cdot 3 = 24$ and $8 + 3 = 11$; choice **c**.

3. $a = 2, b = 13, c = 6$
$a \cdot c = 2 \cdot 6 = 12$
$12 \cdot 1 = 12$ and $12 + 1 = 13$; choice **b**.

4. $a = 4, b = 8, c = 3$
$a \cdot c = 4 \cdot 3 = 12$
$2 \cdot 6 = 12$ and $2 + 6 = 8$; choice **d**.

5. This gives us a four-term polynomial, which may be factored by grouping.

Exercise Set 4.4

1. $$\begin{aligned} x^2 + 3x + 2x + 6 &= x(x+3) + 2(x+3) \\ &= (x+3)(x+2) \end{aligned}$$

3. $$\begin{aligned} y^2 + 8y - 2y - 16 &= y(y+8) - 2(y+8) \\ &= (y+8)(y-2) \end{aligned}$$

5. $$\begin{aligned} 8x^2 - 5x - 24x + 15 &= x(8x-5) - 3(8x-5) \\ &= (8x-5)(x-3) \end{aligned}$$

7. $$\begin{aligned} 5x^4 - 3x^2 + 25x^2 - 15 &= x^2(5x^2 - 3) + 5(5x^2 - 3) \\ &= (5x^2 - 3)(x^2 + 5) \end{aligned}$$

9. a. $9 \cdot 2 = 18$
$9 + 2 = 11$
9 and 2 are numbers whose product is 18 and whose sum is 11.

b. $11x = 9x + 2x$

c. $$\begin{aligned} 6x^2 + 11x + 3 &= 6x^2 + 9x + 2x + 3 \\ &= 3x(2x+3) + 1(2x+3) \\ &= (2x+3)(3x+1) \end{aligned}$$

11. a. $-3 \cdot -20 = 60$
$-3 + (-20) = -23$
-3 and -20 are numbers whose product is 60 and whose sum is -23.

b. $-23x = -3x - 20x$

c. $$\begin{aligned} 15x^2 - 23x + 4 &= 15x^2 - 3x - 20x + 4 \\ &= 3x(5x-1) - 4(5x-1) \\ &= (5x-1)(3x-4) \end{aligned}$$

13. $21 \cdot 2 = 42$
$14 \cdot 3 = 42$
$14 + 3 = 17$

$$\begin{aligned} 21y^2 + 17y + 2 &= 21y^2 + 14y + 3y + 2 \\ &= 7y(3y+2) + 1(3y+2) \\ &= (3y+2)(7y+1) \end{aligned}$$

15. $7 \cdot -11 = -77$
$-11 \cdot 7 = -77$
$-11 + 7 = -4$

$$\begin{aligned} 7x^2 - 4x - 11 &= 7x^2 - 11x + 7x - 11 \\ &= x(7x-11) + 1(7x-11) \\ &= (7x-11)(x+1) \end{aligned}$$

17. $10 \cdot 2 = 20$
$-4 \cdot -5 = 20$
$-4 + (-5) = -9$

$$\begin{aligned} 10x^2 - 9x + 2 &= 10x^2 - 4x - 5x + 2 \\ &= 2x(5x-2) - 1(5x-2) \\ &= (5x-2)(2x-1) \end{aligned}$$

19. $2 \cdot 5 = 10$
$-5 \cdot -2 = 10$
$-5 + (-2) = -7$
$$\begin{aligned}2x^2 - 7x + 5 &= 2x^2 - 5x - 2x + 5\\ &= x(2x-5) - 1(2x-5)\\ &= (2x-5)(x-1)\end{aligned}$$

21. $12x + 4x^2 + 9 = 4x^2 + 12x + 9$
$4 \cdot 9 = 36$
$6 \cdot 6 = 36$
$6 + 6 = 12$
$$\begin{aligned}4x^2 + 12x + 9 &= 4x^2 + 6x + 6x + 9\\ &= 2x(2x+3) + 3(2x+3)\\ &= (2x+3)(2x+3) \text{ or } (2x+3)^2\end{aligned}$$

23. $4 \cdot -21 = -84$
$6 \cdot -14 = -84$
$6 + (-14) = -8$
$$\begin{aligned}4x^2 - 8x - 21 &= 4x^2 + 6x - 14x - 21\\ &= 2x(2x+3) - 7(2x+3)\\ &= (2x+3)(2x-7)\end{aligned}$$

25. $10 \cdot 12 = 120$
$-8 \cdot -15 = 120$
$-8 + (-15) = -23$
$$\begin{aligned}10x^2 - 23x + 12 &= 10x^2 - 8x - 15x + 12\\ &= 2x(5x-4) - 3(5x-4)\\ &= (5x-4)(2x-3)\end{aligned}$$

27. $2x^3 + 13x^2 + 15x = x(2x^2 + 13x + 15)$
$2 \cdot 15 = 30$
$3 \cdot 10 = 30$
$3 + 10 = 13$
$$\begin{aligned}2x^3 + 13x^2 + 15x &= x(2x^2 + 13x + 15)\\ &= x(2x^2 + 3x + 10x + 15)\\ &= x[x(2x+3) + 5(2x+3)]\\ &= x(2x+3)(x+5)\end{aligned}$$

29. $16y^2 - 34y + 18 = 2(8y^2 - 17y + 9)$
$8 \cdot 9 = 72$
$-9 \cdot -8 = 72$
$-9 + (-8) = -17$
$$\begin{aligned}16y^2 - 34y + 18 &= 2(8y^2 - 17y + 9)\\ &= 2(8y^2 - 9y - 8y + 9)\\ &= 2[y(8y-9) - 1(8y-9)]\\ &= 2(8y-9)(y-1)\end{aligned}$$

31. $-13x + 6 + 6x^2 = 6x^2 - 13x + 6$
$6 \cdot 6 = 36$
$-9 \cdot -4 = 36$
$-9 + (-4) = -13$
$$\begin{aligned}6x^2 - 13x + 6 &= 6x^2 - 9x - 4x + 6\\ &= 3x(2x-3) - 2(2x-3)\\ &= (2x-3)(3x-2)\end{aligned}$$

33. $54a^2 - 9a - 30 = 3(18a^2 - 3a - 10)$
$18 \cdot -10 = -180$
$12 \cdot -15 = -180$
$12 + (-15) = -3$
$$\begin{aligned}54a^2 - 9a - 30 &= 3(18a^2 - 3a - 10)\\ &= 3(18a^2 + 12a - 15a - 10)\\ &= 3[6a(3a+2) - 5(3a+2)]\\ &= 3(3a+2)(6a-5)\end{aligned}$$

35. $20a^3 + 37a^2 + 8a = a(20a^2 + 37a + 8)$
$20 \cdot 8 = 160$
$5 \cdot 32 = 160$
$5 + 32 = 37$
$$\begin{aligned}20a^3 + 37a^2 + 8a &= a(20a^2 + 37a + 8)\\ &= a(20a^2 + 5a + 32a + 8)\\ &= a[5a(4a+1) + 8(4a+1)]\\ &= a(4a+1)(5a+8)\end{aligned}$$

37. $12x^3 - 27x^2 - 27x = 3x(4x^2 - 9x - 9)$
$4 \cdot -9 = -36$
$3 \cdot -12 = -36$
$3 + (-12) = -9$
$$\begin{aligned}12x^3 - 27x^2 - 27x &= 3x(4x^2 - 9x - 9)\\ &= 3x(4x^2 + 3x - 12x - 9)\\ &= 3x[x(4x+3) - 3(4x+3)]\\ &= 3x(4x+3)(x-3)\end{aligned}$$

39. $3x^2y + 4xy^2 + y^3 = y(3x^2 + 4xy + y^2)$
$3 \cdot y^2 = 3y^2$
$y \cdot 3y = 3y^2$
$y + 3y = 4y$
$$\begin{aligned}3x^2y + 4xy^2 + y^3 &= y(3x^2 + 4xy + y^2)\\ &= y(3x^2 + xy + 3xy + y^2)\\ &= y[x(3x+y) + y(3x+y)]\\ &= y(3x+y)(x+y)\end{aligned}$$

41. $20 \cdot 1 = 20$
There are no factors of 20 which sum to 7, so $20z^2 + 7z + 1$ is prime.

43. $24a^2 - 6ab - 30b^2 = 6(4a^2 - ab - 5b^2)$
$4 \cdot -5b^2 = -20b^2$
$4b \cdot -5b = -20b^2$
$4b + (-5b) = -b$
$$\begin{aligned} 24a^2 - 6ab - 30b^2 &= 6(4a^2 - ab - 5b^2) \\ &= 6(4a^2 + 4ab - 5ab - 5b^2) \\ &= 6[4a(a+b) - 5b(a+b)] \\ &= 6(a+b)(4a-5b) \end{aligned}$$

45. $15p^4 + 31p^3q + 2p^2q^2$
$= p^2(15p^2 + 31pq + 2q^2)$
$15 \cdot 2q^2 = 30q^2$
$q \cdot 30q = 30q^2$
$q + 30q = 31q$
$15p^4 + 31p^3q + 2p^2q^2$
$= p^2(15p^2 + 31pq + 2q^2)$
$= p^2(15p^2 + pq + 30pq + 2q^2)$
$= p^2[p(15p+q) + 2q(15p+q)]$
$= p^2(15p+q)(p+2q)$

47. $35 + 12x + x^2 = x^2 + 12x + 35$
$1 \cdot 35 = 35$
$7 \cdot 5 = 35$
$7 + 5 = 12$
$$\begin{aligned} x^2 + 12x + 35 &= x^2 + 7x + 5x + 35 \\ &= x(x+7) + 5(x+7) \\ &= (x+7)(x+5) \text{ or } (7+x)(5+x) \end{aligned}$$

49. $6 - 11x + 5x^2 = 5x^2 - 11x + 6$
$5 \cdot 6 = 30$
$-6 \cdot -5 = 30$
$-6 + (-5) = -11$
$$\begin{aligned} 5x^2 - 11x + 6 &= 5x^2 - 6x - 5x + 6 \\ &= x(5x-6) - 1(5x-6) \\ &= (5x-6)(x-1) \text{ or } (6-5x)(1-x) \end{aligned}$$

51. $(x-2)(x+2) = x^2 - 2^2 = x^2 - 4$

53. $$\begin{aligned} (y+4)(y+4) &= (y+4)^2 \\ &= (y)^2 + 2(y)(4) + (4)^2 \\ &= y^2 + 8y + 16 \end{aligned}$$

55. $(9z+5)(9z-5) = (9z)^2 - 5^2 = 81z^2 - 25$

57. $$\begin{aligned} (4x-3)^2 &= (4x)^2 - 2(4x)(3) + (3)^2 \\ &= 16x^2 - 24x + 9 \end{aligned}$$

59. $5(2x^2 + 9x + 9) = 10x^2 + 45x + 45$
$2 \cdot 9 = 18$
$3 \cdot 6 = 18$
$3 + 6 = 9$
$$\begin{aligned} 5(2x^2 + 9x + 9) &= 5(2x^2 + 3x + 6x + 9) \\ &= 5[x(2x+3) + 3(2x+3)] \\ &= 5(2x+3)(x+3) \end{aligned}$$
The perimeter is $10x^2 + 45x + 45$ or $5(2x + 3)(x + 3)$.

61. $$\begin{aligned} x^{2n} + 2x^n + 3x^n + 6 &= x^n(x^n + 2) + 3(x^n + 2) \\ &= (x^n + 2)(x^n + 3) \end{aligned}$$

63. $3 \cdot -35 = -105$
$-5 \cdot 21 = -105$
$-5 + 21 = 16$
$$\begin{aligned} 3x^{2n} + 16x^n - 35 &= 3x^{2n} - 5x^n + 21x^n - 35 \\ &= x^n(3x^n - 5) + 7(3x^n - 5) \\ &= (3x^n - 5)(x^n + 7) \end{aligned}$$

65. answers may vary

Section 4.5 Practice Exercises

1. a. Since $36 = 6^2$ and $12x = 2 \cdot 6 \cdot x$, $x^2 + 12x + 36$ is a perfect square trinomial.

b. Since $100 = 10^2$ and $20x = 2 \cdot 10 \cdot x$, $x^2 + 20x + 100$ is a perfect square trinomial.

2. a. Since $9x^2 = (3x)^2$ and $25 = 5^2$, but $20x \neq 2 \cdot 3x \cdot 5$, the polynomial is not a perfect square trinomial.

b. Since $4x^2 = (2x)^2$, but 11 is not a perfect square, the polynomial is not a perfect square trinomial.

3. a. Since $25x^2 = (5x)^2$ and $1 = 1^2$, and $2 \cdot 5x \cdot 1 = 10x$ which is the opposite of $-10x$, the trinomial is a perfect square trinomial.

b. Since $9x^2 = (3x)^2$ and $49 = 7^2$, and $2 \cdot 3x \cdot 7 = 42x$ which is the opposite of $-42x$, the trinomial is a perfect square trinomial.

4. $$\begin{aligned} x^2 + 16x + 64 &= (x)^2 + 2 \cdot x \cdot 8 + 8^2 \\ &= (x+8)^2 \end{aligned}$$

5. $$\begin{aligned} 9r^2 + 24rs + 16s^2 &= (3r)^2 + 2 \cdot 3r \cdot 4s + (4s)^2 \\ &= (3r+4s)^2 \end{aligned}$$

6. $$\begin{aligned} 9n^4 - 6n^2 + 1 &= (3n^2)^2 - 2 \cdot 3n^2 \cdot 1 + 1^2 \\ &= (3n^2 - 1)^2 \end{aligned}$$

7. Notice that this trinomial is not a perfect square trinomial.
$9x^2 = (3x)^2$ and $4 = 2^2$, but $2 \cdot 3x \cdot 2 = 12x$ and $12x$ is not the middle term $15x$.
Factor by grouping.
$9 \cdot 4 = 36$
$3 \cdot 12 = 36$
$3 + 12 = 15$
$$\begin{aligned} 9x^2 + 15x + 4 &= 9x^2 + 3x + 12x + 4 \\ &= 3x(3x+1) + 4(3x+1) \\ &= (3x+1)(3x+4) \end{aligned}$$

8. a. $$\begin{aligned} 8n^2 + 40n + 50 &= 2(4n^2 + 20n + 25) \\ &= 2[(2n)^2 + 2 \cdot 2n \cdot 5 + 5^2] \\ &= 2(2n+5)^2 \end{aligned}$$

b. $$\begin{aligned} &12x^3 - 84x^2 + 147x \\ &= 3x(4x^2 - 28x + 49) \\ &= 3x[(2x)^2 - 2 \cdot 2x \cdot 7 + 7^2] \\ &= 3x(2x-7)^2 \end{aligned}$$

9. $x^2 - 9 = x^2 - 3^2 = (x-3)(x+3)$

10. $a^2 - 16 = a^2 - 4^2 = (a-4)(a+4)$

11. $c^2 - \frac{9}{25} = c^2 - \left(\frac{3}{5}\right)^2 = \left(c - \frac{3}{5}\right)\left(c + \frac{3}{5}\right)$

12. $s^2 + 9$ is prime since it is the sum of two squares.

13. $9s^2 - 1 = (3s)^2 - 1^2 = (3s-1)(3s+1)$

14. $$\begin{aligned} 16x^2 - 49y^2 &= (4x)^2 - (7y)^2 \\ &= (4x-7y)(4x+7y) \end{aligned}$$

15. $$\begin{aligned} p^4 - 81 &= (p^2)^2 - 9^2 \\ &= (p^2+9)(p^2-9) \\ &= (p^2+9)(p+3)(p-3) \end{aligned}$$

16. $$\begin{aligned} 9x^3 - 25x &= x(9x^2 - 25) \\ &= x[(3x)^2 - 5^2] \\ &= x(3x-5)(3x+5) \end{aligned}$$

17. $$\begin{aligned} 48x^4 - 3 &= 3(16x^4 - 1) \\ &= 3[(4x^2)^2 - 1^2] \\ &= 3(4x^2+1)(4x^2-1) \\ &= 3(4x^2+1)(2x+1)(2x-1) \end{aligned}$$

18. $$\begin{aligned} -9x^2 + 100 &= -1(9x^2 - 100) \\ &= -1[(3x)^2 - 10^2] \\ &= -1(3x-10)(3x+10) \end{aligned}$$

19. $121 - m^2 = 11^2 - m^2 = (11+m)(11-m)$
or
$$\begin{aligned} 121 - m^2 &= -m^2 + 121 \\ &= -1(m^2 - 121) \\ &= -1(m^2 - 11^2) \\ &= -1(m+11)(m-11) \end{aligned}$$

Calculator Explorations

	$x^2 - 2x + 1$	$x^2 - 2x - 1$	$(x-1)^2$
$x = 5$	16	14	16
$x = -3$	16	14	16
$x = 2.7$	2.89	0.89	2.89
$x = -12.1$	171.61	169.61	171.61
$x = 0$	1	1	1

Vocabulary, Readiness & Video Check 4.5

1. A perfect square trinomial is a trinomial that is the square of a binomial.

2. The term $25y^2$ written as a square is $(5y)^2$.

3. The expression $x^2+10x+25y^2$ is called a perfect square trinomial.

4. The expression x^2-49 is called a difference of two squares.

5. The factorization $(x+5y)(x+5y)$ may also be written as $(x+5y)^2$.

6. The factorization $(x-5y)(x+5y)$ may also be written as $(x-5y)^2$. false

7. The trinomial x^2-6x-9 is a perfect square trinomial. false

8. The binomial y^2+9 factors as $(y+3)^2$. false

9. $64=8^2$

10. $9=3^2$

11. $121a^2=(11a)^2$

12. $81b^2=(9b)^2$

13. $36p^4=(6p^2)^2$

14. $4q^4=(2q^2)^2$

15. No, it just means it won't factor into a binomial squared. It may or may not be factorable.

16. The first and last terms are squares, a^2 and b^2, and the middle term is $2\cdot a\cdot b$ or $-2\cdot a\cdot b$.

17. In order to recognize the binomial as a difference of squares and also to identify the terms to use in the special factoring formula.

18. A prime polynomial is one that can't be factored further (using rational real numbers).

Exercise Set 4.5

1. Since $64=8^2$ and $16x=2\cdot 8\cdot x$, $x^2+16x+64$ is a perfect square trinomial.

3. Since $25=5^2$ but $5y\neq 2\cdot 5\cdot y$, $y^2+5y+25$ is not a perfect square trinomial.

5. Since $1=1^2$ and $-2m=-2\cdot 1\cdot m$, m^2-2m+1 is a perfect square trinomial.

7. Since $49=7^2$ but $16a\neq 2\cdot 7\cdot a$, $a^2-16a+49$ is not a perfect square trinomial.

9. $4x^2=(2x)^2$ but $8y^2$ is not a perfect square, so $4x^2+12xy+8y^2$ is not a perfect square trinomial.

11. $25a^2=(5a)^2$, $16b^2=(4b)^2$, and $40ab=2\cdot 5a\cdot 4b$, so $25a^2-40ab+16b^2$ is a perfect square trinomial.

13. $$\begin{aligned}x^2+22x+121&=x^2+2\cdot x\cdot 11+11^2\\&=(x+11)^2\end{aligned}$$

15. $x^2-16x+64=x^2-2\cdot x\cdot 8+8^2=(x-8)^2$

17. $$\begin{aligned}16a^2-24a+9&=(4a)^2-2\cdot 4a\cdot 3+3^2\\&=(4a-3)^2\end{aligned}$$

19. $x^4+4x^2+4=(x^2)^2+2\cdot x^2\cdot 2+2^2=(x^2+2)^2$

21. $$\begin{aligned}2n^2-28n+98&=2(n^2-14n+49)\\&=2[n^2-2\cdot n\cdot 7+7^2]\\&=2(n-7)^2\end{aligned}$$

23. $$\begin{aligned}16y^2+40y+25&=(4y)^2+2\cdot 4y\cdot 5+5^2\\&=(4y+5)^2\end{aligned}$$

25. $$\begin{aligned}x^2y^2-10xy+25&=(xy)^2-2\cdot xy\cdot 5+5^2\\&=(xy-5)^2\end{aligned}$$

27. $$\begin{aligned}m^3+18m^2+81m&=m(m^2+18m+81)\\&=m(m^2+2\cdot m\cdot 9+9^2)\\&=m(m+9)^2\end{aligned}$$

29. Since $1 = 1^2$ and $x^4 = (x^2)^2$, but $6x^2 \neq 2 \cdot 1 \cdot x^2$, the polynomial is not a perfect square trinomial. Since there are no factors of $1 \cdot 1 = 1$ that sum to 6, the trinomial is prime.

31. $9x^2 - 24xy + 16y^2 = (3x)^2 - 2 \cdot 3x \cdot 4y + (4y)^2$
$= (3x - 4y)^2$

33. $x^2 - 4 = x^2 - (2)^2 = (x+2)(x-2)$

35. $81 - p^2 = 9^2 - p^2 = (9+p)(9-p)$

or

$81 - p^2 = -p^2 + 81$
$= -1(p^2 - 81)$
$= -1(p - 9^2)$
$= -1(p+9)(p-9)$

37. $-4r^2 + 1 = -1(4r^2 - 1)$
$= -1[(2r)^2 - 1^2]$
$= -1(2r+1)(2r-1)$

39. $9x^2 - 16 = (3x)^2 - 4^2 = (3x+4)(3x-4)$

41. $16r^2 + 1$ is the sum of two squares, which is prime.

43. $-36 + x^2 = -1(36 - x^2)$
$= -1(6^2 - x^2)$
$= -1(6+x)(6-x)$

or

$-36 + x^2 = x^2 - 36$
$= x^2 - 6^2$
$= (x-6)(x+6)$

45. $m^4 - 1 = (m^2)^2 - 1^2$
$= (m^2 + 1)(m^2 - 1)$
$= (m^2 + 1)(m^2 - 1^2)$
$= (m^2 + 1)(m+1)(m-1)$

47. $x^2 - 169y^2 = x^2 - (13y)^2 = (x^2 + 13y)(x - 13y)$

49. $18r^2 - 8 = 2(9r^2 - 4)$
$= 2[(3r)^2 - 2^2]$
$= 2(3r+2)(3r-2)$

51. $9xy^2 - 4x = x(9y^2 - 4)$
$= x[(3y)^2 - 2^2]$
$= x(3y+2)(3y-2)$

53. $16x^4 - 64x^2 = 16x^2(x^2 - 4)$
$= 16x^2(x^2 - 2^2)$
$= 16x^2(x+2)(x-2)$

55. $xy^3 - 9xyz^2 = xy(y^2 - 9z^2)$
$= xy[y^2 - (3z)^2]$
$= xy(y - 3z)(y + 3z)$

57. $36x^2 - 64y^2 = 4(9x^2 - 16y^2)$
$= 4[(3x)^2 - (4y)^2]$
$= 4(3x+4y)(3x-4y)$

59. $144 - 81x^2 = 9(16 - 9x^2)$
$= 9[4^2 - (3x)^2]$
$= 9(4-3x)(4+3x)$

61. $25y^2 - 9 = (5y)^2 - 3^2 = (5y+3)(5y-3)$

63. $121m^2 - 100n^2 = (11m)^2 - (10n)^2$
$= (11m + 10n)(11m - 10n)$

65. $x^2y^2 - 1 = (xy)^2 - 1^2 = (xy+1)(xy-1)$

67. $x^2 - \frac{1}{4} = x^2 - \left(\frac{1}{2}\right)^2 = \left(x - \frac{1}{2}\right)\left(x + \frac{1}{2}\right)$

69. $49 - \frac{9}{25}m^2 = 7^2 - \left(\frac{3}{5}m\right)^2$
$= \left(7 + \frac{3}{5}m\right)\left(7 - \frac{3}{5}m\right)$

71. $81a^2 - 25b^2 = (9a)^2 - (5b)^2 = (9a+5b)(9a-5b)$

73. $x^2 + 14xy + 49y^2 = x^2 + 2 \cdot x \cdot 7y + (7y)^2$
$= (x+7y)^2$

75. $32n^4 - 112n^2 + 98 = 2(16n^4 - 56n^2 + 49)$
$= 2[(4n^2)^2 - 2 \cdot 4n^2 \cdot 7 + 7^2]$
$= 2(4n^2 - 7)^2$

77. $x^6 - 81x^2 = x^2(x^4 - 81)$
$= x^2[(x^2)^2 - 9^2]$
$= x^2(x^2 + 9)(x^2 - 9)$
$= x^2(x^2 + 9)[(x)^2 - 3^2]$
$= x^2(x^2 + 9)(x + 3)(x - 3)$

79. $64p^3q - 81pq^3 = pq(64p^2 - 81q^2)$
$= pq[(8p)^2 - (9q)^2]$
$= pq(8p - 9q)(8p + 9q)$

81. $x - 6 = 0$
$x - 6 + 6 = 0 + 6$
$x = 6$

83. $2m + 4 = 0$
$2m + 4 - 4 = 0 - 4$
$2m = -4$
$\frac{2m}{2} = \frac{-4}{2}$
$m = -2$

85. $5z - 1 = 0$
$5z - 1 + 1 = 0 + 1$
$5z = 1$
$\frac{5z}{5} = \frac{1}{5}$
$z = \frac{1}{5}$

87. $x^2 - \frac{2}{3}x + \frac{1}{9} = x^2 - 2 \cdot x \cdot \frac{1}{3} + \left(\frac{1}{3}\right)^2 = \left(x - \frac{1}{3}\right)^2$

89. $(x + 2)^2 - y^2 = [(x + 2) + y][(x + 2) - y]$
$= (x + 2 + y)(x + 2 - y)$

91. $a^2(b - 4) - 16(b - 4) = (b - 4)(a^2 - 16)$
$= (b - 4)(a^2 - 4^2)$
$= (b - 4)(a - 4)(a + 4)$

93. $(x^2 + 6x + 9) - 4y^2$
$= (x^2 + 2 \cdot x \cdot 3 + 3^2) - 4y^2$
$= (x + 3)^2 - 4y^2$
$= (x + 3)^2 - (2y)^2$
$= [(x + 3) + 2y][(x + 3) - 2y]$
$= (x + 3 + 2y)(x + 3 - 2y)$

95. $x^{2n} - 100 = (x^n)^2 - 10^2 = (x^n + 10)(x^n - 10)$

97. $x^2 = x^2$ and $16 = 4^2$.
$(x + 4)^2 = x^2 + 2 \cdot x \cdot 4 + 4^2 = x^2 + 8x + 16$, so the number 8 makes the given expression a perfect square trinomial.

99. answers may vary

101. The difference of two squares is the result of multiplying binomials of the form $(x + a)$ and $(x - a)$, so multiplying $(x - 6)$ by $(x + 6)$ results in the difference of two squares.

103. $(a + b)^2 = a^2 + 2 \cdot a \cdot b + b^2 = a^2 + 2ab + b^2$

105. a. $2704 - 16t^2 = 2704 - 16(3)^2$
$= 2704 - 16 \cdot 9$
$= 2704 - 144$
$= 2560$
After 3 seconds, the filter is 2560 feet above the river.

b. $2704 - 16t^2 = 2704 - 16(7)^2$
$= 2704 - 16 \cdot 49$
$= 2704 - 784$
$= 1920$
After 7 seconds, the filter is 1920 feet above the river.

c. The filter lands in the river when its height is 0 feet.
$2704 - 16t^2 = 0$
$(52 + 4t)(52 - 4t) = 0$
$52 + 4t = 0$ or $52 - 4t = 0$
$4t = -52$ $\quad -4t = -52$
$t = -13$ $\quad t = 13$
Discard $t = -13$ since time cannot be negative. The filter lands in the river after 13 seconds.

d. $2704 - 16t^2 = 16(169 - t^2)$
$= 16[(13)^2 - t^2]$
$= 16(13 - t)(13 + t)$

107. a. $2304-16t^2 = 2304-16(3)^2$
$= 2304-16\cdot 9$
$= 2304-144$
$= 2160$
After 3 seconds, the height of the bolt is 2160 feet.

b. $2304-16t^2 = 2304-16(7)^2$
$= 2304-16(49)$
$= 2304-784$
$= 1520$
After 7 seconds, the height of the bolt is 1520 feet.

c. The bolt hits the ground when its height is 0 feet.
$2304-16t^2 = 0$
$(48-4t)(48+4t) = 0$
$48-4t = 0$ or $48+4t = 0$
$-4t = -48$ $\quad 4t = -48$
$t = 12$ $\quad t = -12$
Discard $t = -12$ since time cannot be negative. The bolt hits the ground after 12 seconds.

d. $2304-16t^2 = 16(144-t^2)$
$= 16(12^2-t^2)$
$= 16(12+t)(12-t)$

Integrated Review

1–76. Factoring methods may vary.

1. $x^2+x-12 = (x-3)(x+4)$

2. $x^2-10x+16 = (x-8)(x-2)$

3. $x^2+2x+1 = x^2+2\cdot x\cdot 1+1^2$
$= (x+1)^2$

4. $x^2-6x+9 = x^2-2\cdot x\cdot 3+3^2$
$= (x-3)^2$

5. $x^2-x-6 = (x+2)(x-3)$

6. $x^2+x-2 = (x+2)(x-1)$

7. $x^2+x-6 = (x+3)(x-2)$

8. $x^2+7x+12 = (x+3)(x+4)$

9. $x^2-7x+10 = (x-5)(x-2)$

10. $x^2-x-30 = (x-6)(x+5)$

11. $2x^2-98 = 2(x^2-49)$
$= 2(x^2-7^2)$
$= 2(x-7)(x+7)$

12. $3x^2-75 = 3(x^2-25)$
$= 3(x^2-5^2)$
$= 3(x-5)(x+5)$

13. $x^2+3x+5x+15 = x(x+3)+5(x+3)$
$= (x+3)(x+5)$

14. $3y-21+xy-7x = 3(y-7)+x(y-7)$
$= (y-7)(3+x)$

15. $x^2+6x-16 = (x+8)(x-2)$

16. $x^2-3x-28 = (x-7)(x+4)$

17. $4x^3+20x^2-56x = 4x(x^2+5x-14)$
$= 4x(x+7)(x-2)$

18. $6x^3-6x^2-120x = 6x(x^2-x-20)$
$= 6x(x-5)(x+4)$

19. $12x^2+34x+24 = 2(6x^2+17x+12)$
$= 2(3x+4)(2x+3)$

20. $24a^2+18ab-15b^2 = 3(8a^2+6ab-5b^2)$
$= 3(2a-b)(4a+5b)$

21. $4a^2-b^2 = (2a)^2-b^2 = (2a+b)(2a-b)$

22. $x^2-25y^2 = x^2-(5y)^2 = (x+5y)(x-5y)$

23. $28-13x-6x^2 = (4-3x)(7+2x)$

24. $20-3x-2x^2 = (5-2x)(4+x)$

25. $4-2x+x^2$ is prime.

26. $a+a^2-3$ is prime.

27. $6y^2+y-15=(3y+5)(2y-3)$

28. $4x^2-x-5=(4x-5)(x+1)$

29. $18x^3-63x^2+9x=9x(2x^2-7x+1)$

30. $12a^3-24a^2+4a=4a(3a^2-6a+1)$

31. $16a^2-56a+49=(4a)^2-2\cdot 4a\cdot 7+7^2$
$=(4a-7)^2$

32. $25p^2-70p+49=(5p)^2-2\cdot 5p\cdot 7+7^2$
$=(5p-7)^2$

33. $14+5x-x^2=(7-x)(2+x)$

34. $3-2x-x^2=(3+x)(1-x)$

35. $3x^4y+6x^3y-72x^2y=3x^2y(x^2+2x-24)$
$=3x^2y(x+6)(x-4)$

36. $2x^3y+8x^2y^2-10xy^3=2xy(x^2+4xy-5y^2)$
$=2xy(x+5y)(x-y)$

37. $12x^3y+243xy=3xy(4x^2+81)$

38. $6x^3y^2+8xy^2=2xy^2(3x^2+4)$

39. $2xy-72x^3y=2xy(1-36x^2)$
$=2xy[1-(6x)^2]$
$=2xy(1-6x)(1+6x)$

40. $2x^3-18x=2x(x^2-9)$
$=2x(x^2-3^2)$
$=2x(x-3)(x+3)$

41. $x^3+6x^2-4x-24=x^2(x+6)-4(x+6)$
$=(x+6)(x^2-4)$
$=(x+6)(x+2)(x-2)$

42. $x^3-2x^2-36x+72=x^2(x-2)-36(x-2)$
$=(x-2)(x^2-36)$
$=(x-2)(x-6)(x+6)$

43. $6a^3+10a^2=2a^2(3a+5)$

44. $4n^2-6n=2n(2n-3)$

45. $3x^3-x^2+12x-4=x^2(3x-1)+4(3x-1)$
$=(3x-1)(x^2+4)$

46. $x^3-2x^2+3x-6=x^2(x-2)+3(x-2)$
$=(x-2)(x^2+3)$

47. $6x^2+18xy+12y^2=6(x^2+3xy+2y^2)$
$=6(x+2y)(x+y)$

48. $12x^2+46xy-8y^2=2(6x^2+23xy-4y^2)$
$=2(x+4y)(6x-y)$

49. $5(x+y)+x(x+y)=(x+y)(5+x)$

50. $7(x-y)+y(x-y)=(x-y)(7+y)$

51. $14t^2-9t+1=(7t-1)(2t-1)$

52. $3t^2-5t+1$ is prime.

53. $-3x^2-2x+5=-1(3x^2+2x-5)$
$=-1(3x+5)(x-1)$

54. $-7x^2-19x+6=-1(7x^2+19x-6)$
$=-1(7x-2)(x+3)$

55. $1-8a-20a^2=(1-10a)(1+2a)$

56. $1-7a-60a^2=(1+5a)(1-12a)$

57. $x^4-10x^2+9=(x^2-9)(x^2-1)$
$=(x-3)(x+3)(x-1)(x+1)$

58. $x^4-13x^2+36=(x^2-9)(x^2-4)$
$=(x-3)(x+3)(x-2)(x+2)$

59. $x^2-23x+120=(x-15)(x-8)$

60. $y^2+22y+96=(y+16)(y+6)$

61. $25p^2-70pq+49q^2=(5p)^2-2\cdot 5p\cdot 7q+(7q)^2$
$=(5p-7q)^2$

62. $16a^2-56ab+49b^2=(4a)^2-2\cdot 4a\cdot 7b+(7b)^2$
$=(4a-7b)^2$

63. $x^2-14x-48$ is prime.

64. $7x^2+24xy+9y^2=(7x+3y)(x+3y)$

65. $-x^2-x+30=-1(x^2+x-30)=-1(x-5)(x+6)$

66. $-x^2+6x-8=-1(x^2-6x+8)$
$=-1(x-2)(x-4)$

67. $3rs-s+12r-4=s(3r-1)+4(3r-1)$
$=(3r-1)(s+4)$

68. $x^3-2x^2+x-2=x^2(x-2)+1(x-2)$
$=(x-2)(x^2+1)$

69. $4x^2-8xy-3x+6y=4x(x-2y)-3(x-2y)$
$=(x-2y)(4x-3)$

70. $4x^2-2xy-7yz+14xz=2x(2x-y)-7z(y-2x)$
$=2x(2x-y)+7z(2x-y)$
$=(2x-y)(2x+7z)$

71. $x^2+9xy-36y^2=(x+12y)(x-3y)$

72. $3x^2+10xy-8y^2=(3x-2y)(x+4y)$

73. $x^4-14x^2-32=(x^2+2)(x^2-16)$
$=(x^2+2)(x+4)(x-4)$

74. $x^4-22x^2-75=(x^2+3)(x^2-25)$
$=(x^2+3)(x+5)(x-5)$

75. answers may vary

76. Yes; $9x^2+81y^2=9(x^2+9y^2)$

Section 4.6 Practice Exercises

1. $(x-7)(x+2)=0$
$x-7=0$ or $x+2=0$
$x=7$ $\quad x=-2$
The solutions are 7 and −2.

2. $(x-10)(3x+1)=0$
$x-10=0$ or $3x+1=0$
$x=10$ $\quad 3x=-1$
$x=-\frac{1}{3}$
The solutions are 10 and $-\frac{1}{3}$.

3. a. $y(y+3)=0$
$y=0$ or $y+3=0$
$y=-3$
The solutions are 0 and −3.

b. $x(4x-3)=0$
$x=0$ or $4x-3=0$
$4x=3$
$x=\frac{3}{4}$
The solutions are 0 and $\frac{3}{4}$.

4. $x^2-3x-18=0$
$(x-6)(x+3)=0$
$x-6=0$ or $x+3=0$
$x=6$ $\quad x=-3$
The solutions are 6 and −3.

5. $9x^2-24x=-16$
$9x^2-24x+16=0$
$(3x-4)(3x-4)=0$
$3x-4=0$ or $3x-4=0$
$3x=4$ $\quad 3x=4$
$x=\frac{4}{3}$ $\quad x=\frac{4}{3}$
The solution is $\frac{4}{3}$.

6. a.

$$x(x-4)=5$$
$$x^2-4x=5$$
$$x^2-4x-5=0$$
$$(x-5)(x+1)=0$$
$$x-5=0 \quad \text{or} \quad x+1=0$$
$$x=5 \qquad\qquad x=-1$$

The solutions are 5 and −1.

b.

$$x(3x+7)=6$$
$$3x^2+7x=6$$
$$3x^2+7x-6=0$$
$$(3x-2)(x+3)=0$$
$$3x-2=0 \quad \text{or} \quad x+3=0$$
$$3x=2 \qquad\qquad x=-3$$
$$x=\frac{2}{3}$$

The solutions are $\frac{2}{3}$ and −3.

7.

$$2x^3-18x=0$$
$$2x(x^2-9)=0$$
$$2x(x-3)(x+3)=0$$
$$2x=0 \quad \text{or} \quad x-3=0 \quad \text{or} \quad x+3=0$$
$$x=0 \qquad x=3 \qquad x=-3$$

The solutions are 0, 3, and −3.

8. $(x+3)(3x^2-20x-7)=0$

$$x+3=0 \quad \text{or} \quad 3x^2-20x-7=0$$
$$x=-3 \qquad (3x+1)(x-7)=0$$
$$3x+1=0 \quad \text{or} \quad x-7=0$$
$$3x=-1 \qquad x=7$$
$$x=-\frac{1}{3}$$

The solutions are −3, $-\frac{1}{3}$, and 7.

Vocabulary, Readiness & Video Check 4.6

1. An equation that can be written in the form $ax^2+bx+c=0$, (with $a \neq 0$), is called a quadratic equation.

2. If the product of two numbers is 0, then at least one of the numbers must be 0.

3. The solutions of $(x-3)(x+5)=0$ are 3, −5.

4. If $a \cdot b=0$, then $a=0$ or $b=0$.

5. One side of the equation must be a factored polynomial and the other side must be zero.

6. Because no matter how many factors you have in a multiplication problem, it's still true that for a product to be zero, at least one of the factors must be zero.

Exercise Set 4.6

1. $(x-2)(x+1)=0$

$$x-2=0 \quad \text{or} \quad x+1=0$$
$$x=2 \qquad x=-1$$

The solutions are 2 and −1.

3. $(x-6)(x-7)=0$

$$x-6=0 \quad \text{or} \quad x-7=0$$
$$x=6 \qquad x=7$$

The solutions are 6 and 7.

5. $(x+9)(x+17)=0$

$$x+9=0 \quad \text{or} \quad x+17=0$$
$$x=-9 \qquad x=-17$$

The solutions are −9 and −17.

7. $x(x+6)=0$

$$x=0 \quad \text{or} \quad x+6=0$$
$$x=-6$$

The solutions are 0 and −6.

9. $3x(x-8)=0$

$$3x=0 \quad \text{or} \quad x-8=0$$
$$x=0 \qquad x=8$$

The solutions are 0 and 8.

11. $(2x+3)(4x-5)=0$

$$2x+3=0 \quad \text{or} \quad 4x-5=0$$
$$2x=-3 \qquad 4x=5$$
$$x=-\frac{3}{2} \qquad x=\frac{5}{4}$$

The solutions are $-\frac{3}{2}$ and $\frac{5}{4}$.

13. $(2x-7)(7x+2)=0$

$$2x-7=0 \quad \text{or} \quad 7x+2=0$$
$$2x=7 \qquad 7x=-2$$
$$x=\frac{7}{2} \qquad x=-\frac{2}{7}$$

The solutions are $\frac{7}{2}$ and $-\frac{2}{7}$.

15. $\left(x-\frac{1}{2}\right)\left(x+\frac{1}{3}\right)=0$

$x-\frac{1}{2}=0$ or $x+\frac{1}{3}=0$

$x=\frac{1}{2}$ $\quad$ $x=-\frac{1}{3}$

The solutions are $\frac{1}{2}$ and $-\frac{1}{3}$.

17. $(x + 0.2)(x + 1.5) = 0$

$x+0.2=0$ or $x+1.5=0$

$x=-0.2$ $\quad$ $x=-1.5$

The solutions are −0.2 and −1.5.

19. $x^2-13x+36=0$

$(x-9)(x-4)=0$

$x-9=0$ or $x-4=0$

$x=9$ $\quad$ $x=4$

The solutions are 9 and 4.

21. $x^2+2x-8=0$

$(x+4)(x-2)=0$

$x+4=0$ or $x-2=0$

$x=-4$ $\quad$ $x=2$

The solutions are −4 and 2.

23. $x^2-7x=0$

$x(x-7)=0$

$x=0$ or $x-7=0$

$x=7$

The solutions are 0 and 7.

25. $x^2+20x=0$

$x(x+20)=0$

$x=0$ or $x+20=0$

$x=-20$

The solutions are 0 and −20.

27. $x^2=16$

$x^2-16=0$

$(x-4)(x+4)=0$

$x-4=0$ or $x+4=0$

$x=4$ $\quad$ $x=-4$

The solutions are 4 and −4.

29. $x^2-4x=32$

$x^2-4x-32=0$

$(x-8)(x+4)=0$

$x-8=0$ or $x+4=0$

$x=8$ $\quad$ $x=-4$

The solutions are 8 and −4.

31. $(x+4)(x-9)=4x$

$x^2-5x-36=4x$

$x^2-9x-36=0$

$(x+3)(x-12)=0$

$x+3=0$ or $x-12=0$

$x=-3$ $\quad$ $x=12$

The solutions are −3 and 12.

33. $x(3x-1)=14$

$3x^2-x=14$

$3x^2-x-14=0$

$3x^2-7x+6x-14=0$

$x(3x-7)+2(3x-7)=0$

$(3x-7)(x+2)=0$

$3x-7=0$ or $x+2=0$

$3x=7$ $\quad$ $x=-2$

$x=\frac{7}{3}$

The solutions are $\frac{7}{3}$ and −2.

35. $3x^2+19x-72=0$

$(3x-8)(x+9)=0$

$3x-8=0$ or $x+9=0$

$3x=8$ $\quad$ $x=-9$

$x=\frac{8}{3}$

The solutions are $\frac{8}{3}$ and −9.

37. $4x^3-x=0$

$x(4x^2-1)=0$

$x[(2x)^2-(1)^2]=0$

$x(2x+1)(2x-1)=0$

$x=0$ or $2x+1=0$ or $2x-1=0$
$2x=-1 \qquad 2x=1$
$x=-\frac{1}{2} \qquad x=\frac{1}{2}$

The solutions are 0, $-\frac{1}{2}$, and $\frac{1}{2}$.

39. $4(x-7)=6$
$4x-28=6$
$4x=34$
$x=\frac{34}{4}$
$x=\frac{17}{2}$

The solution is $\frac{17}{2}$.

41. $(4x-3)(16x^2-24x+9)=0$
$(4x-3)[(4x)^2-2\cdot 4x\cdot 3+(3)^2]=0$
$(4x-3)(4x-3)^2=0$
$(4x-3)(4x-3)(4x-3)=0$
$4x-3=0$
$4x=3$
$x=\frac{3}{4}$

The solution is $\frac{3}{4}$.

43. $4y^2-1=0$
$(2y-1)(2y+1)=0$
$2y-1=0$ or $2y+1=0$
$2y=1 \qquad 2y=-1$
$y=\frac{1}{2} \qquad y=-\frac{1}{2}$

The solutions are $\frac{1}{2}$ and $-\frac{1}{2}$.

45. $(2x+3)(2x^2-5x-3)=0$
$(2x+3)(2x^2+x-6x-3)=0$
$(2x+3)[x(2x+1)-3(2x+1)]=0$
$(2x+3)(2x+1)(x-3)=0$
$2x+3=0$ or $2x+1=0$ or $x-3=0$
$2x=-3 \qquad 2x=-1 \qquad x=3$
$x=-\frac{3}{2} \qquad x=-\frac{1}{2}$

The solutions are $-\frac{3}{2}$, $-\frac{1}{2}$, and 3.

47. $x^2-15=-2x$
$x^2+2x-15=0$
$(x+5)(x-3)=0$
$x+5=0$ or $x-3=0$
$x=-5 \qquad x=3$

The solutions are −5 and 3.

49. $30x^2-11x=30$
$30x^2-11x-30=0$
$30x^2+25x-36x-30=0$
$5x(6x+5)-6(6x+5)=0$
$(6x+5)(5x-6)=0$
$6x+5=0$ or $5x-6=0$
$6x=-5 \qquad 5x=6$
$x=-\frac{5}{6} \qquad x=\frac{6}{5}$

The solutions are $-\frac{5}{6}$ and $\frac{6}{5}$.

51. $5x^2-6x-8=0$
$(5x+4)(x-2)=0$
$5x+4=0$ or $x-2=0$
$5x=-4 \qquad x=2$
$x=-\frac{4}{5}$

The solutions are $-\frac{4}{5}$ and 2.

53. $6y^2-22y-40=0$
$2(3y^2-11y-20)=0$
$2(3y^2+4y-15y-20)=0$
$2[y(3y+4)-5(3y+4)]=0$
$2(3y+4)(y-5)=0$
$3y+4=0$ or $y-5=0$
$3y=-4 \qquad y=5$
$y=-\frac{4}{3}$

The solutions are $-\frac{4}{3}$ and 5.

55. $(y-2)(y+3)=6$
$y^2+y-6=6$
$y^2+y-12=0$
$(y+4)(y-3)=0$
$y+4=0$ or $y-3=0$
$y=-4 \qquad y=3$

The solutions are −4 and 3.

57. $x^3-12x^2+32x=0$
$x(x^2-12x+32)=0$
$x(x-8)(x-4)=0$
$x=0$ or $x-8=0$ or $x-4=0$
$x=8$ $\quad x=4$
The solutions are 0, 8, and 4.

59. $x^2+14x+49=0$
$(x+7)(x+7)=0$
$x+7=0$
$x=-7$
The solution is −7.

61. $12y=8y^2$
$0=8y^2-12y$
$0=4y(2y-3)$
$4y=0$ or $2y-3=0$
$y=0$ $\quad 2y=3$
$y=\frac{3}{2}$
The solutions are 0 and $\frac{3}{2}$.

63. $7x^3-7x=0$
$7x(x^2-1)=0$
$7x(x-1)(x+1)=0$
$7x=0$ or $x-1=0$ or $x+1=0$
$x=0$ $\quad x=1$ $\quad x=-1$
The solutions are 0, 1, and −1.

65. $3x^2+8x-11=13-6x$
$3x^2+14x-11=13$
$3x^2+14x-24=0$
$3x^2+18x-4x-24=0$
$3x(x+6)-4(x+6)=0$
$(x+6)(3x-4)=0$

$x+6=0$ or $3x-4=0$
$x=-6$ $\quad 3x=4$
$x=\frac{4}{3}$
The solutions are −6 and $\frac{4}{3}$.

67. $3x^2-20x=-4x^2-7x-6$
$7x^2-13x+6=0$
$(7x-6)(x-1)=0$
$7x-6=0$ or $x-1=0$
$7x=6$ $\quad x=1$
$x=\frac{6}{7}$
The solutions are $\frac{6}{7}$ and 1.

69. $\frac{3}{5}+\frac{4}{9}=\frac{3\cdot 9}{5\cdot 9}+\frac{4\cdot 5}{9\cdot 5}=\frac{27}{45}+\frac{20}{45}=\frac{47}{45}$

71. $\frac{7}{10}-\frac{5}{12}=\frac{7\cdot 6}{10\cdot 6}-\frac{5\cdot 5}{12\cdot 5}=\frac{42}{60}-\frac{25}{60}=\frac{17}{60}$

73. $\frac{4}{5}\cdot\frac{7}{8}=\frac{4\cdot 7}{5\cdot 8}=\frac{4\cdot 7}{5\cdot 4\cdot 2}=\frac{7}{5\cdot 2}=\frac{7}{10}$

75. The equation must first be written in standard form.
$x(x-2)=8$
$x^2-2x=8$
$x^2-2x-8=0$
$(x-4)(x+2)=0$
$x-4=0$ or $x+2=0$
$x=4$ $\quad x=-2$
The solutions are 4 and −2.

77. answers may vary, for example $(x-6)(x+1)=0$

79. answers may vary, for example
$(x-5)(x-7)=0$
$x^2-7x-5x+35=0$
$x^2-12x+35=0$

81. a.

Time, x (seconds)	Height, y (feet)
0	$-16(0)^2+20(0)+300=300$
1	$-16(1)^2+20(1)+300=304$
2	$-16(2)^2+20(2)+300=276$
3	$-16(3)^2+20(3)+300=216$
4	$-16(4)^2+20(4)+300=124$
5	$-16(5)^2+20(5)+300=0$
6	$-16(6)^2+20(6)+300=-156$

b. When the compass strikes the ground, its height is 0 feet, so it strikes the ground after 5 seconds.

c. The maximum height listed in the table is 304 feet, after 1 second.

83.

$$(x-3)(3x+4)=(x+2)(x-6)$$
$$3x^2+4x-9x-12=x^2-6x+2x-12$$
$$3x^2-5x-12=x^2-4x-12$$
$$2x^2-x=0$$
$$x(2x-1)=0$$
$$x=0 \quad \text{or} \quad 2x-1=0$$
$$x=\frac{1}{2}$$

The solutions are 0 and $\frac{1}{2}$.

85.

$$(2x-3)(x+8)=(x-6)(x+4)$$
$$2x^2+16x-3x-24=x^2+4x-6x-24$$
$$2x^2+13x-24=x^2-2x-24$$
$$x^2+15x=0$$
$$x(x+15)=0$$
$$x=0 \quad \text{or} \quad x+15=0$$
$$x=-15$$

The solutions are 0 and -15.

Section 4.7 Practice Exercises

1. The diver will be at a height of 0 when he or she reaches the pool.

$$h=-16t^2+64$$
$$0=-16t^2+64$$
$$0=-16(t^2-4)$$
$$0=-16(t-2)(t+2)$$
$$t-2=0 \quad \text{or} \quad t+2=0$$
$$t=2 \qquad\qquad t=-2$$

It takes the diver 2 seconds to reach the pool.

2. Let x be the number.

$$x^2-2x=63$$
$$x^2-2x-63=0$$
$$(x-9)(x+7)=0$$
$$x-9=0 \quad \text{or} \quad x+7=0$$
$$x=9 \qquad\qquad x=-7$$

The numbers are 9 and -7.

3. Let x be the width. Then $x + 5$ is the length.

$$\text{Area} = \text{length}\cdot\text{width}$$
$$176=(x+5)\cdot x$$
$$176=x^2+5x$$
$$0=x^2+5x-176$$
$$0=(x+16)(x-11)$$
$$x+16=0 \quad \text{or} \quad x-11=0$$
$$x=-16 \qquad\qquad x=11$$

The width is 11 feet and the length is $x + 5 = 11 + 5 = 16$ feet.

4. Let x be the first odd integer. Then $x + 2$ is the next consecutive odd integer.

$$x(x+2)=x+(x+2)+23$$
$$x^2+2x=2x+25$$
$$x^2=25$$
$$x^2-25=0$$
$$(x+5)(x-5)=0$$
$$x+5=0 \quad \text{or} \quad x-5=0$$
$$x=-5 \qquad\qquad x=5$$
$$x+2=-3 \qquad\qquad x+2=7$$

The two consecutive odd integers are -5 and -3 or 5 and 7.

5. Let x be the length of one leg. Then $x-7$ is the length of the other leg.

$$x^2+(x-7)^2=13^2$$
$$x^2+x^2-14x+49=169$$
$$2x^2-14x-120=0$$
$$2(x^2-7x-60)=0$$
$$2(x+5)(x-12)=0$$
$$x+5=0 \quad \text{or} \quad x-12=0$$
$$x=-5 \qquad\qquad x=12$$

Discard $x=-5$ since length cannot be negative. If $x=12$, then $x-7=5$. The lengths of the legs are 5 meters and 12 meters.

Vocabulary, Readiness & Video Check 4.7

1. In applications, the context of the stated application needs to be considered. Each translated equation resulted in both a positive and a negative solution, and a negative solution is not appropriate for any of the stated applications.

Exercise Set 4.7

1. Let x be the width of the rectangle, then the length is $x+4$.

3. Let x be the first odd integer, then the next consecutive odd integer is $x+2$.

5. Let x be the base of the triangle, then the height is $4x+1$.

7. $\text{area}=(\text{side})^2$

$$121=x^2$$
$$0=x^2-121$$
$$0=(x-11)(x+11)$$
$$x-11=0 \quad \text{or} \quad x+11=0$$
$$x=11 \qquad\qquad x=-11$$

Discard $x=-11$ since length cannot be negative. The length of its sides are 11 units.

9. The perimeter is the sum of the lengths of the sides.

$$(x+5)+(x^2-3x)+(3x-8)+(x+3)=120$$
$$x^2+2x=120$$
$$x^2+2x-120=0$$
$$(x+12)(x-10)=0$$
$$x+12=0 \quad \text{or} \quad x-10=0$$
$$x=-12 \qquad\qquad x=10$$

For $x=-12$, $x+5$, $x+3$, and $3x-8$ are negative. Since lengths cannot be negative, $x=10$ is the only solution that works in this context.

$x+3=10+3=13$

$x+5=10+5=15$

$x^2-3x=(10)^2-3(10)=100-30=70$

$3x-8=3(10)-8=30-8=22$

The sides have lengths 13 cm, 15 cm, 70 cm, and 22 cm.

11. $\text{Area}=\text{base}\cdot\text{height}$

$$96=(x+5)(x-5)$$
$$96=x^2-25$$
$$0=x^2-121$$
$$0=(x+11))(x-11)$$
$$x+11=0 \quad \text{or} \quad x-11=0$$
$$x=-11 \qquad\qquad x=11$$

For $x=-11$, $x+5$ and $x-5$ are negative. Since lengths cannot be negative, $x=11$ is the only solution that works in this context.

$x+5=11+5=16$

$x-5=11-5=6$

The base is 16 miles and the height is 6 miles.

13. The object will hit the ground when its height is 0.

$$0=-16t^2+64t+80$$
$$0=-16(t^2-4t-5)$$
$$0=-16(t-5)(t+1)$$
$$0=t-5 \quad \text{or} \quad 0=t+1$$
$$5=t \qquad\qquad -1=t$$

Since the time t cannot be negative, we discard $t=-1$. The object hits the ground after 5 seconds.

15. Let x be the width. Then $2x-7$ is the length.

$\text{Area}=\text{length}\cdot\text{width}$

$$30=(2x-7)\cdot(x)$$
$$30=2x^2-7x$$
$$0=2x^2-7x-30$$
$$0=(2x+5)(x-6)$$
$$2x+5=0 \quad \text{or} \quad x-6=0$$
$$x=-\frac{5}{2} \qquad\qquad x=6$$

Discard $x=-\frac{5}{2}$ since length cannot be negative.

The width is 6 cm and the length is $2x-7=2(6)-7=12-7=5$ cm.

17. $D = \frac{1}{2}n(n-3)$
$= \frac{1}{2}(12)(12-3)$
$= \frac{1}{2}(12)(9)$
$= 54$
A polygon with 12 sides has 54 diagonals.

19. $D = \frac{1}{2}n(n-3)$
$35 = \frac{1}{2}n(n-3)$
$2 \cdot 35 = 2 \cdot \frac{1}{2}n(n-3)$
$70 = n(n-3)$
$70 = n^2 - 3n$
$0 = n^2 - 3n - 70$
$0 = (n-10)(n+7)$
$n - 10 = 0$ or $n + 7 = 0$
$n = 10$ $\quad n = -7$
Discard $n = -7$. The polygon has 10 sides.

21. Let x be the number.
$x + x^2 = 132$
$x^2 + x - 132 = 0$
$(x+12)(x-11) = 0$
$x + 12 = 0$ or $x - 11 = 0$
$x = -12$ $\quad x = 11$
The number is -12 or 11.

23. Let x be the first number. Then $x + 1$ is the next consecutive number.
$x(x+1) = 210$
$x^2 + x = 210$
$x^2 + x - 210 = 0$
$(x-14)(x+15) = 0$
$x - 14 = 0$ or $x + 15 = 0$
$x = 14$ $\quad x = -15$
Discard $x = -15$. The room numbers are 14 and $x + 1 = 14 + 1 = 15$.

25. Use the Pythagorean theorem where x = hypotenuse, $(x - 1)$ = one leg, and 5 = other leg.
$x^2 = (x-1)^2 + 5^2$
$x^2 = x^2 - 2x + 1 + 25$
$x^2 = x^2 - 2x + 26$
$0 = -2x + 26$
$2x = 26$
$x = 13$
The length of the ladder is 13 feet.

27. $(x+3)^2 = 64$
$x^2 + 6x + 9 = 64$
$x^2 + 6x - 55 = 0$
$(x-5)(x+11) = 0$
$x - 5 = 0$ or $x + 11 = 0$
$x = 5$ $\quad x = -11$
Discard $x = -11$. The original square had sides of 5 inches.

29. Let x be the length of the shorter leg. Then the length of the other leg is $x + 4$ and the length of the hypotenuse is $x + 8$.
$(x+8)^2 = x^2 + (x+4)^2$
$x^2 + 16x + 64 = x^2 + x^2 + 8x + 16$
$0 = x^2 - 8x - 48$
$0 = (x-12)(x+4)$
$0 = x - 12$ or $0 = x + 4$
$12 = x$ $\quad -4 = x$
Discard $x = -4$ since length cannot be negative.
$x + 4 = 12 + 4 = 16$
$x + 8 = 12 + 8 = 20$
The lengths of the sides are 12 mm, 16 mm, and 20 mm.

31. $\text{area} = \frac{1}{2} \cdot \text{base} \cdot \text{height}$
$100 = \frac{1}{2} \cdot 2x \cdot x$
$100 = x^2$
$0 = x^2 - 100$
$0 = (x-10)(x+10)$
$x - 10 = 0$ or $x + 10 = 0$
$x = 10$ $\quad x = -10$
Discard $x = -10$. The height is 10 kilometers.

33. Let x be the length of the shorter leg. Then $x + 12$ is the length of the longer leg and $2x - 12$ is the length of the hypotenuse.

$$(2x-12)^2 = x^2 + (x+12)^2$$
$$4x^2 - 48x + 144 = x^2 + x^2 + 24x + 144$$
$$2x^2 - 72x = 0$$
$$2x(x-36) = 0$$
$$2x = 0 \quad \text{or} \quad x - 36 = 0$$
$$x = 0 \qquad\qquad x = 36$$

Discard $x = 0$ since the length must be positive. The shorter leg of the triangle has length 36 feet.

35. When the object reaches the ground, the height is 0.

$$h = -16t^2 + 1444$$
$$0 = -16t^2 + 1444$$
$$0 = -4(4t^2 - 361)$$
$$0 = -4(2t-19)(2t+19)$$
$$2t - 19 = 0 \quad \text{or} \quad 2t + 19 = 0$$
$$t = \frac{19}{2} \qquad\qquad t = -\frac{19}{2}$$

Discard $t = -\frac{19}{2}$ since time must be positive.

The object reaches the ground after $\frac{19}{2}$ or 9.5 seconds.

37. Use $A = P(1+r)^2$ when $P = 100$ and $A = 144$.

$$144 = 100(1+r)^2$$
$$144 = 100(1+2r+r^2)$$
$$144 = 100 + 200r + 100r^2$$
$$0 = 100r^2 + 200r - 44$$
$$0 = 4(25r^2 + 50r - 11)$$
$$0 = 4(5r-1)(5r+11)$$
$$0 = 5r - 1 \quad \text{or} \quad 0 = 5r + 11$$
$$1 = 5r \qquad\qquad -11 = 5r$$
$$\frac{1}{5} = r \qquad\qquad -\frac{11}{5} = r$$

Discard $r = -\frac{11}{5}$ since the interest rate must be positive. The interest rate is $\frac{1}{5} = 0.20 = 20\%$.

39. Let x be the length. Then $x - 7$ is the width.

area = length · width

$$120 = x(x-7)$$
$$0 = x^2 - 7x - 120$$
$$0 = (x-15)(x+8)$$
$$x - 15 = 0 \quad \text{or} \quad x + 8 = 0$$
$$x = 15 \qquad\qquad x = -8$$

Discard $x = -8$ since length is positive. The length is 15 miles and the width is $x - 7 = 15 - 7 = 8$ miles.

41. Let $C = 9500$ in $C = x^2 - 15x + 50$.

$$9500 = x^2 - 15x + 50$$
$$0 = x^2 - 15x - 9450$$
$$0 = (x+90)(x-105)$$
$$0 = x + 90 \quad \text{or} \quad 0 = x - 105$$
$$-90 = x \qquad\qquad 105 = x$$

Discard $x = -90$ since the number of units manufactured cannot be negative.
105 units are manufactured at a cost of \$9500.

43. From the graph, there were 2.2 million, or 2,200,000 visitors to Acadia National Park in 2009.

45. From the graph, there were 2.4 million, or 2,400,000 visitors to Acadia National Park in 2012.

47. From the graph, the lines intersect at approximately 2010.

49. answers may vary

51. $\frac{20}{35} = \frac{4 \cdot 5}{5 \cdot 7} = \frac{4}{7}$

53. $\frac{27}{18} = \frac{3 \cdot 9}{2 \cdot 9} = \frac{3}{2}$

55. $\frac{14}{42} = \frac{1 \cdot 14}{3 \cdot 14} = \frac{1}{3}$

57. Since x is the width of the rectangle, the length is $x + 6$.

$$x^2 + x(x+6) = 176$$
$$x^2 + x^2 + 6x = 176$$
$$2x^2 + 6x - 176 = 0$$
$$2(x^2 + 3x - 88) = 0$$
$$2(x-8)(x+11) = 0$$

$x-8=0$ or $x+11=0$
$x=8$ $\quad x=-11$
Discard $x=-11$ since length cannot be negative. The side of the square is 8 meters.

59. Let x be one of the numbers. Since the numbers sum to 25, the other number is $25-x$.

$$x^2+(25-x)^2=325$$
$$x^2+625-50x+x^2=325$$
$$2x^2-50x+300=0$$
$$2(x^2-25x+150)=0$$
$$2(x-10)(x-15)=0$$

$x-10=0$ or $x-15=0$
$x=10$ $\quad x=15$
If $x = 10$, then $25-x=25-10=15$.
If $x = 15$, then $25-x=25-15=10$.
The numbers are 10 and 15.

61. Dimensions of total area:
length $= x+6+4+4 = x+14$
width $= x+4+4 = x+8$

$$\text{Total area} = \text{pool area}+576$$
$$(x+14)(x+8)=x(x+6)+576$$
$$x^2+22x+112=x^2+6x+576$$
$$22x+112=6x+576$$
$$16x=464$$
$$x=29$$

$x+6=29+6=35$
The pool is 29 meters by 35 meters.

63. answers may vary

Chapter 4 Vocabulary Check

1. An equation that can be written in the form $ax^2+bx+c=0$ (with a not 0) is called a quadratic equation.

2. Factoring is the process of writing an expression as a product.

3. The greatest common factor of a list of terms is the product of all common factors.

4. A trinomial that is the square of some binomial is called a perfect square trinomial.

5. In a right triangle, the side opposite the right angle is called the hypotenuse.

6. In a right triangle, each side adjacent to the right angle is called a leg.

7. The Pythagorean theorem states that $(\text{leg})^2+(\text{leg})^2=(\text{hypotenuse})^2$.

Chapter 4 Review

1. $5m+30=5\cdot m+5\cdot 6=5(m+6)$

2. $6x^2-15x=3x\cdot 2x-3x\cdot 5=3x(2x-5)$

3. $4x^5+2x-10x^4=2x\cdot 2x^4+2x\cdot 1-2x\cdot 5x^3$
$=2x(2x^4+1-5x^3)$

4. $20x^3+12x^2+24x=4x\cdot 5x^2+4x\cdot 3x+4x\cdot 6$
$=4x(5x^2+3x+6)$

5. $3x(2x+3)-5(2x+3)=(2x+3)(3x-5)$

6. $5x(x+1)-(x+1)=5x(x+1)-1\cdot(x+1)$
$=(x+1)(5x-1)$

7. $3x^2-3x+2x-2=3x(x-1)+2(x-1)$
$=(x-1)(3x+2)$

8. $3a^2+9ab+3b^2+ab=3a(a+3b)+b(3b+a)$
$=(a+3b)(3a+b)$

9. $10a^2+5ab+7b^2+14ab$
$=5a(2a+b)+7b(b+2a)$
$=(2a+b)(5a+7b)$

10. $6x^2+10x-3x-5=2x(3x+5)-1(3x+5)$
$=(3x+5)(2x-1)$

11. $x^2+6x+8=(x+4)(x+2)$

12. $x^2-11x+24=(x-8)(x-3)$

13. x^2+x+2 is prime.

14. $x^2-5x-6=(x-6)(x+1)$

15. $x^2+2x-8=(x+4)(x-2)$

16. $x^2+4xy-12y^2=(x+6y)(x-2y)$

17. $x^2+8xy+15y^2=(x+5y)(x+3y)$

18. $72-18x-2x^2=2(36-9x-x^2)$
$=2(3-x)(12+x)$

or

$72-18x-2x^2=-2x^2-18x+72$
$=-2(x^2+9x-36)$
$=-2(x-3)(x+12)$

19. $32+12x-4x^2=4(8+3x-x^2)$

or

$32+12x-4x^2=-4x^2+12x+32$
$=-4(x^2-3x-8)$

20. $5y^3-50y^2+120y=5y(y^2-10y+24)$
$=5y(y-6)(y-4)$

21. To factor $x^2+2x-48$, think of two numbers whose product is $\underline{-48}$ and whose sum is $\underline{2}$.

22. The first step to factor $3x^2+15x+30$ is to factor out the GCF, 3.

23. Factors of $2x^2$: $2x \cdot x$
Factors of 6: $6 = 1 \cdot 6$, $6 = 2 \cdot 3$
$2x^2+13x+6=(2x+1)(x+6)$

24. Factors of $4x^2$: $4x^2=4x \cdot x$, $4x^2=2x \cdot 2x$
Factors of -3: $-3 = -1 \cdot 3$, $-3 = 1 \cdot -3$
$4x^2+4x-3=(2x+3)(2x-1)$

25. Factors of $6x^2$: $6x^2=6x \cdot x$, $6x^2=3x \cdot 2x$
Factors of $-4y^2$: $-4y^2=-4y \cdot y$,
$-4y^2=4y \cdot -y$, $-4y^2=-2y \cdot 2y$
$6x^2+5xy-4y^2=(3x+4y)(2x-y)$

26. x^2-x+2 is prime.

27. $2 \cdot -39 = -78$
$3 \cdot -26 = -78$
$3 + (-26) = -23$
$2x^2-23x-39=2x^2+3x-26x-39$
$=x(2x+3)-13(2x+3)$
$=(2x+3)(x-13)$

28. $18 \cdot -20y^2=-360y^2$
$15y \cdot -24y=-360y^2$
$15y+(-24y)=-9y$
$18x^2-9xy-20y^2=18x^2+15xy-24xy-20y^2$
$=3x(6x+5y)-4y(6x+5y)$
$=(6x+5y)(3x-4y)$

29. $10y^3+25y^2-60y=5y(2y^2+5y-12)$
$2 \cdot -12 = -24$
$-3 \cdot 8 = -24$
$-3 + 8 = 5$
$10y^3+25y^2-60y=5y(2y^2+5y-12)$
$=5y(2y^2-3y+8y-12)$
$=5y[y(2y-3)+4(2y-3)]$
$=5y(2y-3)(y+4)$

30. $60y^3-39y^2+6y=3y(20y^2-13y+2)$
$20 \cdot 2 = 40$
$-5 \cdot -8 = 40$
$-5 + (-8) = -13$
$60y^3-39y^2+6y=3y(20y^2-13y+2)$
$=3y(20y^2-5y-8y+2)$
$=3y[5y(4y-1)-2(4y-1)]$
$=3y(4y-1)(5y-2)$

31. $(x^2-2)+(x^2-4x)+(3x^2-5x)$
$=x^2-2+x^2-4x+3x^2-5x$
$=5x^2-9x-2$
$5 \cdot -2 = -10$
$1 \cdot -10 = -10$
$1 + (-10) = -9$
$5x^2-9x-2=5x^2+x-10x-2$
$=x(5x+1)-2(5x+1)$
$=(5x+1)(x-2)$

The perimeter is $5x^2-9x-2$ or $(5x+1)(x-2)$.

32. $2(2x^2+3)+2(6x^2-14x)=4x^2+6+12x^2-28x$
$=16x^2-28x+6$
$=2(8x^2-14x+3)$

$8 \cdot 3 = 24$
$-2 \cdot -12 = 24$
$-2 + (-12) = -14$

$$2(8x^2-14x+3)=2(8x^2-2x-12x+3)$$
$$=2[2x(4x-1)-3(4x-1)]$$
$$=2(4x-1)(2x-3)$$

The perimeter is $16x^2-28x+6$ or $2(4x-1)(2x-3)$.

33. Since $9=3^2$ and $6x=2\cdot 3\cdot x$, x^2+6x+9 is a perfect square trinomial.

34. Since $64=8^2$, but $8x\neq 2\cdot 8\cdot x$, $x^2+8x+64$ is not a perfect square trinomial.

35. $9m^2=(3m)^2$ and $16=4^2$, but $2\cdot 3m\cdot 4\neq 12m$, so $9m^2-12m+16$ is not a perfect square trinomial.

36. Since $4y^2=(2y)^2$ and $49=7^2$ and $2\cdot 2y\cdot 7=28y$, $4y^2-28y+49$ is a perfect square trinomial.

37. Yes; x^2-9 or x^2-3^2 is the difference of two squares.

38. No; x^2+16 or x^2+4^2 is the sum of two squares.

39. Yes; $4x^2-25y^2$ or $(2x)^2-(5y)^2$ is the difference of two squares.

40. No; $9a^3-1$ is not the difference of two squares [note: $9a^2-1$ or $(3a)^2-1^2$ is the difference of two squares.]

41. $x^2-81=x^2-9^2=(x-9)(x+9)$

42. $x^2+12x+36=x^2+2\cdot x\cdot 6+6^2=(x+6)^2$

43. $4x^2-9=(2x)^2-3^2=(2x-3)(2x+3)$

44. $9t^2-25s^2=(3t)^2-(5s)^2=(3t-5s)(3t+5s)$

45. $16x^2=(4x)^2$ and y^2 are both perfect squares, but they are added instead of subtracted, so the polynomial is prime.

46. $n^2-18n+81=n^2-2\cdot n\cdot 9+9^2=(n-9)^2$

47.
$$3r^2+36r+108=3(r^2+12+36)$$
$$=3(r^2+2\cdot r\cdot 6+6^2)$$
$$=3(r+6)^2$$

48.
$$9y^2-42y+49=(3y)^2-2\cdot 3y\cdot 7+7^2$$
$$=(3y-7)^2$$

49.
$$5m^8-5m^6=5m^6(m^2-1)$$
$$=5m^6(m^2-1^2)$$
$$=5m^6(m+1)(m-1)$$

50.
$$4x^2-28xy+49y^2=(2x)^2-2\cdot 2x\cdot 7y+(7y)^2$$
$$=(2x-7y)^2$$

51.
$$3x^2y+6xy^2+3y^3=3y(x^2+2xy+y^2)$$
$$=3y(x+y)^2$$

52.
$$16x^4-1=(4x^2)^2-1^2$$
$$=(4x^2-1)(4x^2+1)$$
$$=[(2x)^2-1^2](4x^2+1)$$
$$=(2x-1)(2x+1)(4x^2+1)$$

53. $(x+6)(x-2)=0$

$x+6=0$ or $x-2=0$

$x=-6$ $\quad$ $x=2$

The solutions are –6 and 2.

54. $(x-7)(x+11)=0$

$x-7=0$ or $x+11=0$

$x=7$ $\quad$ $x=-11$

The solutions are 7 and –11.

55. $3x(x+1)(7x-2)=0$

$3x=0$ or $x+1=0$ or $7x-2=0$

$x=0$ $\quad$ $x=-1$ $\quad$ $x=\frac{2}{7}$

The solutions are 0, –1, and $\frac{2}{7}$.

56. $4(5x+1)(x+3)=0$

$5x+1=0$ or $x+3=0$

$x=-\frac{1}{5}$ $\quad$ $x=-3$

The solutions are $-\frac{1}{5}$ and –3.

57. $x^2+8x+7=0$
$(x+7)(x+1)=0$
$x+7=0$ or $x+1=0$
$x=-7$ $\quad x=-1$
The solutions are −7 and −1.

58. $x^2-2x-24=0$
$(x+4)(x-6)=0$
$x+4=0$ or $x-6=0$
$x=-4$ $\quad x=6$
The solutions are −4 and 6.

59. $x^2+10x=-25$
$x^2+10x+25=0$
$(x+5)^2=0$
$x+5=0$
$x=-5$
The solution is −5.

60. $x(x-10)=-16$
$x^2-10x=-16$
$x^2-10x+16=0$
$(x-2)(x-8)=0$
$x-2=0$ or $x-8=0$
$x=2$ $\quad x=8$
The solutions are 2 and 8.

61. $(3x-1)(9x^2+3x+1)=0$
$3x-1=0$ or $9x^2+3x+1=0$
$x=\frac{1}{3}$ $\quad 9x^2+3x+1>0$
The solution is $\frac{1}{3}$.

62. $56x^2-5x-6=0$
$(7x+2)(8x-3)=0$
$7x+2=0$ or $8x-3=0$
$x=-\frac{2}{7}$ $\quad x=\frac{3}{8}$
The solutions are $-\frac{2}{7}$ and $\frac{3}{8}$.

63. $m^2=6m$
$m^2-6m=0$
$m(m-6)=0$
$m=0$ or $m-6=0$
$m=6$
The solutions are 0 and 6.

64. $r^2=25$
$r^2-25=0$
$(r-5)(r+5)=0$
$r-5=0$ or $r+5=0$
$r=5$ $\quad r=-5$
The solutions are 5 and −5.

65. $(x-4)(x-5)=0$
$x^2-9x+20=0$

66. $[x-(-1)][x-(-1)]=0$
$(x+1)(x+1)=0$
$x^2+2x+1=0$

67. Let x be the width. Then $2x$ is the length.
$\text{perimeter}=2\cdot\text{length}+2\cdot\text{width}$
$24=2(2x)+2x$
$24=4x+2x$
$24=6x$
$4=x$
$8=2x$
The dimensions are 4 inches by 8 inches. The choice is c.

68. Let x be the width. Then $3x + 1$ is the length.
$\text{area}=\text{length}\cdot\text{width}$
$80=(3x+1)\cdot x$
$80=3x^2+x$
$0=3x^2+x-80$
$0=(3x+16)(x-5)$
$3x+16=0$ or $x-5=0$
$x=-\frac{16}{3}$ $\quad x=5$
Discard $x=-\frac{16}{3}$ since length cannot be negative. The dimensions are 5 meters by $3x + 1 = 3(5) + 1 = 15 + 1 = 16$ meters. The choice is d.

69. area = side^2

$$81 = x^2$$
$$0 = x^2 - 81$$
$$0 = (x-9)(x+9)$$
$$x-9=0 \quad \text{or} \quad x+9=0$$
$$x=9 \qquad\qquad x=-9$$

Discard $x = -9$ since length cannot be negative. The length of each side is 9 units.

70. The perimeter is the sum of the sides.

$$(2x+3)+(3x+1)+(x^2-3x)+(x+3)=47$$
$$x^2+3x+7=47$$
$$x^2+3x-40=0$$
$$(x-5)(x+8)=0$$
$$x-5=0 \quad \text{or} \quad x+8=0$$
$$x=5 \qquad\qquad x=-8$$

Discard $x = -8$ since, for example, $x + 3$ would be negative.
$2x + 3 = 2(5) + 3 = 10 + 3 = 13$
$3x + 1 = 3(5) + 1 = 15 + 1 = 16$
$x^2 - 3x = 5^2 - 3(5) = 25 - 15 = 10$
$x + 3 = 5 + 3 = 8$
The lengths of the sides are 13 units, 16 units, 10 units, and 8 units.

71. Let x be the width. Then $2x - 15$ is the length.

area = length · width

$$500 = (2x-15)\cdot x$$
$$500 = 2x^2 - 15x$$
$$0 = 2x^2 - 15x - 500$$
$$0 = (2x+25)(x-20)$$
$$2x+25=0 \quad \text{or} \quad x-20=0$$
$$x = -\frac{25}{2} \qquad\qquad x=20$$

Discard $x = -\frac{25}{2}$ since length cannot be negative. The width is 20 inches and the length is $2 \cdot 20 - 15 = 40 - 15 = 25$ inches.

72. Let x be the height. Then $4x$ is the base.

$$\text{area} = \frac{1}{2}\cdot \text{base} \cdot \text{height}$$
$$162 = \frac{1}{2}\cdot 4x \cdot x$$
$$162 = 2x^2$$
$$0 = 2x^2 - 162$$
$$0 = 2(x^2 - 81)$$
$$0 = 2(x-9)(x+9)$$
$$x-9=0 \quad \text{or} \quad x+9=0$$
$$x=9 \qquad\qquad x=-9$$

Discard $x = -9$ since length cannot be negative. The height is 9 yards and the base is $4x = 4(9) = 36$ yards.

73. Let x be the first positive integer. Then $x + 1$ is the next consecutive integer.

$$x(x+1) = 380$$
$$x^2 + x = 380$$
$$x^2 + x - 380 = 0$$
$$(x+20)(x-19) = 0$$
$$x+20=0 \quad \text{or} \quad x-19=0$$
$$x=-20 \qquad\qquad x=19$$

Discard $x = -20$ since it is not positive. The integers are 19 and $19 + 1 = 20$.

74. Let x be the first positive even integer. Then $x + 2$ is the next consecutive even integer.

$$x(x+2) = 440$$
$$x^2 + 2x = 440$$
$$x^2 + 2x - 440 = 0$$
$$(x+22)(x-20) = 0$$
$$x+22=0 \quad \text{or} \quad x-20=0$$
$$x=-22 \qquad\qquad x=20$$

Discard $x = -22$ since it is not positive. The integers are 20 and $20 + 2 = 22$.

75. a.

$$h = -16t^2 + 440t$$
$$2800 = -16t^2 + 440t$$
$$16t^2 - 440t + 2800 = 0$$
$$8(2t^2 - 55t + 350) = 0$$
$$8(2t-35)(t-10) = 0$$
$$2t-35=0 \quad \text{or} \quad t-10=0$$
$$t = \frac{35}{2} \text{ or } 17.5 \qquad\qquad t=10$$

The rocket reaches a height of 2800 feet at 10 seconds on the way up and at 17.5 seconds on the way down.

b. The height is 0 when the rocket reaches the ground.

$$h = -16t^2 + 440t$$
$$0 = -16t^2 + 440t$$
$$0 = -8t(2t-55)$$
$$-8t=0 \quad \text{or} \quad 2t-55=0$$
$$t=0 \qquad\qquad 2t=55$$
$$t=27.5$$

The rocket reaches the ground again after 27.5 seconds.

76. Let x be the length of the longer leg. Then $x + 8$ is the length of the hypotenuse and $x - 8$ is the length of the shorter leg.

$$(x+8)^2 = (x-8)^2 + x^2$$
$$x^2 + 16x + 64 = x^2 - 16x + 64 + x^2$$
$$0 = x^2 - 32x$$
$$0 = x(x-32)$$
$$x = 0 \quad \text{or} \quad x - 32 = 0$$
$$x = 32$$

The longer leg is 32 centimeters.

77. $6x + 24 = 6 \cdot x + 6 \cdot 4 = 6(x + 4)$

78. $7x - 63 = 7 \cdot x - 7 \cdot 9 = 7(x - 9)$

79. $11x(4x - 3) - 6(4x - 3) = (4x - 3)(11x - 6)$

80. $2x(x-5)-(x-5) = 2x(x-5)-1(x-5)$
$= (x-5)(2x-1)$

81. $3x^3 - 4x^2 + 6x - 8 = x^2(3x-4) + 2(3x-4)$
$= (3x-4)(x^2+2)$

82. $xy + 2x - y - 2 = x(y+2) - 1(y+2)$
$= (y+2)(x-1)$

83. $2x^2 + 2x - 24 = 2(x^2 + x - 12) = 2(x+4)(x-3)$

84. $3x^3 - 30x^2 + 27x = 3x(x^2 - 10x + 9)$
$= 3x(x-9)(x-1)$

85. $4x^2 - 81 = (2x)^2 - 9^2 = (2x+9)(2x-9)$

86. $2x^2 - 18 = 2(x^2 - 9)$
$= 2(x^2 - 3^2)$
$= 2(x-3)(x+3)$

87. $16x^2 - 24x + 9 = (4x)^2 - 2 \cdot 4x \cdot 3 + 3^2 = (4x-3)^2$

88. $5x^2 + 20x + 20 = 5(x^2 + 4x + 4)$
$= 5(x^2 + 2 \cdot x \cdot 2 + 2^2)$
$= 5(x+2)^2$

89.
$$2x^2 - x - 28 = 0$$
$$(2x+7)(x-4) = 0$$
$$2x + 7 = 0 \quad \text{or} \quad x - 4 = 0$$
$$x = -\frac{7}{2} \qquad x = 4$$

The solutions are $-\frac{7}{2}$ and 4.

90.
$$x^2 - 2x = 15$$
$$x^2 - 2x - 15 = 0$$
$$(x+3)(x-5) = 0$$
$$x + 3 = 0 \quad \text{or} \quad x - 5 = 0$$
$$x = -3 \qquad x = 5$$

The solutions are −3 and 5.

91. $2x(x + 7)(x + 4) = 0$
$$2x = 0 \quad \text{or} \quad x + 7 = 0 \quad \text{or} \quad x + 4 = 0$$
$$x = 0 \qquad x = -7 \qquad x = -4$$

92.
$$x(x-5) = -6$$
$$x^2 - 5x = -6$$
$$x^2 - 5x + 6 = 0$$
$$(x-3)(x-2) = 0$$
$$x - 3 = 0 \quad \text{or} \quad x - 2 = 0$$
$$x = 3 \qquad x = 2$$

The solutions are 3 and 2.

93.
$$x^2 = 16x$$
$$x^2 - 16x = 0$$
$$x(x-16) = 0$$
$$x = 0 \quad \text{or} \quad x - 16 = 0$$
$$x = 16$$

The solutions are 0 and 16.

94. The perimeter is the sum of the sides.
$$(x^2 + 3) + 2x + (4x + 5) = 48$$
$$x^2 + 6x + 8 = 48$$
$$x^2 + 6x - 40 = 0$$
$$(x-4)(x+10) = 0$$
$$x - 4 = 0 \quad \text{or} \quad x + 10 = 0$$
$$x = 4 \qquad x = -10$$

Discard $x = -10$ since length, such as $2x = 2(-10) = -20$ cannot be negative.

$x^2 + 3 = 4^2 + 3 = 16 + 3 = 19$

$2x = 2 \cdot 4 = 8$

$4x + 5 = 4 \cdot 4 + 5 = 16 + 5 = 21$

The lengths are 19 inches, 8 inches, and 21 inches.

95. Let x be the length. Then $x - 4$ is the width.
area = length · width

$$12 = x(x-4)$$
$$12 = x^2 - 4x$$
$$0 = x^2 - 4x - 12$$
$$0 = (x-6)(x+2)$$

$x - 6 = 0$ or $x + 2 = 0$
$x = 6$ $\quad$ $x = -2$

Discard $x = -2$ since length cannot be negative. The length is 6 inches and the width is $x - 4 = 6 - 4 = 2$ inches.

Chapter 4 Test

1. $9x^2 - 3x = 3x(3x-1)$

2. $x^2 + 11x + 28 = (x+7)(x+4)$

3. $49 - m^2 = 7^2 - m^2 = (7-m)(7+m)$

4. $y^2 + 22y + 121 = y^2 + 2 \cdot y \cdot 11 + 11^2$
$= (y+11)^2$

5. $x^4 - 16 = (x^2)^2 - (4)^2$
$= (x^2+4)(x^2-4)$
$= (x^2+4)[(x)^2 - (2)^2]$
$= (x^2+4)(x+2)(x-2)$

6. $4(a+3) - y(a+3) = (a+3)(4-y)$

7. $x^2 + 4$ is prime.

8. $y^2 - 8y - 48 = (y-12)(y+4)$

9. $3a^2 + 3ab - 7a - 7b = 3a(a+b) - 7(a+b)$
$= (a+b)(3a-7)$

10. $3x^2 - 5x + 2 = (3x-2)(x-1)$

11. $180 - 5x^2 = 5(36 - x^2)$
$= 5(6^2 - x^2)$
$= 5(6-x)(6+x)$

12. $3x^3 - 21x^2 + 30x = 3x(x^2 - 7x + 10)$
$= 3x(x-5)(x-2)$

13. $6t^2 - t - 5 = 6t^2 + 5t - 6t - 5$
$= t(6t+5) - 1(6t+5)$
$= (6t+5)(t-1)$

14. $xy^2 - 7y^2 - 4x + 28 = y^2(x-7) - 4(x-7)$
$= (x-7)(y^2-4)$
$= (x-7)(y-2)(y+2)$

15. $x - x^5 = x(1-x^4)$
$= x[1^2 - (x^2)^2]$
$= x(1+x^2)(1-x^2)$
$= x(1+x^2)(1+x)(1-x)$

16. $x^2 + 14xy + 24y^2 = x^2 + 12xy + 2xy + 24y^2$
$= x(x+12y) + 2y(x+12y)$
$= (x+12y)(x+2y)$

17. $(x-3)(x+9) = 0$
$x - 3 = 0$ or $x + 9 = 0$
$x = 3$ $\quad$ $x = -9$
The solutions are 3 and −9.

18. $x^2 + 5x = 14$
$x^2 + 5x - 14 = 0$
$(x+7)(x-2) = 0$
$x + 7 = 0$ or $x - 2 = 0$
$x = -7$ $\quad$ $x = 2$
The solutions are −7 and 2.

19. $x(x+6) = 7$
$x^2 + 6x = 7$
$x^2 + 6x - 7 = 0$
$(x+7)(x-1) = 0$
$x + 7 = 0$ or $x - 1 = 0$
$x = -7$ $\quad$ $x = 1$
The solutions are −7 and 1.

20. $3x(2x-3)(3x+4) = 0$
$3x = 0$ or $2x - 3 = 0$ or $3x + 4 = 0$
$x = 0$ $\quad$ $x = \frac{3}{2}$ $\quad$ $x = -\frac{4}{3}$
The solutions are 0, $\frac{3}{2}$, and $-\frac{4}{3}$.

21. $5t^3 - 45t = 0$
$5t(t^2 - 9) = 0$
$5t[(t)^2 - (3)^2] = 0$
$5t(t+3)(t-3) = 0$
$5t = 0$ or $t+3=0$ or $t-3=0$
$t = 0$ $t=-3$ $t=3$
The solutions are 0, −3, and 3.

22. $t^2 - 2t - 15 = 0$
$(t+3)(t-5) = 0$
$t+3=0$ or $t-5=0$
$t=-3$ $t=5$
The solutions are −3 and 5.

23. $6x^2 = 15x$
$6x^2 - 15x = 0$
$3x(2x-5) = 0$
$3x = 0$ or $2x-5=0$
$x = 0$ $x = \frac{5}{2}$
The solutions are 0 and $\frac{5}{2}$.

24. Let x be the height. Then the base is $x + 9$.
$\text{area} = \frac{1}{2} \cdot \text{base} \cdot \text{height}$
$68 = \frac{1}{2} \cdot (x+9) \cdot x$
$2 \cdot 68 = 2 \cdot \frac{1}{2} \cdot (x+9) \cdot x$
$136 = x(x+9)$
$136 = x^2 + 9x$
$0 = x^2 + 9x - 136$
$0 = (x+17)(x-8)$
$x+17=0$ or $x-8=0$
$x=-17$ $x=8$
Discard $x = -17$ since length cannot be negative.
The height is 8 feet and the base is $x + 9 = 8 + 9 = 17$ feet.

25. $(x-1)(x+2) = 54$
$x^2 + 2x - x - 2 = 54$
$x^2 + x - 56 = 0$
$(x+8)(x-7) = 0$
$x+8=0$ or $x-7=0$
$x=-8$ $x=7$
Discard $x = -8$ since length cannot be negative.
$x - 1 = 7 - 1 = 6$
$x + 2 = 7 + 2 = 9$
The width of the rectangle is 6 units and the length is 9 units.

26. The object is at a height of 0 when it reaches the ground.
$h = -16t^2 + 784$
$0 = -16t^2 + 784$
$0 = -16(t^2 - 49)$
$0 = -16(t-7)(t+7)$
$t-7=0$ or $t+7=0$
$t=7$ $t=-7$
Discard $t = -7$ since time cannot be negative.
The object reaches the ground after 7 seconds.

27. Let x be the length of the shorter leg. Then $x + 10$ is the length of the hypotenuse and $x + 10 - 5 = x + 5$ is the length of the longer leg.
$x^2 + (x+5)^2 = (x+10)^2$
$x^2 + x^2 + 10x + 25 = x^2 + 20x + 100$
$x^2 - 10x - 75 = 0$
$(x-15)(x+5) = 0$
$x-15=0$ or $x+5=0$
$x=15$ $x=-5$
Discard $x = -5$ since length cannot be negative.
The shorter leg is 15 cm, the longer leg is $15 + 5 = 20$ cm, and the hypotenuse is $15 + 10 = 25$ cm.

28. The height of the object is 0 when it reaches the ground.
$h = -16t^2 + 1089$
$0 = -16t^2 + 1089$
$0 = -1(16t^2 - 1089)$
$0 = -1[(4t)^2 - 33^2]$
$0 = -1(4t-33)(4t+33)$
$4t-33=0$ or $4t+33=0$
$t = \frac{33}{4}$ or 8.25 $t = -\frac{33}{4}$
Discard $t = -\frac{33}{4}$ since time cannot be negative.
The object reaches the ground after 8.25 seconds.

Cumulative Review Chapters 1–4

1. a. Nine is less than or equal to eleven is translated as $9 \le 11$.

b. Eight is greater than one is translated as $8 > 1$.

c. Three is not equal to four is translated as $3 \neq 4$.

2. a. $|-5| = 5$
$|-3| = 3$
$|-5| > |-3|$ since $5 > 3$

b. $|0| = 0$
$|-2| = 2$
$|0| < |-2|$ since $0 < 2$

3. Replace x with 2.

$$\begin{aligned} 3x+10 &= 8x \\ 3(2)+10 &\stackrel{?}{=} 8(2) \\ 6+10 &\stackrel{?}{=} 16 \\ 16 &= 16 \quad \text{True} \end{aligned}$$

Since a true statement results, 2 is a solution of the equation.

4. Replace x with 20 and y with 10.

$$\frac{x}{y}+5x = \frac{20}{10}+5(20) = 2+100 = 102$$

5. $-4 - 8 = -4 + (-8) = -12$

6. Replace x with -20 and y with 10.

$$\frac{x}{y}+5x = \frac{-20}{10}+5(-20) = -2+(-100) = -102$$

7. Replace x with -2 and y with -4.

a. $\dfrac{3x}{2y} = \dfrac{3(-2)}{2(-4)} = \dfrac{-6}{-8} = \dfrac{-2\cdot 3}{-2\cdot 4} = \dfrac{3}{4}$

b.
$$\begin{aligned} x^3 - y^2 &= (-2)^3 - (-4)^2 \\ &= -8-16 \\ &= -8+(-16) \\ &= -24 \end{aligned}$$

c. $\dfrac{x-y}{-x} = \dfrac{-2-(-4)}{-(-2)} = \dfrac{-2+4}{2} = \dfrac{2}{2} = 1$

8. Replace x with -20 and y with -10.

$$\frac{x}{y}+5x = \frac{-20}{-10}+5(-20) = 2+(-100) = -98$$

9. $2x + 3x + 5 + 2 = (2x + 3x) + (5 + 2) = 5x + 7$

10.
$$\begin{aligned} 5-2(3x-7) &= 5-2\cdot 3x-2(-7) \\ &= 5-6x+14 \\ &= 19-6x \end{aligned}$$

11. $-5a - 3 + a + 2 = -5a + a - 3 + 2 = -4a - 1$

12.
$$\begin{aligned} 5(x-6)+9(-2x+1) &= 5\cdot x-5\cdot 6+9\cdot(-2x)+9\cdot 1 \\ &= 5x-30-18x+9 \\ &= -13x-21 \end{aligned}$$

13. $2.3x + 5x - 6 = 7.3x - 6$

14.
$$\begin{aligned} 0.8y+0.2(y-1) &= 1.8 \\ 10[0.8y+0.2(y-1)] &= 10(1.8) \\ 8y+2(y-1) &= 18 \\ 8y+2y-2 &= 18 \\ 10y-2 &= 18 \\ 10y-2+2 &= 18+2 \\ 10y &= 20 \\ \frac{10y}{10} &= \frac{20}{10} \\ y &= 2 \end{aligned}$$

The solution is 2.

15.
$$\begin{aligned} -3x &= 33 \\ \frac{-3x}{-3} &= \frac{33}{-3} \\ x &= -11 \end{aligned}$$

The solution is -11.

16.
$$\begin{aligned} \frac{x}{-7} &= -4 \\ -7\cdot\frac{x}{-7} &= -7\cdot -4 \\ x &= 28 \end{aligned}$$

The solution is 28.

17.
$$\begin{aligned} 3(x-4) &= 3x-12 \\ 3x-12 &= 3x-12 \\ -3x+3x-12 &= -3x+3x-12 \\ -12 &= -12 \end{aligned}$$

Since this results in a true statement, the solution is every real number.

18.
$$\begin{aligned} -\frac{2}{3}x &= -22 \\ -\frac{3}{2}\cdot -\frac{2}{3}x &= -\frac{3}{2}\cdot -22 \\ x &= 33 \end{aligned}$$

The solution is 33.

19. $V = lwh$
$$\frac{V}{wh} = \frac{lwh}{wh}$$
$$\frac{V}{wh} = l \text{ or } l = \frac{V}{wh}$$

20.
$$3x + 2y = -7$$
$$-3x + 3x + 2y = -3x - 7$$
$$2y = -3x - 7$$
$$\frac{2y}{2} = \frac{-3x - 7}{2}$$
$$y = \frac{-3x - 7}{2}$$
or
$$y = -\frac{3}{2}x - \frac{7}{2}$$

21. $(5^3)^6 = 5^{3\cdot6} = 5^{18}$

22. $5^2 + 5^1 = 25 + 5 = 30$

23. $(y^8)^2 = y^{8\cdot2} = y^{16}$

24. $y^8 \cdot y^2 = y^{8+2} = y^{10}$

25.
$$\frac{(x^3)^4 x}{x^7} = \frac{x^{3\cdot4} \cdot x^1}{x^7} = \frac{x^{12} \cdot x^1}{x^7} = \frac{x^{12+1}}{x^7} = \frac{x^{13}}{x^7} = x^{13-7} = x^6$$

26. $3^{-2} = \frac{1}{3^2} = \frac{1}{9}$

27. $(y^{-3}z^6)^{-6} = y^{-3(-6)}z^{6(-6)} = y^{18}z^{-36} = \frac{y^{18}}{z^{36}}$

28. $\frac{x^{-3}}{x^{-7}} = x^{-3-(-7)} = x^{-3+7} = x^4$

29. $\frac{x^{-7}}{(x^4)^3} = \frac{x^{-7}}{x^{4\cdot3}} = \frac{x^{-7}}{x^{12}} = \frac{1}{x^{12-(-7)}} = \frac{1}{x^{12+7}} = \frac{1}{x^{19}}$

30. $\frac{(5a^7)^2}{a^5} = \frac{5^2 a^{7\cdot2}}{a^5} = \frac{25a^{14}}{a^5} = 25a^{14-5} = 25a^9$

31. $-3x + 7x = (-3 + 7)x = 4x$

32.
$$\frac{2}{3}x + 23 + \frac{1}{6}x - 100 = \left(\frac{2}{3}x + \frac{1}{6}x\right) + (23 - 100)$$
$$= \left(\frac{4}{6}x + \frac{1}{6}x\right) + (23 - 100)$$
$$= \frac{5}{6}x - 77$$

33.
$$11x^2 + 5 + 2x^2 - 7 = 11x^2 + 2x^2 + 5 - 7$$
$$= 13x^2 - 2$$

34.
$$0.2x - 1.1 + 2.3 - 0.7x = 0.2x - 0.7x - 1.1 + 2.3$$
$$= -0.5x + 1.2$$

35. $(2x - y)^2$
$$= (2x - y)(2x - y)$$
$$= 2x(2x) + 2x(-y) + (-y)(2x) + (-y)(-y)$$
$$= 4x^2 - 2xy - 2xy + y^2$$
$$= 4x^2 - 4xy + y^2$$

36. $(3x - 7y)^2$
$$= (3x - 7y)(3x - 7y)$$
$$= 3x(3x) + 3x(-7y) + (-7y)(3x) + (-7y)(-7y)$$
$$= 9x^2 - 21xy - 21xy + 49y^2$$
$$= 9x^2 - 42xy + 49y^2$$

37. $(t + 2)^2 = t^2 + 2(t)(2) + 2^2 = t^2 + 4t + 4$

38.
$$(x - 13)^2 = x^2 - 2(x)(13) + 13^2$$
$$= x^2 - 26x + 169$$

39.
$$(x^2 - 7y)^2 = (x^2)^2 - 2(x^2)(7y) + (7y)^2$$
$$= x^4 - 14x^2y + 49y^2$$

40.
$$(7x + y)^2 = (7x)^2 + 2(7x)(y) + (y)^2$$
$$= 49x^2 + 14xy + y^2$$

41. $\frac{8x^2y^2 - 16xy + 2x}{4xy} = \frac{8x^2y^2}{4xy} - \frac{16xy}{4xy} + \frac{2x}{4xy}$
$= 2xy - 4 + \frac{1}{2y}$

42. $z^3 + 7z + z^2 + 7 = z(z^2 + 7) + 1(z^2 + 7)$
$= (z^2 + 7)(z + 1)$

43. $5(x + 3) + y(x + 3) = (x + 3)(5 + y)$

44. $2x^3 + 2x^2 - 84x = 2x(x^2 + x - 42)$
$= 2x(x^2 + 7x - 6x - 42)$
$= 2x[x(x + 7) - 6(x + 7)]$
$= 2x(x + 7)(x - 6)$

45. $x^4 + 5x^2 + 6 = x^4 + 2x^2 + 3x^2 + 6$
$= x^2(x^2 + 2) + 3(x^2 + 2)$
$= (x^2 + 2)(x^2 + 3)$

46. $-4x^2 - 23x + 6 = -4x^2 + x - 24x + 6$
$= x(-4x + 1) + 6(-4x + 1)$
$= (-4x + 1)(x + 6)$

or

$-4x^2 - 23x + 6 = -1(4x^2 + 23x - 6)$
$= -1(4x^2 - x + 24x - 6)$
$= -1[x(4x - 1) + 6(4x - 1)]$
$= -1(4x - 1)(x + 6)$

47. $6x^2 - 2x - 20 = 2(3x^2 - x - 10)$
$= 2(3x^2 - 6x + 5x - 10)$
$= 2[3x(x - 2) + 5(x - 2)]$
$= 2(x - 2)(3x + 5)$

48. $9xy^2 - 16x = x(9y^2 - 16)$
$= x[(3y)^2 - 4^2]$
$= x(3y + 4)(3y - 4)$

49. The height is 0 when the diver reaches the ocean.
$h = -16t^2 + 144$
$0 = -16t^2 + 144$
$0 = -16(t^2 - 9)$
$0 = -16(t - 3)(t + 3)$
$t - 3 = 0$ or $t + 3 = 0$
$t = 3$ $t = -3$
Discard $t = -3$ since time cannot be negative.
The diver reaches the ocean after 3 seconds.

50. $x^2 - 13x = -36$
$x^2 - 13x + 36 = 0$
$x^2 - 9x - 4x + 36 = 0$
$x(x - 9) - 4(x - 9) = 0$
$(x - 9)(x - 4) = 0$
$x - 9 = 0$ or $x - 4 = 0$
$x = 9$ $x = 4$
The solutions are 9 and 4.

Chapter 5

Section 5.1 Practice Exercises

1. a. $\frac{x-3}{5x+1}=\frac{4-3}{5(4)+1}=\frac{1}{20+1}=\frac{1}{21}$

b. $\frac{x-3}{5x+1}=\frac{(-3)-3}{5(-3)+1}=\frac{-6}{-15+1}=\frac{-6}{-14}=\frac{3}{7}$

2. a. $x+8=0$
$x=-8$

When $x=-8$, the expression $\frac{x}{x+8}$ is undefined.

b. $x^2+5x+4=0$
$(x+4)(x+1)=0$
$x+4=0$ or $x+1=0$
$x=-4$ $\quad x=-1$
When $x=-4$ or $x=-1$, the expression $\frac{x-3}{x^2+5x+4}$ is undefined.

c. The denominator of $\frac{x^2-3x+2}{5}$ is never 0, so there are no values of x for which this expression is undefined.

3. $\frac{x^4+x^3}{5x+5}=\frac{x^3(x+1)}{5(x+1)}=\frac{x^3}{5}$

4. $\frac{x^2+11x+18}{x^2+x-2}=\frac{(x+9)(x+2)}{(x-1)(x+2)}=\frac{x+9}{x-1}$

5. $\frac{x^2+10x+25}{x^2+5x}=\frac{(x+5)(x+5)}{x(x+5)}=\frac{x+5}{x}$

6. $\frac{x+5}{x^2-25}=\frac{x+5}{(x+5)(x-5)}=\frac{1}{x-5}$

7. a. $\frac{x+4}{4+x}=\frac{x+4}{x+4}=1$

b. $\frac{x-4}{4-x}=\frac{x-4}{(-1)(x-4)}=\frac{1}{-1}=-1$

8.
$$\begin{aligned}\frac{2x^2-5x-12}{16-x^2}&=\frac{(x-4)(2x+3)}{(4-x)(4+x)}\\&=\frac{(x-4)(2x+3)}{(-1)(x-4)(4+x)}\\&=\frac{2x+3}{(-1)(x+4)}\\&=-\frac{2x+3}{x+4}\text{ or }\frac{-2x-3}{x+4}\end{aligned}$$

9. $-\frac{3x+7}{x-6}=\frac{-(3x+7)}{x-6}=\frac{-3x-7}{x-6}$ or
$\frac{3x+7}{-(x-6)}=\frac{3x+7}{-x+6}$

Vocabulary, Readiness & Video Check 5.1

1. A rational expression is an expression that can be written in the form $\frac{P}{Q}$, where P and Q are polynomials and $Q\neq 0$.

2. The expression $\frac{x+3}{3+x}$ simplifies to 1.

3. The expression $\frac{x-3}{3-x}$ simplifies to -1.

4. A rational expression is undefined for values that make the denominator 0.

5. The expression $\frac{7x}{x-2}$ is undefined for $x=2$.

6. The process of writing a rational expression in lowest terms is called simplifying.

7. For a rational expression, $-\frac{a}{b}=\frac{-a}{b}=\frac{a}{-b}$.

8. $\frac{x}{x+7}$ cannot be simplified.

9. Since $3+x=x+3$, $\frac{3+x}{x+3}$ can be simplified.

10. Since $5-x=-1(x-5)$, $\frac{5-x}{x-5}$ can be simplified.

11. $\frac{x+2}{x+8}$ cannot be simplified.

12. replacement values of variables; by evaluating the expression for different replacement values—variables are replaced with these values and the expression is simplified

13. Rational expressions are fractions and are therefore undefined if the denominator is zero; if a denominator contains variables, set it equal to zero and solve.

14. Although x is a factor in the numerator, it is not a factor in the denominator—factor means to write as a product, and the denominator is a difference, not a product.

15. We would need to write parentheses around the numerator or denominator if it had more than one term because the negative sign needs to apply to the entire numerator or denominator.

Exercise Set 5.1

1. $\frac{x+5}{x+2}=\frac{2+5}{2+2}=\frac{7}{4}$

3. $\frac{y^3}{y^2-1}=\frac{(-2)^3}{(-2)^2-1}=\frac{-8}{4-1}=-\frac{8}{3}$

5.
$$\frac{x^2+8x+2}{x^2-x-6}=\frac{2^2+8(2)+2}{2^2-2-6}=\frac{4+16+2}{4-2-6}=\frac{22}{-4}=-\frac{11}{2}$$

7. a. $A=\frac{3x+400}{x}=\frac{3(1)+400}{1}=403$

The cost of producing 1 DVD is \$403.

b. $A=\frac{3(100)+400}{100}=\frac{300+400}{100}=\frac{700}{100}=7$

The average cost for producing 100 DVDs is \$7.

c. decrease; answers may vary

9. $2x=0$

$x=0$

$\frac{7}{2x}$ is undefined for $x=0$.

11. $x+2=0$

$x=-2$

$\frac{x+3}{x+2}$ is undefined for $x=-2$.

13. $2x-5=0$

$2x=5$

$x=\frac{5}{2}$

$\frac{x-4}{2x-5}$ is undefined for $x=\frac{5}{2}$.

15. $15x^2+30x=0$

$15x(x+2)=0$

$15x=0$ or $x+2=0$

$x=0$ $\quad$ $x=-2$

$\frac{9x^3+4}{15x^2+30x}$ is undefined for $x=0$ and $x=-2$.

17. Since $4\neq 0$, there are no values of x for which $\frac{x^2-5x-2}{4}$ is undefined.

19. $x^2-5x-6=0$

$(x-6)(x+1)=0$

$x-6=0$ or $x+1=0$

$x=6$ $\quad$ $x=-1$

$\frac{3x^2+9}{x^2-5x-6}$ is undefined for $x=6$ and $x=-1$.

21. $3x^2+13x+14=0$

$(x+2)(3x+7)=0$

$x+2=0$ or $3x+7=0$

$x=-2$ $\quad$ $3x=-7$

$x=-\frac{7}{3}$

$\frac{x}{3x^2+13x+14}$ is undefined for $x=-2$ and $x=-\frac{7}{3}$.

23. $\frac{x+7}{7+x}=\frac{x+7}{x+7}=1$

25. $\frac{x-7}{7-x}=\frac{-1(-x+7)}{7-x}=\frac{-1(7-x)}{7-x}=-1$

27. $\frac{2}{8x+16}=\frac{2}{8(x+2)}=\frac{1}{4(x+2)}$

29. $\frac{x-2}{x^2-4}=\frac{x-2}{(x+2)(x-2)}=\frac{1}{x+2}$

31. $\frac{2x-10}{3x-30}=\frac{2(x-5)}{3(x-10)}$ can't be simplified.

33. $\frac{-5a-5b}{a+b}=\frac{-5(a+b)}{a+b}=-5$

35. $\frac{7x+35}{x^2+5x}=\frac{7(x+5)}{x(x+5)}=\frac{7}{x}$

37. $\frac{x+5}{x^2-4x-45}=\frac{x+5}{(x-9)(x+5)}=\frac{1}{x-9}$

39. $\frac{5x^2+11x+2}{x+2}=\frac{(5x+1)(x+2)}{x+2}=5x+1$

41. $\frac{x^3+7x^2}{x^2+5x-14}=\frac{x^2(x+7)}{(x-2)(x+7)}=\frac{x^2}{x-2}$

43. $\frac{14x^2-21x}{2x-3}=\frac{7x(2x-3)}{2x-3}=7x$

45. $\frac{x^2+7x+10}{x^2-3x-10}=\frac{(x+5)(x+2)}{(x+2)(x-5)}=\frac{x+5}{x-5}$

47. $\frac{3x^2+7x+2}{3x^2+13x+4}=\frac{(3x+1)(x+2)}{(3x+1)(x+4)}=\frac{x+2}{x+4}$

49. $\frac{2x^2-8}{4x-8}=\frac{2(x^2-4)}{4(x-2)}=\frac{2(x+2)(x-2)}{2\cdot 2(x-2)}=\frac{x+2}{2}$

51. $\frac{4-x^2}{x-2}=\frac{-(x^2-4)}{x-2}=\frac{-(x+2)(x-2)}{x-2}=-(x+2)$

53. $\frac{x^2-1}{x^2-2x+1}=\frac{(x+1)(x-1)}{(x-1)^2}=\frac{x+1}{x-1}$

55.
$$\begin{aligned}\frac{x^2+xy+2x+2y}{x+2}&=\frac{x(x+y)+2(x+y)}{x+2}\\&=\frac{(x+2)(x+y)}{x+2}\\&=x+y\end{aligned}$$

57.
$$\begin{aligned}\frac{5x+15-xy-3y}{2x+6}&=\frac{5(x+3)-y(x+3)}{2(x+3)}\\&=\frac{(x+3)(5-y)}{2(x+3)}\\&=\frac{5-y}{2}\end{aligned}$$

59.
$$\begin{aligned}\frac{2xy+5x-2y-5}{3xy+4x-3y-4}&=\frac{x(2y+5)-1(2y+5)}{x(3y+4)-1(3y+4)}\\&=\frac{(2y+5)(x-1)}{(3y+4)(x-1)}\\&=\frac{2y+5}{3y+4}\end{aligned}$$

61.
$$\begin{aligned}-\frac{x-10}{x+8}&=\frac{-(x-10)}{x+8}\\&=\frac{-x+10}{x+8}\\&=\frac{x-10}{-(x+8)}\\&=\frac{x-10}{-x-8}\end{aligned}$$

63.
$$\begin{aligned}-\frac{5y-3}{y-12}&=\frac{-(5y-3)}{y-12}\\&=\frac{-5y+3}{y-12}\\&=\frac{5y-3}{-(y-12)}\\&=\frac{5y-3}{-y+12}\end{aligned}$$

65. $\frac{9-x^2}{x-3} = \frac{9-x^2}{-(3-x)}$
$= \frac{(3+x)(3-x)}{-(3-x)}$
$= \frac{3+x}{-1}$
$= -3-x$

The given answer is correct.

67. $\frac{7-34x-5x^2}{25x^2-1} = \frac{(7+x)(1-5x)}{(5x+1)(5x-1)}$
$= \frac{(7+x)(1-5x)}{-(5x+1)(1-5x)}$
$= \frac{7+x}{-5x-1}$

The given answer is correct.

69. $\frac{1}{3} \cdot \frac{9}{11} = \frac{1 \cdot 9}{3 \cdot 11} = \frac{1 \cdot 3 \cdot 3}{3 \cdot 11} = \frac{3}{11}$

71. $\frac{1}{3} \div \frac{1}{4} = \frac{1}{3} \cdot \frac{4}{1} = \frac{1 \cdot 4}{3 \cdot 1} = \frac{4}{3}$

73. $\frac{13}{20} \div \frac{2}{9} = \frac{13}{20} \cdot \frac{9}{2} = \frac{13 \cdot 9}{20 \cdot 2} = \frac{117}{40}$

75. $\frac{5a-15}{5} = \frac{5(a-3)}{5} = a-3$

The given answer is correct.

77. $\frac{1+2}{1+3} = \frac{3}{4} \neq \frac{2}{3}$

The given answer is incorrect.

79. answers may vary

81. answers may vary

83. Use $C = \frac{DA}{A+12}$ with $A = 8$ and $D = 1000$.

$C = \frac{DA}{A+12} = \frac{1000 \cdot 8}{8+12} = \frac{8000}{20} = 400$

The child should receive a dose of 400 milligrams.

85. Use $C = \frac{100W}{L}$ with $W = 5$ and $L = 6.4$.

$C = \frac{100W}{L} = \frac{100(5)}{6.4} = \frac{500}{6.4} = 78.125$

The cephalic index is 78.125. The skull is medium.

87. Use $S = \frac{h+d+2t+3r}{b}$ with $h = 130$, $d = 30$, $t = 1$, $r = 37$, and $b = 449$.

$S = \frac{130+30+2(1)+3(37)}{449} = \frac{273}{449} \approx 0.608$

Stanton's slugging percentage was about 60.8%.

Section 5.2 Practice Exercises

1. a. $\frac{16y}{3} \cdot \frac{1}{x^2} = \frac{16y \cdot 1}{3 \cdot x^2} = \frac{16y}{3x^2}$

b. $\frac{-5a^3}{3b^3} \cdot \frac{2b^2}{15a} = \frac{-5a^3 \cdot 2b^2}{3b^3 \cdot 15a}$
$= \frac{-5 \cdot a^3 \cdot 2 \cdot b^2}{3 \cdot b^3 \cdot 3 \cdot 5 \cdot a}$
$= -\frac{2a^2}{9b}$

2. $\frac{3x+6}{14} \cdot \frac{7x^2}{x^3+2x^2} = \frac{3(x+2) \cdot 7x^2}{2 \cdot 7 \cdot x^2(x+2)} = \frac{3}{2}$

3. $\frac{4x+8}{7x^2-14x} \cdot \frac{3x^2-5x-2}{9x^2-1}$
$= \frac{4(x+2)}{7x(x-2)} \cdot \frac{(3x+1)(x-2)}{(3x+1)(3x-1)}$
$= \frac{4(x+2)(3x+1)(x-2)}{7x(x-2)(3x+1)(3x-1)}$
$= \frac{4(x+2)}{7x(3x-1)}$

4. $\frac{7x^2}{6} \div \frac{x}{2y} = \frac{7x^2}{6} \cdot \frac{2y}{x} = \frac{7 \cdot x \cdot x \cdot 2 \cdot y}{2 \cdot 3 \cdot x} = \frac{7xy}{3}$

5. $\frac{(x-4)^2}{6} \div \frac{3x-12}{2} = \frac{(x-4)^2}{6} \cdot \frac{2}{3x-12}$
$= \frac{(x-4)(x-4) \cdot 2}{2 \cdot 3 \cdot 3 \cdot (x-4)}$
$= \frac{x-4}{9}$

6. $\frac{10x+4}{x^2-4} \div \frac{5x^3+2x^2}{x+2} = \frac{10x+4}{x^2-4} \cdot \frac{x+2}{5x^3+2x^2}$
$= \frac{2(5x+2)(x+2)}{(x-2)(x+2) \cdot x^2(5x+2)}$
$= \frac{2}{x^2(x-2)}$

7. $\frac{3x^2-10x+8}{7x-14} \div \frac{9x-12}{21}$
$= \frac{3x^2-10x+8}{7x-14} \cdot \frac{21}{9x-12}$
$= \frac{(3x-4)(x-2) \cdot 3 \cdot 7}{7(x-2) \cdot 3(3x-4)}$
$= 1$

8. a. $\frac{x+3}{x} \cdot \frac{7}{x+3} = \frac{(x+3) \cdot 7}{x \cdot (x+3)} = \frac{7}{x}$

b. $\frac{x+3}{x} \div \frac{7}{x+3} = \frac{x+3}{x} \cdot \frac{x+3}{7} = \frac{(x+3)^2}{7x}$

c. $\frac{3-x}{x^2+6x+5} \cdot \frac{2x+10}{x^2-7x+12}$
$= \frac{(3-x) \cdot 2(x+5)}{(x+5)(x+1) \cdot (x-4)(x-3)}$
$= \frac{-1(x-3) \cdot 2}{(x+1)(x-4)(x-3)}$
$= -\frac{2}{(x+1)(x-4)}$

9. $288 \text{ sq in.} = \frac{288 \text{ sq in.}}{1} \cdot \frac{1 \text{ sq ft}}{144 \text{ sq in.}} = 2 \text{ sq ft}$

10. $3.5 \text{ sq ft} = \frac{3.5 \text{ sq ft}}{1} \cdot \frac{144 \text{ sq in.}}{1 \text{ sq ft}} = 504 \text{ sq in.}$

11. $61{,}000 \text{ sq yd} = 61{,}000 \text{ sq yd} \cdot \frac{9 \text{ sq ft}}{1 \text{ sq yd}}$
$= 549{,}000 \text{ sq ft}$

12. 102.7 feet/second
$= \frac{102.7 \text{ feet}}{1 \text{ second}} \cdot \frac{3600 \text{ seconds}}{1 \text{ hour}} \cdot \frac{1 \text{ mile}}{5280 \text{ feet}}$
$= \frac{102.7 \cdot 3600}{5280} \text{ miles/hour}$
≈ 70.0 miles/hour

Vocabulary, Readiness & Video Check 5.2

1. The expressions $\frac{x}{2y}$ and $\frac{2y}{x}$ are called reciprocals.

2. $\frac{a}{b} \cdot \frac{c}{d} = \frac{a \cdot c}{b \cdot d}$

3. $\frac{a}{b} \div \frac{c}{d} = \frac{a \cdot d}{b \cdot c}$

4. $\frac{x}{7} \cdot \frac{x}{6} = \frac{x^2}{42}$

5. $\frac{x}{7} \div \frac{x}{6} = \frac{6}{7}$

6. Yes, multiplying and simplifying rational expressions often require polynomial factoring. Example 2 alone involves factoring out a GCF, factoring a trinomial with $a \neq 1$, and factoring a difference of squares.

7. Dividing rational expressions is exactly like dividing fractions. Therefore, to divide by a rational expression, multiply by its reciprocal.

8. Multiplication and division of rational expressions are performed similarly—both involve multiplication—but there are important differences. Note the operation first to see whether you multiply by the reciprocal or not.

9. We're converting to cubic feet so we want cubic feet in the numerator. We want cubic yards to divide out so cubic yards is in the denominator.

Exercise Set 5.2

1. $\frac{3x}{y^2} \cdot \frac{7y}{4x} = \frac{3 \cdot 7 \cdot x \cdot y}{4 \cdot x \cdot y \cdot y} = \frac{21}{4y}$

3. $\frac{8x}{2} \cdot \frac{x^5}{4x^2} = \frac{2 \cdot 4 \cdot x \cdot x \cdot x \cdot x \cdot x \cdot x}{2 \cdot 4 \cdot x \cdot x} = x^4$

5. $-\frac{5a^2b}{30a^2b^2}\cdot b^3 = -\frac{5a^2b}{30a^2b^2}\cdot\frac{b^3}{1}$
$= -\frac{5\cdot a^2b\cdot b\cdot b^2}{5\cdot 6\cdot a^2\cdot b^2}$
$= -\frac{b\cdot b}{6}$
$= -\frac{b^2}{6}$

7. $\frac{x}{2x-14}\cdot\frac{x^2-7x}{5} = \frac{x}{2(x-7)}\cdot\frac{x(x-7)}{5}$
$= \frac{x\cdot x}{2\cdot 5}$
$= \frac{x^2}{10}$

9. $\frac{6x+6}{5}\cdot\frac{10}{36x+36} = \frac{6(x+1)}{5}\cdot\frac{10}{36(x+1)}$
$= \frac{6\cdot 10}{5\cdot 36}$
$= \frac{6\cdot 2\cdot 5}{5\cdot 6\cdot 2\cdot 3}$
$= \frac{1}{3}$

11. $\frac{(m+n)^2}{m-n}\cdot\frac{m}{m^2+mn} = \frac{(m+n)(m+n)}{m-n}\cdot\frac{m}{m(m+n)}$
$= \frac{m+n}{m-n}$

13. $\frac{x^2-25}{x^2-3x-10}\cdot\frac{x+2}{x} = \frac{(x+5)(x-5)}{(x+2)(x-5)}\cdot\frac{x+2}{x}$
$= \frac{x+5}{x}$

15. $\frac{x^2+6x+8}{x^2+x-20}\cdot\frac{x^2+2x-15}{x^2+8x+16}$
$= \frac{(x+4)(x+2)}{(x+5)(x-4)}\cdot\frac{(x+5)(x-3)}{(x+4)(x+4)}$
$= \frac{(x+2)(x-3)}{(x-4)(x+4)}$

17. $\frac{5x^7}{2x^5}\div\frac{15x}{4x^3} = \frac{5x^7}{2x^5}\cdot\frac{4x^3}{15x}$
$= \frac{5\cdot x^5\cdot x\cdot x}{2\cdot x^5}\cdot\frac{2\cdot 2\cdot x^3}{5\cdot 3\cdot x}$
$= \frac{2\cdot x\cdot x^3}{3}$
$= \frac{2x^4}{3}$

19. $\frac{8x^2}{y^3}\div\frac{4x^2y^3}{6} = \frac{8x^2}{y^3}\cdot\frac{6}{4x^2y^3}$
$= \frac{2\cdot 4\cdot 6\cdot x\cdot x}{4\cdot x\cdot x\cdot y\cdot y\cdot y\cdot y\cdot y\cdot y}$
$= \frac{2\cdot 6}{y\cdot y\cdot y\cdot y\cdot y\cdot y}$
$= \frac{12}{y^6}$

21. $\frac{(x-6)(x+4)}{4x}\div\frac{2x-12}{8x^2} = \frac{(x-6)(x+4)}{4x}\cdot\frac{8x^2}{2x-12}$
$= \frac{(x-6)(x+4)}{4x}\cdot\frac{4x\cdot 2\cdot x}{2(x-6)}$
$= \frac{x(x+4)}{1}$
$= x(x+4)$

23. $\frac{3x^2}{x^2-1}\div\frac{x^5}{(x+1)^2} = \frac{3x^2}{x^2-1}\cdot\frac{(x+1)^2}{x^5}$
$= \frac{3x^2}{(x-1)(x+1)}\cdot\frac{(x+1)(x+1)}{x^2\cdot x^3}$
$= \frac{3(x+1)}{x^3(x-1)}$

25. $\frac{m^2-n^2}{m+n}\div\frac{m}{m^2+nm} = \frac{m^2-n^2}{m+n}\cdot\frac{m^2+nm}{m}$
$= \frac{(m+n)(m-n)}{m+n}\cdot\frac{m(m+n)}{m}$
$= (m-n)(m+n)$
$= m^2-n^2$

27. $\frac{x+2}{7-x} \div \frac{x^2-5x+6}{x^2-9x+14} = \frac{x+2}{7-x} \cdot \frac{x^2-9x+14}{x^2-5x+6}$
$= \frac{x+2}{7-x} \cdot \frac{(x-7)(x-2)}{(x-3)(x-2)}$
$= \frac{x+2}{-(x-7)} \cdot \frac{(x-7)(x-2)}{(x-3)(x-2)}$
$= -\frac{x+2}{x-3}$

29. $\frac{x^2+7x+10}{x-1} \div \frac{x^2+2x-15}{x-1}$
$= \frac{x^2+7x+10}{x-1} \cdot \frac{x-1}{x^2+2x-15}$
$= \frac{(x+5)(x+2)}{x-1} \cdot \frac{x-1}{(x+5)(x-3)}$
$= \frac{x+2}{x-3}$

31. $\frac{5x-10}{12} \div \frac{4x-8}{8} = \frac{5x-10}{12} \cdot \frac{8}{4x-8}$
$= \frac{5(x-2)}{2 \cdot 6} \cdot \frac{4 \cdot 2}{4(x-2)}$
$= \frac{5}{6}$

33. $\frac{x^2+5x}{8} \cdot \frac{9}{3x+15} = \frac{x(x+5)}{8} \cdot \frac{3 \cdot 3}{3(x+5)} = \frac{3x}{8}$

35. $\frac{7}{6p^2+q} \div \frac{14}{18p^2+3q} = \frac{7}{6p^2+q} \cdot \frac{18p^2+3q}{14}$
$= \frac{7}{6p^2+q} \cdot \frac{3(6p^2+q)}{7 \cdot 2}$
$= \frac{3}{2}$

37. $\frac{3x+4y}{x^2+4xy+4y^2} \cdot \frac{x+2y}{2} = \frac{3x+4y}{(x+2y)^2} \cdot \frac{x+2y}{2}$
$= \frac{3x+4y}{2(x+2y)}$

39. $\frac{(x+2)^2}{x-2} \div \frac{x^2-4}{2x-4} = \frac{(x+2)^2}{x-2} \cdot \frac{2x-4}{x^2-4}$
$= \frac{(x+2)(x+2)}{x-2} \cdot \frac{2(x-2)}{(x-2)(x+2)}$
$= \frac{2(x+2)}{x-2}$

41. $\frac{x^2-4}{24x} \div \frac{2-x}{6xy} = \frac{x^2-4}{24x} \cdot \frac{6xy}{2-x}$
$= \frac{(x+2)(x-2)}{6 \cdot 4 \cdot x} \cdot \frac{6 \cdot x \cdot y}{-1(x-2)}$
$= \frac{y(x+2)}{-4}$
$= -\frac{y(x+2)}{4}$

43. $\frac{a^2+7a+12}{a^2+5a+6} \cdot \frac{a^2+8a+15}{a^2+5a+4}$
$= \frac{(a+4)(a+3)}{(a+3)(a+2)} \cdot \frac{(a+5)(a+3)}{(a+4)(a+1)}$
$= \frac{(a+5)(a+3)}{(a+2)(a+1)}$

45. $\frac{5x-20}{3x^2+x} \cdot \frac{3x^2+13x+4}{x^2-16}$
$= \frac{5(x-4)}{x(3x+1)} \cdot \frac{(x+4)(3x+1)}{(x+4)(x-4)}$
$= \frac{5}{x}$

47. $\frac{8n^2-18}{2n^2-5n+3} \div \frac{6n^2+7n-3}{n^2-9n+8}$
$= \frac{8n^2-18}{2n^2-5n+3} \cdot \frac{n^2-9n+8}{6n^2+7n-3}$
$= \frac{2(4n^2-9)}{(2n-3)(n-1)} \cdot \frac{(n-8)(n-1)}{(3n-1)(2n+3)}$
$= \frac{2(2n-3)(2n+3)}{(2n-3)(n-1)} \cdot \frac{(n-8)(n-1)}{(3n-1)(2n+3)}$
$= \frac{2(n-8)}{3n-1}$

49. 10 square feet
$= \frac{10 \text{ square feet}}{1} \cdot \frac{144 \text{ square inches}}{1 \text{ square foot}}$
$= 1440$ square inches

51. 45 square feet $= \frac{45 \text{ square feet}}{1} \cdot \frac{1 \text{ square yard}}{9 \text{ square feet}}$
$= 5$ square yards

53. 3 cubic yards $= \frac{3 \text{ cubic yards}}{1} \cdot \frac{27 \text{ cubic feet}}{1 \text{ cubic yard}}$
$= 81$ cubic feet

55. $\frac{50 \text{ miles}}{1 \text{ hour}} = \frac{50 \text{ miles}}{1 \text{ hour}} \cdot \frac{5280 \text{ feet}}{1 \text{ mile}} \cdot \frac{1 \text{ hour}}{3600 \text{ seconds}}$
≈ 73 feet per second

57. 6.3 square yards
$= \frac{6.3 \text{ square yards}}{1} \cdot \frac{9 \text{ square feet}}{1 \text{ square yard}}$
$= 56.7$ square feet

59. 133,500 square yards
$= \frac{133{,}500 \text{ square yards}}{1} \cdot \frac{9 \text{ square feet}}{1 \text{ square yard}}$
$= 1{,}201{,}500$ square feet

61. 80.9 feet per second
$= \frac{80.9 \text{ feet}}{1 \text{ second}} \cdot \frac{3600 \text{ seconds}}{1 \text{ hour}} \cdot \frac{1 \text{ mile}}{5280 \text{ feet}}$
≈ 55.2 miles per hour

63. $\frac{1}{5} + \frac{4}{5} = \frac{1+4}{5} = \frac{5}{5} = 1$

65. $\frac{9}{9} - \frac{19}{9} = \frac{9-19}{9} = \frac{-10}{9} = -\frac{10}{9}$

67. $\frac{6}{5} + \left(\frac{1}{5} - \frac{8}{5}\right) = \frac{6}{5} + \left(-\frac{7}{5}\right) = -\frac{1}{5}$

69. $\frac{4}{a} \cdot \frac{1}{b} = \frac{4 \cdot 1}{a \cdot b} = \frac{4}{ab}$
The statement is true.

71. $\frac{x}{5} \cdot \frac{x+3}{4} = \frac{x^2+3x}{20} \neq \frac{2x+3}{20}$
The statement is false.

73. $\frac{2x}{x^2-25} \cdot \frac{x+5}{9x} = \frac{2 \cdot x}{(x+5)(x-5)} \cdot \frac{x+5}{9 \cdot x}$
$= \frac{2}{9(x-5)}$

The area is $\frac{2}{9(x-5)}$ square feet.

75. $\left(\frac{x^2-y^2}{x^2+y^2} \div \frac{x^2-y^2}{3x}\right) \cdot \frac{x^2+y^2}{6}$
$= \frac{x^2-y^2}{x^2+y^2} \cdot \frac{3x}{x^2-y^2} \cdot \frac{x^2+y^2}{6}$
$= \frac{3x}{6}$
$= \frac{x}{2}$

77. $\left(\frac{2a+b}{b^2} \cdot \frac{3a^2-2ab}{ab+2b^2}\right) \div \frac{a^2-3ab+2b^2}{5ab-10b^2}$
$= \frac{2a+b}{b^2} \cdot \frac{3a^2-2ab}{ab+2b^2} \cdot \frac{5ab-10b^2}{a^2-3ab+2b^2}$
$= \frac{2a+b}{b^2} \cdot \frac{a(3a-2b)}{b(a+2b)} \cdot \frac{5 \cdot b(a-2b)}{(a-b)(a-2b)}$
$= \frac{5a(2a+b)(3a-2b)}{b^2(a-b)(a+2b)}$

79. answers may vary

81. $\$2000 \text{ US} = \frac{\$2000 \text{ US}}{1} \cdot \frac{1 \text{ euro}}{\$1.3387 \text{ US}}$
$= \frac{2000}{1.3387}$ euros
≈ 1493.99 euros
On that day, \$2000 US was worth 1493.99 euros.

Section 5.3 Practice Exercises

1. $\frac{8x}{3y} + \frac{x}{3y} = \frac{8x+x}{3y} = \frac{9x}{3y} = \frac{3x}{y}$

2. $\frac{3x}{3x-7} - \frac{7}{3x-7} = \frac{3x-7}{3x-7} = \frac{1}{1}$ or 1

3. $\frac{2x^2+5x}{x+2} - \frac{4x+6}{x+2} = \frac{2x^2+5x-(4x+6)}{x+2}$
$= \frac{2x^2+5x-4x-6}{x+2}$
$= \frac{2x^2+x-6}{x+2}$
$= \frac{(2x-3)(x+2)}{x+2}$
$= 2x-3$

4. a. $\frac{2}{9}, \frac{7}{15}$

$9 = 3^2$ and $15 = 3 \cdot 5$

$\text{LCD} = 3^2 \cdot 5 = 9 \cdot 5 = 45$

b. $\frac{5}{6x^3}, \frac{11}{8x^5}$

$6x^3 = 2 \cdot 3 \cdot x^3$ and $8x^5 = 2^3 \cdot x^5$

$\text{LCD} = 3 \cdot 2^3 \cdot x^5 = 3 \cdot 8 \cdot x^5 = 24x^5$

5. $\frac{3a}{a+5}, \frac{7a}{a-5}$

$\text{LCD} = (a+5)(a-5)$

6. $\frac{7x^2}{(x-4)^2}, \frac{5x}{3x-12}$

$(x-4)^2 = (x-4)(x-4)$ and $3x - 12 = 3(x-4)$

$\text{LCD} = 3(x-4)(x-4) = 3(x-4)^2$

7. $\frac{y+5}{y^2+2y-3}, \frac{y+4}{y^2-3y+2}$

$y^2 + 2y - 3 = (y+3)(y-1)$

$y^2 - 3y + 2 = (y-2)(y-1)$

$\text{LCD} = (y+3)(y-1)(y-2)$

8. $\frac{6}{x-4}, \frac{9}{4-x}$

$(4-x) = -(x-4)$

$\text{LCD} = (x-4)$ or $(4-x)$

9. $\frac{2x}{5y} = \frac{2x}{5y} \cdot 1 = \frac{2x}{5y} \cdot \frac{4x^2y}{4x^2y} = \frac{8x^3y}{20x^2y^2}$

10.
$$\frac{3}{x^2-25} = \frac{3}{(x-5)(x+5)}$$
$$= \frac{3}{(x-5)(x+5)} \cdot \frac{x-3}{x-3}$$
$$= \frac{3(x-3)}{(x-5)(x+5)(x-3)}$$
$$= \frac{3x-9}{(x-5)(x+5)(x-3)}$$

Vocabulary, Readiness & Video Check 5.3

1. $\frac{7}{11} + \frac{2}{11} = \underline{\frac{9}{11}}$

2. $\frac{7}{11} - \frac{2}{11} = \underline{\frac{5}{11}}$

3. $\frac{a}{b} + \frac{c}{b} = \underline{\frac{a+c}{b}}$

4. $\frac{a}{b} - \frac{c}{b} = \underline{\frac{a-c}{b}}$

5. $\frac{5}{x} - \frac{6+x}{x} = \underline{\frac{5-(6+x)}{x}}$

6. In order to carry out the subtraction properly—parentheses make sure each term in the numerator is affected by the subtraction, not just the first term.

7. We completely factor denominators—including coefficients—so we can determine the greatest number of times each unique factor occurs in any one denominator for the LCD.

8. To write an equivalent rational expression, we multiply the <u>numerator</u> of the rational expression by the same expression as the denominator. This means we're multiplying the original rational expression by a factor of <u>one</u> and therefore not changing the <u>value</u> of the original expression.

Exercise Set 5.3

1. $\frac{a}{13} + \frac{9}{13} = \frac{a+9}{13}$

3. $\frac{4m}{3n} + \frac{5m}{3n} = \frac{4m+5m}{3n} = \frac{9m}{3n} = \frac{3m}{n}$

5. $\frac{4m}{m-6} - \frac{24}{m-6} = \frac{4m-24}{m-6} = \frac{4(m-6)}{m-6} = 4$

7. $\frac{9}{3+y} + \frac{y+1}{3+y} = \frac{9+y+1}{3+y} = \frac{y+10}{3+y}$

9. $\frac{5x^2+4x}{x-1}-\frac{2x+3}{x-1}=\frac{5x^2+4x-(2x+3)}{x-1}$
$=\frac{5x^2+2x-3}{x-1}$
$=\frac{(5x-3)(x+1)}{x-1}$

11. $\frac{4a}{a^2+2a-15}-\frac{12}{a^2+2a-15}=\frac{4a-12}{a^2+2a-15}$
$=\frac{4(a-3)}{(a+5)(a-3)}$
$=\frac{4}{a+5}$

13. $\frac{2x+3}{x^2-x-30}-\frac{x-2}{x^2-x-30}=\frac{2x+3-(x-2)}{x^2-x-30}$
$=\frac{x+5}{(x-6)(x+5)}$
$=\frac{1}{x-6}$

15. $\frac{2x+1}{x-3}+\frac{3x+6}{x-3}=\frac{2x+1+3x+6}{x-3}=\frac{5x+7}{x-3}$

17. $\frac{2x^2}{x-5}-\frac{25+x^2}{x-5}=\frac{2x^2-(25+x^2)}{x-5}$
$=\frac{2x^2-25-x^2}{x-5}$
$=\frac{x^2-25}{x-5}$
$=\frac{(x+5)(x-5)}{x-5}$
$=x+5$

19. $\frac{5x+4}{x-1}-\frac{2x+7}{x-1}=\frac{5x+4-(2x+7)}{x-1}$
$=\frac{5x+4-2x-7}{x-1}$
$=\frac{3x-3}{x-1}$
$=\frac{3(x-1)}{x-1}$
$=3$

21. $2x = 2\cdot x$
$4x^3 = 2^2\cdot x^3$
$\text{LCD} = 2^2\cdot x^3 = 4x^3$

23. $8x = 2\cdot 2\cdot 2\cdot x$
$2x + 4 = 2(x + 2)$
$\text{LCD} = 2\cdot 2\cdot 2\cdot x(x + 2) = 8x(x + 2)$

25. $x + 3 = x + 3$
$x - 2 = x - 2$
$\text{LCD} = (x + 3)(x - 2)$

27. $x + 6 = x + 6$
$3x + 18 = 3(x + 6)$
$\text{LCD} = 3(x + 6)$

29. $(x-6)^2=(x-6)(x-6)$
$5x - 30 = 5(x - 6)$
$\text{LCD}=5(x-6)(x-6)=5(x-6)^2$

31. $3x + 3 = 3(x + 1)$
$2x^2+4x+2=2(x^2+2x+1)=2(x+1)^2$
$\text{LCD}=2\cdot 3\cdot (x+1)^2=6(x+1)^2$

33. $8 - x = -(x - 8)$
LCD = $x - 8$ or $8 - x$

35. $x^2+3x-4=(x-1)(x+4)$
$x^2+2x-3=(x-1)(x+3)$
$\text{LCD} = (x - 1)(x + 4)(x + 3)$

37. $3x^2+4x+1=(3x+1)(x+1)$
$2x^2-x-1=(2x+1)(x-1)$
$\text{LCD} = (3x + 1)(x + 1)(2x + 1)(x - 1)$

39. $x^2-16=(x+4)(x-4)$
$2x^3-8x^2=2x^2(x-4)$
$\text{LCD}=2x^2(x+4)(x-4)$

41. $\frac{3}{2x}=\frac{3\cdot 2x}{2x\cdot 2x}=\frac{6x}{4x^2}$

43. $\frac{6}{3a}=\frac{6\cdot 4b^2}{3a\cdot 4b^2}=\frac{24b^2}{12ab^2}$

45. $\frac{9}{2x+6}=\frac{9}{2(x+3)}=\frac{9y}{2(x+3)y}=\frac{9y}{2y(x+3)}$

47. $\frac{9a+2}{5a+10}=\frac{9a+2}{5(a+2)}=\frac{(9a+2)\cdot b}{5(a+2)\cdot b}=\frac{9ab+2b}{5b(a+2)}$

49. $\frac{x}{x^3+6x^2+8x} = \frac{x}{x(x^2+6x+8)}$
$= \frac{x}{x(x+4)(x+2)}$
$= \frac{x(x+1)}{x(x+4)(x+2)(x+1)}$
$= \frac{x^2+x}{x(x+4)(x+2)(x+1)}$

51. $\frac{9y-1}{15x^2-30} = \frac{(9y-1)\cdot 2}{(15x^2-30)\cdot 2} = \frac{18y-2}{30x^2-60}$

53. $\frac{5x}{7}+\frac{9x}{7} = \frac{5x+9x}{7} = \frac{14x}{7} = 2x$

55. $\frac{x+3}{4} \div \frac{2x-1}{4} = \frac{x+3}{4}\cdot\frac{4}{2x-1} = \frac{x+3}{2x-1}$

57. $\frac{x^2}{x-6} - \frac{5x+6}{x-6} = \frac{x^2-(5x+6)}{x-6}$
$= \frac{x^2-5x-6}{x-6}$
$= \frac{(x-6)(x+1)}{x-6}$
$= x+1$

59. $\frac{x^2+5x}{x^2-25}\cdot\frac{3x-15}{x^2} = \frac{x(x+5)}{(x+5)(x-5)}\cdot\frac{3(x-5)}{x^2} = \frac{3}{x}$

61. $\frac{x^3+7x^2}{3x^3-x^2} \div \frac{5x^2+36x+7}{9x^2-1}$
$= \frac{x^3+7x^2}{3x^3-x^2}\cdot\frac{9x^2-1}{5x^2+36x+7}$
$= \frac{x^2(x+7)}{x^2(3x-1)}\cdot\frac{(3x-1)(3x+1)}{(5x+1)(x+7)}$
$= \frac{3x+1}{5x+1}$

63. $\frac{2}{3}+\frac{5}{7} = \frac{2\cdot 7}{3\cdot 7}+\frac{5\cdot 3}{7\cdot 3} = \frac{14}{21}+\frac{15}{21} = \frac{29}{21}$

65. $\frac{2}{6}-\frac{3}{4} = \frac{2\cdot 2}{6\cdot 2}-\frac{3\cdot 3}{4\cdot 3} = \frac{4}{12}-\frac{9}{12} = -\frac{5}{12}$

67. $\frac{1}{12}+\frac{3}{20} = \frac{1\cdot 5}{12\cdot 5}+\frac{3\cdot 3}{20\cdot 3} = \frac{5}{60}+\frac{9}{60} = \frac{14}{60} = \frac{7}{30}$

69. $4a-20 = 4(a-5)$
$(a-5)^2 = (a-5)(a-5)$
$\text{LCD} = 4(a-5)(a-5) = 4(a-5)^2$; d

71. answers may vary

73. The perimeter of a square is 4 times the side length.
$4\cdot\frac{5}{x-2} = \frac{4}{1}\cdot\frac{5}{x-2} = \frac{20}{x-2}$
The perimeter is $\frac{20}{x-2}$ meters.

75. answers may vary

77. $88 = 2^3\cdot 11$
$4332 = 2^2\cdot 3\cdot 19^2$
$\text{LCD} = 2^3\cdot 3\cdot 11\cdot 19^2 = 95,304$ Earth days

79. answers may vary

81. answers may vary

Section 5.4 Practice Exercises

1. a. $\frac{y}{5}-\frac{3y}{15} = \frac{y\cdot 3}{5\cdot 3}-\frac{3y}{15} = \frac{3y-3y}{15} = \frac{0}{15} = 0$

b. $\frac{5}{8x}+\frac{11}{10x^2} = \frac{5\cdot 5x}{8x\cdot 5x}+\frac{11\cdot 4}{10x^2\cdot 4}$
$= \frac{25x}{40x^2}+\frac{44}{40x^2}$
$= \frac{25x+44}{40x^2}$

2. $\frac{10x}{x^2-9}-\frac{5}{x+3} = \frac{10x}{(x-3)(x+3)}-\frac{5(x-3)}{(x+3)(x-3)}$
$= \frac{10x-5(x-3)}{(x+3)(x-3)}$
$= \frac{10x-5x+15}{(x+3)(x-3)}$
$= \frac{5x+15}{(x+3)(x-3)}$
$= \frac{5(x+3)}{(x+3)(x-3)}$
$= \frac{5}{x-3}$

3. $\frac{5}{7x}+\frac{2}{x+1}=\frac{5(x+1)}{7x(x+1)}+\frac{2(7x)}{(x+1)(7x)}$
$=\frac{5(x+1)+2(7x)}{7x(x+1)}$
$=\frac{5x+5+14x}{7x(x+1)}$
$=\frac{19x+5}{7x(x+1)}$

4. $\frac{10}{x-6}-\frac{15}{6-x}=\frac{10}{x-6}-\frac{15}{-(x-6)}$
$=\frac{10}{x-6}-\frac{-15}{x-6}$
$=\frac{10-(-15)}{x-6}$
$=\frac{25}{x-6}$

5. $2+\frac{x}{x+5}=\frac{2}{1}+\frac{x}{x+5}$
$=\frac{2(x+5)}{1(x+5)}+\frac{x}{x+5}$
$=\frac{2x+10+x}{x+5}$
$=\frac{3x+10}{x+5}$

6. $\frac{4}{3x^2+2x}-\frac{3x}{12x+8}=\frac{4}{x(3x+2)}-\frac{3x}{4(3x+2)}$
$=\frac{4(4)}{x(3x+2)(4)}-\frac{3x(x)}{4(3x+2)(x)}$
$=\frac{16-3x^2}{4x(3x+2)}$

7. $\frac{6x}{x^2+4x+4}+\frac{x}{x^2-4}$
$=\frac{6x}{(x+2)(x+2)}+\frac{x}{(x+2)(x-2)}$
$=\frac{6x(x-2)}{(x+2)(x+2)(x-2)}+\frac{x(x+2)}{(x+2)(x-2)(x+2)}$
$=\frac{6x(x-2)+x(x+2)}{(x+2)^2(x-2)}$
$=\frac{6x^2-12x+x^2+2x}{(x+2)^2(x-2)}$
$=\frac{7x^2-10x}{(x+2)^2(x-2)}$
$=\frac{x(7x-10)}{(x+2)^2(x-2)}$

Vocabulary, Readiness & Video Check 5.4

1. $\frac{3}{7x}+\frac{5}{7}=\frac{3}{7x}+\frac{5\cdot x}{7\cdot x}=\frac{3+5x}{7x}$; **b**

2. $\frac{1}{x}+\frac{2}{x^2}=\frac{1\cdot x}{x\cdot x}+\frac{2}{x^2}=\frac{x+2}{x^2}$; **a**

3. The exercise is adding two rational expressions with denominators that are opposites of each other. Recognizing this special case can save us time and effort. If we recognize that one denominator is −1 times the other denominator, we may save many steps.

Exercise Set 5.4

1. $\frac{4}{2x}+\frac{9}{3x}=\frac{4\cdot 3}{2x\cdot 3}+\frac{9\cdot 2}{3x\cdot 2}=\frac{12}{6x}+\frac{18}{6x}=\frac{30}{6x}=\frac{5}{x}$

3. $\frac{15a}{b}+\frac{6b}{5}=\frac{15a\cdot 5}{b\cdot 5}+\frac{6b\cdot b}{5\cdot b}$
$=\frac{75a}{5b}+\frac{6b^2}{5b}$
$=\frac{75a+6b^2}{5b}$

5. $\frac{3}{x}+\frac{5}{2x^2}=\frac{3\cdot 2x}{x\cdot 2x}+\frac{5}{2x^2}=\frac{6x}{2x^2}+\frac{5}{2x^2}=\frac{6x+5}{2x^2}$

7. $$\frac{6}{x+1}+\frac{10}{2x+2}=\frac{6}{x+1}+\frac{10}{2(x+1)}$$
$$=\frac{6}{x+1}+\frac{5}{x+1}$$
$$=\frac{11}{x+1}$$

9. $$\frac{3}{x+2}-\frac{2x}{x^2-4}=\frac{3(x-2)}{(x+2)(x-2)}-\frac{2x}{(x+2)(x-2)}$$
$$=\frac{3x-6-2x}{(x+2)(x-2)}$$
$$=\frac{x-6}{(x+2)(x-2)}$$

11. $$\frac{3}{4x}+\frac{8}{x-2}=\frac{3(x-2)}{4x(x-2)}+\frac{8(4x)}{(x-2)(4x)}$$
$$=\frac{3x-6+32x}{4x(x-2)}$$
$$=\frac{35x-6}{4x(x-2)}$$

13. $$\frac{6}{x-3}+\frac{8}{3-x}=\frac{6}{x-3}+\frac{8}{-(x-3)}$$
$$=\frac{6}{x-3}+\frac{-8}{x-3}$$
$$=\frac{6+(-8)}{x-3}$$
$$=\frac{-2}{x-3}$$
$$=-\frac{2}{x-3}$$

15. $$\frac{9}{x-3}+\frac{9}{3-x}=\frac{9}{x-3}+\frac{9}{-(x-3)}$$
$$=\frac{9}{x-3}+\frac{-9}{x-3}$$
$$=\frac{0}{x-3}$$
$$=0$$

17. $$\frac{-8}{x^2-1}-\frac{7}{1-x^2}=\frac{-8}{x^2-1}-\frac{7}{-(x^2-1)}$$
$$=\frac{-8}{x^2-1}-\frac{-7}{x^2-1}$$
$$=\frac{-8-(-7)}{x^2-1}$$
$$=\frac{-1}{x^2-1}$$
$$=-\frac{1}{x^2-1}$$

19. $$\frac{5}{x}+2=\frac{5}{x}+\frac{2}{1}=\frac{5}{x}+\frac{2\cdot x}{1\cdot x}=\frac{5+2x}{x}$$

21. $$\frac{5}{x-2}+6=\frac{5}{x-2}+\frac{6}{1}$$
$$=\frac{5}{x-2}+\frac{6(x-2)}{1(x-2)}$$
$$=\frac{5+6x-12}{x-2}$$
$$=\frac{6x-7}{x-2}$$

23. $$\frac{y+2}{y+3}-2=\frac{y+2}{y+3}-\frac{2}{1}$$
$$=\frac{y+2}{y+3}-\frac{2(y+3)}{1(y+3)}$$
$$=\frac{y+2}{y+3}-\frac{2y+6}{y+3}$$
$$=\frac{y+2-(2y+6)}{y+3}$$
$$=\frac{y+2-2y-6}{y+3}$$
$$=\frac{-y-4}{y+3}$$
$$=\frac{-(y+4)}{y+3}$$
$$=-\frac{y+4}{y+3}$$

25. $$\frac{-x+2}{x}-\frac{x-6}{4x}=\frac{4(-x+2)}{4x}-\frac{x-6}{4x}$$
$$=\frac{-4x+8-(x-6)}{4x}$$
$$=\frac{-4x+8-x+6}{4x}$$
$$=\frac{-5x+14}{4x} \text{ or } -\frac{5x-14}{4x}$$

27. $\frac{5x}{x+2}-\frac{3x-4}{x+2}=\frac{5x-(3x-4)}{x+2}$
$=\frac{5x-3x+4}{x+2}$
$=\frac{2x+4}{x+2}$
$=\frac{2(x+2)}{x+2}$
$=2$

29. $\frac{3x^4}{7}-\frac{4x^2}{21}=\frac{3x^4\cdot 3}{7\cdot 3}-\frac{4x^2}{21}$
$=\frac{9x^4}{21}-\frac{4x^2}{21}$
$=\frac{9x^4-4x^2}{21}$

31. $\frac{1}{x+3}-\frac{1}{(x+3)^2}=\frac{1\cdot(x+3)}{(x+3)(x+3)}-\frac{1}{(x+3)(x+3)}$
$=\frac{x+3-1}{(x+3)^2}$
$=\frac{x+2}{(x+3)^2}$

33. $\frac{4}{5b}+\frac{1}{b-1}=\frac{4(b-1)}{5b(b-1)}+\frac{1\cdot 5b}{(b-1)(5b)}$
$=\frac{4b-4}{5b(b-1)}+\frac{5b}{5b(b-1)}$
$=\frac{4b-4+5b}{5b(b-1)}$
$=\frac{9b-4}{5b(b-1)}$

35. $\frac{2}{m}+1=\frac{2}{m}+\frac{m}{m}=\frac{2+m}{m}$

37. $\frac{2x}{x-7}-\frac{x}{x-2}=\frac{2x(x-2)}{(x-7)(x-2)}-\frac{x(x-7)}{(x-2)(x-7)}$
$=\frac{2x^2-4x-(x^2-7x)}{(x-7)(x-2)}$
$=\frac{2x^2-4x-x^2+7x}{(x-7)(x-2)}$
$=\frac{x^2+3x}{(x-7)(x-2)}$
$=\frac{x(x+3)}{(x-7)(x-2)}$

39. $\frac{6}{1-2x}-\frac{4}{2x-1}=\frac{6}{1-2x}-\frac{4}{-(1-2x)}$
$=\frac{6}{1-2x}-\frac{-4}{1-2x}$
$=\frac{6+4}{1-2x}$
$=\frac{10}{1-2x}$

41. $\frac{7}{(x+1)(x-1)}+\frac{8}{(x+1)^2}$
$=\frac{7(x+1)}{(x+1)(x-1)(x+1)}+\frac{8(x-1)}{(x+1)^2(x-1)}$
$=\frac{7x+7+8x-8}{(x+1)^2(x-1)}$
$=\frac{15x-1}{(x+1)^2(x-1)}$

43. $\frac{x}{x^2-1}-\frac{2}{x^2-2x+1}$
$=\frac{x(x-1)}{(x-1)(x+1)(x-1)}-\frac{2(x+1)}{(x-1)^2(x+1)}$
$=\frac{x^2-x-2x-2}{(x-1)^2(x+1)}$
$=\frac{x^2-3x-2}{(x-1)^2(x+1)}$

45. $\frac{3a}{2a+6}-\frac{a-1}{a+3}=\frac{3a}{2(a+3)}-\frac{(a-1)(2)}{(a+3)(2)}$
$=\frac{3a-(2a-2)}{2(a+3)}$
$=\frac{3a-2a+2}{2(a+3)}$
$=\frac{a+2}{2(a+3)}$

47. $\frac{y-1}{2y+3}+\frac{3}{(2y+3)^2}=\frac{(y-1)(2y+3)}{(2y+3)^2}+\frac{3}{(2y+3)^2}$
$=\frac{2y^2+3y-2y-3+3}{(2y+3)^2}$
$=\frac{2y^2+y}{(2y+3)^2}$
$=\frac{y(2y+1)}{(2y+3)^2}$

49. $\frac{5}{2-x}+\frac{x}{2x-4}=\frac{5}{-(x-2)}+\frac{x}{2(x-2)}$
$=\frac{-5(2)}{(x-2)(2)}+\frac{x}{2(x-2)}$
$=\frac{-10+x}{2(x-2)}$
$=\frac{x-10}{2(x-2)}$

51. $\frac{15}{x^2+6x-19}+\frac{2}{x+3}=\frac{15}{(x+3)^2}+\frac{2(x+3)}{(x+3)(x+3)}$
$=\frac{15+2x+6}{(x+3)^2}$
$=\frac{2x+21}{(x+3)^2}$

53. $\frac{13}{x^2-5x+6}-\frac{5}{x-3}$
$=\frac{13}{(x-3)(x-2)}-\frac{5(x-2)}{(x-3)(x-2)}$
$=\frac{13-(5x-10)}{(x-3)(x-2)}$
$=\frac{13-5x+10}{(x-3)(x-2)}$
$=\frac{-5x+23}{(x-3)(x-2)}$

55. $\frac{70}{m^2-100}+\frac{7}{2(m+10)}$
$=\frac{70\cdot 2}{(m-10)(m+10)2}+\frac{7(m-10)}{2(m+10)(m-10)}$
$=\frac{140+7m-70}{2(m-10)(m+10)}$
$=\frac{7m+70}{2(m-10)(m+10)}$
$=\frac{7(m+10)}{2(m-10)(m+10)}$
$=\frac{7}{2(m-10)}$

57. $\frac{x+8}{x^2-5x-6}+\frac{x+1}{x^2-4x-5}$
$=\frac{x+8}{(x+1)(x-6)}+\frac{x+1}{(x+1)(x-5)}$
$=\frac{(x+8)(x-5)+(x+1)(x-6)}{(x+1)(x-6)(x-5)}$
$=\frac{x^2+3x-40+x^2-5x-6}{(x+1)(x-6)(x-5)}$
$=\frac{2x^2-2x-46}{(x+1)(x-6)(x-5)}$
$=\frac{2(x^2-x-23)}{(x+1)(x-6)(x-5)}$

59. $\frac{5}{4n^2-12n+8}-\frac{3}{3n^2-6n}$
$=\frac{5\cdot 3n}{4(n-2)(n-1)3n}-\frac{3(n-1)\cdot 4}{3n(n-2)(n-1)4}$
$=\frac{15n-12n+12}{12n(n-2)(n-1)}$
$=\frac{3n+12}{12n(n-2)(n-1)}$
$=\frac{3(n+4)}{12n(n-2)(n-1)}$
$=\frac{n+4}{4n(n-2)(n-1)}$

61. $\frac{15x}{x+8}\cdot\frac{2x+16}{3x}=\frac{5\cdot 3x}{x+8}\cdot\frac{2(x+8)}{3x}=\frac{5\cdot 2}{1}=10$

63. $\frac{8x+7}{3x+5}-\frac{2x-3}{3x+5}=\frac{8x+7-(2x-3)}{3x+5}$
$=\frac{8x+7-2x+3}{3x+5}$
$=\frac{6x+10}{3x+5}$
$=\frac{2(3x+5)}{3x+5}$
$=2$

65. $\frac{5a+10}{18}\div\frac{a^2-4}{10a}=\frac{5a+10}{18}\cdot\frac{10a}{a^2-4}$
$=\frac{5(a+2)}{9\cdot 2}\cdot\frac{2\cdot 5a}{(a+2)(a-2)}$
$=\frac{5\cdot 5a}{9(a-2)}$
$=\frac{25a}{9(a-2)}$

67. $\dfrac{5}{x^2-3x+2}+\dfrac{1}{x-2}=\dfrac{5}{(x-2)(x-1)}+\dfrac{1(x-1)}{(x-2)(x-1)}$
$=\dfrac{5+x-1}{(x-2)(x-1)}$
$=\dfrac{x+4}{(x-2)(x-1)}$

69. $3x+5=7$
$3x=2$
$x=\dfrac{2}{3}$

71. $2x^2-x-1=0$
$(2x+1)(x-1)=0$
$2x+1=0$ or $x-1=0$
$2x=-1$ $\quad x=1$
$x=-\dfrac{1}{2}$

73. $4(x+6)+3=-3$
$4x+24+3=-3$
$4x+27=-3$
$4x=-30$
$x=\dfrac{-30}{4}$
$x=-\dfrac{15}{2}$

75. $\dfrac{3}{x}-\dfrac{2x}{x^2-1}+\dfrac{5}{x+1}=\dfrac{3}{x}-\dfrac{2x}{(x-1)(x+1)}+\dfrac{5}{x+1}$
$=\dfrac{3(x-1)(x+1)}{x(x-1)(x+1)}-\dfrac{2x\cdot x}{(x-1)(x+1)x}+\dfrac{5\cdot x(x-1)}{(x+1)x(x-1)}$
$=\dfrac{3x^2-3-2x^2+5x^2-5x}{x(x-1)(x+1)}$
$=\dfrac{6x^2-5x-3}{x(x-1)(x+1)}$

77. $\frac{5}{x^2-4}+\frac{2}{x^2-4x+4}-\frac{3}{x^2-x-6}=\frac{5}{(x+2)(x-2)}+\frac{2}{(x-2)^2}-\frac{3}{(x+2)(x-3)}$

$$=\frac{5(x-2)(x-3)+2(x+2)(x-3)-3(x-2)^2}{(x+2)(x-2)^2(x-3)}$$

$$=\frac{5(x^2-5x+6)+2(x^2-x-6)-3(x^2-4x+4)}{(x-2)^2(x+2)(x-3)}$$

$$=\frac{5x^2-25x+30+2x^2-2x-12-3x^2+12x-12}{(x-2)^2(x+2)(x-3)}$$

$$=\frac{4x^2-15x+6}{(x-2)^2(x+2)(x-3)}$$

79. $\frac{9}{x^2+9x+14}-\frac{3x}{x^2+10x+21}+\frac{x+4}{x^2+5x+6}=\frac{9}{(x+2)(x+7)}-\frac{3x}{(x+3)(x+7)}+\frac{x+4}{(x+2)(x+3)}$

$$=\frac{9(x+3)}{(x+2)(x+7)(x+3)}-\frac{3x(x+2)}{(x+3)(x+7)(x+2)}+\frac{(x+4)(x+7)}{(x+2)(x+3)(x+7)}$$

$$=\frac{9x+27-3x^2-6x+x^2+7x+4x+28}{(x+2)(x+7)(x+3)}$$

$$=\frac{-2x^2+14x+55}{(x+2)(x+7)(x+3)}$$

81. $\frac{3}{x+4}-\frac{1}{x-4}=\frac{3(x-4)-1(x+4)}{(x+4)(x-4)}$

$$=\frac{3x-12-x-4}{(x+4)(x-4)}$$

$$=\frac{2x-16}{(x+4)(x-4)}$$

$$=\frac{2(x-8)}{(x+4)(x-4)}$$

The other piece of the board measures $\frac{2(x-8)}{(x+4)(x-4)}$ inches.

83. $1-\frac{G}{P}=\frac{P}{P}-\frac{G}{P}=\frac{P-G}{P}$

85. answers may vary

87. $90-\frac{40}{x}=\frac{90}{1}-\frac{40}{x}$

$$=\frac{90x}{x}-\frac{40}{x}$$

$$=\frac{90x-40}{x}$$

The complement measures $\left(\frac{90x-40}{x}\right)^\circ$.

89. answers may vary

Section 5.5 Practice Exercises

1. The LCD is 20.

$$\frac{x}{4}+\frac{4}{5}=\frac{1}{20}$$
$$20\left(\frac{x}{4}+\frac{4}{5}\right)=20\left(\frac{1}{20}\right)$$
$$20\left(\frac{x}{4}\right)+20\left(\frac{4}{5}\right)=20\left(\frac{1}{20}\right)$$
$$5x+16=1$$
$$5x=-15$$
$$x=-3$$

The solution is –3.

2. The LCD is 15.

$$\frac{x+2}{3}-\frac{x-1}{5}=\frac{1}{15}$$
$$15\left(\frac{x+2}{3}-\frac{x-1}{5}\right)=15\left(\frac{1}{15}\right)$$
$$15\left(\frac{x+2}{3}\right)-15\left(\frac{x-1}{5}\right)=15\left(\frac{1}{15}\right)$$
$$5(x+2)-3(x-1)=1$$
$$5x+10-3x+3=1$$
$$2x+13=1$$
$$2x=-12$$
$$x=-6$$

The solution is –6.

3. The LCD is x.

$$2+\frac{6}{x}=x+7$$
$$x\left(2+\frac{6}{x}\right)=x(x+7)$$
$$x(2)+x\left(\frac{6}{x}\right)=x\cdot x+x\cdot 7$$
$$2x+6=x^2+7x$$
$$0=x^2+5x-6$$
$$0=(x+6)(x-1)$$
$$x+6=0 \quad \text{or} \quad x-1=0$$
$$x=-6 \qquad\qquad x=1$$

Neither –6 nor 1 makes the denominator 0 in the original equation. Both –6 and 1 are solutions.

4. The LCD is $(x + 3)(x - 3)$.

$$\frac{2}{x+3}+\frac{3}{x-3}=\frac{-2}{x^2-9}$$
$$(x+3)(x-3)\left(\frac{2}{x+3}+\frac{3}{x-3}\right)=(x+3)(x-3)\left(\frac{-2}{x^2-9}\right)$$
$$(x+3)(x-3)\cdot\frac{2}{x+3}+(x+3)(x-3)\cdot\frac{3}{x-3}=(x+3)(x-3)\left(\frac{-2}{x^2-9}\right)$$
$$2(x-3)+3(x+3)=-2$$
$$2x-6+3x+9=-2$$
$$5x+3=-2$$
$$5x=-5$$
$$x=-1$$

The solution is -1.

5. The LCD is $x - 1$.

$$\frac{5x}{x-1}=\frac{5}{x-1}+3$$
$$(x-1)\left(\frac{5x}{x-1}\right)=(x-1)\left(\frac{5}{x-1}+3\right)$$
$$(x-1)\cdot\frac{5x}{x-1}=(x-1)\cdot\frac{5}{x-1}+(x-1)\cdot 3$$
$$5x=5+3(x-1)$$
$$5x=5+3x-3$$
$$2x=2$$
$$x=1$$

1 makes the denominator 0 in the original equation. Therefore 1 is *not* a solution and this equation has no solution.

6. The LCD is $x + 3$.

$$x-\frac{6}{x+3}=\frac{2x}{x+3}+2$$
$$(x+3)\left(x-\frac{6}{x+3}\right)=(x+3)\left(\frac{2x}{x+3}+2\right)$$
$$(x+3)(x)-(x+3)\left(\frac{6}{x+3}\right)=(x+3)\left(\frac{2x}{x+3}\right)+(x+3)(2)$$
$$(x+3)(x)-6=2x+2(x+3)$$
$$x^2+3x-6=2x+2x+6$$
$$x^2-x-12=0$$
$$(x-4)(x+3)=0$$
$$x-4=0 \quad \text{or} \quad x+3=0$$
$$x=4 \qquad\qquad x=3$$

$x = 3$ can't be a solution of the original equation. The only solution is 4.

7. The LCD is abx.

$$\frac{1}{a}+\frac{1}{b}=\frac{1}{x}$$
$$abx\left(\frac{1}{a}+\frac{1}{b}\right)=abx\left(\frac{1}{x}\right)$$
$$abx\left(\frac{1}{a}\right)+abx\left(\frac{1}{b}\right)=abx\left(\frac{1}{x}\right)$$
$$bx+ax=ab$$
$$bx=ab-ax$$
$$bx=a(b-x)$$
$$\frac{bx}{b-x}=\frac{a(b-x)}{b-x}$$
$$\frac{bx}{b-x}=a$$

Vocabulary, Readiness & Video Check 5.5

1.
$$4\left(\frac{3x}{2}+5\right)=4\left(\frac{1}{4}\right)$$
$$4\left(\frac{3x}{2}\right)+4\cdot 5=4\left(\frac{1}{4}\right)$$
$$6x+20=1$$
The correct choice is **c**.

2.
$$5x\left(\frac{1}{x}-\frac{3}{5x}\right)=5x(2)$$
$$5x\left(\frac{1}{x}\right)-5x\left(\frac{3}{5x}\right)=5x(2)$$
$$5-3=10x$$
The correct choice is **b**.

3. The LCD of $\frac{9}{x}$, $\frac{3}{4}$, and $\frac{1}{12}=\frac{1}{3\cdot 4}$ is $12x$; **b**.

4. The LCD of $\frac{8}{3x}$, $\frac{1}{x}$, and $\frac{7}{9}=\frac{7}{3\cdot 3}$ is $9x$; **d**.

5. The LCD of $\frac{9}{x-1}$ and $\frac{7}{(x-1)^2}$ is $(x-1)^2$; **a**.

6. The LCD of $\frac{1}{x-2}$, $\frac{3}{x^2-4}=\frac{3}{(x+2)(x-2)}$, and $8=\frac{8}{1}$ is (x^2-4); **c**.

7. These equations are solved in very different ways, so we need to determine the next correct step to make. For a linear equation, we first "move" variable terms to one side and numbers to the other; for a quadratic equation, we first set the equation equal to 0.

8. If there are variables in any denominators, we should first check to see if the proposed solutions make these denominators zero in the original equation, giving us an undefined rational expression. If so, that solution is an extraneous solution and is not a solution to the equation.

9. the steps for solving an equation containing rational expressions; as if it's the only variable in the equation

Exercise Set 5.5

1. The LCD is 5.

$$\frac{x}{5}+3=9$$
$$5\left(\frac{x}{5}+3\right)=5(9)$$
$$x+15=45$$
$$x=30$$

Check:
$$\frac{x}{5}+3=9$$
$$\frac{30}{5}+3\stackrel{?}{=}9$$
$$6+3\stackrel{?}{=}9$$
$$9=9 \quad \text{True}$$
The solution is 30.

3. The LCD is 12.

$$\frac{x}{2}+\frac{5x}{4}=\frac{x}{12}$$
$$12\left(\frac{x}{2}+\frac{5x}{4}\right)=12\left(\frac{x}{12}\right)$$
$$6x+15x=x$$
$$21x=x$$
$$20x=0$$
$$x=0$$

Check:
$$\frac{x}{2}+\frac{5x}{4}=\frac{x}{12}$$
$$\frac{0}{2}+\frac{5(0)}{4}\stackrel{?}{=}\frac{0}{12}$$
$$0+0\stackrel{?}{=}0$$
$$0=0 \quad \text{True}$$
The solution is 0.

5. The LCD is x.

$$2-\frac{8}{x}=6$$
$$x\left(2-\frac{8}{x}\right)=x(6)$$
$$2x-8=6x$$
$$-8=4x$$
$$-2=x$$

Check: $2-\frac{8}{x}=6$

$$2-\frac{8}{-2}\stackrel{?}{=}6$$
$$2-(-4)\stackrel{?}{=}6$$
$$2+4=6$$
$$6=6 \quad \text{True}$$

The solution is −2.

7. The LCD is x.

$$2+\frac{10}{x}=x+5$$
$$x\left(2+\frac{10}{x}\right)=x(x+5)$$
$$2x+10=x^2+5x$$
$$0=x^2+3x-10$$
$$0=(x+5)(x-2)$$
$$x+5=0 \quad \text{or} \quad x-2=0$$
$$x=-5 \qquad\qquad x=2$$

Check −5: $2+\frac{10}{x}=x+5$

$$2+\frac{10}{-5}\stackrel{?}{=}-5+5$$
$$2+(-2)\stackrel{?}{=}-5+5$$
$$0=0 \quad \text{True}$$

Check 2: $2+\frac{10}{x}=x+5$

$$2+\frac{10}{2}\stackrel{?}{=}2+5$$
$$2+5\stackrel{?}{=}2+5$$
$$7=7 \quad \text{True}$$

The solutions are −5 and 2.

9. The LCD is $5 \cdot 2 = 10$.

$$\frac{a}{5}=\frac{a-3}{2}$$
$$10\left(\frac{a}{5}\right)=10\left(\frac{a-3}{2}\right)$$
$$2a=5(a-3)$$
$$2a=5a-15$$
$$-3a=-15$$
$$a=5$$

Check: $\frac{a}{5}=\frac{a-3}{2}$

$$\frac{5}{5}\stackrel{?}{=}\frac{5-3}{2}$$
$$1\stackrel{?}{=}\frac{2}{2}$$
$$1=1 \quad \text{True}$$

The solution is 5.

11. The LCD is $5 \cdot 2 = 10$.

$$\frac{x-3}{5}+\frac{x-2}{2}=\frac{1}{2}$$
$$10\left(\frac{x-3}{5}+\frac{x-2}{2}\right)=10\left(\frac{1}{2}\right)$$
$$2(x-3)+5(x-2)=\frac{10}{2}$$
$$2x-6+5x-10=5$$
$$7x-16=5$$
$$7x=21$$
$$x=3$$

Check: $\frac{x-3}{5}+\frac{x-2}{2}=\frac{1}{2}$

$$\frac{3-3}{5}+\frac{3-2}{2}\stackrel{?}{=}\frac{1}{2}$$
$$\frac{0}{5}+\frac{1}{2}\stackrel{?}{=}\frac{1}{2}$$
$$\frac{1}{2}=\frac{1}{2} \quad \text{True}$$

The solution is 3.

13. The LCD is $2a - 5$.

$$\frac{3}{2a-5}=-1$$
$$(2a-5)\left(\frac{3}{2a-5}\right)=(2a-5)(-1)$$
$$3=-2a+5$$
$$-2=-2a$$
$$1=a$$

Check: $\frac{3}{2a-5}=-1$

$$\frac{3}{2(1)-5}\stackrel{?}{=}-1$$
$$\frac{3}{2-5}\stackrel{?}{=}-1$$
$$\frac{3}{-3}\stackrel{?}{=}-1$$
$$-1=-1 \quad \text{True}$$

The solution is 1.

15. The LCD is $y - 4$.

$$\frac{4y}{y-4}+5=\frac{5y}{y-4}$$
$$(y-4)\left(\frac{4y}{y-4}+5\right)=(y-4)\left(\frac{5y}{y-4}\right)$$
$$4y+5(y-4)=5y$$
$$4y+5y-20=5y$$
$$9y-20=5y$$
$$-20=-4y$$
$$5=y$$

Check: $\frac{4y}{y-4}+5=\frac{5y}{y-4}$

$$\frac{4\cdot 5}{5-4}+5\stackrel{?}{=}\frac{5\cdot 5}{5-4}$$
$$\frac{20}{1}+5\stackrel{?}{=}\frac{25}{1}$$
$$25=25 \quad \text{True}$$

The solution is 5.

17. The LCD is $a - 3$.

$$2+\frac{3}{a-3}=\frac{a}{a-3}$$
$$(a-3)\left(2+\frac{3}{a-3}\right)=(a-3)\frac{a}{a-3}$$
$$2(a-3)+3=a$$
$$2a-6+3=a$$
$$2a-3=a$$
$$-3=-a$$
$$3=a$$

Check: $2+\frac{3}{a-3}=\frac{a}{a-3}$

$$2+\frac{3}{3-3}\stackrel{?}{=}\frac{3}{3-3}$$
$$2+\frac{3}{0}\stackrel{?}{=}\frac{3}{0}$$

Since $\frac{3}{0}$ is undefined, $a = 3$ does not check and the equation has no solution.

19. The LCD is $(x+3)(x-3)=x^2-9$.

$$\frac{1}{x+3}+\frac{6}{x^2-9}=1$$
$$(x^2-9)\left(\frac{1}{x+3}+\frac{6}{x^2-9}\right)=1(x^2-9)$$
$$(x-3)+6=x^2-9$$
$$x+3=x^2-9$$
$$0=x^2-x-12$$
$$0=(x-4)(x+3)$$
$$x-4=0 \quad \text{or} \quad x+3=0$$
$$x=4 \qquad\qquad x=-3$$

Check 4: $\frac{1}{x+3}+\frac{6}{x^2-9}=1$

$$\frac{1}{4+3}+\frac{6}{4^2-9}\stackrel{?}{=}1$$
$$\frac{1}{7}+\frac{6}{16-9}\stackrel{?}{=}1$$
$$\frac{1}{7}+\frac{6}{7}\stackrel{?}{=}1$$
$$\frac{7}{7}=1 \quad \text{True}$$

Check −3: $\frac{1}{x+3}+\frac{6}{x^2-9}=1$

$$\frac{1}{-3+3}+\frac{6}{(-3)^2-9}\stackrel{?}{=}1$$
$$\frac{1}{0}+\frac{6}{0}\stackrel{?}{=}1$$

Since $\frac{1}{0}$ and $\frac{6}{0}$ are undefined, $x = -3$ does not check. The solution is 4.

21. The LCD is $y + 4$.

$$\frac{2y}{y+4}+\frac{4}{y+4}=3$$
$$(y+4)\left(\frac{2y}{y+4}+\frac{4}{y+4}\right)=(y+4)(3)$$
$$2y+4=3y+12$$
$$4=y+12$$
$$-8=y$$

Check: $\frac{2y}{y+4}+\frac{4}{y+4}=3$

$$\frac{2(-8)}{-8+4}+\frac{4}{-8+4}\stackrel{?}{=}3$$
$$\frac{-16}{-4}+\frac{4}{-4}\stackrel{?}{=}3$$
$$4-1\stackrel{?}{=}3$$
$$3=3 \quad \text{True}$$

The solution is −8.

23. The LCD is $(x+2)(x-2)=x^2-4$.

$$\frac{2x}{x+2}-2=\frac{x-8}{x-2}$$
$$(x^2-4)\left(\frac{2x}{x+2}-2\right)=(x^2-4)\left(\frac{x-8}{x-2}\right)$$
$$2x(x-2)-2(x^2-4)=(x+2)(x-8)$$
$$2x^2-4x-2x^2+8=x^2-8x+2x-16$$
$$-4x+8=x^2-6x-16$$
$$0=x^2-2x-24$$
$$0=(x-6)(x+4)$$
$$x-6=0 \quad \text{or} \quad x+4=0$$
$$x=6 \qquad\qquad x=-4$$

Check 6: $\dfrac{2x}{x+2}-2=\dfrac{x-8}{x-2}$
$$\frac{2(6)}{6+2}-2\stackrel{?}{=}\frac{6-8}{6-2}$$
$$\frac{12}{8}-2\stackrel{?}{=}\frac{-2}{4}$$
$$\frac{3}{2}-\frac{4}{2}\stackrel{?}{=}-\frac{1}{2}$$
$$-\frac{1}{2}=-\frac{1}{2} \quad \text{True}$$

Check –4: $\dfrac{2x}{x+2}-2=\dfrac{x-8}{x-2}$
$$\frac{2(-4)}{-4+2}-2\stackrel{?}{=}\frac{-4-8}{-4-2}$$
$$\frac{-8}{-2}-2\stackrel{?}{=}\frac{-12}{-6}$$
$$4-2\stackrel{?}{=}2$$
$$2=2 \quad \text{True}$$

The solutions are 6 and –4.

25. The LCD is $2y$.

$$\frac{2}{y}+\frac{1}{2}=\frac{5}{2y}$$
$$2y\left(\frac{2}{y}+\frac{1}{2}\right)=2y\left(\frac{5}{2y}\right)$$
$$2(2)+y(1)=5$$
$$4+y=5$$
$$y=1$$

The solution $y = 1$ checks.

27. The LCD is $(a-6)(a-1)$.

$$\frac{a}{a-6}=\frac{-2}{a-1}$$
$$(a-6)(a-1)\left(\frac{a}{a-6}\right)=(a-6)(a-1)\left(\frac{-2}{a-1}\right)$$
$$a(a-1)=-2(a-6)$$
$$a^2-a=-2a+12$$
$$a^2+a-12=0$$
$$(a+4)(a-3)=0$$
$$a+4=0 \quad \text{or} \quad a-3=0$$
$$a=-4 \qquad\qquad a=3$$

The solutions $a = -4$ and $a = 3$ check.

29. The LCD is $2x \cdot 3 = 6x$.

$$\frac{11}{2x}+\frac{2}{3}=\frac{7}{2x}$$
$$6x\left(\frac{11}{2x}+\frac{2}{3}\right)=6x\left(\frac{7}{2x}\right)$$
$$3(11)+2x(2)=3(7)$$
$$33+4x=21$$
$$4x=-12$$
$$x=-3$$

The solution $x = -3$ checks.

31. The LCD is $(x-2)(x+2)$.

$$\frac{2}{x-2}+1=\frac{x}{x+2}$$
$$(x-2)(x+2)\left(\frac{2}{x-2}+1\right)=(x-2)(x+2)\left(\frac{x}{x+2}\right)$$
$$2(x+2)+1(x-2)(x+2)=x(x-2)$$
$$2x+4+x^2-4=x^2-2x$$
$$4x=0$$
$$x=0$$

The solution $x = 0$ checks.

33. The LCD is 6.

$$\frac{x+1}{3}-\frac{x-1}{6}=\frac{1}{6}$$
$$6\left(\frac{x+1}{3}-\frac{x-1}{6}\right)=6\left(\frac{1}{6}\right)$$
$$2(x+1)-(x-1)=1$$
$$2x+2-x+1=1$$
$$x+3=1$$
$$x=-2$$

The solution $x = -2$ checks.

35. The LCD is $6(t-4)$.

$$\frac{t}{t-4}=\frac{t+4}{6}$$
$$6(t-4)\left(\frac{t}{t-4}\right)=6(t-4)\left(\frac{t+4}{6}\right)$$
$$6t=(t-4)(t+4)$$
$$6t=t^2-16$$
$$0=t^2-6t-16$$
$$0=(t-8)(t+2)$$
$$t-8=0 \quad \text{or} \quad t+2=0$$
$$t=8 \qquad\qquad t=-2$$

The solutions $t = 8$ and $t = -2$ check.

37. $2y + 2 = 2(y + 1)$
$4y + 4 = 4(y + 1) = 2 \cdot 2(y + 1)$
The LCD is $4(y + 1)$.

$$\frac{y}{2y+2}+\frac{2y-16}{4y+4}=\frac{2y-3}{y+1}$$
$$4(y+1)\left(\frac{y}{2(y+1)}+\frac{2y-16}{4(y+1)}\right)=4(y+1)\left(\frac{2y-3}{y+1}\right)$$
$$2y+2y-16=4(2y-3)$$
$$4y-16=8y-12$$
$$-16=4y-12$$
$$-4=4y$$
$$-1=y$$

The solution $y = -1$ makes the denominators $2y + 2$, $4y + 4$, and $y + 1$ zero, so the equation has no solution.

39. $r^2+5r-14=(r+7)(r-2)$
The LCD is $(r + 7)(r - 2)$.

$$\frac{4r-4}{r^2+5r-14}+\frac{2}{r+7}=\frac{1}{r-2}$$
$$(r+7)(r-2)\left(\frac{4r-4}{(r+7)(r-2)}+\frac{2}{r+7}\right)=(r+7)(r-2)\left(\frac{1}{r-2}\right)$$
$$4r-4+2(r-2)=1(r+7)$$
$$4r-4+2r-4=r+7$$
$$6r-8=r+7$$
$$5r=15$$
$$r=3$$

The solution $r = 3$ checks.

41. $x^2+x-6=(x+3)(x-2)$
The LCD is $(x + 3)(x - 2)$.

$$\frac{x+1}{x+3}=\frac{x^2-11x}{x^2+x-6}-\frac{x-3}{x-2}$$
$$(x+3)(x-2)\left(\frac{x+1}{x+3}\right)=(x+3)(x-2)\left(\frac{x^2-11x}{x^2+x-6}-\frac{x-3}{x-2}\right)$$
$$(x-2)(x+1)=x^2-11x-(x+3)(x-3)$$
$$x^2-x-2=x^2-11x-(x^2-9)$$
$$x^2-x-2=x^2-11x-x^2+9$$
$$x^2-x-2=-11x+9$$
$$x^2+10x-11=0$$
$$(x+11)(x-1)=0$$
$$x+11=0 \quad \text{or} \quad x-1=0$$
$$x=-11 \qquad\qquad x=1$$

The solutions $x = -11$ and $x = 1$ check.

43. The LCD is I.

$$R=\frac{E}{I}$$
$$I(R)=I\left(\frac{E}{I}\right)$$
$$IR=E$$
$$I=\frac{E}{R}$$

45. The LCD is $B + E$.

$$T=\frac{2U}{B+E}$$
$$(B+E)T=(B+E)\left(\frac{2U}{B+E}\right)$$
$$BT+ET=2U$$
$$BT=2U-ET$$
$$B=\frac{2U-ET}{T}$$

47. The LCD is h^2.

$$B=\frac{705w}{h^2}$$
$$h^2(B)=h^2\left(\frac{705w}{h^2}\right)$$
$$Bh^2=705w$$
$$\frac{Bh^2}{705}=w$$

49. The LCD is G.

$$N = R + \frac{V}{G}$$
$$G(N) = G\left(R + \frac{V}{G}\right)$$
$$GN = GR + V$$
$$GN - GR = V$$
$$G(N - R) = V$$
$$G = \frac{V}{N - R}$$

51. The LCD is πr.

$$\frac{C}{\pi r} = 2$$
$$\pi r\left(\frac{C}{\pi r}\right) = \pi r(2)$$
$$C = 2\pi r$$
$$\frac{C}{2\pi} = r$$

53. The LCD is $3xy$.

$$\frac{1}{y} + \frac{1}{3} = \frac{1}{x}$$
$$3xy\left(\frac{1}{y} + \frac{1}{3}\right) = 3xy\left(\frac{1}{x}\right)$$
$$3x + xy = 3y$$
$$x(3 + y) = 3y$$
$$x = \frac{3y}{3 + y}$$

55. The reciprocal of x is $\frac{1}{x}$.

57. The reciprocal of x, added to the reciprocal of 2 is $\frac{1}{x} + \frac{1}{2}$.

59. $\frac{1}{3}$ of the tank is filled in 1 hour.

61. answers may vary

63. $\frac{1}{x} + \frac{5}{9}$ is an expression.

$$\frac{1}{x} + \frac{5}{9} = \frac{1 \cdot 9}{x \cdot 9} + \frac{5 \cdot x}{9 \cdot x} = \frac{9}{9x} + \frac{5x}{9x} = \frac{5x + 9}{9x}$$

65. $\frac{5}{x-1} - \frac{2}{x} = \frac{5}{x(x-1)}$ is an equation. The LCD is $x(x - 1)$.

$$\frac{5}{x-1} - \frac{2}{x} = \frac{5}{x(x-1)}$$
$$x(x-1)\left(\frac{5}{x-1} - \frac{2}{x}\right) = x(x-1)\left(\frac{5}{x(x-1)}\right)$$
$$5x - 2(x - 1) = 5$$
$$5x - 2x + 2 = 5$$
$$3x + 2 = 5$$
$$3x = 3$$
$$x = 1$$

Since $\frac{5}{0}$ is undefined, $x = 1$ does not check and the equation has no solution.

67. $\frac{20x}{3} + \frac{32x}{6} = 180$

The LCD is 6.

$$6\left(\frac{20x}{3} + \frac{32x}{6}\right) = 6(180)$$
$$2(20x) + 32x = 1080$$
$$40x + 32x = 1080$$
$$72x = 1080$$
$$x = 15$$
$$\frac{20x}{3} = \frac{20(15)}{3} = 100°$$
$$\frac{32x}{6} = \frac{32(15)}{6} = 80°$$

The angles measure 100° and 80°.

69. $\frac{450}{x} + \frac{150}{x} = 90$

The LCD is x.

$$x\left(\frac{450}{x} + \frac{150}{x}\right) = x(90)$$
$$450 + 150 = 90x$$
$$600 = 90x$$
$$\frac{600}{90} = x$$
$$\frac{20}{3} = x$$
$$\frac{450}{x} = 450 \div \frac{20}{3} = \frac{450}{1} \cdot \frac{3}{20} = \frac{1350}{20} = 67.5$$
$$\frac{150}{x} = 150 \div \frac{20}{3} = \frac{150}{1} \cdot \frac{3}{20} = \frac{450}{20} = 22.5$$

The angles measure 22.5° and 67.5°.

71. $a^2+4a+3=(a+3)(a+1)$

$a^2+a-6=(a+3)(a-2)$

$a^2-a-2=(a+1)(a-2)$

The LCD is $(a + 3)(a + 1)(a - 2)$.

$$\begin{aligned}\frac{4}{a^2+4a+3}+\frac{2}{a^2+a-6}-\frac{3}{a^2-a-2}&=0\\(a+3)(a+1)(a-2)\left(\frac{4}{a^2+4a+3}+\frac{2}{a^2+a-6}-\frac{3}{a^2-a-2}\right)&=(a+3)(a+1)(a-2)(0)\\4(a-2)+2(a+1)-3(a+3)&=0\\4a-8+2a+2-3a-9&=0\\3a-15&=0\\3a&=15\\a&=5\end{aligned}$$

The solution $a = 5$ checks.

Integrated Review

1. $\frac{1}{x}+\frac{2}{3}=\frac{1\cdot 3}{x\cdot 3}+\frac{2\cdot x}{3\cdot x}=\frac{3}{3x}+\frac{2x}{3x}=\frac{3+2x}{3x}$

This is an expression.

2. $\frac{3}{a}+\frac{5}{6}=\frac{3\cdot 6}{a\cdot 6}+\frac{5\cdot a}{6\cdot a}=\frac{18}{6a}+\frac{5a}{6a}=\frac{18+5a}{6a}$

This is an expression.

3.

$$\begin{aligned}\frac{1}{x}+\frac{2}{3}&=\frac{3}{x}\\3x\left(\frac{1}{x}+\frac{2}{3}\right)&=3x\left(\frac{3}{x}\right)\\3+2x&=9\\2x&=6\\x&=3\end{aligned}$$

This is an equation.

4.

$$\begin{aligned}\frac{3}{a}+\frac{5}{6}&=1\\6a\left(\frac{3}{a}+\frac{5}{6}\right)&=6a(1)\\18+5a&=6a\\18&=a\end{aligned}$$

This is an equation.

5. $\frac{2}{x+1}-\frac{1}{x}=\frac{2\cdot x}{(x+1)\cdot x}-\frac{1\cdot(x+1)}{x\cdot(x+1)}$
$$=\frac{2x}{x(x+1)}-\frac{x+1}{x(x+1)}$$
$$=\frac{2x-x-1}{x(x+1)}$$
$$=\frac{x-1}{x(x+1)}$$
This is an expression.

6. $\frac{4}{x-3}-\frac{1}{x}=\frac{4\cdot x}{(x-3)\cdot x}-\frac{1\cdot(x-3)}{x\cdot(x-3)}$
$$=\frac{4x}{x(x-3)}-\frac{x-3}{x(x-3)}$$
$$=\frac{4x-x+3}{x(x-3)}$$
$$=\frac{3x+3}{x(x-3)}$$
$$=\frac{3(x+1)}{x(x-3)}$$
This is an expression.

7.
$$\frac{2}{x+1}-\frac{1}{x}=1$$
$$x(x+1)\left(\frac{2}{x+1}-\frac{1}{x}\right)=x(x+1)(1)$$
$$2x-(x+1)=x(x+1)$$
$$2x-x-1=x^2+x$$
$$x-1=x^2+x$$
$$0=x^2+1$$
There are no solutions. This is an equation.

8.
$$\frac{4}{x-3}-\frac{1}{x}=\frac{6}{x(x-3)}$$
$$x(x-3)\left(\frac{4}{x-3}-\frac{1}{x}\right)=x(x-3)\left(\frac{6}{x(x-3)}\right)$$
$$4x-1(x-3)=6$$
$$4x-x+3=6$$
$$3x=3$$
$$x=1$$
This is an equation.

9. $\frac{15x}{x+8}\cdot\frac{2x+16}{3x}=\frac{3\cdot 5x}{x+8}\cdot\frac{2(x+8)}{3x}=5\cdot 2=10$
This is an expression.

10. $\frac{9z+5}{15}\cdot\frac{5z}{81z^2-25}=\frac{9z+5}{3\cdot 5}\cdot\frac{5z}{(9z+5)(9z-5)}$
$$=\frac{z}{3(9z-5)}$$
This is an expression.

11. $\frac{2x+3}{x-3}+\frac{3x+6}{x-3}=\frac{2x+1+3x+6}{x-3}=\frac{5x+7}{x-3}$
This is an expression.

12. $\frac{4p-3}{2p+7}+\frac{3p+8}{2p+7}=\frac{4p-3+3p+8}{2p+7}=\frac{7p+5}{2p+7}$
This is an expression.

13.
$$\frac{x+5}{7}=\frac{8}{2}$$
$$14\left(\frac{x+5}{7}\right)=14\left(\frac{8}{2}\right)$$
$$2(x+5)=7(8)$$
$$2x+10=56$$
$$2x=46$$
$$x=23$$
This is an equation.

14.
$$\frac{1}{2}=\frac{x+1}{8}$$
$$8\left(\frac{1}{2}\right)=8\left(\frac{x+1}{8}\right)$$
$$4=x+1$$
$$3=x$$
This is an equation.

15. $\frac{5a+10}{18}\div\frac{a^2-4}{10a}=\frac{5a+10}{18}\cdot\frac{10a}{a^2-4}$
$$=\frac{5(a+2)}{2\cdot 9}\cdot\frac{2\cdot 5a}{(a+2)(a-2)}$$
$$=\frac{5\cdot 5a}{9(a-2)}$$
$$=\frac{25a}{9(a-2)}$$
This is an expression.

16. $\frac{9}{x^2-1}\div\frac{12}{3x+3}=\frac{9}{x^2-1}\cdot\frac{3x+3}{12}$
$$=\frac{9}{(x-1)(x+1)}\cdot\frac{3(x+1)}{3\cdot 4}$$
$$=\frac{9}{4(x-1)}$$
This is an expression.

17. $\frac{x+2}{3x-1}+\frac{5}{(3x-1)^2}=\frac{(x+2)(3x-1)}{(3x-1)(3x-1)}+\frac{5}{(3x-1)^2}$
$=\frac{(x+2)(3x-1)+5}{(3x-1)^2}$
$=\frac{3x^2-x+6x-2+5}{(3x-1)^2}$
$=\frac{3x^2+5x+3}{(3x-1)^2}$

This is an expression.

18. $\frac{4}{(2x-5)^2}+\frac{x+1}{2x-5}=\frac{4}{(2x-5)^2}+\frac{(x+1)(2x-5)}{(2x-5)(2x-5)}$
$=\frac{4+(x+1)(2x-5)}{(2x-5)^2}$
$=\frac{4+2x^2-5x+2x-5}{(2x-5)^2}$
$=\frac{2x^2-3x-1}{(2x-5)^2}$

This is an expression.

19. $\frac{x-7}{x}-\frac{x+2}{5x}=\frac{(x-7)\cdot 5}{x\cdot 5}-\frac{x+2}{5x}$
$=\frac{5x-35-x-2}{5x}$
$=\frac{4x-37}{5x}$

This is an expression.

20. $\frac{10x-9}{x}-\frac{x-4}{3x}=\frac{(10x-9)\cdot 3}{x\cdot 3}-\frac{x-4}{3x}$
$=\frac{3(10x-9)-x+4}{3x}$
$=\frac{30x-27-x+4}{3x}$
$=\frac{29x-23}{3x}$

This is an expression.

21. $\frac{3}{x+3}=\frac{5}{x^2-9}-\frac{2}{x-3}$
$(x^2-9)\left(\frac{3}{x+3}\right)=(x^2-9)\left(\frac{5}{x^2-9}-\frac{2}{x-3}\right)$
$3(x-3)=5-2(x+3)$
$3x-9=5-2x-6$
$3x-9=-2x-1$
$5x=8$
$x=\frac{8}{5}$

This is an equation.

22. $\frac{9}{x^2-4}+\frac{2}{x+2}=\frac{-1}{x-2}$
$(x^2-4)\left(\frac{9}{x^2-4}+\frac{2}{x+2}\right)=(x^2-4)\left(\frac{-1}{x-2}\right)$
$9+2(x-2)=-1(x+2)$
$9+2x-4=-x-2$
$2x+5=-x-2$
$3x=-7$
$x=-\frac{7}{3}$

This is an equation.

23. answers may vary

24. answers may vary

Section 5.6 Practice Exercises

1. $\frac{3}{8}=\frac{63}{x}$
$3\cdot x=63\cdot 8$
$3x=504$
$\frac{3x}{3}=\frac{504}{3}$
$x=168$

2. $\frac{2x+1}{7}=\frac{x-3}{5}$
$5(2x+1)=7(x-3)$
$10x+5=7x-21$
$3x+5=-21$
$3x=-26$
$\frac{3x}{3}=\frac{-26}{3}$
$x=-\frac{26}{3}$

3.
$$\frac{26}{250} = \frac{x}{50,000}$$
$$26(50,000) = 250x$$
$$1,300,000 = 250x$$
$$5200 = x$$
5200 people are expected to have a flu shot.

4.
$$\frac{12}{x} = \frac{9}{15}$$
$$12 \cdot 15 = 9 \cdot x$$
$$180 = 9x$$
$$20 = x$$
The missing length is 20 units.

5.
$$\frac{x}{2} - \frac{1}{3} = \frac{x}{6}$$
$$6\left(\frac{x}{2} - \frac{1}{3}\right) = 6\left(\frac{x}{6}\right)$$
$$6\left(\frac{x}{2}\right) - 6\left(\frac{1}{3}\right) = 6\left(\frac{x}{6}\right)$$
$$3x - 2 = x$$
$$-2 = -2x$$
$$\frac{-2}{-2} = \frac{-2x}{-2}$$
$$1 = x$$
The number is 1.

6.

	Hours to Complete Total Job	Part of Job Completed in 1 Hour
Guillaume	2	$\frac{1}{2}$
Greg	3	$\frac{1}{3}$
Together	x	$\frac{1}{x}$

$$\frac{1}{2} + \frac{1}{3} = \frac{1}{x}$$
$$6x\left(\frac{1}{2}\right) + 6x\left(\frac{1}{3}\right) = 6x\left(\frac{1}{x}\right)$$
$$3x + 2x = 6$$
$$5x = 6$$
$$x = \frac{6}{5} = 1\frac{1}{5} \text{ hr}$$

Together they can sort one batch in $1\frac{1}{5}$ hours.

7.

	distance	=	rate	·	time
Car	600		$x + 15$		$\frac{600}{x+15}$
Motorcycle	450		x		$\frac{450}{x}$

$$\frac{600}{x+15} = \frac{450}{x}$$
$$600x = 450(x+15)$$
$$600x = 450x + 6750$$
$$150x = 6750$$
$$x = 45$$
$$x + 15 = 45 + 15 = 60$$
The speed of the motorcycle is 45 mph. The speed of the car is 60 mph.

Vocabulary, Readiness & Video Check 5.6

1. The time to complete the job working together will be less than both of the individual times. The answer is **c**.

2. The time to fill the pond with both pipes on at the same time will be less than both of the individual times. The answer is **a**.

3. A number: x

The reciprocal of the number: $\frac{1}{x}$

The reciprocal of the number, decreased by 3:
$\frac{1}{x} - 3$

4. A number: y

The reciprocal of the number: $\frac{1}{y}$

The reciprocal of the number, increased by 2:
$\frac{1}{y} + 2$

5. A number: z
The sum of the number and 5: $z + 5$
The reciprocal of the sum of the number and 5:
$\frac{1}{z+5}$

6. A number: x
The difference of the number and 1: $x - 1$
The reciprocal of the difference of the number and 1: $\frac{1}{x-1}$

7. A number: y
Twice the number: $2y$
Eleven divided by twice the number: $\frac{11}{2y}$

8. A number: z
Triple the number: $3z$
Negative 10 divided by triple the number: $\frac{-10}{3y}$

9. No. Proportions are actually equations containing rational expressions, so they can also be solved by using the steps to solve those equations.

10. There are also many ways to set up an incorrect proportion, so just checking your solution in your proportion isn't enough. You need to determine if your solution is reasonable from the relationships given in the problem.

11. divided by, quotient

12. Two machines (or people) take different amounts of time to complete the task, one faster and one slower than the other. When working together, they will complete the task in less time than the faster machine, so your answer must be less than the time of the faster machine.

13.

	d	=	r	·	t
Car	325		$x+7$		$\frac{325}{x+7}$
Motorcycle	290		x		$\frac{290}{x}$

$$\frac{325}{x+7}=\frac{290}{x}$$

Exercise Set 5.6

1. The LCD is 6.

$$\frac{2}{3}=\frac{x}{6}$$
$$6\left(\frac{2}{3}\right)=6\left(\frac{x}{6}\right)$$
$$4=x$$

3.
$$\frac{x}{10}=\frac{5}{9}$$
$$9x=50$$
$$x=\frac{50}{9}$$

5.
$$\frac{x+1}{2x+3}=\frac{2}{3}$$
$$3(x+1)=2(2x+3)$$
$$3x+3=4x+6$$
$$3=x+6$$
$$-3=x$$

7.
$$\frac{9}{5}=\frac{12}{3x+2}$$
$$9(3x+2)=12\cdot 5$$
$$27x+18=60$$
$$27x=42$$
$$x=\frac{42}{27}=\frac{14}{9}$$

9. Let x be the elephant's weight on Pluto.
$$\frac{100}{3}=\frac{4100}{x}$$
$$100x=3(4100)$$
$$100x=12,300$$
$$x=123$$
The elephant weighs 123 pounds on Pluto.

11. Let y be the number of calories in 42.6 grams.
$$\frac{110}{28.4}=\frac{y}{42.6}$$
$$42.6(110)=28.4y$$
$$4686=28.4y$$
$$165=y$$
There are 165 calories in 42.6 grams of Crispy Rice.

13.
$$\frac{16}{10}=\frac{34}{y}$$
$$16y=34(10)$$
$$16y=340$$
$$y=\frac{340}{16}$$
$$y=21.25$$

15. $\frac{y}{8}=\frac{20}{28}$
$28y=20(8)$
$28y=160$
$y=\frac{160}{28}=\frac{40}{7}=5\frac{5}{7}$ ft

17. Let x be the number.
$3\left(\frac{1}{x}\right)=9\left(\frac{1}{6}\right)$
$\frac{3}{x}=\frac{9}{6}$
$3(6)=9x$
$18=9x$
$2=x$
The number is 2.

19. Let x be the number.
$\frac{2x+3}{x+1}=\frac{3}{2}$
$2(2x+3)=3(x+1)$
$4x+6=3x+3$
$x=-3$
The number is –3.

21. Let x be the time in hours that it takes them to complete the job working together.
The experienced surveyor completes $\frac{1}{4}$ of the job in 1 hour. The apprentice surveyor completes $\frac{1}{5}$ of the job in 1 hour. Together, they complete $\frac{1}{x}$ of the job in 1 hour.
$\frac{1}{4}+\frac{1}{5}=\frac{1}{x}$
The LCD is $4 \cdot 5 \cdot x = 20x$.
$20x\left(\frac{1}{4}+\frac{1}{5}\right)=20x\left(\frac{1}{x}\right)$
$5x+4x=20$
$9x=20$
$x=\frac{20}{9}$
It takes them $\frac{20}{9}=2\frac{2}{9}$ hours to survey the roadbed together.

23. Let x be the time in minutes that it takes the two belts to complete the job working together. The first belt completes $\frac{1}{2}$ of the job in 1 minute.
The smaller belt completes $\frac{1}{6}$ of the job in 1 minute. Together, they complete $\frac{1}{x}$ of the job in 1 minute.
$\frac{1}{2}+\frac{1}{6}=\frac{1}{x}$
The LCD is $6x$.
$6x\left(\frac{1}{2}+\frac{1}{6}\right)=6x\left(\frac{1}{x}\right)$
$3x+x=6$
$4x=6$
$x=\frac{6}{4}=\frac{3}{2}=1\frac{1}{2}$
It will take $1\frac{1}{2}$ minutes to move the cans to the storage when both belts are used.

25. Let r be her jogging rate.

	distance	=	rate	·	time
Trip to park	12		r		$\frac{12}{r}$
Return trip	18		r		$\frac{18}{r}$

Since the return trip took 1 hour longer,
$\frac{18}{r}=1+\frac{12}{r}$. The LCD is r.
$r\left(\frac{18}{r}\right)=r\left(1+\frac{12}{r}\right)$
$18=r+12$
$6=r$
Her jogging rate is 6 miles per hour.

27. Let x be the speed for the first portion.

	distance	=	rate	·	time
1st portion	20		x		$\frac{20}{x}$
2nd portion	16		$x-2$		$\frac{16}{x-2}$

$$\frac{20}{x}=\frac{16}{x-2}$$

The LCD is $x(x-2)$.

$$x(x-2)\left(\frac{20}{x}\right)=x(x-2)\left(\frac{16}{x-2}\right)$$
$$20(x-2)=16x$$
$$20x-40=16x$$
$$4x=40$$
$$x=10$$

The cyclist's speed for first portion is 10 mph and the speed for the second portion is 8 mph.

29. $40\text{ students}\cdot\frac{9\text{ square feet}}{1\text{ student}}=360\text{ square feet}$

40 students need a minimum of 360 square feet.

31. Let n be the number.

$$\frac{1}{4}=\frac{n}{8}$$
$$8\left(\frac{1}{4}\right)=8\left(\frac{n}{8}\right)$$
$$2=n$$

The number is 2.

33. Let x be the time in hours that it takes Marcus and Tony to do the job working together.

Marcus lays $\frac{1}{6}$ of a slab in 1 hour. Tony lays $\frac{1}{4}$ of a slab in 1 hour. Together, they lay $\frac{1}{x}$ of the slab in 1 hour.

$$\frac{1}{6}+\frac{1}{4}=\frac{1}{x}$$

The LCD is $12x$.

$$12x\left(\frac{1}{6}+\frac{1}{4}\right)=12x\left(\frac{1}{x}\right)$$
$$2x+3x=12$$
$$5x=12$$
$$x=\frac{12}{5}$$

It will take Tony and Marcus $\frac{12}{5}$ hours to lay the slab, so the labor estimate should be $\frac{12}{5}(\$45)=\$108$.

35. Let w be the speed of the wind.

	distance	=	rate	·	time
With wind	400		$230+w$		$\frac{400}{230+w}$
Against wind	336		$230-w$		$\frac{336}{230-w}$

$$\frac{400}{230+w} = \frac{336}{230-w}$$
$$400(230-w) = 336(230+w)$$
$$92{,}000 - 400w = 77{,}280 + 336w$$
$$14{,}720 = 736w$$
$$20 = w$$

The speed of the wind is 20 mph.

37. $\frac{2}{3} = \frac{25}{y}$

$$2y = 3(25)$$
$$2y = 75$$
$$y = \frac{75}{2}$$

The unknown length is $y = \frac{75}{2} = 37.5$ feet.

39. Let x be the speed of the slower train. In 3.5 hours, the slower train travels $3.5x$ miles and the faster train travels $3.5(x + 10)$ miles.

$$3.5x + 3.5(x+10) = 322$$
$$3.5x + 3.5x + 35 = 322$$
$$7x + 35 = 322$$
$$7x = 287$$
$$x = 41$$

The slower train travels 41 mph and the faster train travels 41 + 10 = 51 mph.

41. Let x be the number.

$$\frac{2}{x-3} - \frac{4}{x+3} = 8 \cdot \frac{1}{x^2-9}$$

The LCD is $x^2 - 9 = (x+3)(x-3)$.

$$(x+3)(x-3)\left(\frac{2}{x-3} - \frac{4}{x+3}\right) = (x^2-9)\left(\frac{8}{x^2-9}\right)$$
$$2(x+3) - 4(x-3) = 8$$
$$2x + 6 - 4x + 12 = 8$$
$$-2x + 18 = 8$$
$$-2x = -10$$
$$x = 5$$

The solution $x = 5$ checks, so the number is 5.

43. Let x be the rate in still air.

	distance	=	rate	·	time
With wind	630		$x + 35$		$\frac{630}{x+35}$
Against wind	455		$x - 35$		$\frac{455}{x-35}$

$$\frac{630}{x+35}=\frac{455}{x-35}$$
$$630(x-35)=455(x+35)$$
$$630x-22,050=455x+15,925$$
$$175x=37,975$$
$$x=217$$

The plane flies at a rate of 217 mph in still air.

45. Let x be the number of gallons of water needed.

$$\frac{8\text{ tsp}}{2\text{ gal}}=\frac{36\text{ tsp}}{x\text{ gal}}$$
$$\frac{8}{2}=\frac{36}{x}$$
$$8x=36(2)$$
$$8x=72$$
$$x=9$$

9 gallons of water are needed to mix with a box of weed killer.

47. Let x be the rate of the wind.

	distance	=	rate	·	time
with wind	48		$16+x$		$\frac{48}{16+x}$
into wind	16		$16-x$		$\frac{16}{16-x}$

$$\frac{48}{16+x}=\frac{16}{16-x}$$
$$42(16-x)=16(16+x)$$
$$768-48x=256+16x$$
$$512=64x$$
$$8=x$$

The rate of the wind is 8 mph.

49. Let x be the rate of the slower hiker. Then the rate of the faster hiker is $x + 1.1$. In 2 hours, the slower hiker walks $2x$ miles, while the faster hiker walks $2(x + 1.1)$ miles.

$$2x+2(x+1.1)=11$$
$$2x+2x+2.2=11$$
$$4x+2.2=11$$
$$4x=8.8$$
$$x=2.2$$

$x + 1.1 = 2.2 + 1.1 = 3.3$

The hikers walk 2.2 miles per hour and 3.3 miles per hour.

51. Let x be the time it takes for the second worker to do the same job alone.

$$\frac{1}{3}+\frac{1}{x}=\frac{1}{\frac{3}{2}}$$
$$\frac{1}{3}+\frac{1}{x}=\frac{2}{3}$$
$$3x\left(\frac{1}{3}+\frac{1}{x}\right)=3x\left(\frac{2}{3}\right)$$
$$x+3=2x$$
$$3=x$$

It will take the second worker 3 hours to get the job done.

53. $$\frac{20\text{ feet}}{6\text{ inches}}=\frac{x\text{ feet}}{8\text{ inches}}$$
$$\frac{20}{6}=\frac{x}{8}$$
$$8(20)=6x$$
$$160=6x$$
$$\frac{160}{6}=x$$
$$\frac{80}{3}=x$$

The missing dimension is $\frac{80}{3}=26\frac{2}{3}$ feet.

55. Let x be the number of other nuts.

$$\frac{3}{2}=\frac{324}{x}$$
$$3x=2(324)$$
$$3x=648$$
$$x=216$$

There should be 216 other nuts in the can.

57. Let t be the time in hours that the jet plane travels.

	distance	=	**rate**	·	**time**
jet plane	$500t$		500		t
propeller plane	$200(t+2)$		200		$t+2$

$$500t=200(t+2)$$
$$500t=200t+400$$
$$300t=400$$
$$t=\frac{400}{300}$$
$$t=\frac{4}{3}$$

$$\text{distance}=500t=500\left(\frac{4}{3}\right)=666\frac{2}{3}$$

The planes are $666\frac{2}{3}$ miles from the starting point.

59. Let x be the time that it takes the third pipe to fill the pool alone.

$$\frac{1}{20}+\frac{1}{15}+\frac{1}{x}=\frac{1}{6}$$
$$60x\left(\frac{1}{20}+\frac{1}{15}+\frac{1}{x}\right)=60x\left(\frac{1}{6}\right)$$
$$3x+4x+60=10x$$
$$7x+60=10x$$
$$60=3x$$
$$20=x$$

It will take the third pump 20 hours to do the job alone.

61. Let r be the motorcycle's speed.

	distance	=	**rate**	·	**time**
car	280		$r+10$		$\frac{280}{r+10}$
motorcycle	240		r		$\frac{240}{r}$

$$\frac{280}{r+10}=\frac{240}{r}$$
$$280r=240(r+10)$$
$$280r=240r+2400$$
$$40r=2400$$
$$r=60$$
$$r+10=60+10=70$$

The motorcycle's speed was 60 miles per hour and the car's speed was 70 miles per hour.

63. Let x be the time for the third cook to prepare the same number of pies.

$$\frac{1}{6}+\frac{1}{7}+\frac{1}{x}=\frac{1}{2}$$
$$42x\left(\frac{1}{6}+\frac{1}{7}+\frac{1}{x}\right)=42x\left(\frac{1}{2}\right)$$
$$7x+6x+42=21x$$
$$13x+42=21x$$
$$42=8x$$
$$\frac{42}{8}=x$$
$$\frac{42}{8}=\frac{21}{4}=5\frac{1}{4}$$

It will take the third cook $5\frac{1}{4}$ hours to prepare the pies working alone.

65. Let x be the number.

$$\frac{x}{3}-1=\frac{5}{3}$$
$$3\left(\frac{x}{3}-1\right)=3\left(\frac{5}{3}\right)$$
$$x-3=5$$
$$x=8$$

The number is 8.

67. Let x be the slower speed.

	distance	=	rate	·	time
slower speed	70		x		$\frac{70}{x}$
faster speed	300		$x + 40$		$\frac{300}{x+40}$

$$\frac{300}{x+40}=2\left(\frac{70}{x}\right)$$
$$\frac{300}{x+40}=\frac{140}{x}$$
$$300x=140(x+40)$$
$$300x=140x+5600$$
$$160x=5600$$
$$x=35$$

$x + 40 = 35 + 40 = 75$

The slower speed was 35 miles per hour and the faster speed was 75 miles per hour.

69. Let x be the speed of the plane in still air.

	distance	=	rate	·	time
with wind	2160		$x + 30$		$\frac{2160}{x+30}$
against wind	1920		$x - 30$		$\frac{1920}{x-30}$

$$\frac{2160}{x+30}=\frac{1920}{x-30}$$
$$2160(x-30)=1920(x+30)$$
$$2160x-64,800=1920x+57,600$$
$$240x=122,400$$
$$x=510$$

The speed of the plane in still air is 510 miles per hour.

71. $\frac{x}{3.75}=\frac{12}{9}$

$$9x=12(3.75)$$
$$9x=45$$
$$x=5$$

The missing length is $x = 5$.

73. $\frac{x}{9}=\frac{24}{16}$

$$16x=9(24)$$
$$16x=216$$
$$x=\frac{216}{16}$$
$$x=\frac{27}{2}$$

The missing length is $x=\frac{27}{2}=13.5$.

75. $\frac{\frac{3}{4}+\frac{1}{4}}{\frac{3}{8}+\frac{13}{8}}=\frac{\frac{4}{4}}{\frac{16}{8}}=\frac{1}{2}$

77. $\frac{\frac{2}{5}+\frac{1}{5}}{\frac{7}{10}+\frac{7}{10}}=\frac{\frac{3}{5}}{\frac{14}{10}}=\frac{3}{5}\div\frac{14}{10}=\frac{3}{5}\cdot\frac{10}{14}=\frac{3\cdot 2\cdot 5}{5\cdot 2\cdot 7}=\frac{3}{7}$

79. Let x be the time in minutes that it takes for the faster pump to fill the tank, so it fills $\frac{1}{x}$ of the tank in 1 minute. It takes the slower pump $3x$ minutes to fill the tank, so the slower pump fills $\frac{1}{3x}$ of the tank in 1 minute. Together, the pumps fill $\frac{1}{21}$ of the tank in 1 minute.

$$\frac{1}{x}+\frac{1}{3x}=\frac{1}{21}$$
$$21x\left(\frac{1}{x}+\frac{1}{3x}\right)=21x\left(\frac{1}{21}\right)$$
$$21+7=x$$
$$28=x$$

$3x = 3(28) = 84$

The faster pump fills the tank in 28 minutes, while the slower pump takes 84 minutes.

81. answers may vary

83. $D = RT$

$\frac{D}{T} = \frac{RT}{T}$

$\frac{D}{T} = R$ or $R = \frac{D}{T}$

85. Let t be the time it takes for the hyena to overtake the giraffe.

$0.5 + 32t = 40t$

$0.5 = 8t$

$0.0625 = t$

$0.0625 \text{ hr} \cdot \frac{60 \text{ min}}{1 \text{ hr}} = 3.75 \text{ min}$

It will take the hyena 3.75 minutes to overtake the giraffe.

Section 5.7 Practice Exercises

1. $\frac{\frac{3}{7}}{\frac{5}{9}} = \frac{3}{7} \div \frac{5}{9} = \frac{3}{7} \cdot \frac{9}{5} = \frac{27}{35}$

2. $\frac{\frac{3}{4} - \frac{2}{3}}{\frac{1}{2} + \frac{3}{8}} = \frac{\frac{3(3)}{4(3)} - \frac{2(4)}{3(4)}}{\frac{1(4)}{2(4)} + \frac{3}{8}}$

$= \frac{\frac{9}{12} - \frac{8}{12}}{\frac{4}{8} + \frac{3}{8}}$

$= \frac{\frac{1}{12}}{\frac{7}{8}}$

$= \frac{1}{12} \cdot \frac{8}{7}$

$= \frac{1 \cdot 2 \cdot 4}{3 \cdot 4 \cdot 7}$

$= \frac{2}{21}$

3. $\frac{\frac{2}{5} - \frac{1}{x}}{\frac{2x}{15} - \frac{1}{3}} = \frac{\frac{2(x)}{5(x)} - \frac{1(5)}{x(5)}}{\frac{2x}{15} - \frac{1(5)}{3(5)}}$

$= \frac{\frac{2x}{5x} - \frac{5}{5x}}{\frac{2x}{15} - \frac{5}{15}}$

$= \frac{\frac{2x-5}{5x}}{\frac{2x-5}{15}}$

$= \frac{2x-5}{5x} \cdot \frac{15}{2x-5}$

$= \frac{2x-15}{5x} \cdot \frac{3 \cdot 5}{2x-15}$

$= \frac{3}{x}$

4. The LCD is 24.

$\frac{\frac{3}{4} - \frac{2}{3}}{\frac{1}{2} + \frac{3}{8}} = \frac{24\left(\frac{3}{4} - \frac{2}{3}\right)}{24\left(\frac{1}{2} + \frac{3}{8}\right)}$

$= \frac{24\left(\frac{3}{4}\right) - 24\left(\frac{2}{3}\right)}{24\left(\frac{1}{2}\right) + 24\left(\frac{3}{8}\right)}$

$= \frac{18-16}{12+9}$

$= \frac{2}{21}$

5. The LCD is y.

$\frac{1 + \frac{x}{y}}{\frac{2x+1}{y}} = \frac{y\left(1 + \frac{x}{y}\right)}{y\left(\frac{2x+1}{y}\right)} = \frac{y(1) + y\left(\frac{x}{y}\right)}{y\left(\frac{2x+1}{y}\right)} = \frac{y+x}{2x+1}$

6. The LCD is $6xy$.

$\frac{\frac{5}{6y} + \frac{y}{x}}{\frac{y}{3} - x} = \frac{6xy\left(\frac{5}{6y} + \frac{y}{x}\right)}{6xy\left(\frac{y}{3} - x\right)}$

$= \frac{6xy\left(\frac{5}{6y}\right) + 6xy\left(\frac{y}{x}\right)}{6xy\left(\frac{y}{3}\right) - 6xy(x)}$

$= \frac{5x + 6y^2}{2xy^2 - 6x^2y}$

$= \frac{5x + 6y^2}{2xy(y - 3x)}$

Vocabulary, Readiness & Video Check 5.7

1. The LCD for $\frac{1}{4}=\frac{1}{2\cdot 2}$, $\frac{1}{2}$, $\frac{1}{3}$, and $\frac{1}{2}$ is $2\cdot 2\cdot 3=12$; **c.**

2. The LCD for $\frac{3}{5}$, $\frac{2}{3}$, $\frac{1}{10}=\frac{1}{2\cdot 5}$, and $\frac{1}{6}=\frac{1}{2\cdot 3}$ is $2\cdot 3\cdot 5=30$; **b.**

3. The LCD for $\frac{5}{2x^2}$, $\frac{3}{16x}=\frac{3}{2\cdot 2\cdot 2\cdot 2\cdot x}$, $\frac{x}{8}=\frac{x}{2\cdot 2\cdot 2}$, and $\frac{3}{4x}=\frac{3}{2\cdot 2\cdot x}$ is $2\cdot 2\cdot 2\cdot 2\cdot x^2=16x^2$; **a.**

4. The LCD for $\frac{11}{6}=\frac{11}{2\cdot 3}$, $\frac{10}{x^2}$, $\frac{7}{9}=\frac{7}{3\cdot 3}$, and $\frac{5}{x}$ is $2\cdot 3\cdot 3\cdot x^2=18x^2$; **c.**

5. a single fraction in the numerator and in the denominator

6. In Method 2, we find the LCD of all fractions in the complex fraction, then multiply both the numerator and the denominator by this LCD so that both will no longer contain fractions; in Method 1, we find the LCD of the fractions only in the numerator and then the LCD of the fractions only in the denominator in order to perform the addition and/or subtraction and get single fractions in the numerator and in the denominator.

Exercise Set 5.7

1. $\frac{\frac{1}{2}}{\frac{3}{4}}=\frac{1}{2}\div\frac{3}{4}=\frac{1}{2}\cdot\frac{4}{3}=\frac{1\cdot 4}{2\cdot 3}=\frac{2}{3}$

3. $\frac{-\frac{4x}{9}}{-\frac{2x}{3}}=-\frac{4x}{9}\div -\frac{2x}{3}=-\frac{4x}{9}\cdot\frac{3}{-2x}=\frac{4x\cdot 3}{9\cdot 2x}=\frac{2}{3}$

5. $\frac{\frac{1+x}{6}}{\frac{1+x}{3}}=\frac{1+x}{6}\div\frac{1+x}{3}=\frac{1+x}{6}\cdot\frac{3}{1+x}=\frac{3(1+x)}{6(1+x)}=\frac{1}{2}$

7. $\frac{\frac{1}{2}+\frac{2}{3}}{\frac{5}{9}-\frac{5}{6}}=\frac{18\left(\frac{1}{2}+\frac{2}{3}\right)}{18\left(\frac{5}{9}-\frac{5}{6}\right)}=\frac{9+12}{10-15}=\frac{21}{-5}=-\frac{21}{5}$

9. $\frac{2+\frac{7}{10}}{1+\frac{3}{5}}=\frac{10\left(2+\frac{7}{10}\right)}{10\left(1+\frac{3}{5}\right)}=\frac{20+7}{10+6}=\frac{27}{16}$

11. $\frac{\frac{1}{3}}{\frac{1}{2}-\frac{1}{4}}=\frac{12\left(\frac{1}{3}\right)}{12\left(\frac{1}{2}-\frac{1}{4}\right)}=\frac{4}{6-3}=\frac{4}{3}$

13. $\frac{-\frac{2}{9}}{-\frac{14}{3}}=-\frac{2}{9}\div\left(-\frac{14}{3}\right)=-\frac{2}{9}\cdot\left(-\frac{3}{14}\right)=\frac{2\cdot 3}{9\cdot 14}=\frac{1}{21}$

15.
$$\frac{-\frac{5}{12x^2}}{\frac{25}{16x^3}}=\frac{-5}{12x^2}\div\frac{25}{16x^3}=-\frac{5}{12x^2}\cdot\frac{16x^3}{25}=-\frac{5\cdot 16x^3}{12x^2\cdot 25}=-\frac{4x}{15}$$

17. $\frac{\frac{m}{n}-1}{\frac{m}{n}+1}=\frac{n\left(\frac{m}{n}-1\right)}{n\left(\frac{m}{n}+1\right)}=\frac{m-n}{m+n}$

19.
$$\frac{\frac{1}{5}-\frac{1}{x}}{\frac{7}{10}+\frac{1}{x^2}}=\frac{10x^2\left(\frac{1}{5}-\frac{1}{x}\right)}{10x^2\left(\frac{7}{10}+\frac{1}{x^2}\right)}=\frac{2x^2-10x}{7x^2+10}=\frac{2x(x-5)}{7x^2+10}$$

21.
$$\frac{1+\frac{1}{y-2}}{y+\frac{1}{y-2}}=\frac{(y-2)\left(1+\frac{1}{y-2}\right)}{(y-2)\left(y+\frac{1}{y-2}\right)}=\frac{y-2+1}{(y-2)y+1}=\frac{y-1}{y^2-2y+1}=\frac{y-1}{(y-1)^2}=\frac{1}{y-1}$$

23. $\dfrac{\frac{4y-8}{16}}{\frac{6y-12}{4}} = \dfrac{16\left(\frac{4y-8}{16}\right)}{16\left(\frac{6y-12}{4}\right)} = \dfrac{4y-8}{24y-48} = \dfrac{4(y-2)}{24(y-2)} = \dfrac{1}{6}$

25. $\dfrac{\frac{x}{y}+1}{\frac{x}{y}-1} = \dfrac{y\left(\frac{x}{y}+1\right)}{y\left(\frac{x}{y}-1\right)} = \dfrac{x+y}{x-y}$

27. $\dfrac{1}{2+\frac{1}{3}} = \dfrac{3(1)}{3\left(2+\frac{1}{3}\right)} = \dfrac{3}{6+1} = \dfrac{3}{7}$

29. $\dfrac{\frac{ax+ab}{x^2-b^2}}{\frac{x+b}{x-b}} = \dfrac{ax+ab}{x^2-b^2} \div \dfrac{x+b}{x-b}$

$= \dfrac{ax+ab}{x^2-b^2} \cdot \dfrac{x-b}{x+b}$

$= \dfrac{a(x+b)}{(x+b)(x-b)} \cdot \dfrac{x-b}{x+b}$

$= \dfrac{a}{x+b}$

31. $\dfrac{-\frac{3+y}{4}}{\frac{8+y}{28}} = \dfrac{28\left(\frac{-3+y}{4}\right)}{28\left(\frac{8+y}{28}\right)}$

$= \dfrac{-21+7y}{8+y}$

$= \dfrac{7y-21}{8+y}$

$= \dfrac{7(y-3)}{8+y}$

33. $\dfrac{3+\frac{12}{x}}{1-\frac{16}{x^2}} = \dfrac{x^2\left(3+\frac{12}{x}\right)}{x^2\left(1-\frac{16}{x^2}\right)}$

$= \dfrac{3x^2+12x}{x^2-16}$

$= \dfrac{3x(x+4)}{(x+4)(x-4)}$

$= \dfrac{3x}{x-4}$

35. $\dfrac{\frac{8}{x+4}+2}{\frac{12}{x+4}-2} = \dfrac{(x+4)\left(\frac{8}{x+4}+2\right)}{(x+4)\left(\frac{12}{x+4}-2\right)}$

$= \dfrac{8+2x+8}{12-2x-8}$

$= \dfrac{2x+16}{-2x+4}$

$= \dfrac{2(x+8)}{2(-x+2)}$

$= -\dfrac{x+8}{x-2}$

37. $\dfrac{\frac{s}{r}+\frac{r}{s}}{\frac{s}{r}-\frac{r}{s}} = \dfrac{rs\left(\frac{s}{r}+\frac{r}{s}\right)}{rs\left(\frac{s}{r}-\frac{r}{s}\right)} = \dfrac{s^2+r^2}{s^2-r^2}$

39. $\dfrac{\frac{6}{x-5}+\frac{x}{x-2}}{\frac{3}{x-6}-\frac{2}{x-5}} = \dfrac{\frac{6(x-2)}{(x-5)(x-2)}+\frac{x(x-5)}{(x-2)(x-5)}}{\frac{3(x-5)}{(x-6)(x-5)}-\frac{2(x-6)}{(x-5)(x-6)}}$

$= \dfrac{\frac{6x-12+x^2-5x}{(x-5)(x-2)}}{\frac{3x-15-2x+12}{(x-6)(x-5)}}$

$= \dfrac{\frac{x^2+x-12}{(x-5)(x-2)}}{\frac{x-3}{(x-6)(x-5)}}$

$= \dfrac{(x+4)(x-3)}{(x-5)(x-2)} \cdot \dfrac{(x-6)(x-5)}{x-3}$

$= \dfrac{(x-6)(x+4)}{x-2}$

41. The longest bar corresponds to Serena Williams, so Serena Williams has won the most prize money in her career.

43. $30.0 - 24.4 = 5.6$
The approximate spread in lifetime prize money between Kim Clijsters and Venus Williams is \$5.6 million.

45. answers may vary

47. $\dfrac{\frac{1}{3}+\frac{3}{4}}{2} = \dfrac{\frac{1\cdot 4}{3\cdot 4}+\frac{3\cdot 3}{4\cdot 3}}{2} = \dfrac{\frac{4}{12}+\frac{9}{12}}{2} = \dfrac{13}{12}\cdot\dfrac{1}{2} = \dfrac{13}{24}$

49. The distance is the average of the two given marked distances.

$\dfrac{3\frac{1}{2}+5}{2} = \dfrac{\frac{7}{2}+5}{2} = \dfrac{\frac{7}{2}+\frac{10}{2}}{2} = \dfrac{\frac{17}{2}}{2} = \dfrac{17}{2}\cdot\dfrac{1}{2} = \dfrac{17}{4} = 4\dfrac{1}{4}$

He should drill $4\dfrac{1}{4}$ (or 4.25) feet from the left side of the board.

51. $\dfrac{1}{\frac{1}{R_1}+\frac{1}{R_2}} = \dfrac{R_1R_2(1)}{R_1R_2\left(\frac{1}{R_1}+\frac{1}{R_2}\right)} = \dfrac{R_1R_2}{R_2+R_1}$

53.
$$\frac{x^{-1}+2^{-1}}{x^{-2}-4^{-1}} = \frac{\frac{1}{x}+\frac{1}{2}}{\frac{1}{x^2}-\frac{1}{4}}$$
$$= \frac{\frac{1\cdot 2}{x\cdot 2}+\frac{1\cdot x}{2\cdot x}}{\frac{1\cdot 4}{x^2\cdot 4}-\frac{1\cdot x^2}{4\cdot x^2}}$$
$$= \frac{\frac{2+x}{2x}}{\frac{4-x^2}{4x^2}}$$
$$= \frac{2+x}{2x}\cdot\frac{4x^2}{4-x^2}$$
$$= \frac{2+x}{2x}\cdot\frac{4x^2}{(2-x)(2+x)}$$
$$= \frac{2x}{2-x}$$

55. $\dfrac{y^{-2}}{1-y^{-2}} = \dfrac{\frac{1}{y^2}}{1-\frac{1}{y^2}} = \dfrac{y^2\left(\frac{1}{y^2}\right)}{y^2\left(1-\frac{1}{y^2}\right)} = \dfrac{1}{y^2-1}$

57. $t = \dfrac{d}{r} = \dfrac{\frac{20x}{3}}{\frac{5x}{9}} = \dfrac{20x}{3}\cdot\dfrac{9}{5x} = 12$

The time is 12 hours.

Chapter 5 Vocabulary Check

1. A <u>rational expression</u> is an expression that can be written in the form $\frac{P}{Q}$, where P and Q are polynomials and Q is not 0.

2. In a <u>complex fraction</u>, the numerator or denominator or both may contain fractions.

3. For a rational expression, $-\dfrac{a}{b} = \dfrac{-a}{\underline{b}} = \dfrac{a}{\underline{-b}}$.

4. A rational expression is undefined when the <u>denominator</u> is 0.

5. The process of writing a rational expression in lowest terms is called <u>simplifying</u>.

6. The expressions $\frac{2x}{7}$ and $\frac{7}{2x}$ are called <u>reciprocals</u>.

7. The <u>least common denominator</u> of a list of rational expressions is a polynomial of least degree whose factors include all factors of the denominators in the list.

8. A <u>unit</u> fraction is a fraction that equals 1.

Chapter 5 Review

1.
$$x^2-4=0$$
$$(x-2)(x+2)=0$$
$$x-2=0 \quad\text{or}\quad x+2=0$$
$$x=2 \qquad\qquad x=-2$$

The expression $\dfrac{x+5}{x^2-4}$ is undefined for $x = 2$ and $x = -2$.

2.
$$4x^2-4x-15=0$$
$$(2x-5)(2x+3)=0$$
$$2x-5=0 \quad\text{or}\quad 2x+3=0$$
$$2x=5 \qquad\qquad 2x=3$$
$$x=\frac{5}{2} \qquad\qquad x=-\frac{3}{2}$$

The expression $\dfrac{5x+9}{4x^2-4x-15}$ is undefined for $x=\frac{5}{2}$ and $x=-\frac{3}{2}$.

3. Replace z with -2.
$$\frac{2-z}{z+5} = \frac{2-(-2)}{-2+5} = \frac{4}{3}$$

4. Replace x with 5 and y with 7.
$$\frac{x^2+xy-y^2}{x+y} = \frac{(5)^2+(5)(7)-(7)^2}{5+7}$$
$$= \frac{25+35-49}{12}$$
$$= \frac{11}{12}$$

5. $\dfrac{2x+6}{x^2+3x} = \dfrac{2(x+3)}{x(x+3)} = \dfrac{2}{x}$

6. $\dfrac{3x-12}{x^2-4x} = \dfrac{3(x-4)}{x(x-4)} = \dfrac{3}{x}$

7. $\dfrac{x+2}{x^2-3x-10}=\dfrac{x+2}{(x-5)(x+2)}=\dfrac{1}{x-5}$

8. $\dfrac{x+4}{x^2+5x+4}=\dfrac{x+4}{(x+4)(x+1)}=\dfrac{1}{x+1}$

9. $\dfrac{x^3-4x}{x^2+3x+2}=\dfrac{x(x^2-4)}{(x+2)(x+1)}$
$=\dfrac{x(x-2)(x+2)}{(x+2)(x+1)}$
$=\dfrac{x(x-2)}{x+1}$

10. $\dfrac{5x^2-125}{x^2+2x-15}=\dfrac{5(x^2-25)}{(x+5)(x-3)}$
$=\dfrac{5(x-5)(x+5)}{(x+5)(x-3)}$
$=\dfrac{5(x-5)}{x-3}$

11. $\dfrac{x^2-x-6}{x^2-3x-10}=\dfrac{(x-3)(x+2)}{(x-5)(x+2)}=\dfrac{x-3}{x-5}$

12. $\dfrac{x^2-2x}{x^2+2x-8}=\dfrac{x(x-2)}{(x+4)(x-2)}=\dfrac{x}{x+4}$

13. $\dfrac{x^2+xa+xb+ab}{x^2-xc+bx-bc}=\dfrac{(x^2+xa)+(xb+ab)}{(x^2-xc)+(bx-bc)}$
$=\dfrac{x(x+a)+b(x+a)}{x(x-c)+b(x-c)}$
$=\dfrac{(x+b)(x+a)}{(x+b)(x-c)}$
$=\dfrac{x+a}{x-c}$

14. $\dfrac{x^2+5x-2x-10}{x^2-3x-2x+6}=\dfrac{(x^2+5x)+(-2x-10)}{(x^2-3x)+(-2x+6)}$
$=\dfrac{x(x+5)-2(x+5)}{x(x-3)-2(x-3)}$
$=\dfrac{(x-2)(x+5)}{(x-2)(x-3)}$
$=\dfrac{x+5}{x-3}$

15. $\dfrac{15x^3y^2}{z}\cdot\dfrac{z}{5xy^3}=\dfrac{3\cdot5\cdot x\cdot x^2\cdot y^2\cdot z}{z\cdot5\cdot x\cdot y\cdot y^2}=\dfrac{3x^2}{y}$

16. $\dfrac{-y^3}{8}\cdot\dfrac{9x^2}{y^3}=-\dfrac{y^3\cdot9x^2}{8\cdot y^3}=-\dfrac{9x^2}{8}$

17. $\dfrac{x^2-9}{x^2-4}\cdot\dfrac{x-2}{x+3}=\dfrac{(x-3)(x+3)}{(x-2)(x+2)}\cdot\dfrac{x-2}{x+3}=\dfrac{x-3}{x+2}$

18. $\dfrac{2x+5}{x-6}\cdot\dfrac{2x}{-x+6}=\dfrac{2x+5}{x-6}\cdot\dfrac{2x}{-(x-6)}=\dfrac{-2x(2x+5)}{(x-6)^2}$

19. $\dfrac{x^2-5x-24}{x^2-x-12}\div\dfrac{x^2-10x+16}{x^2+x-6}$
$=\dfrac{x^2-5x-24}{x^2-x-12}\cdot\dfrac{x^2+x-6}{x^2-10x+16}$
$=\dfrac{(x-8)(x+3)}{(x-4)(x+3)}\cdot\dfrac{(x+3)(x-2)}{(x-8)(x-2)}$
$=\dfrac{x+3}{x-4}$

20. $\dfrac{4x+4y}{xy^2}\div\dfrac{3x+3y}{x^2y}=\dfrac{4x+4y}{xy^2}\cdot\dfrac{x^2y}{3x+3y}$
$=\dfrac{4(x+y)}{xy^2}\cdot\dfrac{x^2y}{3(x+y)}$
$=\dfrac{4x}{3y}$

21. $\dfrac{x^2+x-42}{x-3}\cdot\dfrac{(x-3)^2}{x+7}=\dfrac{(x+7)(x-6)}{x-3}\cdot\dfrac{(x-3)^2}{x+7}$
$=(x-6)(x-3)$

22. $\dfrac{2a+2b}{3}\cdot\dfrac{a-b}{a^2-b^2}=\dfrac{2(a+b)}{3}\cdot\dfrac{a-b}{(a-b)(a+b)}$
$=\dfrac{2}{3}$

23. $\dfrac{2x^2-9x+9}{8x-12}\div\dfrac{x^2-3x}{2x}=\dfrac{2x^2-9x+9}{8x-12}\cdot\dfrac{2x}{x^2-3x}$
$=\dfrac{(2x-3)(x-3)}{4(2x-3)}\cdot\dfrac{2x}{x(x-3)}$
$=\dfrac{1}{2}$

24. $\frac{x^2-y^2}{x^2+xy} \div \frac{3x^2-2xy-y^2}{3x^2+6x}$
$= \frac{x^2-y^2}{x^2+xy} \cdot \frac{3x^2+6x}{3x^2-2xy-y^2}$
$= \frac{(x-y)(x+y)}{x(x+y)} \cdot \frac{3x(x+2)}{(3x+y)(x-y)}$
$= \frac{3(x+2)}{3x+y}$

25. $\frac{x}{x^2+9x+14} + \frac{7}{x^2+9x+14} = \frac{x+7}{x^2+9x+14}$
$= \frac{x+7}{(x+7)(x+2)}$
$= \frac{1}{x+2}$

26. $\frac{x}{x^2+2x-15} + \frac{5}{x^2+2x-15} = \frac{x+5}{x^2+2x-15}$
$= \frac{x+5}{(x+5)(x-3)}$
$= \frac{1}{x-3}$

27. $\frac{4x-5}{3x^2} - \frac{2x+5}{3x^2} = \frac{4x-5-2x-5}{3x^2}$
$= \frac{2x-10}{3x^2}$
$= \frac{2(x-5)}{3x^2}$

28. $\frac{9x+7}{6x^2} - \frac{3x+4}{6x^2} = \frac{9x+7-3x-4}{6x^2}$
$= \frac{6x+3}{6x^2}$
$= \frac{3(2x+1)}{6x^2}$
$= \frac{2x+1}{2x^2}$

29. The LCD is $2 \cdot 7 \cdot x$ or $14x$.

30. $x^2-5x-24 = (x-8)(x+3)$
$x^2+11x+24 = (x+8)(x+3)$
The LCD is $(x+3)(x+8)(x-8)$.

31. $\frac{5}{7x} = \frac{5 \cdot 2x^2y}{7x \cdot 2x^2y} = \frac{10x^2y}{14x^3y}$

32. $\frac{9}{4y} = \frac{9 \cdot 4y^2x}{4y \cdot 4y^2x} = \frac{36y^2x}{16y^3x}$

33. $\frac{x+2}{x^2+11x+18} = \frac{x+2}{(x+2)(x+9)}$
$= \frac{(x+2)(x-5)}{(x+2)(x+9)(x-5)}$
$= \frac{x^2-3x-10}{(x+2)(x-5)(x+9)}$

34. $\frac{3x-5}{x^2+4x+4} = \frac{3x-5}{(x+2)^2}$
$= \frac{(3x-5)(x+3)}{(x+2)^2(x+3)}$
$= \frac{3x^2+4x-15}{(x+2)^2(x+3)}$

35. $\frac{4}{5x^2} + \frac{6}{y} = \frac{4y}{5x^2y} + \frac{6 \cdot 5x^2}{y \cdot 5x^2} = \frac{4y+30x^2}{5x^2y}$

36. $\frac{2}{x-3} - \frac{4}{x-1} = \frac{2(x-1)-4(x-3)}{(x-3)(x-1)}$
$= \frac{2x-2-4x+12}{(x-3)(x-1)}$
$= \frac{-2x+10}{(x-3)(x-1)}$

37. $\frac{4}{x+3} - 2 = \frac{4-2(x+3)}{x+3} = \frac{4-2x-6}{x+3} = \frac{-2x-2}{x+3}$

38. $\frac{3}{x^2+2x-8} + \frac{2}{x^2-3x+2}$
$= \frac{3}{(x+4)(x-2)} + \frac{2}{(x-2)(x-1)}$
$= \frac{3(x-1)+2(x+4)}{(x+4)(x-2)(x-1)}$
$= \frac{3x-3+2x+8}{(x+4)(x-2)(x-1)}$
$= \frac{5x+5}{(x+4)(x-2)(x-1)}$
$= \frac{5(x+1)}{(x+4)(x-2)(x-1)}$

39.
$$\frac{2x-5}{6x+9}-\frac{4}{2x^2+3x}=\frac{2x-5}{3(2x+3)}-\frac{4}{x(2x+3)}$$
$$=\frac{x(2x-5)-4(3)}{3x(2x+3)}$$
$$=\frac{2x^2-5x-12}{3x(2x+3)}$$
$$=\frac{(2x+3)(x-4)}{3x(2x+3)}$$
$$=\frac{x-4}{3x}$$

40.
$$\frac{x-1}{x^2-2x+1}-\frac{x+1}{x-1}=\frac{x-1}{(x-1)(x-1)}-\frac{x+1}{x-1}$$
$$=\frac{1}{x-1}-\frac{x+1}{x-1}$$
$$=\frac{1-x-1}{x-1}$$
$$=-\frac{x}{x-1}$$

41.
$$\frac{n}{10}=9-\frac{n}{5}$$
$$10\left(\frac{n}{10}\right)=10\left(9-\frac{n}{5}\right)$$
$$n=90-2n$$
$$3n=90$$
$$n=30$$
The solution is 30.

42.
$$\frac{2}{x+1}-\frac{1}{x-2}=-\frac{1}{2}$$
$$2(x+1)(x-2)\left(\frac{2}{x+1}-\frac{1}{x-2}\right)=2(x+1)(x-2)\left(-\frac{1}{2}\right)$$
$$2\cdot 2(x-2)-2(x+1)=-1(x+1)(x-2)$$
$$4x-8-2x-2=-(x^2-x-2)$$
$$2x-10=-x^2+x+2$$
$$x^2+x-12=0$$
$$(x+4)(x-3)=0$$
$x+4=0$ or $x-3=0$
$x=-4$ $x=3$
The solutions are –4 and 3.

43.
$$\frac{y}{2y+2}+\frac{2y-16}{4y+4}=\frac{y-3}{y+1}$$
$$\frac{y}{2(y+1)}+\frac{2y-16}{4(y+1)}=\frac{y-3}{y+1}$$
$$4(y+1)\left(\frac{y}{2(y+1)}+\frac{2y-16}{4(y+1)}\right)=4(y+1)\left(\frac{y-3}{y+1}\right)$$
$$2y+2y-16=4(y-3)$$
$$4y-16=4y-12$$
$$0y=4$$
$$0=4;\text{ no solution}$$

44.
$$\frac{2}{x-3}-\frac{4}{x+3}=\frac{8}{x^2-9}$$
$$(x^2-9)\left(\frac{2}{x-3}-\frac{4}{x+3}\right)=(x^2-9)\left(\frac{8}{x^2-9}\right)$$
$$2(x+3)-4(x-3)=8$$
$$2x+6-4x+12=8$$
$$-2x+18=8$$
$$-2x=-10$$
$$x=5$$
The solution is 5.

45.
$$\frac{x-3}{x+1}-\frac{x-6}{x+5}=0$$
$$(x+1)(x+5)\left(\frac{x-3}{x+1}-\frac{x-6}{x+5}\right)=(x+1)(x+5)0$$
$$(x+5)(x-3)-(x+1)(x-6)=0$$
$$x^2+2x-15-x^2+5x+6=0$$
$$7x-9=0$$
$$7x=9$$
$$x=\frac{9}{7}$$
The solution is $\frac{9}{7}$.

46.
$$x+5=\frac{6}{x}$$
$$x(x+5)=x\left(\frac{6}{x}\right)$$
$$x^2+5x=6$$
$$x^2+5x-6=0$$
$$(x+6)(x-1)=0$$
$$x+6=0 \quad\text{or}\quad x-1=0$$
$$x=-6 \qquad\qquad x=1$$
The solutions are −6 and 1.

47.
$$\frac{2}{x-1}=\frac{3}{x+3}$$
$$2(x+3)=3(x-1)$$
$$2x+6=3x-3$$
$$9=x$$
The solution is 9.

48.
$$\frac{4}{y-3}=\frac{2}{y-3}$$
$$4(y-3)=2(y-3)$$
$$4y-12=2y-6$$
$$2y=6$$
$$y=3$$
$y=3$ does not check. There is no solution.

49.
$$\frac{300}{20}=\frac{x}{45}$$
$$300(45)=20x$$
$$13{,}500=20x$$
$$675=x$$
The machine will process 675 parts in 45 minutes.

50.
$$\frac{90}{8}=\frac{x}{3}$$
$$90(3)=8x$$
$$270=8x$$
$$33.75=x$$
Mr. Visconti charges \$33.75 for 3 hours of consulting.

51. Let n be the number.
$$5\left(\frac{1}{n}\right)=\frac{3}{2}\left(\frac{1}{n}\right)+\frac{7}{6}$$
$$\frac{5}{n}=\frac{3}{2n}+\frac{7}{6}$$
$$6n\left(\frac{5}{n}\right)=6n\left(\frac{3}{2n}+\frac{7}{6}\right)$$
$$30=9+7n$$
$$21=7n$$
$$3=n$$
The number is 3.

52. Let n be the number.

$$\frac{1}{n}=\frac{1}{4-n}$$
$$n(4-n)\left(\frac{1}{n}\right)=n(4-n)\left(\frac{1}{4-n}\right)$$
$$4-n=n$$
$$4=2n$$
$$2=n$$

The number is 2.

53. Let x be the speed of the faster car.

	distance	=	rate	·	time
Faster car	90		x		$\frac{90}{x}$
Slower car	60		$x-10$		$\frac{60}{x-10}$

The times are equal.

$$\frac{90}{x}=\frac{60}{x-10}$$
$$90(x-10)=60x$$
$$90x-900=60x$$
$$30x=900$$
$$x=30$$
$$x-10=20$$

The faster car is traveling at 30 mph and the slower car is traveling at 20 mph.

54. Let x be the speed of the boat in still water.

	distance	=	rate	·	time
Upstream	48		$x-4$		$\frac{48}{x-4}$
Downstream	72		$x+4$		$\frac{72}{x+4}$

The times are equal.

$$\frac{48}{x-4}=\frac{72}{x+4}$$
$$48(x+4)=72(x-4)$$
$$48x+192=72x-288$$
$$480=24x$$
$$20=x$$

The speed of the boat in still water is 20 mph.

55. Let x be the time for Maria alone. Together, Mark and Maria complete $\frac{1}{5}$ of the job in 1 hour. Individually, they complete $\frac{1}{7}$ and $\frac{1}{x}$ of the job in 1 hour.

$$\frac{1}{5}=\frac{1}{7}+\frac{1}{x}$$
$$35x\left(\frac{1}{5}\right)=35x\left(\frac{1}{7}+\frac{1}{x}\right)$$
$$7x=5x+35$$
$$2x=35$$
$$x=\frac{35}{2}=17\frac{1}{2}$$

It will take Maria $17\frac{1}{2}$ hours to manicure Mr. Sturgeon's lawn alone.

56. Let x be the time for the pipes to fill the pond working together. Pipe A fills $\frac{1}{20}$ of the pond in 1 day, pipe B fills $\frac{1}{15}$ of the pond in 1 day, and together they fill $\frac{1}{x}$ of the pond in one day.

$$\frac{1}{20}+\frac{1}{15}=\frac{1}{x}$$
$$60x\left(\frac{1}{20}+\frac{1}{15}\right)=60x\left(\frac{1}{x}\right)$$
$$3x+4x=60$$
$$7x=60$$
$$x=\frac{60}{7}=8\frac{4}{7}$$

It takes $8\frac{4}{7}$ days to fill the pond using both pipes.

57.
$$\frac{2}{3}=\frac{10}{x}$$
$$2\cdot x=3\cdot 10$$
$$2x=30$$
$$x=15$$

The missing length is $x = 15$.

58. $\frac{12}{4} = \frac{18}{x}$
$12 \cdot x = 4 \cdot 18$
$12x = 72$
$x = 6$
The missing length is $x = 6$.

59. $\frac{\frac{5x}{27}}{-\frac{10xy}{21}} = \frac{5x}{27} \div -\frac{10xy}{21}$
$= \frac{5x}{27} \cdot \frac{-21}{10xy}$
$= -\frac{5 \cdot x \cdot 3 \cdot 7}{3 \cdot 9 \cdot 2 \cdot 5 \cdot x \cdot y}$
$= -\frac{7}{18y}$

60. $\frac{\frac{3}{5}+\frac{2}{7}}{\frac{1}{5}+\frac{5}{6}} = \frac{\frac{3 \cdot 7}{5 \cdot 7}+\frac{2 \cdot 5}{7 \cdot 5}}{\frac{1 \cdot 6}{5 \cdot 6}+\frac{5 \cdot 5}{6 \cdot 5}}$
$= \frac{\frac{21+10}{35}}{\frac{6+25}{30}}$
$= \frac{\frac{31}{35}}{\frac{31}{30}}$
$= \frac{31}{35} \div \frac{31}{30}$
$= \frac{31}{35} \cdot \frac{30}{31}$
$= \frac{31 \cdot 5 \cdot 6}{5 \cdot 7 \cdot 31}$
$= \frac{6}{7}$

61. $\frac{3-\frac{1}{y}}{2-\frac{1}{y}} = \frac{y\left(3-\frac{1}{y}\right)}{y\left(2-\frac{1}{y}\right)} = \frac{3y-1}{2y-1}$

62. $\frac{\frac{6}{x+2}+4}{\frac{8}{x+2}-4} = \frac{(x+2)\left(\frac{6}{x+2}+4\right)}{(x+2)\left(\frac{8}{x+2}-4\right)}$
$= \frac{6+4(x+2)}{8-4(x+2)}$
$= \frac{6+4x+8}{8-4x-8}$
$= \frac{4x+14}{-4x}$
$= -\frac{2(2x+7)}{2 \cdot 2x}$
$= -\frac{2x+7}{2x}$

63. $\frac{4x+12}{8x^2+24x} = \frac{4(x+3)}{8x(x+3)} = \frac{1}{2x}$

64. $\frac{x^3-6x^2+9x}{x^2+4x-21} = \frac{x(x^2-6x+9)}{(x+7)(x-3)}$
$= \frac{x(x-3)(x-3)}{(x+7)(x-3)}$
$= \frac{x(x-3)}{x+7}$

65. $\frac{x^2+9x+20}{x^2-25} \cdot \frac{x^2-9x+20}{x^2+8x+16}$
$= \frac{(x+4)(x+5)}{(x-5)(x+5)} \cdot \frac{(x-4)(x-5)}{(x+4)(x+4)}$
$= \frac{x-4}{x+4}$

66. $\frac{x^2-x-72}{x^2-x-30} \div \frac{x^2+6x-27}{x^2-9x+18}$
$= \frac{x^2-x-72}{x^2-x-30} \cdot \frac{x^2-9x+18}{x^2+6x-27}$
$= \frac{(x-9)(x+8)}{(x-6)(x+5)} \cdot \frac{(x-6)(x-3)}{(x+9)(x-3)}$
$= \frac{(x-9)(x+8)}{(x+5)(x+9)}$

67. $\frac{x}{x^2-36} + \frac{6}{x^2-36} = \frac{x+6}{x^2-36}$
$= \frac{x+6}{(x+6)(x-6)}$
$= \frac{1}{x-6}$

68. $\frac{5x-1}{4x}-\frac{3x-2}{4x}=\frac{5x-1-3x+2}{4x}=\frac{2x+1}{4x}$

69. $\frac{4}{3x^2+8x-3}+\frac{2}{3x^2-7x+2}$
$=\frac{4}{(3x-1)(x+3)}+\frac{2}{(3x-1)(x-2)}$
$=\frac{4(x-2)}{(3x-1)(x+3)(x-2)}+\frac{2(x+3)}{(3x-1)(x-2)(x+3)}$
$=\frac{4x-8+2x+6}{(3x-1)(x+3)(x-2)}$
$=\frac{6x-2}{(3x-1)(x+3)(x-2)}$
$=\frac{2(3x-1)}{(3x-1)(x+3)(x-2)}$
$=\frac{2}{(x+3)(x-2)}$

70. $\frac{3x}{x^2+9x+14}-\frac{6x}{x^2+4x-21}$
$=\frac{3x}{(x+7)(x+2)}-\frac{6x}{(x+7)(x-3)}$
$=\frac{3x(x-3)}{(x+7)(x+2)(x-3)}-\frac{6x(x+2)}{(x+7)(x-3)(x+2)}$
$=\frac{3x^2-9x-6x^2-12x}{(x+7)(x+2)(x-3)}$
$=\frac{-3x^2-21x}{(x+7)(x+2)(x-3)}$
$=\frac{-3x(x+7)}{(x+7)(x+2)(x-3)}$
$=-\frac{3x}{(x+2)(x-3)}$

71. $\frac{4}{a-1}+2=\frac{3}{a-1}$
$(a-1)\left(\frac{4}{a-1}+2\right)=(a-1)\left(\frac{3}{a-1}\right)$
$4+2(a-1)=3$
$4+2a-2=3$
$2+2a=3$
$2a=1$
$a=\frac{1}{2}$
The solution is $\frac{1}{2}$.

72. $\frac{x}{x+3}+4=\frac{x}{x+3}$
$(x+3)\left(\frac{x}{x+3}+4\right)=(x+3)\left(\frac{x}{x+3}\right)$
$x+4(x+3)=x$
$x+4x+12=x$
$5x+12=x$
$4x+12=0$
$4x=-12$
$x=-3$
-3 makes the denominator $x+3$ zero. Therefore there is no solution.

73. Let n be the number.
$\frac{2n}{3}-\frac{1}{6}=\frac{n}{2}$
$6\left(\frac{2n}{3}-\frac{1}{6}\right)=6\left(\frac{n}{2}\right)$
$4n-1=3n$
$n-1=0$
$n=1$
The number is 1.

74. Let x be the time to paint the shed together. Mr. Crocker can paint $\frac{1}{3}$ of the shed in one day, and his son can paint $\frac{1}{4}$. Together, they can paint $\frac{1}{x}$.
$\frac{1}{3}+\frac{1}{4}=\frac{1}{x}$
$12x\left(\frac{1}{3}+\frac{1}{4}\right)=12x\left(\frac{1}{x}\right)$
$4x+3x=12$
$7x=12$
$x=\frac{12}{7}$ or $1\frac{5}{7}$
It will take $1\frac{5}{7}$ days to paint the shed when they work together.

75. $\frac{x}{10}=\frac{3}{5}$
$5x=3(10)$
$5x=30$
$x=6$
The missing length is $x=6$.

76. $\frac{x}{4}=\frac{18}{6}$
$6x=4(18)$
$6x=72$
$x=12$
The missing length is $x = 12$.

77. $\frac{\frac{1}{4}}{\frac{1}{3}+\frac{1}{2}}=\frac{12\left(\frac{1}{4}\right)}{12\left(\frac{1}{3}+\frac{1}{2}\right)}=\frac{3}{4+6}=\frac{3}{10}$

78. $\frac{4+\frac{2}{x}}{6+\frac{3}{x}}=\frac{x\left(4+\frac{2}{x}\right)}{x\left(6+\frac{3}{x}\right)}=\frac{4x+2}{6x+3}=\frac{2(2x+1)}{3(2x+1)}=\frac{2}{3}$

Chapter 5 Test

1. $x^2+4x+3=0$
$(x+1)(x+3)=0$
$x+1=0$ or $x+3=0$
$x=-1$ $\quad x=-3$
The expression $\frac{x+5}{x^2+4x+3}$ is undefined for $x=-1$ and $x=-3$.

2. a. $C=\frac{100x+3000}{x}=\frac{100(200)+3000}{200}=\$115$

b. $C=\frac{100x+3000}{x}$
$=\frac{100(1000)+3000}{1000}$
$=\$103$

3. $\frac{3x-6}{5x-10}=\frac{3(x-2)}{5(x-2)}=\frac{3}{5}$

4. $\frac{x+6}{x^2+12x+36}=\frac{x+6}{(x+6)(x+6)}=\frac{1}{x+6}$

5. $\frac{7-x}{x-7}=\frac{-(x-7)}{x-7}=-1$

6. $\frac{y-x}{x^2-y^2}=\frac{-(x-y)}{(x-y)(x+y)}=-\frac{1}{x+y}$

7. $\frac{2m^3-2m^2-12m}{m^2-5m+6}=\frac{2m(m^2-m-6)}{(m-3)(m-2)}$
$=\frac{2m(m-3)(m+2)}{(m-3)(m-2)}$
$=\frac{2m(m+2)}{m-2}$

8. $\frac{ay+3a+2y+6}{ay+3a+5y+15}=\frac{(ay+3a)+(2y+6)}{(ay+3a)+(5y+15)}$
$=\frac{a(y+3)+2(y+3)}{a(y+3)+5(y+3)}$
$=\frac{(a+2)(y+3)}{(a+5)(y+3)}$
$=\frac{a+2}{a+5}$

9. $\frac{x^2-13x+42}{x^2+10x+21}\div\frac{x^2-4}{x^2+x-6}$
$=\frac{x^2-13x+42}{x^2+10x+21}\cdot\frac{x^2+x-6}{x^2-4}$
$=\frac{(x-6)(x-7)}{(x+3)(x+7)}\cdot\frac{(x+3)(x-2)}{(x+2)(x-2)}$
$=\frac{(x-6)(x-7)}{(x+7)(x+2)}$

10. $\frac{3}{x-1}\cdot(5x-5)=\frac{3}{x-1}\cdot5(x-1)=15$

11. $\frac{y^2-5y+6}{2y+4}\cdot\frac{y+2}{2y-6}=\frac{(y-3)(y-2)}{2(y+2)}\cdot\frac{y+2}{2(y-3)}$
$=\frac{y-2}{4}$

12. $\frac{5}{2x+5}-\frac{6}{2x+5}=\frac{5-6}{2x+5}=-\frac{1}{2x+5}$

13. $\dfrac{5a}{a^2-a-6}-\dfrac{2}{a-3}$
$=\dfrac{5a}{(a-3)(a+2)}-\dfrac{2}{a-3}$
$=\dfrac{5a}{(a-3)(a+2)}-\dfrac{2(a+2)}{(a-3)(a+2)}$
$=\dfrac{5a-2(a+2)}{(a-3)(a+2)}$
$=\dfrac{5a-2a-4}{(a-3)(a+2)}$
$=\dfrac{3a-4}{(a-3)(a+2)}$

14. $\dfrac{6}{x^2-1}+\dfrac{3}{x+1}=\dfrac{6}{(x-1)(x+1)}+\dfrac{3}{x+1}$
$=\dfrac{6}{(x-1)(x+1)}+\dfrac{3(x-1)}{(x+1)(x-1)}$
$=\dfrac{6+3x-3}{(x+1)(x-1)}$
$=\dfrac{3x+3}{(x+1)(x-1)}$
$=\dfrac{3(x+1)}{(x+1)(x-1)}$
$=\dfrac{3}{x-1}$

15. $\dfrac{x^2-9}{x^2-3x}\div\dfrac{x^2+4x+1}{2x+10}=\dfrac{x^2-9}{x^2-3x}\cdot\dfrac{2x+10}{x^2+4x+1}$
$=\dfrac{(x-3)(x+3)}{x(x-3)}\cdot\dfrac{2(x+5)}{x^2+4x+1}$
$=\dfrac{2(x+3)(x+5)}{x(x^2+4x+1)}$

16. $\dfrac{x+2}{x^2+11x+18}+\dfrac{5}{x^2-3x-10}$
$=\dfrac{x+2}{(x+9)(x+2)}+\dfrac{5}{(x-5)(x+2)}$
$=\dfrac{(x+2)(x-5)}{(x+9)(x+2)(x-5)}+\dfrac{5(x+9)}{(x-5)(x+2)(x+9)}$
$=\dfrac{x^2-3x-10+5x+45}{(x+9)(x+2)(x-5)}$
$=\dfrac{x^2+2x+35}{(x+9)(x+2)(x-5)}$

17. $\dfrac{4y}{y^2+6y+5}-\dfrac{3}{y^2+5y+4}$
$=\dfrac{4y}{(y+5)(y+1)}-\dfrac{3}{(y+1)(y+4)}$
$=\dfrac{4y(y+4)-3(y+5)}{(y+1)(y+5)(y+4)}$
$=\dfrac{4y^2+16y-3y-15}{(y+1)(y+5)(y+4)}$
$=\dfrac{4y^2+13y-15}{(y+1)(y+5)(y+4)}$

18. The LCD is $3\cdot 5\cdot y=15y$.
$\dfrac{4}{y}-\dfrac{5}{3}=-\dfrac{1}{5}$
$15y\left(\dfrac{4}{y}-\dfrac{5}{3}\right)=15y\left(-\dfrac{1}{5}\right)$
$60-25y=-3y$
$60=22y$
$\dfrac{60}{22}=y$
$\dfrac{30}{11}=y$

The solution is $\dfrac{30}{11}$.

19. $\dfrac{5}{y+1}=\dfrac{4}{y+2}$
$5(y+2)=4(y+1)$
$5y+10=4y+4$
$y=-6$

The solution is -6.

20. The LCD is $2(a-3)$.
$\dfrac{a}{a-3}=\dfrac{3}{a-3}-\dfrac{3}{2}$
$2(a-3)\left(\dfrac{a}{a-3}\right)=2(a-3)\left(\dfrac{3}{a-3}-\dfrac{3}{2}\right)$
$2a=6-3(a-3)$
$2a=6-3a+9$
$5a=15$
$a=3$

Since $a=3$ causes the denominator $a-3$ to be 0, the equation has no solution.

21. The LCD is $x^2-25=(x+5)(x-5)$.

$$\frac{10}{x^2-25}=\frac{3}{x+5}+\frac{1}{x-5}$$
$$(x^2-25)\left(\frac{10}{x^2-25}\right)=(x+5)(x-5)\left(\frac{3}{x+5}+\frac{1}{x-5}\right)$$
$$10=3(x-5)+1(x+5)$$
$$10=3x-15+x+5$$
$$10=4x-10$$
$$20=4x$$
$$5=x$$

Since $x = 5$ causes the denominators x^2-25 and $x-5$ to be 0, the equation has no solution.

22.
$$x-\frac{14}{x-1}=4-\frac{2x}{x-1}$$
$$(x-1)\left(x-\frac{14}{x-1}\right)=(x-1)\left(4-\frac{2x}{x-1}\right)$$
$$x(x-1)-14=4(x-1)-2x$$
$$x^2-x-14=4x-4-2x$$
$$x^2-x-14=2x-4$$
$$x^2-3x-10=0$$
$$(x-5)(x+2)=0$$
$$x-5=0 \quad \text{or} \quad x+2=0$$
$$x=5 \qquad\qquad x=-2$$

The solutions are 5 and −2.

23.
$$\frac{\frac{5x^2}{yz^2}}{\frac{10x}{z^3}}=\frac{5x^2}{yz^2}\div\frac{10x}{z^3}$$
$$=\frac{5x^2}{yz^2}\cdot\frac{z^3}{10x}$$
$$=\frac{5\cdot x\cdot x\cdot z\cdot z^2}{y\cdot z^2\cdot 2\cdot 5\cdot x}$$
$$=\frac{xz}{2y}$$

24.
$$\frac{\frac{b}{a}-\frac{a}{b}}{\frac{1}{b}+\frac{1}{a}}=\frac{\left(\frac{b}{a}-\frac{a}{b}\right)ab}{\left(\frac{1}{b}+\frac{1}{a}\right)ab}$$
$$=\frac{b^2-a^2}{a+b}$$
$$=\frac{(b-a)(b+a)}{a+b}$$
$$=b-a$$

25.
$$\frac{5-\frac{1}{y^2}}{\frac{1}{y}+\frac{2}{y^2}}=\frac{y^2\left(5-\frac{1}{y^2}\right)}{y^2\left(\frac{1}{y}+\frac{2}{y^2}\right)}=\frac{5y^2-1}{y+2}$$

26. Let n be the number.

$$n+5\left(\frac{1}{n}\right)=6$$
$$n+\frac{5}{n}=6$$
$$n\left(n+\frac{5}{n}\right)=n(6)$$
$$n^2+5=6n$$
$$n^2-6n+5=0$$
$$(n-5)(n-1)=0$$
$$n-5=0 \quad \text{or} \quad n-1=0$$
$$n=5 \qquad\qquad n=1$$

The number is 1 or 5.

27. Let x be the speed of the boat in still water. Let $x + 2$ be the speed of the boat going downstream. Let $x - 2$ be the speed of the boat going upstream.

	distance	=	**rate**	·	**time**
Upstream	14		$x-2$		$\frac{14}{x-2}$
Downstream	16		$x+2$		$\frac{16}{x+2}$

$$\frac{14}{x-2}=\frac{16}{x+2}$$
$$14(x+2)=16(x-2)$$
$$14x+28=16x-32$$
$$60=2x$$
$$30=x$$

The speed of the boat in still water is 30 mph.

28. Let x be the time in hours that it takes for both inlet pipes together to fill the tank.

The first pipe fills $\frac{1}{12}$ of the tank in 1 hour, the second pipe fills $\frac{1}{15}$ of the tank in 1 hour, and the two pipes fill $\frac{1}{x}$ of the tank in 1 hour.

$$\frac{1}{12}+\frac{1}{15}=\frac{1}{x}$$

The LCD is $60x$.

$$60x\left(\frac{1}{12}+\frac{1}{15}\right)=60x\left(\frac{1}{x}\right)$$
$$5x+4x=60$$
$$9x=60$$
$$x=\frac{60}{9}$$
$$x=\frac{20}{3}$$

It takes both pipes $\frac{20}{3}=6\frac{2}{3}$ hours to fill the tank.

29. $\frac{8}{x}=\frac{10}{15}$
$$8(15)=10x$$
$$120=10x$$
$$12=x$$

30. $\frac{3}{85}=\frac{x}{510}$
$$3(510)=85x$$
$$1530=85x$$
$$18=x$$
18 defective bulbs should be found in 510 bulbs.

Cumulative Review Chapters 1–5

1. a. The quotient of 15 and a number is 4 is written as $\frac{15}{x}=4$.

b. Three subtracted from 12 is a number is written as $12-3=x$.

c. 17 added to four times a number is 21 is written as $4x+17=21$.

2. a. The difference of 12 and a number is –45 is written as $12-x=-45$.

b. The product of 12 and a number is –45 is written as $12x=-45$.

c. A number less 10 is twice the number is written as $x-10=2x$.

3. a. $3+(-7)+(-8)=(-4)+(-8)=-12$

b. $[7+(-10)]+[-2+(-4)]=(-3)+(-6)=-9$

4. a. $28-6-30=22-30=-8$

b. $7-2-22=5-22=-17$

5. $3(x+y)=3\cdot x+3\cdot y$ illustrates the distributive property.

6. $3+y=y+3$ illustrates the commutative property of addition.

7. $(x+7)+9=x+(7+9)$ illustrates the associative property of addition.

8. $(x\cdot 7)\cdot 9=x\cdot(7\cdot 9)$ illustrates the associative property of multiplication.

9. $3-x=7$
$$2-x-3=7-3$$
$$-x=4$$
$$(-1)(-x)=(-1)4$$
$$x=-4$$
The solution is –4.

10. $7x-6=6x-6$
$$7x-6+6=6x-6+6$$
$$7x=6x$$
$$7x-6x=6x-6x$$
$$x=0$$
The solution is 0.

11. Let x be the length of the shorter piece. Then $4x$ is the length of the longer piece.
$$x+4x=10$$
$$5x=10$$
$$x=2$$
$4x=4(2)=8$
The shorter piece is 2 feet. The longer piece is 8 feet.

12. Let n be the first even integer. Then $n+2$ is the second.
$$n+(n+2)=382$$
$$2n+2=382$$
$$2n=380$$
$$n=190$$
$n+2=192$
The two consecutive even integers are 190 and 192.

13. $y=mx+b$
$$y-b=mx$$
$$\frac{y-b}{m}=x$$

14. $3x-2y=6$
$$3x=6+2y$$
$$x=\frac{6+2y}{3}$$

15. $x+4\le -6$
$$x+4-4\le -6-4$$
$$x\le -10$$

−12 −10 −8 −6 −4 −2 0 2 4 6 8

16. $-3x+7>-x+9$
$$7>2x+9$$
$$-2>2x$$
$$-1>x$$
$$\{x|x<-1\}$$

17. $\dfrac{x^5}{x^2}=x^{5-2}=x^3$

18. $\dfrac{y^{14}}{y^{14}}=1$

19. $\dfrac{4^7}{4^3}=4^{7-3}=4^4=256$

20. $(x^5y^2)^3=x^{5\cdot 3}y^{2\cdot 3}=x^{15}y^6$

21. $\dfrac{(-3)^5}{(-3)^2}=(-3)^{5-2}=(-3)^3=-27$

22. $\dfrac{x^{19}y^5}{xy}=\dfrac{x^{19}}{x}\cdot\dfrac{y^5}{y}=x^{19-1}\cdot y^{5-1}=x^{18}y^4$

23. $\dfrac{2x^5y^2}{xy}=\dfrac{2}{1}\cdot\dfrac{x^5}{x}\cdot\dfrac{y^2}{y}=2x^{5-1}y^{2-1}=2x^4y$

24. $(-3a^2b)(5a^3b)=-3\cdot 5\cdot a^2\cdot a^3\cdot b\cdot b$
$$=-15a^{2+3}b^{1+1}$$
$$=-15a^5b^2$$

25. $2x^{-3}=2\cdot\dfrac{1}{x^3}=\dfrac{2}{x^3}$

26. $7^{-2}=\dfrac{1}{7^2}=\dfrac{1}{49}$

27. $(-2)^{-4}=\dfrac{1}{(-2)^4}=\dfrac{1}{16}$

28. $5z^{-7}=5\cdot\dfrac{1}{z^7}=\dfrac{5}{z^7}$

29. $5x(2x^3+6)=5x(2x^3)+5x(6)=10x^4+30x$

30. $(x+9)^2=(x)^2+2(x)(9)+(9)^2=x^2+18x+81$

31. $-3x^2(5x^2+6x-1)$
$$=-3x^2(5x^2)+(-3x^2)(6x)-(-3x^2)(1)$$
$$=-15x^4-18x^3+3x^2$$

32. $(2x+1)(2x-1)=(2x)^2-(1)^2=4x^2-1$

33. $\dfrac{4x^2+7+8x^3}{2x+3}=4x^2-4x+6-\dfrac{11}{2x+3}$

$$\begin{array}{r|l} & 4x^2-4x+6 \\ 2x+3 & 8x^3+4x^2+0x^1+7 \\ & \underline{8x^3+12x^2} \\ & -8x^2+0x^1 \\ & \underline{-8x^2-12x} \\ & 12x+7 \\ & \underline{12x+18} \\ & -11 \end{array}$$

34. $\dfrac{4x^3-9x+2}{x-4}=4x^2+16x+55+\dfrac{222}{x-4}$

$$\begin{array}{r|l} & 4x^2+16x+55 \\ x-4 & 4x^3-0x^2-9x+2 \\ & \underline{4x^3-16x^2} \\ & 16x^2-9x \\ & \underline{16x^2-64x} \\ & 55x+2 \\ & \underline{55x-220} \\ & 222 \end{array}$$

35. $x^2+7x+12=(x+4)(x+3)$

36. $-2a^2+10a+12=-2(a^2-5a-6)$
$$=-2(a-6)(a+1)$$

37. $25x^2+20xy+4y^2=(5x+2y)(5x+2y)$
$$=(5x+2y)^2$$

38. $x^2 - 4 = (x-2)(x+2)$

39. $x^2 - 9x - 22 = 0$
$(x-11)(x+2) = 0$
$x - 11 = 0$ or $x + 2 = 0$
$x = 11$ $\quad x = -2$
The solutions are 11 and −2.

40. $3x^2 + 5x = 2$
$3x^2 + 5x - 2 = 0$
$(3x-1)(x+2) = 0$
$3x - 1 = 0$ or $x + 2 = 0$
$3x = 1$ $\quad x = -2$
$x = \frac{1}{3}$
The solutions are $\frac{1}{3}$ and −2.

41. $\frac{x^2+x}{3x} \cdot \frac{6}{5x+5} = \frac{x(x+1)}{3x} \cdot \frac{2 \cdot 3}{5(x+1)} = \frac{2}{5}$

42. $\frac{2x^2-50}{4x^4-20x^3} = \frac{2(x^2-25)}{4x^3(x-5)}$
$= \frac{2(x-5)(x+5)}{2 \cdot 2x^3(x-5)}$
$= \frac{x+5}{2x^3}$

43. $\frac{3x^2+2x}{x-1} - \frac{10x-5}{x-1} = \frac{3x^2+2x-10x+5}{x-1}$
$= \frac{3x^2-8x+5}{x-1}$
$= \frac{(3x-5)(x-1)}{x-1}$
$= 3x - 5$

44. $7x^6 - 7x^5 + 7x^4 = 7x^4(x^2 - x + 1)$

45. $\frac{6x}{x^2-4} - \frac{3}{x+2} = \frac{6x}{(x+2)(x-2)} - \frac{3(x-2)}{(x+2)(x-2)}$
$= \frac{6x-3x+6}{(x+2)(x-2)}$
$= \frac{3x+6}{(x+2)(x-2)}$
$= \frac{3(x+2)}{(x+2)(x-2)}$
$= \frac{3}{x-2}$

46. $4x^2 + 12x + 9 = (2x+3)(2x+3) = (2x+3)^2$

47. $\frac{t-4}{2} - \frac{t-3}{9} = \frac{5}{18}$
$18\left(\frac{t-4}{2} - \frac{t-3}{9}\right) = 18\left(\frac{5}{18}\right)$
$9(t-4) - 2(t-3) = 5$
$9t - 36 - 2t + 6 = 5$
$7t - 30 = 5$
$7t = 35$
$t = 5$
The solution is 5.

48. $\frac{6x^2-18x}{3x^2-2x} \cdot \frac{15x-10}{x^2-9} = \frac{6x(x-3)}{x(3x-2)} \cdot \frac{5(3x-2)}{(x-3)(x+3)}$
$= \frac{30}{x+3}$

49. Let x be the time in hours for Sam and Frank to complete the tour together. Sam completes $\frac{1}{3}$ of a tour in 1 hour, and Frank completes $\frac{1}{7}$ of a tour in 1 hour. Together, they complete $\frac{1}{x}$ of a tour in 1 hour.
$\frac{1}{3} + \frac{1}{7} = \frac{1}{x}$
$21x\left(\frac{1}{3} + \frac{1}{7}\right) = 21x\left(\frac{1}{x}\right)$
$7x + 3x = 21$
$10x = 21$
$x = \frac{21}{10} = 2\frac{1}{10}$
Sam and Frank can complete a quality control tour together in $2\frac{1}{10}$ hours.

50. $\frac{\frac{m}{3}+\frac{n}{6}}{\frac{m+n}{12}} = \frac{12\left(\frac{m}{3}+\frac{n}{6}\right)}{12\left(\frac{m+n}{12}\right)} = \frac{4m+2n}{m+n} = \frac{2(2m+n)}{m+n}$

Chapter 6

Section 6.1 Practice Exercises

1. a. The height of the bar for birds is 78, so approximately 78 endangered species are birds.

b. The shortest bar corresponds to arachnids, so arachnids have the fewest endangered species.

2. a. Locate 40 along the time axis and move vertically upward to the line. Then move horizontally to the left until the pulse rate axis is reached. The pulse rate is 70 beats per minute 40 minutes after lighting a cigarette.

b. Read the pulse rate for time = 0. The pulse rate is 60 beats per minute when the cigarette in being lit.

c. Read the time when the pulse rate is the highest, i.e., when the line is at the peak. The pulse rate is the highest 5 minutes after lighting the cigarette.

3. a. Point (4, 2) lies in quadrant I.

b. Point (−1, −3) lies in quadrant III.

c. Point (2, −2) lies in quadrant IV.

d. Point (−5, 1) lies in quadrant II.

e. Point (0, 3) lies on the y-axis.

f. Point (3, 0) lies on the x-axis.

g. Point (0, −4) lies on the y-axis.

h. Point $\left(-2\frac{1}{2}, 0\right)$ lies on the x-axis.

i. Point $\left(1, -3\frac{3}{4}\right)$ lies in quadrant IV.

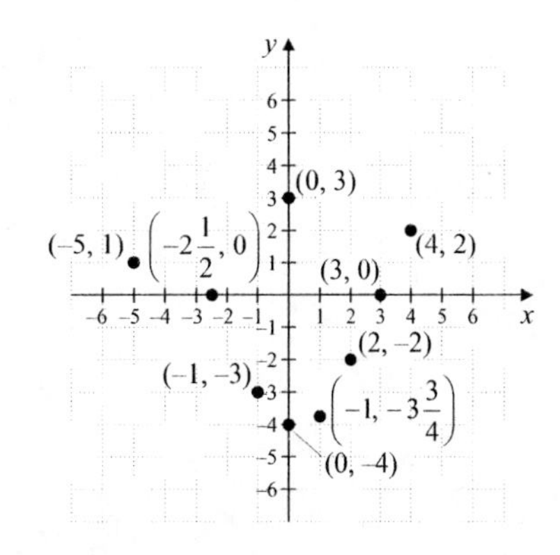

4. a. (2008, 1692), (2009, 1156), (2010, 1282), (2011, 1693), (2012, 939), (2013, 908)

b. U.S. Tornadoes

Number of Tornadoes: 0, 200, 400, 600, 800, 1000, 1200, 1400, 1600, 1800

Year: 2009, 2011, 2013

c. The number of tornadoes varies greatly from year to year.

5. a. $x + 2y = 8$
In (0,), the x-coordinate is 0.
$0 + 2y = 8$
$2y = 8$
$y = 4$
The ordered pair is (0, 4).

b. $x + 2y = 8$
In (, 3) the y-coordinate is 3.
$x + 2(3) = 8$
$x + 6 = 8$
$x = 2$
The ordered pair is (2, 3).

c. $x + 2y = 8$
In (−4,), the x-coordinate is −4.
$-4 + 2y = 8$
$2y = 12$
$y = 6$
The ordered pair is (−4, 6).

6. a. Let $x = -3$
$y = -2x$
$y = -2(-3)$
$y = 6$

b. Let $y = 0$.
$y = -2x$
$0 = -2x$
$0 = x$

c. Let $y = 10$.
$y = -2x$
$10 = -2x$
$-5 = x$

The ordered pairs are (−3, 6), (0, 0), and (−5, 10).

	x	y
a.	−3	6
b.	0	0
c.	−5	10

7. a. Let $x = -3$.
$y = \frac{1}{3}x - 1$
$y = \frac{1}{3}(-3) - 1$
$y = -1 - 1$
$y = -2$

b. Let $x = 0$.
$y = \frac{1}{3}x - 1$
$y = \frac{1}{3}(0) - 1$
$y = 0 - 1$
$y = -1$

c. Let $y = 0$.
$0 = \frac{1}{3}x - 1$
$1 = \frac{1}{3}x$
$3(1) = x$
$3 = x$

The ordered pairs are (−3, −2), (0, −1), and (3, 0).

	x	y
a.	−3	−2
b.	0	−1
c.	3	0

8. When $x = 1$,
$y = -50x + 400$
$y = -50(1) + 400$
$y = -50 + 400$
$y = 350$

When $x = 2$,
$y = -50x + 400$
$y = -50(2) + 400$
$y = -100 + 400$
$y = 300$

When $x = 3$,
$y = -50x + 400$
$y = -50(3) + 400$
$y = -150 + 400$
$y = 250$

When $x = 4$,
$y = -50x + 400$
$y = -50(4) + 400$
$y = -200 + 400$
$y = 200$

When $x = 5$,
$y = -50x + 400$
$y = -50(5) + 400$
$y = -250 + 400$
$y = 150$

When $x = 6$,
$y = -50x + 400$
$y = -50(6) + 400$
$y = -300 + 400$
$y = 100$

When $x = 7$,
$y = -50x + 400$
$y = -50(7) + 400$
$y = -350 + 400$
$y = 50$

The completed table is shown.

x	1	2	3	4	5	6	7
y	350	300	250	200	150	100	50

Vocabulary, Readiness & Video Check 6.1

1. The horizontal axis is called the x-axis.

2. The vertical axis is called the y-axis.

3. The intersection of the horizontal axis and the vertical axis is a point called the origin.

4. The axes divide the plane into regions, called quadrants. There are four of these regions.

5. In the ordered pair of numbers (−2, 5), the number −2 is called the x-coordinate and the number 5 is called the y-coordinate.

6. Each ordered pair of numbers corresponds to one point in the plane.

7. An ordered pair is a solution of an equation in two variables if replacing the variables by the coordinates of the ordered pair results in a true statement.

8. The graph of paired data as points in a rectangular coordinate system is called a scatter diagram.

9. 2003; 9.4

10. origin; left or right; up or down

11. Paired data, which can be written as ordered pairs and graphed.

12. Replace both values of the ordered pair in the linear equation and see if a true statement results.

Exercise Set 6.1

1. The longest bar corresponds to September, so the month in which most hurricanes made landfall is September.

3. The length of the bar for August is 77, so approximately 77 hurricanes made landfall in August.

5. Two of the 77 hurricanes that made landfall in August did so in 2008. The fraction is $\frac{2}{77}$.

7. The longest bar corresponds to Tokyo, Japan, and the length of the bar is about 39.4 million. So, the city with the largest population is Tokyo, Japan and its population is about 39.4 million or 39,400,000.

9. The longest bar corresponding to a city in Mexico is the bar for Mexico City. The population is approximately 22.2 million or 22,200,000.

11. The bar corresponding to Seoul, South Korea has length about 24.2 and the bar corresponding to Delhi, India has length about 25.2. Thus, Delhi, India is about $25.2 - 24.2 \approx 1$ million larger than Seoul, South Korea.

13. The point on the graph corresponding to 2013 is 7.6, so the average number of goals per game in 2013 was 7.6.

15. The highest point on the graph corresponds to 2003, so the average number of goals per game was the greatest in 2003.

17.

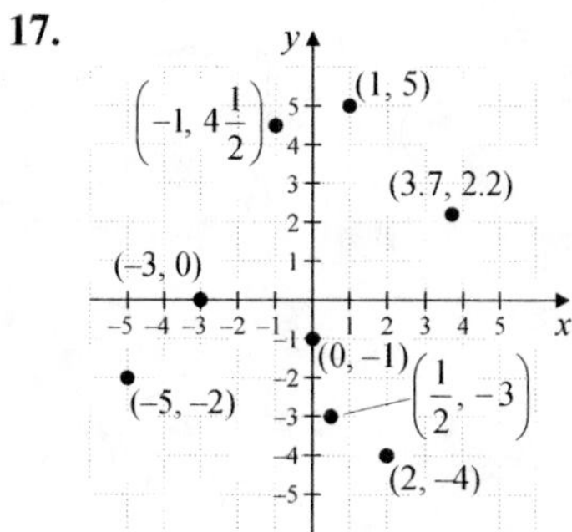

(1, 5) and (3.7, 2.2) are in quadrant I.

$\left(-1, 4\frac{1}{2}\right)$ is in quadrant II.

(−5, −2) is in quadrant III.

(2, −4) and $\left(\frac{1}{2}, -3\right)$ are in quadrant IV.

(−3, 0) lies on the x-axis.
(0, −1) lies on the y-axis.

19. Point A is at the origin, so its coordinates are (0, 0).

21. Point C is 3 units right and 2 units up from the origin, so its coordinates are (3, 2).

23. Point E is 2 units left and 2 units down from the origin, so its coordinates are (−2, −2).

25. Point G is 2 units right and 1 unit down from the origin, so its coordinates are (2, −1).

27. Point B is on the y-axis and 3 units down from the origin, so its coordinates are (0, −3).

29. Point D is 1 unit right and 3 units up from the origin, so its coordinates are (1, 3).

31. Point F is 3 units left and 1 unit down from the origin, so its coordinates are (−3, −1).

33. a. The ordered pairs are (2007, 9.6), (2008, 9.6), (2009, 10.6), (2010, 10.6), (2011, 10.2), (2012, 10.8).

b. The ordered pair (2010, 10.6) indicates that the domestic box office in 2010 was $10.6 billion.

c.

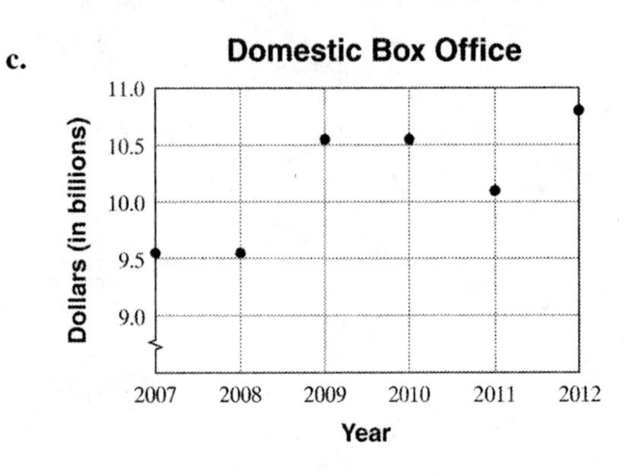

d. answers may vary

35. a. The ordered pairs are (0.50, 10), (0.75, 12), (1.00, 15), (1.25, 16), (1.50, 18), (1.50, 19), (1.75, 19), and (2.00, 20).

b. The ordered pair (1.25, 16) indicates that when Minh studied 1.25 hours, her quiz score was 16.

c.

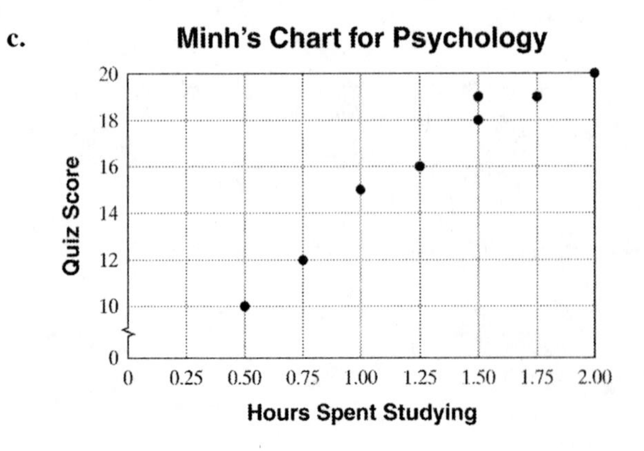

d. answers may vary

37. $x - 4y = 4$

In (, −2), the y-coordinate is −2.

$x - 4(-2) = 4$
$x + 8 = 4$
$x = -4$

In (4,), the x-coordinate is 4.

$4 - 4y = 4$
$-4y = 0$
$y = 0$

The completed coordinates are (−4, −2) and (4, 0).

39. $y = \frac{1}{4}x - 3$

In (−8,), the x-coordinate is −8.

$y = \frac{1}{4}(-8) - 3 = -2 - 3 = -5$

In (, 1), the y-coordinate is 1.

$1 = \frac{1}{4}x - 3$
$4 = \frac{1}{4}x$
$16 = x$

The completed coordinates are (−8, −5) and (16, 1).

41. $y = -7x$
$-\frac{1}{7}y = x$

$x = -\frac{1}{7}y$	$y = -7x$
0	$-7(0) = 0$
−1	$-7(-1) = 7$
$-\frac{1}{7}(2) = -\frac{2}{7}$	2

43. $-y + 2 = x$
$-y = x - 2$
$y = -x + 2$

$x = -y + 2$	$y = -x + 2$
0	$-0 + 2 = 2$
$-0 + 2 = 2$	0
−3	$-(-3) + 2 = 3 + 2 = 5$

45. $y = \frac{1}{2}x$
$2y = x$

$x = 2y$	$y = \frac{1}{2}x$
0	$\frac{1}{2}(0) = 0$
−6	$\frac{1}{2}(-6) = -3$
$2(1) = 2$	1

47. $x + 3y = 6$
$x = -3y + 6$

$3y = -x + 6$
$y = -\frac{1}{3}x + 2$

$x = -3y + 6$	$y = -\frac{1}{3}x + 2$
0	$-\frac{1}{3}(0) + 2 = 0 + 2 = 2$
$-3(0) + 6 = 0 + 6 = 6$	0
$-3(1) + 6 = -3 + 6 = 3$	1

49. $y = 2x - 12$
$y + 12 = 2x$
$\frac{1}{2}(y + 12) = x$

$x = \frac{1}{2}(y + 12)$	$y = 2x - 12$
0	$2(0) - 12 = 0 - 12 = -12$
$\frac{1}{2}(-2 + 12) = \frac{1}{2}(10) = 5$	−2
3	$2(3) - 12 = 6 - 12 = -6$

51. $2x + 7y = 5$
$2x = -7y + 5$
$x = \frac{-7y + 5}{2}$

$7y = -2x + 5$
$y = \frac{-2x + 5}{7}$

$x = \frac{-7y+5}{2}$	$y = \frac{-2x+5}{7}$
0	$\frac{-2(0)+5}{7} = \frac{5}{7}$
$\frac{-7(0)+5}{2} = \frac{5}{2}$	0
$\frac{-7(1)+5}{2} = \frac{-2}{2} = -1$	1

53. $x = -5y$
$y = 0$: $x = -5(0) = 0$
$y = 1$: $x = -5(1) = -5$
$x = 10$: $10 = -5y$
$-2 = y$
The ordered pairs are (0, 0), (−5, 1), and (10, −2).

x	y
0	0
−5	1
10	−2

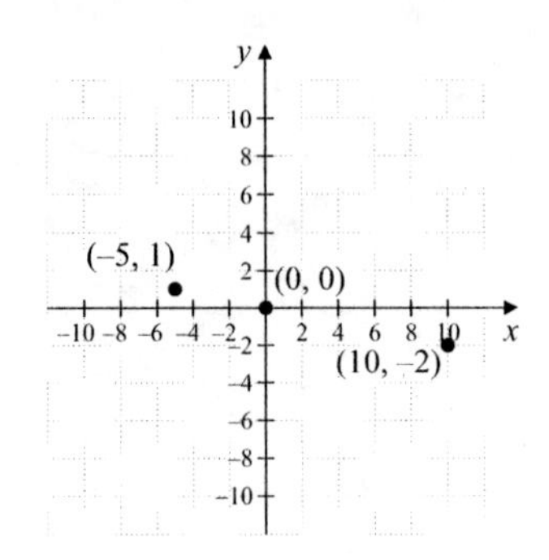

55. $y = \frac{1}{3}x + 2$

$x = 0$: $y = \frac{1}{3}(0) + 2 = 0 + 2 = 2$

$x = -3$: $y = \frac{1}{3}(-3) + 2 = -1 + 2 = 1$

$y = 0$: $0 = \frac{1}{3}x + 2$

$-2 = \frac{1}{3}x$

$-6 = x$

The ordered pairs are (0, 2), (−3, 1), and (−6, 0).

x	y
0	2
−3	1
−6	0

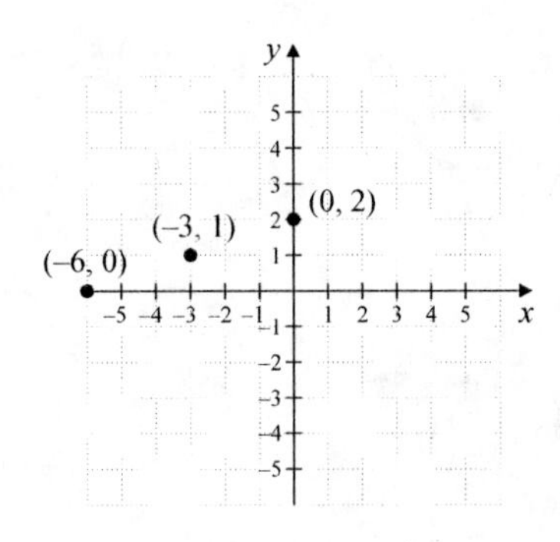

57. a.

x	100	200	300
$y = 80x + 5000$	$80(100) + 5000$ $= 8000 + 5000$ $= 13{,}000$	$80(200) + 5000$ $= 16{,}000 + 5000$ $= 21{,}000$	$80(300) + 5000$ $= 24{,}000 + 5000$ $= 29{,}000$

b. Find x when $y = 8600$.

$$8600 = 80x + 5000$$
$$3600 = 80x$$
$$45 = x$$

45 computer desks can be produced for $8600.

59. a.

x	2	5	8
$y = 0.24x + 5.96$	$0.24(2) + 5.96$ $= 0.48 + 5.96$ $= 6.44$	$0.24(5) + 5.96$ $= 1.20 + 5.96$ $= 7.16$	$0.24(8) + 5.96$ $= 1.92 + 5.96$ $= 7.88$

b. Find x when $y = 7.40$.

$$7.40 = 0.24x + 5.96$$
$$1.44 = 0.24x$$
$$6 = x$$

2003 + 6 = 2009

The average cinema admission price was $7.40 in 2009.

c. Find x when $y = 9.00$.

$$9.00 = 0.24x + 5.96$$
$$3.04 = 0.24x$$
$$13 \approx x$$

2003 + 13 = 2016

The average cinema admission price is predicted to be $9.00 in 2016.

61. $x + y = 5$

$$y = 5 - x$$

63. $2x + 4y = 5$

$$4y = 5 - 2x$$
$$y = \frac{5 - 2x}{4}$$

65. $10x = -5y$
$-2x = y$
$y = -2x$

67. False; point (−1, 5) lies in quadrant II.

69. True

71. Points in quadrant III are to the left and down from the origin, so the x- and y-coordinates are negative. (negative, negative) corresponds to quadrant III.

73. Points in quadrant IV are to the right and down from the origin, so the x-coordinate is positive and the y-coordinate is negative. (positive, negative) corresponds to quadrant IV.

75. The origin corresponds to (0, 0).

77. If the x-coordinate of a point is 0, the point is neither to the left nor to the right of the origin, so it is on the y-axis.

79. no; answers may vary

81. answers may vary

83. answers may vary

85. A point four units to right of the y-axis and seven units below the x-axis has coordinates (4, −7).

87. The length of the rectangle is $3 - (-1) = 4$ and the width of the rectangle is $5 - (-4) = 9$.

$$\begin{aligned}\text{Perimeter} &= 2(\text{length}) + 2(\text{width})\\ &= 2(4) + 2(9)\\ &= 8 + 18\\ &= 26\end{aligned}$$

The perimeter is 26 units.

89. The numbers of members are approximately 21 million, 23 million, 25 million, and 27 million.

91. The point on the high temperature graph corresponding to Thursday is 83, so the high temperature reading on Thursday was 83°F.

93. The lowest point on the graph of low temperatures corresponds to Sunday. The low temperature on Sunday was 68°F.

95. The difference between the graphs is the greatest for Tuesday. The high temperature was 86°F and the low temperature was 73°F, so the difference is $86 - 73 = 13$°F.

Section 6.2 Practice Exercises

1. $x + 3y = 6$

$x = 0$: $0 + 3y = 6$
$3y = 6$
$y = 2$

$x = 3$: $3 + 3y = 6$
$3y = 3$
$y = 1$

$y = 0$: $x + 3(0) = 6$
$x + 0 = 6$
$x = 6$

The ordered pairs are (0, 2), (3, 1), and (6, 0).

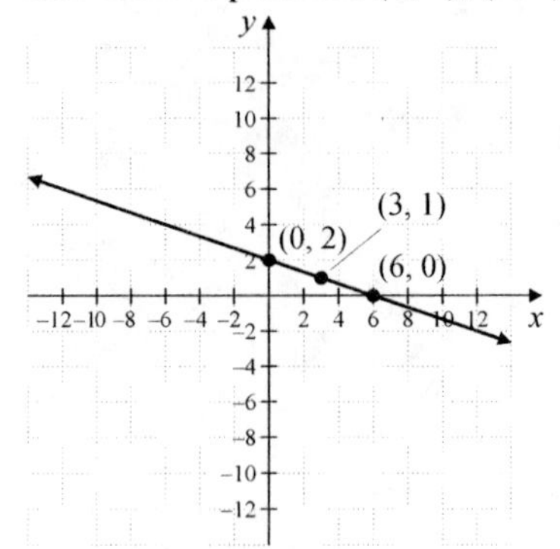

2. $-2x + 4y = 8$

$x = -2$: $-2(-2) + 4y = 8$
$4 + 4y = 8$
$4y = 4$
$y = 1$

$x = 0$: $-2(0) + 4y = 8$
$0 + 4y = 8$
$4y = 8$
$y = 2$

$x = 2$: $-2(2) + 4y = 8$
$-4 + 4y = 8$
$4y = 12$
$y = 3$

The ordered pairs are (−2, 1), (0, 2), and (2, 3).

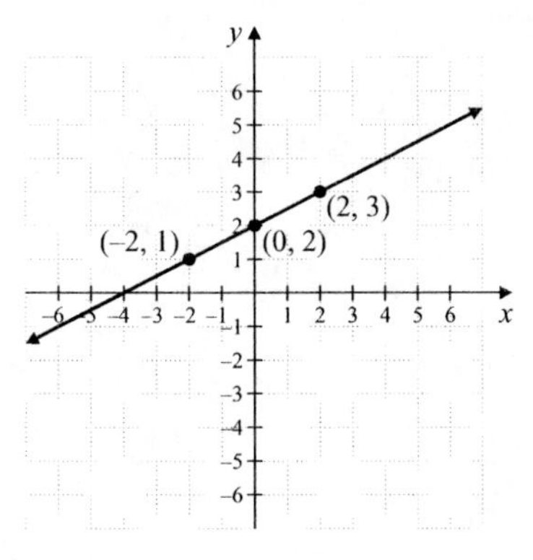

3. $y = 2x$
$x = -2$: $y = 2(-2) = -4$
$x = 0$: $y = 2(0) = 0$
$x = 3$: $y = 2(3) = 6$
The ordered pairs are $(-2, -4)$, $(0, 0)$, and $(3, 6)$.

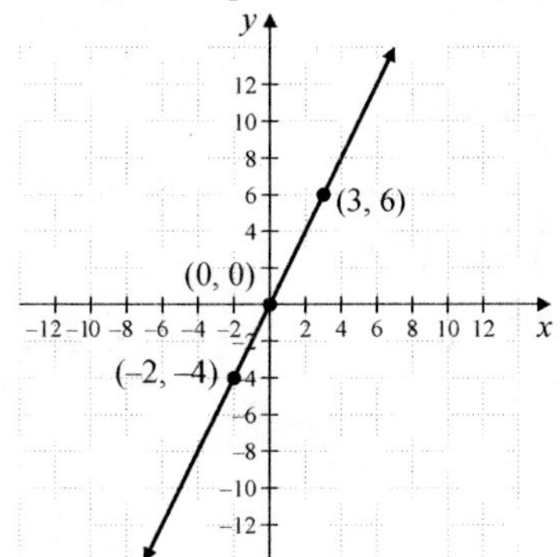

4. $y = -\frac{1}{2}x + 4$

$x = -6$: $y = -\frac{1}{2}(-6) + 4 = 3 + 4 = 7$

$x = 0$: $-\frac{1}{2}(0) + 4 = 0 + 4 = 4$

$x = 4$: $y = -\frac{1}{2}(4) + 4 = -2 + 4 = 2$

The ordered pairs are $(-6, 7)$, $(0, 4)$, and $(4, 2)$.

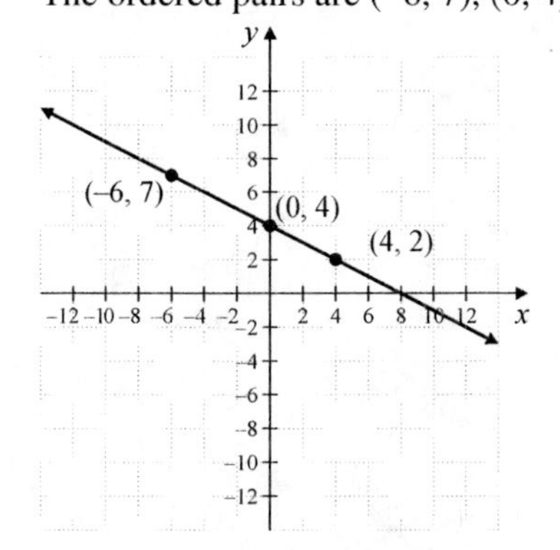

5. The equation $x = 3$ can be written in standard form as $x + 0y = 3$. No matter what value replaces y, x is always 3. It is a vertical line. Plot points $(3, 2)$, $(3, 0)$, and $(3, -4)$, for example.

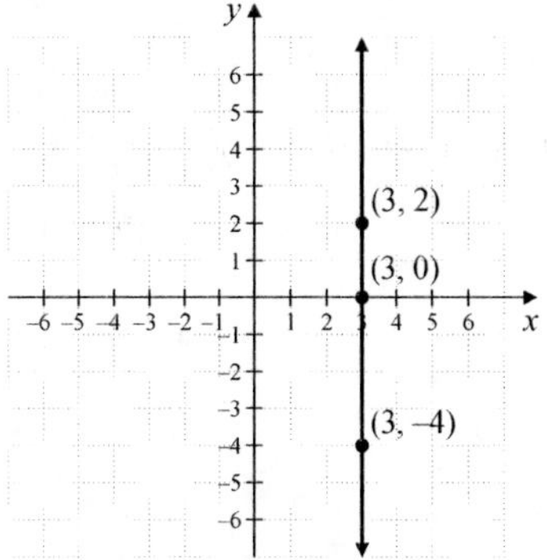

6. $x = 2022 - 2010 = 12$
Find 12 on the x-axis. Move vertically upward to the line and then horizontally to the left. In 2022, we predict that there will be 3600 thousand registered nurses.

Calculator Explorations

1. $y = -3x + 7$

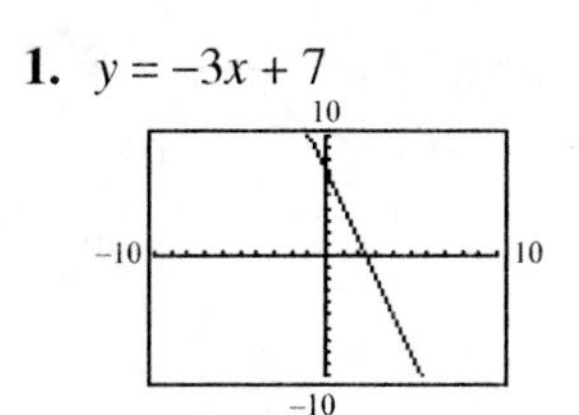

2. $y = -x + 5$

3. $y = 2.5x - 7.9$

4. $y = -1.3x + 5.2$

5. $y = -\dfrac{3}{10}x + \dfrac{32}{5}$

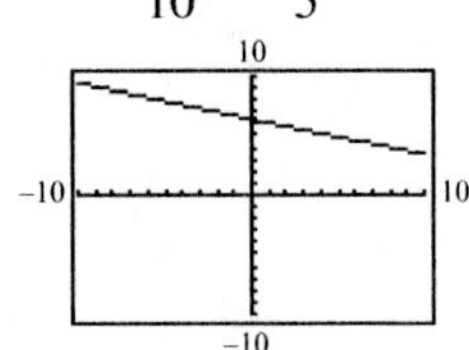

6. $y = \dfrac{2}{9}x - \dfrac{22}{3}$

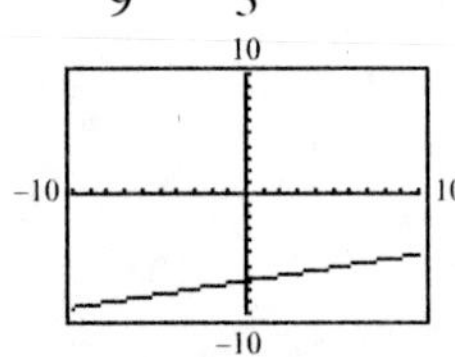

Vocabulary, Readiness & Video Check 6.2

1. It is always good practice to use a third point as a check to see that our points lie along a straight line.

2. An infinite number of points make up the line, and every point corresponds to an ordered pair that is a solution of the linear equation in two variables.

Exercise Set 6.2

1. $x - y = 6$

$y = 0$: $x - 0 = 6$
$x = 6$

$x = 4$: $4 - y = 6$
$-y = 2$
$y = -2$

$y = -1$: $x - (-1) = 6$
$x + 1 = 6$
$x = 5$

The ordered pairs are (6, 0), (4, −2), and (5, −1).

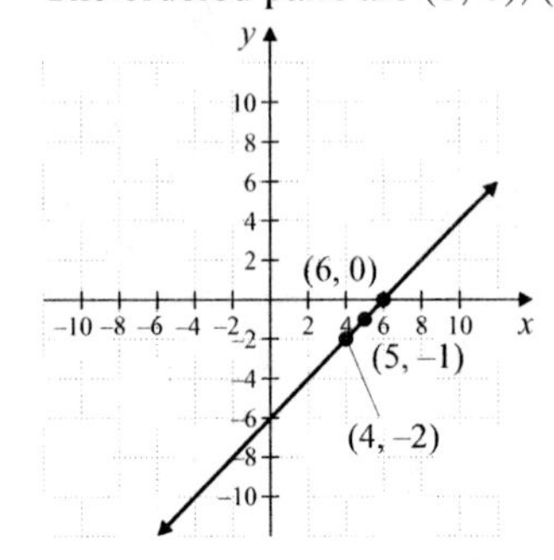

3. $y = -4x$

$x = 1$: $y = -4(1) = -4$
$x = 0$: $y = 4(0) = 0$
$x = -1$: $y = -4(-1) = 4$

The ordered pairs are (1, −4), (0, 0), and (−1, 4).

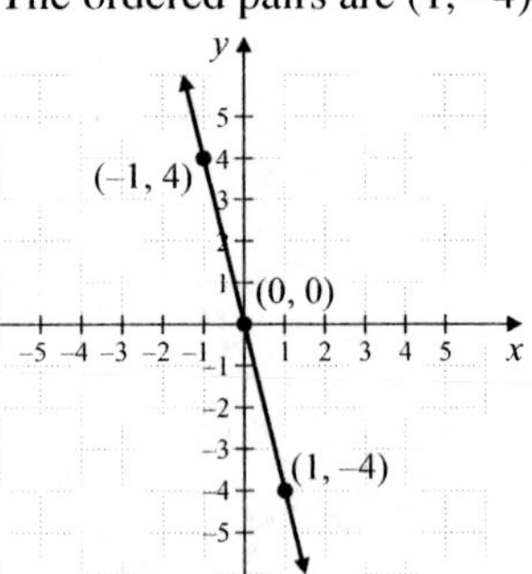

5. $y = \dfrac{1}{3}x$

$x = 0$: $y = \dfrac{1}{3}(0) = 0$

$x = 6$: $y = \dfrac{1}{3}(6) = 2$

$x = -3$: $y = \dfrac{1}{3}(-3) = -1$

The ordered pairs are (0, 0), (6, 2), and (−3, −1).

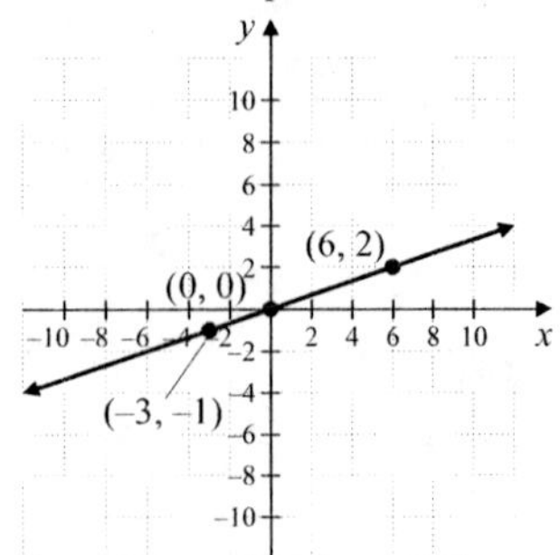

7. $y = -4x + 3$

$x = 0$: $y = -4(0) + 3 = 0 + 3 = 3$
$x = 1$: $y = -4(1) + 3 = -4 + 3 = -1$
$x = 2$: $y = -4(2) + 3 = -8 + 3 = -5$

The ordered pairs are (0, 3), (1, −1), and (2, −5).

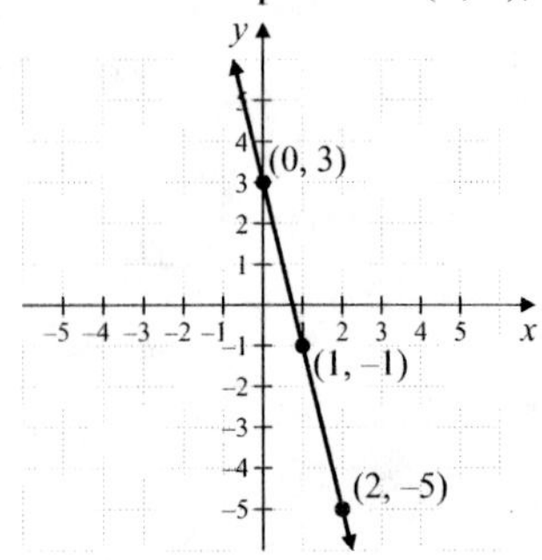

9. $x + y = 1$

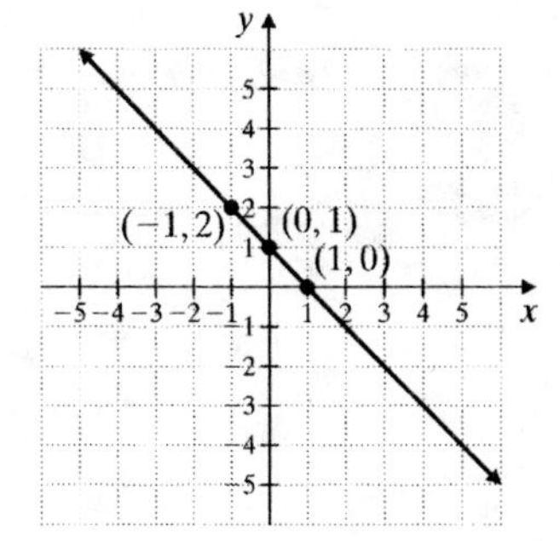

11. $x - y = -2$

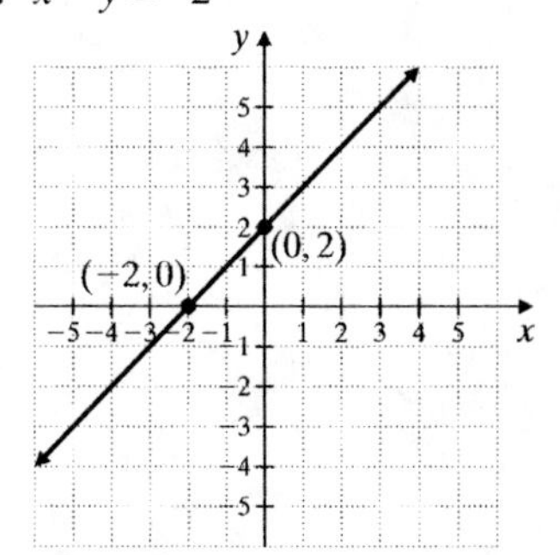

13. $x - 2y = 6$

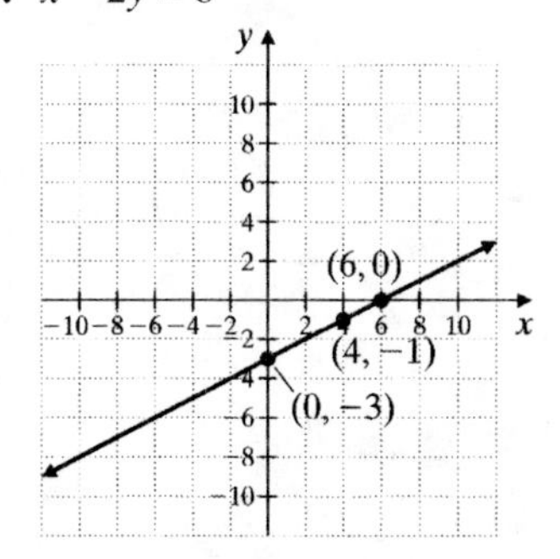

15. $y = 6x + 3$

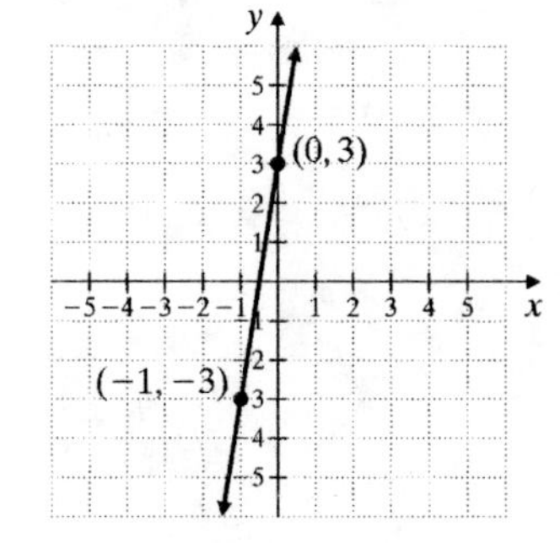

17. $x = -4$

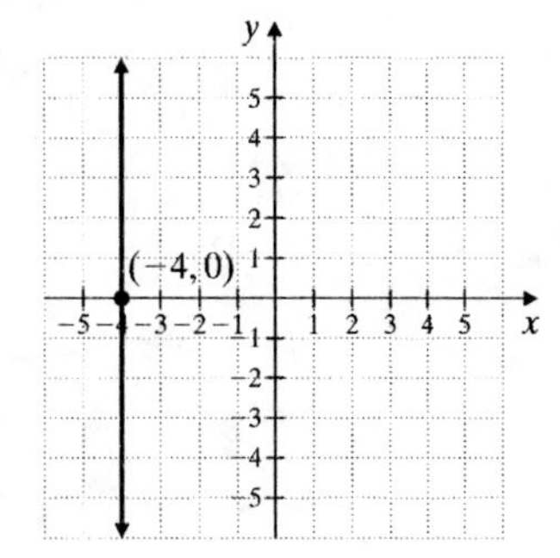

19. $y = 3$

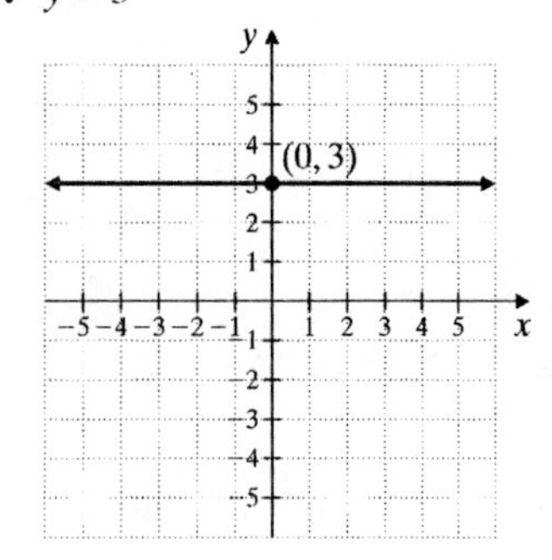

21. $y = x$

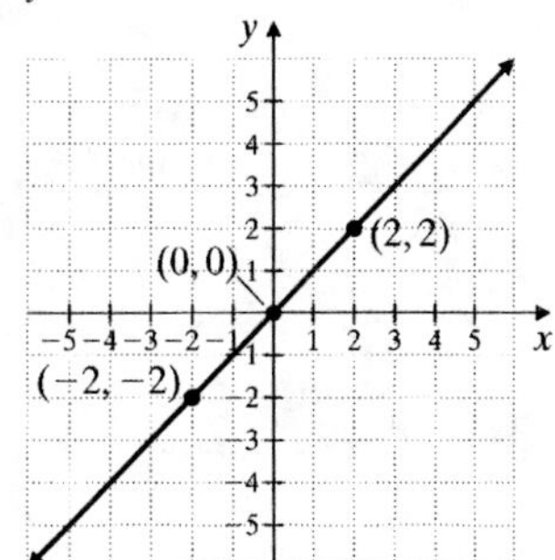

23. $x = -3y$

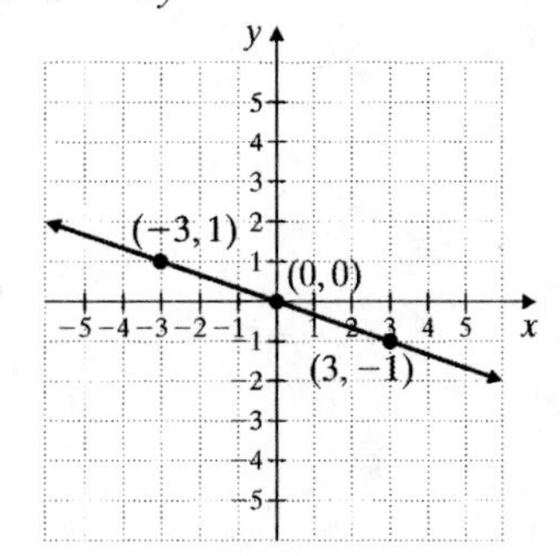

25. $x + 3y = 9$

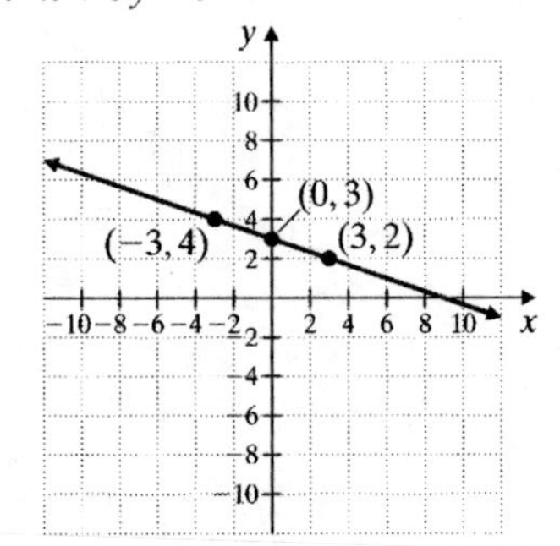

27. $y = \frac{1}{2}x + 2$

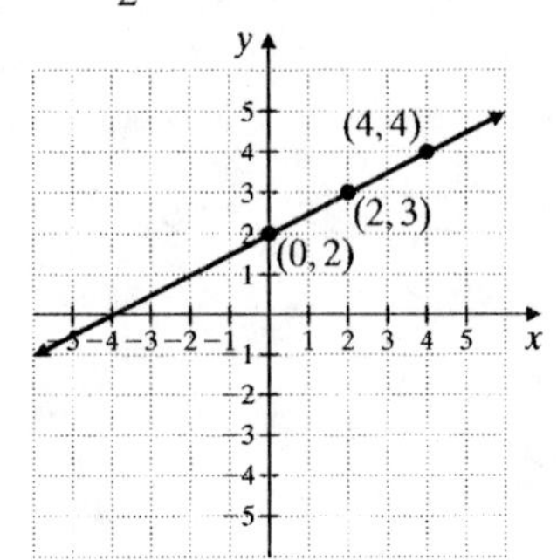

29. $3x - 2y = 12$

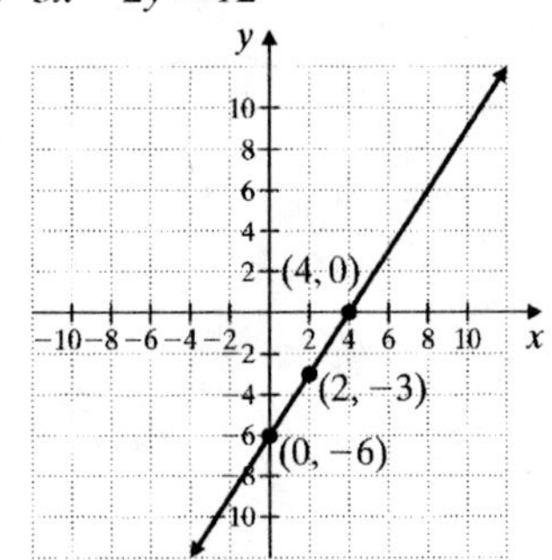

31. $y = -3.5x + 4$

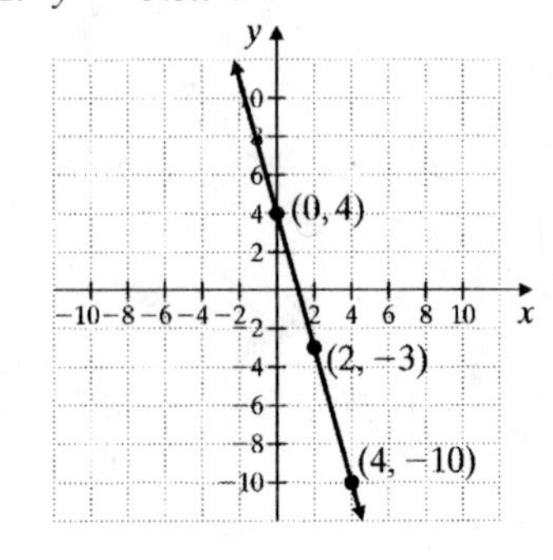

33. a.

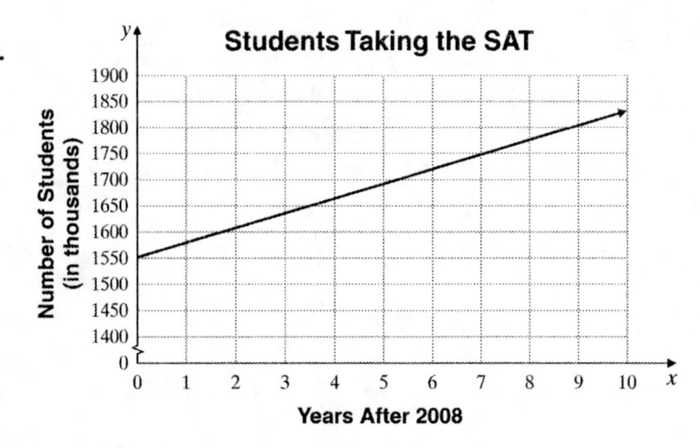

b. $x = 7$: $y = 28(7) + 1552 = 196 + 1552 = 1748$
Yes, the point (7, 1748) lies on the line. answers may vary

35. a.

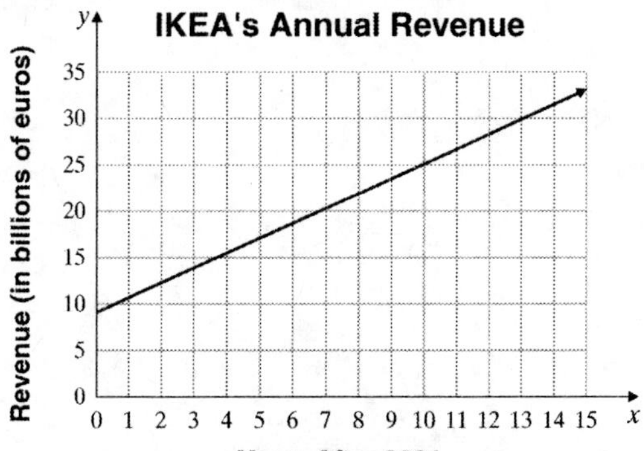

b. $x = 11$: $y = 1.6(11) + 9.1 = 17.6 + 9.1 = 26.7$
The ordered pair is (11, 26.7).

c. Year $= x + 2001 = 11 + 2001 = 2012$
In 2012, IKEA's total annual revenue was 26.7 billion euros.

37. The fourth vertex is the bottom-right corner of the rectangle. The x-coordinate must line the point up with the top-right corner, and the y-coordinate must line the point up with the bottom-left corner. The coordinates are $(4, -1)$.

39. $x - y = -3$
$x = 0$: $0 - y = -3$
$y = 3$
$y = 0$: $x - 0 = -3$
$x = -3$

x	y
0	3
−3	0

41. $y = 2x$
$x = 0$: $y = 2(0) = 0$
$y = 0$: $0 = 2x$
$0 = x$

x	y
0	0
0	0

43. $y = 5x$
$y = 5x + 4$

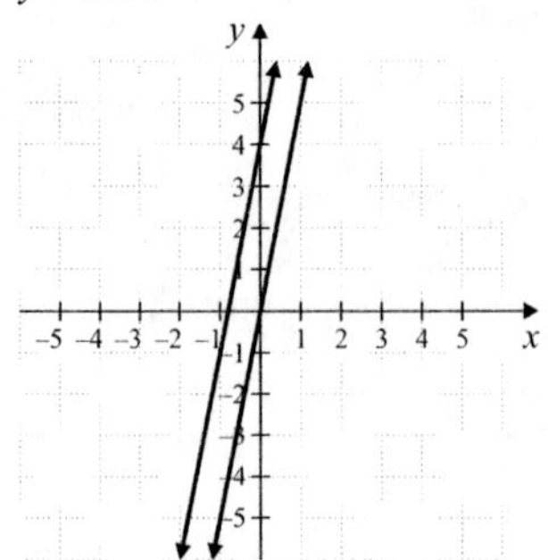

45. $y = -2x$
$y = -2x - 3$

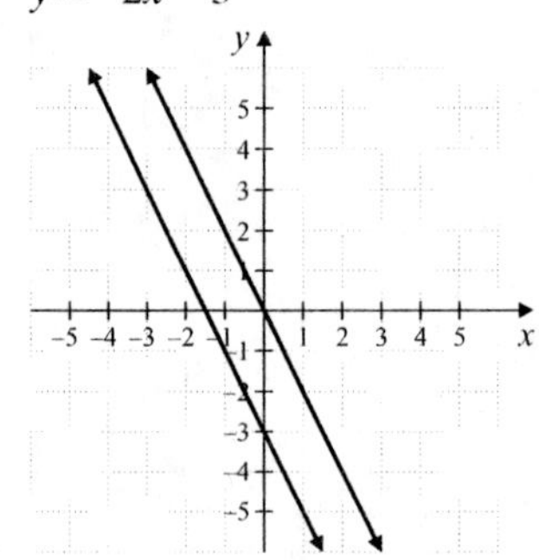

47. $y = x^2$

$x = 0$: $y = 0^2 = 0$

$x = 1$: $y = 1^2 = 1$

$x = -1$: $y = (-1)^2 = 1$

$x = 2$: $y = 2^2 = 4$

$x = -2$: $y = (-2)^2 = 4$

x	y
0	0
1	1
−1	1
2	4
−2	4

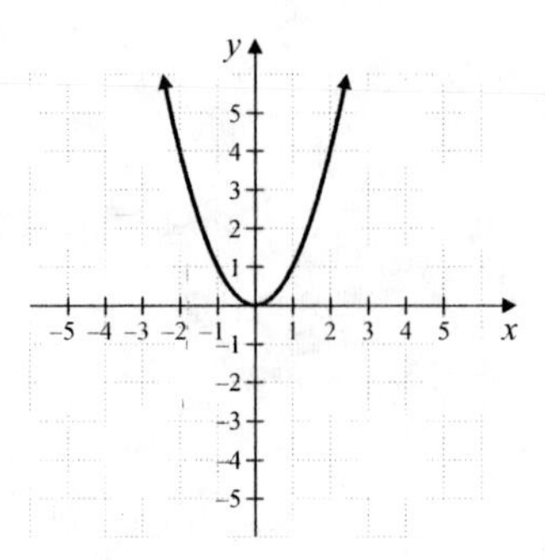

49. The perimeter is the distance around.
$x + 5 + y + 5 = 22$
$x + y + 10 = 22$
$x + y = 12$
$x = 3$: $3 + y = 12$
$y = 9$
If x is 3 centimeters, then y is 9 centimeters.

51. yes; answers may vary

Section 6.3 Practice Exercises

1. x-intercept: (2, 0)
y-intercept: (0, −4)

2. x-intercepts: (−4, 0), (2, 0)
y-intercept: (0, 2)

3. x-intercept and y-intercept: (0, 0)

4. $2x - y = 4$
$y = 0$: $2x - 0 = 4$
$2x = 4$
$x = 2$
x-intercept: (2, 0)
$x = 0$: $2(0) - y = 4$
$0 - y = 4$
$-y = 4$
$y = -4$
y-intercept: (0, −4)

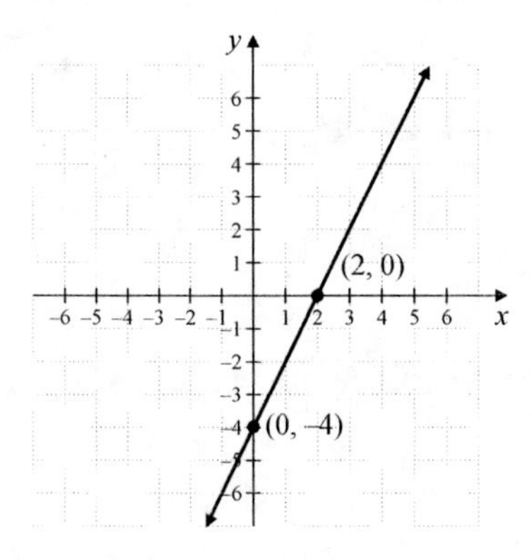

5. $y = 3x$
$y = 0$: $0 = 3x$
$0 = x$
x-intercept: (0, 0)
$x = 0$: $y = 3(0)$
$y = 0$
y-intercept: (0, 0)
Let $x = 1$ to find a second point.
$x = 1$: $y = 3$
(1, 3) is another point on the line.

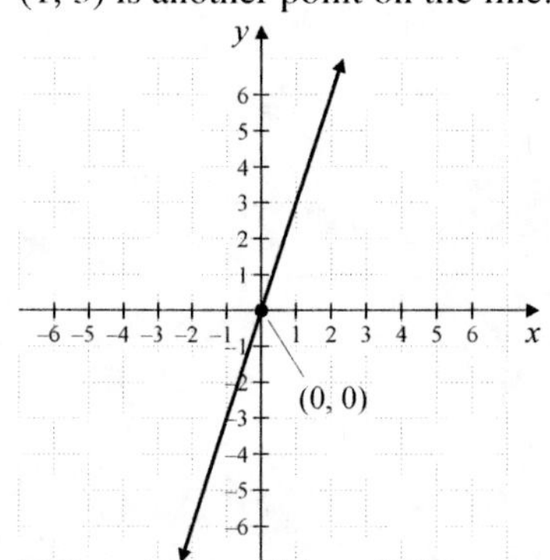

6. $x = -3$
The graph of $x = -3$ is a vertical line with x-intercept $(-3, 0)$.

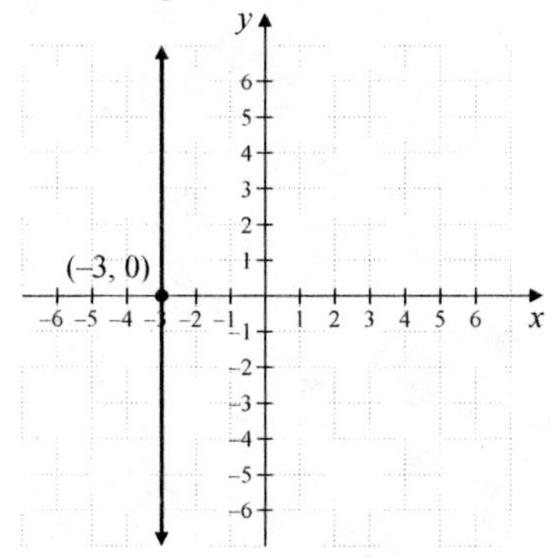

7. $y = 4$
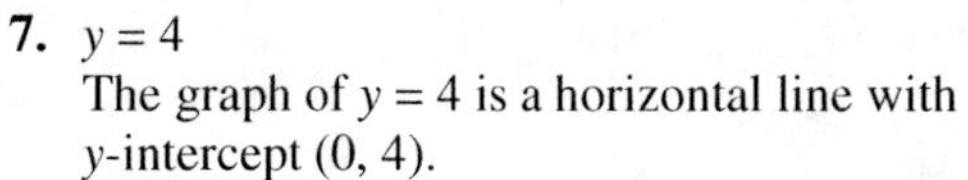
The graph of $y = 4$ is a horizontal line with y-intercept (0, 4).

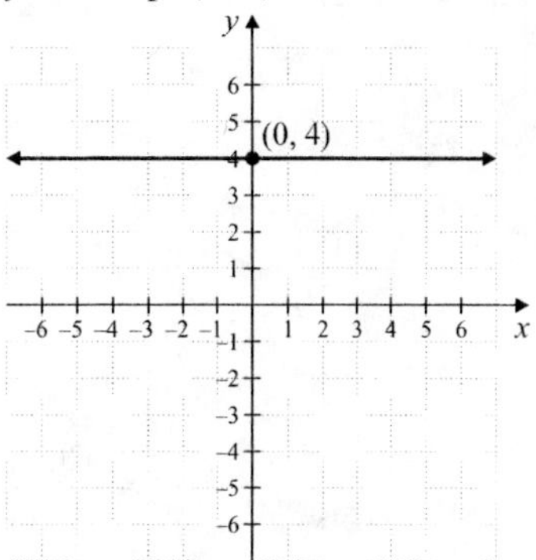

Calculator Explorations

1. $x = 3.78y$

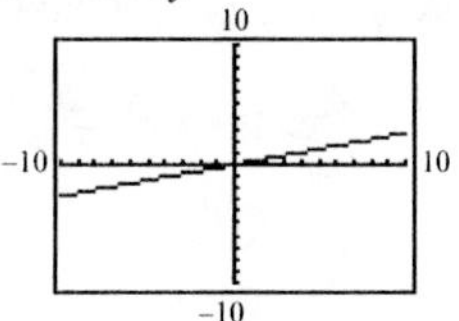

2. $-2.61y = x$

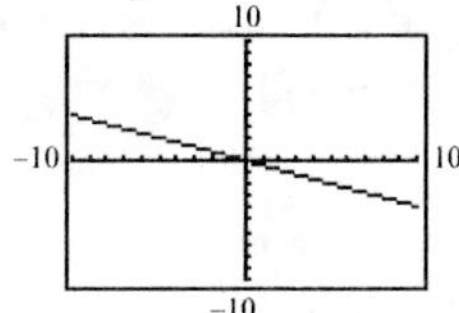

3. $3x + 7y = 21$

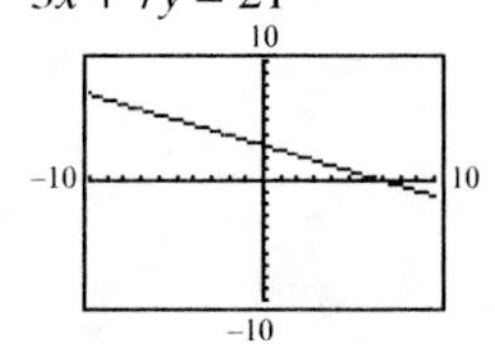

4. $-4x + 6y = 12$

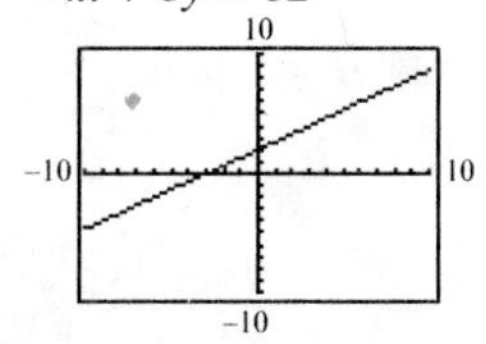

5. $-2.2x + 6.8y = 15.5$

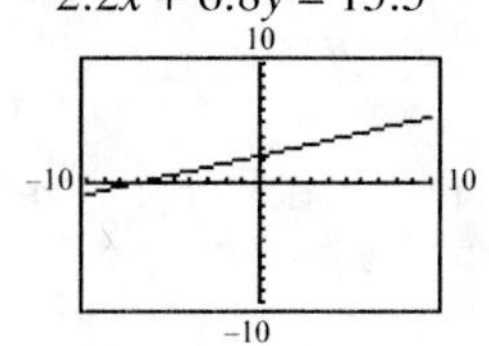

6. $5.9x - 0.8y = -10.4$

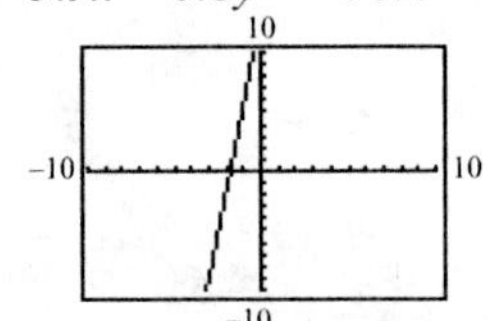

Vocabulary, Readiness & Video Check 6.3

1. An equation that can be written in the form $Ax + By = C$ is called a <u>linear</u> equation in two variables.

2. The form $Ax + By = C$ is called <u>standard</u> form.

3. The graph of the equation $y = -1$ is a <u>horizontal</u> line.

4. The graph of the equation $x = 5$ is a <u>vertical</u> line.

5. A point where a graph crosses the y-axis is called a <u>y-intercept</u>.

6. A point where a graph crosses the x-axis is called an <u>x-intercept</u>.

7. Given an equation of a line, to find the x-intercept (if there is one), let <u>y</u> = 0 and solve for <u>x</u>.

8. Given an equation of a line, to find the y-intercept (if there is one), let <u>x</u> = 0 and solve for <u>y</u>.

9. False; a horizontal line (other than $y = 0$) has no x-intercept and a vertical line (other than $x = 0$) has no y-intercept.

10. True

11. True

12. False; $x = 5$: $y = 5(5) = 25$, so the point (5, 1) is not on the graph of $y = 5x$.

13. because x-intercepts lie on the x-axis; because y-intercepts lie on the y-axis

14. It is always good practice to use a third point as a check to see that our points lie along a straight line.

15. For a horizontal line, the coefficient of x will be 0; for a vertical line, the coefficient of y will be 0.

Exercise Set 6.3

1. x-intercept: (–1, 0)
y-intercept: (0, 1)

3. x-intercepts: (–2, 0), (2, 0)
y-intercept: (0, –2)

5. x-intercepts: (–2, 0), (1, 0), (3, 0)
y-intercept: (0, 3)

7. x-intercepts: (–1, 0), (1, 0)
y-intercepts: (0, 1), (0, –2)

9. $x - y = 3$
$y = 0$: $x - 0 = 0$
$x = 3$
x-intercept: (3, 0)
$x = 0$: $0 - y = 3$
$-y = 3$
$y = -3$
y-intercept: (0, –3)

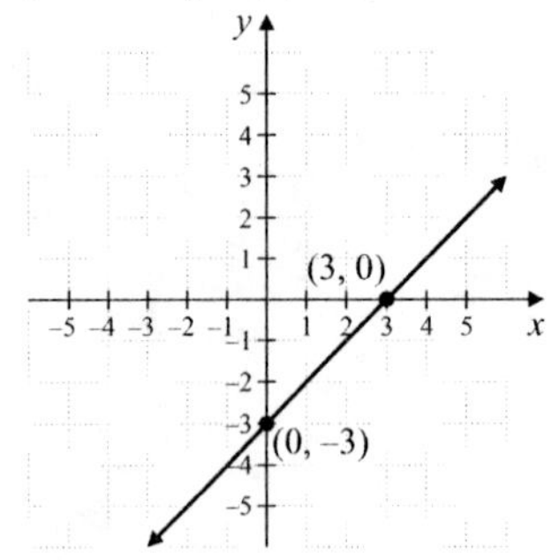

11. $x = 5y$
$y = 0$: $x = 5(0) = 0$
x-intercept: (0, 0)
$x = 0$: $0 = 5y$
$0 = y$
y-intercept: (0, 0)
Let $y = 1$ to find a second point.
$y = 1$: $x = 5$
(5, 1) is another point on the line.

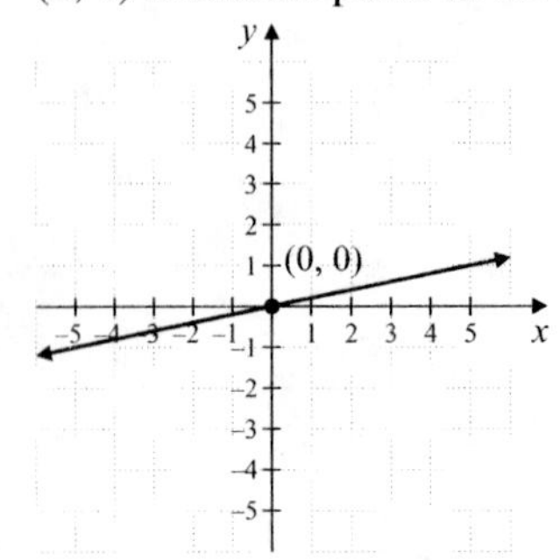

13. $-x + 2y = 6$
$y = 0$: $-x + 2(0) = 6$
$-x = 6$
$x = -6$
x-intercept: $(-6, 0)$
$x = 0$: $-0 + 2y = 6$
$2y = 6$
$y = 3$
y-intercept: $(0, 3)$

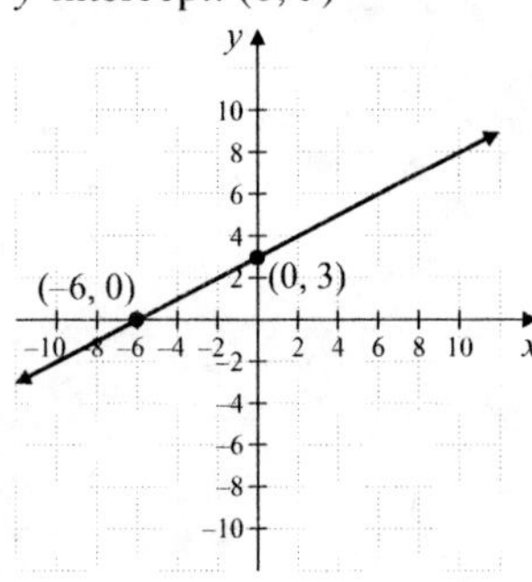

15. $2x - 4y = 8$
$y = 0$: $2x - 4(0) = 8$
$2x - 0 = 8$
$2x = 8$
$x = 4$
x-intercept: $(4, 0)$
$x = 0$: $2(0) - 4y = 8$
$0 - 4y = 8$
$-4y = 8$
$y = -2$
y-intercept: $(0, -2)$

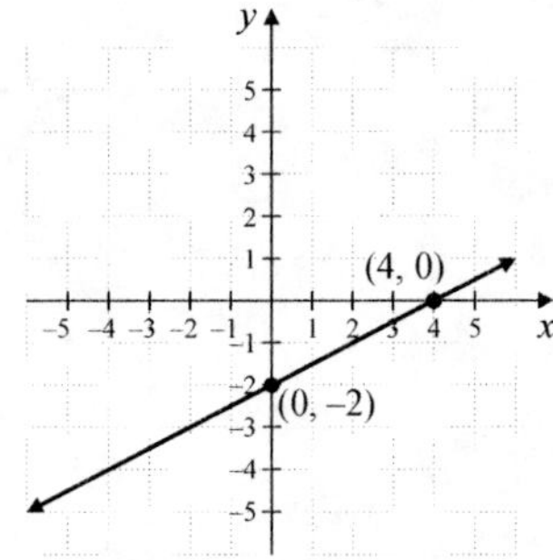

17. $y = 2x$
$y = 0$: $0 = 2x$
$0 = x$
x-intercept: $(0, 0)$
$(0, 0)$ is also the y-intercept. Let $x = 1$ to find a second point.
$x = 1$: $y = 2(1)$
$y = 2$
$(1, 2)$ is another point on the line.

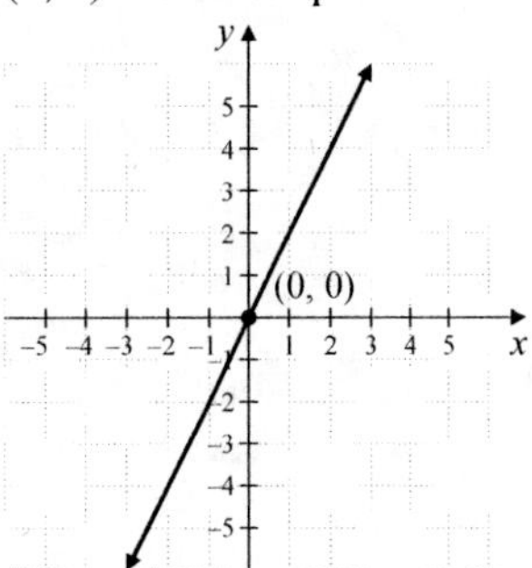

19. $y = 3x + 6$
$y = 0$: $0 = 3x + 6$
$-6 = 3x$
$-2 = x$
x-intercept: $(-2, 0)$
$x = 0$: $y = 3(0) + 6$
$y = 0 + 6$
$y = 6$
y-intercept: $(0, 6)$

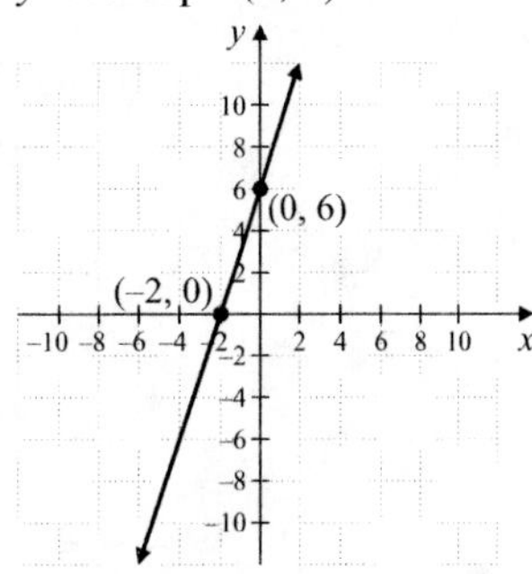

21. The graph of $x = -1$ is a vertical line with x-intercept $(-1, 0)$.

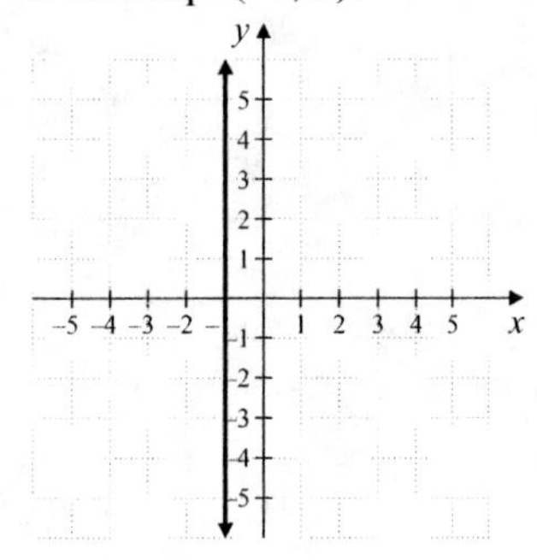

23. The graph of $y = 0$ is a horizontal line with y-intercept (0, 0).

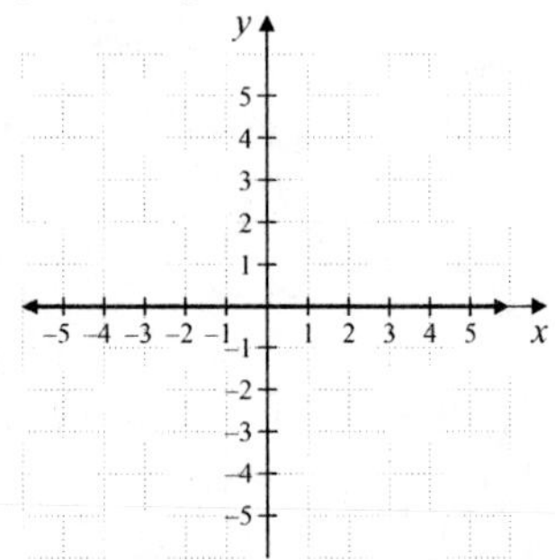

25. $y+7=0$

$y=-7$

The graph of $y = -7$ is a horizontal line with y-intercept (0, −7).

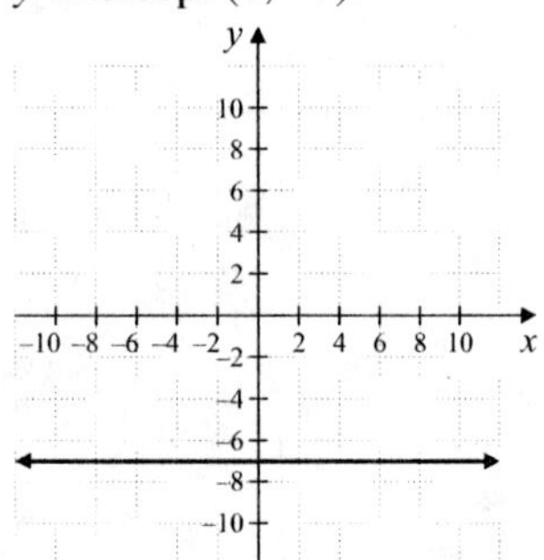

27. $x+3=0$

$x=-3$

The graph of $x = -3$ is a vertical line with x-intercept (0, −3).

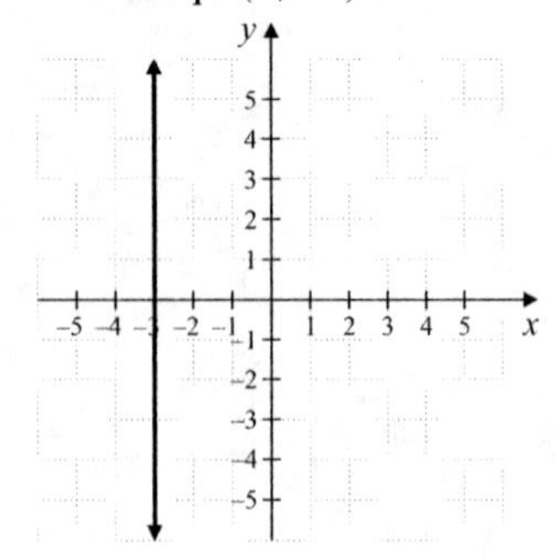

29. $x = y$

$y = 0$: $x = 0$

x-intercept: (0, 0)

(0, 0) is also the y-intercept. Let $x = 3$ to find a second point.

$x = 3$: $3 = y$

(3, 3) is another point on the line.

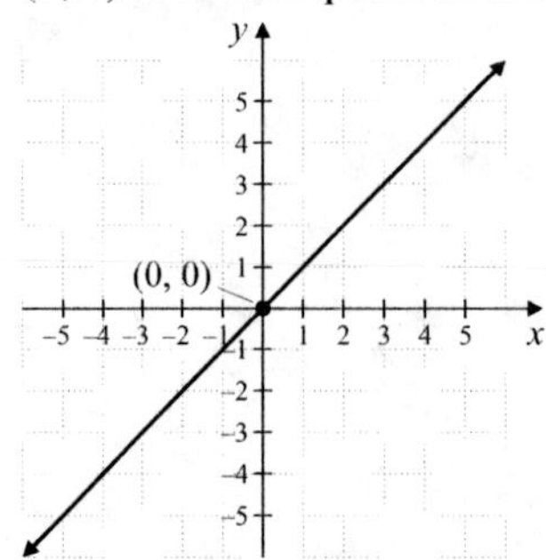

31. $x + 8y = 8$

$y = 0$: $x+8(0)=8$

$x+0=8$

$x=8$

x-intercept: (8, 0)

$x = 0$: $0+8y=8$

$8y=8$

$y=1$

y-intercept: (0, 1)

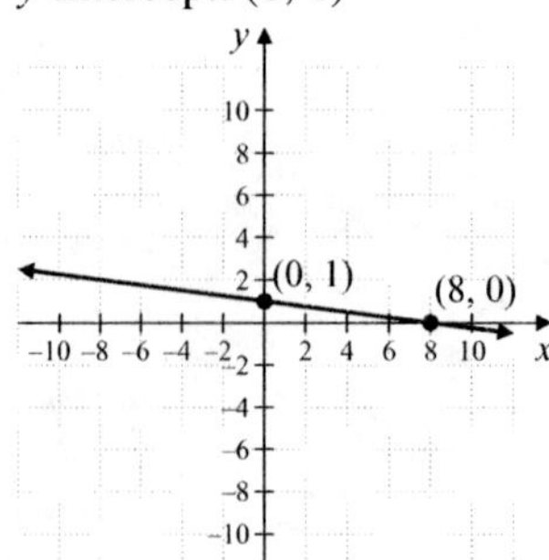

33. $5 = 6x - y$

$y = 0$: $5=6x-0$

$5=6x$

$\frac{5}{6}=x$

x-intercept: $\left(\frac{5}{6}, 0\right)$

$x = 0$: $5=6(0)-y$

$5=-y$

$-5=y$

y-intercept: (0, −5)

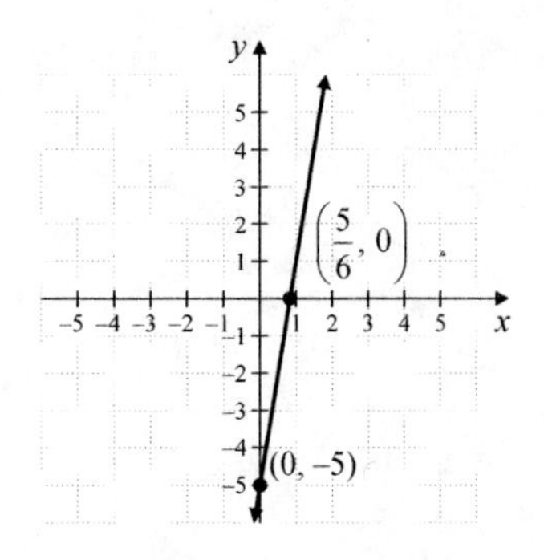

35. $-x + 10y = 11$

$y = 0$: $-x + 10(0) = 11$

$$-x + 0 = 11$$
$$-x = 11$$
$$x = -11$$

x-intercept: $(-11, 0)$

$x = 0$: $-0 + 10y = 11$

$$10y = 11$$
$$y = \frac{11}{10}$$

y-intercept: $\left(0, \frac{11}{10}\right)$

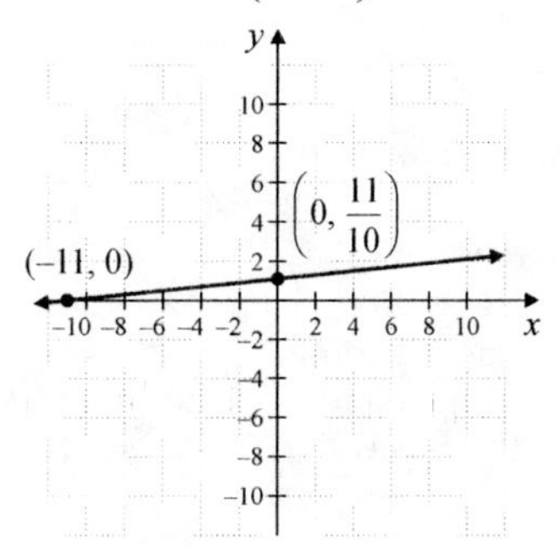

37. $x = -4\frac{1}{2}$

This is a vertical line with x-intercept $\left(-4\frac{1}{2}, 0\right)$.

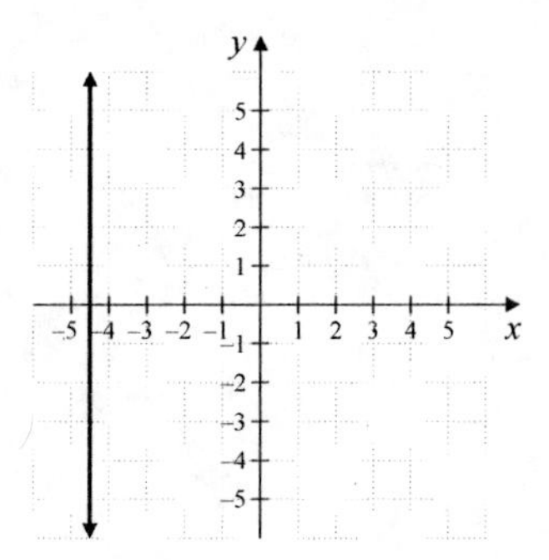

39. $y = 3\frac{1}{4}$

This is a horizontal line with y-intercept $\left(0, 3\frac{1}{4}\right)$.

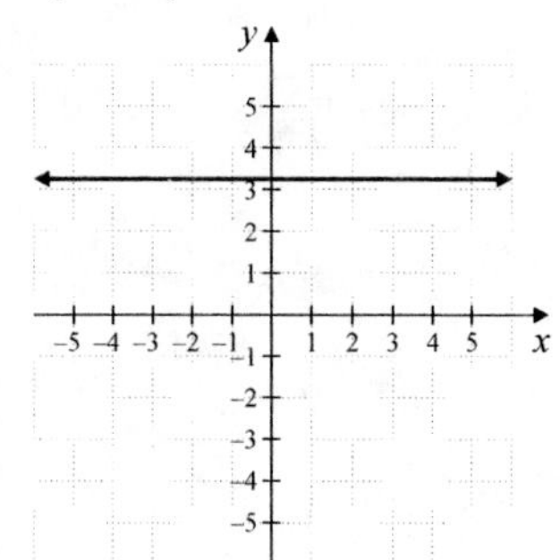

41. $y = -\frac{2}{3}x + 1$

$y = 0$: $0 = -\frac{2}{3}x + 1$

$$\frac{2}{3}x = 1$$
$$x = \frac{3}{2}$$

x-intercept: $\left(\frac{3}{2}, 0\right)$

$x = 0$: $y = -\frac{2}{3}(0) + 1$

$$y = 1$$

y-intercept: $(0, 1)$

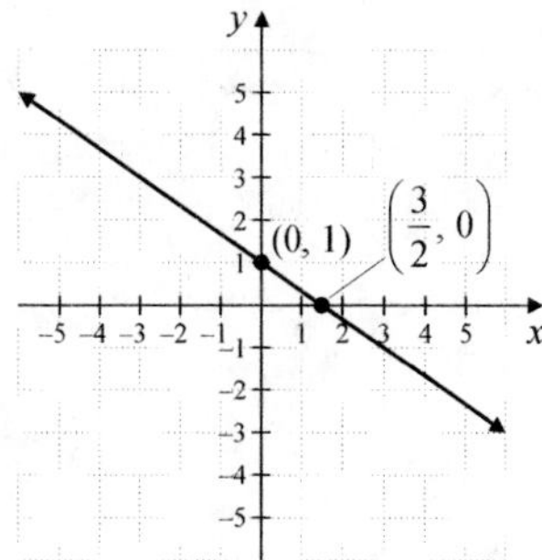

43. $4x - 6y + 2 = 0$

$y = 0$: $4x - 6(0) + 2 = 0$

$$4x - 0 + 2 = 0$$
$$4x = -2$$
$$x = -\frac{1}{2}$$

x-intercept: $\left(-\frac{1}{2}, 0\right)$

$$x = 0:\ 4(0) - 6y + 2 = 0$$
$$0 - 6y + 2 = 0$$
$$-6y = -2$$
$$y = \frac{1}{3}$$

y-intercept: $\left(0, \frac{1}{3}\right)$

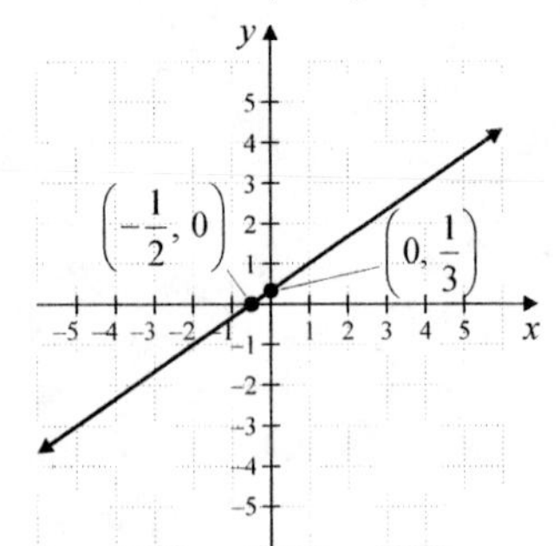

45. $\frac{-6-3}{2-8} = \frac{-9}{-6} = \frac{3}{2}$

47. $\frac{-8-(-2)}{-3-(-2)} = \frac{-8+2}{-3+2} = \frac{-6}{-1} = 6$

49. $\frac{0-6}{5-0} = \frac{-6}{5} = -\frac{6}{5}$

51. The graph of $y = 3$ is a horizontal line with y-intercept (0, 3). This is graph **c**.

53. The graph of $x = 3$ is a vertical line with x-intercept (3, 0). This is graph **a**.

55. The line can be on the x-axis or the y-axis so it can have infinitely many x- and y-intercepts.

57. A circle can have no x- and y-intercepts. That is, it does not have to intersect the axes.

59. answers may vary

61. $3x + 6y = 1200$
$$x = 0:\ 3(0) + 6y = 1200$$
$$6y = 1200$$
$$y = 200$$
The ordered pair (0, 200) corresponds to manufacturing 0 chairs and 200 desks.

63. Manufacturing 50 desks corresponds to $y = 50$.
$3x + 6y = 1200$
$$y = 50:\ 3x + 6(50) = 1200$$
$$3x + 300 = 1200$$
$$3x = 900$$
$$x = 300$$
When 50 desks are manufactured, 300 chairs can be manufactured.

65.

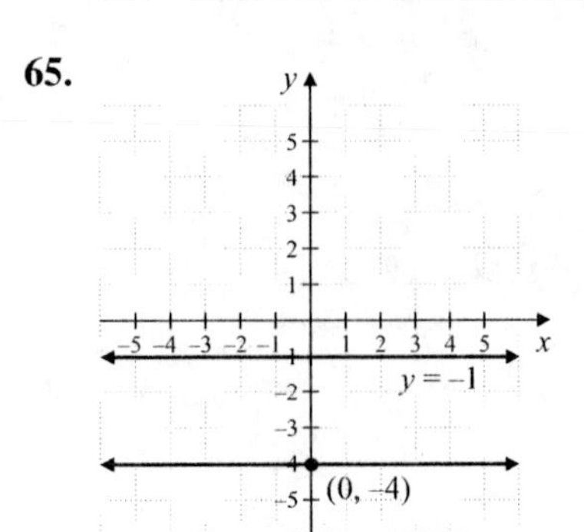

The equation of the line is $y = -4$.

67. a. $y = -1.7x + 52$
$$y = 0:\quad 0 = -1.7x + 52$$
$$1.7x = 52$$
$$x \approx 30.6$$
The x-intercept is (30.6, 0).

b. The x-intercept of (30.6, 0) means that 30.6 years after 2006, there may be no newspaper circulation.

Section 6.4 Practice Exercises

1. $m = \frac{y_2 - y_1}{x_2 - x_1} = \frac{-1-3}{4-(-2)} = \frac{-4}{6} = -\frac{2}{3}$

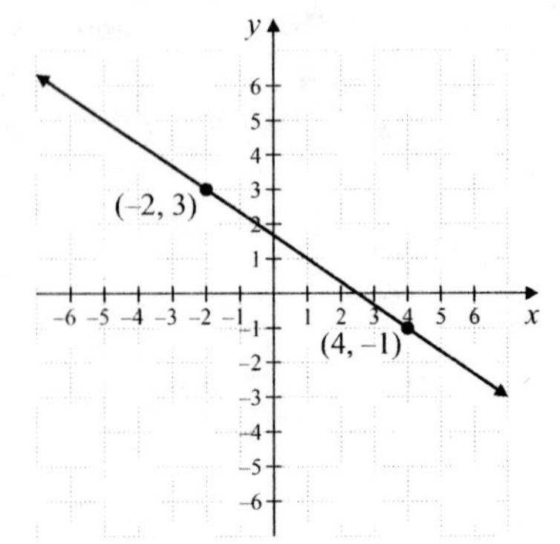

2. 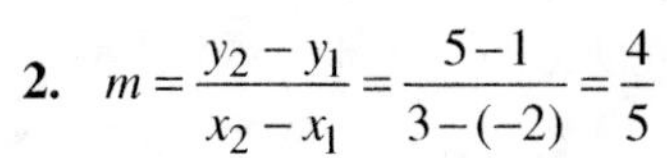
$m=\frac{y_2-y_1}{x_2-x_1}=\frac{5-1}{3-(-2)}=\frac{4}{5}$

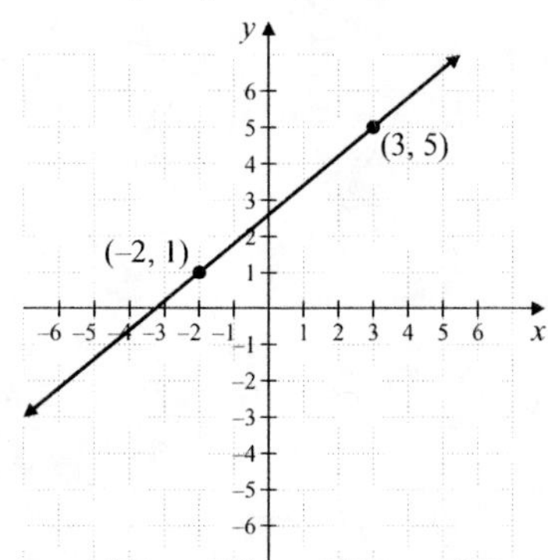

3. $5x+4y=10$
$4y=-5x+10$
$y=-\frac{5}{4}x+\frac{10}{4}$

The slope is $m=-\frac{5}{4}$.

4. $-y=-2x+7$
$\frac{-y}{-1}=\frac{-2x}{-1}+\frac{7}{-1}$
$y=2x-7$

The slope is $m=2$.

5. $y=3$ is a horizontal line. Horizontal lines have a slope of $m=0$.

6. $x=-2$ is a vertical line. Vertical lines have undefined slopes.

7. a. $x+y=5$ $2x+y=5$
$y=-x+5$ $y=-2x+5$
slope $=-1$ slope $=-2$
The slopes are not the same, so the lines are not parallel. The product, $(-1)(-2)=2$, is not -1, so the lines are not perpendicular.

b. $5y=2x-3$ $5x+2y=1$
$y=\frac{2}{5}x-\frac{3}{5}$ $2y=-5x+1$
$y=-\frac{5}{2}x+\frac{1}{2}$

slope $=\frac{2}{5}$ slope $=-\frac{5}{2}$
The slopes are not the same, so the lines are not parallel. The product, $\left(\frac{2}{5}\right)\left(-\frac{5}{2}\right)=-1$, is -1, so the lines are perpendicular.

c. $y=2x+1$ $4x-2y=8$
$-2y=-4x+8$
$y=\frac{-4x}{-2}+\frac{8}{-2}$
$y=2x-4$
slope $=2$ slope $=2$
The slopes are the same, so the lines are parallel.

8. grade $=\frac{\text{rise}}{\text{run}}=\frac{3}{20}=0.15=15\%$
The grade is 15%.

9. $m=\frac{13.1-11.6}{2013-2003}=\frac{1.5}{10}=0.15$
Each year the number of workers employed in the U.S. restaurant industry increases by 0.15 million, or 150,000, workers per year.

Calculator Explorations

1.

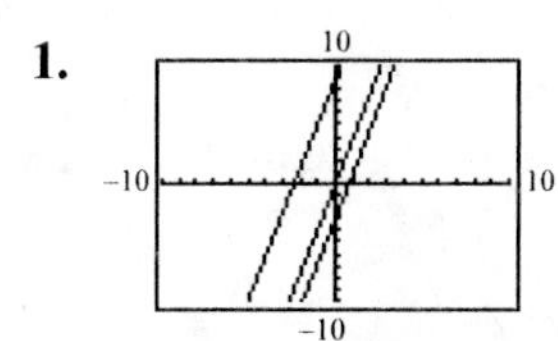

The lines are parallel since they all have a slope of 3.8. The graph of $y=3.8x-3$ is the graph of $y=3.8x$ moved 3 units down with a y-intercept of -3. The graph of $y=3.8x+9$ is the graph of $y=3.8x$ moved 9 units up with a y-intercept of 9.

2.

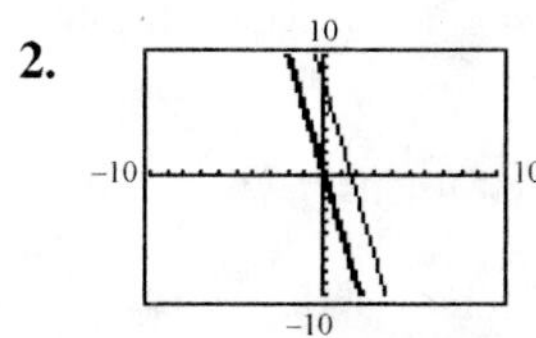

The lines are parallel since they all have a slope of -4.9. The graph of $y=-4.9x+1$ is the graph of $y=-4.9x$ moved 1 unit up with a y-intercept of 1. The graph of $y=-4.9x+8$ is the graph of $y=-4.9x$ moved 8 units up with a y-intercept of 8.

3.

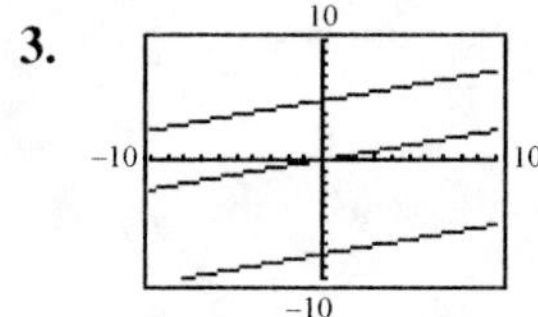

The lines are parallel since they all have a slope of $\frac{1}{4}$. The graph of $y=\frac{1}{4}x+5$ is the graph of

$y=\frac{1}{4}x$ moved 5 units up with a y-intercept of

5. The graph of $y=\frac{1}{4}x-8$ is the graph of $y=\frac{1}{4}x$ moved 8 units down with a y-intercept of -8.

4. 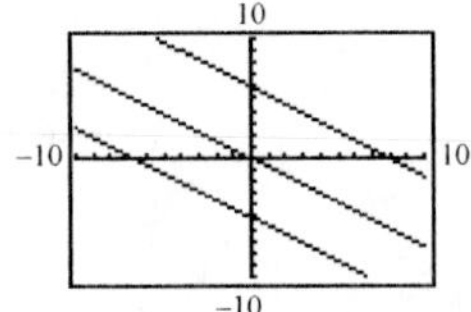

The lines are parallel since they all have a slope of $-\frac{3}{4}$. The graph of $y=-\frac{3}{4}x-5$ is the graph of $y=-\frac{3}{4}x$ moved 5 units down with a y-intercept of -5. The graph of $y=-\frac{3}{4}x+6$ is the graph of $y=-\frac{3}{4}x$ moved 6 units up with a y-intercept of 6.

Vocabulary, Readiness & Video Check 6.4

1. The measure of the steepness or tilt of a line is called slope.
2. If an equation is written in the form $y = mx + b$, the value of the letter m is the value of the slope of the graph.
3. The slope of a horizontal line is 0.
4. The slope of a vertical line is undefined.
5. If the graph of a line moves upward from left to right, the line has positive slope.
6. If the graph of a line moves downward from left to right, the line has negative slope.
7. Given two points of a line, slope $=\frac{\text{change in } y}{\text{change in } x}$.
8. The line slants downward, so its slope is negative.
9. The line slants upward, so its slope is positive.
10. The line is vertical, so its slope is undefined.
11. The line is horizontal, so its slope is 0.
12. Since $m=\frac{7}{6}$ is positive, the line slants upward.
13. Since $m = -3$ is negative, the line slants downward.
14. Since $m = 0$, the line is horizontal.
15. Since m is undefined, the line is vertical.
16. Whatever y-value we decide to start with in the numerator, we *must* start with the corresponding x-value in the denominator.
17. Solve the equation for y; the slope is the coefficient of x.
18. Zero slope indicates $m = 0$ and a horizontal line; undefined slope indicates m is undefined and a vertical line; "no slope" refers to an undefined slope.
19. Slope-intercept form; this form makes the slope easy to see, and we need to compare slopes to determine if two lines are parallel or perpendicular.
20. Step 4: INTERPRET the results.

Exercise Set 6.4

1. $m=\frac{y_2-y_1}{x_2-x_1}=\frac{-2-5}{6-(-1)}=\frac{-7}{7}=-1$

3. $m=\frac{y_2-y_1}{x_2-x_1}=\frac{3-4}{5-1}=\frac{-1}{4}=-\frac{1}{4}$

5. $m=\frac{y_2-y_1}{x_2-x_1}=\frac{1-1}{-2-5}=\frac{0}{-7}=0$

7. $m=\frac{y_2-y_1}{x_2-x_1}=\frac{5-3}{-4-(-4)}=\frac{5-3}{-4+4}=\frac{2}{0}$
   The slope is undefined.

9. $(x_1, y_1)=(-1, 2), (x_2, y_2)=(2, -2)$
   $m=\frac{y_2-y_1}{x_2-x_1}=\frac{-2-2}{2-(-1)}=\frac{-4}{3}=-\frac{4}{3}$

11. $(x_1, y_1) = (1, -2), (x_2, y_2) = (3, 3)$

$$m = \frac{y_2 - y_1}{x_2 - x_1} = \frac{3-(-2)}{3-1} = \frac{3+2}{3-1} = \frac{5}{2}$$

13. Line 1 has a positive slope and line 2 has a negative slope, so line 1 has the greater slope.

15. Line 2 is steeper, so it has the greater slope.

17. $y = 5x - 2$
The slope is $m = 5$.

19. $y = -0.3x + 2.5$
The slope is $m = -0.3$.

21. $2x + y = 7$

$y = -2x + 7$

The slope is $m = -2$.

23. The line is vertical, so it has an undefined slope.

25. $2x - 3y = 10$

$-3y = -2x + 10$

$y = \frac{2}{3}x - \frac{10}{3}$

The slope is $m = \frac{2}{3}$.

27. $x = 1$ is a vertical line, so its slope is undefined.

29. $x = 2y$

$\frac{1}{2}x = y$

$y = \frac{1}{2}x$

The slope is $m = \frac{1}{2}$.

31. $y = -3$ is a horizontal line, so its slope is 0.

33. $-3x - 4y = 6$

$-4y = 3x + 6$

$y = -\frac{3}{4}x - \frac{3}{2}$

The slope is $m = -\frac{3}{4}$.

35. $20x - 5y = 1.2$

$-5y = -20x + 1.2$

$y = \frac{-20x}{-5} + \frac{1.2}{-5}$

$y = 4x - 0.24$

The slope is $m = 4$.

37. $y = \frac{2}{9}x + 3$

$y = -\frac{2}{9}x$

$\frac{2}{9} \neq -\frac{2}{9}$, so the lines are not parallel.

$\left(\frac{2}{9}\right)\left(-\frac{2}{9}\right) = -\frac{4}{81} \neq -1$, so the lines are not perpendicular.
The lines are neither parallel nor perpendicular.

39. $x - 3y = -6$

$-3y = -x - 6$

$y = \frac{1}{3}x + 2$

$y = 3x - 9$

$\frac{1}{3} \neq 3$, so the lines are not parallel.

$\left(\frac{1}{3}\right)(3) = 1 \neq -1$, so the lines are not perpendicular. The lines are neither parallel nor perpendicular.

41. $6x = 5y + 1$

$-5y = -6x + 1$

$y = \frac{6}{5}x - \frac{1}{5}$

$-12x + 10y = 1$

$10y = 12x + 1$

$y = \frac{6}{5}x + \frac{1}{10}$

Both lines have slope $\frac{6}{5}$ and the y-intercepts are different, so they are parallel.

43. $6 + 4x = 3y$

$2+\frac{4}{3}x = y$ or $y=\frac{4}{3}x+2$

$3x+4y=8$

$4y=-3x+8$

$y=-\frac{3}{4}x+2$

$\left(\frac{4}{3}\right)\left(-\frac{3}{4}\right)=-1$, so the lines are perpendicular.

45. $m=\frac{y_2-y_1}{x_2-x_1}=\frac{0-(-3)}{0-(-3)}=\frac{3}{3}=1$

a. The slope of a parallel line is 1.

b. The slope of a perpendicular line is $-\frac{1}{1}=-1$.

47. $m=\frac{y_2-y_1}{x_2-x_1}=\frac{5-(-4)}{3-(-8)}=\frac{5+4}{3+8}=\frac{9}{11}$

a. The slope of a parallel line is $\frac{9}{11}$.

b. The slope of a perpendicular line is $-\frac{1}{\frac{9}{11}}=-\frac{11}{9}$.

49. $\text{slope}=\frac{\text{rise}}{\text{run}}=\frac{6\text{ feet}}{10\text{ feet}}=\frac{3}{5}$

The pitch of the roof is $\frac{3}{5}$.

51. $\text{slope}=\frac{\text{rise}}{\text{run}}=\frac{2}{16}=0.125=12.5\%$

The grade of the road is 12.5%

53. $\text{grade}=\frac{\text{rise}}{\text{run}}=\frac{2580\text{ meters}}{6450\text{ meters}}=0.40=40\%$

The grade of the track is 40%.

55. Canton Avenue:

$\text{grade}=\frac{\text{rise}}{\text{run}}=\frac{11\text{ meters}}{30\text{ meters}}\approx 0.37=37\%$

The grade of Canton Avenue is 37%.

Baldwin Street:

$\text{grade}=\frac{\text{rise}}{\text{run}}=\frac{1\text{ meter}}{2.86\text{ meters}}\approx 0.35=35\%$

The grade of Baldwin Street is 35%.

57. $m=\frac{y_2-y_1}{x_2-x_1}=\frac{115.6-108.4}{2013-2003}=\frac{7.2}{10}=0.72$

Every year there were 0.72 million more U.S. households with televisions.

59. $m=\frac{y_2-y_1}{x_2-x_1}=\frac{11.2-10.3}{2012-2006}=\frac{0.9}{6}=0.15$

Every year, the median age of automobiles in the United States increases by 0.15 year.

61. $y-(-6)=2(x-4)$

$y+6=2x-8$

$y=2x-14$

63. $y-1=-6(x-(-2))$

$y-1=-6(x+2)$

$y-1=-6x-12$

$y=-6x-11$

65. $(x_1, y_1)=(0, 0), (x_2, y_2)=(1, 1)$

$m=\frac{y_2-y_1}{x_2-x_1}=\frac{1-0}{1-0}=\frac{1}{1}=1$

The slope is $m = 1$; **d**.

67. The line is vertical, so its slope is undefined; **b**.

69. $(x_1, y_1)=(2, 0), (x_2, y_2)=(4, -1)$

$m=\frac{y_2-y_1}{x_2-x_1}=\frac{-1-0}{4-2}=\frac{-1}{2}=-\frac{1}{2}$

The slope is $m=-\frac{1}{2}$; **e**.

71. $m=\frac{y_2-y_1}{x_2-x_1}=\frac{0-1}{0-2}=\frac{-1}{-2}=\frac{1}{2}$

$\frac{-1-1}{-2-2}=\frac{-2}{-4}=\frac{1}{2}$; $\frac{-2-1}{-4-2}=\frac{-3}{-6}=\frac{1}{2}$

$\frac{-1-0}{-2-0}=\frac{-1}{-2}=\frac{1}{2}$; $\frac{-2-0}{-4-0}=\frac{-2}{-4}=\frac{1}{2}$

$\frac{-2-(-1)}{-4-(-2)}=\frac{-2+1}{-4+2}=\frac{-1}{-2}=\frac{1}{2}$

73. answers may vary

75. From the graph, the average miles per gallon was 31.5 for the 2008 model year.

77. The lowest points on the graph correspond to 2003 and 2004. The average fuel economy for those model years was 29.5 miles per gallon.

79. Of the line segments listed, the line from 2011 to 2012 is the steepest and therefore has the greatest slope.

81. $\text{pitch} = \frac{\text{rise}}{\text{run}}$

$\frac{2}{5} = \frac{4}{\frac{x}{2}}$

$2 \cdot \frac{x}{2} = 5 \cdot 4$

$x = 20$

83. a. (2007, 2209) and (2012, 2378)

b. $m = \frac{y_2 - y_1}{x_2 - x_1}$

$= \frac{2378 - 2209}{2012 - 2007}$

$= \frac{169}{5}$

$= 33.8$

The slope is 33.8.

c. For the years 2007 through 2012, the number of heart transplants increased at a rate of 33.8 per year.

85. Slope through (1, 3) and (2, 1):

$m = \frac{1-3}{2-1} = \frac{-2}{1} = -2$

Slope through (−4, 0) and (−3, −2):

$m = \frac{-2-0}{-3-(-4)} = \frac{-2}{-3+4} = \frac{-2}{1} = -2$

Slope through (1, 3) and (−4, 0):

$m = \frac{0-3}{-4-1} = \frac{-3}{-5} = \frac{3}{5}$

Slope through (2, 1) and (−3, −2):

$m = \frac{-2-1}{-3-2} = \frac{-3}{-5} = \frac{3}{5}$

Opposite sides are parallel and their slopes are equal.

87. $m = \frac{y_2 - y_1}{x_2 - x_1} = \frac{4.5-1.2}{-2.2-(-3.8)} = \frac{3.3}{1.6} = 2.0625$

89. $m = \frac{y_2 - y_1}{x_2 - x_1} = \frac{-2.9-(-10.1)}{9.8-14.3} = \frac{7.2}{-4.5} = -1.6$

91.

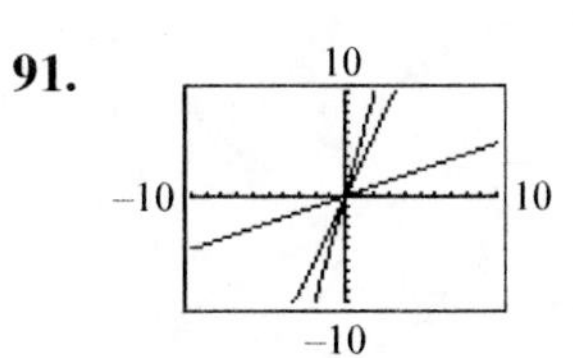

As the slope becomes larger, the line becomes steeper.

Section 6.5 Practice Exercises

1. From $y = \frac{2}{3}x - 4$, the y-intercept is (0, −4). The slope is $\frac{2}{3}$, so another point on the graph is (0 + 3, −4 + 2) or (3, −2).

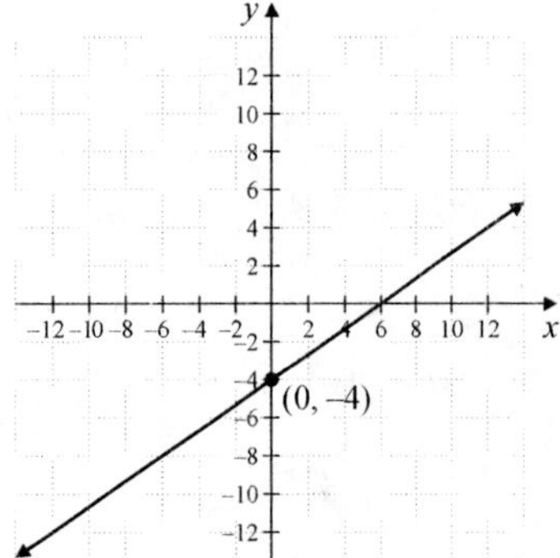

2. $3x + y = 2$

$y = -3x + 2$

The slope is $-3 = \frac{-3}{1}$ and the y-intercept is (0, 2). Another point on the graph is (0 + 1, 2 − 3) or (1, −1).

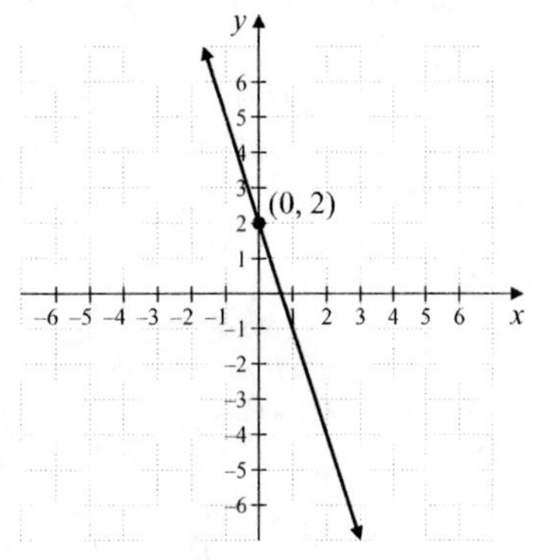

3. $y = mx + b$

$y = \frac{3}{5}x + (-2)$

$y = \frac{3}{5}x - 2$

4. $y - y_1 = m(x - x_1)$

$y - (-4) = -3(x - 2)$

$y + 4 = -3(x - 2)$

$y + 4 = -3x + 6$

$y = -3x + 2$

5. $m = \frac{-2-3}{5-1} = \frac{-5}{4}$

$y - y_1 = m(x - x_1)$

$y - 3 = \frac{-5}{4}(x - 1)$

$4(y - 3) = 4\left[\frac{-5}{4}(x - 1)\right]$

$4(y - 3) = -5(x - 1)$

$4y - 12 = -5x + 5$

$5x + 4y = 17$

6. a. (10, 200), (9, 250)

$m = \frac{y_2 - y_1}{x_2 - x_1} = \frac{250 - 200}{9 - 10} = \frac{50}{-1} = -50$

$y - y_1 = m(x - x_1)$

$y - 200 = -50(x - 10)$

$y - 200 = -50x + 500$

$y = -50x + 700$

b. Let $x = 7.50$.

$y = -50x + 700$

$y = -50(7.50) + 700$

$y = -375 + 700$

$y = 325$

The predicted weekly sales is 325.

Calculator Explorations

1.

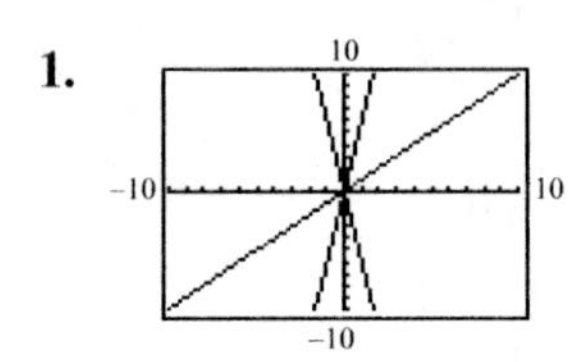

2.

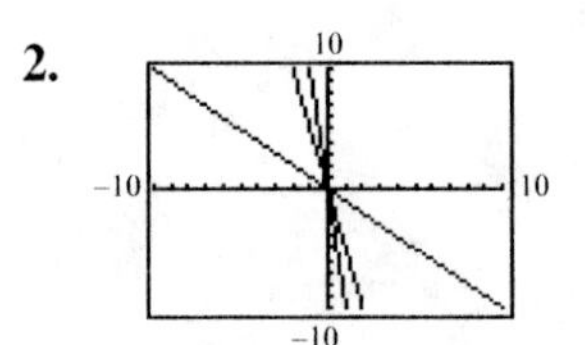

3.

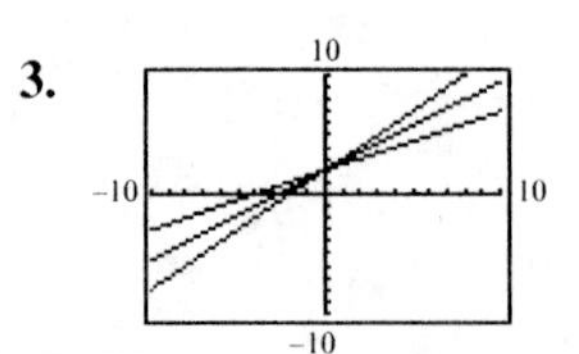

4.

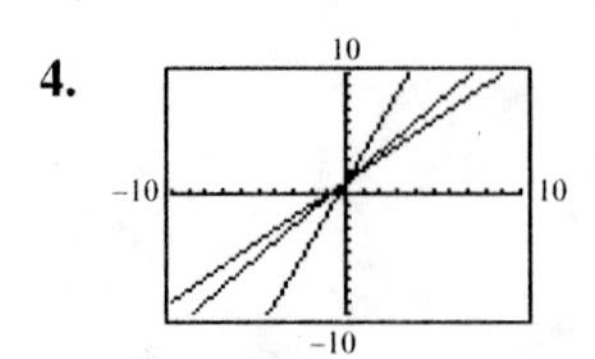

Vocabulary, Readiness & Video Check 6.5

1. The form $y = mx + b$ is called <u>slope-intercept</u> form. When a linear equation in two variables is written in this form, <u>m</u> is the slope of its graph and (0, <u>b</u>) is its y-intercept.

2. The form $y - y_1 = m(x - x_1)$ is called <u>point-slope</u> form. When a linear equation in two variables is written in this form, <u>m</u> is the slope of its graph and <u>(x_1, y_1)</u> is a point on the graph.

3. $y - 7 = 4(x + 3)$; <u>point-slope</u> form

4. $5x - 9y = 11$; <u>standard</u> form

5. $y = \frac{3}{4}x - \frac{1}{3}$; <u>slope-intercept</u> form

6. $y + 2 = \frac{-1}{3}(x - 2)$; <u>point-slope form</u>

7. $y = \frac{1}{2}$; <u>horizontal</u> line

8. $x = -17$; <u>vertical</u> line

9. Start by graphing the <u>y-intercept</u>. From this point, find another point by applying the slope—if necessary, rewrite the slope as a <u>fraction</u>.

10. $\left(0, -\frac{1}{6}\right)$

11. Write the equation with x- and y-terms on one side of the equal sign and a constant on the other side.

12. Yes, if one of the points given is the y-intercept. We will still need to use the slope formula to find the slope, but then we'll have the slope and y-intercept for the slope-intercept form.

13. We need to know what our variables stand for in order to solve part (b) of the example, and that depends on how we set up our ordered pairs in part (a).

Exercise Set 6.5

1. From $y = 2x + 1$, the y-intercept is (0, 1). The slope is 2 or $\frac{2}{1}$, so another point on the graph is $(0 + 1, 1 + 2)$ or (1, 3).

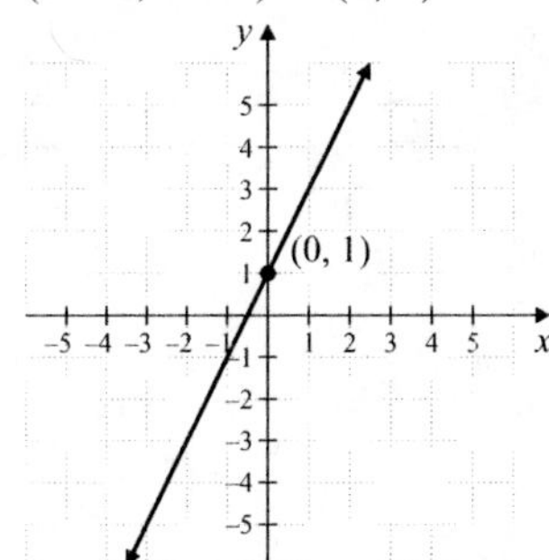

3. From $y = \frac{2}{3}x + 5$, the y-intercept is (0, 5). The slope is $\frac{2}{3}$, so another point on the graph is $(0 + 3, 5 + 2)$ or (3, 7).

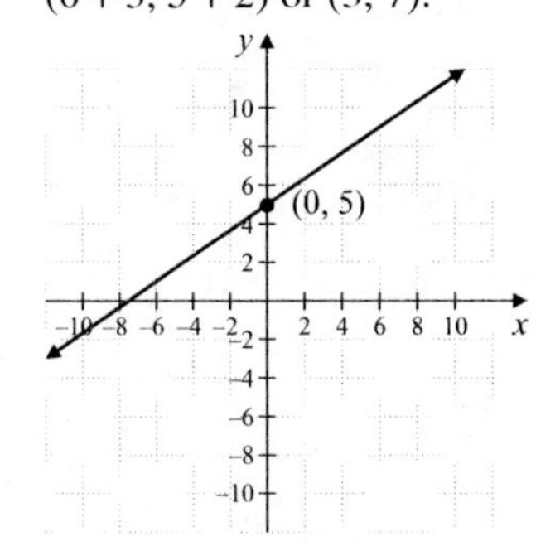

5. From $y = -5x$, the y-intercept is (0, 0). The slope is -5 or $\frac{-5}{1}$, so another point on the graph is $(0 + 1, 0 + (-5))$ or $(1, -5)$.

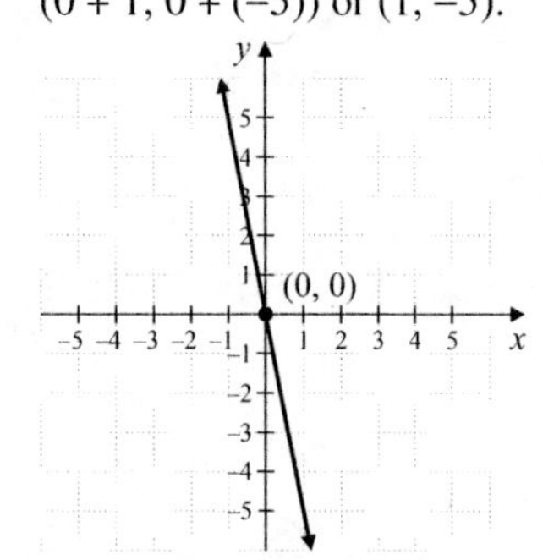

7.
$$4x + y = 6$$
$$y = -4x + 6$$

The slope is $-4 = \frac{-4}{1}$ and the y-intercept is (0, 6). Another point on the graph is $(0 + 1, 6 - 4)$ or (1, 2).

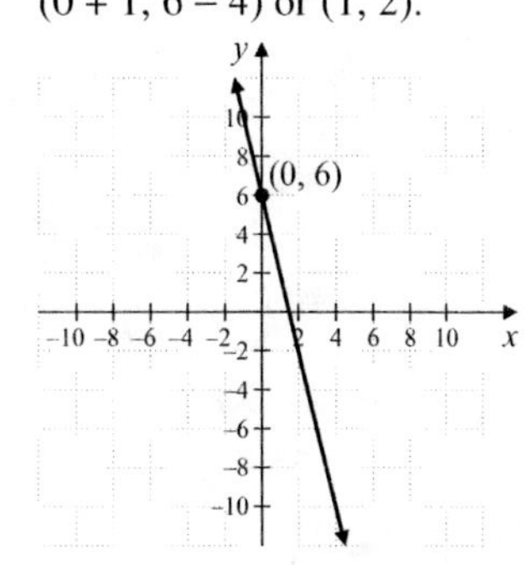

9.
$$4x - 7y = -14$$
$$-7y = -4x - 14$$
$$y = \frac{4}{7}x + 2$$

The slope is $\frac{4}{7}$ and the y-intercept is (0, 2). Another point on the graph is $(0 + 7, 2 + 4)$ or (7, 6).

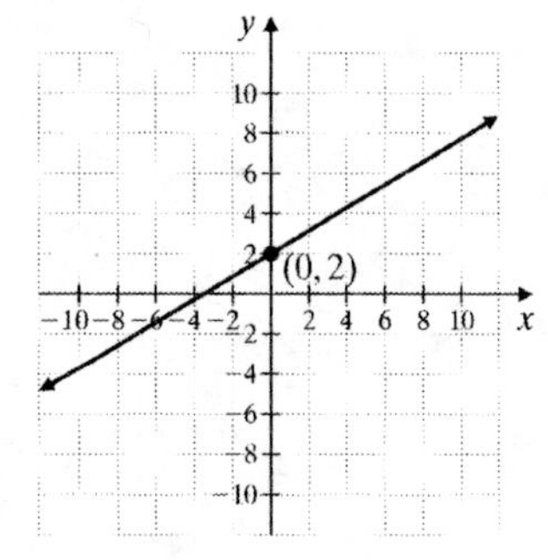

11. $x = \frac{5}{4}y$

$\frac{4}{5}x = y$

$y = \frac{4}{5}x + 0$

The slope is $\frac{4}{5}$ and the y-intercept is (0, 0).

Another point on the graph is (0 + 5, 0 + 4) or (5, 4).

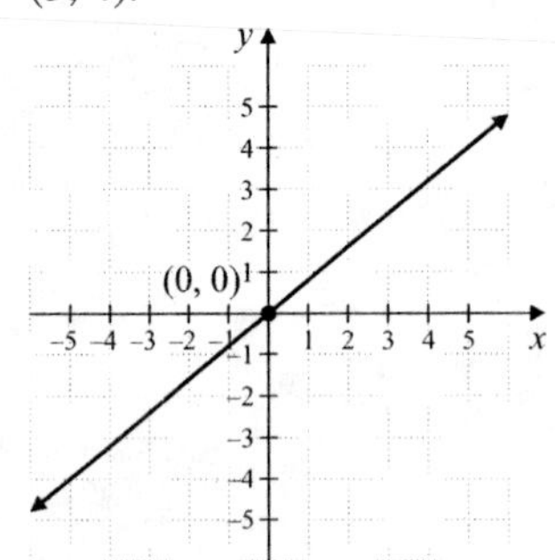

13. $y = mx + b$
$y = 5x + 3$

15. $y = mx + b$

$y = -4x + \left(-\frac{1}{6}\right)$

$y = -4x - \frac{1}{6}$

17. $y = mx + b$

$y = \frac{2}{3}x + 0$

$y = \frac{2}{3}x$

19. $y = mx + b$
$y = 0x + (-8)$
$y = -8$

21. $y = mx + b$

$y = -\frac{1}{5}x + \frac{1}{9}$

23. $y - y_1 = m(x - x_1)$
$y - 2 = 6(x - 2)$
$y - 2 = 6x - 12$
$y = 6x - 10$
$-6x + y = -10$

25. $y - y_1 = m(x - x_1)$
$y - (-5) = -8[x - (-1)]$
$y + 5 = -8(x + 1)$
$y + 5 = -8x - 8$
$y = -8x - 13$
$8x + y = -13$

27. $y - y_1 = m(x - x_1)$

$y - (-6) = \frac{3}{2}(x - 5)$

$y + 6 = \frac{3}{2}(x - 5)$

$2(y + 6) = 2\left[\frac{3}{2}(x - 5)\right]$

$2y + 12 = 3x - 15$
$2y = 3x - 27$
$-3x + 2y = -27$ or $3x - 2y = 27$

29. $y - y_1 = m(x - x_1)$

$y - 0 = -\frac{1}{2}[x - (-3)]$

$y = -\frac{1}{2}(x + 3)$

$2y = -1(x + 3)$
$2y = -x - 3$
$x + 2y = -3$

31. $m = \frac{y_2 - y_1}{x_2 - x_1} = \frac{6 - 2}{5 - 3} = \frac{4}{2} = 2$

$y - y_1 = m(x - x_1)$
$y - 2 = 2(x - 3)$
$y - 2 = 2x - 6$
$y + 4 = 2x$
$4 = 2x - y$
$2x - y = 4$

33. $m = \frac{y_2 - y_1}{x_2 - x_1} = \frac{-5 - 3}{-2 - (-1)} = \frac{-8}{-1} = 8$

$y - y_1 = m(x - x_1)$
$y - 3 = 8[x - (-1)]$
$y - 3 = 8(x + 1)$
$y - 3 = 8x + 8$
$y - 11 = 8x$
$-11 = 8x - y$
$8x - y = -11$

35. $m = \frac{y_2 - y_1}{x_2 - x_1} = \frac{-1-3}{-1-2} = \frac{-4}{-3} = \frac{4}{3}$

$$\begin{aligned} y - y_1 &= m(x - x_1) \\ y - 3 &= \frac{4}{3}(x-2) \\ 3(y-3) &= 4(x-2) \\ 3y - 9 &= 4x - 8 \\ 3y - 1 &= 4x \\ -1 &= 4x - 3y \\ 4x - 3y &= -1 \end{aligned}$$

37. $m = \frac{y_2 - y_1}{x_2 - x_1} = \frac{\frac{1}{13} - 0}{-\frac{1}{8} - 0} = \frac{\frac{1}{13}}{-\frac{1}{8}} = \frac{1}{13}\left(-\frac{8}{1}\right) = -\frac{8}{13}$

$$\begin{aligned} y - y_1 &= m(x - x_1) \\ y - 0 &= -\frac{8}{13}(x-0) \\ y &= -\frac{8}{13}x \\ 13y &= -8x \\ 8x + 13y &= 0 \end{aligned}$$

39. $y = mx + b$

$y = -\frac{1}{2}x + \frac{5}{3}$

41. $m = \frac{y_2 - y_1}{x_2 - x_1} = \frac{10-7}{7-10} = \frac{3}{-3} = -1$

$$\begin{aligned} y - y_1 &= m(x - x_1) \\ y - 7 &= -1(x - 10) \\ y - 7 &= -x + 10 \\ y &= -x + 17 \end{aligned}$$

43. A line with undefined slope is a vertical line. This one has an x-intercept of $\left(-\frac{3}{4}, 0\right)$. The equation is $x = -\frac{3}{4}$.

45.

$$\begin{aligned} y - y_1 &= m(x - x_1) \\ y - 9 &= 1[x - (-7)] \\ y - 9 &= x + 7 \\ y &= x + 16 \end{aligned}$$

47. $y = mx + b$

$y = -5x + 7$

49. A line parallel to $y = 5$ is a horizontal line.

$y = 2$

51. $m = \frac{y_2 - y_1}{x_2 - x_1} = \frac{3-0}{2-0} = \frac{3}{2}$

$$\begin{aligned} y - y_1 &= m(x - x_1) \\ y - 0 &= \frac{3}{2}(x - 0) \\ y &= \frac{3}{2}x \end{aligned}$$

53. A line perpendicular to the y-axis is a horizontal line.

$y = -3$

55.

$$\begin{aligned} y - y_1 &= m(x - x_1) \\ y - (-2) &= -\frac{4}{7}[x - (-1)] \\ y + 2 &= -\frac{4}{7}(x+1) \\ y + 2 &= -\frac{4}{7}x - \frac{4}{7} \\ y &= -\frac{4}{7}x - \frac{4}{7} - \frac{14}{7} \\ y &= -\frac{4}{7}x - \frac{18}{7} \end{aligned}$$

57. a. The ordered pairs are (0, 370) and (5, 312).

b. $m = \frac{y_2 - y_1}{x_2 - x_1} = \frac{312 - 370}{5 - 0} = \frac{-58}{5} = -11.6$

$$\begin{aligned} y - y_1 &= m(x - x_1) \\ y - 370 &= -11.6(x - 0) \\ y - 370 &= -11.6x \\ y &= -11.6x + 370 \end{aligned}$$

c. 2010 corresponds to $x = 3$.

$$\begin{aligned} x = 3:\ y &= -11.6(3) + 370 \\ &= -34.8 + 370 \\ &= 335.2 \end{aligned}$$

There were approximately 335.2 million magazine subscriptions in 2010.

59. a. The ordered pairs are (1, 32) and (3, 96).

$m = \frac{s_2 - s_1}{t_2 - t_1} = \frac{96 - 32}{3 - 1} = \frac{64}{2} = 32$

$$\begin{aligned} s - s_1 &= m(t - t_1) \\ s - 32 &= 32(t - 1) \\ s - 32 &= 32t - 32 \\ s &= 32t \end{aligned}$$

b. $t = 4$: $s = 32(4)$
$s = 128$
The speed of the rock 4 seconds after it was dropped is 128 feet per second.

61. a. The ordered pairs are (0, 314,000) and (4, 434,000).

$$m = \frac{y_2 - y_1}{x_2 - x_1} = \frac{434,000 - 314,000}{4 - 0} = \frac{120,000}{4} = 30,000$$

$$\begin{aligned} y - y_1 &= m(x - x_1) \\ y - 314,000 &= 30,000(x - 0) \\ y - 314,000 &= 30,000x \\ y &= 30,000x + 314,000 \end{aligned}$$

b. The year 2014 is 6 years past 2008.
$x = 2014 - 2008 = 6$
$y = 30,000(6) + 314,000$
$y = 180,000 + 314,000$
$y = 494,000$
494,000 vehicles are predicted for 2014.

63. a. The ordered pairs are (0, 5545) and (5, 5320).

$$m = \frac{y_2 - y_1}{x_2 - x_1} = \frac{5320 - 5545}{5 - 0} = \frac{-225}{5} = -45$$

$$\begin{aligned} y - y_1 &= m(x - x_1) \\ y - 5545 &= -45(x - 0) \\ y - 5545 &= -45x \\ y &= -45x + 5545 \end{aligned}$$

b. The year 2015 is 8 years past 2007, so it corresponds to $x = 8$.
$x = 8$: $y = -45(8) + 5545$
$y = -360 + 5545$
$y = 5185$
5185 indoor cinema sites are predicted for 2015.

65. a. The ordered pairs are (3, 10,000) and (5, 8000).

$$m = \frac{S_2 - S_1}{p_2 - p_1} = \frac{8000 - 10,000}{5 - 3} = \frac{-2000}{2} = -1000$$

$$\begin{aligned} S - S_1 &= m(p - p_1) \\ S - 10,000 &= -1000(p - 3) \\ S - 10,000 &= -1000p + 3000 \\ S &= -1000p + 13,000 \end{aligned}$$

b. $p = 3.50$: $S = -1000(3.50) + 13,000$
$S = -3500 + 13,000$
$S = 9500$
9500 Fun Noodles will be sold when the price is \$3.50 each.

67. $x = 2$: $x^2 - 3x + 1 = 2^2 - 3(2) + 1 = 4 - 6 + 1 = -1$

69. $x = -1$: $x^2 - 3x + 1 = (-1)^2 - 3(-1) + 1$
$= 1 + 3 + 1$
$= 5$

71. The graph of $y = 2x + 1$ has slope $m = 2$ and y-intercept (0, 1). This is graph **b**.

73. The graph of $y = -3x - 2$ has slope $m = -3$ and y-intercept (0, −2). This is graph **d**.

75. The slope of the line $y = 2x + 5$ is $m = 2$.

$$\begin{aligned} y - y_1 &= m(x - x_1) \\ y - 4 &= 2[x - (-2)] \\ y - 4 &= 2(x + 2) \\ y - 4 &= 2x + 4 \\ -8 &= 2x - y \\ 2x - y &= -8 \end{aligned}$$

77. a. A line parallel to the line $y = 3x - 1$ will have slope 3.

$$\begin{aligned} y - y_1 &= m(x - x_1) \\ y - 2 &= 3[x - (-1)] \\ y - 2 &= 3(x + 1) \\ y - 2 &= 3x + 3 \\ y &= 3x + 5 \\ -5 &= 3x - y \\ 3x - y &= -5 \end{aligned}$$

b. A line perpendicular to the line $y = 3x - 1$ will have slope $-\frac{1}{3}$.

$$y - y_1 = m(x - x_1)$$
$$y - 2 = -\frac{1}{3}[x - (-1)]$$
$$y - 2 = -\frac{1}{3}(x + 1)$$
$$3y - 6 = -1(x + 1)$$
$$3y - 6 = -x - 1$$
$$3y = -x + 5$$
$$x + 3y = 5$$

Integrated Review

1. Select two points on the line, such as (0, 0) and (1, 2).
$$m = \frac{y_2 - y_1}{x_2 - x_1} = \frac{2 - 0}{1 - 0} = \frac{2}{1} = 2$$

2. Horizontal lines have slopes of $m = 0$.

3. Select two points on the line, such as (0, 1) and (−3, 3).
$$m = \frac{y_2 - y_1}{x_2 - x_1} = \frac{3 - 1}{-3 - 0} = \frac{2}{-3} = -\frac{2}{3}$$

4. Vertical lines have undefined slopes.

5. $y = -2x$
$y = 0$: $0 = -2x$
$0 = x$
The x-intercept is (0, 0).
$x = 0$: $y = -2(0) = 0$
The y-intercept is (0, 0).
Find another point, for example let $x = 1$.
$y = -2(1) = -2$
Another point on the line is (1, −2).

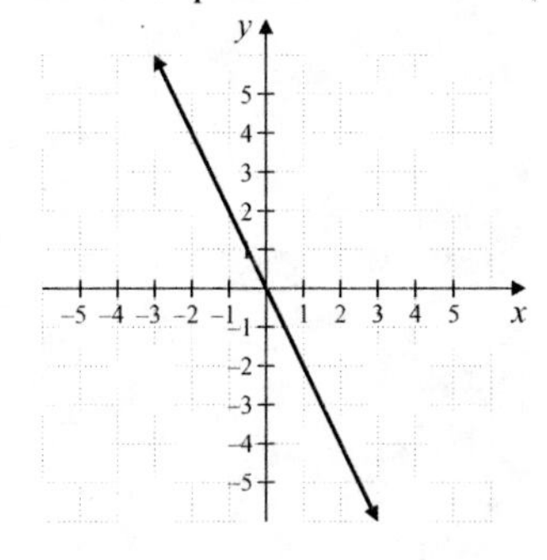

6. $x + y = 3$
$y = 0$: $x + 0 = 3$
$x = 3$
The x-intercept is (3, 0).
$x = 0$: $0 + y = 3$
$y = 3$
The y-intercept is (0, 3).

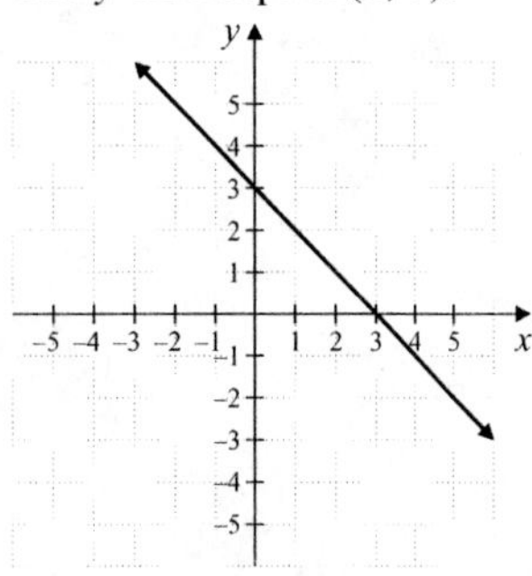

7. The graph of $x = -1$ is a vertical line with an x-intercept of (−1, 0).

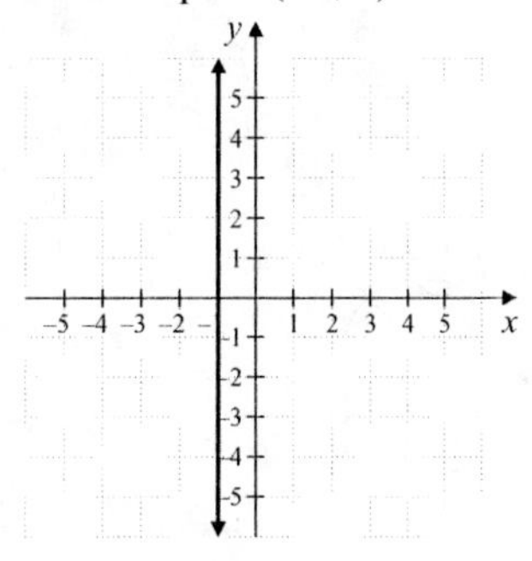

8. The graph of $y = 4$ is a horizontal line with a y-intercept of (0, 4).

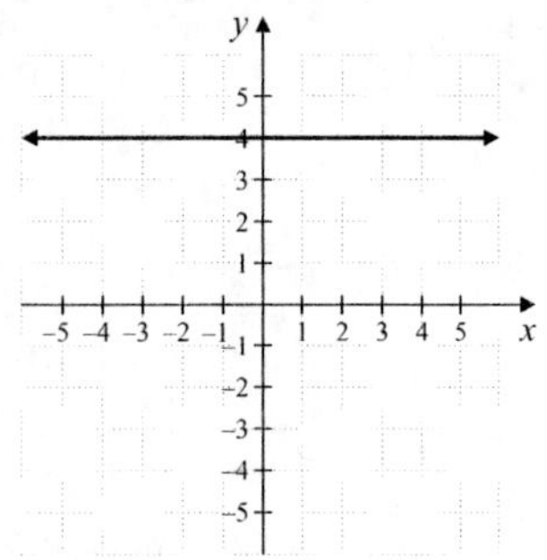

9. $x - 2y = 6$
$y = 0$: $x - 2(0) = 6$
$x - 0 = 6$
$x = 6$
The x-intercept is (6, 0).
$x = 0$: $0 - 2y = 6$
$-2y = 6$
$y = -3$
The y-intercept is (0, −3).

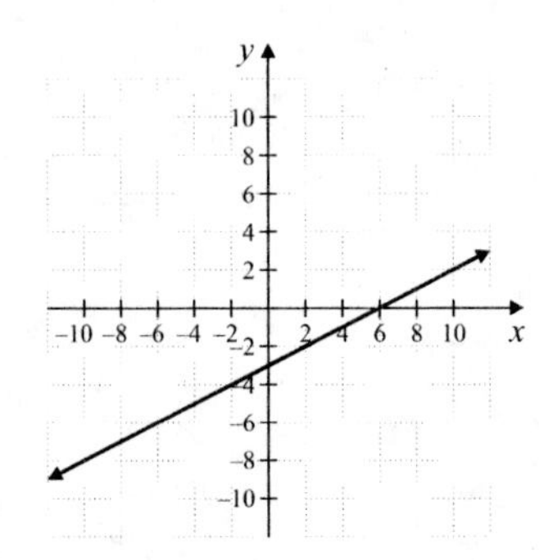

10. $y = 3x + 2$

$y = 0$: $\quad 0 = 3x + 2$

$-2 = 3x$

$-\frac{2}{3} = x$

The x-intercept is $\left(-\frac{2}{3}, 0\right)$.

$x = 0$: $\quad y = 3(0) + 2$

$y = 0 + 2$

$y = 2$

The y-intercept is (0, 2).

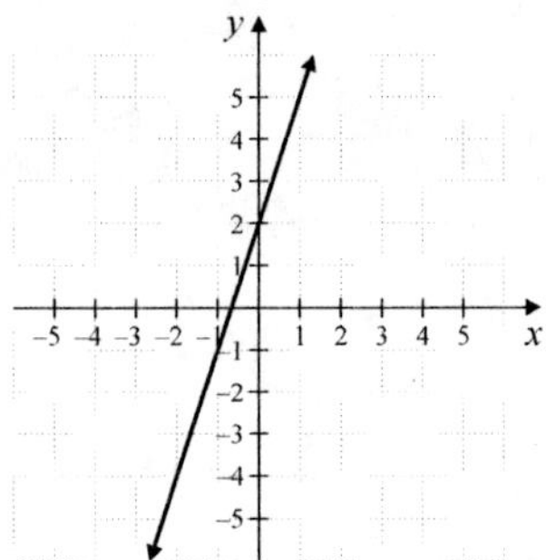

11. $y = -\frac{3}{4}x + 3$

$y = 0$: $\quad 0 = -\frac{3}{4}x + 3$

$-3 = -\frac{3}{4}x$

$4 = x$

The x-intercept is (4, 0).

$x = 0$: $\quad y = -\frac{3}{4}(0) + 3$

$y = 0 + 3$

$y = 3$

The y-intercept is (0, 3).

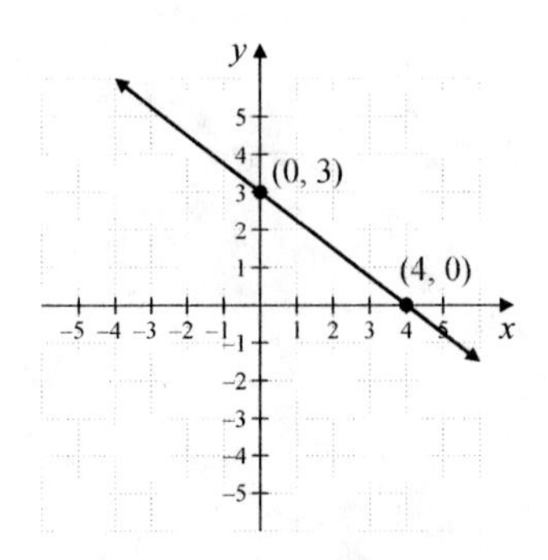

12. $5x - 2y = 8$

$y = 0$: $\quad 5x - 2(0) = 8$

$5x - 0 = 8$

$5x = 8$

$x = \frac{8}{5}$

The x-intercept is $\left(\frac{8}{5}, 0\right)$.

$x = 0$: $\quad 5(0) - 2y = 8$

$0 - 2y = 8$

$-2y = 8$

$y = -4$

The y-intercept is (0, –4).

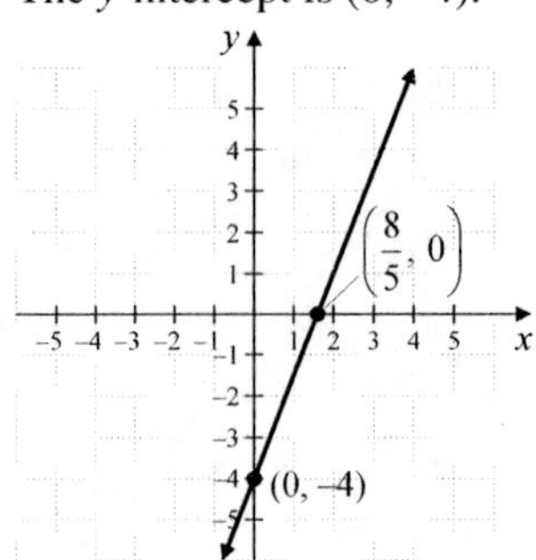

13. $y = mx + b$

$y = 3x - 1$

The slope is $m = 3$.

14. $y = mx + b$

$y = -6x + 2$

The slope is $m = -6$.

15. $y = mx + b$

$7x + 2y = 11$

$2y = -7x + 11$

$y = -\frac{7}{2}x + \frac{11}{2}$

The slope is $m = -\frac{7}{2}$.

16. $y = mx + b$
$2x - y = 0$
$-y = -2x$
$y = 2x$
$y = 2x + 0$
The slope is $m = 2$.

17. The graph of $x = 2$ is a vertical line. Vertical lines have undefined slopes.

18. The graph of $y = -4$ is a horizontal line. Horizontal lines have slopes of $m = 0$.

19. $y = mx + b$
$y = 2x + \left(-\frac{1}{3}\right)$
$y = 2x - \frac{1}{3}$

20. $y - y_1 = m(x - x_1)$
$y - 3 = -4[x - (-1)]$
$y - 3 = -4(x + 1)$
$y - 3 = -4x - 4$
$y = -4x - 1$

21. $m = \frac{y_2 - y_1}{x_2 - x_1} = \frac{-3 - 0}{-1 - 2} = \frac{-3}{-3} = 1$
$y - y_1 = m(x - x_1)$
$y - 0 = 1(x - 2)$
$y = x - 2$
$-x + y = -2$ or $x - y = 2$

22. $6x - y = 7$
$-y = -6x + 7$
$y = 6x - 7$
$m = 6$

$2x + 3y = 4$
$3y = -2x + 4$
$y = -\frac{2}{3}x + \frac{4}{3}$
$m = -\frac{2}{3}$

Since $6 \neq -\frac{2}{3}$, the lines are not parallel. Since $6\left(-\frac{2}{3}\right) = -4 \neq -1$, the lines are not perpendicular. The lines are neither parallel nor perpendicular.

23. $3x - 6y = 4$
$-6y = -3x + 4$
$y = \frac{1}{2}x - \frac{2}{3}$
$m = \frac{1}{2}$

$y = -2x$
$m = -2$

Since $\left(\frac{1}{2}\right)(-2) = -1$, the lines are perpendicular.

24. **a.** The ordered pairs are (2008, 3600) and (2012, 4416).

b. $m = \frac{y_2 - y_1}{x_2 - x_1} = \frac{4416 - 3600}{2012 - 2008} = \frac{816}{4} = 204$

c. For the years 2008 through 2012, the amount of yogurt produced increased at a rate of 204 million pounds per year.

Section 6.6 Practice Exercises

1. The domain is the set of x-coordinates: $\{-3, 4, 7\}$.
The range is the set of y-coordinates: $\{0, 1, 5, 6\}$.

2. **a.** Each x-value is only assigned to one y-value, so the relation is a function.

b. The x-value 1 is paired with two y-values, 4 and -3, so this set of ordered pairs is not a function.

3. **a.** This is the graph of the relation $\{(-3, -2), (-1, -1), (0, 0), (1, 1)\}$. Each x-coordinate has exactly one y-value, so this is the graph of a function.

b. This is the graph of the relation $\{(-1, -1), (-1, 2), (1, 0), (3, 1)\}$. The x-value -1 is paired with two y-values, -1 and 2, so this is not the graph of a function.

4. **a.** No vertical line will intersect the graph more than once, so the graph is the graph of a function.

b. No vertical line will intersect the graph more than once, so the graph is the graph of a function.

c. Vertical lines can be drawn that intersect the graph in two points, so the graph is not the graph of a function.

d. A vertical line can be drawn that intersects this line at every point, so the graph is not the graph of a function.

5. **a, b,** and **c** are functions because their graphs are nonvertical lines. **d** is not a function because its graph is a vertical line.

6. a. According to the graph, the time of the sunrise on March 1st is 6:30 A.M.

b. According to the graph, the sun rises at 6 A.M. in the middle of March and the middle of September.

7. $f(x) = x^2 + 1$

a. $f(1) = 1^2 + 1 = 1 + 1 = 2$
Ordered pair: (1, 2)

b. $f(-3) = (-3)^2 + 1 = 9 + 1 = 10$
(−3, 10)

c. $f(0) = 0^2 + 1 = 0 + 1 = 1$
(0, 1)

8. a. In this function, x can be any real number. The domain of $h(x)$ is the set of all real numbers.

b. Since we cannot divide by 0, the domain of $f(x)$ is the set of all real numbers except 0.

9. a. The x-values go from −4 to 6, so the domain is $-4 \le x \le 6$.
The y-values go from −2 to 3, so the range is $-2 \le y \le 3$.

b. There are no restrictions on the x-values, so the domain is all real numbers.
The y-values are all less than or equal to 3, so the range is $y \le 3$.

Vocabulary, Readiness & Video Check 6.6

1. A set of ordered pairs is called a relation.

2. A set of ordered pairs that assigns to each x-value exactly one y-value is called a function.

3. The set of all y-coordinates of a relation is called the range.

4. The set of all x-coordinates of a relation is called the domain.

5. All linear equations are functions except those whose graphs are vertical lines.

6. All linear equations are functions except those whose equations are of the form $x = c$.

7. If $f(3) = 7$, the corresponding ordered pair is (3, 7).

8. The domain of $f(x) = x + 5$ is all real numbers.

9. For the function $y = mx + b$, the dependent variable is y and the independent variable is x.

10. A relation is a set of ordered pairs and an equation in two variables defines a set of ordered pairs. Therefore, an equation in two variables defines a relation.

11. Yes, this is a function. The definition restricts x-values to be assigned to exactly one y-value, but it makes no such restriction of the y-values.

12. A vertical line represents one x-value paired with many y-values. A function only allows an x-value paired with exactly one y-value, so if a vertical line intersects a graph more than once, there's an x-value paired with more than one y-value and we don't have a function.

13. $f(-2) = 6$ corresponds to (−2, 6) and $f(3) = 11$ corresponds to (3, 11).

Exercise Set 6.6

1. The domain is the set of x-coordinates: $\{-7, 0, 2, 10\}$.
The range is the set of y-coordinates: $\{-7, 0, 4, 10\}$.

3. The domain is the set of x-coordinates: $\{0, 1, 5\}$
The range is the set of y-coordinates: $\{-2\}$

5. Each x-value is only assigned to one y-value, so the relation is a function.

7. The x-value −1 is paired with more than one y-value, 0, 6, and 8, so the relation is not a function.

9. The vertical line $x = 1$ will intersect the graph in two points, so the graph is not the graph of a function.

11. No vertical line will intersect the graph more than once, so the graph is the graph of a function.

13. No vertical line will intersect the graph more than once, so the graph is the graph of a function.

15. Vertical lines can be drawn that intersect the graph in two points, so the graph is not the graph of a function.

17. If $x = -1$, the relation is not also a function; a.

19. $y - x = 7$
The graph of this linear equation is not a vertical line, so the equation describes a function.

21. $y = 6$
The graph of this linear equation is not a vertical line, so the equation describes a function.

23. $x = -2$
The graph of this linear equation is a vertical line, so the equation does not describe a function.

25. $x = y^2$

$y = 1$: $x = (1)^2 = 1$

$y = -1$: $x = (-1)^2 = 1$

Since there is an x-value that is paired with two y-values, the equation does not describe a function.

27. On June 1, the graph shows sunset to be at approximately 9:30 P.M.

29. At 3 P.M., the graph shows this happens on January 1 and December 1.

31. The graph passes the vertical line test, so it is the graph of a function.

33. Before October 1996, the graph shows the minimum wage was $4.25 per hour.

35. According to the graph, the minimum wage increased to over $7.00 in 2009.

37. yes; answers may vary

39. According to the graph, the postage would be $1.72.

41. From the graph, it would cost $1.12 to mail a large envelope that weighs more than 1 ounce and less than or equal to 2 ounces.

43. yes; answers may vary

45. $f(x) = 2x - 5$
$f(-2) = 2(-2) - 5 = -4 - 5 = -9$
$f(0) = 2(0) - 5 = 0 - 5 = -5$
$f(3) = 2(3) - 5 = 6 - 5 = 1$

47. $f(x) = x^2 + 2$

$f(-2) = (-2)^2 + 2 = 4 + 2 = 6$

$f(0) = 0^2 + 2 = 0 + 2 = 2$

$f(3) = 3^2 + 2 = 9 + 2 = 11$

49. $f(x) = 3x$
$f(-2) = 3(-2) = -6$
$f(0) = 3(0) = 0$
$f(3) = 3(3) = 9$

51. $f(x) = |x|$
$f(-2) = |-2| = 2$
$f(0) = |0| = 0$
$f(3) = |3| = 3$

53. $h(x) = -5x$
$h(-1) = -5(-1) = 5$
$h(0) = -5(0) = 0$
$h(4) = -5(4) = -20$

55. $h(x) = 2x^2 + 3$

$h(-1) = 2(-1)^2 + 3 = 2(1) + 3 = 2 + 3 = 5$

$h(0) = 2(0)^2 + 3 = 2(0) + 3 = 0 + 3 = 3$

$h(4) = 2(4)^2 + 3 = 2(16) + 3 = 32 + 3 = 35$

57. The ordered pair solution corresponding to $f(3) = 6$ is $(3, 6)$.

59. The ordered pair solution corresponding to $g(0) = -\frac{1}{2}$ is $\left(0, -\frac{1}{2}\right)$.

61. The ordered pair solution corresponding to $h(-2) = 9$ is $(-2, 9)$.

63. The domain of $f(x)$ is all real numbers.

65. $x + 5$ cannot be 0.
$$x+5=0$$
$$x=-5$$
The domain of $f(x)$ is all real numbers except −5.

67. The domain is all real numbers. The range is $y \geq -4$.

69. The domain is all real numbers. The range is all real numbers.

71. The domain is all real numbers. The range is $\{2\}$.

73. When $x = 0$, $y = -1$, so the ordered pair solution is $(0, -1)$.

75. When $x = 0$, $y = -1$, so $f(0) = -1$.

77. When $y = 0$, $x = -1$ and $x = 5$.

79.
$$2x+5<7$$
$$2x<2$$
$$x<1$$

81.
$$-x+6\leq 9$$
$$-x\leq 3$$
$$x\geq -3$$

83.
$$\frac{3}{x}+\frac{3}{2x}+\frac{5}{x}=\frac{3\cdot 2}{x\cdot 2}+\frac{3}{2x}+\frac{5\cdot 2}{x\cdot 2}$$
$$=\frac{6}{2x}+\frac{3}{2x}+\frac{10}{2x}$$
$$=\frac{6+3+10}{2x}$$
$$=\frac{19}{2x}$$
The perimeter is $\frac{19}{2x}$ meters.

85. A function f evaluated at −5 as 12 is written as $f(-5) = 12$.

87. answers may vary

89. $y = x + 7$ written in function notation is $f(x) = x + 7$.

91. $f(x)=\frac{136}{25}x$

a. $f(35)=\frac{136}{25}(35)=\frac{4760}{25}=\frac{952}{5}=190.4$
The proper dosage for a 35-pound dog is 190.4 milligrams.

b. $f(70)=\frac{136}{25}(70)=\frac{9520}{25}=\frac{1904}{5}=380.8$
The proper dosage for a 70-pound dog is 380.8 milligrams.

Section 6.7 Practice Exercises

1. $x - 4y > 8$

a.
$$(-3, 2):\ -3-4(2)>8$$
$$-3-8>8$$
$$-11>8 \quad \text{False}$$
$(-3, 2)$ is not a solution of the inequality.

b.
$$(9, 0):\ 9-4(0)>8$$
$$9-0>8$$
$$9>8 \quad \text{True}$$
$(9, 0)$ is a solution of the inequality.

2. Graph the boundary line, $x - y = 3$, with a dashed line.
Test (0, 0):
$$x-y>3$$
$$0-0>3$$
$$0>3 \quad \text{False}$$
Shade the half-plane not containing (0, 0).

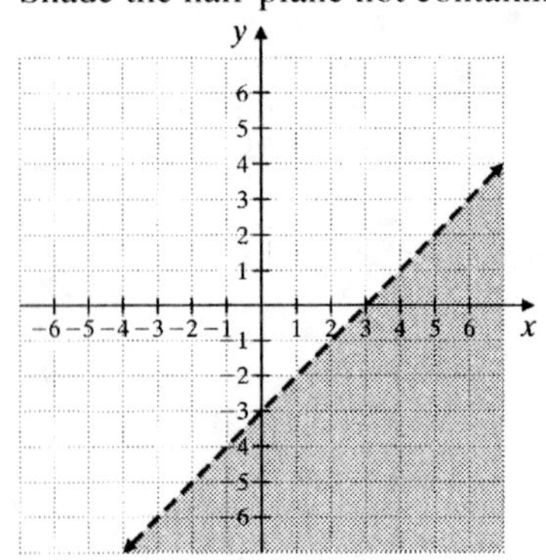

3. Graph the boundary line, $x - 4y = 4$, with a solid line.
Test (0, 0):
$$x-4y\leq 4$$
$$0-4(0)\leq 4$$
$$0-0\leq 4$$
$$0\leq 4 \quad \text{True}$$
Shade the half-plane containing (0, 0).

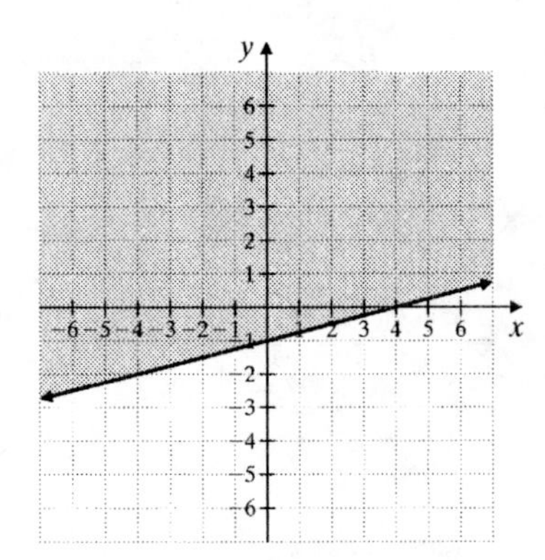

4. Graph the boundary line, $y = 3x$, with a dashed line.
Test (1, 1): $y < 3x$
$1 < 3(1)$
$1 < 3$ True
Shade the half-plane containing (1, 1).

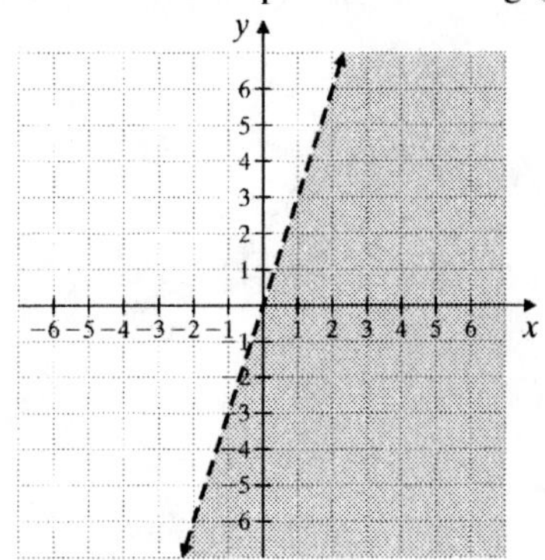

5. Graph the boundary line, $3x + 2y = 12$, with a solid line.
Test (0, 0): $3x + 2y \geq 12$
$3(0) + 2(0) \geq 12$
$0 + 0 \geq 12$
$0 \geq 12$ False
Shade the half-plane not containing (0, 0).

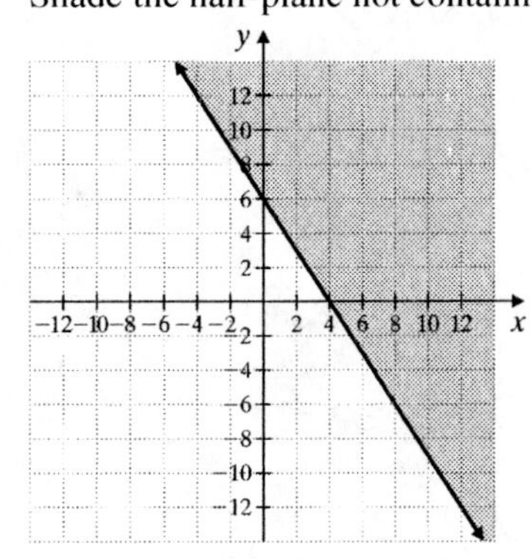

6. Graph the boundary line, $x = 2$, with a dashed line.
Test (0, 0): $x < 2$
$0 < 2$ True
Shade the half-plane containing (0, 0).

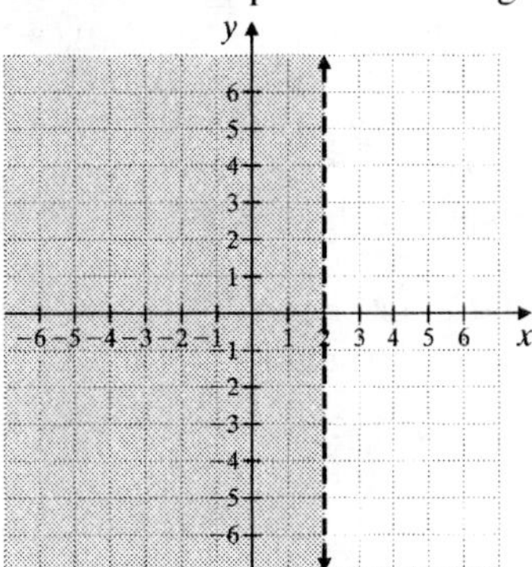

7. Graph the boundary line, $y = \frac{1}{4}x + 3$, with a solid line.

Test (0, 0): $y \geq \frac{1}{4}x + 3$

$0 \geq \frac{1}{4}(0) + 3$

$0 \geq 0 + 3$

$0 \geq 3$ False

Shade the half-plane not containing (0, 0).

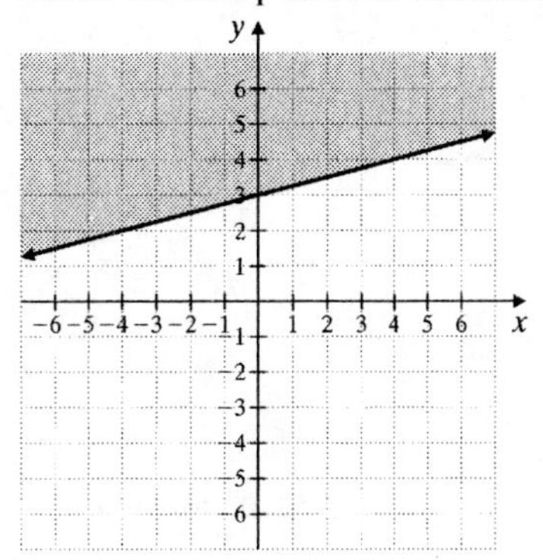

Vocabulary, Readiness & Video Check 6.7

1. The statement $5x - 6y < 7$ is an example of a linear inequality in two variables.

2. A boundary line divides a plane into two regions called half-planes.

3. The graph of $5x - 6y < 7$ includes its corresponding boundary line. false

4. When graphing a linear inequality, to determine which side of the boundary line to shade, choose a point *not* on the boundary line. true

5. The boundary line for the inequality $5x - 6y < 7$ is the graph of $5x - 6y = 7$. true

6. The graph shown is $y < 2$.

7. An ordered pair is a solution of an inequality if replacing the variables with the coordinates of the ordered pair results in a true statement.

8. We find the boundary line equation by replacing the inequality symbol with =. The points on this line are solutions (line is solid) if the inequality is ≥ or ≤; they are not solutions (line is dashed) if the inequality is > or <.

Exercise Set 6.7

1. $x - y > 3$
   $(0, 3)$: $0 - 3 > 3$
   $-3 > 3$ False
   $(0, 3)$ is not a solution of the inequality.
   $(2, -1)$: $2 - (-1) > 3$
   $2 + 1 > 3$
   $3 > 3$ False
   $(2, -1)$ is not a solution of the inequality.

3. $3x - 5y \le -4$
   $(2, 3)$: $3(2) - 5(3) \le -4$
   $6 - 15 \le -4$
   $-9 \le -4$ True
   $(2, 3)$ is a solution of the inequality.
   $(-1, -1)$: $3(-1) - 5(-1) \le -4$
   $-3 + 5 \le -4$
   $2 \le -4$ False
   $(-1, -1)$ is not a solution of the inequality.

5. $x < -y$
   $(0, 2)$: $0 < -2$ False
   $(0, 2)$ is not a solution of the inequality.
   $(-5, 1)$: $-5 < -1$ True
   $(-5, 1)$ is a solution of the inequality.

7. Graph the boundary line, $x + y = 1$, with a solid line.
   Test $(0, 0)$: $x + y \le 1$
   $0 + 0 \le 1$
   $0 \le 1$ True
   Shade the half-plane containing $(0, 0)$.

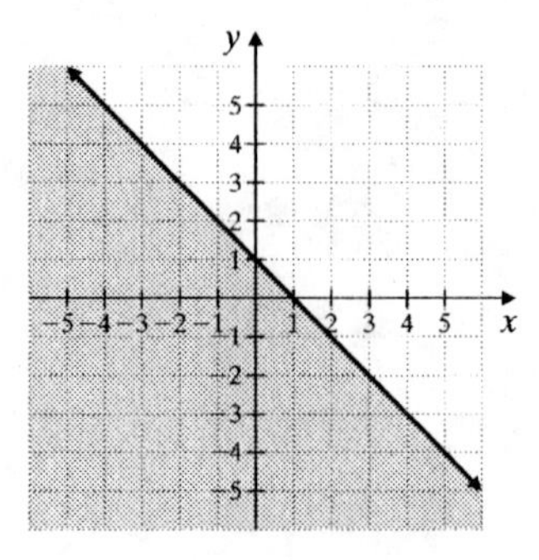

9. Graph the boundary line, $2x - y = -4$, with a dashed line.
   Test $(0, 0)$: $2x - y > -4$
   $2(0) - 0 > -4$
   $0 - 0 > -4$
   $0 > -4$ True
   Shade the half-plane containing $(0, 0)$.

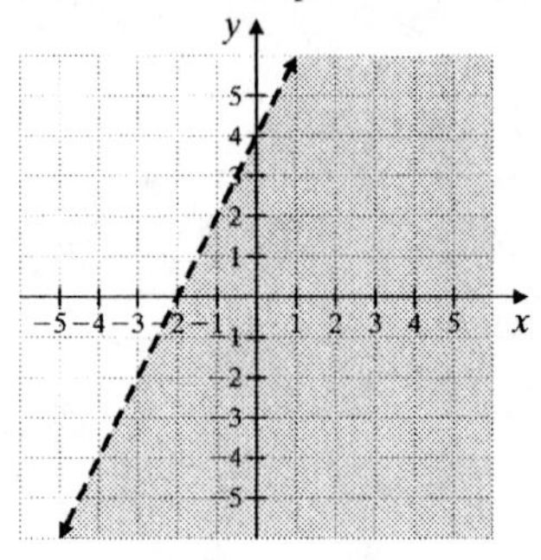

11. Graph the boundary line, $y = 2x$, with a solid line.
   Test $(1, 1)$: $y \ge 2x$
   $1 \ge 2(1)$
   $1 \ge 2$ False
   Shade the half-plane not containing $(1, 1)$.

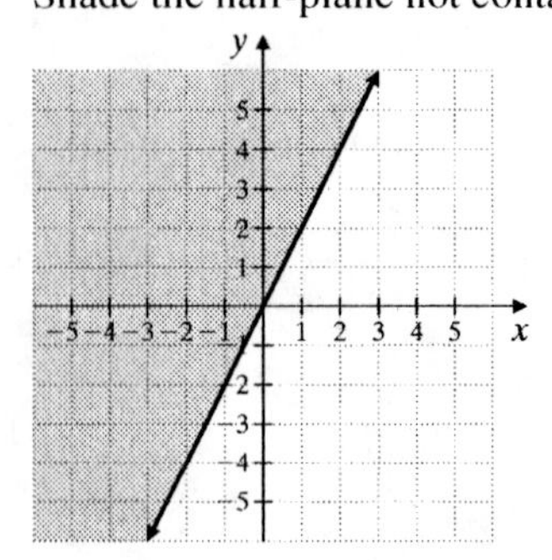

13. Graph the boundary line, $x = -3y$, with a dashed line.
Test (1, 1): $x < -3y$
$1 < -3(1)$
$1 < -3$ False
Shade the half-plane not containing (1, 1).

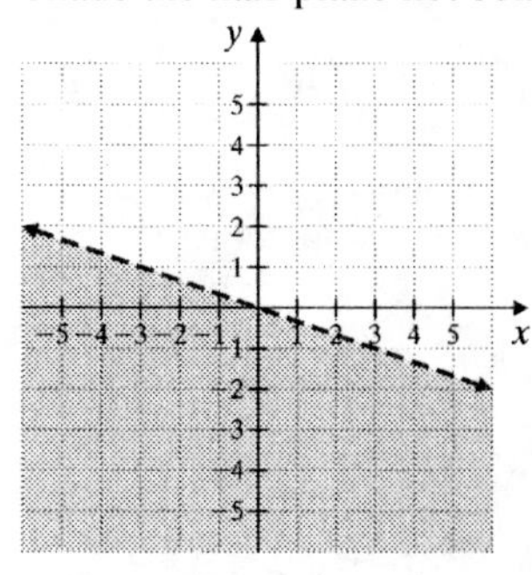

15. Graph the boundary line, $y = x + 5$, with a solid line.
Test (0, 0): $y \geq x + 5$
$0 \geq 0 + 5$
$0 \geq 5$ False
Shade the half-plane not containing (0, 0).

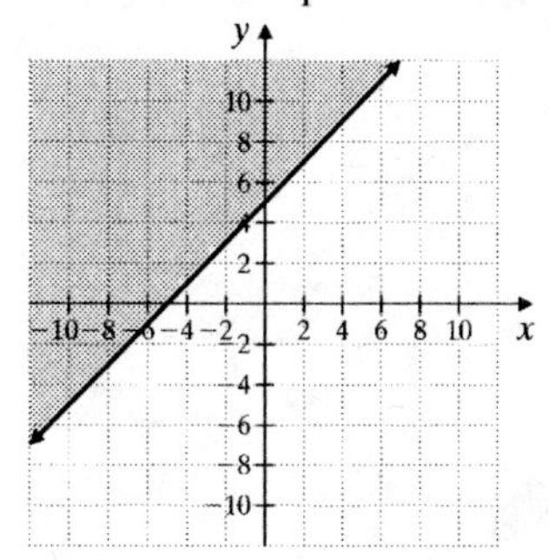

17. Graph the boundary line, $y = 4$, with a dashed line.
Test (0, 0): $y < 4$
$0 < 4$ True
Shade the half-plane containing (0, 0).

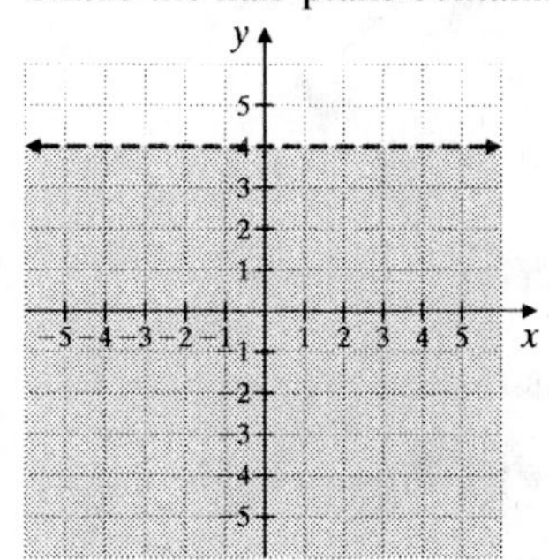

19. Graph the boundary line, $x = -3$, with a solid line.
Test (0, 0): $x \geq -3$
$0 \geq -3$ True
Shade the half-plane containing (0, 0).

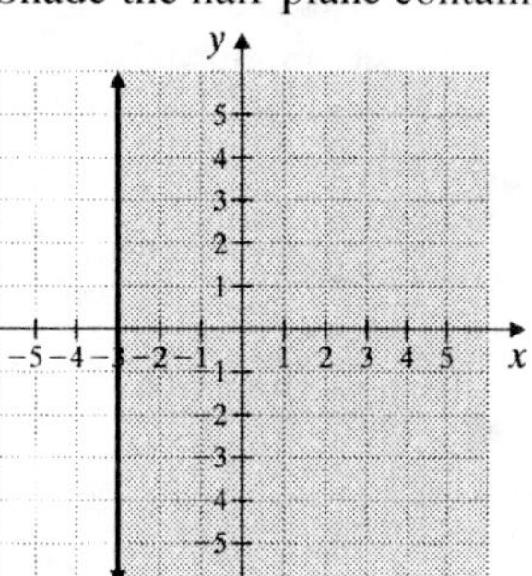

21. Graph the boundary line, $5x + 2y = 10$, with a solid line.
Test (0, 0): $5x + 2y \leq 10$
$5(0) + 2(0) \leq 10$
$0 + 0 \leq 10$
$0 \leq 10$ True
Shade the half-plane containing (0, 0).

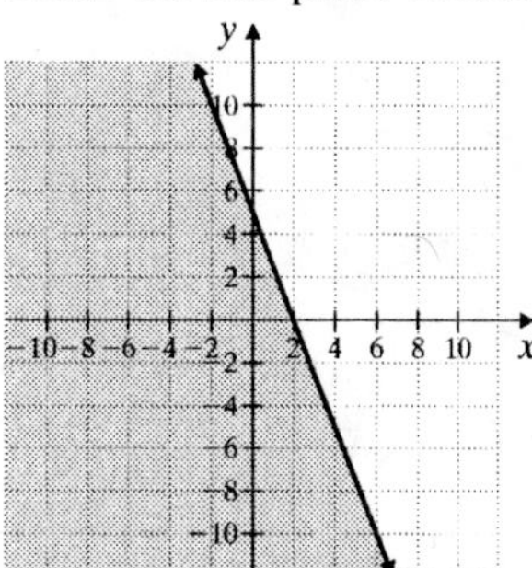

23. Graph the boundary line, $x = y$, with a dashed line.
Test (1, 4): $x > y$
$1 > 4$ False
Shade the half-plane not containing (1, 4).

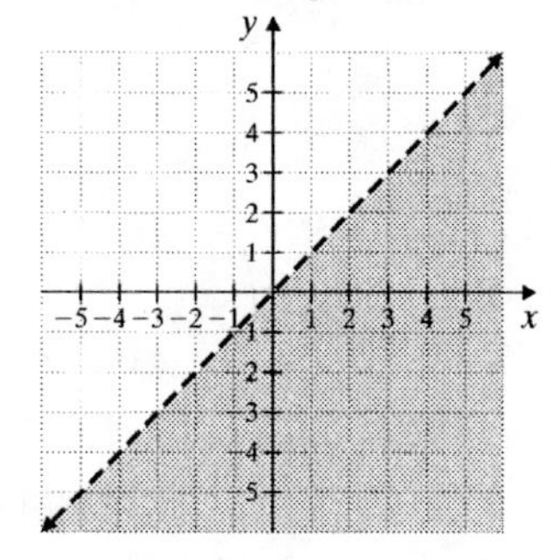

25. Graph the boundary line, $x - y = 6$, with a solid line.
Test (0, 0): $x - y \le 6$
$0 - 0 \le 6$
$0 \le 6$ True
Shade the half-plane containing (0, 0).

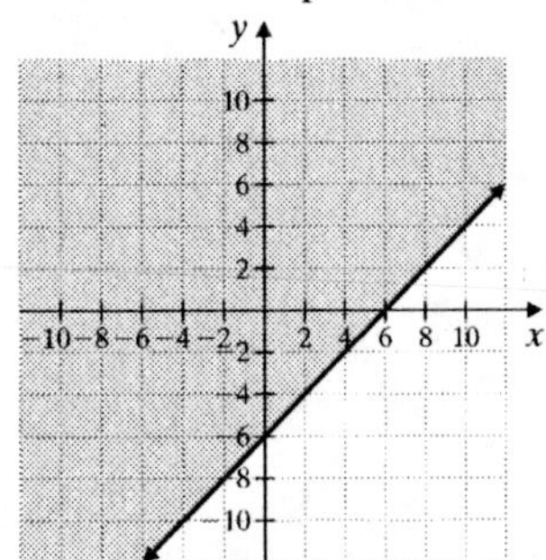

27. Graph the boundary line, $x = 0$, with a solid line.
Test (1, 1): $x \ge 0$
$1 \ge 0$ True
Shade the half-plane containing (1, 1).

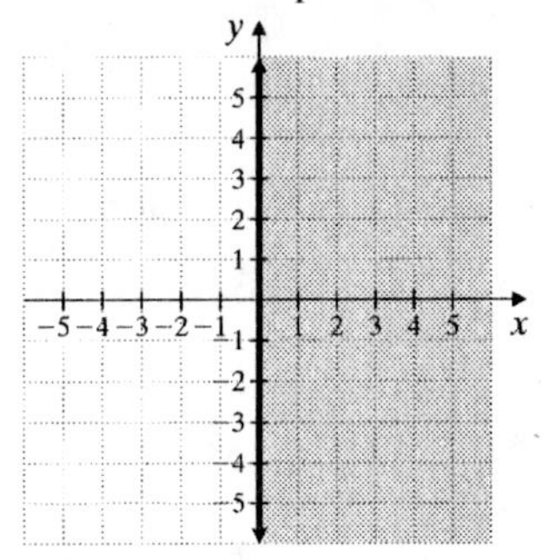

29. Graph the boundary line, $2x + 7y = 5$, with a dashed line.
Test (0, 0): $2x + 7y > 5$
$2(0) + 7(0) > 5$
$0 + 0 > 5$
$0 > 5$ False
Shade the half-plane not containing (0, 0).

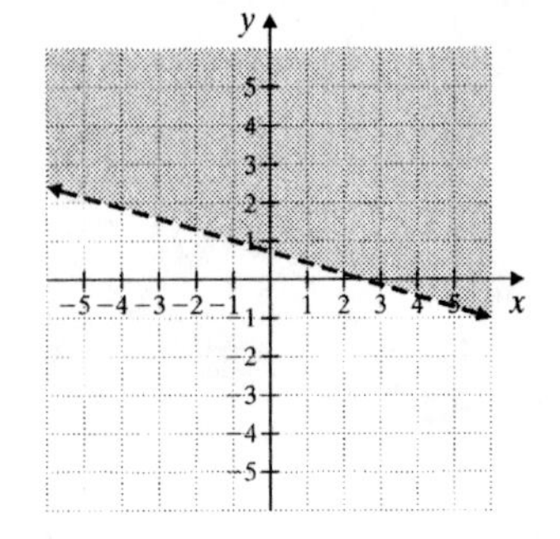

31. Graph the boundary line, $y = \frac{1}{2}x - 4$, with a solid line.
Test (0, 0): $y \ge \frac{1}{2}x - 4$
$0 \ge \frac{1}{2}(0) - 4$
$0 \ge 0 - 4$
$0 \ge -4$ True
Shade the half-plane containing (0, 0).

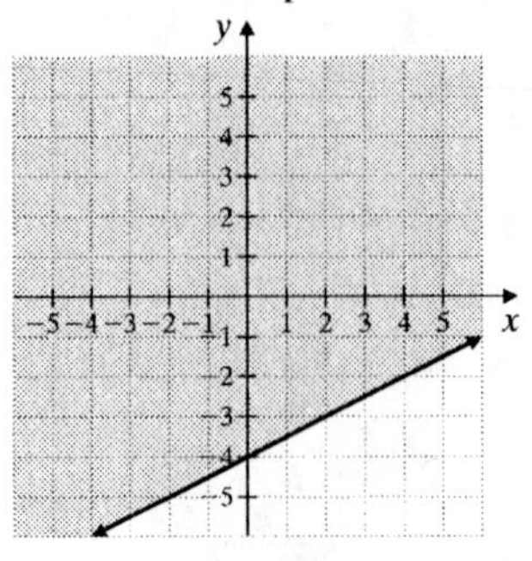

33. Graph the boundary line, $y = -\frac{3}{4}x + 2$, with a dashed line.
Test (0, 0): $y < -\frac{3}{4}x + 2$
$0 < -\frac{3}{4}(0) + 2$
$0 < 0 + 2$
$0 < 2$ True
Shade the half-plane containing (0, 0).

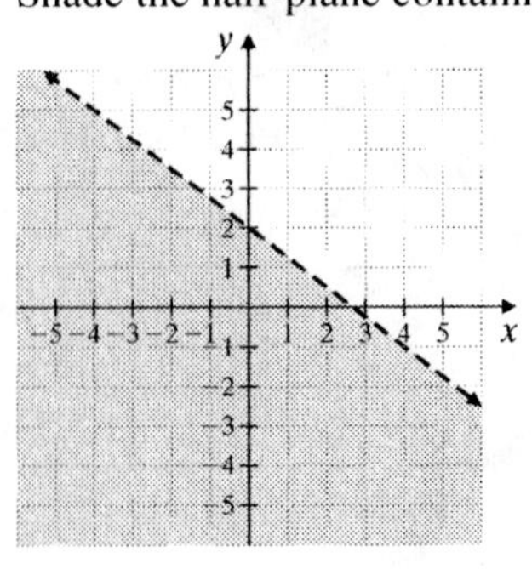

35. The point of intersection appears to be $(-2, 1)$.

37. The point of intersection appears to be $(-3, -1)$.

39. The graph is the half-plane with the dashed boundary line $x = 2$. The choice is **a**.

41. The graph is the half-plane with the dashed boundary line $y = 2$. The choice is **b**.

43. answers may vary

45. Test (1, 1): $3x+4y<8$

$$3(1)+4(1)<8$$
$$3+4<8$$
$$7<8 \quad \text{True}$$

(1, 1) is included in the graph of $3x + 4y < 8$.

47. Test (1, 1): $y \ge -\frac{1}{2}x$

$$1 \ge -\frac{1}{2}(1)$$
$$1 \ge -\frac{1}{2} \quad \text{True}$$

(1, 1) is included in the graph of $y \ge -\frac{1}{2}x$.

49. a. The sum of the number of days, x, times \$30, and the number of miles, y, times \$0.15, must be at most \$500.
$30x + 0.15y \le 500$

b.

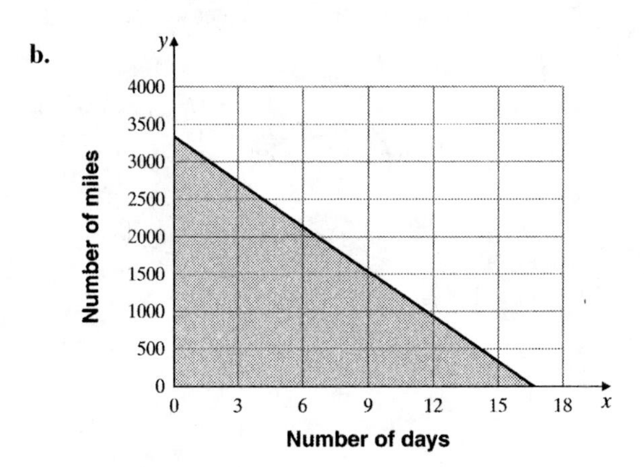

c. answers may vary

Section 6.8 Practice Exercises

1. Use (4, 8).

$$y = kx$$
$$8 = k \cdot 4$$
$$\frac{8}{4} = \frac{k \cdot 4}{4}$$
$$2 = k$$

Since $k = 2$, the equation is $y = 2x$.

2. Let $y = 15$ and $x = 45$.

$$y = kx$$
$$15 = k(45)$$
$$\frac{15}{45} = \frac{k(45)}{45}$$
$$\frac{1}{3} = k$$

The equation is $y = \frac{1}{3}x.$

Let $x = 3$.

$y = \frac{1}{3}x$

$y = \frac{1}{3} \cdot 3$

$y = 1$

Thus, when x is 3, y is 1.

3. Use (−1, −2) and (0, 0).

$\text{slope} = \frac{0-(-2)}{0-(-1)} = \frac{2}{1} = 2$

Thus, $k = 2$ and the variation equation is $y = 2x$.

4. Use (4, 5).

$y = \frac{k}{x}$

$5 = \frac{k}{4}$

$4 \cdot 5 = 4 \cdot \frac{k}{4}$

$20 = k$

Since $k = 20$, the equation is $y = \frac{20}{x}.$

5. Let $y = 4$ and $x = 0.8$.

$y = \frac{k}{x}$

$4 = \frac{k}{0.8}$

$0.8(4) = 0.8\left(\frac{k}{0.8}\right)$

$3.2 = k$

The equation is $y = \frac{3.2}{x}.$

Let $x = 20$.

$y = \frac{3.2}{20}$

$y = 0.16$

Thus, when x is 20, y is 0.16.

6. $A = kr^2$

$49\pi = k(7)^2$

$49\pi = 49k$

$\pi = k$

The formula for the area of a circle is $A = \pi r^2.$

Let $r = 4$.

$A = \pi r^2$

$A = \pi \cdot 4^2$

$A = 16\pi$

The area is 16π square feet.

7. $d = kt^2$

$144 = k(3)^2$

$144 = 9k$

$16 = k$

The equation is $d = 16t^2.$

Let $t = 5$.

$d = 16t^2$

$d = 16 \cdot 5^2$

$d = 16 \cdot 25$

$d = 400$

The object will fall 400 feet in 5 seconds.

Vocabulary, Readiness & Video Check 6.8

1. $y = \frac{k}{x}$, where k is a constant. inverse

2. $y = kx$, where k is a constant. direct

3. $y = 5x$ direct

4. $y = \frac{5}{x}$ inverse

5. $y = \frac{7}{x^2}$ inverse

6. $y = 6.5x^4$ direct

7. $y = \frac{11}{x}$ inverse

8. $y = 18x$ direct

9. $y = 12x^2$ direct

10. $y = \frac{20}{x^3}$ inverse

11. linear; slope

12. With inverse variation, we know that $y = \frac{k}{x}$ or $yx = k$, so we just need to multiply the values of x and y together to find k.

13. No; the direct relationship is the power of x times a constant, and the inverse relationship is the reciprocal of the power of x times a constant.

14. This is a direct variation problem, $y = kx$ (with k positive), so as either amount (y or x) increases, the other amount also increases. We're asked to find the new distance given a weight increase, so we know that our answer will also show a distance increase.

Exercise Set 6.8

1.
$$y = kx$$
$$3 = k(6)$$
$$\frac{1}{2} = k$$
$$y = \frac{1}{2}x$$

3.
$$y = kx$$
$$12 = k(2)$$
$$6 = k$$
$$y = 6x$$

5. $m = \frac{y_2 - y_1}{x_2 - x_1} = \frac{3-0}{1-0} = \frac{3}{1} = 3$

$y = 3x$

7. $m = \frac{y_2 - y_1}{x_2 - x_1} = \frac{2-0}{3-0} = \frac{2}{3}$

$y = \frac{2}{3}x$

9.
$$y = \frac{k}{x}$$
$$7 = \frac{k}{1}$$
$$7 = k$$
$$y = \frac{7}{x}$$

11.
$$y = \frac{k}{x}$$
$$0.05 = \frac{k}{10}$$
$$0.5 = k$$
$$y = \frac{0.5}{x}$$

13. y varies directly as x is written as $y = kx$.

15. h varies inversely as t is written as $h = \frac{k}{t}$.

17. z varies directly as x^2 is written as $z = kx^2$.

19. y varies inversely as z^3 is written as $y = \frac{k}{z^3}$.

21. x varies inversely as $\sqrt{y}$ is written as $x = \frac{k}{\sqrt{y}}$.

23. $y = kx$

$y = 20$ when $x = 5$: $20 = k(5)$

$4 = k$

$y = 4x$

$x = 10$: $y = 4(10) = 40$

$y = 40$ when $x = 10$.

25. $y = \frac{k}{x}$

$y = 5$ when $x = 60$: $5 = \frac{k}{60}$

$300 = k$

$y = \frac{300}{x}$

$x = 100$: $y = \frac{300}{100} = 3$

$y = 3$ when $x = 100$.

27. $z = kx^2$

$z = 96$ when $x = 4$: $96 = k(4)^2$

$96 = 16k$

$6 = k$

$z = 6x^2$

$x = 3$: $z = 6(3)^2 = 6(9) = 54$

$z = 54$ when $x = 3$.

29. $a = \frac{k}{b^3}$

$a = \frac{3}{2}$ when $b = 2$: $\frac{3}{2} = \frac{k}{2^3}$

$\frac{3}{2} = \frac{k}{8}$

$12 = k$

$a = \frac{12}{b^3}$

$b = 3$: $a = \frac{12}{3^3} = \frac{12}{27} = \frac{4}{9}$

$a = \frac{4}{9}$ when $b = 3$.

31. Let p be the paycheck amount when h hours are worked.

$p = kh$

$p = 166.50$ when $h = 18$: $166.50 = k(18)$

$9.25 = k$

$p = 9.25h$

$h = 10$: $p = 9.25(10) = 92.50$

The pay is $92.50 for 10 hours.

33. Let c be the cost per headphone when h headphones are manufactured.

$c = \frac{k}{h}$

$c = 9$ when $h = 5000$: $9 = \frac{k}{5000}$

$45,000 = k$

$c = \frac{45,000}{h}$

$h = 7500$: $c = \frac{45,000}{7500} = 6$

The cost to manufacture 7500 headphones is $6 per headphone.

35. Let d be the distance when a weight of w is attached.

$d = kw$

$d = 4$ when $w = 60$: $4 = k(60)$

$\frac{1}{15} = k$

$d = \frac{1}{15}w$

$w = 80$: $d = \frac{1}{15}(80) = 5\frac{1}{3}$

The spring stretches $5\frac{1}{3}$ inches when 80 pounds is attached to the spring.

37. Let w be the weight of an object when it is d miles from the center of the Earth.

$w = \frac{k}{d^2}$

$w = 180$ when $d = 4000$:

$180 = \frac{k}{4000^2}$

$180 = \frac{k}{16,000,000}$

$2,880,000,000 = k$

$w = \frac{2,880,000,000}{d^2}$

$d = 4010$: $w = \frac{2,880,000,000}{4010^2}$

$= \frac{2,880,000,000}{16,080,100}$

≈ 179.1

The man will weigh about 179.1 pounds when he is 10 miles above the surface of the Earth.

39. $d = kt^2$

$d = 64$ when $t = 2$: $64 = k(2)^2$

$64 = 4k$

$16 = k$

$d = 16t^2$

$t = 10$: $d = 16(10)^2 = 16(100) = 1600$

He will fall 1600 feet in 10 seconds.

41.
$$\begin{array}{r} -3x + 4y = 7 \\ 3x - 2y = 9 \\ \hline 2y = 16 \end{array}$$

43.
$$\begin{array}{r} 5x - 0.4y = 0.7 \\ -9x + 0.4y = -0.2 \\ \hline -4x \qquad = 0.5 \end{array}$$

45. If y varies directly as x, then $y = kx$. If x is tripled, to become $3x$, then $y = k(3x) = 3(kx)$, and y is multiplied by 3.

47. If p varies directly with the square root of l, then $p = k\sqrt{l}$. If l is quadrupled, to become $4l$, then $k\sqrt{4l} = 2\left(k\sqrt{l}\right)$, and p is doubled.

Chapter 6 Vocabulary Check

1. An ordered pair is a <u>solution</u> of an equation in two variables if replacing the variables by the coordinates of the ordered pair results in a true statement.

2. The vertical number line in the rectangular coordinate system is called the <u>y-axis</u>.

3. A <u>linear</u> equation can be written in the form $Ax + By = C$.

4. A(n) <u>x-intercept</u> is a point of the graph where the graph crosses the x-axis.

5. The form $Ax + By = C$ is called <u>standard</u> form.

6. A(n) <u>y-intercept</u> is a point of the graph where the graph crosses the y-axis.

7. A set of ordered pairs that assigns to each x-value exactly one y-value is called a <u>function</u>.

8. The equation $y = 7x - 5$ is written in <u>slope-intercept</u> form.

9. The set of all x-coordinates of a relation is called the <u>domain</u> of the relation.

10. The set of all y-coordinates of a relation is called the <u>range</u> of the relation.

11. The set of ordered pairs is called a <u>relation</u>.

12. The equation $y + 1 = 7(x - 2)$ is written in <u>point-slope</u> form.

13. To find an x-intercept of a graph, let <u>y</u> = 0.

14. The horizontal number line in the rectangular coordinate system is called the <u>x-axis</u>.

15. To find a y-intercept of a graph, let <u>x</u> = 0.

16. The <u>slope</u> of a line measures the steepness or tilt of a line.

17. The equation $y = kx$ is an example of <u>direct</u> variation.

18. The equation $y = \frac{k}{x}$ is an example of <u>inverse</u> variation.

Chapter 6 Review

1–6.

(−6, 4), $\left(0, 4\frac{4}{5}\right)$, (−7, 0), (0.7, 0.7), (1, −3), (−2, −5)

7. $-2 + y = 6x$

In (7,), the x-coordinate is 7.

$$-2 + y = 6(7)$$
$$-2 + y = 42$$
$$y = 44$$

The ordered pair solution is (7, 44).

8. $y = 3x + 5$

In (, −8), the y-coordinate is −8.

$$-8 = 3x + 5$$
$$-13 = 3x$$
$$-\frac{13}{3} = x$$

The ordered pair solution is $\left(-\frac{13}{3}, -8\right)$.

9. $9 = -3x + 4y$

$y = 0$: $9 = -3x + 4(0)$
$$9 = -3x + 0$$
$$9 = -3x$$
$$-3 = x$$

$y = 3$: $9 = -3x + 4(3)$
$$9 = -3x + 12$$
$$-3 = -3x$$
$$1 = x$$

$x = 9$: $9 = -3(9) + 4y$
$$9 = -27 + 4y$$
$$36 = 4y$$
$$9 = y$$

x	y
−3	0
1	3
9	9

10. $y = 5$ for each value of x.

x	y
7	5
−7	5
0	5

11. $x = 2y$

$y = 0$: $x = 2(0)$
$x = 0$

$y = 5$: $x = 2(5)$
$x = 10$

$y = -5$: $x = 2(-5)$
$x = -10$

x	y
0	0
10	5
−10	−5

12. a. $y = 5x + 2000$

$x = 1$: $y = 5(1) + 2000 = 5 + 2000 = 2005$

$x = 100$: $y = 5(100) + 2000$
$= 500 + 2000$
$= 2500$

$x = 1000$: $y = 5(1000) + 2000$
$= 5000 + 2000$
$= 7000$

x	1	100	1000
y	2005	2500	7000

b. Let $y = 6430$ and solve for x.

$6430 = 5x + 2000$
$4430 = 5x$
$886 = x$

886 compact disc holders can be produced for $6430.

13. $x - y = 1$

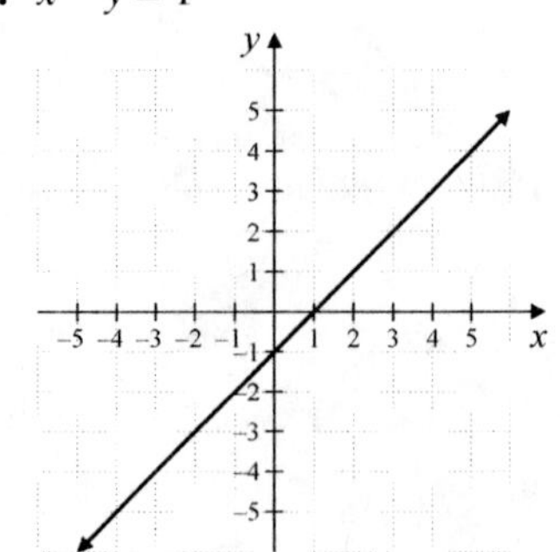

14. $x + y = 6$

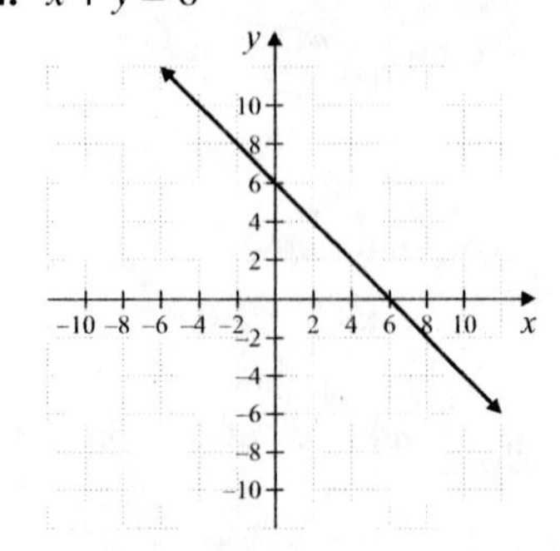

15. $x - 3y = 12$

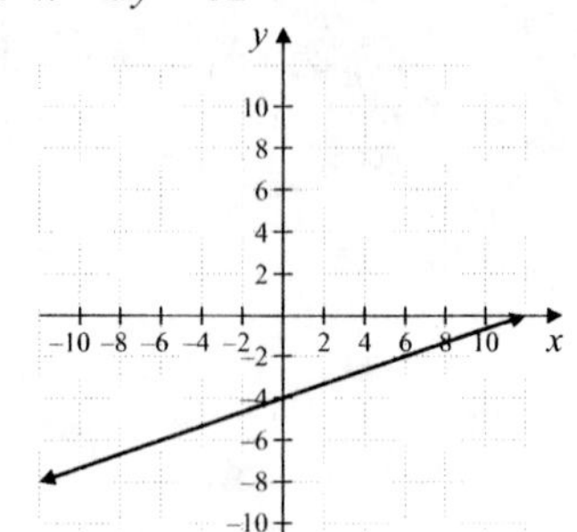

16. $5x - y = -8$

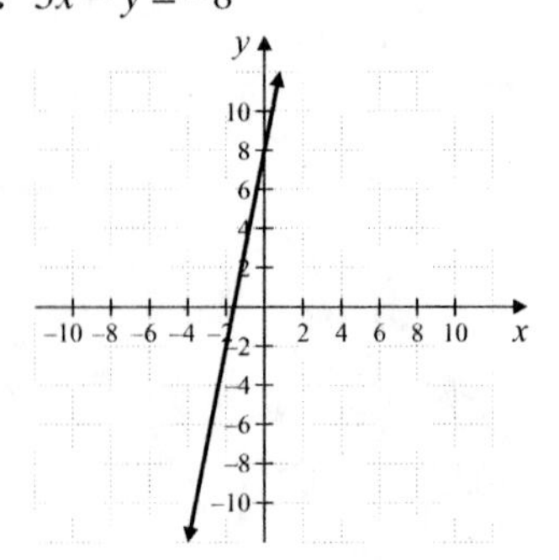

17. $x = 3y$

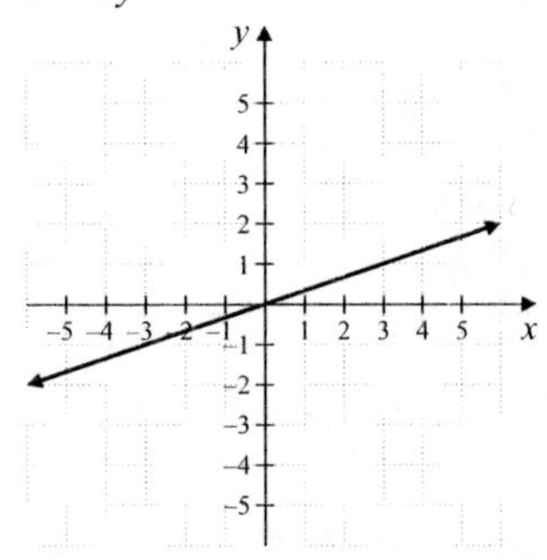

18. $y = -2x$

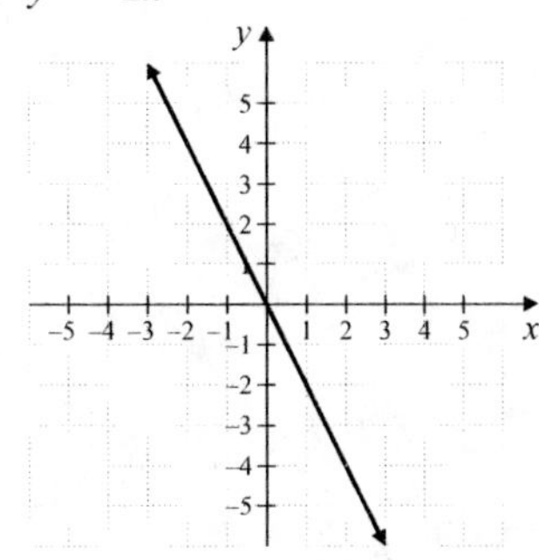

19. The x-intercept is (4, 0).
The y-intercept is (0, −2).

20. The x-intercepts are (−2, 0) and (2, 0).
The y-intercepts are (0, 2) and (0, −2).

21. $y = -3$ is a horizontal line with y-intercept (0, −3).

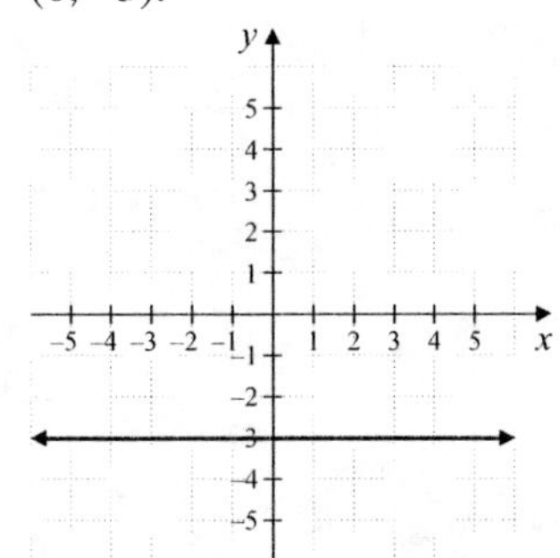

22. $x = 5$ is a vertical line with x-intercept (5, 0).

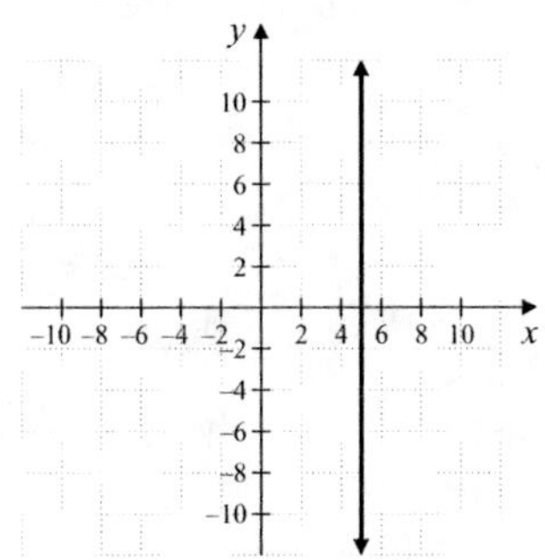

23. $x - 3y = 12$

$$\begin{aligned} y = 0:\ x - 3(0) &= 12 \\ x - 0 &= 12 \\ x &= 12 \end{aligned}$$

x-intercept: (12, 0)

$$\begin{aligned} x = 0:\ 0 - 3y &= 12 \\ -3y &= 12 \\ y &= -4 \end{aligned}$$

y-intercept: (0, −4)

24. $-4x + y = 8$

$$\begin{aligned} y = 0:\ -4x + 0 &= 8 \\ -4x &= 8 \\ x &= -2 \end{aligned}$$

x-intercept: (−2, 0)

$$\begin{aligned} x = 0:\ -4(0) + y &= 8 \\ 0 + y &= 8 \\ y &= 8 \end{aligned}$$

y-intercept: (0, 8)

25. $(x_1, y_1) = (-1, 2),\ (x_2, y_2) = (3, -1)$

$$m = \frac{y_2 - y_1}{x_2 - x_1} = \frac{-1-2}{3-(-1)} = \frac{-3}{3+1} = -\frac{3}{4}$$

26. $(x_1, y_1) = (-2, -2), (x_2, y_2) = (3, -1)$

$$m = \frac{y_2 - y_1}{x_2 - x_1} = \frac{-1-(-2)}{3-(-2)} = \frac{-1+2}{3+2} = \frac{1}{5}$$

27. When $m = 0$, the line is horizontal. The choice is **d**.

28. The slope is $m = -1$. The choice is **b**.

29. When the slope is undefined, the line is vertical. The choice is **c**.

30. The slope is $m = 4$. The choice is **a**.

31. $m = \frac{y_2 - y_1}{x_2 - x_1} = \frac{8-5}{6-2} = \frac{3}{4}$

32. $m = \frac{y_2 - y_1}{x_2 - x_1} = \frac{2-7}{1-4} = \frac{-5}{-3} = \frac{5}{3}$

33. $m = \frac{y_2 - y_1}{x_2 - x_1} = \frac{-9-3}{-2-1} = \frac{-12}{-3} = 4$

34. $m = \frac{y_2 - y_1}{x_2 - x_1} = \frac{-6-1}{3-(-4)} = \frac{-6-1}{3+4} = \frac{-7}{7} = -1$

35. $y = mx + b$
$y = 3x + 7$
The slope is $m = 3$.

36. $y = mx + b$
$x - 2y = 4$
$-2y = -x + 4$
$y = \frac{1}{2}x - 2$
The slope is $m = \frac{1}{2}$.

37. $y = mx + b$
$y = -2$
$y = 0x - 2$
The slope is $m = 0$.

38. $x = 0$ is a vertical line. The slope is undefined.

39. $x - y = -6$
$-y = -x - 6$
$y = x + 6$
$m = 1$

$x + y = 3$
$y = -x + 3$
$m = -1$

Since $(1)(-1) = -1$, the lines are perpendicular.

40. $3x + y = 7$
$y = -3x + 7$
$m = -3$

$-3x - y = 10$
$-y = 3x + 10$
$y = -3x - 10$
$m = -3$

Since the slopes are equal, the lines are parallel.

41. $y = 4x + \frac{1}{2}$
$m = 4$

$4x + 2y = 1$
$2y = -4x + 1$
$y = -2x + \frac{1}{2}$
$m = -2$

Since $4 \neq -2$ and $(4)(-2) = -8 \neq -1$, the lines are neither parallel nor perpendicular.

42. $y = 6x - \frac{1}{3}$
$m = 6$

$x + 6y = 6$
$6y = -x + 6$
$y = -\frac{1}{6}x + 1$
$m = -\frac{1}{6}$

Since $(6)\left(-\frac{1}{6}\right) = -1$, the lines are perpendicular.

43. $m = \frac{y_2 - y_1}{x_2 - x_1} = \frac{7390 - 6730}{2012 - 2006} = \frac{660}{6} = 110$
The total number of U.S. magazines in print increases by 110 magazines per year.

44. $m = \frac{y_2 - y_1}{x_2 - x_1} = \frac{1750 - 1470}{2012 - 2006} = \frac{280}{6} \approx 47$
The number of U.S. lung transplants increases by about 47 transplants per year.

45. $y = mx + b$
$x - 6y = -1$
$-6y = -x - 1$
$y = \frac{1}{6}x + \frac{1}{6}$
$m = \frac{1}{6}$; *y*-intercept $\left(0, \frac{1}{6}\right)$

46. $y = mx + b$
$3x + y = 7$
$y = -3x + 7$
$m = -3$; *y*-intercept $(0, 7)$

47. $y = mx + b$
$y = -5x + \frac{1}{2}$

48. $y = mx + b$
$y = \frac{2}{3}x + 6$

49. $y = mx + b$
$y = 2x + 1$
$m = 2$, *y*-intercept $(0, 1)$
The choice is **d**.

50. $y = mx + b$
$y = -4x$
$y = -4x + 0$
$m = -4$; *y*-intercept $(0, 0)$
The choice is **c**.

51. $y = mx + b$
$y = 2x$
$y = 2x + 0$
$m = 2$; *y*-intercept $(0, 0)$
The choice is **a**.

52. $y = mx + b$
$y = 2x - 1$
$m = 2$; *y*-intercept $(0, -1)$
The choice is **b**.

53. $y - y_1 = m(x - x_1)$
$y - 0 = 4(x - 2)$
$y = 4x - 8$
$-4x + y = -8$

54. $y - y_1 = m(x - x_1)$
$y - (-5) = -3(x - 0)$
$y + 5 = -3x$
$y = -3x - 5$
$3x + y = -5$

55. $y - y_1 = m(x - x_1)$
$y - 4 = \frac{3}{5}(x - 1)$
$5(y - 4) = 5 \cdot \frac{3}{5}(x - 1)$
$5y - 20 = 3(x - 1)$
$5y - 20 = 3x - 3$
$5y = 3x + 17$
$-3x + 5y = 17$

56. $y - y_1 = m(x - x_1)$
$y - 3 = -\frac{1}{3}[x - (-3)]$
$3y - 9 = -1[x - (-3)]$
$3y - 9 = -(x + 3)$
$3y - 9 = -x - 3$
$3y = -x + 6$
$x + 3y = 6$

57. $m = \frac{y_2 - y_1}{x_2 - x_1} = \frac{-7 - 7}{2 - 1} = \frac{-14}{1} = -14$
$y - y_1 = m(x - x_1)$
$y - 7 = -14(x - 1)$
$y - 7 = -14x + 14$
$y = -14x + 21$

58. $m = \frac{y_2 - y_1}{x_2 - x_1} = \frac{6 - 5}{-4 - (-2)} = \frac{6 - 5}{-4 + 2} = \frac{1}{-2} = -\frac{1}{2}$
$y - y_1 = m(x - x_1)$
$y - 5 = -\frac{1}{2}[x - (-2)]$
$y - 5 = -\frac{1}{2}(x + 2)$
$2(y - 5) = 2\left(-\frac{1}{2}\right)(x + 2)$
$2y - 10 = -1(x + 2)$
$2y - 10 = -x - 2$
$2y = -x + 8$
$y = -\frac{1}{2}x + 4$

59. The x-value 7 is paired with two y-values, 1 and 5, so the relation is not a function.

60. Each x-value is only assigned to one y-value, so the relation is a function.

61. No vertical line will intersect the graph more than once, so the graph is the graph of a function.

62. No vertical line will intersect the graph more than once, so the graph is the graph of a function.

63. The vertical line $x = 3$ will intersect the graph at more than one point, so the graph is not the graph of a function.

64. No vertical line will intersect the graph more than once, so the graph is the graph of a function.

65. $f(x) = -2x + 6$
$f(0) = -2(0) + 6 = 0 + 6 = 6$

66. $f(x) = -2x + 6$
$f(-2) = -2(-2) + 6 = 4 + 6 = 10$

67. $f(x) = -2x + 6$
$f\left(\frac{1}{2}\right) = -2\left(\frac{1}{2}\right) + 6 = -1 + 6 = 5$

68. $f(x) = -2x + 6$
$f\left(-\frac{1}{2}\right) = -2\left(-\frac{1}{2}\right) + 6 = 1 + 6 = 7$

69. Graph the boundary line, $x + 6y = 6$, with a dashed line.
Test (0, 0): $x + 6y < 6$
$0 + 6(0) < 6$
$0 + 0 < 6$
$0 < 6$ True
Shade the half-plane containing (0, 0).

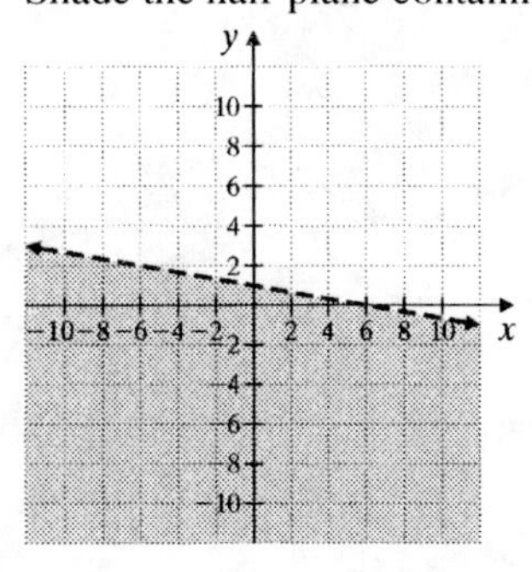

70. Graph the boundary line, $x + y = -2$, with a dashed line.
Test (0, 0): $x + y > -2$
$0 + 0 > -2$
$0 > -2$ True
Shade the half-plane containing (0, 0).

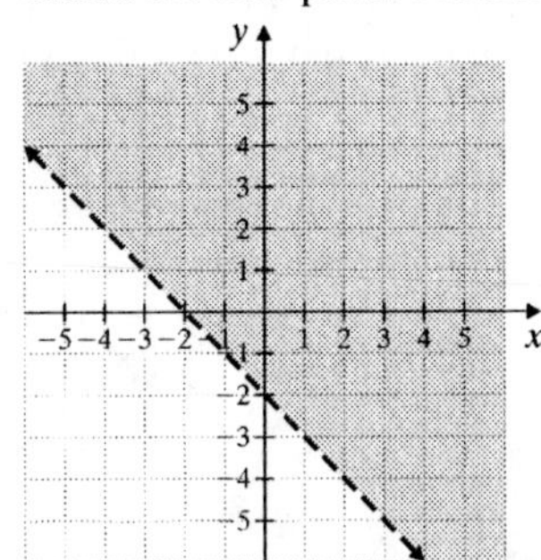

71. Graph the boundary line, $y = -7$, with a solid line.
Test (0, 0): $y \geq -7$
$0 \geq -7$ True
Shade the half-plane containing (0, 0).

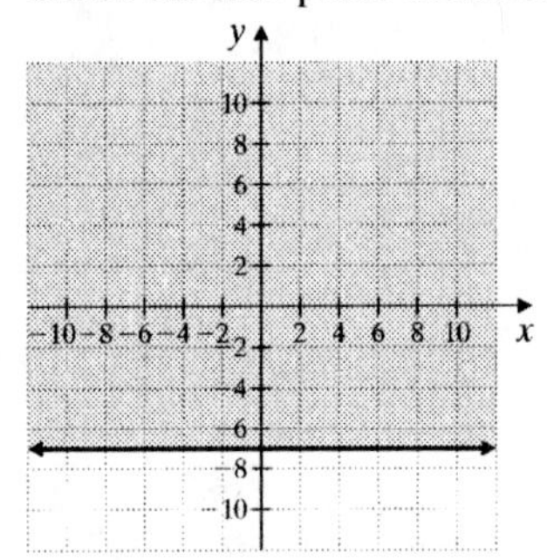

72. Graph the boundary line, $y = -4$, as a solid line.
Test (0, 0): $y \leq -4$
$0 \leq -4$ False
Shade the half-plane not containing (0, 0).

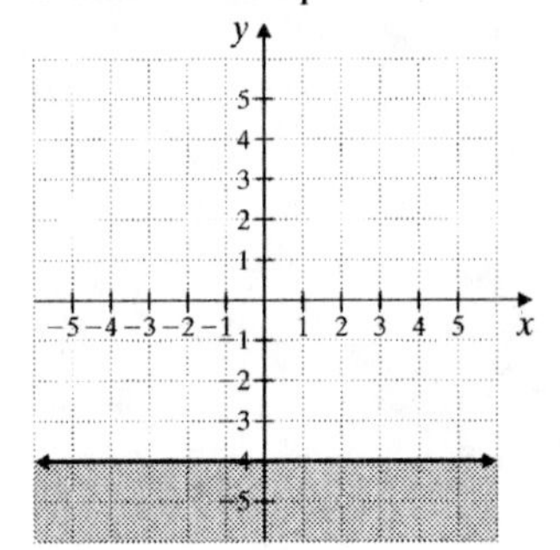

73. Graph the boundary line, $-x = y$, as a solid line.
Test (1, 1): $-x \leq y$
$-1 \leq 1$ True
Shade the half-plane containing (1, 1).

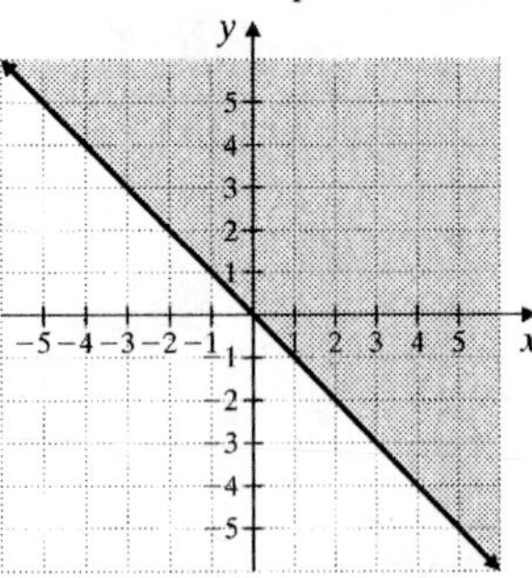

74. Graph the boundary line, $x = -y$, as a solid line.
Test (1, 1): $x \geq -y$
$1 \geq -1$ True
Shade the half-plane containing (1, 1).

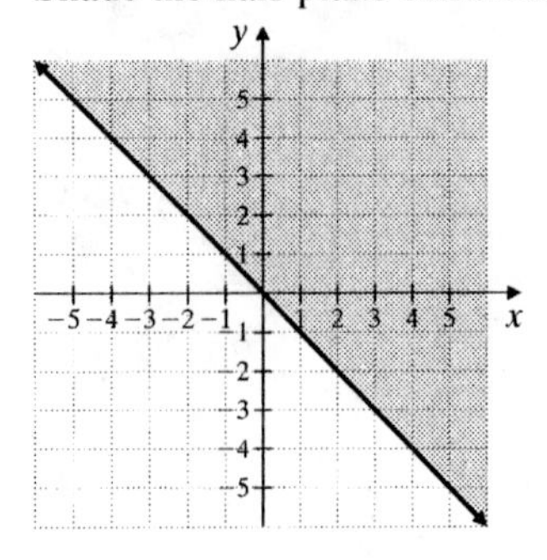

75. $y = kx$
$y = 40$ when $x = 4$: $40 = k(4)$
$10 = k$

$y = 10x$
$x = 11$: $y = 10(11) = 110$
$y = 110$ when $x = 11$.

76. $y = \frac{k}{x}$

$y = 4$ when $x = 6$: $4 = \frac{k}{6}$
$24 = k$

$y = \frac{24}{x}$

$x = 48$: $y = \frac{24}{48} = \frac{1}{2}$

$y = \frac{1}{2}$ when $x = 48$.

77. $y = \frac{k}{x^3}$

$y = 12.5$ when $x = 2$: $12.5 = \frac{k}{2^3}$

$12.5 = \frac{k}{8}$

$100 = k$

$y = \frac{100}{x^3}$

$x = 3$: $y = \frac{100}{3^3} = \frac{100}{27}$

$y = \frac{100}{27}$ when $x = 3$.

78. $y = kx^2$

$y = 175$ when $x = 5$: $175 = k(5)^2$

$175 = 25k$

$7 = k$

$y = 7x^2$

$x = 10$: $y = 7(10)^2 = 7(100) = 700$

$y = 700$ when $x = 10$.

79. Let c be the cost for manufacturing m milliliters.

$c = \frac{k}{m}$

$c = 6600$ when $m = 3000$: $6600 = \frac{k}{3000}$

$19{,}800{,}000 = k$

$c = \frac{19{,}800{,}000}{m}$

Let $m = 5000$: $c = \frac{19{,}800{,}000}{5000} = 3960$

It costs \$3960 to manufacture 5000 milliliters.

80. Let d be the distance when a weight of w is attached.

$d = kw$

$d = 8$ when $w = 150$: $8 = k(150)$

$\frac{4}{75} = k$

$d = \frac{4}{75}w$

Let $w = 90$: $d = \frac{4}{75} \cdot 90 = 4\frac{4}{5}$

The spring stretches $4\frac{4}{5}$ inches when 90 pounds is attached.

81. $2x - 5y = 9$

$y = 1$: $2x - 5(1) = 9$

$2x - 5 = 9$

$2x = 14$

$x = 7$

$x = 2$: $2(2) - 5y = 9$

$4 - 5y = 9$

$-5y = 5$

$y = -1$

$y = -3$: $2x - 5(-3) = 9$

$2x + 15 = 9$

$2x = -6$

$x = -3$

x	y
7	1
2	−1
−3	−3

82. $x = -3y$

$x = 0$: $0 = -3y$

$0 = y$

$y = 1$: $x = -3(1)$

$x = -3$

$x = 6$: $6 = -3y$

$-2 = y$

x	y
0	0
−3	1
6	−2

83. $2x - 3y = 6$

$y = 0$: $2x - 3(0) = 6$

$2x - 0 = 6$

$2x = 6$

$x = 3$

x-intercept: (3, 0)

$x = 0$: $2(0) - 3y = 6$

$0 - 3y = 6$

$-3y = 6$

$y = -2$

y-intercept: (0, −2)

84. $-5x + y = 10$

$y = 0$: $-5x + 0 = 10$

$-5x = 10$

$x = -2$

x-intercept: $(-2, 0)$

$x = 0$: $-5(0) + y = 10$

$0 + y = 10$

$y = 10$

y-intercept: $(0, 10)$

85. $x - 5y = 10$

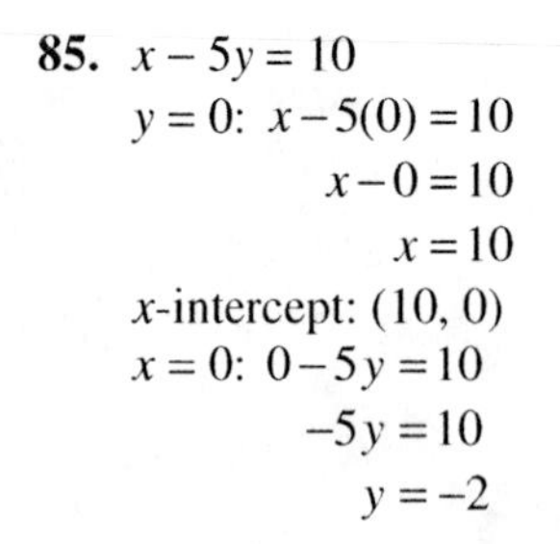

$y = 0$: $x - 5(0) = 10$

$x - 0 = 10$

$x = 10$

x-intercept: $(10, 0)$

$x = 0$: $0 - 5y = 10$

$-5y = 10$

$y = -2$

y-intercept: $(0, -2)$

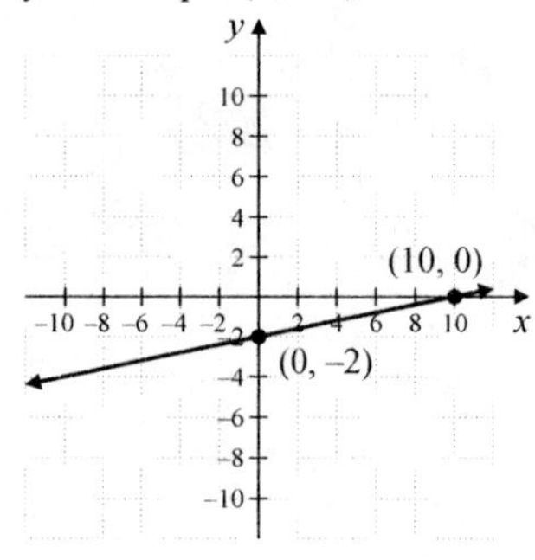

86. $x + y = 4$

$y = 0$: $x + 0 = 4$

$x = 4$

x-intercept: $(4, 0)$

$x = 0$: $0 + y = 4$

$y = 4$

y-intercept: $(0, 4)$

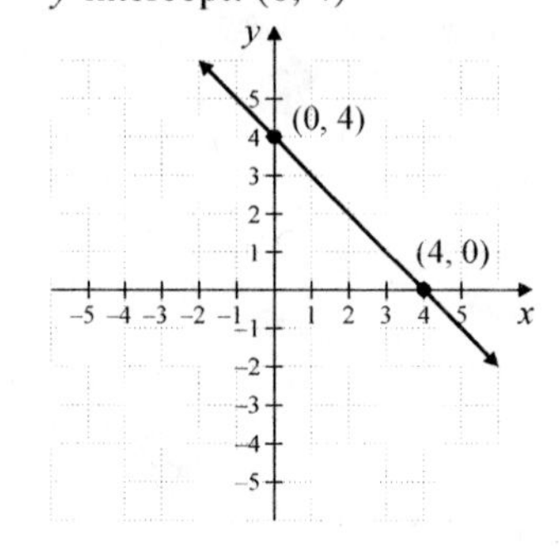

87. $y = -4x$

$y = 0$: $0 = -4x$

$0 = x$

x-intercept: $(0, 0)$

The y-intercept is also $(0, 0)$. Find another point.

Let $x = 1$.

$y = -4(1) = -4$

Another point is $(1, -4)$.

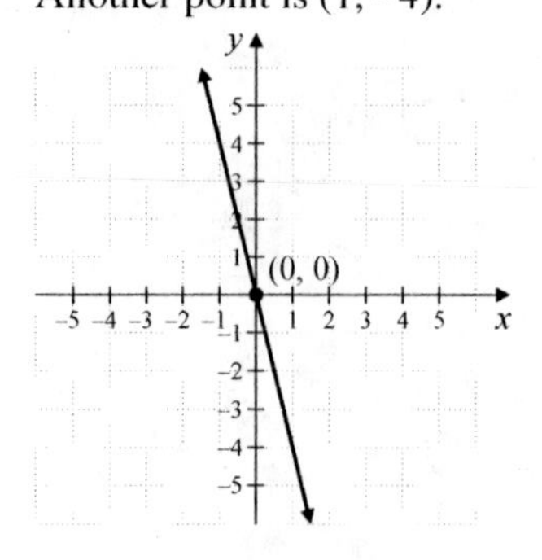

88. $2x + 3y = -6$

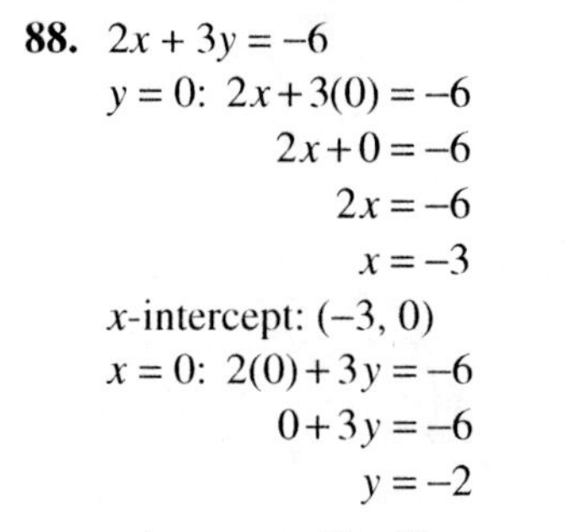

$y = 0$: $2x + 3(0) = -6$

$2x + 0 = -6$

$2x = -6$

$x = -3$

x-intercept: $(-3, 0)$

$x = 0$: $2(0) + 3y = -6$

$0 + 3y = -6$

$y = -2$

y-intercept: $(0, -2)$

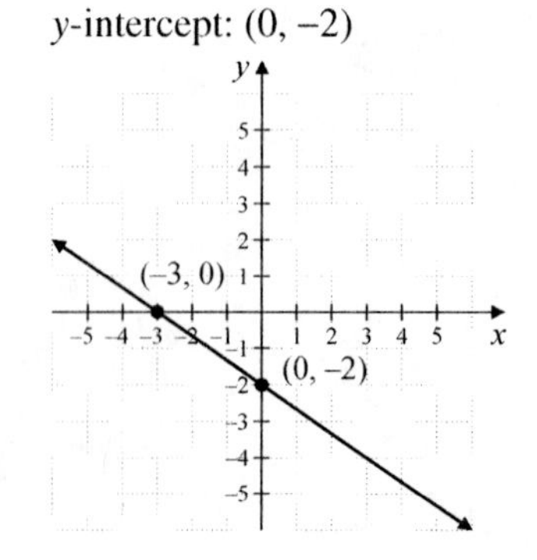

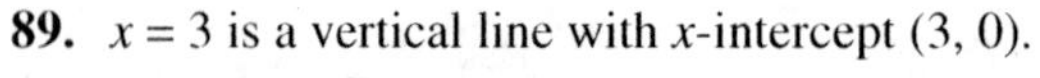

89. $x = 3$ is a vertical line with x-intercept $(3, 0)$.

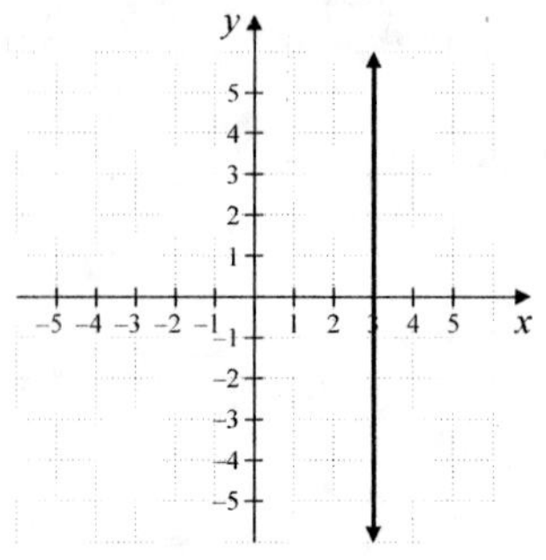

90. $y = -2$ is a horizontal line with y-intercept $(0, -2)$.

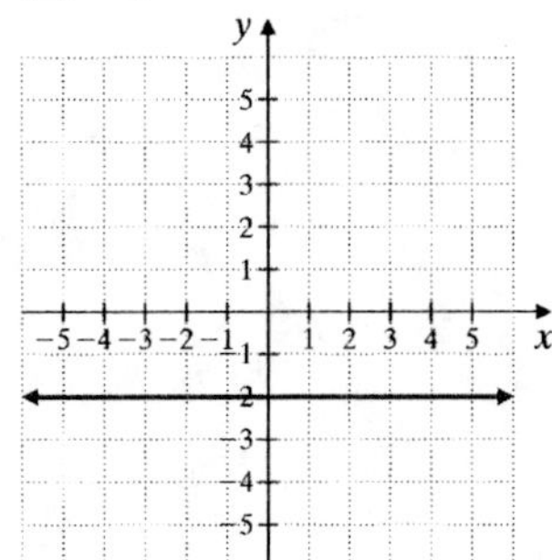

91. $m = \frac{y_2 - y_1}{x_2 - x_1} = \frac{2-(-5)}{-4-3} = \frac{2+5}{-4-3} = \frac{7}{-7} = -1$

92. $m = \frac{y_2 - y_1}{x_2 - x_1} = \frac{-8-3}{-6-1} = \frac{-11}{-7} = \frac{11}{7}$

93. $(x_1, y_1) = (0, -4), (x_2, y_2) = (2, 0)$

$m = \frac{y_2 - y_1}{x_2 - x_1} = \frac{0-(-4)}{2-0} = \frac{4}{2} = 2$

94. $(x_1, y_1) = (0, 2), (x_2, y_2) = (6, 0)$

$m = \frac{y_2 - y_1}{x_2 - x_1} = \frac{0-2}{6-0} = \frac{-2}{6} = -\frac{1}{3}$

95. $y = mx + b$

$$-2x + 3y = -15$$
$$3y = 2x - 15$$
$$y = \frac{2}{3}x - 5$$

$m = \frac{2}{3}$; y-intercept: $(0, -5)$

96. $y = mx + b$

$$6x + y - 2 = 0$$
$$6x + y = 2$$
$$y = -6x + 2$$

$m = -6$; y-intercept: $(0, 2)$

97.
$$y - y_1 = m(x - x_1)$$
$$y - (-7) = -5(x - 3)$$
$$y + 7 = -5x + 15$$
$$y = -5x + 8$$
$$5x + y = 8$$

98.
$$y - y_1 = m(x - x_1)$$
$$y - 6 = 3(x - 0)$$
$$y - 6 = 3x$$
$$-6 = 3x - y$$
$$3x - y = -6$$

99. $m = \frac{y_2 - y_1}{x_2 - x_1} = \frac{5-9}{-2-(-3)} = \frac{5-9}{-2+3} = \frac{-4}{1} = -4$

$$y - y_1 = m(x - x_1)$$
$$y - 9 = -4[x - (-3)]$$
$$y - 9 = -4(x + 3)$$
$$y - 9 = -4x - 12$$
$$y = -4x - 3$$
$$4x + y = -3$$

100. $m = \frac{y_2 - y_1}{x_2 - x_1} = \frac{-9-1}{5-3} = \frac{-10}{2} = -5$

$$y - y_1 = m(x - x_1)$$
$$y - 1 = -5(x - 3)$$
$$y - 1 = -5x + 15$$
$$y = -5x + 16$$
$$5x + y = 16$$

101. The tallest bar in the graph corresponds to France, so France has the most tourist arrivals.

102. The shortest bar on the graph corresponds to Malaysia, so Malaysia has the fewest tourist arrivals.

103. The bars for France, U.S., Spain, and China extend above 50, so these four countries have more than 50 million tourist arrivals each year.

104. The bars on the graph that do not extend above 30 are U.K., Malaysia, and Russian Federation, so those 3 countries have fewer than 30 million tourist arrivals per year.

105. The height of the bar for Germany appears to be 30, so Germany has 30 million tourist arrivals per year.

106. The height of the bar for Malaysia appears to be 25, so Malaysia has 25 million tourist arrivals per year.

Chapter 6 Test

1. $12y - 7x = 5$
$x = 1$: $12y - 7(1) = 5$
$12y - 7 = 5$
$12y = 12$
$y = 1$
The ordered pair is (1, 1).

2. $y = 17$ for each value of x. The ordered pair is (−4, 17).

3. $(x_1, y_1) = (-1, -1), (x_2, y_2) = (4, 1)$
$$m = \frac{y_2 - y_1}{x_2 - x_1} = \frac{1-(-1)}{4-(-1)} = \frac{1+1}{4+1} = \frac{2}{5}$$

4. The slope of a horizontal line is $m = 0$.

5. $$m = \frac{y_2 - y_1}{x_2 - x_1} = \frac{2-(-5)}{-1-6} = \frac{7}{-7} = -1$$

6. $$m = \frac{y_2 - y_1}{x_2 - x_1} = \frac{-1-(-8)}{-1-0} = \frac{-1+8}{-1} = \frac{7}{-1} = -7$$

7. $y = mx + b$
$-3x + y = 5$
$y = 3x + 5$
$m = 3$

8. $x = 6$ is a vertical line. The slope of a vertical line is undefined.

9. $2x + y = 8$
$y = 0$: $2x + 0 = 8$
$2x = 8$
$x = 4$
x-intercept: (4, 0)
$x = 0$: $2(0) + y = 8$
$y = 8$
y-intercept: (0, 8)

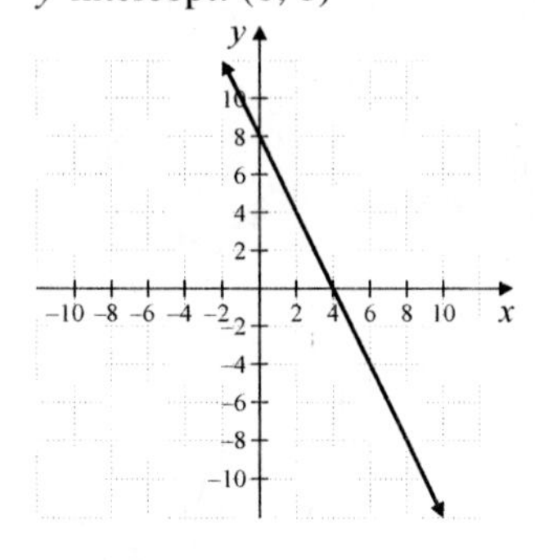

10. $-x + 4y = 5$
$y = 0$: $-x + 4(0) = 5$
$-x + 0 = 5$
$-x = 5$
$x = -5$
x-intercept: (−5, 0)
$x = 0$: $-0 + 4y = 5$
$4y = 5$
$y = \frac{5}{4}$
y-intercept: $\left(0, \frac{5}{4}\right)$

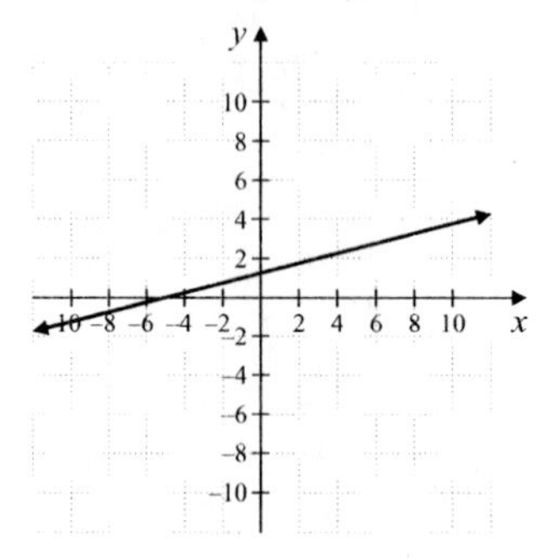

11. Graph the boundary line, $x - y = -2$, with a solid line.
Test (0, 0): $x - y \geq -2$
$0 - 0 \geq -2$
$0 \geq -2$ True
Shade the half-plane containing (0, 0).

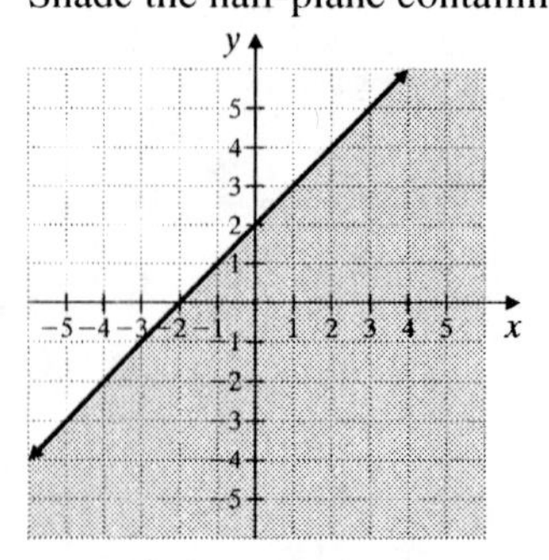

12. Graph the boundary line, $y = -4x$, with a solid line.
Test (1, 1): $y \geq -4x$
$1 \geq -4(1)$
$1 \geq -4$ True
Shade the half-plane containing (1, 1).

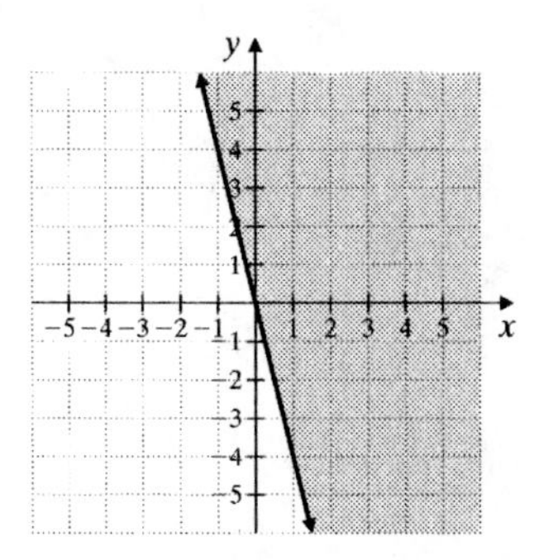

13. $5x - 7y = 10$

$y = 0$: $5x - 7(0) = 10$
$5x - 0 = 10$
$5x = 10$
$x = 2$

x-intercept: (2, 0)

$x = 0$: $5(0) - 7y = 10$
$0 - 7y = 10$
$-7y = 10$
$y = -\frac{10}{7}$

y-intercept: $\left(0, -\frac{10}{7}\right)$

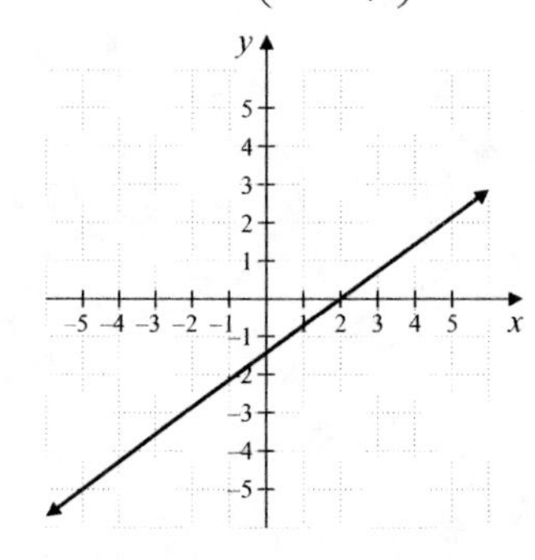

14. Graph the boundary line, $2x - 3y = -6$, with a dashed line.

Test (0, 0): $2x - 3y > -6$
$2(0) - 3(0) > -6$
$0 - 0 > -6$
$0 > -6$ True

Shade the half-plane containing (0, 0).

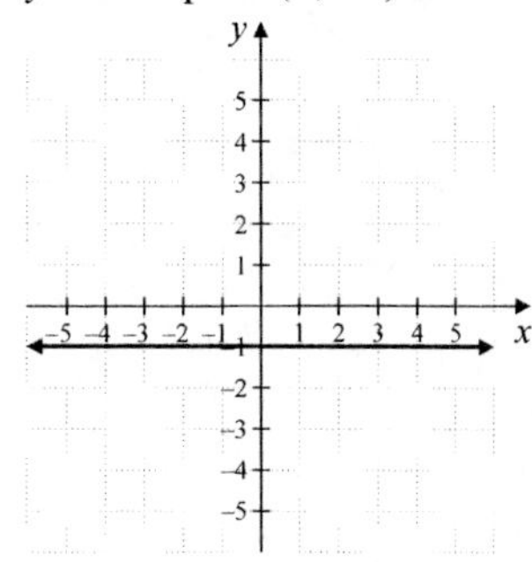

15. Graph the boundary line, $6x + y = -1$, with a dashed line.

Test (0, 0): $6x + y > -1$
$6(0) + 0 > -1$
$0 + 0 > -1$
$0 > -1$ True

Shade the half-plane containing (0, 0).

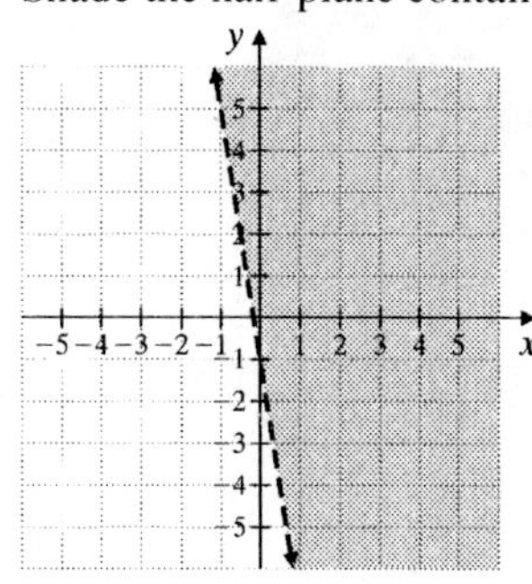

16. The graph of $y = -1$ is a horizontal line with a y-intercept of (0, −1).

17. $y = 2x - 6$

The slope is $m = 2$ and the y-intercept is (0, −6).

$-4x = 2y$
$-2x = y$
$y = -2x + 0$

The slope is $m = -2$ and the y-intercept is (0, 0). Since the slopes are different, the lines are not parallel. Since $2(-2) = -4 \neq -1$, the lines are not perpendicular. The lines are neither parallel nor perpendicular.

18.

$$y - y_1 = m(x - x_1)$$
$$y - 2 = -\frac{1}{4}(x - 2)$$
$$4(y - 2) = 4\left(-\frac{1}{4}\right)(x - 2)$$
$$4y - 8 = -1(x - 2)$$
$$4y - 8 = -x + 2$$
$$4y = -x + 10$$
$$x + 4y = 10$$

19. $m = \frac{y_2 - y_1}{x_2 - x_1} = \frac{-7-0}{6-0} = \frac{-7}{6} = -\frac{7}{6}$

$y - y_1 = m(x - x_1)$
$y - 0 = -\frac{7}{6}(x - 0)$
$y = -\frac{7}{6}x$
$6y = 6\left(-\frac{7}{6}x\right)$
$6y = -7x$
$7x + 6y = 0$

20. $m = \frac{y_2 - y_1}{x_2 - x_1} = \frac{3-(-5)}{1-2} = \frac{3+5}{-1} = -8$

$y - y_1 = m(x - x_1)$
$y - (-5) = -8(x - 2)$
$y + 5 = -8x + 16$
$y = -8x + 11$
$8x + y = 11$

21. $m = \frac{1}{8}$; $b = 12$

$y = \frac{1}{8}x + 12$
$8y = x + 8(12)$
$8y = x + 96$
$x - 8y = -96$

22. Each x-value is only assigned to one y-value, so the relation is a function.

23. The x-value -3 is assigned to two y-values, -3 and 2, so the relation is not a function. Note that the x-value 0 is also assigned to two y-values, 5 and 0.

24. No vertical line will intersect the graph more than once, so the graph is the graph of a function.

25. No vertical line will intersect the graph more than once, so the graph is the graph of a function.

26. $f(x) = 2x - 4$

a. $f(-2) = 2(-2) - 4 = -4 - 4 = -8$

b. $f(0.2) = 2(0.2) - 4 = 0.4 - 4 = -3.6$

c. $f(0) = 2(0) - 4 = 0 - 4 = -4$

27. $f(x) = x^3 - x$

a. $f(-1) = (-1)^3 - (-1) = -1 + 1 = 0$

b. $f(0) = 0^3 - 0 = 0 - 0 = 0$

c. $f(4) = 4^3 - 4 = 64 - 4 = 60$

28. $2x + 2(2y) = 42$
$2(x + 2y) = 2(21)$
$x + 2y = 21$

Let $y = 8$: $x + 2(8) = 21$
$x + 16 = 21$
$x = 5$

When $y = 8$ meters, $x = 5$ meters.

29. a. The ordered pairs are (2008, 19.6), (2009, 22.2), (2010, 23.9), (2011, 25.3), (2012, 26.7).

b.

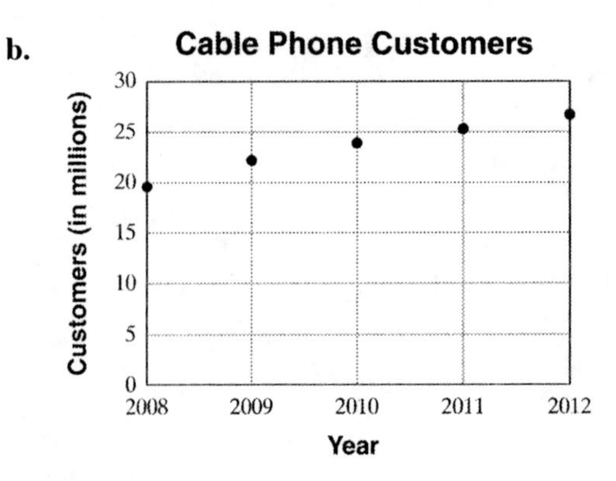

30. $m = \frac{y_2 - y_1}{x_2 - x_1} = \frac{1360 - 1484}{2012 - 2004} = \frac{-124}{8} = -15.5$

Every 1 year, 15.5 million fewer movie tickets are sold.

31. $y = kx$

$y = 10$ when $x = 15$: $10 = k(15)$
$\frac{2}{3} = k$

$y = \frac{2}{3}x$

Let $x = 42$: $y = \frac{2}{3}(42) = 28$

When $x = 42$, $y = 28$.

32. $y = \frac{k}{x^2}$

$y = 8$ when $x = 5$: $8 = \frac{k}{5^2}$

$8 = \frac{k}{25}$

$200 = k$

$y = \frac{200}{x^2}$

Let $x = 15$: $y = \frac{200}{15^2} = \frac{200}{225} = \frac{8}{9}$

When $x = 15$, $y = \frac{8}{9}$.

Cumulative Review Chapters 1–6

1. $6 \div 3 + 5^2 = 6 \div 3 + 25 = 2 + 25 = 27$

2. $\frac{10}{3} + \frac{5}{21} = \frac{10}{3} \cdot \frac{7}{7} + \frac{5}{21} = \frac{70}{21} + \frac{5}{21} = \frac{75}{21} = \frac{25}{7}$

3.
$$\begin{aligned} 1 + 2[5(2 \cdot 3 + 1) - 10] &= 1 + 2[5(6 + 1) - 10] \\ &= 1 + 2[5(7) - 10] \\ &= 1 + 2(35 - 10) \\ &= 1 + 2(25) \\ &= 1 + 50 \\ &= 51 \end{aligned}$$

4.
$$\begin{aligned} 16 - 3 \cdot 3 + 2^4 &= 16 - 3 \cdot 3 + 16 \\ &= 16 - 9 + 16 \\ &= 7 + 16 \\ &= 23 \end{aligned}$$

5. $20{,}320 - (-282) = 20{,}320 + 282 = 20{,}602$
The difference in elevation is 20,602 feet.

6.
$$\begin{aligned} 1.7x - 11 - 0.9x - 25 &= (1.7x - 0.9x) + (-11 - 25) \\ &= 0.8x + (-36) \\ &= 0.8x - 36 \end{aligned}$$

7. Twice a number, plus 6 is written as $2x + 6$.

8. The product of -15 and the sum of a number and $\frac{2}{3}$ is written as $-15\left(x + \frac{2}{3}\right)$.

9. The difference of a number and 4, divided by 7 is written as $(x - 4) \div 7$ or $\frac{x-4}{7}$.

10. The quotient of -9 and twice a number is written as $\frac{-9}{2x}$.

11. Five plus the sum of a number and 1 is written as $5 + (x + 1) = 5 + x + 1 = 6 + x$.

12. A number subtracted from -86 is written as $-86 - x$.

13.
$$\begin{aligned} \frac{5}{2}x &= 15 \\ \frac{2}{5} \cdot \frac{5}{2}x &= \frac{2}{5} \cdot 15 \\ x &= 6 \end{aligned}$$

14.
$$\begin{aligned} \frac{x}{4} - 1 &= -7 \\ \frac{x}{4} - 1 + 1 &= -7 + 1 \\ \frac{x}{4} &= -6 \\ 4 \cdot \frac{x}{4} &= 4(-6) \\ x &= -24 \end{aligned}$$

15.
$$\begin{aligned} 2x &< -4 \\ \frac{2x}{2} &< \frac{-4}{2} \\ x &< -2 \end{aligned}$$

$\{x | x < -2\}$

−5 −4 −3 −2 −1 0 1 2 3 4 5

16.
$$\begin{aligned} 5(x + 4) &\ge 4(2x + 3) \\ 5x + 20 &\ge 8x + 12 \\ 5x + 20 - 20 &\ge 8x + 12 - 20 \\ 5x &\ge 8x - 8 \\ 5x - 8x &\ge 8x - 8 - 8x \\ -3x &\ge -8 \\ \frac{-3x}{-3} &\le \frac{-8}{-3} \\ x &\le \frac{8}{3} \end{aligned}$$

$\left\{x \middle| x \le \frac{8}{3}\right\}$

17. a. The degree of the trinomial $-2t^2 + 3t + 6$ is 2, the greatest degree of any of its terms.

b. The degree of the binomial $15x - 10$ or $15x^1 - 10$ is 1.

c. The degree of the polynomial $7x + 3x^3 + 2x^2 - 1$ is 3. It is not a monomial, binomial, nor a trinomial, so the answer is none of these.

18.
$$x + 2y = 6$$
$$x + 2y - x = 6 - x$$
$$2y = 6 - x$$
$$\frac{2y}{2} = \frac{6-x}{2}$$
$$y = \frac{6-x}{2}$$

19. $(-2x^2 + 5x - 1) + (-2x^2 + x + 3)$
$$= -2x^2 + 5x - 1 - 2x^2 + x + 3$$
$$= (-2x^2 - 2x^2) + (5x + x) + (-1 + 3)$$
$$= -4x^2 + 6x + 2$$

20. $(-2x^2 + 5x - 1) - (-2x^2 + x + 3)$
$$= -2x^2 + 5x - 1 + 2x^2 - x - 3$$
$$= (-2x^2 + 2x^2) + (5x - x) + (-1 - 3)$$
$$= 0x^2 + 4x + (-4)$$
$$= 4x - 4$$

21. $(3y + 1)^2 = (3y + 1)(3y + 1)$
$$= (3y)(3y) + (3y)(1) + 1(3y) + 1(1)$$
$$= 9y^2 + 3y + 3y + 1$$
$$= 9y^2 + 6y + 1$$

22. $(x - 12)^2 = x^2 - 2(x)(12) + (12)^2$
$$= x^2 - 24x + 144$$

23. $-9a^5 + 18a^2 - 3a = 3a(-3a^4) + 3a(6a) + 3a(-1)$
$$= 3a(-3a^4 + 6a - 1)$$

24. $4x^2 - 36 = 4(x^2 - 9)$
$$= 4(x^2 - 3^2)$$
$$= 4(x + 3)(x - 3)$$

25. $x^2 + 4x - 12$

Look for two numbers whose product is -12 and whose sum is 4.

$x^2 + 4x - 12 = (x - 2)(x + 6)$

26. $3x^2 - 20xy - 7y^2 = (3x + y)(x - 7y)$

27. Factors of $8x^2$: $8x^2 = 8x \cdot x$, $8x^2 = 4x \cdot 2x$

Factors of 5: $5 = -1 \cdot -5$

$8x^2 - 22x + 5 = (4x - 1)(2x - 5)$

28. Factors of $18x^2$: $18x^2 = 18x \cdot x$, $18x^2 = 9x \cdot 2x$, $18x^2 = 6x \cdot 3x$

Factors of -2: $-2 = -1 \cdot 2$, $-2 = 1 \cdot -2$

$18x^2 + 35x - 2 = (18x - 1)(x + 2)$

29.
$$x^2 - 9x - 22 = 0$$
$$(x - 11)(x + 2) = 0$$
$$x - 11 = 0 \quad \text{or} \quad x + 2 = 0$$
$$x = 11 \quad\quad x = -2$$

The solutions are 11 and -2.

30.
$$x^2 = x$$
$$x^2 - x = 0$$
$$x(x - 1) = 0$$
$$x = 0 \quad \text{or} \quad x - 1 = 0$$
$$x = 1$$

The solutions are 0 and 1.

31.
$$\frac{2x^2 - 11x + 5}{5x - 25} \div \frac{4x - 2}{10} = \frac{2x^2 - 11x + 5}{5x - 25} \cdot \frac{10}{4x - 2}$$
$$= \frac{(2x - 1)(x - 5) \cdot 2 \cdot 5}{5(x - 5) \cdot 2(2x - 1)}$$
$$= \frac{1}{1} \text{ or } 1$$

32.
$$\frac{2x^2 - 50}{4x^4 - 20x^3} = \frac{2(x^2 - 25)}{4x^3(x - 5)}$$
$$= \frac{2(x - 5)(x + 5)}{2 \cdot 2x^3(x - 5)}$$
$$= \frac{x + 5}{2x^3}$$

33. $\frac{4b}{9a} = \frac{4b}{9a} \cdot 1 = \frac{4b}{9a} \cdot \frac{3ab}{3ab} = \frac{4b(3ab)}{9a(3ab)} = \frac{12ab^2}{27a^2b}$

34. $\frac{1}{2x} = \frac{1}{2x} \cdot 1 = \frac{1}{2x} \cdot \frac{7x^2}{7x^2} = \frac{1(7x^2)}{2x(7x^2)} = \frac{7x^2}{14x^3}$

35. $1+\frac{m}{m+1}=\frac{1}{1}+\frac{m}{m+1}$
$=\frac{1(m+1)}{1(m+1)}+\frac{m}{m+1}$
$=\frac{m+1+m}{m+1}$
$=\frac{2m+1}{m+1}$

36. $\frac{2x+1}{x-6}-\frac{x-4}{x-6}=\frac{(2x+1)-(x-4)}{x-6}$
$=\frac{2x+1-x+4}{x-6}$
$=\frac{x+5}{x-6}$

37. $3-\frac{6}{x}=x+8$
$x\left(3-\frac{6}{x}\right)=x(x+8)$
$3x-6=x^2+8x$
$0=x^2+5x+6$
$0=(x+3)(x+2)$
$x+3=0$ or $x+2=0$
$x=-3$ $\quad x=-2$
The solutions are −3 and −2.

38. $3x^2+5x=2$
$3x^2+5x-2=0$
$(3x-1)(x+2)=0$
$3x-1=0$ or $x+2=0$
$x=\frac{1}{3}$ $\quad x=-2$

The solutions are −2 and $\frac{1}{3}$.

39. $\frac{\frac{x+1}{y}}{\frac{x}{y}+2}=\frac{y\left(\frac{x+1}{y}\right)}{y\left(\frac{x}{y}+2\right)}=\frac{y\left(\frac{x+1}{y}\right)}{y\left(\frac{x}{y}\right)+y(2)}=\frac{x+1}{x+2y}$

40. $\frac{\frac{x}{2}-\frac{y}{6}}{\frac{x}{12}-\frac{y}{3}}=\frac{12\left(\frac{x}{2}-\frac{y}{6}\right)}{12\left(\frac{x}{12}-\frac{y}{3}\right)}$
$=\frac{12\left(\frac{x}{2}\right)-12\left(\frac{y}{6}\right)}{12\left(\frac{x}{12}\right)-12\left(\frac{y}{3}\right)}$
$=\frac{6x-2y}{x-4y}$ or $\frac{2(3x-y)}{x-4y}$

41. $3x+y=12$

a. In (0,), the x-coordinate is 0.
$x=0$: $3(0)+y=12$
$0+y=12$
$y=12$
The ordered-pair solution is (0, 12).

b. In (, 6), the y-coordinate is 6.
$y=6$: $3x+6=12$
$3x=6$
$x=2$
The ordered-pair solution is (2, 6).

c. In (−1,), the x-coordinate is −1.
$x=-1$: $3(-1)+y=12$
$-3+y=12$
$y=15$
The ordered-pair solution is (−1, 15).

42. $y=-5x$
$x=0$: $y=-5(0)=0$
$x=-1$: $y=-5(-1)=5$
$y=-10$: $-10=-5x$
$2=x$

x	y
0	0
−1	5
2	−10

43. $2x+y=5$
$y=0$: $2x+0=5$
$2x=5$
$x=\frac{5}{2}$

The x-intercept is $\left(\frac{5}{2}, 0\right)$.

$x = 0$: $2(0) + y = 5$
$0 + y = 5$
$y = 5$

The y-intercept is (0, 5).

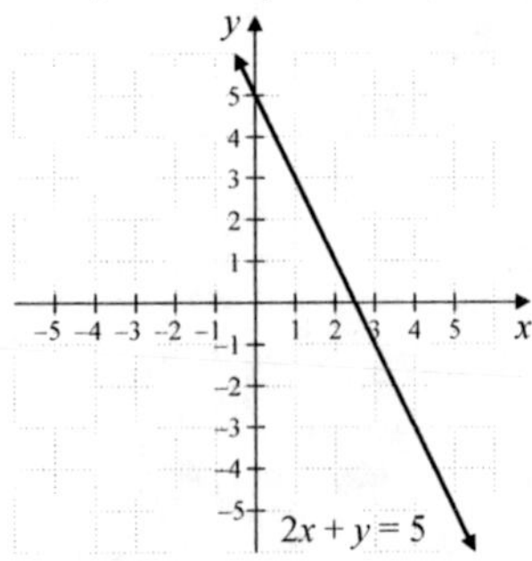

44. $(x_1, y_1) = (0, 5), (x_2, y_2) = (-5, 4)$

$$m = \frac{y_2 - y_1}{x_2 - x_1} = \frac{4-5}{-5-0} = \frac{-1}{-5} = \frac{1}{5}$$

45. $y = mx + b$

$$-2x + 3y = 11$$
$$3y = 2x + 11$$
$$y = \frac{2}{3}x + \frac{11}{3}$$

The slope is $m = \frac{2}{3}$.

46. The graph of $x = -10$ is a vertical line. Slopes of vertical lines are undefined.

47.
$$y - y_1 = m(x - x_1)$$
$$y - 5 = -2[x - (-1)]$$
$$y - 5 = -2(x + 1)$$
$$y - 5 = -2x - 2$$
$$y = -2x + 3$$
$$2x + y = 3$$

48. $y = mx + b$

$$2x - 5y = 10$$
$$-5y = -2x + 10$$
$$y = \frac{2}{5}x - 2$$

The slope is $m = \frac{2}{5}$ and the y-intercept is (0, −2).

49. $g(x) = x^2 - 3$

a. $g(2) = 2^2 - 3 = 4 - 3 = 1$
The ordered pair is (2, 1).

b. $g(-2) = (-2)^2 - 3 = 4 - 3 = 1$
The ordered pair is (−2, 1).

c. $g(0) = 0^2 - 3 = 0 - 3 = -3$
The ordered pair is (0, −3).

50. $(x_1, y_1) = (2, 3), (x_2, y_2) = (0, 0)$

$$m = \frac{y_2 - y_1}{x_2 - x_1} = \frac{0-3}{0-2} = \frac{-3}{-2} = \frac{3}{2}$$
$$y - y_1 = m(x - x_1)$$
$$y - 3 = \frac{3}{2}(x - 2)$$
$$2(y - 3) = 2\left(\frac{3}{2}\right)(x - 2)$$
$$2(y - 3) = 3(x - 2)$$
$$2y - 6 = 3x - 6$$
$$2y = 3x$$
$$0 = 3x - 2y$$
$$3x - 2y = 0$$

Chapter 7

Section 7.1 Practice Exercises

1.
$$\begin{aligned} 5x-2y&=-3 \\ 5(3)-2(9)&=-3 \\ 15-18&=-3 \\ -3&=-3 \end{aligned} \qquad \begin{aligned} y&=3x \\ 9&=3(3) \\ 9&=9 \end{aligned}$$

(3, 9) is a solution of the system.

2.
$$\begin{aligned} 2x-y&=8 \\ 2(3)-(-2)&=8 \\ 6+2&=8 \\ 8&=8 \end{aligned} \qquad \begin{aligned} x+3y&=4 \\ 3+3(-2)&=4 \\ 3+-6&=4 \\ -3&\neq 4 \end{aligned}$$

(3, −2) is not a solution of the system.

3.

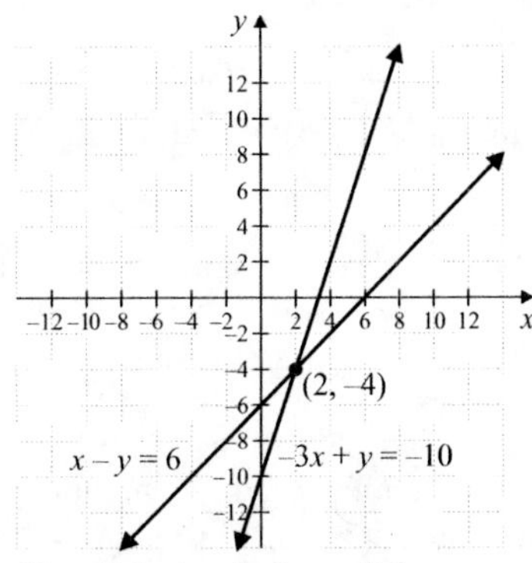

(2, −4) is a solution of the system.

4.

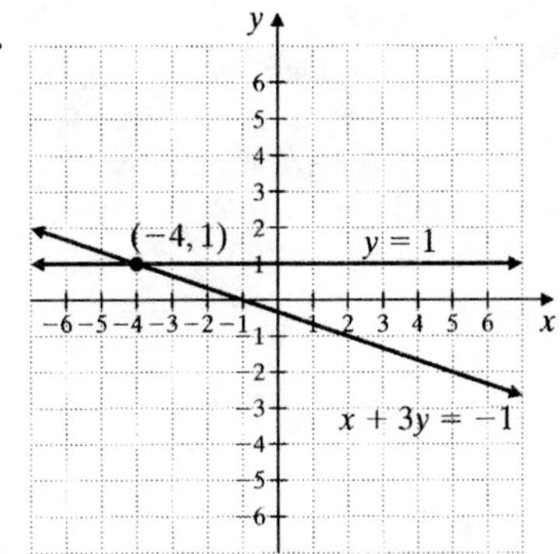

(−4, 1) is a solution of the system.

5.

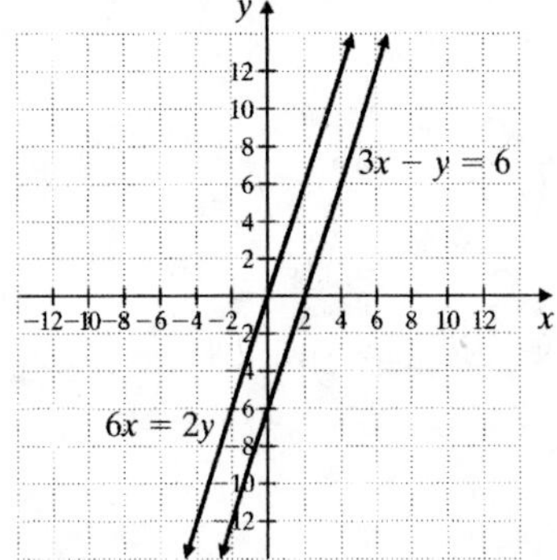

The system has no solution because the lines are parallel.

6.

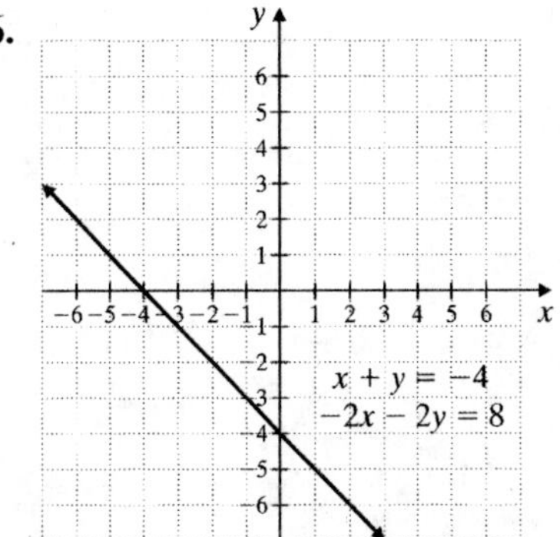

There are an infinite number of solutions because it is the same line.

7.
$$\begin{aligned} 5x+4y&=6 \\ 4y&=-5x+6 \\ y&=-\frac{5}{4}x+\frac{3}{2} \end{aligned} \qquad \begin{aligned} x-y&=3 \\ -y&=-x+3 \\ y&=x-3 \end{aligned}$$

The slope of the first line is $-\frac{5}{4}$, while the slope of the second line is 1. Since the slopes are not equal, the system has one solution.

8.
$$\begin{aligned} -\frac{2}{3}x+y&=6 \\ y&=\frac{2}{3}x+6 \end{aligned} \qquad \begin{aligned} 3y&=2x+5 \\ y&=\frac{2}{3}x+\frac{5}{3} \end{aligned}$$

Both lines have slope $\frac{2}{3}$, but the *y*-intercepts are different. The lines are parallel, so the system has no solution.

Calculator Explorations

1. $y = -2.68x + 1.21$
$y = 5.22x - 1.68$

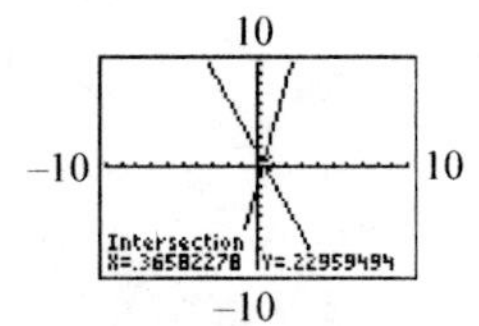

(0.37, 0.23) is the approximate point of intersection.

2. $y = 4.25x + 3.89$
$y = -1.88x + 3.21$

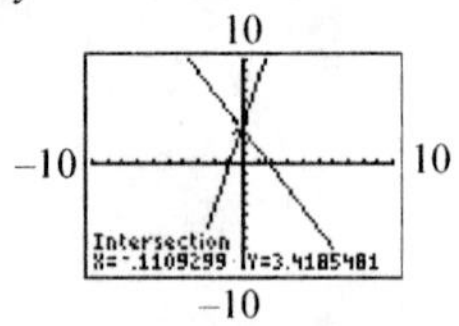

(−0.11, 3.42) is the approximate point of intersection.

3. $4.3x - 2.9y = 5.6 \rightarrow y = -(5.6 - 4.3x)/2.9$
$8.1x + 7.6y = -14.1 \rightarrow y = (-14.1 - 8.1x)/7.6$

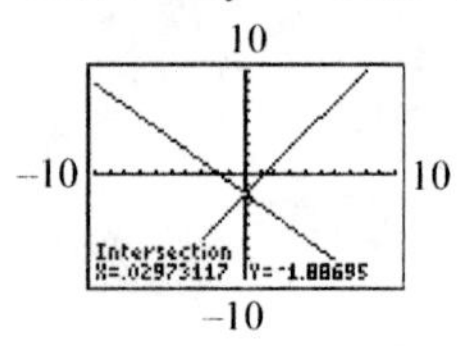

(0.03, −1.89) is the approximate point of intersection.

4. $-3.6x - 8.6y = 10 \rightarrow y = -(10 + 3.6x)/8.6$
$-4.5x + 9.6y = -7.7 \rightarrow y = (-7.7 + 4.5x)/9.6$

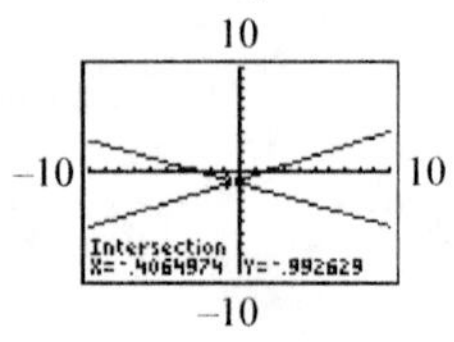

(−0.41, −0.99) is the approximate point of intersection.

Vocabulary, Readiness & Video Check 7.1

1. In a system of linear equations in two variables, if the graphs of the equations are the same, the equations are dependent equations.

2. Two or more linear equations are called a system of linear equations.

3. A system of equations that has at least one solution is called a consistent system.

4. A solution of a system of two equations in two variables is an ordered pair of numbers that is a solution of both equations in the system.

5. A system of equations that has no solution is called an inconsistent system.

6. In a system of linear equations in two variables, if the graphs of the equations are different, the equations are independent equations.

7. The lines intersect at (−1, 3); therefore there is one solution.

8. Since the lines are parallel and do not intersect, there is no solution.

9. Since the lines are the same, there is an infinite number of solutions.

10. The lines intersect at (3, 4); therefore there is one solution.

11. The ordered pair must satisfy all equations of the system in order to be a solution of the system, so we must check that the ordered pair is a solution of both equations.

12. Graphing is not the most accurate method, especially if your graph is off just slightly, or the point of intersection does not have integer coordinates.

13. Writing the equations of a system in slope-intercept form lets us see and compare their slopes and *y*-intercepts. Different slopes mean one solution; same slopes with different *y*-intercepts mean no solution; same slopes with same *y*-intercepts mean an infinite number of solutions.

Exercise Set 7.1

1. a. First equation:
$x + y = 8$
$2 + 4 \stackrel{?}{=} 8$
$6 = 8$ False
(2, 4) is not a solution of the first equation, so it is not a solution of the system.

b. First equation:
$x + y = 8$
$5 + 3 \stackrel{?}{=} 8$
$8 = 8$ True
Second equation:
$3x + 2y = 21$
$3(5) + 2(3) \stackrel{?}{=} 21$
$15 + 6 \stackrel{?}{=} 21$
$21 = 21$ True
Since (5, 3) is a solution of both equations, it is a solution of the system.

3. a. First equation:
$3x - y = 5$
$3(3) - 4 \stackrel{?}{=} 5$
$9 - 4 \stackrel{?}{=} 5$
$5 = 5$ True
Second equation:
$x + 2y = 11$
$3 + 2(4) \stackrel{?}{=} 11$
$3 + 8 \stackrel{?}{=} 11$
$11 = 11$ True
Since (3, 4) is a solution of both equations, it is a solution of the system.

b. First equation:
$3x - y = 5$
$3(0) - (-5) \stackrel{?}{=} 5$
$0 + 5 \stackrel{?}{=} 5$
$5 = 5$ True
Second equation:
$x + 2y = 11$
$0 + 2(-5) \stackrel{?}{=} 11$
$0 + (-10) \stackrel{?}{=} 11$
$-10 = 11$ False
(0, −5) is not a solution of the second equation, so it is not a solution of the system.

5. a. First equation:
$2y = 4x + 6$
$2(-3) \stackrel{?}{=} 4(-3) + 6$
$-6 \stackrel{?}{=} -12 + 6$
$-6 = -6$ True
Second equation:
$2x - y = -3$
$2(-3) - (-3) \stackrel{?}{=} -3$
$-6 + 3 \stackrel{?}{=} -3$
$-3 = -3$ True
(−3, −3) is a solution of both equations, it is a solution of the system.

b. First equation:
$2y = 4x + 6$
$2(3) \stackrel{?}{=} 4(0) + 6$
$6 \stackrel{?}{=} 0 + 6$
$6 = 6$ True
Second equation:
$2x - y = -3$
$2(0) - 3 \stackrel{?}{=} -3$
$0 - 3 \stackrel{?}{=} -3$
$-3 = -3$ True
Since (0, 3) is a solution of both equations, it is a solution of the system.

7. a. First equation:
$-2 = x - 7y$
$-2 \stackrel{?}{=} -2 - 7(0)$
$-2 \stackrel{?}{=} -2 - 0$
$-2 = -2$ True
Second equation:
$6x - y = 13$
$6(-2) - 0 \stackrel{?}{=} 13$
$-12 - 0 \stackrel{?}{=} 13$
$-12 = 13$ False
(−2, 0) is not a solution of the second equation, so it is not a solution of the system.

b. First equation:
$-2 = x - 7y$
$-2 \stackrel{?}{=} \frac{1}{2} - 7\left(\frac{5}{14}\right)$
$-2 \stackrel{?}{=} \frac{1}{2} - \frac{5}{2}$
$-2 = -\frac{4}{2}$
$-2 = -2$ True
Second equation:
$6x - y = 13$
$6\left(\frac{1}{2}\right) - \frac{5}{14} \stackrel{?}{=} 13$
$\frac{42}{14} - \frac{5}{14} \stackrel{?}{=} 13$
$\frac{37}{14} = 13$ False
$\left(\frac{1}{2}, \frac{5}{14}\right)$ is not a solution of the second equation, so it is not a solution of the system.

9.

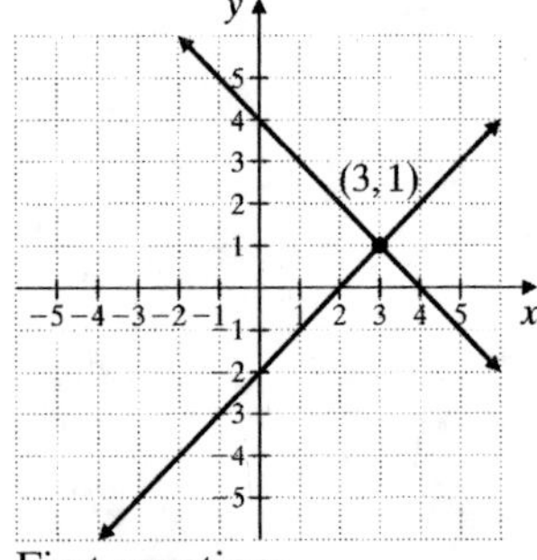

First equation:

$x + y = 4$

$3 + 1 \stackrel{?}{=} 4$

$4 = 4$ True

Second equation:

$x - y = 2$

$3 - 1 \stackrel{?}{=} 2$

$2 = 2$ True

The solution of the system is (3, 1).

11.

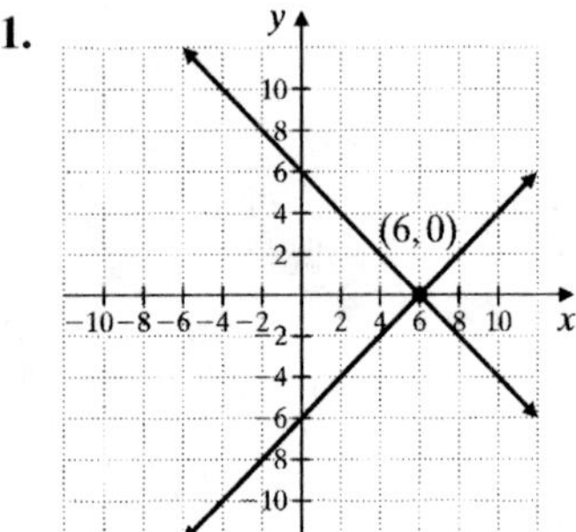

First equation:

$x + y = 6$

$6 + 0 \stackrel{?}{=} 6$

$6 = 6$ True

Second equation:

$-x + y = -6$

$-6 + 0 = -6$

$-6 = -6$ True

The solution of the system is (6, 0).

13.

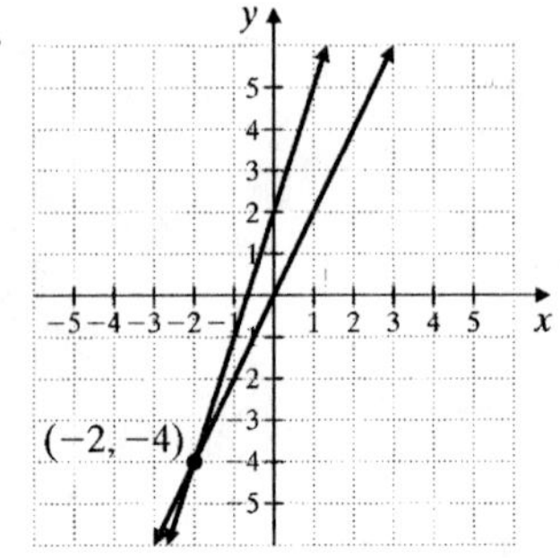

First equation:

$y = 2x$

$-4 \stackrel{?}{=} 2(-2)$

$-4 = -4$ True

Second equation:

$3x - y = -2$

$3(-2) - (-4) \stackrel{?}{=} -2$

$-6 + 4 \stackrel{?}{=} -2$

$-2 = -2$ True

The solution of the system is (−2, −4).

15.

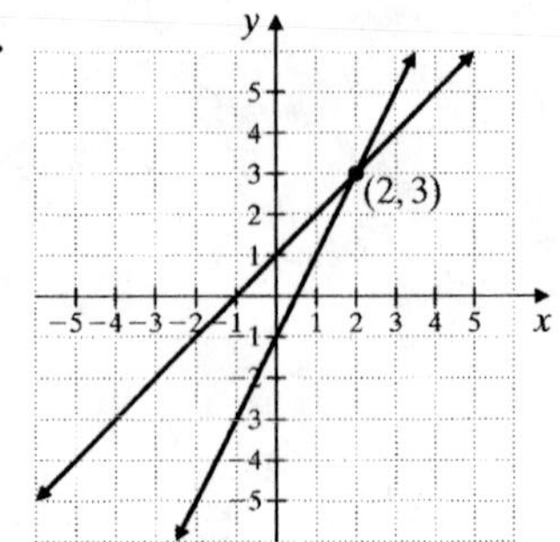

First equation:

$y = x + 1$

$3 \stackrel{?}{=} 2 + 1$

$3 = 3$ True

Second equation:

$y = 2x - 1$

$3 \stackrel{?}{=} 2(2) - 1$

$3 \stackrel{?}{=} 4 - 1$

$3 = 3$ True

The solution of the system is (2, 3).

17.

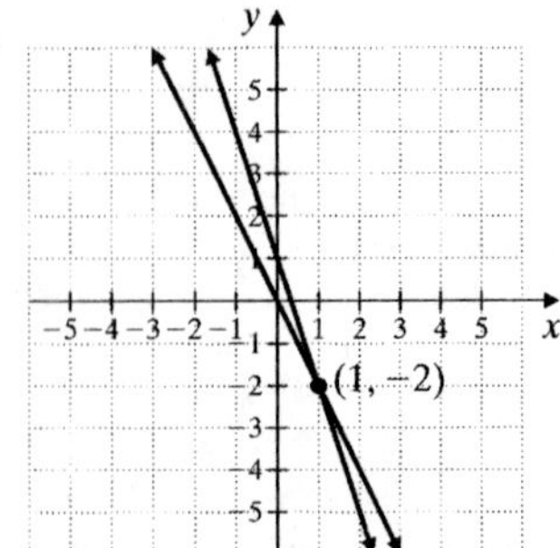

First equation:

$2x + y = 0$

$2(1) + (-2) \stackrel{?}{=} 0$

$2 - 2 \stackrel{?}{=} 0$

$0 = 0$ True

Second equation:

$3x + y = 1$

$3(1) + (-2) \stackrel{?}{=} 1$

$3 - 2 \stackrel{?}{=} 1$

$1 = 1$ True

The solution of the system is (1, −2).

19.

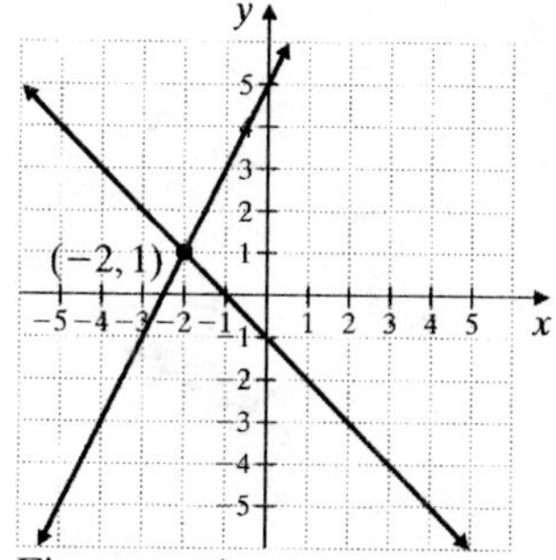

First equation:

$y = -x - 1$

$1 \overset{?}{=} -(-2) - 1$

$1 \overset{?}{=} 2 - 1$

$1 = 1$ True

Second equation:

$y = 2x + 5$

$1 \overset{?}{=} 2(-2) + 5$

$1 \overset{?}{=} -4 + 5$

$1 = 1$ True

The solution of the system is (−2, 1).

21.

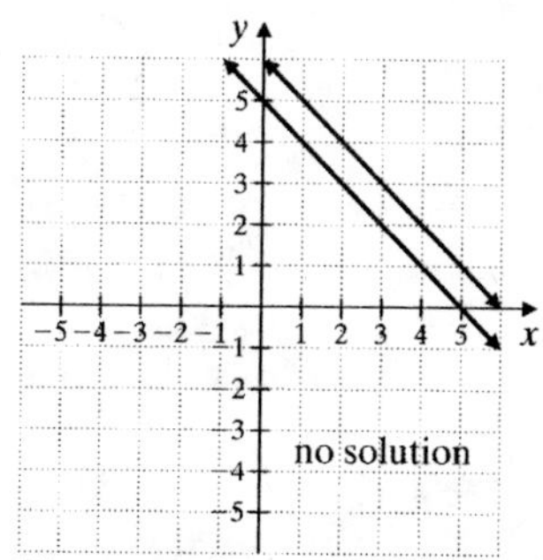

First equation:

$x + y = 5$

$y = -x + 5$

Second equation:

$x + y = 6$

$y = -x + 6$

The lines have the same slope, but different y-intercepts, so they are parallel. The system has no solution.

23.

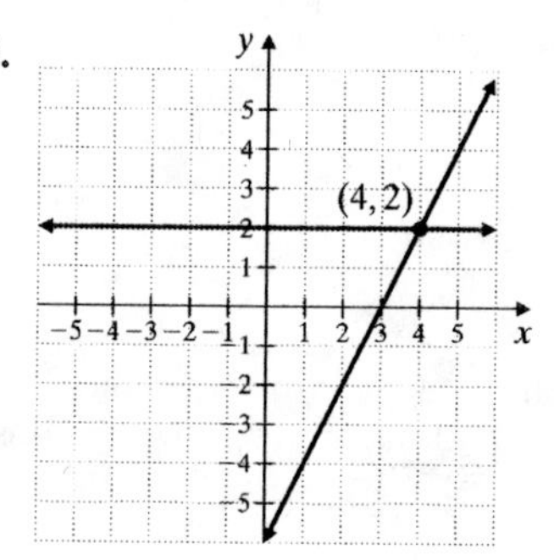

First equation:

$2x - y = 6$

$2(4) - 2 \overset{?}{=} 6$

$8 - 2 \overset{?}{=} 6$

$6 = 6$ True

Second equation:

$y = 2$

$2 = 2$ True

The solution of the system is (4, 2).

25.

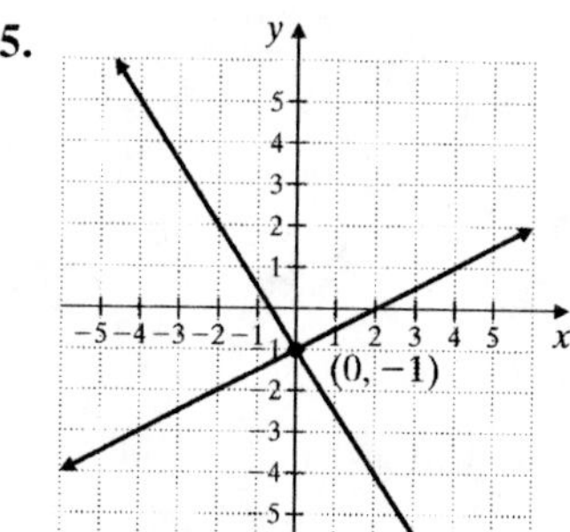

First equation:

$x - 2y = 2$

$0 - 2(-1) \overset{?}{=} 2$

$0 + 2 \overset{?}{=} 2$

$2 = 2$ True

Second equation:

$3x + 2y = -2$

$3(0) + 2(-1) \overset{?}{=} -2$

$0 - 2 \overset{?}{=} -2$

$-2 = -2$ True

The solution of the system is (0, −1).

27.

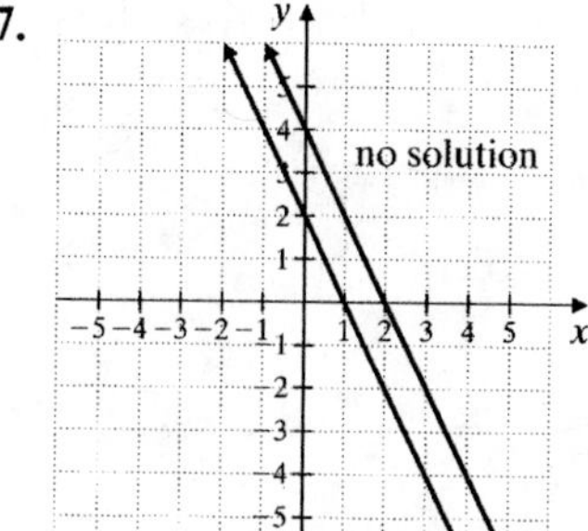

First equation:

$2x + y = 4$

$y = -2x + 4$

Second equation:

$6x = -3y + 6$

$3y = -6x + 6$

$y = -2x + 2$

The lines have the same slope, but different y-intercepts, so they are parallel. The system has no solution.

29.

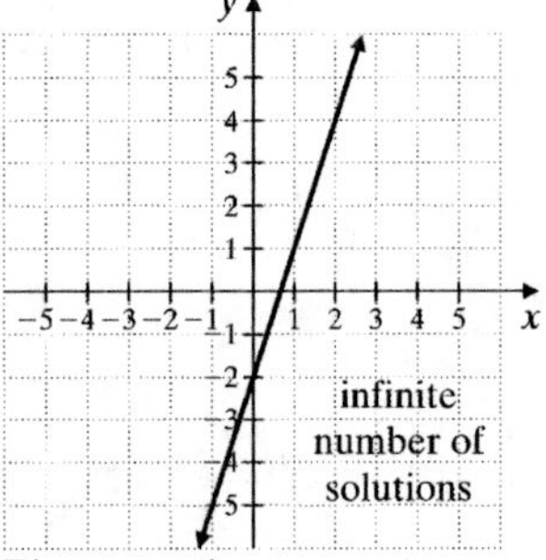

First equation:

$y - 3x = -2$

$y = 3x - 2$

Second equation:

$6x - 2y = 4$

$-2y = -6x + 4$

$y = 3x - 2$

The graphs of the equations are the same line, so the system has an infinite number of solutions.

31.

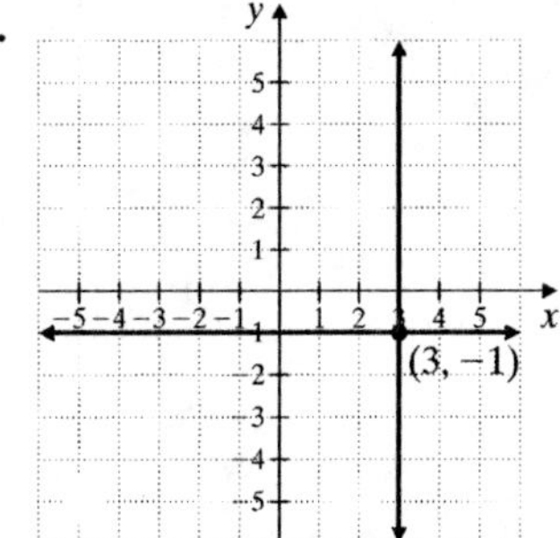

First equation:

$x = 3$

$3 = 3$ True

Second equation:

$y = -1$

$-1 = -1$ True

The solution of the system is (3, −1).

33.

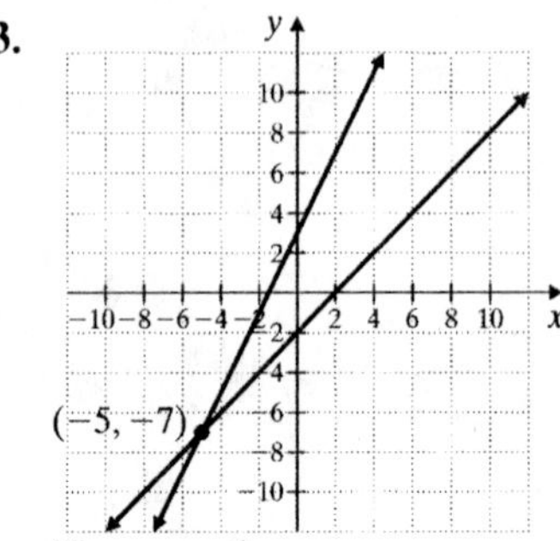

First equation:

$y = x - 2$

$-7 \stackrel{?}{=} -5 - 2$

$-7 = -7$ True

Second equation:

$y = 2x + 3$

$-7 \stackrel{?}{=} 2(-5) + 3$

$-7 \stackrel{?}{=} -10 + 3$

$-7 = -7$ True

The solution of the system is (−5, −7).

35.

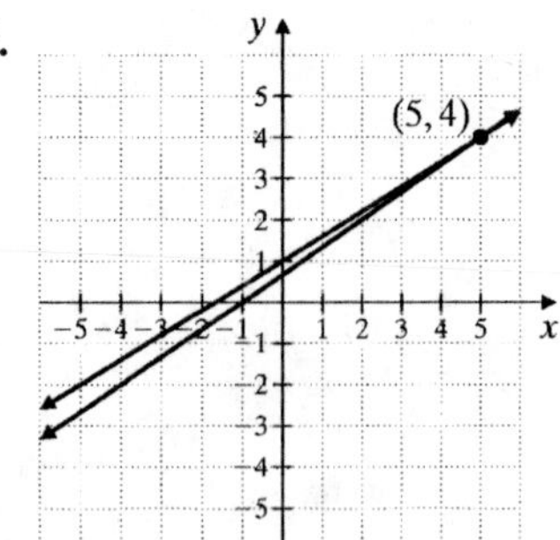

First equation:

$2x - 3y = -2$

$2(5) - 3(4) \stackrel{?}{=} -2$

$10 - 12 \stackrel{?}{=} -2$

$-2 = -2$ True

Second equation:

$-3x + 5y = 5$

$-3(5) + 5(4) \stackrel{?}{=} 5$

$-15 + 20 \stackrel{?}{=} 5$

$5 = 5$ True

The solution of the system is (5, 4).

37.

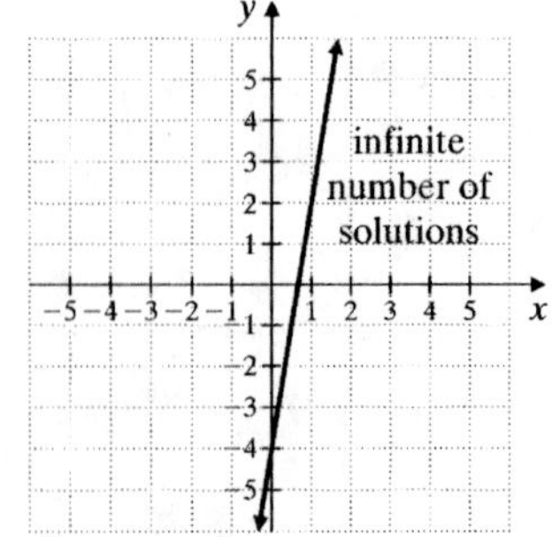

First equation:

$6x - y = 4$

$-y = -6x + 4$

$y = 6x - 4$

Second equation:

$\frac{1}{2}y = -2 + 3x$

$y = -4 + 6x$

$y = 6x - 4$

The graphs of the equations are the same line, so the system has an infinite number of solutions.

39. $4x + y = 24$ $\quad x + 2y = 2$

$y = -4x + 24$ $\quad 2y = -x + 2$

$y = -\frac{1}{2}x + 1$

a. The slopes of the lines are different, so the lines intersect.

b. The system has one solution.

41. $2x + y = 0$ $\quad 2y = 6 - 4x$

$y = -2x$ $\quad y = -2x + 3$

a. The lines have the same slope, but different y-intercepts, so the lines are parallel.

b. The system has no solution.

43. $6x - y = 4$ $\quad \frac{1}{2}y = -2 + 3x$

$6x - 4 = y$ $\quad y = 6x - 4$

a. The lines are identical.

b. The system has an infinite number of solutions.

45. $x = 5$ $\quad y = -2$

vertical line $\quad$ horizontal line

a. The slopes of the lines are different, so the lines intersect.

b. The system has one solution.

47. $3y - 2x = 3$ $\quad x + 2y = 9$

$3y = 2x + 3$ $\quad 2y = -x + 9$

$y = \frac{2}{3}x + 1$ $\quad y = -\frac{1}{2}x + \frac{9}{2}$

a. The slopes of the lines are different, so the lines intersect.

b. The system has one solution.

49. $6y + 4x = 6$ $\quad 3y - 3 = -2x$

$6y = -4x + 6$ $\quad 3y = -2x + 3$

$y = -\frac{2}{3}x + 1$ $\quad y = -\frac{2}{3}x + 1$

a. The lines are identical.

b. The system has an infinite number of solutions.

51. $x + y = 4$ $\quad x + y = 3$

$y = -x + 4$ $\quad y = -x + 3$

a. The lines have the same slope, but different y-intercepts, so the lines are parallel.

b. The system has no solution.

53.
$$\begin{aligned} 5(x-3) + 3x &= 1 \\ 5x - 15 + 3x &= 1 \\ 8x &= 16 \\ x &= 2 \end{aligned}$$

55.
$$\begin{aligned} 4\left(\frac{y+1}{2}\right) + 3y &= 0 \\ 2(y+1) + 3y &= 0 \\ 2y + 2 + 3y &= 0 \\ 5y + 2 &= 0 \\ 5y &= -2 \\ y &= -\frac{2}{5} \end{aligned}$$

57.
$$\begin{aligned} 8a - 2(3a - 1) &= 6 \\ 8a - 6a + 2 &= 6 \\ 2a &= 4 \\ a &= 2 \end{aligned}$$

59. Answers may vary. Possible answer:

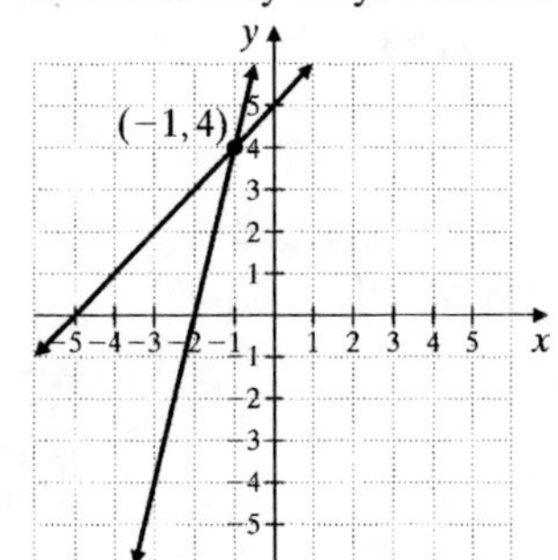

61. Answers may vary. Possible answer:

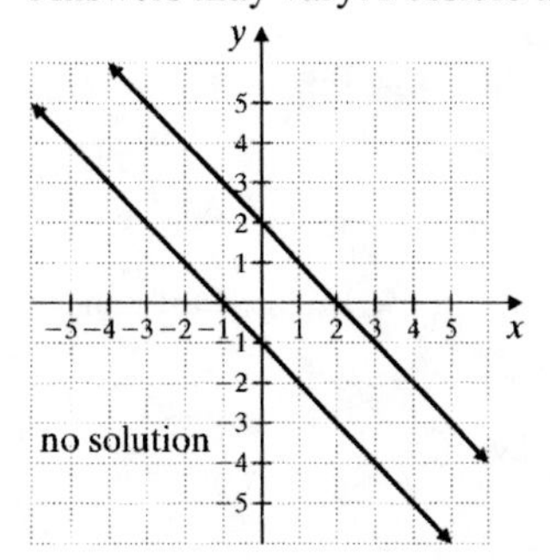

63. The lines cross at a point between 2010 and 2011. The number of digital cinema screens equaled the number of analog cinema screens between 2010 and 2011.

65. The average attendance per game for the Pittsburgh Pirates was greater than the average attendance per game for the Cleveland Indians in 2010, 2011, 2012, and 2013.

67. answers may vary

69. answers may vary

71. **a.** (4, 9) appears in both tables, so it is a solution of the system.

b.

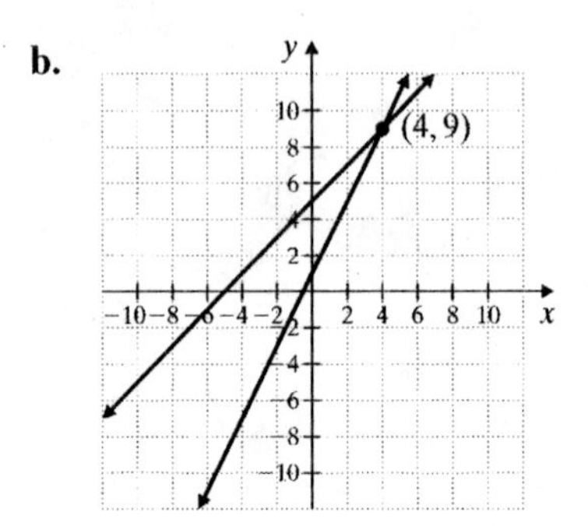

c. Yes; the two lines intersect at (4, 9).

Section 7.2 Practice Exercises

1. $\begin{cases} 2x+3y=13 \\ x=y+4 \end{cases}$

Substitute $y + 4$ for x in the first equation and solve for y.

$$2(y+4)+3y=13$$
$$2y+8+3y=13$$
$$5y=5$$
$$y=1$$

Solve for x.

$x = y + 4 = 1 + 4 = 5$

The solution of the system is (5, 1).

2. $\begin{cases} 4x-y=2 \\ y=5x \end{cases}$

Substitute $5x$ for y in the first equation and solve for x.

$$4x-5x=2$$
$$-x=2$$
$$x=-2$$

Solve for y.

$y = 5x = 5(-2) = -10$

The solution of the system is (−2, −10).

3. $\begin{cases} 3x+y=5 \\ 3x-2y=-7 \end{cases}$

Solve the first equation for y.

$$3x+y=5$$
$$y=-3x+5$$

Substitute $-3x + 5$ for y in the second equation.

$$3x-2y=-7$$
$$3x-2(-3x+5)=-7$$
$$3x+6x-10=-7$$
$$9x=3$$
$$x=\frac{3}{9}=\frac{1}{3}$$

Solve for y.

$$y=-3x+5=-3\left(\frac{1}{3}\right)+5=-1+5=4$$

The solution of the system is $\left(\frac{1}{3}, 4\right)$.

4. $\begin{cases} 5x-2y=6 \\ -3x+y=-3 \end{cases}$

Solve the second equation for y.

$$-3x+y=-3$$
$$y=3x-3$$

Substitute $3x - 3$ for y in the first equation.

$$5x-2y=6$$
$$5x-2(3x-3)=6$$
$$5x-6x+6=6$$
$$-x=0$$
$$x=0$$

Solve for y.

$y = 3x - 3 = 3(0) - 3 = 0 - 3 = -3$

The solution of the system is (0, −3).

5. $\begin{cases} -x+3y=6 \\ y=\frac{1}{3}x+2 \end{cases}$

Substitute $\frac{1}{3}x+2$ for y in the first equation.

$$-x+3y=6$$
$$-x+3\left(\frac{1}{3}x+2\right)=6$$
$$-x+x+6=6$$
$$0x=0$$
$$0=0$$

The statement $0 = 0$ indicates that this system has an infinite number of solutions. It is a graph of the same line.

6. $\begin{cases} 2x-3y=6 \\ -4x+6y=12 \end{cases}$

$$\begin{aligned} 2x-3y&=6 \\ -3y&=-2x+6 \\ y&=\frac{2}{3}x-2 \end{aligned}$$

Substitute $\frac{2}{3}x-2$ for y in the second equation.

$$\begin{aligned} -4x+6y&=12 \\ -4x+6\left(\frac{2}{3}x-2\right)&=12 \\ -4x+4x-12&=12 \\ -12&=12 \end{aligned}$$

The statement $-12 = 12$ indicates that this system has no solution. It is a graph of parallel lines.

Vocabulary, Readiness & Video Check 7.2

1. $x = 1$: $y = 4(1) = 4$
The solution is (1, 4).

2. $0 = 34$ is a false statement. The system has no solution.

3. $0 = 0$ is a true statement. The system has an infinite number of solutions.

4. $y = 0$: $x = 0 + 5 = 5$
The solution is (5, 0).

5. $x = 0$: $0+y=0$
$y=0$
The solution is (0, 0).

6. $0 = 0$ is a true statement. The system has an infinite number of solutions.

7. We solved one equation for a variable. Next, be sure to substitute this expression for the variable into the *other* equation.

Exercise Set 7.2

1. $\begin{cases} x+y=3 \\ x=2y \end{cases}$

Substitute $2y$ for x in the first equation and solve for y.

$$\begin{aligned} x+y&=3 \\ 2y+y&=3 \\ 3y&=3 \\ y&=1 \end{aligned}$$

Now solve for x.
$x = 2y = 2(1) = 2$
The solution of the system is (2, 1).

3. $\begin{cases} x+y=6 \\ y=-3x \end{cases}$

Substitute $-3x$ for y in the first equation and solve for x.

$$\begin{aligned} x+y&=6 \\ x+(-3x)&=6 \\ -2x&=6 \\ x&=-3 \end{aligned}$$

Now solve for y.
$y = -3x = -3(-3) = 9$
The solution of the system is (−3, 9).

5. $\begin{cases} y=3x+1 \\ 4y-8x=12 \end{cases}$

Substitute $3x + 1$ for y in the second equation and solve for x.

$$\begin{aligned} 4y-8x&=12 \\ 4(3x+1)-8x&=12 \\ 12x+4-8x&=12 \\ 4x+4&=12 \\ 4x&=8 \\ x&=2 \end{aligned}$$

Now solve for y.
$y = 3x + 1 = 3(2) + 1 = 6 + 1 = 7$
The solution of the system is (2, 7).

7. $\begin{cases} y=2x+9 \\ y=7x+10 \end{cases}$

Substitute $2x + 9$ for y in the second equation and solve for x.

$$\begin{aligned} y&=7x+10 \\ 2x+9&=7x+10 \\ -5x&=1 \\ y&=-\frac{1}{5} \end{aligned}$$

Now solve for y.

$$y=2x+9=2\left(-\frac{1}{5}\right)+9=\frac{-2}{5}+\frac{45}{5}=\frac{43}{5}$$

The solution of the system is $\left(-\frac{1}{5}, \frac{43}{5}\right)$.

9. $\begin{cases} 3x-4y=10 \\ y=x-3 \end{cases}$

Substitute $x - 3$ for y in the first equation and solve for x.

$$\begin{aligned} 3x-4y&=10\\ 3x-4(x-3)&=10\\ 3x-4x+12&=10\\ -x+12&=10\\ -x&=-2\\ x&=2 \end{aligned}$$

Now solve for y.
$y = x - 3 = 2 - 3 = -1$
The solution of the system is (2, −1).

11. $\begin{cases} x+2y=6 \\ 2x+3y=8 \end{cases}$

Solve the first equation for x.
$x+2y=6$
$x=-2y+6$

Substitute $-2y + 6$ for x in the second equation and solve for y.

$$\begin{aligned} 2x+3y&=8\\ 2(-2y+6)+3y&=8\\ -4y+12+3y&=8\\ -y&=-4\\ y&=4 \end{aligned}$$

Now solve for x.
$x = -2y + 6 = -2(4) + 6 = -8 + 6 = -2$
The solution of the system is (−2, 4).

13. $\begin{cases} 3x+2y=16 \\ x=3y-2 \end{cases}$

Substitute $3y - 2$ for x in the first equation and solve for y.

$$\begin{aligned} 3x+2y&=16\\ 3(3y-2)+2y&=16\\ 9y-6+2y&=16\\ 11y-6&=16\\ 11y&=22\\ y&=2 \end{aligned}$$

Now solve for x.
$x = 3y - 2 = 3(2) - 2 = 6 - 2 = 4$
The solution of the system is (4, 2).

15. $\begin{cases} 2x-5y=1 \\ 3x+y=-7 \end{cases}$

Solve the second equation for y.
$3x+y=-7$
$y=-3x-7$

Substitute $-3x - 7$ for y in the first equation and solve for x.

$$\begin{aligned} 2x-5y&=1\\ 2x-5(-3x-7)&=1\\ 2x+15x+35&=1\\ 17x&=-34\\ x&=-2 \end{aligned}$$

Now solve for y.
$y = -3x - 7 = -3(-2) - 7 = 6 - 7 = -1$
The solution of the system is (−2, −1).

17. $\begin{cases} 4x+2y=5 \\ -2x=y+4 \end{cases}$

Solve the second equation for x.
$-2x=y+4$
$x=-\frac{1}{2}y-2$

Substitute $-\frac{1}{2}y-2$ for x in the first equation and solve for y.

$$\begin{aligned} 4x+2y&=5\\ 4\left(-\frac{1}{2}y-2\right)+2y&=5\\ -2y-8+2y&=5\\ -8&=5 \quad \text{False} \end{aligned}$$

Since the statement $-8 = 5$ is false, the system has no solution.

19. $\begin{cases} 4x+y=11 \\ 2x+5y=1 \end{cases}$

Solve the first equation for y.
$4x+y=11$
$y=-4x+11$

Substitute $-4x + 11$ for y in the second equation and solve for x.

$$\begin{aligned} 2x+5y&=1\\ 2x+5(-4x+11)&=1\\ 2x+(-20x)+55&=1\\ -18x&=-54\\ x&=3 \end{aligned}$$

Now solve for y.
$y = -4x + 11 = -4(3) + 11 = -12 + 11 = -1$
The solution of the system is (3, −1).

21. $\begin{cases} x+2y+5=-4+5y-x \\ 2x+x=y+4 \end{cases}$

Simplify each equation.
$\begin{cases} 2x+9=3y \\ 3x=y+4 \end{cases}$

Solve the second simplified equation for y.
$3x=y+4$
$3x-4=y$

Substitute $3x - 4$ for y in the first simplified equation and solve for x.

$$\begin{aligned} 2x+9 &= 3y \\ 2x+9 &= 3(3x-4) \\ 2x+9 &= 9x-12 \\ 2x+21 &= 9x \\ 21 &= 7x \\ 3 &= x \end{aligned}$$

Now solve for y.

$y = 3x - 4 = 3(3) - 4 = 9 - 4 = 5$

The solution of the system is (3, 5).

23. $\begin{cases} 6x-3y=5 \\ x+2y=0 \end{cases}$

Solve the second equation for x.

$$\begin{aligned} x+2y &= 0 \\ x &= -2y \end{aligned}$$

Substitute $-2y$ for x in the first equation and solve for y.

$$\begin{aligned} 6x-3y &= 5 \\ 6(-2y)-3y &= 5 \\ -12y-3y &= 5 \\ -15y &= 5 \\ y &= -\frac{5}{15} = -\frac{1}{3} \end{aligned}$$

Now solve for x.

$$x = -2y = -2\left(-\frac{1}{3}\right) = \frac{2}{3}$$

The solution of the system is $\left(\frac{2}{3}, -\frac{1}{3}\right)$.

25. $\begin{cases} 3x-y=1 \\ 2x-3y=10 \end{cases}$

Solve the first equation for y.

$$\begin{aligned} 3x-y &= 1 \\ -y &= -3x+1 \\ y &= 3x-1 \end{aligned}$$

Substitute $3x - 1$ for y in the second equation and solve for x.

$$\begin{aligned} 2x-3y &= 10 \\ 2x-3(3x-1) &= 10 \\ 2x-9x+3 &= 10 \\ -7x+3 &= 10 \\ -7x &= 7 \\ x &= -1 \end{aligned}$$

Now solve for y.

$y = 3x - 1 = 3(-1) - 1 = -3 - 1 = -4$

The solution of the system is (−1, −4).

27. $\begin{cases} -x+2y=10 \\ -2x+3y=18 \end{cases}$

Solve the first equation for x.

$$\begin{aligned} -x+2y &= 10 \\ 2y-10 &= x \end{aligned}$$

Substitute $2y - 10$ for x in the second equation and solve for y.

$$\begin{aligned} -2x+3y &= 18 \\ -2(2y-10)+3y &= 18 \\ -4y+20+3y &= 18 \\ -y+20 &= 18 \\ 2 &= y \end{aligned}$$

Now solve for x.

$x = 2y - 10 = 2(2) - 10 = 4 - 10 = -6$

The solution of the system is (−6, 2).

29. $\begin{cases} 5x+10y=20 \\ 2x+6y=10 \end{cases}$

Solve the first equation for x. (Note that the second equation could also be easily solved for x.)

$$\begin{aligned} 5x+10y &= 20 \\ 5x &= -10y+20 \\ x &= -2y+4 \end{aligned}$$

Substitute $-2y + 4$ for x in the second equation and solve for y.

$$\begin{aligned} 2x+6y &= 10 \\ 2(-2y+4)+6y &= 10 \\ -4y+8+6y &= 10 \\ 2y+8 &= 10 \\ 2y &= 2 \\ y &= 1 \end{aligned}$$

Now solve for x.

$x = -2y + 4 = -2(1) + 4 = -2 + 4 = 2$

The solution of the system is (2, 1).

31. $\begin{cases} 3x+6y=9 \\ 4x+8y=16 \end{cases}$

Solve the first equation for x.

$$\begin{aligned} 3x+6y &= 9 \\ 3x &= -6y+9 \\ x &= -2y+3 \end{aligned}$$

Substitute $-2y + 3$ for x in the second equation and solve for y.

$$\begin{aligned} 4x+8y &= 16 \\ 4(-2y+3)+8y &= 16 \\ -8y+12+8y &= 16 \\ 0 &= 4 \quad \text{False} \end{aligned}$$

Since the statement $0 = 4$ is false, the system has no solution.

33. $\begin{cases} \frac{1}{3}x - y = 2 \\ x - 3y = 6 \end{cases}$

Solve the second equation for x.

$$x - 3y = 6$$
$$x = 3y + 6$$

Substitute $3y + 6$ for x in the first equation and solve for y.

$$\frac{1}{3}x - y = 2$$
$$\frac{1}{3}(3y+6) - y = 2$$
$$y + 2 - y = 2$$
$$2 = 2$$

Since 2 = 2 is a true statement, the two equations in the original system are equivalent. The system has an infinite number of solutions.

35. $\begin{cases} x = \frac{3}{4}y - 1 \\ 8x - 5y = -6 \end{cases}$

Substitute $\frac{3}{4}y - 1$ for x in the second equation and solve for y.

$$8x - 5y = -6$$
$$8\left(\frac{3}{4}y - 1\right) - 5y = -6$$
$$6y - 8 - 5y = -6$$
$$y = 2$$

Now solve for x.

$$x = \frac{3}{4}y - 1 = \frac{3}{4}(2) - 1 = \frac{3}{2} - 1 = \frac{1}{2}$$

The solution of the system is $\left(\frac{1}{2}, 2\right)$.

37.

$$3x + 2y = 6$$
$$-2(3x + 2y) = -2(6)$$
$$-6x - 4y = -12$$

39.

$$-4x + y = 3$$
$$3(-4x + y) = 3(3)$$
$$-12x + 3y = 9$$

41.

$$\begin{array}{r} 3n + 6m \\ +\ 2n - 6m \\ \hline 5n \end{array}$$

43.

$$\begin{array}{r} -5a - 7b \\ 5a - 8b \\ \hline -15b \end{array}$$

45. $\begin{cases} -5y + 6y = 3x + 2(x-5) - 3x + 5 \\ 4(x+y) - x + y = -12 \end{cases}$

Simplify each equation.

$\begin{cases} y = 2x - 5 \\ 3x + 5y = -12 \end{cases}$

Substitute $2x - 5$ for y in the second simplified equation and solve for x.

$$3x + 5y = -12$$
$$3x + 5(2x - 5) = -12$$
$$3x + 10x - 25 = -12$$
$$13x - 25 = -12$$
$$13x = 13$$
$$x = 1$$

Now solve for y.

$y = 2x - 5 = 2(1) - 5 = 2 - 5 = -3$

The solution of the system is (1, −3).

47. answers may vary

49. no

51. **c**; answers may vary

53. Using a graphing calculator, the solution of the system is (−2.6, 1.3).

55. Using a graphing calculator, the solution of the system is (3.28, 2.1).

57. **a.** $\begin{cases} y = -1.6x + 20.9 \\ y = 0.6x - 0.2 \end{cases}$

Substitute $-1.6x + 20.9$ for y in the second equation and solve for x.

$$y = 0.6x - 0.2$$
$$-1.6 + 20.9 = 0.6x - 0.2$$
$$-1.6x + 21.1 = 0.6x$$
$$21.1 = 2.2x$$
$$9.6 \approx x$$

Now solve for y.

$$\begin{aligned} y &= -1.6x + 20.9 \\ &= -1.6(9.6) + 20.9 \\ &= -15.36 + 20.9 \\ &= 5.54 \\ &\approx 6 \end{aligned}$$

The rounded solution is (9.6, 6).

b. In about 9.6 years after 2006, U.S. consumer spending on DVD- and Blu-ray-format home entertainment will be approximately $6 billion for each.

c.

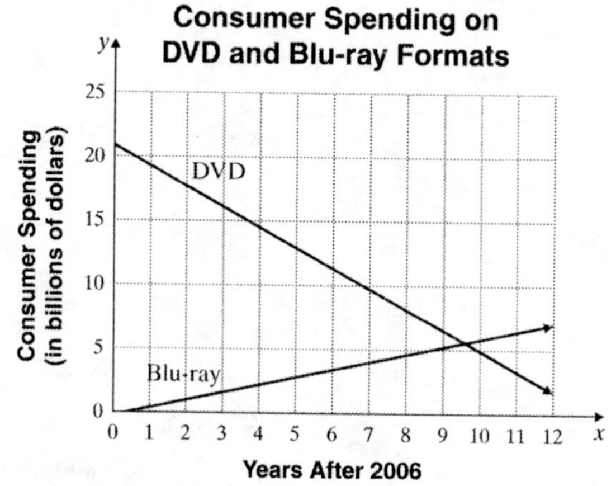

answers may vary

Section 7.3 Practice Exercises

1. $\begin{cases} x+y=13 \\ x-y=5 \end{cases}$

Add the equations to eliminate y; then solve for x.

$$\begin{array}{l} x+y=13 \\ \underline{x-y=5} \\ 2x \quad =18 \\ x \quad =9 \end{array}$$

Now solve for y.

$$\begin{aligned} x+y&=13 \\ 9+y&=13 \\ y&=4 \end{aligned}$$

The solution of the system is (9, 4).

2. $\begin{cases} 2x-y=-6 \\ -x+4y=17 \end{cases}$

Multiply the first equation by 4.

$$\begin{cases} 4(2x-y)=4(-6) \\ -x+4y=17 \end{cases} \Rightarrow \begin{cases} 8x-4y=-24 \\ -x+4y=17 \end{cases}$$

Add the equations to eliminate y; then solve for x.

$$\begin{array}{l} 8x-4y=-24 \\ \underline{-x+4y=17} \\ 7x \quad =-7 \\ x=-1 \end{array}$$

Now solve for y.

$$\begin{aligned} 2x-y&=-6 \\ 2(-1)-y&=-6 \\ -2-y&=-6 \\ -y&=-4 \\ y&=4 \end{aligned}$$

The solution of the system is (−1, 4).

3. $\begin{cases} x-3y=-2 \\ -3x+9y=5 \end{cases}$

Multiply the first equation by 3.

$\begin{cases} 3(x-3y)=3(-2) \\ -3x+9y=5 \end{cases} \Rightarrow \begin{cases} 3x-9y=-6 \\ -3x+9y=5 \end{cases}$

Add the equations to eliminate y; then solve for x.

$$\begin{array}{r} 3x-9y=-6 \\ \underline{-3x+9y=5} \\ 0=-1 \end{array}$$

Since the statement $0=-1$ is false, there is no solution to the system.

4. $\begin{cases} 2x+5y=1 \\ -4x-10y=-2 \end{cases}$

Multiply the first equation by 2.

$\begin{cases} 2(2x+5y)=2(1) \\ -4x-10y=-2 \end{cases} \Rightarrow \begin{cases} 4x+10y=2 \\ -4x-10y=-2 \end{cases}$

Add the equations to eliminate y; then solve for x.

$$\begin{array}{r} 4x+10y=2 \\ \underline{-4x-10y=-2} \\ 0=0 \end{array}$$

Since the statement $0=0$ is true, there are an infinite number of solutions.

5. $\begin{cases} 4x+5y=14 \\ 3x-2y=-1 \end{cases}$

Multiply the first equation by 2 and multiply the second equation by 5.

$\begin{cases} 2(4x+5y)=2(14) \\ 5(3x-2y)=5(-1) \end{cases} \Rightarrow \begin{cases} 8x+10y=28 \\ 15x-10y=-5 \end{cases}$

Add the equations to eliminate y; then solve for x.

$$\begin{array}{r} 8x+10y=28 \\ \underline{15x-10y=-5} \\ 23x \qquad =23 \\ x=1 \end{array}$$

Now solve for y.

$$\begin{aligned} 4x+5y&=14 \\ 4(1)+5y&=14 \\ 5y&=10 \\ y&=2 \end{aligned}$$

The solution of the system is (1, 2).

6. $\begin{cases} -\frac{x}{3}+y=\frac{4}{3} \\ \frac{x}{2}-\frac{5}{2}y=-\frac{1}{2} \end{cases}$

Multiply the first equation by 3 and the second equation by 2.

$\begin{cases} 3\left(-\frac{x}{3}+y\right)=3\left(\frac{4}{3}\right) \\ 2\left(\frac{x}{2}-\frac{5}{2}y\right)=2\left(-\frac{1}{2}\right) \end{cases} \Rightarrow \begin{cases} -x+3y=4 \\ x-5y=-1 \end{cases}$

Add the equations to eliminate x; then solve for y.

$$\begin{array}{r} -x+3y=4 \\ \underline{x-5y=-1} \\ -2y=3 \\ y=-\frac{3}{2} \end{array}$$

Now solve for y through elimination. Multiply the first equation by 15 and the second equation by 6.

$\begin{cases} 15\left(-\frac{x}{3}+y\right)=15\left(\frac{4}{3}\right) \\ 6\left(\frac{x}{2}-\frac{5}{2}y\right)=6\left(-\frac{1}{2}\right) \end{cases} \Rightarrow \begin{cases} -5x+15y=20 \\ 3x-15y=-3 \end{cases}$

Add the equations to eliminate y; then solve for x.

$$\begin{array}{r} -5x+15y=20 \\ \underline{3x-15y=-3} \\ -2x \qquad =17 \\ x=-\frac{17}{2} \end{array}$$

The solution to the system is $\left(-\frac{17}{2}, -\frac{3}{2}\right)$.

Vocabulary, Readiness & Video Check 7.3

1. $\begin{cases} 3x-2y=-9 \\ x+5y=14 \end{cases}$

Multiply the second equation by −3, then add the resulting equations.

$$\begin{array}{r} 3x-2y=-9 \\ \underline{-3x-15y=-42} \\ -17y=-51 \end{array}$$

The y's are not eliminated; the statement is false.

2. $\begin{cases} 3x-2y=-9 \\ x+5y=14 \end{cases}$

Multiply the second equation by −3, then add the resulting equations.

$$\begin{array}{r} 3x-2y=-9 \\ \underline{-3x-15y=-42} \\ -17y=-51 \end{array}$$

The statement is true.

3. $\begin{cases} 3x-2y=-9 \\ x+5y=14 \end{cases}$

Multiply the first equation by 5 and the second equation by 2, then add the two new equations.

$$\begin{aligned} 15x-10y&=-45 \\ 2x+10y&=\ \ 28 \\ \hline 17x\qquad&=-17 \end{aligned}$$

The statement is true.

4. $\begin{cases} 3x-2y=-9 \\ x+5y=14 \end{cases}$

Multiply the first equation by 5 and the second equation by −2, then add the two new equations.

$$\begin{aligned} 15x-10y&=-45 \\ -2x-10y&=-28 \\ \hline 13x-20y&=-73 \end{aligned}$$

The *y*'s are not eliminated; the statement is false.

5. The multiplication property of equality; be sure to multiply *both* sides of the equation by the nonzero number chosen.

Exercise Set 7.3

1. $\begin{cases} 3x+y=5 \\ 6x-y=4 \end{cases}$

Add the equations to eliminate *y*; then solve for *x*.

$$\begin{aligned} 3x+y&=5 \\ 6x-y&=4 \\ \hline 9x\quad&=9 \\ x&=1 \end{aligned}$$

Now solve for *y*.

$$\begin{aligned} 3x+y&=5 \\ 3(1)+y&=5 \\ 3+y&=5 \\ y&=2 \end{aligned}$$

The solution of the system is (1, 2).

3. $\begin{cases} x-2y=8 \\ -x+5y=-17 \end{cases}$

Add the equations to eliminate *x*; then solve for *y*.

$$\begin{aligned} x-2y&=8 \\ -x+5y&=-17 \\ \hline 3y&=-9 \\ y&=-3 \end{aligned}$$

Now solve for *x*.

$$\begin{aligned} x-2y&=8 \\ x-2(-3)&=8 \\ x+6&=8 \\ x&=2 \end{aligned}$$

The solution of the system is (2, −3).

5. $\begin{cases} 3x+y=-11 \\ 6x-2y=-2 \end{cases}$

Multiply the first equation by 2.

$\begin{cases} 2(3x+y)=2(-11) \\ 6x-2y=-2 \end{cases} \rightarrow \begin{cases} 6x+2y=-22 \\ 6x-2y=-2 \end{cases}$

Add the equations to eliminate *y*; then solve for *x*.

$$\begin{aligned} 6x+2y&=-22 \\ 6x-2y&=-2 \\ \hline 12x\quad&=-24 \\ x&=-2 \end{aligned}$$

Now solve for *y*.

$$\begin{aligned} 3x+y&=-11 \\ 3(-2)+y&=-11 \\ -6+y&=-11 \\ y&=-5 \end{aligned}$$

The solution of the system is (−2, −5).

7. $\begin{cases} 3x+2y=11 \\ 5x-2y=29 \end{cases}$

Add the equations to eliminate *y*; then solve for *x*.

$$\begin{aligned} 3x+2y&=11 \\ 5x-2y&=29 \\ \hline 8x&=40 \\ x&=5 \end{aligned}$$

Now solve for *y*.

$$\begin{aligned} 3x+2y&=11 \\ 3(5)+2y&=11 \\ 15+2y&=11 \\ 2y&=-4 \\ y&=-2 \end{aligned}$$

The solution of the system is (5, −2).

9. $\begin{cases} x+5y=18 \\ 3x+2y=-11 \end{cases}$

Multiply the first equation by −3.

$\begin{cases} -3(x+5y)=-3(18) \\ 3x+2y=-11 \end{cases} \rightarrow \begin{cases} -3x-15y=-54 \\ 3x+2y=-11 \end{cases}$

Add the equations to eliminate x; then solve for y.

$$\begin{array}{r} -3x-15y=-54 \\ 3x+2y=-11 \\ \hline -13y=-65 \\ y=5 \end{array}$$

Now solve for x.

$$\begin{aligned} x+5y&=18 \\ x+5(5)&=18 \\ x+25&=18 \\ x&=-7 \end{aligned}$$

The solution of the system is $(-7, 5)$.

11. $\begin{cases} x+y=6 \\ x-y=6 \end{cases}$

Add the equations to eliminate y; then solve for x.

$$\begin{array}{r} x+y=6 \\ x-y=6 \\ \hline 2x\quad =12 \\ x=6 \end{array}$$

Now solve for y.

$$\begin{aligned} x+y&=6 \\ 6+y&=6 \\ y&=0 \end{aligned}$$

The solution of the system is $(6, 0)$.

13. $\begin{cases} 2x+3y=0 \\ 4x+6y=3 \end{cases}$

Multiply the first equation by -2.

$$\begin{cases} -2(2x+3y)=-2(0) \\ 4x+6y=3 \end{cases} \rightarrow \begin{cases} -4x-6y=0 \\ 4x+6y=3 \end{cases}$$

Add the equations to eliminate x.

$$\begin{array}{r} -4x-6y=0 \\ 4x+6y=3 \\ \hline 0=3 \end{array}$$

Since the statement $0 = 3$ is false, the system has no solution.

15. $\begin{cases} -x+5y=-1 \\ 3x-15y=3 \end{cases}$

Multiply the first equation by 3.

$$\begin{cases} 3(-x+5y)=3(-1) \\ 3x-15y=3 \end{cases} \rightarrow \begin{cases} -3x+15y=-3 \\ 3x-15y=3 \end{cases}$$

Add the equations to eliminate x.

$$\begin{array}{r} -3x+15y=-3 \\ 3x-15y=3 \\ \hline 0=0 \end{array}$$

Since the statement $0 = 0$ is true, the system has an infinite number of solutions.

17. $\begin{cases} 3x-2y=7 \\ 5x+4y=8 \end{cases}$

Multiply the first equation by 2.

$$\begin{cases} 2(3x-2y)=2(7) \\ 5x+4y=8 \end{cases} \rightarrow \begin{cases} 6x-4y=14 \\ 5x+4y=8 \end{cases}$$

Add the equations to eliminate y; then solve for x.

$$\begin{array}{r} 6x-4y=14 \\ 5x+4y=8 \\ \hline 11x\quad =22 \\ x=2 \end{array}$$

Now solve for y.

$$\begin{aligned} 3x-2y&=7 \\ 3(2)-2y&=7 \\ 6-2y&=7 \\ -2y&=1 \\ y&=-\frac{1}{2} \end{aligned}$$

The solution of the system is $\left(2, -\frac{1}{2}\right)$.

19. $\begin{cases} 8x=-11y-16 \\ 2x+3y=-4 \end{cases} \rightarrow \begin{cases} 8x+11y=-16 \\ 2x+3y=-4 \end{cases}$

Multiply the second equation by -4.

$$\begin{cases} 8x+11y=-16 \\ -4(2x+3y)=-4(-4) \end{cases} \rightarrow \begin{cases} 8x+11y=-16 \\ -8x-12y=16 \end{cases}$$

Add the equations to eliminate x; then solve for y.

$$\begin{array}{r} 8x+11y=-16 \\ -8x-12y=16 \\ \hline -y=0 \\ y=0 \end{array}$$

Now solve for x.

$$\begin{aligned} 8x+11y&=-16 \\ 8x+11(0)&=-16 \\ 8x&=-16 \\ x&=-2 \end{aligned}$$

The solution of the system is $(-2, 0)$.

21. $\begin{cases} 4x-3y=7 \\ 7x+5y=2 \end{cases}$

Multiply the first equation by 5 and the second equation by 3.

$$\begin{cases} 5(4x-3y)=5(7) \\ 3(7x+5y)=3(2) \end{cases} \rightarrow \begin{cases} 20x-15y=35 \\ 21x+15y=6 \end{cases}$$

Add the equations to eliminate y; then solve for x.

$$\begin{array}{l} 20x-15y=35 \\ \underline{21x+15y=6} \\ 41x \quad\quad = 41 \\ \quad\quad x=1 \end{array}$$

Now solve for y.

$$\begin{aligned} 4x-3y&=7 \\ 4(1)-3y&=7 \\ 4-3y&=7 \\ -3y&=3 \\ y&=-1 \end{aligned}$$

The solution of the system is $(1, -1)$.

23. $\begin{cases} 4x-6y=8 \\ 6x-9y=12 \end{cases}$

Multiply the first equation by –3 and the second equation by 2.

$$\begin{cases} -3(4x-6y)=-3(8) \\ 2(6x-9y)=2(12) \end{cases} \rightarrow \begin{cases} -12x+18y=-24 \\ 12x-18y=24 \end{cases}$$

Add the equations to eliminate x.

$$\begin{array}{r} -12x+18y=-24 \\ \underline{12x-18y=24} \\ 0=0 \end{array}$$

Since the statement $0 = 0$ is true, the system has an infinite number of solutions.

25. $\begin{cases} 2x-5y=4 \\ 3x-2y=4 \end{cases}$

Multiply the first equation by –3 and the second equation by 2.

$$\begin{cases} -3(2x-5y)=-3(4) \\ 2(3x-2y)=2(4) \end{cases} \rightarrow \begin{cases} -6x+15y=-12 \\ 6x-4y=8 \end{cases}$$

Add the equations to eliminate x; then solve for y.

$$\begin{array}{r} -6x+15y=-12 \\ \underline{6x-4y=8} \\ 11y=-4 \\ y=-\frac{4}{11} \end{array}$$

Multiply the first original equation by –2 and the second original equation by 5.

$$\begin{cases} -2(2x-5y)=-2(4) \\ 5(3x-2y)=5(4) \end{cases} \rightarrow \begin{cases} -4x+10y=-8 \\ 15x-10y=20 \end{cases}$$

Add the equations to eliminate y; then solve for x.

$$\begin{array}{l} -4x+10y=-8 \\ \underline{15x-10y=20} \\ 11x \quad\quad =12 \\ \quad\quad x=\frac{12}{11} \end{array}$$

The solution of the system is $\left(\frac{12}{11}, -\frac{4}{11}\right)$.

27. $\begin{cases} \frac{x}{3}+\frac{y}{6}=1 \\ \frac{x}{2}-\frac{y}{4}=0 \end{cases}$

Multiply the first equation by 6 and the second equation by 4.

$$\begin{cases} 6\left(\frac{x}{3}+\frac{y}{6}\right)=6(1) \\ 4\left(\frac{x}{2}-\frac{y}{4}\right)=4(0) \end{cases} \rightarrow \begin{cases} 2x+y=6 \\ 2x-y=0 \end{cases}$$

Add the equations to eliminate y; then solve for x.

$$\begin{array}{l} 2x+y=6 \\ \underline{2x-y=0} \\ 4x=6 \\ x=\frac{6}{4}=\frac{3}{2} \end{array}$$

Multiply the original first equation by 6 and the second original equation by –4.

$$\begin{cases} 6\left(\frac{x}{3}+\frac{y}{6}\right)=6(1) \\ -4\left(\frac{x}{2}-\frac{y}{4}\right)=-4(0) \end{cases} \rightarrow \begin{cases} 2x+y=6 \\ -2x+y=0 \end{cases}$$

Add the equations to eliminate x; then solve for y.

$$\begin{array}{r} 2x+y=6 \\ \underline{-2x+y=0} \\ 2y=6 \\ y=3 \end{array}$$

The solution of the system is $\left(\frac{3}{2}, 3\right)$.

29. $\begin{cases} \frac{10}{3}x+4y=-4 \\ 5x+6y=-6 \end{cases}$

Multiply the first equation by 3 and the second equation by –2.

$$\begin{cases} 3\left(\frac{10}{3}x+4y\right)=3(-4) \\ -2(5x+6y)=-2(-6) \end{cases} \rightarrow \begin{cases} 10x+12y=-12 \\ -10x-12y=12 \end{cases}$$

Add the equations to eliminate x.

$$\begin{array}{r} 10x+12y=-12 \\ \underline{-10x-12y=12} \\ 0=0 \end{array}$$

Since the statement $0 = 0$ is true, the system has an infinite number of solutions.

31. $\begin{cases} x - \dfrac{y}{3} = -1 \\ -\dfrac{x}{2} + \dfrac{y}{8} = \dfrac{1}{4} \end{cases}$

Multiply the first equation by 3 and the second equation by 8.

$$\begin{cases} 3\left(x - \dfrac{y}{3}\right) = 3(-1) \\ 8\left(-\dfrac{x}{2} + \dfrac{y}{8}\right) = 8\left(\dfrac{1}{4}\right) \end{cases} \rightarrow \begin{cases} 3x - y = -3 \\ -4x + y = 2 \end{cases}$$

Add the equations to eliminate y; then solve for x.

$$\begin{array}{r} 3x - y = -3 \\ \underline{-4x + y = 2} \\ -x \quad = -1 \\ x = 1 \end{array}$$

Now solve for y.

$$\begin{aligned} x - \frac{y}{3} &= -1 \\ 1 - \frac{y}{3} &= -1 \\ -\frac{y}{3} &= -2 \\ y &= 6 \end{aligned}$$

The solution of the system is (1, 6).

33. $\begin{cases} -4(x+2) = 3y \\ 2x - 2y = 3 \end{cases}$

Rewrite the first equation.

$$\begin{aligned} -4(x+2) &= 3y \\ -4x - 8 &= 3y \\ -4x &= 3y + 8 \\ -4x - 3y &= 8 \end{aligned}$$

$$\begin{cases} -4x - 3y = 8 \\ 2x - 2y = 3 \end{cases}$$

Multiply the second equation by 2.

$$\begin{cases} -4x - 3y = 8 \\ 2(2x - 2y) = 2(3) \end{cases} \rightarrow \begin{cases} -4x - 3y = 8 \\ 4x - 4y = 6 \end{cases}$$

Add the equations to eliminate x; then solve for y.

$$\begin{array}{r} -4x - 3y = 8 \\ \underline{4x - 4y = 6} \\ -7y = 14 \\ y = -2 \end{array}$$

Now solve for x.

$$\begin{aligned} 2x - 2y &= 3 \\ 2x - 2(-2) &= 3 \\ 2x + 4 &= 3 \\ 2x &= -1 \\ x &= -\frac{1}{2} \end{aligned}$$

The solution of the system is $\left(-\dfrac{1}{2}, -2\right)$.

35. $\begin{cases} \dfrac{x}{3} - y = 2 \\ -\dfrac{x}{2} + \dfrac{3y}{2} = -3 \end{cases}$

Multiply the first equation by 3 and the second by 2.

$$\begin{cases} 3\left(\dfrac{x}{3} - y\right) = 3(2) \\ 2\left(-\dfrac{x}{2} + \dfrac{3y}{2}\right) = 2(-3) \end{cases} \rightarrow \begin{cases} x - 3y = 6 \\ -x + 3y = -6 \end{cases}$$

Add the equations to eliminate y.

$$\begin{array}{r} x - 3y = 6 \\ \underline{-x + 3y = -6} \\ 0 = 0 \end{array}$$

Since the statement $0 = 0$ is true, the system has an infinite number of solutions.

37. $\begin{cases} \dfrac{3}{5}x - y = -\dfrac{4}{5} \\ 3x + \dfrac{y}{2} = -\dfrac{9}{5} \end{cases}$

Multiply the first equation by 5 and the second equation by 10 to eliminate fractions.

$$\begin{cases} 5\left(\dfrac{3}{5}x - y\right) = 5\left(-\dfrac{4}{5}\right) \\ 10\left(3x + \dfrac{y}{2}\right) = 10\left(-\dfrac{9}{5}\right) \end{cases} \rightarrow \begin{cases} 3x - 5y = -4 \\ 30x + 5y = -18 \end{cases}$$

Add the equations to eliminate y; then solve for x.

$$\begin{array}{r} 3x - 5y = -4 \\ \underline{30x + 5y = -18} \\ 33x \quad = -22 \\ x = -\dfrac{22}{33} = -\dfrac{2}{3} \end{array}$$

Now use the equation $3x - 5y = -4$ to solve for y.

$$3x - 5y = -4$$
$$3\left(-\frac{2}{3}\right) - 5y = -4$$
$$-2 - 5y = -4$$
$$-5y = -2$$
$$y = \frac{2}{5}$$

The solution of the system is $\left(-\frac{2}{3}, \frac{2}{5}\right)$.

39. $\begin{cases} 3.5x + 2.5y = 17 \\ -1.5x - 7.5y = -33 \end{cases}$

Multiply the first equation by 30 and the second equation by 10.

$\begin{cases} 30(3.5x + 2.5y) = 30(17) \\ 10(-1.5x - 7.5y) = 10(-33) \end{cases}$

$\rightarrow \begin{cases} 105x + 75y = 510 \\ -15x - 75y = -330 \end{cases}$

Add the equations to eliminate y; then solve for x.

$$\begin{array}{r} 105x + 75y = 510 \\ -15x - 75y = -330 \\ \hline 90x \qquad = 180 \\ x = 2 \end{array}$$

Now solve for y.

$$3.5x + 2.5y = 17$$
$$3.5(2) + 2.5y = 17$$
$$7 + 2.5y = 17$$
$$2.5y = 10$$
$$y = 4$$

The solution of the system is (2, 4).

41. $\begin{cases} 0.02x + 0.04y = 0.09 \\ -0.1x + 0.3y = 0.8 \end{cases}$

Multiply the first equation by 100 and the second equation by 10 to eliminate decimals.

$\begin{cases} 100(0.02x + 0.04y) = 100(0.09) \\ 10(-0.1x + 0.3y) = 10(0.8) \end{cases}$

$\rightarrow \begin{cases} 2x + 4y = 9 \\ -x + 3y = 8 \end{cases}$

Multiply the second equation by 2.

$\begin{cases} 2x + 4y = 9 \\ 2(-x + 3y) = 2(8) \end{cases} \rightarrow \begin{cases} 2x + 4y = 9 \\ -2x + 6y = 16 \end{cases}$

Add the equations to eliminate x; then solve for y.

$$\begin{array}{r} 2x + 4y = 9 \\ -2x + 6y = 16 \\ \hline 10y = 25 \\ y = 2.5 \end{array}$$

Use the equation $-x + 3y = 8$ to solve for x.

$$-x + 3y = 8$$
$$-x + 3(2.5) = 8$$
$$-x + 7.5 = 8$$
$$-x = 0.5$$
$$x = -0.5$$

The solution of the system is (−0.5, 2.5).

43. Twice a number, added to 6, is 3 less than the number is written as $2x + 6 = x - 3$.

45. Three times a number, subtracted from 20, is 2 is written as $20 - 3x = 2$.

47. The product of 4 and the sum of a number and 6 is twice the number is written as $4(x + 6) = 2x$.

49. To eliminate the variable y, multiply the second equation by 2.

$$3x - y = -12$$
$$2(3x - y) = 2(-12)$$
$$6x - 2y = -24$$

51. **b**; answers may vary

53. answers may vary

55. **a.** When $b = 15$, the system has an infinite number of solutions.

b. When b is any real number except 15, the system has no solutions.

57. $\begin{cases} 2x + 3y = 14 \\ 3x - 4y = -69.1 \end{cases}$

Multiply the first equation by −3 and the second equation by 2.

$\begin{cases} -3(2x + 3y) = -3(14) \\ 2(3x - 4y) = 2(-69.1) \end{cases}$

$\rightarrow \begin{cases} -6x - 9y = -42 \\ 6x - 8y = -138.2 \end{cases}$

Add the equations to eliminate x; then solve for y.

$$\begin{array}{r} -6x - 9y = -42 \\ 6x - 8y = -138.2 \\ \hline -17y = -180.2 \\ y = 10.6 \end{array}$$

Now solve for x.

$$\begin{aligned} 2x+3y &= 14 \\ 2x+3(10.6) &= 14 \\ 2x+31.8 &= 14 \\ 2x &= -17.8 \\ x &= -8.9 \end{aligned}$$

The solution of the system is (−8.9, 10.6).

59. **a.** $\begin{cases} 38x+10y=3167 \\ 117x-10y=-2827 \end{cases}$

Add the equations to eliminate y; then solve for x.

$$\begin{aligned} 38x+10y &= 3167 \\ 117x-10y &= -2827 \\ \hline 155x \qquad &= 340 \\ x &\approx 2 \end{aligned}$$

Now solve for y.

$$\begin{aligned} 38x+10y &= 3167 \\ 38(2)+10y &= 3167 \\ 76+10y &= 3167 \\ 10y &= 3091 \\ y &\approx 309 \end{aligned}$$

The rounded solution is (2, 309). (If the second equation is used to find y, the rounded solution is (2, 306).)

b. In about 2012 (2010 + 2), the number of mail carrier jobs was approximately equal to the number of market research analyst jobs.

c. There were 309 thousand (or 306 thousand) mail carrier and market research analyst jobs in 2012.

Integrated Review

1. $\begin{cases} 2x-3y=-11 \\ y=4x-3 \end{cases}$

Substitute $4x - 3$ for y in the first equation and solve for x.

$$\begin{aligned} 2x-3y &= -11 \\ 2x-3(4x-3) &= -11 \\ 2x-12x+9 &= -11 \\ -10x &= -20 \\ x &= 2 \end{aligned}$$

Now solve for y.

$y = 4x - 3 = 4(2) - 3 = 8 - 3 = 5$

The solution to the system is (2, 5).

2. $\begin{cases} 4x-5y=6 \\ y=3x-10 \end{cases}$

Substitute $3x - 10$ for y in the first equation and solve for x.

$$\begin{aligned} 4x-5y &= 6 \\ 4x-5(3x-10) &= 6 \\ 4x-15x+50 &= 6 \\ -11x &= -44 \\ x &= 4 \end{aligned}$$

Now solve for y.

$y = 3x - 10 = 3(4) - 10 = 12 - 10 = 2$

The solution of the system is (4, 2).

3. $\begin{cases} x+y=3 \\ x-y=7 \end{cases}$

Add the equations to eliminate y; then solve for x.

$$\begin{aligned} x+y &= 3 \\ x-y &= 7 \\ \hline 2x \quad &= 10 \\ x &= 5 \end{aligned}$$

Now solve for y.

$$\begin{aligned} x+y &= 3 \\ 5+y &= 3 \\ y &= -2 \end{aligned}$$

The solution of the system is (5, −2).

4. $\begin{cases} x-y=20 \\ x+y=-8 \end{cases}$

Add the equations to eliminate y; then solve for x.

$$\begin{aligned} x-y &= 20 \\ x+y &= -8 \\ \hline 2x &= 12 \\ x &= 6 \end{aligned}$$

Now solve for y.

$$\begin{aligned} x+y &= -8 \\ 6+y &= -8 \\ y &= -14 \end{aligned}$$

The solution of the system is (6, −14).

5. $\begin{cases} x+2y=1 \\ 3x+4y=-1 \end{cases}$

Solve the first equation for x.

$$\begin{aligned} x+2y &= 1 \\ x &= -2y+1 \end{aligned}$$

Substitute $-2y + 1$ for x in the second equation and solve for y.

$$\begin{aligned} 3x+4y &= -1 \\ 3(-2y+1)+4y &= -1 \\ -6y+3+4y &= -1 \\ -2y &= -4 \\ y &= 2 \end{aligned}$$

Now solve for x.

$x = -2y + 1 = -2(2) + 1 = -4 + 1 = -3$

The solution of the system is (−3, 2).

6. $\begin{cases} x+3y=5 \\ 5x+6y=-2 \end{cases}$

Solve the first equation for x.

$x+3y=5$
$x=-3y+5$

Substitute $-3y+5$ for x in the second equation and solve for y.

$5x+6y=-2$
$5(-3y+5)+6y=-2$
$-15y+25+6y=-2$
$-9y=-27$
$y=3$

Now solve for x.

$x=-3y+5=-3(3)+5=-9+5=-4$

The solution of the system is $(-4, 3)$.

7. $y=x+3$
$3x=2y-6$

Substitute $x+3$ for y in the second equation and solve for x.

$3x=2y-6$
$3x=2(x+3)-6$
$3x=2x+6-6$
$x=0$

Now solve for y.

$y=x+3=0+3=3$

The solution of the system is $(0, 3)$.

8. $\begin{cases} y=-2x \\ 2x-3y=-16 \end{cases}$

Substitute $-2x$ for y in the second equation and solve for x.

$2x-3y=-16$
$2x-3(-2x)=-16$
$2x+6x=-16$
$8x=-16$
$x=-2$

Now solve for y.

$y=-2x=-2(-2)=4$

The solution of the system is $(-2, 4)$.

9. $\begin{cases} y=2x-3 \\ y=5x-18 \end{cases}$

Substitute $2x-3$ for y in the second equation and solve for x.

$y=5x-18$
$2x-3=5x-18$
$15=3x$
$5=x$

Now solve for y.

$y=2x-3=2(5)-3=10-3=7$

The solution of the system is $(5, 7)$.

10. $\begin{cases} y=6x-5 \\ y=4x-11 \end{cases}$

Substitute $6x-5$ for y in the second equation and solve for x.

$y=4x-11$
$6x-5=4x-11$
$2x=-6$
$x=-3$

Now solve for y.

$y=6x-5=6(-3)-5=-18-5=-23$

The solution of the system is $(-3, -23)$.

11. $\begin{cases} x+\frac{1}{6}y=\frac{1}{2} \\ 3x+2y=3 \end{cases}$

Multiply the first equation by 6 to eliminate fractions.

$6\left(x+\frac{1}{6}y\right)=6\left(\frac{1}{2}\right)$
$6x+y=3$

Now solve for y.

$y=-6x+3$

Substitute $-6x+3$ for y in the second equation and solve for x.

$3x+2y=3$
$3x+2(-6x+3)=3$
$3x-12x+6=3$
$-9x=-3$
$x=\frac{3}{9}=\frac{1}{3}$

Now solve for y.

$y=-6x+3=-6\left(\frac{1}{3}\right)+3=-2+3=1$

The solution of the system is $\left(\frac{1}{3}, 1\right)$.

12. $x+\frac{1}{3}y=\frac{5}{12}$
$8x+3y=4$

Multiply the first equation by 12 to eliminate fractions.

$12\left(x+\frac{1}{3}y\right)=12\left(\frac{5}{12}\right)$
$12x+4y=5$

Multiply the revised first equation by -3 and the second original equation by 4.

$\begin{cases} -3(12x+4y)=-3(5) \\ 4(8x+3y)=4(4) \end{cases} \rightarrow \begin{cases} -36x-12y=-15 \\ 32x+12y=16 \end{cases}$

Add the equations to eliminate y; then solve for x.

$$\begin{array}{r} -36x-12y=-15 \\ \underline{32x+12y=16} \\ -4x\qquad=1 \\ x=-\frac{1}{4} \end{array}$$

Now solve for y.

$$\begin{aligned} 8x+3y&=4 \\ 8\left(-\frac{1}{4}\right)+3y&=4 \\ -2+3y&=4 \\ 3y&=6 \\ y&=2 \end{aligned}$$

The solution of the system is $\left(-\frac{1}{4}, 2\right)$.

13. $\begin{cases} x-5y=1 \\ -2x+10y=3 \end{cases}$

Solve the first equation for x.

$$\begin{aligned} x-5y&=1 \\ x&=5y+1 \end{aligned}$$

Substitute $5y + 1$ for x in the second equation and then solve for y.

$$\begin{aligned} -2x+10y&=3 \\ -2(5y+1)+10y&=3 \\ -10y-2+10y&=3 \\ -2&=3 \end{aligned}$$

Since the statement is false, the system has no solution.

14. $\begin{cases} -x+2y=3 \\ 3x-6y=-9 \end{cases}$

Solve the first equation for x.

$$\begin{aligned} -x+2y&=3 \\ x&=2y-3 \end{aligned}$$

Substitute $2y - 3$ for x in the second equation and solve for y.

$$\begin{aligned} 3x-6y&=-9 \\ 3(2y-3)-6y&=-9 \\ 6y-9-6y&=-9 \\ 0&=0 \end{aligned}$$

Since the statement $0 = 0$ is true, the system has an infinite number of solutions.

15. $\begin{cases} 0.2x-0.3y=-0.95 \\ 0.4x+0.1y=0.55 \end{cases}$

Multiply both equations by 100 to eliminate decimals.

$$\begin{cases} 100(0.2x-0.3y)=100(-0.95) \\ 100(0.4x+0.1y)=100(0.55) \end{cases}$$

$$\rightarrow \begin{cases} 20x-30y=-95 \\ 40x+10y=55 \end{cases}$$

Multiply the second revised equation by 3.

$$\begin{cases} 20x-30y=-95 \\ 3(40x+10y)=3(55) \end{cases} \rightarrow \begin{cases} 20x-30y=-95 \\ 120x+30y=165 \end{cases}$$

Add the equations to eliminate y; then solve for x.

$$\begin{array}{r} 20x-30y=-95 \\ \underline{120x+30y=165} \\ 140x\qquad=70 \\ x=0.5 \end{array}$$

Now solve for y.

$$\begin{aligned} 0.4x+0.1y&=0.55 \\ 0.4(0.5)+0.1y&=0.55 \\ 0.2+0.1y&=0.55 \\ 0.1y&=0.35 \\ y&=3.5 \end{aligned}$$

The solution of the system is (0.5, 3.5).

16. $\begin{cases} 0.08x-0.04y=-0.11 \\ 0.02x-0.06y=-0.09 \end{cases}$

Multiply both equations by 100 to eliminate decimals.

$$\begin{cases} 100(0.08x-0.04y)=100(-0.11) \\ 100(0.02x-0.06y=100(-0.09) \end{cases}$$

$$\rightarrow \begin{cases} 8x-4y=-11 \\ 2x-6y=-9 \end{cases}$$

Multiply the second revised equation by –4.

$$\begin{cases} 8x-4y=-11 \\ -4(2x-6y)=-4(-9) \end{cases} \rightarrow \begin{cases} 8x-4y=-11 \\ -8x+24y=36 \end{cases}$$

Add the equations to eliminate x; then solve for y.

$$\begin{array}{r} 8x-4y=-11 \\ \underline{-8x+24y=36} \\ 20y=25 \\ y=1.25 \end{array}$$

Now solve for x.

$$\begin{aligned} 0.08x-0.04y&=-0.11 \\ 0.08x-0.04(1.25)&=-0.11 \\ 0.08x-0.05&=-0.11 \\ 0.08x&=-0.06 \\ x&=-0.75 \end{aligned}$$

The solution of the system is (–0.75, 1.25).

17. $\begin{cases} x = 3y - 7 \\ 2x - 6y = -14 \end{cases}$

Substitute $3y - 7$ for x in the second equation and solve for y.

$$\begin{aligned} 2x - 6y &= -14 \\ 2(3y - 7) - 6y &= -14 \\ 6y - 14 - 6y &= -14 \\ 0 &= 0 \end{aligned}$$

Since the statement $0 = 0$ is true, the system has an infinite number of solutions.

18. $\begin{cases} y = \dfrac{x}{2} - 3 \\ 2x - 4y = 0 \end{cases}$

Substitute $\dfrac{x}{2} - 3$ for y in the second equation and solve for x.

$$\begin{aligned} 2x - 4y &= 0 \\ 2x - 4\left(\frac{x}{2} - 3\right) &= 0 \\ 2x - 2x + 12 &= 0 \\ 12 &= 0 \end{aligned}$$

Since the statement $12 = 0$ is false, the system has no solution.

19. $\begin{cases} 2x + 5y = -1 \\ 3x - 4y = 33 \end{cases}$

Multiply the first equation by 3 and the second equation by -2.

$$\begin{cases} 3(2x + 5y) = 3(-1) \\ -2(3x - 4y) = -2(33) \end{cases} \rightarrow \begin{cases} 6x + 15y = -3 \\ -6x + 8y = -66 \end{cases}$$

Add the equations to eliminate x; then solve for y.

$$\begin{aligned} 6x + 15y &= -3 \\ -6x + 8y &= -66 \\ \hline 23y &= -69 \\ y &= -3 \end{aligned}$$

Now solve for x.

$$\begin{aligned} 2x + 5y &= -1 \\ 2x + 5(-3) &= -1 \\ 2x - 15 &= -1 \\ 2x &= 14 \\ x &= 7 \end{aligned}$$

The solution of the system is $(7, -3)$.

20. $\begin{cases} 7x - 3y = 2 \\ 6x + 5y = -21 \end{cases}$

Multiply the first equation by 5 and the second equation by 3.

$$\begin{cases} 5(7x - 3y) = 5(2) \\ 3(6x + 5y) = 3(-21) \end{cases} \rightarrow \begin{cases} 35x - 15y = 10 \\ 18x + 15y = -63 \end{cases}$$

Add the equations to eliminate y; then solve for x.

$$\begin{aligned} 35x - 15y &= 10 \\ 18x + 15y &= -63 \\ \hline 53x \quad &= -53 \\ x &= -1 \end{aligned}$$

Now solve for y.

$$\begin{aligned} 7x - 3y &= 2 \\ 7(-1) - 3y &= 2 \\ -7 - 3y &= 2 \\ -3y &= 9 \\ y &= -3 \end{aligned}$$

The solution of the system is $(-1, -3)$.

21. answers may vary

22. answers may vary

Section 7.4 Practice Exercises

1. $\begin{cases} x + y = 50 \\ x - y = 22 \end{cases}$

Add the equations to eliminate y; then solve for x.

$$\begin{aligned} x + y &= 50 \\ x - y &= 22 \\ \hline 2x \quad &= 72 \\ x &= 36 \end{aligned}$$

Now solve for y.

$$\begin{aligned} x + y &= 50 \\ 36 + y &= 50 \\ y &= 14 \end{aligned}$$

The two numbers are 36 and 14.

2. $\begin{cases} A + C = 587 \\ 7A + 5C = 3379 \end{cases}$

Multiply the first equation by -5.

$$\begin{cases} -5(A + C) = -5(587) \\ 7A + 5C = 3379 \end{cases} \rightarrow \begin{cases} -5A - 5C = -2935 \\ 7A + 5C = 3379 \end{cases}$$

Add the equations to eliminate C; then solve for A.

$$\begin{aligned} -5A - 5C &= -2935 \\ 7A + 5C &= 3379 \\ \hline 2A \quad &= 444 \\ A &= 222 \end{aligned}$$

Now solve for C.

$$\begin{aligned} A + C &= 587 \\ 222 + C &= 587 \\ C &= 365 \end{aligned}$$

There were 222 adults and 365 children.

3.

	r ·	t =	d
Faster car	x	3	$3x$
Slower car	y	3	$3y$

$3x+3y=440$
$x=y+10$

Substitute $y + 1$ for x in the first equation and solve for y.

$$3x+3y=440$$
$$3(y+10)+3y=440$$
$$3y+30+3y=440$$
$$6y=410$$
$$y=68\frac{1}{3}$$

Now solve for x.

$$x=y+10=68\frac{1}{3}+10=78\frac{1}{3}$$

One car's speed is $68\frac{1}{3}$ mph and the other car's speed is $78\frac{1}{3}$ mph.

4. Let x be the liters of 20% solution.
Let y be the liters of 70% solution.

$$\begin{cases} x+y=50 \\ 0.2x+0.7y=0.6(50) \end{cases}$$

Multiply the first equation by −2 and the second equation by 10.

$$\begin{cases} -2(x+y)=-2(50) \\ 10(0.2x+0.7y)=10(30) \end{cases}$$

$$\rightarrow \begin{cases} -2x-2y=-100 \\ 2x+7y=300 \end{cases}$$

Add the equations to eliminate x; then solve for y.

$$\begin{aligned} -2x-2y&=-100 \\ 2x+7y&=300 \\ \hline 5y&=200 \\ y&=40 \end{aligned}$$

Now solve for x.

$$x+y=50$$
$$x+40=50$$
$$x=10$$

10 liters of the 20% alcohol solution and 40 liters of the 70% alcohol solution make 50 liters of the 60% alcohol solution.

Vocabulary, Readiness & Video Check 7.4

1. Up to now we've been working with one variable/unknown and one equation. Because these systems involve two equations with two unknowns, for these applications we need to choose two variables to represent two unknowns and translate the problem into two equations.

Exercise Set 7.4

1. In choice **b**, the length is not 3 feet longer than the width. In choice **a**, the perimeter is $2(8+5)=2(13)=26$ feet, not 30 feet.
Choice **c** gives the solution, since $9=6+3$ and $2(9+6)=2(15)=30$.

3. In choice **a**, the total cost is $2(3)+3(4)=6+12=\$18$, not \$17.
In choice **c**, the total cost is $2(2)+3(5)=4+15=\$19$, not \$17.
Choice **b** gives the solution, since $2(4)+3(3)=8+9=\$17$ and $5(4)+4(3)=20+12=\$32$.

5. In choice **b**, the total number of coins is $20+44=64$, not 100.
In choice **c**, the total value of the coins is $60(0.10)+40(0.25)=6.00+10.00=\$16.00$, not \$13.00.
Choice **a** gives the solution, since $80+20=100$ and $80(0.10)+20(0.25)=8.00+5.00=\$13.00$.

7. Let x be the first number and y the second.

$$\begin{cases} x+y=15 \\ x-y=7 \end{cases}$$

9. Let x be the amount in the larger account and y be the amount in the smaller account.

$$\begin{cases} x+y=6500 \\ x=y+800 \end{cases}$$

11. Let x be the first number and y be the second.

$$x+y=83$$
$$x-y=17$$

Add the equations to eliminate y then solve for x.

$$\begin{aligned} x+y&=83 \\ x-y&=17 \\ \hline 2x\quad&=100 \\ x&=50 \end{aligned}$$

Now solve for y.

$$x+y=83$$
$$50+y=83$$
$$y=33$$

The numbers are 50 and 33.

13. Let x be the first number and y the second.

$$\begin{cases} x+2y=8 \\ 2x+y=25 \end{cases}$$

Solve the first equation for x.

$$x+2y=8$$
$$x=8-2y$$

Substitute $8-2y$ for x in the second equation and solve for y.

$$2x+y=25$$
$$2(8-2y)+y=25$$
$$16-4y+y=25$$
$$16-3y=25$$
$$-3y=9$$
$$y=-3$$

Now solve for x.

$x=8-2y=8-2(-3)=8+6=14$

The numbers are 14 and -3.

15. Let x be the number of runs that Miguel Cabrera batted in and y be the number that Josh Hamilton batted in.

$$\begin{cases} y=x-11 \\ x+y=267 \end{cases}$$

Substitute $x-11$ for y in the second equation and solve for x.

$$x+y=267$$
$$x+x-11=267$$
$$2x=278$$
$$x=139$$

Now solve for y.

$y=x-11=139-11=128$

Miguel Cabrera batted in 139 runs and Josh Hamilton batted in 128 runs.

17. Let a be the price of an adult's ticket and c be the price of a child's ticket.

$$\begin{cases} 3a+4c=159 \\ 2a+3c=112 \end{cases}$$

Multiply the first equation by -2 and the second equation by 3.

$$\begin{cases} -2(3a+4c)=-2(159) \\ 3(2a+3c)=3(112) \end{cases} \rightarrow \begin{cases} -6a-8c=-318 \\ 6a+9c=336 \end{cases}$$

Add the equations to eliminate a and solve for c.

$$\begin{array}{r} -6a-8c=-318 \\ 6a+9c=336 \\ \hline c=18 \end{array}$$

Now solve for a.

$$2a+3c=112$$
$$2a+3(18)=112$$
$$2a+54=112$$
$$2a=58$$
$$a=29$$

The price of an adult's ticket is \$29 and the price of a child's ticket is \$18.

19. Let x be quarters and y be nickels.

$$\begin{cases} x+y=80 \\ 0.25x+0.05y=14.60 \end{cases}$$

Solve the first equation in terms of y.

$$x+y=80$$
$$y=-x+80$$

Substitute $-x+80$ for y in the second equation and solve for x.

$$0.25x+0.05y=14.60$$
$$0.25x+0.05(-x+80)=14.60$$
$$0.25x-0.05x+4=14.60$$
$$0.20x=10.60$$
$$x=53$$

Now solve for y.

$y=-x+80=-53+80=27$

There are 53 quarters and 27 nickels.

21. Let x be the value of one share of McDonald's stock and let y be the value of one share of Ford Motor Co. stock.

$$\begin{cases} 30x+68y=4107 \\ x=y+80.55 \end{cases}$$

Substitute $y+80.55$ for x in the first equation and solve for y.

$$30x+68y=4107$$
$$30(y+80.55)+68y=4107$$
$$30y+2416.5+68y=4107$$
$$98y+2416.5=4107$$
$$98y=1690.5$$
$$y=17.25$$

Now solve for x.

$x=y+80.55=17.25+80.55=97.8$

The value of one share of McDonald's stock was \$97.80 and the value of one share of Ford Motor Co. stock was \$17.25.

23. Let x be the daily fee and y be the mileage charge.

$$\begin{cases} 4x+450y=240.50 \\ 3x+200y=146.00 \end{cases}$$

Multiply the first equation by 3 and the second by -4.

$$\begin{cases} 3(4x+450y)=3(240.50) \\ -4(3x+200y)=-4(146.00) \end{cases}$$

$$\rightarrow \begin{cases} 12x+1350y=721.5 \\ -12x-800y=-584 \end{cases}$$

Add the equations to eliminate x and solve for y.

$$\begin{aligned} 12x+1350y&=721.5 \\ -12x-800y&=-584 \\ \hline 550y&=137.5 \\ y&=0.25 \end{aligned}$$

Now solve for x.

$$\begin{aligned} 3x+200y&=146 \\ 3x+200(0.25)&=146 \\ 3x+50&=146 \\ 3x&=96 \\ x&=32 \end{aligned}$$

There is a \$32 daily fee and a \$0.25 per mile mileage charge.

25. $\begin{cases} 18=2(x+y) \\ 18=\frac{9}{2}(x-y) \end{cases}$

Multiply the first equation by $\frac{1}{2}$ and the second equation by $\frac{2}{9}$.

$$\begin{cases} \frac{1}{2}(18)=\frac{1}{2}[2(x+y)] \\ \frac{2}{9}(18)=\frac{2}{9}\left[\frac{9}{2}(x-y)\right] \end{cases} \rightarrow \begin{cases} 9=x+y \\ 4=x-y \end{cases}$$

Add the equations to eliminate y; then solve for x.

$$\begin{aligned} 9&=x+y \\ 4&=x-y \\ \hline 13&=2x \\ 6.5&=x \end{aligned}$$

Now solve for y.

$$\begin{aligned} 9&=x+y \\ 9&=6.5+y \\ 2.5&=y \end{aligned}$$

The rate that Pratap can row in still water is 6.5 miles per hour and the rate of the current is 2.5 miles per hour.

27. Let x = rate of flight in still wind. Then y = rate of wind.

$$\begin{cases} 780=\frac{3}{2}(x+y) \\ 780=2(x-y) \end{cases}$$

Multiply the first equation by $\frac{2}{3}$ and the second equation by $\frac{1}{2}$.

$$\begin{cases} \frac{2}{3}(780)=\frac{2}{3}\left[\frac{3}{2}(x+y)\right] \\ \frac{1}{2}(780)=\frac{1}{2}[2(x-y)] \end{cases} \rightarrow \begin{cases} 520=x+y \\ 390=x-y \end{cases}$$

Add the equations to eliminate y; then solve for x.

$$\begin{aligned} 520&=x+y \\ 390&=x-y \\ \hline 910&=2x \\ 455&=x \end{aligned}$$

Now solve for y.

$$\begin{aligned} 780&=2(x-y) \\ 780&=2(455)-2y \\ 780&=910-2y \\ -130&=-2y \\ 65&=y \end{aligned}$$

The speed of the plane in still air is 455 mph and the speed of the wind is 65 mph.

29. Let x be the number of hours that Kevin spent on his bicycle and y be the number of hours he spent walking. Then the distance he rode was $40x$ miles and the distance he walked was $4y$ miles.

$$\begin{cases} x+y=6 \\ 40x+4y=186 \end{cases}$$

Solve the first equation for y.

$$\begin{aligned} x+y&=6 \\ y&=6-x \end{aligned}$$

Substitute $6-x$ for y in the second equation and solve for x.

$$\begin{aligned} 40x+4y&=186 \\ 40x+4(6-x)&=186 \\ 40x+24-4x&=186 \\ 36x+24&=186 \\ 36x&=162 \\ x&=4.5 \end{aligned}$$

Kevin spent 4.5 hours on his bike.

31. Let x be ounces of 4% solution, and let y be ounces of 12% solution.

$$\begin{cases} x+y=12 \\ 0.04x+0.12y=0.09(12) \end{cases}$$

Solve the first equation in terms of x.

$$\begin{aligned} x+y&=12 \\ x&=-y+12 \end{aligned}$$

Substitute $-y+12$ for x in the second equation and solve for y.

$$0.04x + 0.12y = 1.08$$
$$0.04(-y + 12) + 0.12y = 1.08$$
$$-0.04y + 0.48 + 0.12y = 1.08$$
$$0.08y = 0.60$$
$$y = 7.5$$

Now solve for x.

$x = -y + 12 = -7.5 + 12 = 4.5$

Darren will need 4.5 ounces of the 4% solution and 7.5 ounces of the 12% solution to make 12 ounces of the 9% solution.

33. Let x be the number of pounds of high-quality coffee, and let y be the number of pounds of the cheaper coffee.

$$\begin{cases} x + y = 200 \\ 4.95x + 2.65y = 200(3.95) \end{cases}$$

Solve the first equation for x.

$$x + y = 200$$
$$x = 200 - y$$

Substitute $200 - y$ for x in the second equation and solve for y.

$$4.95x + 2.65y = 200(3.95)$$
$$4.95(200 - y) + 2.65y = 790$$
$$990 - 4.95y + 2.65y = 790$$
$$990 - 2.30y = 790$$
$$-2.30y = -200$$
$$y = \frac{200}{2.30}$$
$$y \approx 87$$

$x = 200 - y \approx 200 - 87 = 113$

Wayne should blend 113 pounds of the coffee that sells for \$4.95 per pound with 87 pounds of the cheaper coffee.

35. Let x be one angle, and let y be the other angle.

$$x + y = 90$$
$$y = 2x$$

Substitute $2x$ for y in the first equation and solve for x.

$$x + y = 90$$
$$x + 2x = 90$$
$$3x = 90$$
$$x = 30$$

Now solve for y.

$y = 2x = 2(30) = 60$

One angle is 30° and the other is 60°.

37. Let x be the measure of one angle and y be the measure of the other.

$$\begin{cases} x + y = 90 \\ x = 10 + 3y \end{cases}$$

Substitute $10 + 3y$ for x in the first equation and solve for y.

$$x + y = 90$$
$$10 + 3y + y = 90$$
$$10 + 4y = 90$$
$$4y = 80$$
$$y = 20$$

$x = 10 + 3y = 10 + 3(20) = 10 + 60 = 70$

The angles measure 20° and 70°.

39. Let x be the number of pieces sold at the original price, and let y be the number of pieces sold at the discounted price.

$$\begin{cases} x + y = 90 \\ 9.5x + 7.5y = 721 \end{cases}$$

Solve the first equation in terms of x.

$$x + y = 90$$
$$x = -y + 90$$

Substitute $-y + 90$ for x in the second equation and solve for y.

$$9.5x + 7.5y = 721$$
$$9.5(-y + 90) + 7.5y = 721$$
$$-9.5y + 855 + 7.5y = 721$$
$$-2.0y = -134$$
$$y = 67$$

Now solve for x.

$x = -y + 90 = -67 + 90 = 23$

They sold 23 pieces at \$9.50 each and 67 pieces at \$7.50 each.

41. Let x be the rate of the faster group and y be the rate of the slower group.

$$\begin{cases} 240x + 240y = 1200 \\ y = x - \frac{1}{2} \end{cases}$$

Substitute $x - \frac{1}{2}$ for y in the fist equation and solve for x.

$$240x + 240y = 1200$$
$$240x + 240\left(x - \frac{1}{2}\right) = 1200$$
$$240x + 240x - 120 = 1200$$
$$480x - 120 = 1200$$
$$480x = 1320$$
$$x = 2.75$$

$$y = x - \frac{1}{2} = x - 0.5 = 2.75 - 0.5 = 2.25$$

The hiking rates are $2.75 = 2\frac{3}{4}$ miles per hour and $2.25 = 2\frac{1}{4}$ miles per hour.

43. Let x be the number of gallons of 30% solution and y be the number of gallons of 60% solution.

$$\begin{cases} x+y=150 \\ 0.3x+0.6y=0.5(150) \end{cases}$$

Solve the first equation in terms of x.

$x = 150 - y$

Substitute $150 - y$ for x in the second equation and solve for y.

$$\begin{aligned} 0.3x+0.6y &= 0.5(150) \\ 0.3x+0.6y &= 75 \\ 0.3(150-y)+0.6y &= 75 \\ 45-0.3y+0.6y &= 75 \\ 0.3y &= 30 \\ y &= 100 \end{aligned}$$

Now solve for x.

$x = 150 - y = 150 - 100 = 50$

Combining 50 gallons of the 30% solution and 100 gallons of the 60% solution is necessary to create 150 gallons of a 50% solution.

45. Let x be the length and y the width.

$$\begin{cases} 2(x+y)=144 \\ x=y+12 \end{cases}$$

Substitute $y + 12$ for x in the first equation and solve for y.

$$\begin{aligned} 2(x+y) &= 144 \\ 2(y+12+y) &= 144 \\ 2(2y+12) &= 144 \\ 4y+24 &= 144 \\ 4y &= 120 \\ y &= 30 \end{aligned}$$

$x = y + 12 = 30 + 12 = 42$

The length is 42 inches and the width is 30 inches.

47. $4^2 = 4 \cdot 4 = 16$

49. $(6x)^2 = (6x)(6x) = 36x^2$

51. $(10y^3)^2 = (10y^3)(10y^3) = 100y^6$

53. The price of the result must be between \$0.49 and \$0.65, so choice **a** is the only possibility.

55. $y+2x=33$

$y=2x-3$

Substitute $2x - 3$ for y in the first equation and solve for y.

$$\begin{aligned} y+2x &= 33 \\ 2x-3+2x &= 33 \\ 4x &= 36 \\ x &= 9 \end{aligned}$$

Now solve for y.

$y = 2x - 3 = 2(9) - 3 = 18 - 3 = 15$

The width is 9 feet and the length is 15 feet.

Chapter 7 Vocabulary Check

1. In a system of linear equations in two variables, if the graphs of the equations are the same, the equations are dependent equations.
2. Two or more linear equations are called a system of linear equations.
3. A system of equations that has at least one solution is called a(n) consistent system.
4. A solution of a system of two equations in two variables is an ordered pair of numbers that is a solution of both equations in the system.
5. Two algebraic methods for solving systems of equations are addition and substitution.
6. A system of equations that has no solution is called a(n) inconsistent system.
7. In a system of linear equations in two variables, if the graphs of the equations are different, the equations are independent equations.

Chapter 7 Review

1. a. $\begin{cases} 2x-3y=12 \\ 3x+4y=1 \end{cases}$

First equation:

$$\begin{aligned} 2x-3y &= 12 \\ 2(12)-3(4) &\stackrel{?}{=} 12 \\ 24-12 &\stackrel{?}{=} 12 \\ 12 &= 12 \quad \text{True} \end{aligned}$$

Second equation:
$$3x+4y=1$$
$$3(12)-4(4) \stackrel{?}{=} 1$$
$$36-16 \stackrel{?}{=} 1$$
$$20=1 \quad \text{False}$$
(12, 4) is not a solution of the system.

b. First equation:
$$2x-3y=12$$
$$2(3)-3(-2) \stackrel{?}{=} 12$$
$$6+6 \stackrel{?}{=} 12$$
$$12=12 \quad \text{True}$$
Second equation:
$$3x+4y=1$$
$$3(3)+4(-2) \stackrel{?}{=} 1$$
$$9-8 \stackrel{?}{=} 1$$
$$1=1 \quad \text{True}$$
(3, −2) is a solution of the system.

2. a. $\begin{cases} 2x+3y=1 \\ 3y-x=4 \end{cases}$

First equation:
$$2x+3y=1$$
$$2(2)+3(2) \stackrel{?}{=} 1$$
$$4+6 \stackrel{?}{=} 1$$
$$10=1 \quad \text{False}$$
(2, 2) is not a solution of the system.

b. First equation:
$$2x+3y=1$$
$$2(-1)+3(1) \stackrel{?}{=} 1$$
$$-2+3 \stackrel{?}{=} 1$$
$$1=1 \quad \text{True}$$
Second equation:
$$3y-x=4$$
$$3(1)-(-1) \stackrel{?}{=} 4$$
$$3+1 \stackrel{?}{=} 4$$
$$4=4 \quad \text{True}$$
(−1, 1) is a solution of the system.

3. a. $\begin{cases} 5x-6y=18 \\ 2y-x=-4 \end{cases}$

First equation:
$$5x-6y=18$$
$$5(-6)-6(-8) \stackrel{?}{=} 18$$
$$-30+48 \stackrel{?}{=} 18$$
$$18=18 \quad \text{True}$$
Second equation:
$$2y-x=-4$$
$$2(-8)-(-6) \stackrel{?}{=} -4$$
$$-16+6 \stackrel{?}{=} -4$$
$$-10=-4 \quad \text{False}$$
(−6, −8) is not a solution of the system.

b. First equation:
$$5x-6y=18$$
$$5(3)-6\left(\frac{5}{2}\right) \stackrel{?}{=} 18$$
$$15-15 \stackrel{?}{=} 18$$
$$0=18 \quad \text{False}$$
$\left(3, \frac{5}{2}\right)$ is not a solution of the system.

4. a. $\begin{cases} 4x+y=0 \\ -8x-5y=9 \end{cases}$

First equation:
$$4x+y=0$$
$$4\left(\frac{3}{4}\right)+(-3) \stackrel{?}{=} 0$$
$$3-3 \stackrel{?}{=} 0$$
$$0=0 \quad \text{True}$$
Second equation:
$$-8x-5y=9$$
$$-8\left(\frac{3}{4}\right)-5(-3) \stackrel{?}{=} 9$$
$$-6+15 \stackrel{?}{=} 9$$
$$9=9 \quad \text{True}$$
$\left(\frac{3}{4}, -3\right)$ is a solution of the system.

b. First equation:
$$4x+y=0$$
$$4(-2)+8 \stackrel{?}{=} 0$$
$$-8+8 \stackrel{?}{=} 0$$
$$0=0 \quad \text{True}$$
Second equation:
$$-8x-5y=9$$
$$-8(-2)-5(8) \stackrel{?}{=} 9$$
$$16-40 \stackrel{?}{=} 9$$
$$-24=9 \quad \text{False}$$
(−2, 8) is not a solution of the system.

5. $\begin{cases} x+y=5 \\ x-y=1 \end{cases}$

The solution of the system is (3, 2).

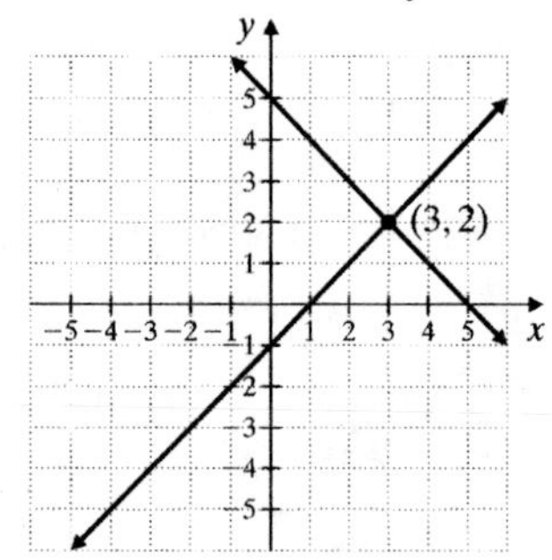

6. $\begin{cases} x+y=3 \\ x-y=-1 \end{cases}$

The solution of the system is (1, 2).

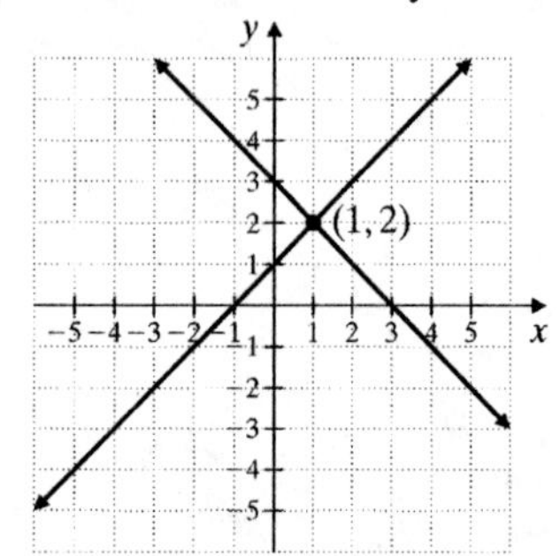

7. $\begin{cases} x=5 \\ y=-1 \end{cases}$

The solution of the system is (5, −1).

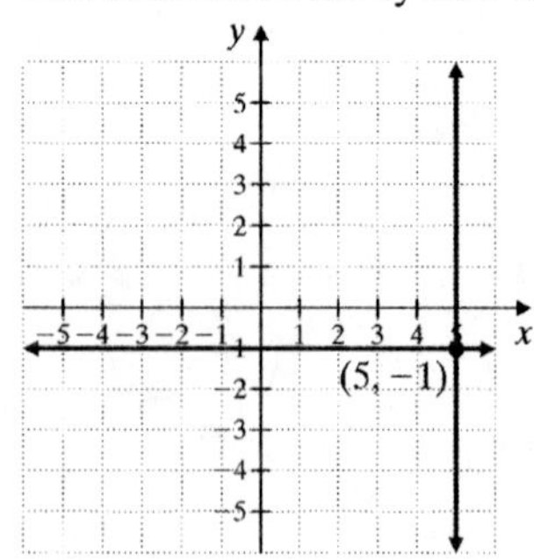

8. $\begin{cases} x=-3 \\ y=2 \end{cases}$

The solution of the system is (−3, 2).

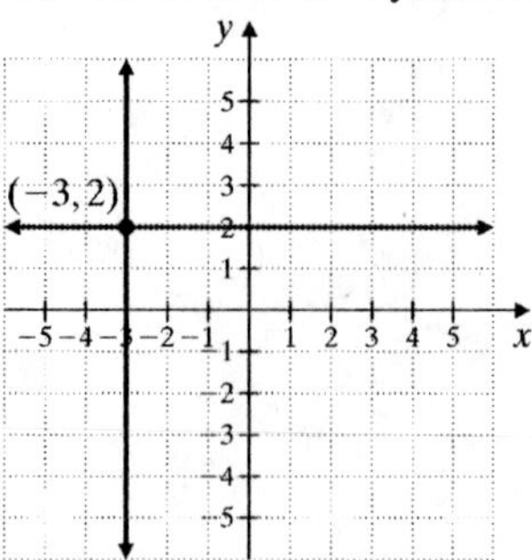

9. $\begin{cases} 2x+y=5 \\ x=-3y \end{cases}$

The solution of the system is (3, −1).

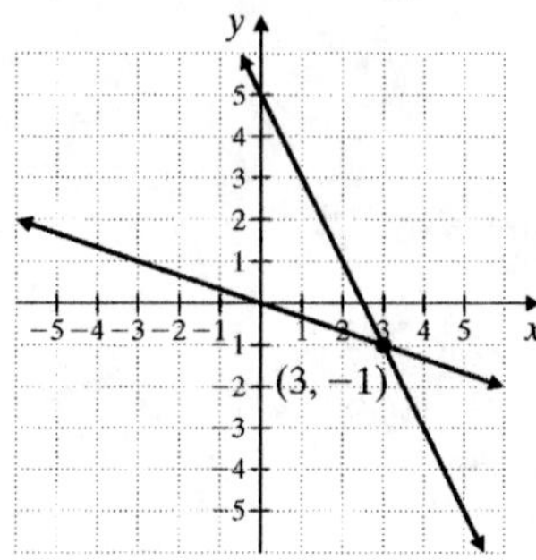

10. $\begin{cases} 3x+y=-2 \\ y=-5x \end{cases}$

The solution of the system is (1, −5).

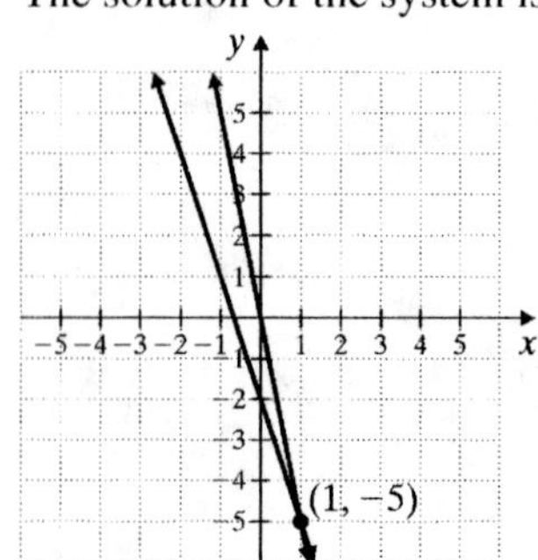

11. $\begin{cases} y=3x \\ -6x+2y=6 \end{cases}$

There are no solutions to the system because the lines are parallel.

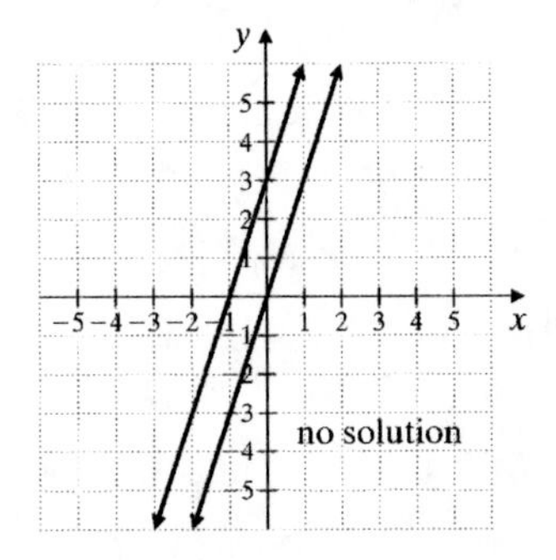

12. $\begin{cases} x-2y=2 \\ -2x+4y=-4 \end{cases}$

Since they are the same lines, there are an infinite number of solutions.

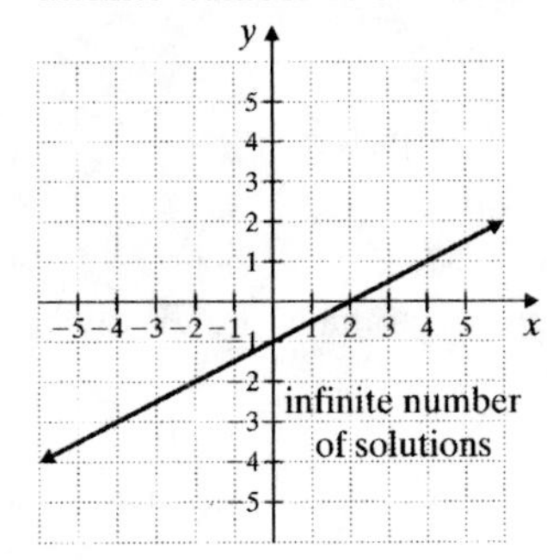

13. $\begin{cases} y=2x+6 \\ 3x-2y=-11 \end{cases}$

Substitute $2x + 6$ for y in the second equation and solve for x.

$$3x-2y=-11$$
$$3x-2(2x+6)=-11$$
$$3x-4x-12=-11$$
$$-x=1$$
$$x=-1$$

Now solve for y.

$y = 2x + 6 = 2(-1) + 6 = -2 + 6 = 4$

The solution of the system is $(-1, 4)$.

14. $\begin{cases} y=3x-7 \\ 2x-3y=7 \end{cases}$

Substitute $3x - 7$ for y in the second equation and solve for x.

$$2x-3y=7$$
$$2x-3(3x-7)=7$$
$$2x-9x+21=7$$
$$-7x=-14$$
$$x=2$$

Now solve for y.

$y = 3x - 7 = 3(2) - 7 = 6 - 7 = -1$

The solution of the system is $(2, -1)$.

15. $\begin{cases} x+3y=-3 \\ 2x+y=4 \end{cases}$

Solve the second equation in terms of y.

$$2x+y=4$$
$$y=-2x+4$$

Substitute $-2x + 4$ for y in the first equation and solve for x.

$$x+3y=-3$$
$$x+3(-2x+4)=-3$$
$$x-6x+12=-3$$
$$-5x=-15$$
$$x=3$$

Now solve for y.

$y = -2x + 4 = -2(3) + 4 = -6 + 4 = -2$

The solution of the system is $(3, -2)$.

16. $\begin{cases} 3x+y=11 \\ x+2y=12 \end{cases}$

Solve the first equation in terms of y.

$$3x+y=11$$
$$y=-3x+11$$

Substitute $-3x + 11$ for y in the second equation and solve for x.

$$x+2y=12$$
$$x+2(-3x+11)=12$$
$$x-6x+22=12$$
$$-5x=-10$$
$$x=2$$

Now solve for y.

$y = -3x + 11 = -3(2) + 11 = -6 + 11 = 5$

The solution of the system is $(2, 5)$.

17. $\begin{cases} 4y=2x+6 \\ x-2y=-3 \end{cases}$

Solve the second equation in terms of x.

$$x-2y=-3$$
$$x=2y-3$$

Substitute $2y - 3$ for x in the first equation and solve for y.

$$4y=2x+6$$
$$4y=2(2y-3)+6$$
$$4y=4y-6+6$$
$$0=0$$

Since the statement $0 = 0$ is true, there are an infinite number of solutions for the system.

18. $\begin{cases} 9x=6y+3 \\ 6x-4y=2 \end{cases}$

Solve the first equation in terms of x.

$9x = 6y + 3$
$x = \frac{6}{9}y + \frac{3}{9}$
$x = \frac{2}{3}y + \frac{1}{3}$

Substitute $\frac{2}{3}y + \frac{1}{3}$ for x in the second equation and solve for y.

$6x - 4y = 2$
$6\left(\frac{2}{3}y + \frac{1}{3}\right) - 4y = 2$
$4y + 2 - 4y = 2$
$0 = 0$

Since the statement 0 = 0 is true, there are an infinite number of solutions for the system.

19. $\begin{cases} x + y = 6 \\ y = -x - 4 \end{cases}$

Substitute $-x - 4$ for y in the first equation and solve for x.

$x + y = 6$
$x + (-x - 4) = 6$
$0 = 10$ False

Since the statement 0 = 10 is false, there is no solution for the system.

20. $\begin{cases} -3x + y = 6 \\ y = 3x + 2 \end{cases}$

Substitute $3x + 2$ for y in the first equation and solve for x.

$-3x + y = 6$
$-3x + 3x + 2 = 6$
$2 = 6$ False

Since the statement 2 = 6 is false, there is no solution for the system.

21. $\begin{cases} 2x + 3y = -6 \\ x - 3y = -12 \end{cases}$

Add the equations to eliminate y and solve for x.

$2x + 3y = -6$
$x - 3y = -12$
$3x = -18$
$x = -6$

Now solve for y.

$x - 3y = -12$
$-6 - 3y = -12$
$-3y = -6$
$y = 2$

The solution for the system is (−6, 2).

22. $\begin{cases} 4x + y = 15 \\ -4x + 3y = -19 \end{cases}$

Add the equations to eliminate x and solve for y.

$4x + y = 15$
$-4x + 3y = -19$
$4y = -4$
$y = -1$

Now solve for x.

$4x + y = 15$
$4x - 1 = 15$
$4x = 16$
$x = 4$

The solution for the system is (4, −1).

23. $\begin{cases} 2x - 3y = -15 \\ x + 4y = 31 \end{cases}$

Multiply the second equation by −2.

$\begin{cases} 2x - 3y = -15 \\ -2(x + 4y) = -2(31) \end{cases} \rightarrow \begin{cases} 2x - 3y = -15 \\ -2x - 8y = -62 \end{cases}$

Add the equations to eliminate x and solve for y.

$2x - 3y = -15$
$-2x - 8y = -62$
$-11y = -77$
$y = 7$

Now solve for x.

$x + 4y = 31$
$x + 4(7) = 31$
$x + 28 = 31$
$x = 3$

The solution of the system is (3, 7).

24. $\begin{cases} x - 5y = -22 \\ 4x + 3y = 4 \end{cases}$

Multiply the first equation by −4.

$\begin{cases} -4(x - 5y) = -4(-22) \\ 4x + 3y = 4 \end{cases} \rightarrow \begin{cases} -4x + 20y = 88 \\ 4x + 3y = 4 \end{cases}$

Add the equations to eliminate x and solve for y.

$-4x + 20y = 88$
$4x + 3y = 4$
$23y = 92$
$y = 4$

Now solve for x.

$x - 5y = -22$
$x - 5(4) = -22$
$x - 20 = -22$
$x = -2$

The solution of the system is (−2, 4).

25. $\begin{cases} 2x-6y=-1 \\ -x+3y=\frac{1}{2} \end{cases}$

Multiply the second equation by 2.

$$\begin{cases} 2x-6y=-1 \\ 2(-x+3y)=2\left(\frac{1}{2}\right) \end{cases} \rightarrow \begin{cases} 2x-6y=-1 \\ -2x+6y=1 \end{cases}$$

Add the equations to eliminate x.

$$\begin{array}{r} 2x-6y=-1 \\ -2x+6y=1 \\ \hline 0=0 \end{array}$$

Since the statement $0 = 0$ is true, there are an infinite number of solutions for the system.

26. $\begin{cases} 0.6x-0.3y=-1.5 \\ 0.04x-0.02y=-0.1 \end{cases}$

Multiply the first equation by 10 and the second by 100 to eliminate decimals.

$$\begin{cases} 10(0.6x-0.3y)=10(-1.5) \\ 100(0.04x-0.02y)=100(-0.1) \end{cases}$$

$$\rightarrow \begin{cases} 6x-3y=-15 \\ 4x-2y=-10 \end{cases}$$

Multiply the first revised equation by 2 and the second revised equation by −3.

$$\begin{cases} 2(6x-3y)=2(-15) \\ -3(4x-2y)=-3(-10) \end{cases} \rightarrow \begin{cases} 12x-6y=-30 \\ -12x+6y=30 \end{cases}$$

Add the equations to eliminate y.

$$\begin{array}{r} 12x-6y=-30 \\ -12x+6y=30 \\ \hline 0=0 \end{array}$$

Since the statement $0 = 0$ is true, there are an infinite number of solutions for the system.

27. $\begin{cases} \frac{3}{4}x+\frac{2}{3}y=2 \\ x+\frac{y}{3}=6 \end{cases}$

Multiply the first equation by 12 and the second by 3 to eliminate fractions.

$$\begin{cases} 12\left(\frac{3}{4}x+\frac{2}{3}y\right)=12(2) \\ 3\left(x+\frac{y}{3}\right)=3(6) \end{cases} \rightarrow \begin{cases} 9x+8y=24 \\ 3x+y=18 \end{cases}$$

Multiply the second revised equation by −3.

$$\begin{cases} 9x+8y=24 \\ -3(3x+y)=-3(18) \end{cases} \rightarrow \begin{cases} 9x+8y=24 \\ -9x-3y=-54 \end{cases}$$

Add the equations to eliminate x and solve for y.

$$\begin{array}{r} 9x+8y=24 \\ -9x-3y=-54 \\ \hline 5y=-30 \\ y=-6 \end{array}$$

Now solve for x.

$$x+\frac{y}{3}=6$$
$$x+\left(\frac{-6}{3}\right)=6$$
$$x+(-2)=6$$
$$x=8$$

The solution of the system is $(8, -6)$.

28. $\begin{cases} 10x+2y=0 \\ 3x+5y=33 \end{cases}$

Multiply the first equation by 5 and the second by −2.

$$\begin{cases} 5(10x+2y)=5(0) \\ -2(3x+5y)=-2(33) \end{cases} \rightarrow \begin{cases} 50x+10y=0 \\ -6x-10y=-66 \end{cases}$$

Add the equations to eliminate y and solve for x.

$$\begin{array}{r} 50x+10y=0 \\ -6x-10y=-66 \\ \hline 44x \quad =-66 \end{array}$$

$$x=\frac{-66}{44}=-\frac{3}{2}$$

Now solve for y.

$$10x+2y=0$$
$$10\left(-\frac{3}{2}\right)+2y=0$$
$$-15+2y=0$$
$$2y=15$$
$$y=\frac{15}{2}$$

The solution of the system is $\left(-\frac{3}{2}, \frac{15}{2}\right)$.

29. Let x be the smaller number and y be the larger.

$$\begin{cases} x+y=16 \\ 3y-x=72 \end{cases}$$

Solve the first equation in terms of y.

$$x+y=16$$
$$y=-x+16$$

Substitute $-x + 16$ for y in the second equation and solve for x.

$$3y-x=72$$
$$3(-x+16)-x=72$$
$$-3x+48-x=72$$
$$-4x=24$$
$$x=-6$$

Now solve for y.
$y = -x + 16 = -(-6) + 16 = 6 + 16 = 22$
The two numbers are -6 and 22.

30. Let x be the number of orchestra seats and y be the number of balcony seats.

$$\begin{cases} x + y = 360 \\ 45x + 35y = 15{,}150 \end{cases}$$

Solve the first equation in terms of x.
$x = -y + 360$
Substitute $-y + 360$ for x in the second equation and solve for y.

$$\begin{aligned} 45x + 35y &= 15{,}150 \\ 45(-y + 360) + 35y &= 15{,}150 \\ -45y + 16{,}200 + 35y &= 15{,}150 \\ -10y &= -1050 \\ y &= 105 \end{aligned}$$

Now solve for x.

$$\begin{aligned} x + y &= 360 \\ x + 105 &= 360 \\ x &= 255 \end{aligned}$$

There are 255 orchestra seats and 105 balcony seats.

31. Let x be the rate of the current and y be the speed in still water.

$$19(x - y) = 340$$
$$14(x + y) = 340$$

Multiply the first equation by $\frac{1}{19}$ and the second by $\frac{1}{14}$.

$$\begin{cases} \frac{1}{19}[19(x - y)] = \frac{1}{19}(340) \\ \frac{1}{14}[14(x + y)] = \frac{1}{14}(340) \end{cases} \rightarrow \begin{cases} x - y = 17.9 \\ x + y = 24.3 \end{cases}$$

Add the equations to eliminate y and solve for x.

$$\begin{aligned} x - y &= 17.9 \\ x + y &= 24.3 \\ \hline 2x &= 42.2 \\ x &= 21.1 \end{aligned}$$

Now solve for y.

$$\begin{aligned} x + y &= 24.3 \\ 21.1 + y &= 24.3 \\ y &= 3.2 \end{aligned}$$

The speed in still water is 21.1 mph and the current of the river is 3.2 mph.

32. Let x = number of cc of 6% acid solution and y = number of cc of 14% acid solution.

$$x + y = 50$$
$$0.06x + 0.14y = 0.12(50) \rightarrow 0.06x + 0.14y = 6$$

Solve the first equation in terms of x.

$$\begin{aligned} x + y &= 50 \\ x &= -y + 50 \end{aligned}$$

Substitute $-y + 50$ for x in the second equation and solve for y.

$$\begin{aligned} 0.06x + 0.14y &= 6 \\ 0.06(-y + 50) + 0.14y &= 6 \\ -0.06y + 3 + 0.14y &= 6 \\ 0.08y &= 3 \\ y &= 37.5 \end{aligned}$$

Now solve for x.
$x = -y + 50 = -37.5 + 50 = 12.5$
12.5 cc of the 6% solution and 37.5 cc of the 14% solution are needed to make 50 cc of the 12% solution.

33. Let x be the cost of an egg and y be the cost of a strip of bacon.

$$\begin{cases} 3x + 4y = 3.80 \\ 2x + 3y = 2.75 \end{cases}$$

Multiply the first equation by 2 and the second by -3.

$$\begin{cases} 2(3x + 4y) = 2(3.80) \\ -3(2x + 3y) = -3(2.75) \end{cases} \rightarrow \begin{cases} 6x + 8y = 7.60 \\ -6x - 9y = -8.25 \end{cases}$$

Add the equations to eliminate x and solve for y.

$$\begin{aligned} 6x + 8y &= 7.60 \\ -6x - 9y &= -8.25 \\ \hline -y &= -0.65 \\ y &= 0.65 \end{aligned}$$

Now solve for x.

$$\begin{aligned} 2x + 3y &= 2.75 \\ 2x + 3(0.65) &= 2.75 \\ 2x + 1.95 &= 2.75 \\ 2x &= 0.80 \\ x &= 0.40 \end{aligned}$$

Each egg costs \$0.40 and each strip of bacon costs \$0.65.

34. Let x be time spent jogging, and let y be the time spent walking.

$$\begin{cases} x + y = 3 \\ 7.5x + 4y = 15 \end{cases}$$

Solve the first equation in terms of x.

$$\begin{aligned} x + y &= 3 \\ x &= -y + 3 \end{aligned}$$

Substitute $-y + 3$ for x in the second equation and solve for y.

$$\begin{aligned}7.5x+4y&=15\\7.5(-y+3)+4y&=15\\-7.5y+22.5+4y&=15\\-3.5y&=-7.5\\y&=2.14\end{aligned}$$

Now solve for x.

$$\begin{aligned}x+y&=3\\x+2.14&=3\\x&=0.86\end{aligned}$$

He spent 0.86 hour jogging and 2.14 hours walking.

35. $\begin{cases}x-2y=1\\2x+3y=-12\end{cases}$

The solution to the system is (−3, −2).

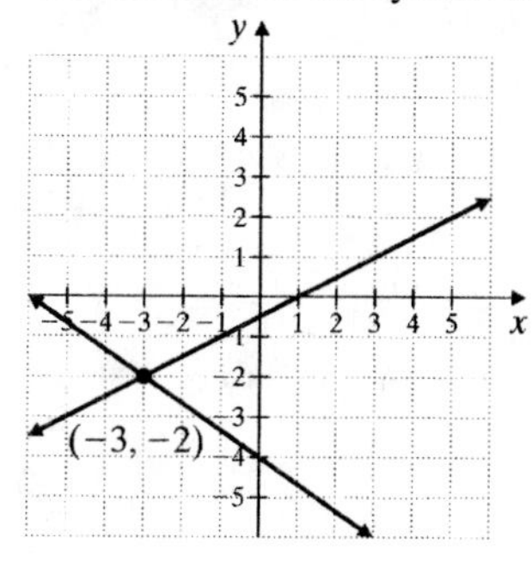

36. $\begin{cases}3x-y=-4\\6x-2y=-8\end{cases}$

The system has an infinite number of solutions because it is the same line.

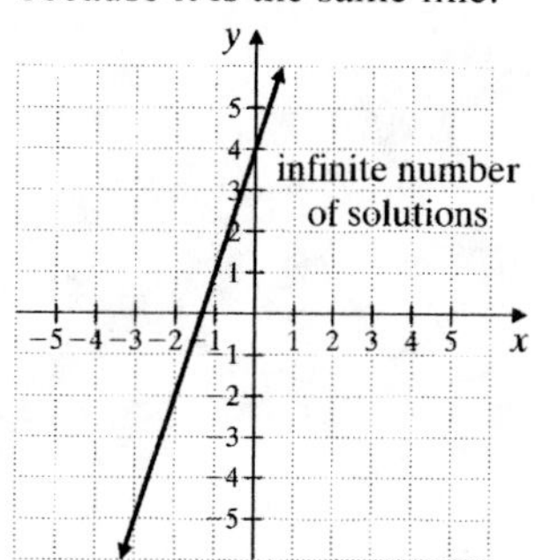

37. $\begin{cases}x+4y=11\\5x-9y=-3\end{cases}$

Solve the first equation in terms of x.

$$\begin{aligned}x+4y&=11\\x&=-4y+11\end{aligned}$$

Substitute $-4y+11$ for x in the second equation and solve for y.

$$\begin{aligned}5x-9y&=-3\\5(-4y+11)-9y&=-3\\-20y+55-9y&=-3\\-29y&=-58\\y&=2\end{aligned}$$

Now solve for x.

$x=-4y+11=-4(2)+11=-8+11=3$

The solution of the system is (3, 2).

38. $\begin{cases}x+9y=16\\3x-8y=13\end{cases}$

Solve the first equation in terms of x.

$$\begin{aligned}x+9y&=16\\x&=-9y+16\end{aligned}$$

Substitute $-9y+16$ for x in the second equation and solve for y.

$$\begin{aligned}3x-8y&=13\\3(-9y+16)-8y&=13\\-27y+48-8y&=13\\-35y&=-35\\y&=1\end{aligned}$$

Now solve for x.

$x=-9y+16=-9(1)+16=-9+16=7$

The solution of the system is (7, 1).

39. $y=-2x$

$4x+7y=-15$

Substitute $-2x$ for y in the second equation and solve for x.

$$\begin{aligned}4x+7y&=-15\\4x+7(-2x)&=-15\\4x-14x&=-15\\-10x&=-15\\x&=\frac{-15}{-10}=\frac{3}{2}\end{aligned}$$

Now solve for y.

$$y=-2x=-2\left(\frac{3}{2}\right)=-3$$

The solution of the system is $\left(\frac{3}{2}, -3\right)$.

40. $\begin{cases}3y=2x+15\\-2x+3y=21\end{cases}\rightarrow\begin{cases}2x-3y=-15\\-2x+3y=21\end{cases}$

Add the equations to eliminate x.

$$\begin{aligned}2x-3y&=-15\\-2x+3y&=21\\\hline 0&=6\quad\text{False}\end{aligned}$$

Since the statement $0=6$ in false, there is no solution for the system.

41. $\begin{cases} 3x - y = 4 \\ 4y = 12x - 16 \end{cases}$

Solve the first equation in terms of y.

$$\begin{aligned} 3x - y &= 4 \\ 3x - 4 &= y \end{aligned}$$

Substitute $3x - 4$ for y in the second equation and solve for x.

$$\begin{aligned} 4y &= 12x - 16 \\ 4(3x - 4) &= 12x - 16 \\ 12x - 16 &= 12x - 16 \\ 0 &= 0 \end{aligned}$$

Since the statement $0 = 0$ is true, there are an infinite number of solutions for the system.

42. $\begin{cases} x + y = 19 \\ x - y = -3 \end{cases}$

Add the equations to eliminate y and solve for x.

$$\begin{aligned} x + y &= 19 \\ x - y &= -3 \\ \hline 2x \quad &= 16 \\ x &= 8 \end{aligned}$$

Now solve for y.

$$\begin{aligned} x + y &= 19 \\ 8 + y &= 19 \\ y &= 11 \end{aligned}$$

The solution of the system is (8, 11).

43. $\begin{cases} x - 3y = -11 \\ 4x + 5y = -10 \end{cases}$

Solve the first equation in terms of x.

$$\begin{aligned} x - 3y &= -11 \\ x &= 3y - 11 \end{aligned}$$

Substitute $3y - 11$ for x in the second equation and solve for y.

$$\begin{aligned} 4x + 5y &= -10 \\ 4(3y - 11) + 5y &= -10 \\ 12y - 44 + 5y &= -10 \\ 17y &= 34 \\ y &= 2 \end{aligned}$$

Now solve for x.

$x = 3y - 11 = 3(2) - 11 = 6 - 11 = -5$

The solution of the system is (–5, 2).

44. $\begin{cases} -x - 15y = 44 \\ 2x + 3y = 20 \end{cases}$

Solve the first equation in terms of x.

$$\begin{aligned} -x - 15y &= 44 \\ -15y - 44 &= x \end{aligned}$$

Substitute $-15y - 44$ for x in the second equation and solve for y.

$$\begin{aligned} 2x + 3y &= 20 \\ 2(-15y - 44) + 3y &= 20 \\ -30y - 88 + 3y &= 20 \\ -27y &= 108 \\ y &= -4 \end{aligned}$$

Now solve for x.

$x = -15y - 44 = -15(-4) - 44 = 60 - 44 = 16$

The solution of the system is (16, –4).

45. $\begin{cases} 2x + y = 3 \\ 6x + 3y = 9 \end{cases}$

Solve the first equation in terms of y.

$$\begin{aligned} 2x + y &= 3 \\ y &= -2x + 3 \end{aligned}$$

Substitute $-2x + 3$ for y in the second equation and solve for x.

$$\begin{aligned} 6x + 3y &= 9 \\ 6x + 3(-2x + 3) &= 9 \\ 6x - 6x + 9 &= 9 \\ 0 &= 0 \end{aligned}$$

Since the statement $0 = 0$ is true, there are an infinite number of solutions for the system.

46. $\begin{cases} -3x + y = 5 \\ -3x + y = -2 \end{cases}$

Solve the first equation in terms of y.

$$\begin{aligned} -3x + y &= 5 \\ y &= 3x + 5 \end{aligned}$$

Substitute $3x + 5$ for y in the second equation and solve for x.

$$\begin{aligned} -3x + y &= -2 \\ -3x + 3x + 5 &= -2 \\ 5 &= -2 \quad \text{False} \end{aligned}$$

Since the statement $5 = -2$ is false, there is no solution for the system.

47. Let x be the smaller number and y be the larger number.

$\begin{cases} x + y = 12 \\ 3x + y = 20 \end{cases}$

Solve the first equation in terms of y.

$$\begin{aligned} x + y &= 12 \\ y &= -x + 12 \end{aligned}$$

Substitute $-x + 12$ for y in the second equation and solve for x.

$$\begin{aligned} 3x + y &= 20 \\ 3x + (-x + 12) &= 20 \\ 2x + 12 &= 20 \\ 2x &= 8 \\ x &= 4 \end{aligned}$$

Now solve for y.
$y = -x + 12 = -4 + 12 = 8$
The two numbers are 4 and 8.

48. Let x be the smaller number and y be the larger number.
$$x - y = -18$$
$$2x - y = -23$$
Solve the first equation in terms of x.
$$x - y = -18$$
$$x = y - 18$$
Substitute $y - 18$ for x in the second equation and solve for y.
$$2x - y = -23$$
$$2(y-18) - y = -23$$
$$2y - 36 - y = -23$$
$$y = 13$$
Now solve for x.
$x = y - 18 = 13 - 18 = -5$
The two numbers are −5 and 13.

49. Let x be nickels and y be dimes.
$$x + y = 65$$
$$0.05x + 0.1y = 5.30$$
Solve the first equation in terms of x.
$$x + y = 65$$
$$x = -y + 65$$
Substitute $-y + 65$ for x in the second equation and solve for y.
$$0.05x + 0.1y = 5.30$$
$$0.05(-y + 65) + 0.1y = 5.30$$
$$-0.05y + 3.25 + 0.1y = 5.30$$
$$0.05y = 2.05$$
$$y = 41$$
Now solve for x.
$x = -y + 65 = -41 + 65 = 24.$
There are 24 nickels and 41 dimes.

50. Let x be the number of 13¢ stamps and y be the number of 22¢ stamps.
$$\begin{cases} x + y = 26 \\ 0.13x + 0.22y = 4.19 \end{cases}$$
Solve the first equation in terms of x.
$$x + y = 26$$
$$x = -y + 26$$
Substitute $-y + 26$ for x in the second equation and solve for y.
$$0.13x + 0.22y = 4.19$$
$$0.13(-y + 26) + 0.22y = 4.19$$
$$-0.13y + 3.38 + 0.22y = 4.19$$
$$0.09y = 0.81$$
$$y = 9$$
Now solve for x.
$x = -y + 26 = -9 + 26 = 17$
They purchased 17 13¢ stamps and 9 22¢ stamps.

Chapter 7 Test

1. False; a system of two linear equations can have no solutions, exactly one solution, or infinitely many solutions.

2. False; a solution has to be a solution to both equations to be a solution of the system.

3. True; when the resulting statement is false the system has no solutions.

4. False; when $3x = 0 \rightarrow x = 0$; the system does have a solution.

5. First equation:
$$2x - 3y = 5$$
$$2(1) - 3(-1) \stackrel{?}{=} 5$$
$$2 + 3 \stackrel{?}{=} 5$$
$$5 = 5 \quad \text{True}$$
Second equation:
$$6x + y = 1$$
$$6(1) + (-1) \stackrel{?}{=} 1$$
$$6 - 1 \stackrel{?}{=} 1$$
$$5 = 1 \quad \text{False}$$
Since the statement $5 = 1$ is false, $(1, -1)$ is not a solution of the system.

6. $$\begin{cases} 4x - 3y = 24 \\ 4x + 5y = -8 \end{cases}$$
First equation:
$$4x - 3y = 24$$
$$4(3) - 3(-4) \stackrel{?}{=} 24$$
$$12 + 12 \stackrel{?}{=} 24$$
$$24 = 24 \quad \text{True}$$
Second equation:
$$4x + 5y = -8$$
$$4(3) + 5(-4) \stackrel{?}{=} -8$$
$$12 - 20 \stackrel{?}{=} -8$$
$$-8 = -8 \quad \text{True}$$
$(3, -4)$ is a solution of the system.

7. $\begin{cases} x - y = 2 \\ 3x - y = -2 \end{cases}$

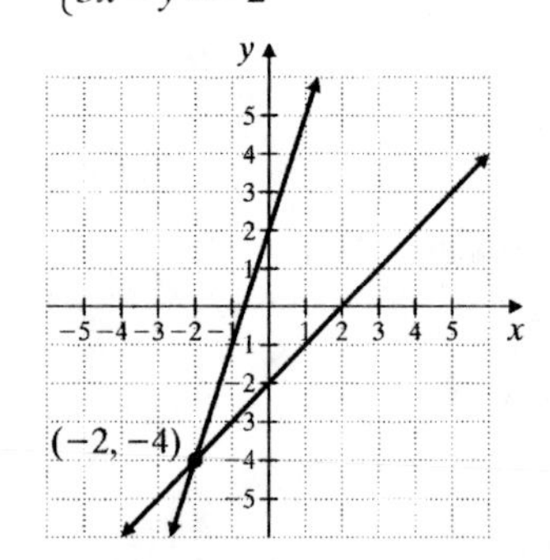

8. $\begin{cases} y = -3x \\ 3x + y = 6 \end{cases}$

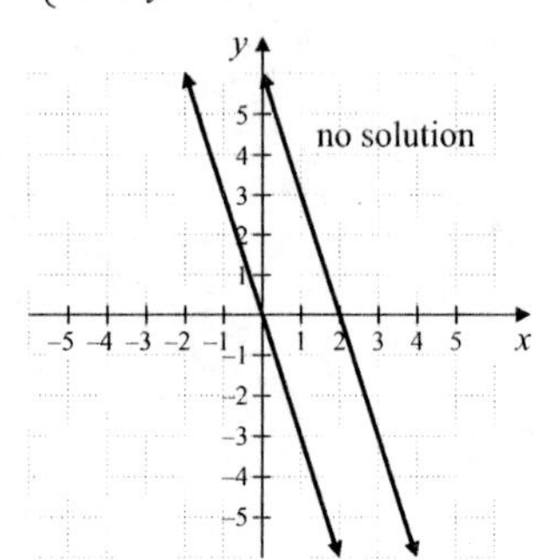

9. $\begin{cases} 3x - 2y = -14 \\ y = x + 5 \end{cases}$

Substitute $x + 5$ for y in the first equation and solve for x.

$$3x - 2y = -14$$
$$3x - 2(x + 5) = -14$$
$$3x - 2x - 10 = -14$$
$$x - 10 = -14$$
$$x = -4$$

$y = x + 5 = -4 + 5 = 1$

The solution of the system is (−4, 1).

10. $\begin{cases} \frac{1}{2}x + 2y = -\frac{15}{4} \\ 4x = -y \end{cases}$

Multiply the second equation by −1.

$(-1)4x = (-1)(-y)$

$-4x = y$

Substitute $-4x$ for y in the first equation.

$$\frac{1}{2}x + 2y = -\frac{15}{4}$$
$$\frac{1}{2}x + 2(-4x) = -\frac{15}{4}$$
$$\frac{1}{2}x - 8x = -\frac{15}{4}$$
$$4\left(\frac{1}{2}x - 8x\right) = 4\left(-\frac{15}{4}\right)$$
$$2x - 32x = -15$$
$$-30x = -15$$
$$x = \frac{15}{30} = \frac{1}{2}$$

Now solve for y.

$y = -4x = -4\left(\frac{1}{2}\right) = -2$

The solution of the system is $\left(\frac{1}{2}, -2\right)$.

11. $\begin{cases} x + y = 28 \\ x - y = 12 \end{cases}$

Add the equations to eliminate y.

$$\begin{aligned} x + y &= 28 \\ x - y &= 12 \\ \hline 2x \quad &= 40 \\ x &= 20 \end{aligned}$$

Now solve for y.

$$x + y = 28$$
$$20 + y = 28$$
$$y = 8$$

The solution of the system is (20, 8).

12. $\begin{cases} 4x - 6y = 7 \\ -2x + 3y = 0 \end{cases}$

Multiply the second equation by 2.

$\begin{cases} 4x - 6y = 7 \\ 2(-2x + 3y) = 2(0) \end{cases} \rightarrow \begin{cases} 4x - 6y = 7 \\ -4x + 6y = 0 \end{cases}$

Add the equations to eliminate x.

$$\begin{aligned} 4x - 6y &= 7 \\ -4x + 6y &= 0 \\ \hline 0 &= 7 \end{aligned}$$

Since the statement $0 = 7$ is false, the system has no solution.

13. $\begin{cases} 3x + y = 7 \\ 4x + 3y = 1 \end{cases}$

Solve the first equation for y.

$3x + y = 7$

$y = 7 - 3x$

Substitute $7 - 3x$ for y in the second equation

and solve for x.

$$\begin{aligned} 4x+3y&=1 \\ 4x+3(7-3x)&=1 \\ 4x+21-9x&=1 \\ 21-5x&=1 \\ -5x&=-20 \\ x&=4 \end{aligned}$$

$y = 7 - 3x = 7 - 3(4) = 7 - 12 = -5$

The solution of the system is (4, −5).

14. $\begin{cases} 3(2x+y)=4x+20 \\ x-2y=3 \end{cases}$

Simplify the first equation.

$$\begin{aligned} 3(2x+y)&=4x+20 \\ 6x+3y&=4x+20 \\ 2x+3y&=20 \end{aligned}$$

$$\begin{cases} 2x+3y=20 \\ x-2y=3 \end{cases}$$

Multiply the second equation by −2.

$$\begin{cases} 2x+3y=20 \\ -2(x-2y)=-2(3) \end{cases} \rightarrow \begin{cases} 2x+3y=20 \\ -2x+4y=-6 \end{cases}$$

Add the equations to eliminate x; then solve for y.

$$\begin{aligned} 2x+3y&=20 \\ -2x+4y&=-6 \\ \hline 7y&=14 \\ y&=2 \end{aligned}$$

Now solve for x.

$$\begin{aligned} x-2y&=3 \\ x-2(2)&=3 \\ x-4&=3 \\ x&=7 \end{aligned}$$

The solution of the system is (7, 2).

15. $\begin{cases} \dfrac{x-3}{2}=\dfrac{2-y}{4} \\ \dfrac{7-2x}{3}=\dfrac{y}{2} \end{cases}$

Multiply the first equation by 4 and the second equation by 6 to eliminate fractions and simplify.

$$\begin{cases} 4\left(\dfrac{x-3}{2}\right)=4\left(\dfrac{2-y}{4}\right) \\ 6\left(\dfrac{7-2x}{3}\right)=6\left(\dfrac{y}{2}\right) \end{cases} \rightarrow \begin{cases} 2x-6=2-y \\ 14-4x=3y \end{cases}$$

$$\rightarrow \begin{cases} 2x+y=8 \\ 4x+3y=14 \end{cases}$$

Multiply the first revised equation by −3.

$$\begin{cases} -3(2x+y)=-3(8) \\ 4x+3y=14 \end{cases} \rightarrow \begin{cases} -6x-3y=-24 \\ 4x+3y=14 \end{cases}$$

Add the equations to eliminate y.

$$\begin{aligned} -6x-3y&=-24 \\ 4x+3y&=14 \\ \hline -2x\quad&=-10 \\ x&=5 \end{aligned}$$

Now solve for y.

$$\begin{aligned} 2x+y&=8 \\ 2(5)+y&=8 \\ 10+y&=8 \\ y&=-2 \end{aligned}$$

The solution of the system is (5, −2).

16. $\begin{cases} 8x-4y=12 \\ y=2x-3 \end{cases}$

Substitute $2x - 3$ for y in the first equation and solve for x.

$$\begin{aligned} 8x-4y&=12 \\ 8x-4(2x-3)&=12 \\ 8x-8x+12&=12 \\ 0&=0 \end{aligned}$$

Since the statement 0 = 0 is true, the system has an infinite number of solutions.

17. $\begin{cases} 0.01x-0.06y=-0.23 \\ 0.2x+0.4y=0.2 \end{cases}$

Multiply the first equation by 100 and the second equation by 10 to eliminate decimals.

$$\begin{cases} 100(0.01x-0.06y)=100(-0.23) \\ 10(0.2x+0.4y)=10(0.2) \end{cases}$$

$$\rightarrow \begin{cases} x-6y=-23 \\ 2x+4y=2 \end{cases}$$

Multiply the first equation by −2.

$$\begin{cases} -2(x-6y)=-2(-23) \\ 2x+4y=2 \end{cases} \rightarrow \begin{cases} -2x+12y=46 \\ 2x+4y=2 \end{cases}$$

Add the equations to eliminate x; then solve for y.

$$\begin{aligned} -2x+12y&=46 \\ 2x+4y&=2 \\ \hline 16y&=48 \\ y&=3 \end{aligned}$$

Now solve for x.

$$\begin{aligned} x-6y&=-23 \\ x-6(3)&=-23 \\ x-18&=-23 \\ x&=-5 \end{aligned}$$

The solution of the system is (−5, 3).

18. $\begin{cases} x-\frac{2}{3}y=3 \\ -2x+3y=10 \end{cases}$

Solve the first equation in terms of x.

$$x-\frac{2}{3}y=3$$
$$x=\frac{2}{3}y+3$$

Substitute $\frac{2}{3}y+3$ for x in the second equation and solve for y.

$$-2x+3y=10$$
$$-2\left(\frac{2}{3}y+3\right)+3y=10$$
$$-\frac{4}{3}y-6+3y=10$$
$$-4y-18+9y=30$$
$$5y=48$$
$$y=\frac{48}{5}$$

Now solve for x.

$$x=\frac{2}{3}y+3=\frac{2}{3}\left(\frac{48}{5}\right)+3=\frac{32}{5}+\frac{15}{5}=\frac{47}{5}$$

The solution of the system is $\left(\frac{47}{5}, \frac{48}{5}\right)$.

19. Let x be the first number and the y be the second.

$\begin{cases} x+y=124 \\ x-y=32 \end{cases}$

Add the equations to eliminate y then solve for x.

$$x+y=124$$
$$x-y=32$$
$$2x=156$$
$$x=78$$

Now solve for y.

$$x+y=124$$
$$78+y=124$$
$$y=46$$

The numbers are 78 and 46.

20. Let x = number of cc of 12% solution and y = number of cc of 16% solution.

$\begin{cases} x+80=y \\ 0.12x+0.22(80)=0.16y \end{cases}$

Substitute $x + 80$ for y in the second equation; then solve for x.

$$0.12x+17.6=0.16y$$
$$0.12x+17.6=0.16(x+80)$$
$$0.12x+17.6=0.16x+12.8$$
$$4.8=0.04x$$
$$120=x$$

120 cc of the 12% saline is needed to make the 16% solution.

21. Let t be the number of farms in Texas (in thousands) and let m be the number of farms in Missouri (in thousands).

$\begin{cases} t+m=351 \\ t=m+139 \end{cases}$

Substitute $m + 139$ for t in the first equation and solve for m.

$$t+m=351$$
$$m+139+m=351$$
$$2m+139=351$$
$$2m=212$$
$$m=106$$

$t = m + 139 = 106 + 139 = 245$

There were 245 thousand farms in Texas and 106 thousand in Missouri.

22. Let x be speed of one hiker and y be the speed of the other hiker.

$\begin{cases} 4x+4y=36 \\ y=2x \end{cases}$

Substitute $2x$ for y in the first equation; then solve for x.

$$4x+4y=36$$
$$4x+4(2x)=36$$
$$4x+8x=36$$
$$12x=36$$
$$x=3$$

Now solve for y.

$y = 2x = 2(3) = 6$

One hiker hikes at 3 mph and the other hikes at 6 mph.

23. The lines cross between 2008 and 2009, and again between 2011 and 2012, so those were the times when album sales of R&B music were equal to album sales of alternative music.

24. The line for alternative music is above the line for R&B music for 2008 and 2012, so the album sales of alternative music were greater than the album sales of R&B music in 2008 and 2012.

Cumulative Review Chapters 1–7

1. a. $-14-8+10-(-6)=-6$

b. $1.6-(-10.3)+(-5.6)=6.3$

2. a. $5^2=25$

b. $2^5=32$

3. reciprocal of $22=\frac{1}{22}$

4. opposite of $22=-22$

5. reciprocal of $\frac{3}{16}=\frac{16}{3}$

6. opposite of $\frac{3}{16}=-\frac{3}{16}$

7. reciprocal of $-10=-\frac{1}{10}$

8. opposite of $-10=10$

9. reciprocal of $-\frac{9}{13}=-\frac{13}{9}$

10. opposite of $-\frac{9}{13}=\frac{9}{13}$

11. reciprocal of $1.7=\frac{1}{1.7}$

12. opposite of $1.7=-1.7$

13. a.
$$x+y=8$$
$$x+3=8$$
$$x=5$$

b.
$$x+y=8$$
$$y=8-x$$

14.
$$5(x-1)=6x$$
$$5x-5=6x$$
$$-5=x$$

15.
$$-2(x-5)+10=-3(x+2)+x$$
$$-2x+10+10=-3x-6+x$$
$$-2x+20=-2x-6$$
$$0x=-26$$
$$0=-26 \quad \text{False}$$
Since the statement $0=-26$ is false, there is no solution.

16.
$$5(y-5)=5y+10$$
$$5y-25=5y+10$$
$$0y=35$$
$$0=35 \quad \text{False}$$
Since the statement $0=35$ is false, there is no solution.

17.
$$\frac{x}{2}-1=\frac{2}{3}x-3$$
$$6\left(\frac{x}{2}-1\right)=6\left(\frac{2}{3}x-3\right)$$
$$3x-6=4x-18$$
$$12=x$$

18.
$$7(x-2)-6(x+1)=20$$
$$7x-14-6x-6=20$$
$$x-20=20$$
$$x=40$$

19.
$$-5x+7<2(x-3)$$
$$-5x+7<2x-6$$
$$13<7x$$
$$\frac{13}{7}<x$$
$$\left\{x \middle| x>\frac{13}{7}\right\}$$

(Number line from −5 to 5 with an open circle at $\frac{13}{7}$ and shading to the right.)

20. $P=a+b+c$ for b.
$b=P-a-c$

21. $\left(\frac{m}{n}\right)^7=\frac{m^7}{n^7}$ where $n\neq 0$

22. $\frac{a^7b^{10}}{ab^{15}}=\frac{a^6}{b^5}$ where $b\neq 0$

23. $\left(\frac{2x^4}{3y^5}\right)^4=\frac{2^4x^{4\cdot 4}}{3^4y^{5\cdot 4}}=\frac{16x^{16}}{81y^{20}}$ where $y\neq 0$

24. $(7a^2b^{-3})^2 = 7^2a^{2\cdot 2}b^{-3\cdot 2} = 49a^4b^{-6} = \dfrac{49a^4}{b^6}$

where $b \neq 0$.

25. $(2x^3 + 8x^2 - 6x) - (2x^3 - x^2 + 1)$
$= 2x^3 + 8x^2 - 6x - 2x^3 + x^2 - 1$
$= 9x^2 - 6x - 1$

26. $\left(5x^2 + 6x + \dfrac{1}{2}\right) + \left(x^2 - \dfrac{4}{3}x - \dfrac{10}{21}\right)$
$= 6x^2 + \dfrac{14}{3}x + \dfrac{1}{42}$

27.
$$\begin{array}{r} 2x+4 \\ 3x-1 \overline{)6x^2+10x-5} \\ \underline{6x^2 - 2x} \\ 12x-5 \\ \underline{12x-4} \\ -1 \end{array}$$

$\dfrac{6x^2 + 10x - 1}{3x - 1} = 2x + 4 - \dfrac{1}{3x - 1}$

28. $9x^2 = 3 \cdot 3 \cdot x \cdot x$
$6x^3 = 2 \cdot 3 \cdot x \cdot x \cdot x$
$21x^5 = 3 \cdot 7 \cdot x \cdot x \cdot x \cdot x \cdot x$
The common factors are 3 and $x \cdot x = x^2$, so the GCF is $3x^2$.

29.
$x(2x - 7) = 4$
$2x^2 - 7x = 4$
$2x^2 - 7x - 4 = 0$
$(2x + 1)(x - 4) = 0$
$2x + 1 = 0$ or $x - 4 = 0$
$2x = -1$ $\quad x = 4$
$x = -\dfrac{1}{2}$

The solutions are $x = -\dfrac{1}{2}$ or $x = 4$.

30. $x(2x - 7) = 0$
$x = 0$ or $2x - 7 = 0$
$2x = 7$
$x = \dfrac{7}{2}$

The solutions are $x = 0$ or $x = \dfrac{7}{2}$.

31. x = one leg
$x + 2$ = other leg
$x + 4$ = hypotenuse
$x^2 + (x + 2)^2 = (x + 4)^2$
$x^2 + x^2 + 4x + 4 = x^2 + 8x + 16$
$x^2 - 4x - 12 = 0$
$(x - 6)(x + 2) = 0$
$x - 6 = 0$ or $x + 2 = 0$
$x = 6$ $\quad x = -2$
Discard $x = -2$ because length cannot be negative.
one leg: $x = 6$
other leg: $x + 2 = 6 + 2 = 8$
hypotenuse: $x + 4 = 6 + 4 = 10$
The lengths of the sides of the right triangle are 6, 8, and 10 units.

32. $A = bh$
Let x = base; then $h = 3x + 5$.
$A = x(3x + 5) = 182$
$3x^2 + 5x - 182 = 0$
$(3x + 26)(x - 7) = 0$
$3x + 26 = 0$ or $x - 7 = 0$
$3x = -26$ $\quad x = 7$
$x = -\dfrac{26}{3}$

Discard $x = -\dfrac{26}{3}$ because length cannot be negative.
base = 7 ft
height = $3x + 5 = 3(7) + 5 = 21 + 5 = 26$ ft

33. $\dfrac{2y}{2y - 7} - \dfrac{7}{2y - 7} = \dfrac{2y - 7}{2y - 7} = 1$

34. $\dfrac{2}{x - 6} + \dfrac{3}{x + 1} = \dfrac{2(x + 1)}{(x - 6)(x + 1)} + \dfrac{3(x - 6)}{(x + 1)(x - 6)}$
$= \dfrac{2x + 2 + 3x - 18}{(x - 6)(x + 1)}$
$= \dfrac{5x - 16}{(x - 6)(x + 1)}$

35. $\dfrac{\frac{x}{y} + \frac{3}{2x}}{\frac{x}{2} + y} = \dfrac{(2xy)\left(\frac{x}{y} + \frac{3}{2x}\right)}{(2xy)\left(\frac{x}{2} + y\right)} = \dfrac{2x^2 + 3y}{x^2y + 2xy^2}$

36. $m = \dfrac{y_2 - y_1}{x_2 - x_1} = \dfrac{3 - (-8)}{-1 - 2} = -\dfrac{11}{3}$

37. $y = mx + b$
If $y = -1$, then $m = 0$.

38. Pick any 2 points on the line $x = 2$.
Let $(x_1, y_1) = (2, 0)$ and $(x_2, y_2) = (2, 5)$.

$$m = \frac{y_2 - y_1}{x_2 - x_1} = \frac{5-0}{2-2} = \frac{5}{0} = \text{undefined}$$

39. $m = \frac{y_2 - y_1}{x_2 - x_1} = \frac{4-5}{-3-2} = \frac{-1}{-5} = \frac{1}{5}$

$$\begin{aligned} y - y_1 &= m(x - x_1) \\ y - 5 &= \frac{1}{5}(x-2) \\ y - 5 &= \frac{1}{5}x - \frac{2}{5} \\ 5(y-5) &= 5\left(\frac{1}{5}x - \frac{2}{5}\right) \\ 5y - 25 &= x - 2 \\ -x + 5y &= 23 \end{aligned}$$

40.
$$\begin{aligned} y - y_1 &= m(x - x_1) \\ y - 3 &= -5(x-(-2)) \\ y - 3 &= -5x - 10 \\ y &= -5x - 7 \end{aligned}$$

41. $\{(0, 2), (3, 3), (-1, 0), (3, -2)\}$
Domain: $\{-1, 0, 3\}$
Range: $\{-2, 0, 2, 3\}$

42. $f(x) = 5x^2 - 6$

$f(0) = 5(0)^2 - 6 = -6$

$f(-2) = 5(-2)^2 - 6 = 5(4) - 6 = 20 - 6 = 14$

43. $\begin{cases} 2x - 3y = 6 \\ x = 2y \end{cases}$

First equation:
$$\begin{aligned} 2x - 3y &= 6 \\ 2(12) - 3(6) &\stackrel{?}{=} 6 \\ 24 - 18 &\stackrel{?}{=} 6 \\ 6 &= 6 \quad \text{True} \end{aligned}$$
Second equation:
$$\begin{aligned} x &= 2y \\ 12 &\stackrel{?}{=} 2(6) \\ 12 &= 12 \quad \text{True} \end{aligned}$$
$(12, 6)$ is a solution of the system.

44. a. $\begin{cases} 2x - y = 6 \\ 3x + 2y = -5 \end{cases}$

First equation:
$$\begin{aligned} 2(1) - (-4) &\stackrel{?}{=} 6 \\ 2 + 4 &\stackrel{?}{=} 6 \\ 6 &= 6 \quad \text{True} \end{aligned}$$
Second equation:
$$\begin{aligned} 3x + 2y &= -5 \\ 3(1) + 2(-4) &\stackrel{?}{=} -5 \\ 3 + (-8) &\stackrel{?}{=} -5 \\ -5 &= -5 \quad \text{True} \end{aligned}$$
$(1, -4)$ is a solution of the system.

b. First equation:
$$\begin{aligned} 2x - y &= 6 \\ 2(0) - (6) &\stackrel{?}{=} 6 \\ -6 &= 6 \quad \text{False} \end{aligned}$$
$(0, 6)$ is not a solution of the system.

c. First equation:
$$\begin{aligned} 2x - y &= 6 \\ 2(3) - 0 &= 6 \\ 6 &= 6 \quad \text{True} \end{aligned}$$
Second equation:
$$\begin{aligned} 3x + 2y &= -5 \\ 3(3) + 2(0) &\stackrel{?}{=} -5 \\ 9 + 0 &\stackrel{?}{=} -5 \\ 9 &= -5 \quad \text{False} \end{aligned}$$
$(3, 0)$ is not a solution of the system.

45. $\begin{cases} x + 2y = 7 \\ 2x + 2y = 13 \end{cases}$

Multiply the first equation by -1.

$$\begin{cases} -1(x + 2y) = -1(7) \\ 2x + 2y = 13 \end{cases} \rightarrow \begin{cases} -x - 2y = -7 \\ 2x + 2y = 13 \end{cases}$$

Add the equations to eliminate y; then solve for x.

$$\begin{array}{r} -x - 2y = -7 \\ \underline{2x + 2y = 13} \\ x \qquad = 6 \end{array}$$

Now solve for y.
$$\begin{aligned} x + 2y &= 7 \\ 6 + 2y &= 7 \\ 2y &= 1 \\ y &= \frac{1}{2} \end{aligned}$$

The solution of the system is $\left(6, \frac{1}{2}\right)$.

46. $\begin{cases} 3x-4y=10 \\ y=2x \end{cases}$

Substitute $2x$ for y in the first equation and solve for x.

$$\begin{aligned} 3x-4y&=10 \\ 3x-4(2x)&=10 \\ 3x-8x&=10 \\ -5x&=10 \\ x&=-2 \end{aligned}$$

Now solve for y.

$y=2x=2(-2)=-4$

The solution of the system is $(-2, -4)$.

47. $\begin{cases} -x-\dfrac{y}{2}=\dfrac{5}{2} \\ \dfrac{x}{6}-\dfrac{y}{2}=0 \end{cases}$

Multiply the first equation by -6 and the second equation by 6.

$$\begin{cases} -6\left(-x-\dfrac{y}{2}\right)=-6\left(\dfrac{5}{2}\right) \\ 6\left(\dfrac{x}{6}-\dfrac{y}{2}\right)=6(0) \end{cases} \rightarrow \begin{cases} 6x+3y=-15 \\ x-3y=0 \end{cases}$$

Add the equations to eliminate y; then solve for x.

$$\begin{aligned} 6x+3y&=-15 \\ x-3y&=0 \\ \hline 7x\quad&=-15 \\ x&=-\frac{15}{7} \end{aligned}$$

Now solve for y.

$$\begin{aligned} -x-\frac{y}{2}&=\frac{5}{2} \\ -\left(-\frac{15}{7}\right)-\frac{y}{2}&=\frac{5}{2} \\ 14\left(\frac{15}{7}-\frac{y}{2}\right)&=14\left(\frac{5}{2}\right) \\ 30-7y&=35 \\ -7y&=5 \\ y&=-\frac{5}{7} \end{aligned}$$

The solution of the system is $\left(-\dfrac{15}{7}, -\dfrac{5}{7}\right)$.

48. $\begin{cases} x=5y-3 \\ x=8y+4 \end{cases}$

Substitute $5y-3$ for x in the second equation and solve for y.

$$\begin{aligned} x&=8y+4 \\ 5y-3&=8y+4 \\ -7&=3y \\ -\frac{7}{3}&=y \end{aligned}$$

Now solve for x.

$$x=5y-3=5\left(-\frac{7}{3}\right)-3=-\frac{35}{3}-\frac{9}{3}=-\frac{44}{3}$$

The solution of the system is $\left(-\dfrac{44}{3}, -\dfrac{7}{3}\right)$.

49. Let x be the first number and y be the second.

$\begin{cases} x+y=37 \\ x-y=21 \end{cases}$

Add the equations to eliminate y; then solve for x.

$$\begin{aligned} x+y&=37 \\ x-y&=21 \\ \hline 2x\quad&=58 \\ x&=29 \end{aligned}$$

Now solve for y.

$$\begin{aligned} x+y&=37 \\ 29+y&=37 \\ y&=8 \end{aligned}$$

The two numbers are 29 and 8.

50. **a.** It is not a function because each x-coordinate (except 0) has more than one y-coordinate.

b. It is a function because each x-coordinate has exactly one y-coordinate.

c. It is not a function because all but two x-coordinates have more than one y-coordinate.

Chapter 8

Section 8.1 Practice Exercises

1. $\sqrt{100} = 10$, because $10^2 = 100$ and 10 is positive.

2. $-\sqrt{81} = -9$. The negative sign in front of the radical indicates the negative square root of 81.

3. $\sqrt{\frac{25}{81}} = \frac{5}{9}$, because $\left(\frac{5}{9}\right)^2 = \frac{25}{81}$ and $\frac{5}{9}$ is positive.

4. $\sqrt{1} = 1$, because $1^2 = 1$ and 1 is positive.

5. $\sqrt{0.81} = 0.9$ because $(0.9)^2 = 0.81$ and 0.9 is positive.

6. $\sqrt[3]{27} = 3$ because $3^3 = 27$.

7. $\sqrt[3]{-8} = -2$ because $(-2)^3 = -8$.

8. $\sqrt[3]{\frac{1}{64}} = \frac{1}{4}$ because $\left(\frac{1}{4}\right)^3 = \frac{1}{64}$.

9. $\sqrt[4]{-16}$ is not a real number since the index 4 is even and the radicand -16 is negative.

10. $\sqrt[5]{-1} = -1$ because $(-1)^5 = -1$.

11. $\sqrt[4]{256} = 4$ because $4^4 = 256$ and 4 is positive.

12. $\sqrt[6]{-1}$ is not a real number because the index 6 is even and the radicand -1 is negative.

13. To three decimal places, $\sqrt{22} \approx 4.690$.

14. $\sqrt{z^8} = z^4$ because $(z^4)^2 = z^8$.

15. $\sqrt{x^{20}} = x^{10}$ because $(x^{10})^2 = x^{20}$.

16. $\sqrt{4x^6} = 2x^3$ because $(2x^3)^2 = 4x^6$.

17. $\sqrt[3]{8y^{12}} = 2y^4$ because $(2y^4)^3 = 8y^{12}$.

18. $\sqrt[3]{-64x^9y^{24}} = -4x^3y^8$ because $(-4x^3y^8)^3 = -64x^9y^{24}$.

19. $\sqrt[3]{-64x^9y^{24}} = -4x^3y^8$ because $(-4x^3y^8)^3 = -64x^9y^{24}$.

Calculator Explorations

1. $\sqrt{6} \approx 2.449$; since 6 is between perfect squares 4 and 9, $\sqrt{6}$ is between $\sqrt{4} = 2$ and $\sqrt{9} = 3$.

2. $\sqrt{14} \approx 3.742$; since 14 is between perfect squares 9 and 16, $\sqrt{14}$ is between $\sqrt{9} = 3$ and $\sqrt{16} = 4$.

3. $\sqrt{11} \approx 3.317$; since 11 is between perfect squares 9 and 16, $\sqrt{11}$ is between $\sqrt{9} = 3$ and $\sqrt{16} = 4$.

4. $\sqrt{200} \approx 14.142$; since 200 is between perfect squares 196 and 225, $\sqrt{200}$ is between $\sqrt{196} = 14$ and $\sqrt{225} = 15$.

5. $\sqrt{82} \approx 9.055$; since 82 is between perfect squares 81 and 100, $\sqrt{82}$ is between $\sqrt{81} = 9$ and $\sqrt{100} = 10$.

6. $\sqrt{46} \approx 6.782$; since 46 is between perfect squares 36 and 49, $\sqrt{46}$ is between $\sqrt{36} = 6$ and $\sqrt{49} = 7$.

7. $\sqrt[3]{40} \approx 3.420$

8. $\sqrt[3]{71} \approx 4.141$

9. $\sqrt[4]{20} \approx 2.115$

10. $\sqrt[4]{15} \approx 1.968$

11. $\sqrt[5]{18} \approx 1.783$

12. $\sqrt[6]{2} \approx 1.122$

Vocabulary, Readiness & Video Check 8.1

1. The symbol $\sqrt{}$ is used to denote the positive, or principal, square root.

2. In the expression $\sqrt[4]{16}$, the number 4 is called the index, the number 16 is called the radicand, and $\sqrt{}$ is called the radical sign.

3. The reverse operation of squaring a number is finding a square root of a number.

4. For a positive number a, $-\sqrt{a}$ is the negative square root of a and $\sqrt{a}$ is the positive square root of a.

5. An nth root of a number a is a number whose nth power is a.

6. $\sqrt{4} = 2$; the statement is false.

7. $\sqrt{-9}$ is not a real number; the statement is false.

8. $\sqrt{1000} \approx 31.623$; the statement is false.

9. True

10. True

11. The radical sign, $\sqrt{}$, indicates a positive square root only. A negative sign before the radical sign, $-\sqrt{}$, indicates a negative square root.

12. The square root of a negative number *is not* a real number; the cube root of a negative number *is* a real number.

13. an odd-numbered index

14. Take the two integers that your answer falls between and square them; then check to make sure that the radicand falls between these two squares.

15. Divide the index into each exponent in the radicand—but still check by raising your answer to a power equal to the index.

Exercise Set 8.1

1. $\sqrt{16} = 4$, because $4^2 = 16$ and 4 is positive.

3. $\sqrt{\frac{1}{25}} = \frac{1}{5}$, because $\left(\frac{1}{5}\right)^2 = \frac{1}{25}$ and $\frac{1}{5}$ is positive.

5. $-\sqrt{100} = -10$. The negative sign indicates the negative square root of 100.

7. $\sqrt{-4}$ is not a real number, because there is no real number whose square is -4.

9. $-\sqrt{121} = -11$. The negative sign indicates the negative square root of 121.

11. $\sqrt{\frac{9}{25}} = \frac{3}{5}$, because $\left(\frac{3}{5}\right)^2 = \frac{9}{5}$ and $\frac{3}{5}$ is positive.

13. $\sqrt{900} = 30$, because $30^2 = 900$ and 30 is positive.

15. $\sqrt{144} = 12$, because $12^2 = 144$ and 12 is positive.

17. $\sqrt{\frac{1}{100}} = \frac{1}{10}$, because $\left(\frac{1}{10}\right)^2 = \frac{1}{100}$ and $\frac{1}{10}$ is positive.

19. $\sqrt{0.25} = 0.5$, because $0.5^2 = 0.25$ and 0.5 is positive.

21. $\sqrt[3]{125} = 5$, because $5^3 = 125$.

23. $\sqrt[3]{-64} = -4$, because $(-4)^3 = -64$.

25. $-\sqrt[3]{8} = -2$, because $2^3 = 8$.

27. $\sqrt[3]{\frac{1}{8}} = \frac{1}{2}$, because $\left(\frac{1}{2}\right)^3 = \frac{1}{8}$.

29. $\sqrt[3]{-125} = -5$, because $(-5)^3 = -125$.

31. $\sqrt[5]{32} = 2$, because $2^5 = 32$.

33. $\sqrt{81} = 9$, because $9^2 = 81$ and 9 is positive.

35. $\sqrt[4]{-16}$ is not a real number since the index 4 is even and the radicand -16 is negative.

37. $\sqrt[3]{-\frac{27}{64}} = -\frac{3}{4}$, because $\left(-\frac{3}{4}\right)^3 = -\frac{27}{64}$.

39. $-\sqrt[4]{625} = -5$, because $5^4 = 625$.

41. $\sqrt[6]{1} = 1$, because $1^6 = 1$ and 1 is positive.

43. $\sqrt{7} \approx 2.646$

45. $\sqrt{37} \approx 6.083$

47. $\sqrt{136} \approx 11.662$

49. $\sqrt{2} \approx 1.41$
$90\sqrt{2} \approx 90(1.41) = 126.90$
The distance from home plate to second base is approximately 126.90 feet.

51. $\sqrt{m^2} = m$, because $m^2 = m^2$.

53. $\sqrt{x^4} = x^2$, because $(x^2)^2 = x^4$.

55. $\sqrt{9x^8} = 3x^4$, because $(3x^4)^2 = 9x^8$.

57. $\sqrt{81x^2} = 9x$ because $(9x)^2 = 81x^2$.

59. $\sqrt{a^2b^4} = ab^2$, because $(ab^2)^2 = a^2b^4$.

61. $\sqrt{16a^6b^4} = 4a^3b^2$, because $(4a^3b^2)^2 = 16a^6b^4$

63. $\sqrt[3]{a^6b^{18}} = a^2b^6$, because $(a^2b^6)^3 = a^6b^{18}$.

65. $\sqrt[3]{-8x^3y^{27}} = -2xy^9$, because
$(-2xy^9)^3 = -8x^3y^{27}$

67. $\sqrt{\frac{x^6}{36}} = \frac{x^3}{6}$, because $\left(\frac{x^3}{6}\right)^2 = \frac{x^6}{36}$.

69. $\sqrt{\frac{25y^2}{9}} = \frac{5y}{3}$, because $\left(\frac{5y}{3}\right)^2 = \frac{25y^2}{9}$.

71. $50 = 25 \cdot 2$

73. $32 = 16 \cdot 2$ or
$32 = 4 \cdot 8$

75. $28 = 4 \cdot 7$

77. $27 = 9 \cdot 3$

79. a. $\sqrt[7]{-1}$ is a real number because the index is odd.

b. $\sqrt[3]{-125}$ is a real number because the index is odd.

c. $\sqrt[6]{-128}$ is not a real number because the index is even and the radicand is negative.

d. $\sqrt[8]{-1}$ is not a real number because the index is even and the radicand is negative.

81. The length of the side is $\sqrt{49}$. Since $7^2 = 49$, $\sqrt{49} = 7$ and the sides of the square have length 7 miles.

83. The length of a side is $\sqrt{100}$ millimeters. Since $10^2 = 100$, $\sqrt{100} = 10$. The length of a side is 10 millimeters.

85. $\sqrt{\sqrt{81}} = \sqrt{9} = 3$, since $3^2 = 9$ and $9^2 = 81$.

87. $\sqrt{\sqrt{10,000}} = 10$ since $10^2 = 100$ and
$100^2 = 10,000$.

89. Since $\sqrt{18}$ is between $\sqrt{16}$ and $\sqrt{25}$, then $\sqrt{18}$ is between 4 and 5.

91. Since $\sqrt{80}$ is between $\sqrt{64}$ and $\sqrt{81}$, then $\sqrt{80}$ is between 8 and 9.

93. $T = 2\pi\sqrt{\frac{L}{g}} = 2\pi\sqrt{\frac{30}{32}} \approx 2(3.14)(0.968) \approx 6.1$
The period of the pendulum is about 6.1 seconds.

95. answers may vary

97.

x	$y=\sqrt{x}$
0	$\sqrt{0}=0$
1	$\sqrt{1}=1$
3	$\sqrt{3}\approx 1.7$
4	$\sqrt{4}=2$
9	$\sqrt{9}=3$

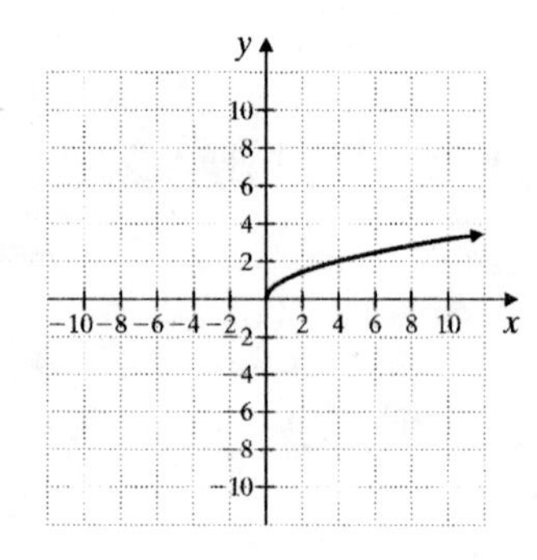

99. $\sqrt{x^2}=|x|$

101. $\sqrt{(x+2)^2}=|x+2|$

103. The graph of $y=\sqrt{x-2}$ 'starts' at (2, 0).

105. The graph of $y=\sqrt{x+4}$ 'starts' at (−4, 0).

Section 8.2 Practice Exercises

1. $\sqrt{40}=\sqrt{4\cdot 10}=\sqrt{4}\cdot\sqrt{10}=2\sqrt{10}$

2. $\sqrt{18}=\sqrt{9\cdot 2}=\sqrt{9}\cdot\sqrt{2}=3\sqrt{2}$

3. $\sqrt{500}=\sqrt{100\cdot 5}=\sqrt{100}\cdot\sqrt{5}=10\sqrt{5}$

4. $\sqrt{15}$
The radicand 15 contains no perfect square factors other than 1. Thus $\sqrt{15}$ is in simplest form.

5.
$$\begin{aligned}7\sqrt{75}&=7\cdot\sqrt{25\cdot 3}\\&=7\cdot\sqrt{25}\cdot\sqrt{3}\\&=7\cdot 5\cdot\sqrt{3}\\&=35\sqrt{3}\end{aligned}$$

6. $\sqrt{\dfrac{16}{81}}=\dfrac{\sqrt{16}}{\sqrt{81}}=\dfrac{4}{9}$

7. $\sqrt{\dfrac{2}{25}}=\dfrac{\sqrt{2}}{\sqrt{25}}=\dfrac{\sqrt{2}}{5}$

8. $\sqrt{\dfrac{45}{49}}=\dfrac{\sqrt{45}}{\sqrt{49}}=\dfrac{\sqrt{9}\cdot\sqrt{5}}{7}=\dfrac{3\sqrt{5}}{7}$

9. $\sqrt{x^{11}}=\sqrt{x^{10}\cdot x}=\sqrt{x^{10}}\cdot\sqrt{x}=x^5\sqrt{x}$

10.
$$\begin{aligned}\sqrt{18x^4}&=\sqrt{9\cdot 2\cdot x^4}\\&=\sqrt{9x^4\cdot 2}\\&=\sqrt{9x^4}\cdot\sqrt{2}\\&=3x^2\sqrt{2}\end{aligned}$$

11. $\sqrt{\dfrac{27}{x^8}}=\dfrac{\sqrt{27}}{\sqrt{x^8}}=\dfrac{\sqrt{9\cdot 3}}{x^4}=\dfrac{\sqrt{9}\cdot\sqrt{3}}{x^4}=\dfrac{3\sqrt{3}}{x^4}$

12.
$$\begin{aligned}\sqrt{\frac{7y^7}{25}}&=\frac{\sqrt{7y^7}}{\sqrt{25}}\\&=\frac{\sqrt{y^6\cdot 7y}}{5}\\&=\frac{\sqrt{y^6}\cdot\sqrt{7}}{5}\\&=\frac{y^3\sqrt{7y}}{5}\end{aligned}$$

13. $\sqrt[3]{88}=\sqrt[3]{8\cdot 11}=\sqrt[3]{8}\cdot\sqrt[3]{11}=2\sqrt[3]{11}$

14. $\sqrt[3]{50}$
The number 50 contains no perfect cube factors, so $\sqrt[3]{50}$ cannot be simplified further.

15. $\sqrt[3]{\dfrac{10}{27}}=\dfrac{\sqrt[3]{10}}{\sqrt[3]{27}}=\dfrac{\sqrt[3]{10}}{3}$

16. $\sqrt[3]{\dfrac{81}{8}}=\dfrac{\sqrt[3]{81}}{\sqrt[3]{8}}=\dfrac{\sqrt[3]{27\cdot 3}}{2}=\dfrac{\sqrt[3]{27}\cdot\sqrt[3]{3}}{2}=\dfrac{3\sqrt[3]{3}}{2}$

Vocabulary, Readiness & Video Check 8.2

1. If $\sqrt{a}$ and $\sqrt{b}$ are real numbers, then $\sqrt{a\cdot b}=\underline{\sqrt{a}\cdot\sqrt{b}}$.

2. If $\sqrt{a}$ and $\sqrt{b}$ are real numbers, then $\sqrt{\frac{a}{b}} = \underline{\frac{\sqrt{a}}{\sqrt{b}}}.$

3. $\sqrt{16\cdot 25} = \sqrt{\underline{16}}\cdot\sqrt{\underline{25}} = \underline{4}\cdot\underline{5} = \underline{20}$

4. $\sqrt{36\cdot 3} = \sqrt{\underline{36}}\cdot\sqrt{\underline{3}} = \underline{6}\cdot\sqrt{\underline{3}} = \underline{6\sqrt{3}}$

5. $\sqrt{48} = 2\sqrt{12}$
$= 2\sqrt{4\cdot 3}$
$= 2\sqrt{4}\cdot\sqrt{3}$
$= 2\cdot 2\sqrt{3}$
$= 4\sqrt{3}$
The statement is false.

6. True; 6 has no perfect cube factors.

7. Factor until we have a product of primes. A repeated prime factor means a perfect square—if more than one factor is repeated, we can multiply all the repeated factors together to get one larger perfect square factor.

8. In words, the quotient rule for square roots says that the square root of a quotient is equal to the square root of the numerator over the square root of the denominator.

9. The power must be 1. Any even power is a perfect square and can be simplified; any higher odd power is the product of an even power times the variable with a power of 1.

10. If a factor is repeated the same number of times as the index, then we have a perfect root and the product rule can be applied.

Exercise Set 8.2

1. $\sqrt{20} = \sqrt{4\cdot 5} = \sqrt{4}\sqrt{5} = 2\sqrt{5}$

3. $\sqrt{50} = \sqrt{25\cdot 2} = \sqrt{25}\cdot\sqrt{2} = 5\sqrt{2}$

5. $\sqrt{33}$ is in simplest form.

7. $\sqrt{98} = \sqrt{49\cdot 2} = \sqrt{49}\cdot\sqrt{2} = 7\sqrt{2}$

9. $\sqrt{60} = \sqrt{4\cdot 15} = \sqrt{4}\cdot\sqrt{15} = 2\sqrt{15}$

11. $\sqrt{180} = \sqrt{36\cdot 5} = \sqrt{36}\cdot\sqrt{5} = 6\sqrt{5}$

13. $\sqrt{52} = \sqrt{4\cdot 13} = \sqrt{4}\cdot\sqrt{13} = 2\sqrt{13}$

15. $3\sqrt{25} = 3\cdot 5 = 15$

17. $7\sqrt{63} = 7\sqrt{9\cdot 7} = 7\cdot\sqrt{9}\cdot\sqrt{7} = 7\cdot 3\cdot\sqrt{7} = 21\sqrt{7}$

19. $-5\sqrt{27} = -5\cdot\sqrt{9\cdot 3}$
$= -5\cdot\sqrt{9}\cdot\sqrt{3}$
$= -5\cdot 3\cdot\sqrt{3}$
$= -15\sqrt{3}$

21. $\sqrt{\frac{8}{25}} = \frac{\sqrt{8}}{\sqrt{25}} = \frac{\sqrt{4\cdot 2}}{5} = \frac{\sqrt{4}\sqrt{2}}{5} = \frac{2\sqrt{2}}{5}$

23. $\sqrt{\frac{27}{121}} = \frac{\sqrt{27}}{\sqrt{121}} = \frac{\sqrt{9\cdot 3}}{11} = \frac{\sqrt{9}\cdot\sqrt{3}}{11} = \frac{3\sqrt{3}}{11}$

25. $\sqrt{\frac{9}{4}} = \frac{\sqrt{9}}{\sqrt{4}} = \frac{3}{2}$

27. $\sqrt{\frac{125}{9}} = \frac{\sqrt{125}}{\sqrt{9}} = \frac{\sqrt{25\cdot 5}}{3} = \frac{\sqrt{25}\cdot\sqrt{5}}{3} = \frac{5\sqrt{5}}{3}$

29. $\sqrt{\frac{11}{36}} = \frac{\sqrt{11}}{\sqrt{36}} = \frac{\sqrt{11}}{6}$

31. $-\sqrt{\frac{27}{144}} = -\frac{\sqrt{27}}{\sqrt{144}}$
$= -\frac{\sqrt{9\cdot 3}}{12}$
$= -\frac{\sqrt{9}\cdot\sqrt{3}}{12}$
$= -\frac{3\sqrt{3}}{12}$
$= -\frac{\sqrt{3}}{4}$

33. $\sqrt{x^7} = \sqrt{x^6\cdot x} = \sqrt{x^6}\sqrt{x} = x^3\sqrt{x}$

35. $\sqrt{x^{13}} = \sqrt{x^{12}\cdot x} = \sqrt{x^{12}}\cdot\sqrt{x} = x^6\sqrt{x}$

37. $\sqrt{36a^3} = \sqrt{36a^2\cdot a} = \sqrt{36a^2}\sqrt{a} = 6a\sqrt{a}$

39. $\sqrt{96x^4} = \sqrt{16x^4\cdot 6} = \sqrt{16x^4}\cdot\sqrt{6} = 4x^2\sqrt{6}$

41. $\sqrt{\frac{12}{m^2}} = \frac{\sqrt{12}}{\sqrt{m^2}} = \frac{\sqrt{4\cdot 3}}{m} = \frac{\sqrt{4}\sqrt{3}}{m} = \frac{2\sqrt{3}}{m}$

43. $\sqrt{\frac{9x}{y^{10}}} = \frac{\sqrt{9x}}{\sqrt{y^{10}}} = \frac{\sqrt{9}\sqrt{x}}{y^5} = \frac{3\sqrt{x}}{y^5}$

45. $\sqrt{\frac{88}{x^{12}}} = \frac{\sqrt{88}}{\sqrt{x^{12}}} = \frac{\sqrt{4\cdot 22}}{x^6} = \frac{\sqrt{4}\sqrt{22}}{x^6} = \frac{2\sqrt{22}}{x^6}$

47. $8\sqrt{4} = 8\cdot 2 = 16$

49. $\sqrt{\frac{36}{121}} = \frac{\sqrt{36}}{\sqrt{121}} = \frac{6}{11}$

51. $\sqrt{175} = \sqrt{25\cdot 7} = \sqrt{25}\cdot\sqrt{7} = 5\sqrt{7}$

53. $\sqrt{\frac{20}{9}} = \frac{\sqrt{20}}{\sqrt{9}} = \frac{\sqrt{4\cdot 5}}{3} = \frac{\sqrt{4}\sqrt{5}}{3} = \frac{2\sqrt{5}}{3}$

55. $\sqrt{24m^7} = \sqrt{4m^6\cdot 6m} = \sqrt{4m^6}\cdot\sqrt{6m} = 2m^3\sqrt{6m}$

57. $$\sqrt{\frac{23y^3}{4x^6}} = \frac{\sqrt{23y^3}}{\sqrt{4x^6}} = \frac{\sqrt{y^2\cdot 23y}}{2x^3} = \frac{\sqrt{y^2}\sqrt{23y}}{2x^3} = \frac{y\sqrt{23y}}{2x^3}$$

59. $\sqrt[3]{24} = \sqrt[3]{8\cdot 3} = \sqrt[3]{8}\cdot\sqrt[3]{3} = 2\sqrt[3]{3}$

61. $\sqrt[3]{250} = \sqrt[3]{125\cdot 2} = \sqrt[3]{125}\cdot\sqrt[3]{2} = 5\sqrt[3]{2}$

63. $\sqrt[3]{\frac{5}{64}} = \frac{\sqrt[3]{5}}{\sqrt[3]{64}} = \frac{\sqrt[3]{5}}{4}$

65. $\sqrt[3]{\frac{23}{8}} = \frac{\sqrt[3]{23}}{\sqrt[3]{8}} = \frac{\sqrt[3]{23}}{2}$

67. $\sqrt[3]{\frac{15}{64}} = \frac{\sqrt[3]{15}}{\sqrt[3]{64}} = \frac{\sqrt[3]{15}}{4}$

69. $\sqrt[3]{80} = \sqrt[3]{8\cdot 10} = \sqrt[3]{8}\cdot\sqrt[3]{10} = 2\sqrt[3]{10}$

71. $6x + 8x = 14x$

73. $$\begin{aligned}(2x+3)(x-5) &= 2x^2 - 10x + 3x - 15\\ &= 2x^2 - 7x - 15\end{aligned}$$

75. $9y^2 - 9y^2 = 0$

77. $\sqrt{x^6y^3} = \sqrt{x^6y^2y} = \sqrt{x^6}\cdot\sqrt{y^2}\cdot\sqrt{y} = x^3y\sqrt{y}$

79. $$\begin{aligned}\sqrt{98x^5y^4} &= \sqrt{49x^4y^4\cdot 2x}\\ &= \sqrt{49x^4y^4}\cdot\sqrt{2x}\\ &= 7x^2y^2\sqrt{2x}\end{aligned}$$

81. $\sqrt[3]{-8x^6} = \sqrt[3]{-8}\cdot\sqrt[3]{x^6} = -2x^3$

83. $\sqrt[3]{80} = \sqrt[3]{8\cdot 10} = \sqrt[3]{8}\cdot\sqrt[3]{10} = 2\sqrt[3]{10}$

Each side length is $2\sqrt[3]{10}$ inches.

85. answers may vary; possible answer: let $a = 1$ and $b = 1$ so $\sqrt{a^2+b^2} = \sqrt{2} \neq a+b = 2$.

87. $\sqrt{31,329} = 177$

The roof of the Water Cube is 177 meters by 177 meters.

89. $$\frac{\sqrt{6A}}{6} = \frac{\sqrt{6\cdot 120}}{6} = \frac{\sqrt{720}}{6} = \frac{\sqrt{144\cdot 5}}{6} = \frac{\sqrt{144}\cdot\sqrt{5}}{6} = \frac{12\sqrt{5}}{6} = 2\sqrt{5}$$

The length of a side is $2\sqrt{5}$ inches.

91. Use $\frac{\sqrt{6A}}{6}$ with $A = 30.375$.

$$\frac{\sqrt{6\cdot 30.375}}{6} = \frac{\sqrt{182.25}}{6} = \frac{13.5}{6} = 2.25$$

The length of one side is 2.25 inches.

93. $C = 100\sqrt[3]{n} + 700$
$= 100\sqrt[3]{1000} + 700$
$= 100 \cdot 10 + 700$
$= 1000 + 700$
$= 1700$
The cost is \$1700.

95. $h = 169$ and $w = 64$.

$$B = \sqrt{\frac{hw}{3600}} = \sqrt{\frac{169 \cdot 64}{3600}} = \frac{\sqrt{169 \cdot 64}}{\sqrt{3600}} = \frac{\sqrt{169}\sqrt{64}}{60} = \frac{13 \cdot 8}{60} = \frac{104}{60} = \frac{26}{15} \approx 1.7$$

The body surface area is about 1.7 square meters.

Section 8.3 Practice Exercises

1. $6\sqrt{11} + 9\sqrt{11} = (6+9)\sqrt{11} = 15\sqrt{11}$

2. $\sqrt{7} - 3\sqrt{7} = 1\sqrt{7} - 3\sqrt{7} = (1-3)\sqrt{7} = -2\sqrt{7}$

3. $\sqrt{2} + \sqrt{2} - \sqrt{15} = 1\sqrt{2} + 1\sqrt{2} - \sqrt{15}$
$= (1+1)\sqrt{2} - \sqrt{15}$
$= 2\sqrt{2} - \sqrt{15}$

4. $3\sqrt{3} - 3\sqrt{2}$ cannot be simplified further since the radicands are not the same.

5. $\sqrt{27} + \sqrt{75} = \sqrt{9 \cdot 3} + \sqrt{25 \cdot 3}$
$= \sqrt{9} \cdot \sqrt{3} + \sqrt{25} \cdot \sqrt{3}$
$= 3\sqrt{3} + 5\sqrt{3}$
$= 8\sqrt{3}$

6. $3\sqrt{20} - 7\sqrt{45} = 3\sqrt{4 \cdot 5} - 7\sqrt{9 \cdot 5}$
$= 3\sqrt{4} \cdot \sqrt{5} - 7\sqrt{9} \cdot \sqrt{5}$
$= 3 \cdot 2\sqrt{5} - 7 \cdot 3\sqrt{5}$
$= 6\sqrt{5} - 21\sqrt{5}$
$= -15\sqrt{5}$

7. $\sqrt{36} - \sqrt{48} - 4\sqrt{3} - \sqrt{9} = 6 - \sqrt{16 \cdot 3} - 4\sqrt{3} - 3$
$= 6 - \sqrt{16} \cdot \sqrt{3} - 4\sqrt{3} - 3$
$= 6 - 4\sqrt{3} - 4\sqrt{3} - 3$
$= 3 - 8\sqrt{3}$

8. $\sqrt{9x^4} - \sqrt{36x^3} + \sqrt{x^3}$
$= 3x^2 - \sqrt{36x^2 \cdot x} + \sqrt{x^2 \cdot x}$
$= 3x^2 - \sqrt{36x^2} \cdot \sqrt{x} + \sqrt{x^2} \cdot \sqrt{x}$
$= 3x^2 - 6x\sqrt{x} + x\sqrt{x}$
$= 3x^2 - 5x\sqrt{x}$

9. $10\sqrt[3]{81p^6} - \sqrt[3]{24p^6} = 10\sqrt[3]{27p^6 \cdot 3} - \sqrt[3]{8p^6 \cdot 3}$
$= 10\sqrt[3]{27p^6} \cdot \sqrt[3]{3} - \sqrt[3]{8p^6} \cdot \sqrt[3]{3}$
$= 10 \cdot 3p^2\sqrt[3]{3} - 2p^2\sqrt[3]{3}$
$= 30p^2\sqrt[3]{3} - 2p^2\sqrt[3]{3}$
$= 28p^2\sqrt[3]{3}$

Vocabulary, Readiness & Video Check 8.3

1. Radicals that have the same index and same radicand are called <u>like radicals</u>.

2. The expressions $7\sqrt[3]{2x}$ and $-\sqrt[3]{2x}$ are called <u>like radicals</u>.

3. $11\sqrt{2} + 6\sqrt{2} = \underline{17\sqrt{2}}$

4. $\sqrt{5}$ is the same as $\underline{1\sqrt{5}}$.

5. $\sqrt{5} + \sqrt{5} = \underline{2\sqrt{5}}$

6. $9\sqrt{7} - \sqrt{7} = \underline{8\sqrt{7}}$

7. Both like terms and like radicals are combined using the distributive property; also, only like (vs. unlike) terms can be combined, as with like radicals (same index and same radicand).

8. Sometimes we can't see until we simplify that there are like radicals to combine, so we may incorrectly think we cannot add or subtract unless we simplify first.

9. the product rule for radicals

Exercise Set 8.3

1. $4\sqrt{3}-8\sqrt{3}=(4-8)\sqrt{3}=-4\sqrt{3}$

3. $3\sqrt{6}+8\sqrt{6}-2\sqrt{6}-5=(3+8-2)\sqrt{6}-5$
$=9\sqrt{6}-5$

5. $6\sqrt{5}-5\sqrt{5}+\sqrt{2}=(6-5)\sqrt{5}+\sqrt{2}=\sqrt{5}+\sqrt{2}$

7. $2\sqrt{3}+5\sqrt{3}-\sqrt{2}=(2+5)\sqrt{3}-\sqrt{2}=7\sqrt{3}-\sqrt{2}$

9. $2\sqrt{2}-7\sqrt{2}-6=(2-7)\sqrt{2}-6=-5\sqrt{2}-6$

11. $\sqrt{12}+\sqrt{27}=\sqrt{4\cdot 3}+\sqrt{9\cdot 3}$
$=\sqrt{4}\sqrt{3}+\sqrt{9}\sqrt{3}$
$=2\sqrt{3}+3\sqrt{3}$
$=(2+3)\sqrt{3}$
$=5\sqrt{3}$

13. $\sqrt{45}+3\sqrt{20}=\sqrt{9\cdot 5}+3\sqrt{4\cdot 5}$
$=\sqrt{9}\sqrt{5}+3\sqrt{4}\sqrt{5}$
$=3\sqrt{5}+3\cdot 2\sqrt{5}$
$=3\sqrt{5}+6\sqrt{5}$
$=(3+6)\sqrt{5}$
$=9\sqrt{5}$

15. $2\sqrt{54}-\sqrt{20}+\sqrt{45}-\sqrt{24}$
$=2\sqrt{9\cdot 6}-\sqrt{4\cdot 5}+\sqrt{9\cdot 5}-\sqrt{4\cdot 6}$
$=2\sqrt{9}\sqrt{6}-\sqrt{4}\sqrt{5}+\sqrt{9}\sqrt{5}-\sqrt{4}\sqrt{6}$
$=2\cdot 3\sqrt{6}-2\sqrt{5}+3\sqrt{5}-2\sqrt{6}$
$=6\sqrt{6}-2\sqrt{5}+3\sqrt{5}-2\sqrt{6}$
$=(6-2)\sqrt{6}+(3-2)\sqrt{5}$
$=4\sqrt{6}+1\sqrt{5}$
$=4\sqrt{6}+\sqrt{5}$

17. $4x-3\sqrt{x^2}+\sqrt{x}=4x-3x+\sqrt{x}=x+\sqrt{x}$

19. $\sqrt{25x}+\sqrt{36x}-11\sqrt{x}=\sqrt{25}\sqrt{x}+\sqrt{36}\sqrt{x}-11\sqrt{x}$
$=5\sqrt{x}+6\sqrt{x}-11\sqrt{x}$
$=(5+6-11)\sqrt{x}$
$=0$

21. $\sqrt{\frac{5}{9}}+\sqrt{\frac{5}{81}}=\frac{\sqrt{5}}{\sqrt{9}}+\frac{\sqrt{5}}{\sqrt{81}}$
$=\frac{\sqrt{5}}{3}+\frac{\sqrt{5}}{9}$
$=\frac{3\sqrt{5}}{9}+\frac{\sqrt{5}}{9}$
$=\left(\frac{3}{9}+\frac{1}{9}\right)\sqrt{5}$
$=\frac{4}{9}\sqrt{5}$
$=\frac{4\sqrt{5}}{9}$

23. $\sqrt{\frac{3}{4}}-\sqrt{\frac{3}{64}}=\frac{\sqrt{3}}{\sqrt{4}}-\frac{\sqrt{3}}{\sqrt{64}}$
$=\frac{\sqrt{3}}{2}-\frac{\sqrt{3}}{8}$
$=\frac{4\sqrt{3}}{8}-\frac{\sqrt{3}}{8}$
$=\left(\frac{4}{8}-\frac{1}{8}\right)\sqrt{3}$
$=\frac{3}{8}\sqrt{3}$
$=\frac{3\sqrt{3}}{8}$

25. $12\sqrt{5}-\sqrt{5}-4\sqrt{5}=(12-1-4)\sqrt{5}=7\sqrt{5}$

27. $\sqrt{75}+\sqrt{48}=\sqrt{25\cdot 3}+\sqrt{16\cdot 3}$
$=\sqrt{25}\sqrt{3}+\sqrt{16}\sqrt{3}$
$=5\sqrt{3}+4\sqrt{3}$
$=(5+4)\sqrt{3}$
$=9\sqrt{3}$

29. $\sqrt{5}+\sqrt{15}$ is in simplest form.

31. $3\sqrt{x^3}-x\sqrt{4x}=3\sqrt{x^2\cdot x}-x\sqrt{4\cdot x}$
$=3\sqrt{x^2}\sqrt{x}-x\sqrt{4}\sqrt{x}$
$=3x\sqrt{x}-2x\sqrt{x}$
$=(3x-2x)\sqrt{x}$
$=x\sqrt{x}$

33. $\sqrt{8}+\sqrt{9}+\sqrt{18}+\sqrt{81}=\sqrt{4\cdot 2}+3+\sqrt{9\cdot 2}+9$
$=\sqrt{4}\sqrt{2}+3+\sqrt{9}\sqrt{2}+9$
$=2\sqrt{2}+3+3\sqrt{2}+9$
$=(2+3)\sqrt{2}+3+9$
$=5\sqrt{2}+12$

35. $4+8\sqrt{2}-9=8\sqrt{2}-5$

37. $2\sqrt{45}-2\sqrt{20}=2\sqrt{9\cdot 5}-2\sqrt{4\cdot 5}$
$=2\sqrt{9}\sqrt{5}-2\sqrt{4}\sqrt{5}$
$=2\cdot 3\sqrt{5}-2\cdot 2\sqrt{5}$
$=6\sqrt{5}-4\sqrt{5}$
$=(6-4)\sqrt{5}$
$=2\sqrt{5}$

39. $\sqrt{35}-\sqrt{140}=\sqrt{35}-\sqrt{4\cdot 35}$
$=\sqrt{35}-\sqrt{4}\sqrt{35}$
$=\sqrt{35}-2\sqrt{35}$
$=(1-2)\sqrt{35}$
$=-1\sqrt{35}$
$=-\sqrt{35}$

41. $6-2\sqrt{3}-\sqrt{3}=6+(-2-1)\sqrt{3}=6-3\sqrt{3}$

43. $3\sqrt{9x}+2\sqrt{x}=3\sqrt{9}\sqrt{x}+2\sqrt{x}$
$=3\cdot 3\sqrt{x}+2\sqrt{x}$
$=9\sqrt{x}+2\sqrt{x}$
$=(9+2)\sqrt{x}$
$=11\sqrt{x}$

45. $\sqrt{9x^2}+\sqrt{81x^2}-11\sqrt{x}=3x+9x-11\sqrt{x}$
$=12x-11\sqrt{x}$

47. $\sqrt{3x^3}+3x\sqrt{x}=\sqrt{x^2\cdot 3x}+3x\sqrt{x}$
$=\sqrt{x^2}\sqrt{3x}+3x\sqrt{x}$
$=x\sqrt{3x}+3x\sqrt{x}$

49. $\sqrt{32x^2}+\sqrt{32x^2}+\sqrt{4x^2}$
$=\sqrt{16x^2\cdot 2}+\sqrt{16x^2\cdot 2}+2x$
$=\sqrt{16x^2}\sqrt{2}+\sqrt{16x^2}\sqrt{2}+2x$
$=4x\sqrt{2}+4x\sqrt{2}+2x$
$=(4x+4x)\sqrt{2}+2x$
$=8x\sqrt{2}+2x$

51. $\sqrt{40x}+\sqrt{40x^4}-2\sqrt{10x}-\sqrt{5x^4}$
$=\sqrt{4\cdot 10x}+\sqrt{4x^4\cdot 10}-2\sqrt{10x}-\sqrt{x^4\cdot 5}$
$=\sqrt{4}\sqrt{10x}+\sqrt{4x^4}\sqrt{10}-2\sqrt{10x}-\sqrt{x^4}\sqrt{5}$
$=2\sqrt{10x}+2x^2\sqrt{10}-2\sqrt{10x}-x^2\sqrt{5}$
$=(2-2)\sqrt{10x}+2x^2\sqrt{10}-x^2\sqrt{5}$
$=0\sqrt{10x}+2x^2\sqrt{10}-x^2\sqrt{5}$
$=2x^2\sqrt{10}-x^2\sqrt{5}$

53. $2\sqrt[3]{9}+5\sqrt[3]{9}-\sqrt[3]{25}=(2+5)\sqrt[3]{9}-\sqrt[3]{25}$
$=7\sqrt[3]{9}-\sqrt[3]{25}$

55. $2\sqrt[3]{2}-7\sqrt[3]{2}-6=(2-7)\sqrt[3]{2}-6=-5\sqrt[3]{2}-6$

57. $\sqrt[3]{81}+\sqrt[3]{24}=\sqrt[3]{27\cdot 3}+\sqrt[3]{8\cdot 3}$
$=\sqrt[3]{27}\sqrt[3]{3}+\sqrt[3]{8}\sqrt[3]{3}$
$=3\sqrt[3]{3}+2\sqrt[3]{3}$
$=(3+2)\sqrt[3]{3}$
$=5\sqrt[3]{3}$

59. $\sqrt[3]{8}+\sqrt[3]{54}-5=2+\sqrt[3]{27}\sqrt[3]{2}-5=-3+3\sqrt[3]{2}$

61. $2\sqrt[3]{8x^3}+2\sqrt[3]{16x^3}=2\cdot 2x+2\sqrt[3]{8x^3\cdot 2}$
$=4x+2\sqrt[3]{8x^3}\sqrt[3]{2}$
$=4x+2\cdot 2x\sqrt[3]{2}$
$=4x+4x\sqrt[3]{2}$

63. $12\sqrt[3]{y^7}-y^2\sqrt[3]{8y}=12\sqrt[3]{y^6\cdot y}-y^2\sqrt[3]{8}\sqrt[3]{y}$
$=12\sqrt[3]{y^6}\sqrt[3]{y}-2y^2\sqrt[3]{y}$
$=12y^2\sqrt[3]{y}-2y^2\sqrt[3]{y}$
$=(12y^2-2y^2)\sqrt[3]{y}$
$=10y^2\sqrt[3]{y}$

65. $\sqrt{40x}+x\sqrt[3]{40}-2\sqrt{10x}-x\sqrt[3]{5}$
$=\sqrt{4\cdot 10x}+x\sqrt[3]{8\cdot 5}-2\sqrt{10x}-x\sqrt[3]{5}$
$=\sqrt{4}\sqrt{10x}+x\sqrt[3]{8}\sqrt[3]{5}-2\sqrt{10x}-x\sqrt[3]{5}$
$=2\sqrt{10x}+2x\sqrt[3]{5}-2\sqrt{10x}-x\sqrt[3]{5}$
$=(2-2)\sqrt{10x}+(2x-x)\sqrt[3]{5}$
$=x\sqrt[3]{5}$

67. $(x+6)^2=(x)^2+2(x)(6)+(6)^2=x^2+12x+36$

69. $(2x-1)^2=(2x)^2-2(2x)(1)+(1)^2=4x^2-4x+1$

71. answers may vary

73. $P = 2l + 2w$
$= 2 \cdot 3\sqrt{5} + 2 \cdot \sqrt{5}$
$= 6\sqrt{5} + 2\sqrt{5}$
$= (6+2)\sqrt{5}$
$= 8\sqrt{5}$
The perimeter is $8\sqrt{5}$ inches.

75. Two triangular end pieces and two rectangular side panels are needed. Each side panel has area $8 \cdot 3 = 24$ square feet.
$$2 \cdot 24 + 2 \cdot \frac{3\sqrt{27}}{4} = 48 + \frac{3\sqrt{9 \cdot 3}}{2} = 48 + \frac{3\sqrt{9}\sqrt{3}}{2} = 48 + \frac{3 \cdot 3\sqrt{3}}{2} = 48 + \frac{9\sqrt{3}}{2}$$
The total area of wood needed is $\left(48 + \frac{9\sqrt{3}}{2}\right)$ square feet.

77. The expression can be simplified.
$4\sqrt{2} + 3\sqrt{2} = (4+3)\sqrt{2} = 7\sqrt{2}$

79. The expression $6 + 7\sqrt{6}$ cannot be simplified.

81. The expression can be simplified.
$\sqrt{7} + \sqrt{7} + \sqrt{7} = (1+1+1)\sqrt{7} = 3\sqrt{7}$

83.
$$\sqrt{\frac{x^3}{16}} - x\sqrt{\frac{9x}{25}} + \frac{\sqrt{81x^3}}{2}$$
$$= \frac{\sqrt{x^3}}{\sqrt{16}} - x\frac{\sqrt{9x}}{\sqrt{25}} + \frac{\sqrt{81x^3}}{2}$$
$$= \frac{\sqrt{x^2 \cdot x}}{4} - x\frac{\sqrt{9 \cdot x}}{5} + \frac{\sqrt{81x^2 \cdot x}}{2}$$
$$= \frac{\sqrt{x^2}\sqrt{x}}{4} - x\frac{\sqrt{9}\sqrt{x}}{5} + \frac{\sqrt{81x^2}\sqrt{x}}{2}$$
$$= \frac{x\sqrt{x}}{4} - \frac{3x\sqrt{x}}{5} + \frac{9x\sqrt{x}}{2}$$
$$= \left(\frac{1}{4} - \frac{3}{5} + \frac{9}{2}\right)x\sqrt{x}$$
$$= \left(\frac{5}{20} - \frac{12}{20} + \frac{90}{20}\right)x\sqrt{x}$$
$$= \frac{83}{20}x\sqrt{x}$$
$$= \frac{83x\sqrt{x}}{20}$$

Section 8.4 Practice Exercises

1. $\sqrt{5} \cdot \sqrt{2} = \sqrt{5 \cdot 2} = \sqrt{10}$

2. $\sqrt{7} \cdot \sqrt{7} = \sqrt{7 \cdot 7} = \sqrt{49} = 7$

3. $\sqrt{6} \cdot \sqrt{3} = \sqrt{18} = \sqrt{9 \cdot 2} = \sqrt{9} \cdot \sqrt{2} = 3\sqrt{2}$

4. $\sqrt{10x} \cdot \sqrt{2x} = \sqrt{10x \cdot 2x}$
$= \sqrt{20x^2}$
$= \sqrt{4x^2 \cdot 5}$
$= \sqrt{4x^2} \cdot \sqrt{5}$
$= 2x\sqrt{5}$

5. a. $\sqrt{7}\left(\sqrt{7} - \sqrt{3}\right) = \sqrt{7} \cdot \sqrt{7} - \sqrt{7} \cdot \sqrt{3} = 7 - \sqrt{21}$

b. $\sqrt{5x}\left(\sqrt{x} - 3\sqrt{5}\right) = \sqrt{5x} \cdot \sqrt{x} - \sqrt{5x} \cdot 3\sqrt{5}$
$= \sqrt{5x \cdot x} - 3\sqrt{5x \cdot 5}$
$= \sqrt{5 \cdot x^2} - 3\sqrt{25 \cdot x}$
$= \sqrt{5} \cdot \sqrt{x^2} - 3 \cdot \sqrt{25} \cdot \sqrt{x}$
$= x\sqrt{5} - 3 \cdot 5 \cdot \sqrt{x}$
$= x\sqrt{5} - 15\sqrt{x}$

c. $\left(\sqrt{x}+\sqrt{5}\right)\left(\sqrt{x}-\sqrt{3}\right)$
$= \sqrt{x}\cdot\sqrt{x}-\sqrt{x}\cdot\sqrt{3}+\sqrt{5}\cdot\sqrt{x}-\sqrt{5}\cdot\sqrt{3}$
$= x-\sqrt{3x}+\sqrt{5x}-\sqrt{15}$

6. a. $\left(\sqrt{3}+8\right)\left(\sqrt{3}-8\right)=\left(\sqrt{3}\right)^2-8^2$
$=3-64$
$=-61$

b. $\left(\sqrt{5x}+4\right)^2=\left(\sqrt{5x}\right)^2+2\left(\sqrt{5x}\right)(4)+(4)^2$
$=5x+8\sqrt{5x}+16$

7. $\frac{\sqrt{15}}{\sqrt{3}}=\sqrt{\frac{15}{3}}=\sqrt{5}$

8. $\frac{\sqrt{90}}{\sqrt{2}}=\sqrt{\frac{90}{2}}=\sqrt{45}=\sqrt{9\cdot 5}=\sqrt{9}\cdot\sqrt{5}=3\sqrt{5}$

9. $\frac{\sqrt{125x^3}}{\sqrt{5x}}=\sqrt{\frac{125x^3}{5x}}=\sqrt{25x^2}=5x$

10. $\frac{5}{\sqrt{3}}=\frac{5}{\sqrt{3}}\cdot\frac{\sqrt{3}}{\sqrt{3}}=\frac{5\cdot\sqrt{3}}{\sqrt{3}\cdot\sqrt{3}}=\frac{5\sqrt{3}}{3}$

11. $\frac{\sqrt{7}}{\sqrt{20}}=\frac{\sqrt{7}}{\sqrt{4\cdot 5}}$
$=\frac{\sqrt{7}}{2\sqrt{5}}\cdot\frac{\sqrt{5}}{\sqrt{5}}$
$=\frac{\sqrt{7}\cdot\sqrt{5}}{2\sqrt{5}\cdot\sqrt{5}}$
$=\frac{\sqrt{35}}{2\cdot 5}$
$=\frac{\sqrt{35}}{10}$

12. $\sqrt{\frac{2}{45x}}=\frac{\sqrt{2}}{\sqrt{45x}}$
$=\frac{\sqrt{2}}{\sqrt{9}\cdot\sqrt{5x}}$
$=\frac{\sqrt{2}}{3\sqrt{5x}}\cdot\frac{\sqrt{5x}}{\sqrt{5x}}$
$=\frac{\sqrt{2}\cdot\sqrt{5x}}{3\sqrt{5x}\cdot\sqrt{5x}}$
$=\frac{\sqrt{10x}}{3\cdot 5x}$
$=\frac{\sqrt{10x}}{15x}$

13. $\frac{3}{2+\sqrt{7}}=\frac{3\left(2-\sqrt{7}\right)}{\left(2+\sqrt{7}\right)\left(2-\sqrt{7}\right)}$
$=\frac{3\left(2-\sqrt{7}\right)}{2^2-\left(\sqrt{7}\right)^2}$
$=\frac{3\left(2-\sqrt{7}\right)}{4-7}$
$=\frac{3\left(2-\sqrt{7}\right)}{-3}$
$=-\frac{3\left(2-\sqrt{7}\right)}{3}$
$=-1\left(2-\sqrt{7}\right)$
$=-2+\sqrt{7}$

14. $\frac{\sqrt{2}+5}{\sqrt{2}-1}=\frac{\left(\sqrt{2}+5\right)\left(\sqrt{2}+1\right)}{\left(\sqrt{2}-1\right)\left(\sqrt{2}+1\right)}$
$=\frac{2+\sqrt{2}+5\sqrt{2}+5}{2-1}$
$=\frac{7+6\sqrt{2}}{1}$
$=7+6\sqrt{2}$

15. $\frac{7}{2-\sqrt{x}}=\frac{7\left(2+\sqrt{x}\right)}{\left(2-\sqrt{x}\right)\left(2+\sqrt{x}\right)}=\frac{7\left(2+\sqrt{x}\right)}{4-x}$

Vocabulary, Readiness & Video Check 8.4

1. $\sqrt{7}\cdot\sqrt{3}=\underline{\sqrt{21}}$

2. $\sqrt{10}\cdot\sqrt{10}=\underline{\sqrt{100}\text{ or }10}$

3. $\dfrac{\sqrt{15}}{\sqrt{3}}=\underline{\sqrt{\dfrac{15}{3}}\text{ or }\sqrt{5}}$

4. The process of eliminating the radical in the denominator of a radical expression is called rationalizing the denominator.

5. The conjugate of $2+\sqrt{3}$ is $\underline{2-\sqrt{3}}$.

6. In each example, the product rule is first used to multiply the radicals and then later used to simplify the radical.

7. The square root of a positive number times the square root of the same positive number (or the square root of a positive number squared) is that positive number.

8. If we notice that some simplifying can be done to the fraction if both radicands are under one radical.

9. To write an equivalent expression without a radical in the denominator.

10. Using the FOIL order to multiply, the Outer product and the Inner product are the only terms with radicals and they will subtract out.

Exercise Set 8.4

1. $\sqrt{8}\cdot\sqrt{2}=\sqrt{16}=4$

3. $\sqrt{10}\cdot\sqrt{5}=\sqrt{50}=\sqrt{25\cdot 2}=\sqrt{25}\sqrt{2}=5\sqrt{2}$

5. $\left(\sqrt{6}\right)^2=6$

7. $\sqrt{2x}\cdot\sqrt{2x}=\left(\sqrt{2x}\right)^2=2x$

9. $\left(2\sqrt{5}\right)^2=\left(2\sqrt{5}\right)\left(2\sqrt{5}\right)=4\left(\sqrt{5}\right)^2=4\cdot 5=20$

11. $\left(6\sqrt{x}\right)^2=\left(6\sqrt{x}\right)\left(6\sqrt{x}\right)=36\left(\sqrt{x}\right)^2=36x$

13. $$\begin{aligned}\sqrt{3x^5}\cdot\sqrt{6x}&=\sqrt{3x^5\cdot 6x}\\&=\sqrt{18x^6}\\&=\sqrt{9x^6\cdot 2}\\&=\sqrt{9x^6}\sqrt{2}\\&=3x^3\sqrt{2}\end{aligned}$$

15. $$\begin{aligned}\sqrt{2xy^2}\cdot\sqrt{8xy}&=\sqrt{2xy^2\cdot 8xy}\\&=\sqrt{16x^2y^3}\\&=\sqrt{16x^2y^2\cdot y}\\&=\sqrt{16x^2y^2}\sqrt{y}\\&=4xy\sqrt{y}\end{aligned}$$

17. $\sqrt{6}\left(\sqrt{5}+\sqrt{7}\right)=\sqrt{6}\cdot\sqrt{5}+\sqrt{6}\cdot\sqrt{7}=\sqrt{30}+\sqrt{42}$

19. $$\begin{aligned}\sqrt{10}\left(\sqrt{2}+\sqrt{5}\right)&=\sqrt{10}\cdot\sqrt{2}+\sqrt{10}\cdot\sqrt{5}\\&=\sqrt{20}+\sqrt{50}\\&=\sqrt{4\cdot 5}+\sqrt{25\cdot 2}\\&=\sqrt{4}\sqrt{5}+\sqrt{25}\sqrt{2}\\&=2\sqrt{5}+5\sqrt{2}\end{aligned}$$

21. $$\begin{aligned}\sqrt{7y}\left(\sqrt{y}-2\sqrt{7}\right)&=\sqrt{7y}\cdot\sqrt{y}-\sqrt{7y}\cdot 2\sqrt{7}\\&=\sqrt{7y\cdot y}-2\sqrt{7y\cdot 7}\\&=\sqrt{7y^2}-2\sqrt{49y}\\&=\sqrt{y^2\cdot 7}-2\sqrt{49\cdot y}\\&=\sqrt{y^2}\sqrt{7}-2\sqrt{49}\sqrt{y}\\&=y\sqrt{7}-2\cdot 7\sqrt{y}\\&=y\sqrt{7}-14\sqrt{y}\end{aligned}$$

23. $\left(\sqrt{3}+6\right)\left(\sqrt{3}-6\right)=\left(\sqrt{3}\right)^2-6^2=3-36=-33$

25. $$\begin{aligned}&\left(\sqrt{3}+\sqrt{5}\right)\left(\sqrt{2}-\sqrt{5}\right)\\&=\sqrt{3}\cdot\sqrt{2}-\sqrt{3}\sqrt{5}+\sqrt{5}\cdot\sqrt{2}-\sqrt{5}\cdot\sqrt{5}\\&=\sqrt{6}-\sqrt{15}+\sqrt{10}-\sqrt{25}\\&=\sqrt{6}-\sqrt{15}+\sqrt{10}-5\end{aligned}$$

27. $(2\sqrt{11}+1)(\sqrt{11}-6)$
$= 2\sqrt{11}\cdot\sqrt{11} - 2\sqrt{11}\cdot 6 + 1\cdot\sqrt{11} - 1\cdot 6$
$= 2\cdot 11 - 12\sqrt{11} + \sqrt{11} - 6$
$= 22 - 11\sqrt{11} - 6$
$= 16 - 11\sqrt{11}$

29. $(\sqrt{x}+6)(\sqrt{x}-6) = (\sqrt{x})^2 - (6)^2 = x - 36$

31. $(\sqrt{x}-7)^2 = (\sqrt{x})^2 - 2(\sqrt{x})(7) + (7)^2$
$= x - 14\sqrt{x} + 49$

33. $(\sqrt{6y}+1)^2 = (\sqrt{6y})^2 + 2(\sqrt{6y})(1) + (1)^2$
$= 6y + 2\sqrt{6y} + 1$

35. $\frac{\sqrt{32}}{\sqrt{2}} = \sqrt{\frac{32}{2}} = \sqrt{16} = 4$

37. $\frac{\sqrt{21}}{\sqrt{3}} = \sqrt{\frac{21}{3}} = \sqrt{7}$

39. $\frac{\sqrt{90}}{\sqrt{5}} = \sqrt{\frac{90}{5}} = \sqrt{18} = \sqrt{9\cdot 2} = \sqrt{9}\sqrt{2} = 3\sqrt{2}$

41. $\frac{\sqrt{75y^5}}{\sqrt{3y}} = \sqrt{\frac{75y^5}{3y}} = \sqrt{25y^4} = 5y^2$

43. $\frac{\sqrt{150}}{\sqrt{2}} = \sqrt{\frac{150}{2}} = \sqrt{75} = \sqrt{25\cdot 3} = \sqrt{25}\sqrt{3} = 5\sqrt{3}$

45. $\frac{\sqrt{72y^5}}{\sqrt{3y^3}} = \sqrt{\frac{72y^5}{3y^3}}$
$= \sqrt{24y^2}$
$= \sqrt{4y^2\cdot 6}$
$= \sqrt{4y^2}\sqrt{6}$
$= 2y\sqrt{6}$

47. $\frac{\sqrt{24x^3y^4}}{\sqrt{2xy}} = \sqrt{\frac{24x^3y^4}{2xy}}$
$= \sqrt{12x^2y^3}$
$= \sqrt{4x^2y^2\cdot 3y}$
$= \sqrt{4x^2y^2}\sqrt{3y}$
$= 2xy\sqrt{3y}$

49. $\frac{\sqrt{3}}{\sqrt{5}} = \frac{\sqrt{3}}{\sqrt{5}}\cdot\frac{\sqrt{5}}{\sqrt{5}} = \frac{\sqrt{15}}{5}$

51. $\frac{7}{\sqrt{2}} = \frac{7}{\sqrt{2}}\cdot\frac{\sqrt{2}}{\sqrt{2}} = \frac{7\sqrt{2}}{2}$

53. $\frac{1}{\sqrt{6y}} = \frac{1}{\sqrt{6y}}\cdot\frac{\sqrt{6y}}{\sqrt{6y}} = \frac{\sqrt{6y}}{6y}$

55. $\sqrt{\frac{5}{18}} = \frac{\sqrt{5}}{\sqrt{18}}$
$= \frac{\sqrt{5}}{\sqrt{9\cdot 2}}$
$= \frac{\sqrt{5}}{3\sqrt{2}}\cdot\frac{\sqrt{2}}{\sqrt{2}}$
$= \frac{\sqrt{5}\sqrt{2}}{3\sqrt{2}\cdot\sqrt{2}}$
$= \frac{\sqrt{10}}{3\cdot 2}$
$= \frac{\sqrt{10}}{6}$

57. $\sqrt{\frac{3}{x}} = \frac{\sqrt{3}}{\sqrt{x}} = \frac{\sqrt{3}}{\sqrt{x}}\cdot\frac{\sqrt{x}}{\sqrt{x}} = \frac{\sqrt{3x}}{x}$

59. $\sqrt{\frac{1}{8}} = \frac{\sqrt{1}}{\sqrt{8}}$
$= \frac{1}{\sqrt{4\cdot 2}}$
$= \frac{1}{2\sqrt{2}}$
$= \frac{1}{2\sqrt{2}}\cdot\frac{\sqrt{2}}{\sqrt{2}}$
$= \frac{\sqrt{2}}{2\cdot 2}$
$= \frac{\sqrt{2}}{4}$

61. $\sqrt{\frac{2}{15}}=\frac{\sqrt{2}}{\sqrt{15}}=\frac{\sqrt{2}}{\sqrt{15}}\cdot\frac{\sqrt{15}}{\sqrt{15}}=\frac{\sqrt{30}}{15}$

63. $$\begin{aligned}\sqrt{\frac{3}{20}}&=\frac{\sqrt{3}}{\sqrt{20}}\\&=\frac{\sqrt{3}}{\sqrt{4\cdot 5}}\\&=\frac{\sqrt{3}}{2\sqrt{5}}\\&=\frac{\sqrt{3}}{2\sqrt{5}}\cdot\frac{\sqrt{5}}{\sqrt{5}}\\&=\frac{\sqrt{15}}{2\cdot 5}\\&=\frac{\sqrt{15}}{10}\end{aligned}$$

65. $\frac{3x}{\sqrt{2x}}=\frac{3x}{\sqrt{2x}}\cdot\frac{\sqrt{2x}}{\sqrt{2x}}=\frac{3x\sqrt{2x}}{2x}=\frac{3\sqrt{2x}}{2}$

67. $\frac{8y}{\sqrt{5}}=\frac{8y}{\sqrt{5}}\cdot\frac{\sqrt{5}}{\sqrt{5}}=\frac{8y\sqrt{5}}{5}$

69. $$\begin{aligned}\sqrt{\frac{x}{36y}}&=\frac{\sqrt{x}}{\sqrt{36y}}\\&=\frac{\sqrt{x}}{\sqrt{36}\sqrt{y}}\\&=\frac{\sqrt{x}}{6\sqrt{y}}\\&=\frac{\sqrt{x}}{6\sqrt{y}}\cdot\frac{\sqrt{y}}{\sqrt{y}}\\&=\frac{\sqrt{xy}}{6\cdot y}\\&=\frac{\sqrt{xy}}{6y}\end{aligned}$$

71. $$\begin{aligned}\sqrt{\frac{y}{12x}}&=\frac{\sqrt{y}}{\sqrt{12x}}\\&=\frac{\sqrt{y}}{\sqrt{4}\sqrt{3x}}\\&=\frac{\sqrt{y}}{2\sqrt{3x}}\\&=\frac{\sqrt{y}}{2\sqrt{3x}}\cdot\frac{\sqrt{3x}}{\sqrt{3x}}\\&=\frac{\sqrt{3xy}}{2\cdot 3x}\\&=\frac{\sqrt{3xy}}{6x}\end{aligned}$$

73. $$\begin{aligned}\frac{3}{\sqrt{2}+1}&=\frac{3}{\sqrt{2}+1}\cdot\frac{\sqrt{2}-1}{\sqrt{2}-1}\\&=\frac{3\left(\sqrt{2}-1\right)}{\left(\sqrt{2}\right)^2-1^2}\\&=\frac{3\sqrt{2}-3}{2-1}\\&=\frac{3\sqrt{2}-3}{1}\\&=3\sqrt{2}-3\end{aligned}$$

75. $$\begin{aligned}\frac{4}{2-\sqrt{5}}&=\frac{4}{2-\sqrt{5}}\cdot\frac{2+\sqrt{5}}{2+\sqrt{5}}\\&=\frac{4\left(2+\sqrt{5}\right)}{2^2-\left(\sqrt{5}\right)^2}\\&=\frac{8+4\sqrt{5}}{4-5}\\&=\frac{8+4\sqrt{5}}{-1}\\&=-8-4\sqrt{5}\end{aligned}$$

77. $$\begin{aligned}\frac{\sqrt{5}+1}{\sqrt{6}-\sqrt{5}}&=\frac{\sqrt{5}+1}{\sqrt{6}-\sqrt{5}}\cdot\frac{\sqrt{6}+\sqrt{5}}{\sqrt{6}+\sqrt{5}}\\&=\frac{\left(\sqrt{5}+1\right)\left(\sqrt{6}+\sqrt{5}\right)}{\left(\sqrt{6}\right)^2-\left(\sqrt{5}\right)^2}\\&=\frac{\sqrt{5}\sqrt{6}+\sqrt{5}\sqrt{5}+1\cdot\sqrt{6}+1\cdot\sqrt{5}}{6-5}\\&=\frac{\sqrt{30}+5+\sqrt{6}+\sqrt{5}}{1}\\&=\sqrt{30}+5+\sqrt{6}+\sqrt{5}\end{aligned}$$

79. $\frac{\sqrt{3}+1}{\sqrt{2}-1} = \frac{\sqrt{3}+1}{\sqrt{2}-1} \cdot \frac{\sqrt{2}+1}{\sqrt{2}+1}$
$= \frac{(\sqrt{3}+1)(\sqrt{2}+1)}{(\sqrt{2})^2 - 1}$
$= \frac{\sqrt{3}\sqrt{2}+\sqrt{3}\cdot 1+1\cdot\sqrt{2}+1^2}{2-1}$
$= \frac{\sqrt{6}+\sqrt{3}+\sqrt{2}+1}{1}$
$= \sqrt{6}+\sqrt{3}+\sqrt{2}+1$

81. $\frac{5}{2+\sqrt{x}} = \frac{5}{2+\sqrt{x}} \cdot \frac{2-\sqrt{x}}{2-\sqrt{x}}$
$= \frac{5(2-\sqrt{x})}{2^2-(\sqrt{x})^2}$
$= \frac{10-5\sqrt{x}}{4-x}$

83. $\frac{3}{\sqrt{x}-4} = \frac{3}{\sqrt{x}-4} \cdot \frac{\sqrt{x}+4}{\sqrt{x}+4}$
$= \frac{3(\sqrt{x}+4)}{(\sqrt{x})^2-(4)^2}$
$= \frac{3\sqrt{x}+12}{x-16}$

85. $x+5=7^2$
$x=49-5$
$x=44$

87. $4z^2+6z-12=(2z)^2$
$4z^2+6z-12=4z^2$
$6z-12=0$
$6z=12$
$z=2$

89. $9x^2+5x+4=(3x+1)^2$
$9x^2+5x+4=9x^2+6x+1$
$5x+4=6x+1$
$4=x+1$
$3=x$

91. Area = (length)(width)
$13\sqrt{2}\cdot 5\sqrt{6} = 13\cdot 5\cdot\sqrt{2}\cdot\sqrt{6}$
$= 65\sqrt{12}$
$= 65\sqrt{4}\sqrt{3}$
$= 65\cdot 2\sqrt{3}$
$= 130\sqrt{3}$

The area is $130\sqrt{3}$ square meters.

93. $\sqrt{\frac{A}{\pi}} = \frac{\sqrt{A}}{\sqrt{\pi}} = \frac{\sqrt{A}}{\sqrt{\pi}} \cdot \frac{\sqrt{\pi}}{\sqrt{\pi}} = \frac{\sqrt{A\pi}}{\pi}$

95. $\sqrt{5}\cdot\sqrt{5} = (\sqrt{5})^2 = 5$

The statement is true.

97. $\sqrt{3x}\cdot\sqrt{3x} = (\sqrt{3x})^2 = 3x$

The statement is false.

99. $\sqrt{11}+\sqrt{2}$ cannot be simplified because the radicands are different. The statement is false.

101. answers may vary

103. answers may vary

105. $\frac{\sqrt{3}+1}{\sqrt{2}-1} = \frac{\sqrt{3}+1}{\sqrt{2}-1} \cdot \frac{\sqrt{3}-1}{\sqrt{3}-1}$
$= \frac{(\sqrt{3})^2-1^2}{(\sqrt{2}-1)(\sqrt{3}-1)}$
$= \frac{3-1}{\sqrt{2}\sqrt{3}-1\cdot\sqrt{2}-1\cdot\sqrt{3}+1\cdot 1}$
$= \frac{2}{\sqrt{6}-\sqrt{2}-\sqrt{3}+1}$

Integrated Review

1. $\sqrt{36}=6$, because $6^2=36$ and 6 is positive.

2. $\sqrt{48}=\sqrt{16\cdot 3}=\sqrt{16}\cdot\sqrt{3}=4\sqrt{3}$

3. $\sqrt{x^4}=x^2$, because $(x^2)^2=x^4$.

4. $\sqrt{y^7}=\sqrt{y^6\cdot y}=\sqrt{y^6}\sqrt{y}=y^3\sqrt{y}$

5. $\sqrt{16x^2}=4x$, because $(4x)^2=16x^2$.

6. $\sqrt{18x^{11}} = \sqrt{9x^{10} \cdot 2x} = \sqrt{9x^{10}}\sqrt{2x} = 3x^5\sqrt{2x}$

7. $\sqrt[3]{8} = 2$, because $2^3 = 8$.

8. $\sqrt[4]{81} = 3$, because $3^4 = 81$.

9. $\sqrt[3]{-27} = -3$, because $(-3)^3 = -27$.

10. $\sqrt{-4}$ is not a real number.

11. $\sqrt{\frac{11}{9}} = \frac{\sqrt{11}}{\sqrt{9}} = \frac{\sqrt{11}}{3}$

12. $\sqrt[3]{\frac{7}{64}} = \frac{\sqrt[3]{7}}{\sqrt[3]{64}} = \frac{\sqrt[3]{7}}{4}$

13. $-\sqrt{16} = -4$. The negative sign indicates the negative square root of 16.

14. $-\sqrt{25} = -5$. The negative sign indicates the negative square root of 25.

15. $\sqrt{\frac{9}{49}} = \frac{\sqrt{9}}{\sqrt{49}} = \frac{3}{7}$

16. $\sqrt{\frac{1}{64}} = \frac{\sqrt{1}}{\sqrt{64}} = \frac{1}{8}$

17. $\sqrt{a^8 a^2} = \sqrt{a^8}\sqrt{b^2} = a^4 b$

18. $\sqrt{x^{10} y^{20}} = \sqrt{x^{10}}\sqrt{y^{20}} = x^5 y^{10}$

19. $\sqrt{25m^6} = \sqrt{25}\sqrt{m^6} = 5m^3$

20. $\sqrt{9n^{16}} = \sqrt{9}\sqrt{n^{16}} = 3n^8$

21. $5\sqrt{7} + \sqrt{7} = (5+1)\sqrt{7} = 6\sqrt{7}$

22.
$$\begin{aligned}\sqrt{50} - \sqrt{8} &= \sqrt{25 \cdot 2} - \sqrt{4 \cdot 2}\\ &= \sqrt{25}\sqrt{2} - \sqrt{4}\sqrt{2}\\ &= 5\sqrt{2} - 2\sqrt{2}\\ &= (5-2)\sqrt{2}\\ &= 3\sqrt{2}\end{aligned}$$

23. $5\sqrt{2} - 5\sqrt{3}$ cannot be simplified.

24.
$$\begin{aligned}&2\sqrt{x} + \sqrt{25x} - \sqrt{36x} + 3x\\ &= 2\sqrt{x} + \sqrt{25}\sqrt{x} - \sqrt{36}\sqrt{x} + 3x\\ &= 2\sqrt{x} + 5\sqrt{x} - 6\sqrt{x} + 3x\\ &= (2+5-6)\sqrt{x} + 3x\\ &= \sqrt{x} + 3x\end{aligned}$$

25. $\sqrt{2} \cdot \sqrt{15} = \sqrt{2 \cdot 15} = \sqrt{30}$

26. $\sqrt{3} \cdot \sqrt{3} = \sqrt{3 \cdot 3} = \sqrt{9} = 3$

27. $\left(2\sqrt{7}\right)^2 = \left(2\sqrt{7}\right)\left(2\sqrt{7}\right) = 4\left(\sqrt{7}\right)^2 = 4 \cdot 7 = 28$

28. $\left(3\sqrt{5}\right)^2 = \left(3\sqrt{5}\right)\left(3\sqrt{5}\right) = 9\left(\sqrt{5}\right)^2 = 9 \cdot 5 = 45$

29. $\sqrt{3}\left(\sqrt{11} + 1\right) = \sqrt{3} \cdot \sqrt{11} + \sqrt{3} \cdot 1 = \sqrt{33} + \sqrt{3}$

30.
$$\begin{aligned}\sqrt{6}\left(\sqrt{3} - 2\right) &= \sqrt{6} \cdot \sqrt{3} - \sqrt{6} \cdot 2\\ &= \sqrt{18} - 2\sqrt{6}\\ &= \sqrt{9 \cdot 2} - 2\sqrt{6}\\ &= 3\sqrt{2} - 2\sqrt{6}\end{aligned}$$

31. $\sqrt{8y}\sqrt{2y} = \sqrt{8y \cdot 2y} = \sqrt{16y^2} = 4y$

32.
$$\begin{aligned}\sqrt{15x^2} \cdot \sqrt{3x^2} &= \sqrt{15x^2 \cdot 3x^2}\\ &= \sqrt{45x^4}\\ &= \sqrt{9x^4 \cdot 5}\\ &= 3x^2\sqrt{5}\end{aligned}$$

33.
$$\begin{aligned}\left(\sqrt{x} - 5\right)\left(\sqrt{x} + 2\right) &= \sqrt{x} \cdot \sqrt{x} + 2\sqrt{x} - 5\sqrt{x} - 5 \cdot 2\\ &= x - 3\sqrt{x} - 10\end{aligned}$$

34.
$$\begin{aligned}\left(3 + \sqrt{2}\right)^2 &= (3)^2 + 2(3)\left(\sqrt{2}\right) + \left(\sqrt{2}\right)^2\\ &= 9 + 6\sqrt{2} + 2\\ &= 11 + 6\sqrt{2}\end{aligned}$$

35. $\frac{\sqrt{8}}{\sqrt{2}} = \sqrt{\frac{8}{2}} = \sqrt{4} = 2$

36. $\frac{\sqrt{45}}{\sqrt{15}} = \sqrt{\frac{45}{15}} = \sqrt{3}$

37. $\dfrac{\sqrt{24x^5}}{\sqrt{2x}} = \sqrt{\dfrac{24x^5}{2x}} = \sqrt{12x^4} = \sqrt{4x^4 \cdot 3} = 2x^2\sqrt{3}$

38. $\dfrac{\sqrt{75a^4b^5}}{\sqrt{5ab}} = \sqrt{\dfrac{75a^4b^5}{5ab}}$
$= \sqrt{15a^3b^4}$
$= \sqrt{a^2b^4 \cdot 15a}$
$= ab^2\sqrt{15a}$

39. $\sqrt{\dfrac{1}{6}} = \dfrac{\sqrt{1}}{\sqrt{6}} = \dfrac{1}{\sqrt{6}} = \dfrac{1}{\sqrt{6}} \cdot \dfrac{\sqrt{6}}{\sqrt{6}} = \dfrac{\sqrt{6}}{6}$

40. $\dfrac{x}{\sqrt{20}} = \dfrac{x}{\sqrt{4 \cdot 5}} = \dfrac{x}{2\sqrt{5}} = \dfrac{x}{2\sqrt{5}} \cdot \dfrac{\sqrt{5}}{\sqrt{5}} = \dfrac{x\sqrt{5}}{2 \cdot 5} = \dfrac{x\sqrt{5}}{10}$

41. $\dfrac{4}{\sqrt{6}+1} = \dfrac{4}{\sqrt{6}+1} \cdot \dfrac{\sqrt{6}-1}{\sqrt{6}-1}$
$= \dfrac{4(\sqrt{6}-1)}{(\sqrt{6})^2 - 1^2}$
$= \dfrac{4\sqrt{6}-4}{6-1}$
$= \dfrac{4\sqrt{6}-4}{5}$

42. $\dfrac{\sqrt{2}+1}{\sqrt{x}-5} = \dfrac{\sqrt{2}+1}{\sqrt{x}-5} \cdot \dfrac{\sqrt{x}+5}{\sqrt{x}+5}$
$= \dfrac{(\sqrt{2}+1)(\sqrt{x}+5)}{(\sqrt{x})^2 - 5^2}$
$= \dfrac{\sqrt{2}\sqrt{x} + 5\sqrt{2} + 1\sqrt{x} + 1 \cdot 5}{x-25}$
$= \dfrac{\sqrt{2x} + 5\sqrt{2} + \sqrt{x} + 5}{x-25}$

Section 8.5 Practice Exercises

1. $\sqrt{x-2} = 7$
$(\sqrt{x-2})^2 = 7^2$
$x - 2 = 49$
$x = 51$

2. $\sqrt{6x-1} = \sqrt{x}$
$(\sqrt{6x-1})^2 = (\sqrt{x})^2$
$6x - 1 = x$
$5x - 1 = 0$
$5x = 1$
$x = \dfrac{1}{5}$

3. $\sqrt{x} + 9 = 2$
$\sqrt{x} = -7$
$\sqrt{x}$ cannot equal −7. Thus, the equation has no solution.

4. $\sqrt{9y^2 + 2y - 10} = 3y$
$\left(\sqrt{9y^2 + 2y - 10}\right)^2 = (3y)^2$
$9y^2 + 2y - 10 = 9y^2$
$2y - 10 = 0$
$2y = 10$
$y = 5$

5. $\sqrt{x+1} - x = -5$
$\sqrt{x+1} = x - 5$
$(\sqrt{x+1})^2 = (x-5)^2$
$x + 1 = x^2 - 10x + 25$
$0 = x^2 - 11x + 24$
$0 = (x-8)(x-3)$
$0 = x - 8$ or $0 = x - 3$
$8 = x$ $\quad$ $3 = x$
Replacing x with 3 results in a false statement. 3 is an extraneous solution. The only solution is 8.

6. $\sqrt{x} + 3 = \sqrt{x+15}$
$(\sqrt{x}+3)^2 = (\sqrt{x+15})^2$
$x + 6\sqrt{x} + 9 = x + 15$
$6\sqrt{x} = 6$
$\sqrt{x} = 1$
$x = 1$

Vocabulary, Readiness & Video Check 8.5

1. The squaring property can result in extraneous solutions, so we need to check our solutions in the original equation—before the squaring property was applied—to make sure they are actual solutions.

2. No; if the first squaring leaves a radical term, this new equation can be thought of as an equation that needs the property applied once.

Exercise Set 8.5

1.
$$\sqrt{x}=9$$
$$\left(\sqrt{x}\right)^2=9^2$$
$$x=81$$

3.
$$\sqrt{x+5}=2$$
$$\left(\sqrt{x+5}\right)^2=2^2$$
$$x+5=4$$
$$x=-1$$

5.
$$\sqrt{x}-2=5$$
$$\sqrt{x}=7$$
$$\left(\sqrt{x}\right)^2=7^2$$
$$x=49$$

7.
$$3\sqrt{x}+5=2$$
$$3\sqrt{x}=-3$$
$$\sqrt{x}=-1$$

$\sqrt{x}$ cannot equal -1. Thus, the equation has no solution.

9.
$$\sqrt{x}=\sqrt{3x-8}$$
$$\left(\sqrt{x}\right)^2=\left(\sqrt{3x-8}\right)^2$$
$$x=3x-8$$
$$-2x=-8$$
$$x=4$$

11.
$$\sqrt{4x-3}=\sqrt{x+3}$$
$$\left(\sqrt{4x-3}\right)^2=\left(\sqrt{x+3}\right)^2$$
$$4x-3=x+3$$
$$3x-3=3$$
$$3x=6$$
$$x=2$$

13.
$$\sqrt{9x^2+2x-4}=3x$$
$$\left(\sqrt{9x^2+2x-4}\right)^2=(3x)^2$$
$$9x^2+2x-4=9x^2$$
$$2x-4=0$$
$$2x=4$$
$$x=2$$

15.
$$\sqrt{x}=x-6$$
$$\left(\sqrt{x}\right)^2=(x-6)^2$$
$$x=x^2-12x+36$$
$$0=x^2-13x+36$$
$$0=(x-9)(x-4)$$
$$0=x-9 \quad \text{or} \quad 0=x-4$$
$$9=x \qquad\qquad 4=x$$

$x=4$ does not check, so the solution is $x=9$.

17.
$$\sqrt{x+7}=x+5$$
$$\left(\sqrt{x+7}\right)^2=(x+5)^2$$
$$x+7=x^2+10x+25$$
$$0=x^2+9x+18$$
$$0=(x+6)(x+3)$$
$$x+6=0 \quad \text{or} \quad x+3=0$$
$$x=-6 \qquad\qquad x=-3$$

$x=-6$ does not check, so the solution is $x=-3$.

19.
$$\sqrt{3x+7}-x=3$$
$$\sqrt{3x+7}=x+3$$
$$\left(\sqrt{3x+7}\right)^2=(x+3)^2$$
$$3x+7=x^2+6x+9$$
$$0=x^2+3x+2$$
$$0=(x+2)(x+1)$$
$$x+2=0 \quad \text{or} \quad x+1=0$$
$$x=-2 \qquad\qquad x=-1$$

21.
$$\sqrt{16x^2+2x+2}=4x$$
$$\left(\sqrt{16x^2+2x+2}\right)^2=(4x)^2$$
$$16x^2+2x+2=16x^2$$
$$2x+2=0$$
$$2x=-2$$
$$x=-1$$

$x=-1$ does not check, so the equation has no solution.

23. $\sqrt{2x^2+6x+9}=3$
$\left(\sqrt{2x^2+6x+9}\right)^2=3^2$
$2x^2+6x+9=9$
$2x^2+6x=0$
$2x(x+3)=0$
$2x=0$ or $x+3=0$
$x=0$ $\quad x=-3$

25. $\sqrt{x-7}=\sqrt{x}-1$
$\left(\sqrt{x-7}\right)^2=\left(\sqrt{x}-1\right)^2$
$x-7=x-2\sqrt{x}+1$
$2\sqrt{x}=8$
$\sqrt{x}=4$
$\left(\sqrt{x}\right)^2=4^2$
$x=16$

27. $\sqrt{x}+2=\sqrt{x+24}$
$\left(\sqrt{x}+2\right)^2=\left(\sqrt{x+24}\right)^2$
$x+4\sqrt{x}+4=x+24$
$4\sqrt{x}=20$
$\sqrt{x}=5$
$\left(\sqrt{x}\right)^2=5^2$
$x=25$

29. $\sqrt{x+8}=\sqrt{x}+2$
$\left(\sqrt{x+8}\right)^2=\left(\sqrt{x}+2\right)^2$
$x+8=x+4\sqrt{x}+4$
$4=4\sqrt{x}$
$1=\sqrt{x}$
$1^2=\left(\sqrt{x}\right)^2$
$1=x$

31. $\sqrt{2x+6}=4$
$\left(\sqrt{2x+6}\right)^2=4^2$
$2x+6=16$
$2x=10$
$x=5$

33. $\sqrt{x+6}+1=3$
$\sqrt{x+6}=2$
$\left(\sqrt{x+6}\right)^2=2^2$
$x+6=4$
$x=-2$

35. $\sqrt{x+6}+5=3$
$\sqrt{x+6}=-2$
$\sqrt{x+6}$ cannot equal -2. Thus, the equation has no solution.

37. $\sqrt{16x^2-3x+6}=4x$
$\left(\sqrt{16x^2-3x+6}\right)^2=(4x)^2$
$16x^2-3x+6=16x^2$
$-3x+6=0$
$-3x=-6$
$x=2$

39. $-\sqrt{x}=-6$
$\sqrt{x}=6$
$\left(\sqrt{x}\right)^2=6^2$
$x=36$

41. $\sqrt{x+9}=\sqrt{x}-3$
$\left(\sqrt{x+9}\right)^2=\left(\sqrt{x}-3\right)^2$
$x+9=x-6\sqrt{x}+9$
$0=-6\sqrt{x}$
$0=\sqrt{x}$
$0^2=\left(\sqrt{x}\right)^2$
$0=x$
$x=0$ does not check, so the equation has no solution.

43. $\sqrt{2x+1}+3=5$
$\sqrt{2x+1}=2$
$\left(\sqrt{2x+1}\right)^2=2^2$
$2x+1=4$
$2x=3$
$x=\frac{3}{2}$

45. $\sqrt{x}+3=7$
$\sqrt{x}=4$
$\left(\sqrt{x}\right)^2=4^2$
$x=16$

47. $\sqrt{4x}=\sqrt{2x+6}$
$\left(\sqrt{4x}\right)^2=\left(\sqrt{2x+6}\right)^2$
$4x=2x+6$
$2x=6$
$x=3$

49. $\sqrt{2x+1}=x-7$
$\left(\sqrt{2x+1}\right)^2=(x-7)^2$
$2x+1=x^2-14x+49$
$0=x^2-16x+48$
$0=(x-4)(x-12)$
$x-4=0$ or $x-12=0$
$x=4$ $\quad x=12$
$x=4$ does not check, so the only solution is $x=12$.

51. $x=\sqrt{2x-2}+1$
$x-1=\sqrt{2x-2}$
$(x-1)^2=\left(\sqrt{2x-2}\right)^2$
$x^2-2x+1=2x-2$
$x^2-4x+3=0$
$(x-3)(x-1)=0$
$x-3=0$ or $x-1=0$
$x=3$ $\quad x=1$

53. $\sqrt{1-8x}-x=4$
$\sqrt{1-8x}=x+4$
$\left(\sqrt{1-8x}\right)^2=(x+4)^2$
$1-8x=x^2+8x+16$
$0=x^2+16x+15$
$0=(x+15)(x+1)$
$x+15=0$ or $x+1=0$
$x=-15$ $\quad x=-1$
$x=-15$ does not check, so the solution is $x=-1$.

55. $3x-8=19$
$3x=27$
$x=9$

57. Let x be the width of the rectangle, then the length is $2x$.
$2(2x)+2x=24$
$4x+2x=24$
$6x=24$
$x=4$
$2x=2(4)=8$
The length of the rectangle is 8 inches.

59. $\sqrt{x-3}+3=\sqrt{3x+4}$
$\left(\sqrt{x-3}+3\right)^2=\left(\sqrt{3x+4}\right)^2$
$x-3+6\sqrt{x-3}+9=3x+4$
$6\sqrt{x-3}=2x-2$
$\left(6\sqrt{x-3}\right)^2=(2x-2)^2$
$36(x-3)=4x^2-8x+4$
$36x-108=4x^2-8x+4$
$0=4x^2-44x+112$
$0=4(x-7)(x-4)$
$x-7=0$ or $x-4=0$
$x=7$ $\quad x=4$

61. answers may vary

63. a. For $V=20$, $b=\sqrt{\frac{20}{2}}=\sqrt{10}\approx 3.2$.

For $V=200$, $b=\sqrt{\frac{200}{2}}=\sqrt{100}=10$.

For $V=2000$, $b=\sqrt{\frac{2000}{2}}=\sqrt{1000}\approx 31.6$.

V	20	200	2000
b	3.2	10	31.6

b. No, the volume increases by a factor of $\sqrt{10}$.

65.

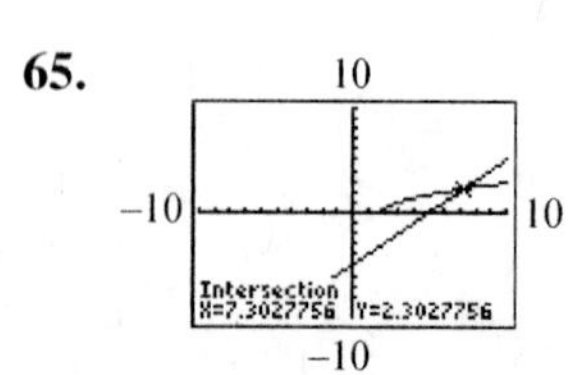

The solution of $\sqrt{x-2}=x-5$ is $x\approx 7.30$.

67.

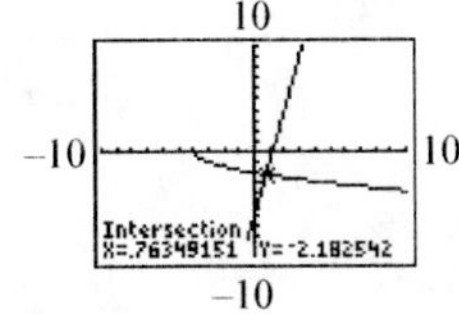

The solution of $-\sqrt{x+4}=5x-6$ is $x\approx 0.76$.

Section 8.6 Practice Exercises

1.
$$\begin{aligned} a^2+b^2&=c^2\\ 3^2+4^2&=c^2\\ 9+16&=c^2\\ 25&=c^2\\ \sqrt{25}&=c\\ 5&=c \end{aligned}$$
The hypotenuse has a length of 5 centimeters.

2.
$$\begin{aligned} a^2+b^2&=c^2\\ 5^2+3^2&=c^2\\ 25+9&=c^2\\ 34&=c^2\\ \sqrt{34}&=c\\ 5.83&\approx c \end{aligned}$$

3. The property owner is using a right triangle with one leg measuring 40 feet and hypotenuse measuring 65 feet to find the unknown distance.
$$\begin{aligned} a^2+b^2&=c^2\\ 40^2+b^2&=65^2\\ 1600+b^2&=4225\\ b^2&=2625\\ b&=\sqrt{2625}=5\sqrt{105} \end{aligned}$$
The distance across the pond is $5\sqrt{105}$ feet or approximately 51.2 feet.

4.
$$\begin{aligned} v&=\sqrt{2gh}\\ &=\sqrt{2\cdot 32\cdot 20}\\ &=\sqrt{1280}\\ &=16\sqrt{5} \end{aligned}$$
The velocity of the object after falling 20 feet is exactly $16\sqrt{5}$ feet per second or approximately 35.8 feet per second.

Vocabulary, Readiness & Video Check 8.6

1. The Pythagorean theorem applies to right triangles only, and in the formula $a^2+b^2+c^2$, c is the length of the hypotenuse.

2. Our answer is a number that when squared equals another value. We're looking for a distance, which must be positive, so our answer must be positive.

3. Both examples ask for an answer rounded to a given place, meaning an estimated answer is expected rather than an exact answer. An exact answer would be given in radical form.

Exercise Set 8.6

1.
$$\begin{aligned} a^2+b^2&=c^2\\ 2^2+3^2&=c^2\\ 4+9&=c^2\\ 13&=c^2\\ \sqrt{13}&=\sqrt{c^2}\\ \sqrt{13}&=c\\ c=\sqrt{13}&\approx 3.61 \end{aligned}$$

3.
$$\begin{aligned} a^2+b^2&=c^2\\ 3^2+b^2&=6^2\\ 9+b^2&=36\\ b^2&=27\\ \sqrt{b^2}&=\sqrt{27}\\ b&=3\sqrt{3}\approx 5.20 \end{aligned}$$

5.
$$\begin{aligned} a^2+b^2&=c^2\\ 7^2+24^2&=c^2\\ 49+576&=c^2\\ 625&=c^2\\ \sqrt{625}&=\sqrt{c^2}\\ 25&=c \end{aligned}$$

7.
$$a^2+b^2=c^2$$
$$\left(\sqrt{3}\right)^2+b^2=5^2$$
$$3+b^2=25$$
$$b^2=22$$
$$\sqrt{b^2}=\sqrt{22}$$
$$b=\sqrt{22}\approx 4.69$$

9.
$$a^2+b^2=c^2$$
$$4^2+b^2=13^2$$
$$16+b^2=169$$
$$b^2=153$$
$$\sqrt{b^2}=\sqrt{153}$$
$$b=3\sqrt{17}$$
$$b=3\sqrt{17}\approx 12.37$$

11.
$$a^2+b^2=c^2$$
$$4^2+5^2=c^2$$
$$16+25=c^2$$
$$41=c^2$$
$$\sqrt{41}=\sqrt{c^2}$$
$$c=\sqrt{41}\approx 6.40$$

13.
$$a^2+b^2=c^2$$
$$a^2+2^2=6^2$$
$$a^2+4=36$$
$$a^2=32$$
$$\sqrt{a^2}=\sqrt{32}$$
$$a=4\sqrt{2}$$
$$a=4\sqrt{2}\approx 5.66$$

15.
$$a^2+b^2=c^2$$
$$\left(\sqrt{10}\right)^2+b^2=10^2$$
$$10+b^2=100$$
$$b^2=90$$
$$\sqrt{b^2}=\sqrt{90}$$
$$b=3\sqrt{10}\approx 9.49$$

17. The pole, wire, and ground form a right triangle with legs of 5 feet and 20 feet.
$$a^2+b^2=c^2$$
$$5^2+20^2=c^2$$
$$25+400=c^2$$
$$425=c^2$$
$$\sqrt{425}=\sqrt{c^2}$$
$$\sqrt{425}=c$$
$$c=\sqrt{425}\approx 20.6$$
The length of the wire is 20.6 feet.

19. The diagonal brace is the hypotenuse of a right triangle with legs measuring 6 feet and 10 feet.
$$a^2+b^2=c^2$$
$$6^2+10^2=c^2$$
$$36+100=c^2$$
$$136=c^2$$
$$\sqrt{136}=\sqrt{c^2}$$
$$c=\sqrt{136}\approx 11.7$$
The brace needs to be 11.7 feet.

21.
$$b=\sqrt{\frac{3V}{h}}$$
$$6=\sqrt{\frac{3V}{2}}$$
$$6^2=\left(\sqrt{\frac{3V}{2}}\right)^2$$
$$36=\frac{3V}{2}$$
$$72=3V$$
$$24=V$$
The volume is 24 cubic feet.

23.
$$s=\sqrt{30fd}$$
$$s=\sqrt{30\cdot 0.35\cdot 280}$$
$$s=\sqrt{2940}$$
$$s\approx 54$$
The car was moving at a speed of 54 miles per hour.

25.
$$v=\sqrt{2.5r}$$
$$v=\sqrt{2.5(300)}$$
$$v=\sqrt{750}$$
$$v\approx 27.4$$
The maximum safe speed is 27 miles per hour.

27. $d = 3.5\sqrt{h}$
$d = 3.5\sqrt{276.9}$
$d \approx 58.2$
You can see a distance of 58.2 kilometers.

29. $d = 3.5\sqrt{h}$
$d = 3.5\sqrt{541.3}$
$d \approx 81.4$
You can see a distance of 81.4 kilometers.

31. $\sqrt{9} = 3$ and $-\sqrt{9} = -3$, so -3 and 3 are numbers whose square is 9.

33. $\sqrt{100} = 10$ and $-\sqrt{100} = -10$, so -10 and 10 are numbers whose square is 100.

35. $\sqrt{64} = 8$ and $-\sqrt{64} = -8$, so -8 and 8 are numbers whose square is 64.

37. First find y.
$a^2 + b^2 = c^2$
$3^2 + y^2 = 7^2$
$9 + y^2 = 49$
$y^2 = 40$
$y = \sqrt{40}$
$y = \sqrt{4 \cdot 10}$
$y = 2\sqrt{10}$
Let b be the second leg of the right triangle with hypotenuse 5.
$a^2 + b^2 = c^2$
$3^2 + b^2 = 5^2$
$9 + b^2 = 25$
$b^2 = 16$
$\sqrt{b^2} = \sqrt{16}$
$b = 4$
Now find x.
$x = y - 4$
$x = 2\sqrt{10} - 4$

39. The distance is the length of the hypotenuse of a right triangle. One leg has length $3 \cdot 30 = 90$ miles, and the other leg has length $3 \cdot 60 = 180$ miles.
$a^2 + b^2 = c^2$
$90^2 + 180^2 = c^2$
$8100 + 32,400 = c^2$
$40,500 = c^2$
$\sqrt{40,500} = \sqrt{c^2}$
$201 \approx c$
They are 201 miles apart.

41. answers may vary

Chapter 8 Vocabulary Check

1. The expressions $5\sqrt{x}$ and $7\sqrt{x}$ are examples of like radicals.

2. In the expression $\sqrt[3]{45}$ the number 3 is the index, the number 45 is the radicand, and $\sqrt{}$ is called the radical sign.

3. The conjugate of $a + b$ is $a - b$.

4. The principal square root of 25 is 5.

5. The process eliminating the radical in the denominator of a radical expression is called rationalizing the denominator.

6. The Pythagorean theorem states that for a right triangle, $(\text{leg})^2 + (\text{leg})^2 = (\text{hypotenuse})^2$.

Chapter 8 Review

1. $\sqrt{81} = 9$, because $9^2 = 81$ and 9 is positive.

2. $-\sqrt{49} = -7$. The negative indicates the negative square root of 49.

3. $\sqrt[3]{27} = 3$, because $3^3 = 27$.

4. $\sqrt[4]{81} = 3$, because $3^4 = 81$.

5. $-\sqrt{\frac{9}{64}} = -\frac{3}{8}$ because $\left(\frac{3}{8}\right)^2 = \frac{9}{64}$.

6. $\sqrt{\frac{36}{81}} = \frac{6}{9}$ because $\left(\frac{6}{9}\right)^2 = \frac{36}{81}$, and $\frac{6}{9} = \frac{2}{3}$.

7. $\sqrt[4]{16} = 2$ because $2^4 = 16$.

8. $\sqrt[3]{-8} = -2$ because $(-2)^3 = -8$.

9. **c**; $\sqrt{-4}$ is not a real number because the radicand is negative and the index is even.

10. **a, c**; $\sqrt{-5}$ and $\sqrt[4]{-5}$ are not real numbers because the radicands are negatives and the indexes are even.

11. $\sqrt{x^{12}} = x^6$, because $(x^6)^2 = x^{12}$.

12. $\sqrt{x^8} = x^4$, because $(x^4)^2 = x^8$.

13. $\sqrt{9y^2} = 3y$, because $(3y)^2 = 9y^2$.

14. $\sqrt{25x^4} = 5x^2$, because $(5x^2)^2 = 25x^4$.

15. $\sqrt{40} = \sqrt{4 \cdot 10} = \sqrt{4}\sqrt{10} = 2\sqrt{10}$

16. $\sqrt{24} = \sqrt{4 \cdot 6} = \sqrt{4}\sqrt{6} = 2\sqrt{6}$

17. $\sqrt{54} = \sqrt{9 \cdot 6} = \sqrt{9}\sqrt{6} = 3\sqrt{6}$

18. $\sqrt{88} = \sqrt{4 \cdot 22} = \sqrt{4}\sqrt{22} = 2\sqrt{22}$

19. $\sqrt{x^5} = \sqrt{x^4 \cdot x} = \sqrt{x^4}\sqrt{x} = x^2\sqrt{x}$

20. $\sqrt{y^7} = \sqrt{y^6 \cdot y} = \sqrt{y^6}\sqrt{y} = y^3\sqrt{y}$

21. $\sqrt{20x^2} = \sqrt{4x^2 \cdot 5} = \sqrt{4x^2}\sqrt{5} = 2x\sqrt{5}$

22. $\sqrt{50y^4} = \sqrt{25y^4 \cdot 2} = \sqrt{25y^4}\sqrt{2} = 5y^2\sqrt{2}$

23. $\sqrt[3]{54} = \sqrt[3]{27 \cdot 2} = \sqrt[3]{27}\sqrt[3]{2} = 3\sqrt[3]{2}$

24. $\sqrt[3]{88} = \sqrt[3]{8 \cdot 11} = \sqrt[3]{8}\sqrt[3]{11} = 2\sqrt[3]{11}$

25. $\sqrt{\frac{18}{25}} = \frac{\sqrt{18}}{\sqrt{25}} = \frac{\sqrt{9 \cdot 2}}{5} = \frac{\sqrt{9}\sqrt{2}}{5} = \frac{3\sqrt{2}}{5}$

26. $\sqrt{\frac{75}{64}} = \frac{\sqrt{75}}{\sqrt{64}} = \frac{\sqrt{25 \cdot 3}}{8} = \frac{\sqrt{25}\sqrt{3}}{8} = \frac{5\sqrt{3}}{8}$

27. $-\sqrt{\frac{50}{9}} = -\frac{\sqrt{50}}{\sqrt{9}} = -\frac{\sqrt{25 \cdot 2}}{3} = -\frac{\sqrt{25}\sqrt{2}}{3} = -\frac{5\sqrt{2}}{3}$

28. $-\sqrt{\frac{12}{49}} = -\frac{\sqrt{12}}{\sqrt{49}} = -\frac{\sqrt{4 \cdot 3}}{7} = -\frac{\sqrt{4}\sqrt{3}}{7} = -\frac{2\sqrt{3}}{7}$

29. $\sqrt{\frac{11}{x^2}} = \frac{\sqrt{11}}{\sqrt{x^2}} = \frac{\sqrt{11}}{x}$

30. $\sqrt{\frac{7}{y^4}} = \frac{\sqrt{7}}{\sqrt{y^4}} = \frac{\sqrt{7}}{y^2}$

31. $\sqrt{\frac{y^5}{100}} = \frac{\sqrt{y^5}}{\sqrt{100}} = \frac{\sqrt{y^4 \cdot y}}{10} = \frac{\sqrt{y^4}\sqrt{y}}{10} = \frac{y^2\sqrt{y}}{10}$

32. $\sqrt{\frac{x^3}{81}} = \frac{\sqrt{x^3}}{\sqrt{81}} = \frac{\sqrt{x^2 \cdot x}}{9} = \frac{\sqrt{x^2}\sqrt{x}}{9} = \frac{x\sqrt{x}}{9}$

33. $5\sqrt{8} - 8\sqrt{2} = (5-8)\sqrt{2} = -3\sqrt{2}$

34. $\sqrt{3} - 6\sqrt{3} = (1-6)\sqrt{3} = -5\sqrt{3}$

35. $6\sqrt{5} + 3\sqrt{6} - 2\sqrt{5} + \sqrt{6} = (6-2)\sqrt{5} + (3+1)\sqrt{6}$
$= 4\sqrt{5} + 4\sqrt{6}$

36. $-\sqrt{7} + 8\sqrt{2} - \sqrt{7} - 6\sqrt{2} = (-1-1)\sqrt{7} + (8-6)\sqrt{2}$
$= -2\sqrt{7} + 2\sqrt{2}$

37. $\sqrt{28} + \sqrt{63} + \sqrt{56} = \sqrt{4 \cdot 7} + \sqrt{9 \cdot 7} + \sqrt{4 \cdot 14}$
$= \sqrt{4}\sqrt{7} + \sqrt{9}\sqrt{7} + \sqrt{4}\sqrt{14}$
$= 2\sqrt{7} + 3\sqrt{7} + 2\sqrt{14}$
$= (2+3)\sqrt{7} + 2\sqrt{14}$
$= 5\sqrt{7} + 2\sqrt{14}$

38. $\sqrt{75} + \sqrt{48} - \sqrt{16} = \sqrt{25 \cdot 3} + \sqrt{16 \cdot 3} - 4$
$= \sqrt{25}\sqrt{3} + \sqrt{16}\sqrt{3} - 4$
$= 5\sqrt{3} + 4\sqrt{3} - 4$
$= (5+4)\sqrt{3} - 4$
$= 9\sqrt{3} - 4$

39. $\sqrt{\frac{5}{9}}-\sqrt{\frac{5}{36}}=\frac{\sqrt{5}}{\sqrt{9}}-\frac{\sqrt{5}}{\sqrt{36}}$
$=\frac{\sqrt{5}}{3}-\frac{\sqrt{5}}{6}$
$=\frac{2\sqrt{5}}{6}-\frac{\sqrt{5}}{6}$
$=\frac{\sqrt{5}}{6}$

40. $\sqrt{\frac{11}{25}}+\sqrt{\frac{11}{16}}=\frac{\sqrt{11}}{\sqrt{25}}+\frac{\sqrt{11}}{\sqrt{16}}$
$=\frac{\sqrt{11}}{5}+\frac{\sqrt{11}}{4}$
$=\frac{4\sqrt{11}}{20}+\frac{5\sqrt{11}}{20}$
$=\frac{9\sqrt{11}}{20}$

41. $\sqrt{45x^2}+3\sqrt{5x^2}-7x\sqrt{5}+10$
$=\sqrt{9x^2\cdot 5}+3\sqrt{x^2\cdot 5}-7x\sqrt{5}+10$
$=\sqrt{9x^2}\sqrt{5}+3\sqrt{x^2}\sqrt{5}-7x\sqrt{5}+10$
$=3x\sqrt{5}+3x\sqrt{5}-7x\sqrt{5}+10$
$=(3x+3x-7x)\sqrt{5}+10$
$=-x\sqrt{5}+10$ or $10-x\sqrt{5}$

42. $\sqrt{50x}-9\sqrt{2x}+\sqrt{72x}-\sqrt{3x}$
$=\sqrt{25\cdot 2x}-9\sqrt{2x}+\sqrt{36\cdot 2x}-\sqrt{3x}$
$=\sqrt{25}\sqrt{2x}-9\sqrt{2x}+\sqrt{36}\sqrt{2x}-\sqrt{3x}$
$=5\sqrt{2x}-9\sqrt{2x}+6\sqrt{2x}-\sqrt{3x}$
$=(5-9+6)\sqrt{2x}-\sqrt{3x}$
$=2\sqrt{2x}-\sqrt{3x}$

43. $\sqrt{3}\cdot\sqrt{6}=\sqrt{3\cdot 6}=\sqrt{18}=\sqrt{9\cdot 2}=\sqrt{9}\cdot\sqrt{2}=3\sqrt{2}$

44. $\sqrt{5}\cdot\sqrt{15}=\sqrt{5\cdot 15}$
$=\sqrt{75}$
$=\sqrt{25\cdot 3}$
$=\sqrt{25}\cdot\sqrt{3}$
$=5\sqrt{3}$

45. $\sqrt{2}\left(\sqrt{5}-\sqrt{7}\right)=\sqrt{2}\sqrt{5}-\sqrt{2}\sqrt{7}=\sqrt{10}-\sqrt{14}$

46. $\sqrt{5}\left(\sqrt{11}+\sqrt{3}\right)=\sqrt{5}\sqrt{11}+\sqrt{5}\sqrt{3}=\sqrt{55}+\sqrt{15}$

47. $\left(\sqrt{3}+2\right)\left(\sqrt{6}-5\right)=\sqrt{3}\sqrt{6}-5\sqrt{3}+2\sqrt{6}-2\cdot 5$
$=\sqrt{18}-5\sqrt{3}+2\sqrt{6}-10$
$=\sqrt{9\cdot 2}-5\sqrt{3}+2\sqrt{6}-10$
$=3\sqrt{2}-5\sqrt{3}+2\sqrt{6}-10$

48. $\left(\sqrt{5}+1\right)\left(\sqrt{5}-3\right)=\sqrt{5}\sqrt{5}-3\sqrt{5}+1\sqrt{5}-1\cdot 3$
$=5+(-3+1)\sqrt{5}-3$
$=2-2\sqrt{5}$

49. $\left(\sqrt{x}-2\right)^2=\left(\sqrt{x}\right)^2-2\left(\sqrt{x}\right)(2)+2^2$
$=x-4\sqrt{x}+4$

50. $\left(\sqrt{y}+4\right)^2=\left(\sqrt{y}\right)^2+2\left(\sqrt{y}\right)(4)+4^2$
$=y+8\sqrt{y}+16$

51. $\frac{\sqrt{27}}{\sqrt{3}}=\sqrt{\frac{27}{3}}=\sqrt{9}=3$

52. $\frac{\sqrt{20}}{\sqrt{5}}=\sqrt{\frac{20}{5}}=\sqrt{4}=2$

53. $\frac{\sqrt{160}}{\sqrt{8}}=\sqrt{\frac{160}{8}}=\sqrt{20}=\sqrt{4\cdot 5}=\sqrt{4}\sqrt{5}=2\sqrt{5}$

54. $\frac{\sqrt{96}}{\sqrt{3}}=\sqrt{\frac{96}{3}}=\sqrt{32}=\sqrt{16\cdot 2}=\sqrt{16}\sqrt{2}=4\sqrt{2}$

55. $\frac{\sqrt{30x^6}}{\sqrt{2x^3}}=\sqrt{\frac{30x^6}{2x^3}}$
$=\sqrt{15x^3}$
$=\sqrt{x^2\cdot 15x}$
$=\sqrt{x^2}\sqrt{15x}$
$=x\sqrt{15x}$

56. $\frac{\sqrt{54x^5y^2}}{\sqrt{3xy^2}}=\sqrt{\frac{54x^5y^2}{3xy^2}}$
$=\sqrt{18x^4}$
$=\sqrt{9x^4\cdot 2}$
$=\sqrt{9x^4}\sqrt{2}$
$=3x^2\sqrt{2}$

57. $\frac{\sqrt{2}}{\sqrt{11}} = \frac{\sqrt{2}}{\sqrt{11}} \cdot \frac{\sqrt{11}}{\sqrt{11}} = \frac{\sqrt{2 \cdot 11}}{11} = \frac{\sqrt{22}}{11}$

58. $\frac{\sqrt{3}}{\sqrt{13}} = \frac{\sqrt{3}}{\sqrt{13}} \cdot \frac{\sqrt{13}}{\sqrt{13}} = \frac{\sqrt{3 \cdot 13}}{13} = \frac{\sqrt{39}}{13}$

59. $\sqrt{\frac{5}{6}} = \frac{\sqrt{5}}{\sqrt{6}} = \frac{\sqrt{5}}{\sqrt{6}} \cdot \frac{\sqrt{6}}{\sqrt{6}} = \frac{\sqrt{5 \cdot 6}}{6} = \frac{\sqrt{30}}{6}$

60. $\sqrt{\frac{7}{10}} = \frac{\sqrt{7}}{\sqrt{10}} = \frac{\sqrt{7}}{\sqrt{10}} \cdot \frac{\sqrt{10}}{\sqrt{10}} = \frac{\sqrt{7 \cdot 10}}{10} = \frac{\sqrt{70}}{10}$

61. $\frac{1}{\sqrt{5x}} = \frac{1}{\sqrt{5x}} \cdot \frac{\sqrt{5x}}{\sqrt{5x}} = \frac{\sqrt{5x}}{5x}$

62. $\frac{5}{\sqrt{3y}} = \frac{5}{\sqrt{3y}} \cdot \frac{\sqrt{3y}}{\sqrt{3y}} = \frac{5\sqrt{3y}}{3y}$

63. $\sqrt{\frac{3}{x}} = \frac{\sqrt{3}}{\sqrt{x}} = \frac{\sqrt{3}}{\sqrt{x}} \cdot \frac{\sqrt{x}}{\sqrt{x}} = \frac{\sqrt{3x}}{x}$

64. $\sqrt{\frac{6}{y}} = \frac{\sqrt{6}}{\sqrt{y}} = \frac{\sqrt{6}}{\sqrt{y}} \cdot \frac{\sqrt{y}}{\sqrt{y}} = \frac{\sqrt{6y}}{y}$

65.
$$\begin{aligned} \frac{3}{\sqrt{5}-2} &= \frac{3}{\sqrt{5}-2} \cdot \frac{\sqrt{5}+2}{\sqrt{5}+2} \\ &= \frac{3(\sqrt{5}+2)}{5-4} \\ &= \frac{3(\sqrt{5}+2)}{1} \\ &= 3(\sqrt{5}+2) \text{ or } 3\sqrt{5}+6 \end{aligned}$$

66.
$$\begin{aligned} \frac{8}{\sqrt{10}-3} &= \frac{8}{\sqrt{10}-3} \cdot \frac{\sqrt{10}+3}{\sqrt{10}+3} \\ &= \frac{8(\sqrt{10}+3)}{10-9} \\ &= \frac{8(\sqrt{10}+3)}{1} \\ &= 8(\sqrt{10}+3) \text{ or } 8\sqrt{10}+24 \end{aligned}$$

67.
$$\begin{aligned} \frac{\sqrt{2}+1}{\sqrt{3}-1} &= \frac{\sqrt{2}+1}{\sqrt{3}-1} \cdot \frac{\sqrt{3}+1}{\sqrt{3}+1} \\ &= \frac{\sqrt{2}\sqrt{3}+1 \cdot \sqrt{2}+1 \cdot \sqrt{3}+1 \cdot 1}{3-1} \\ &= \frac{\sqrt{6}+\sqrt{2}+\sqrt{3}+1}{2} \end{aligned}$$

68.
$$\begin{aligned} \frac{\sqrt{3}-2}{\sqrt{5}+2} &= \frac{\sqrt{3}-2}{\sqrt{5}+2} \cdot \frac{\sqrt{5}-2}{\sqrt{5}-2} \\ &= \frac{\sqrt{3}\sqrt{5}-2\sqrt{3}-2\sqrt{5}+2 \cdot 2}{5-4} \\ &= \frac{\sqrt{15}-2\sqrt{3}-2\sqrt{5}+4}{1} \\ &= \sqrt{15}-2\sqrt{3}-2\sqrt{5}+4 \end{aligned}$$

69.
$$\begin{aligned} \frac{10}{\sqrt{x}+5} &= \frac{10}{\sqrt{x}+5} \cdot \frac{\sqrt{x}-5}{\sqrt{x}-5} \\ &= \frac{10(\sqrt{x}-5)}{x-25} \\ &= \frac{10\sqrt{x}-50}{x-25} \end{aligned}$$

70. $\frac{8}{\sqrt{x}-1} = \frac{8}{\sqrt{x}-1} \cdot \frac{\sqrt{x}+1}{\sqrt{x}+1} = \frac{8(\sqrt{x}+1)}{x-1} = \frac{8\sqrt{x}+8}{x-1}$

71.
$$\begin{aligned} \sqrt{2x} &= 6 \\ \left(\sqrt{2x}\right)^2 &= 6^2 \\ 2x &= 36 \\ x &= 18 \end{aligned}$$

72.
$$\begin{aligned} \sqrt{x+3} &= 4 \\ \left(\sqrt{x+3}\right)^2 &= 4^2 \\ x+3 &= 16 \\ x &= 13 \end{aligned}$$

73.
$$\begin{aligned} \sqrt{x}+3 &= 8 \\ \sqrt{x} &= 5 \\ \left(\sqrt{x}\right)^2 &= 5^2 \\ x &= 25 \end{aligned}$$

74.
$$\begin{aligned} \sqrt{x}+8 &= 3 \\ \sqrt{x} &= -5 \end{aligned}$$

$\sqrt{x}$ cannot equal -5. Thus, the equation has no solution.

75. $\sqrt{2x+1} = x-7$

$\left(\sqrt{2x+1}\right)^2 = (x-7)^2$

$2x+1 = x^2 - 14x + 49$

$0 = x^2 - 16x + 48$

$0 = (x-12)(x-4)$

$x-12 = 0$ or $x-4 = 0$

$x = 12$ $\quad x = 4$

$x = 4$ does not check, so the solution is $x = 12$.

76. $\sqrt{3x+1} = x-1$

$\left(\sqrt{3x+1}\right)^2 = (x-1)^2$

$3x+1 = x^2 - 2x + 1$

$0 = x^2 - 5x$

$0 = x(x-5)$

$x = 0$ or $x-5 = 0$

$x = 5$

$x = 0$ does not check, so the solution is $x = 5$.

77. $\sqrt{x}+3 = \sqrt{x+15}$

$\left(\sqrt{x}+3\right)^2 = \left(\sqrt{x+15}\right)^2$

$x + 6\sqrt{x} + 9 = x + 15$

$6\sqrt{x} = 6$

$\sqrt{x} = 1$

$\left(\sqrt{x}\right)^2 = 1^2$

$x = 1$

78. $\sqrt{x-5} = \sqrt{x}-1$

$\left(\sqrt{x-5}\right)^2 = \left(\sqrt{x}-1\right)^2$

$x-5 = x - 2\sqrt{x} + 1$

$-6 = -2\sqrt{x}$

$3 = \sqrt{x}$

$3^2 = \left(\sqrt{x}\right)^2$

$9 = x$

79. $a^2 + b^2 = c^2$

$5^2 + b^2 = 9^2$

$25 + b^2 = 81$

$b^2 = 56$

$\sqrt{b^2} = \sqrt{56}$

$b = 2\sqrt{14} \approx 7.48$

80. $a^2 + b^2 = c^2$

$6^2 + 9^2 = c^2$

$36 + 81 = c^2$

$117 = c^2$

$\sqrt{117} = \sqrt{c^2}$

$c = 3\sqrt{13} \approx 10.82$

81. The distance between Romeo and Juliet is the length of the hypotenuse of a right triangle with legs of length 20 feet and 12 feet.

$a^2 + b^2 = c^2$

$20^2 + 12^2 = c^2$

$400 + 144 = c^2$

$544 = c^2$

$\sqrt{544} = \sqrt{c^2}$

$c = 4\sqrt{34} \approx 23.32$

The distance is exactly $4\sqrt{34}$ feet or approximately 23.32 feet.

82. The diagonal of a rectangle forms right triangles with the sides of the rectangle.

$a^2 + b^2 = c^2$

$5^2 + b^2 = 10^2$

$25 + b^2 = 100$

$b^2 = 75$

$\sqrt{b^2} = \sqrt{75}$

$b = 5\sqrt{3} \approx 8.66$

The length of the rectangle is exactly $5\sqrt{3}$ inches or approximately 8.66 inches.

83. $r = \sqrt{\frac{S}{4\pi}}$

$r = \sqrt{\frac{72}{4\pi}}$

$r \approx 2.4$

The radius is 2.4 inches.

84.
$$r=\sqrt{\frac{S}{4\pi}}$$
$$6=\sqrt{\frac{S}{4\pi}}$$
$$6^2=\left(\sqrt{\frac{S}{4\pi}}\right)^2$$
$$36=\frac{S}{4\pi}$$
$$144\pi=S$$
The surface area is 144π square inches.

85. $\sqrt{144}=12$, because $12^2=144$ and 12 is positive.

86. $-\sqrt[3]{64}=-4$, because $4^3=64$.

87. $\sqrt{16x^{16}}=4x^8$, because $(4x^8)^2=16x^{16}$.

88. $\sqrt{4x^{24}}=2x^{12}$, because $(2x^{12})^2=4x^{24}$.

89. $\sqrt{18x^7}=\sqrt{9x^6\cdot 2x^1}=\sqrt{9x^6}\sqrt{2x}=3x^3\sqrt{2x}$

90. $\sqrt{48y^6}=\sqrt{16y^6\cdot 3}=\sqrt{16y^6}\sqrt{3}=4y^3\sqrt{3}$

91. $\sqrt{\frac{y^4}{81}}=\frac{\sqrt{y^4}}{\sqrt{81}}=\frac{y^2}{9}$

92. $\sqrt{\frac{x^9}{9}}=\frac{\sqrt{x^9}}{\sqrt{9}}=\frac{\sqrt{x^8\cdot x}}{3}=\frac{\sqrt{x^8}\sqrt{x}}{3}=\frac{x^4\sqrt{x}}{3}$

93.
$$\sqrt{12}+\sqrt{75}=\sqrt{4\cdot 3}+\sqrt{25\cdot 3}$$
$$=\sqrt{4}\sqrt{3}+\sqrt{25}\sqrt{3}$$
$$=2\sqrt{3}+5\sqrt{3}$$
$$=(2+5)\sqrt{3}$$
$$=7\sqrt{3}$$

94.
$$\sqrt{63}+\sqrt{28}-\sqrt{9}=\sqrt{9\cdot 7}+\sqrt{4\cdot 7}-3$$
$$=\sqrt{9}\sqrt{7}+\sqrt{4}\sqrt{7}-3$$
$$=3\sqrt{7}+2\sqrt{7}-3$$
$$=(3+2)\sqrt{7}-3$$
$$=5\sqrt{7}-3$$

95.
$$\sqrt{\frac{3}{16}}-\sqrt{\frac{3}{4}}=\frac{\sqrt{3}}{\sqrt{16}}-\frac{\sqrt{3}}{\sqrt{4}}$$
$$=\frac{\sqrt{3}}{4}-\frac{\sqrt{3}}{2}$$
$$=\frac{\sqrt{3}}{4}-\frac{2\sqrt{3}}{4}$$
$$=-\frac{\sqrt{3}}{4}$$

96.
$$\sqrt{45x^3}+x\sqrt{20x}-\sqrt{5x^3}$$
$$=\sqrt{9x^2\cdot 5x}+x\sqrt{4\cdot 5x}-\sqrt{x^2\cdot 5x}$$
$$=\sqrt{9x^2}\sqrt{5x}+x\sqrt{4}\sqrt{5x}-\sqrt{x^2}\sqrt{5x}$$
$$=3x\sqrt{5x}+2x\sqrt{5x}-x\sqrt{5x}$$
$$=(3x+2x-x)\sqrt{5x}$$
$$=4x\sqrt{5x}$$

97.
$$\sqrt{7}\cdot\sqrt{14}=\sqrt{7\cdot 14}$$
$$=\sqrt{98}$$
$$=\sqrt{49\cdot 2}$$
$$=\sqrt{49}\sqrt{2}$$
$$=7\sqrt{2}$$

98.
$$\sqrt{3}\left(\sqrt{9}-\sqrt{2}\right)=\sqrt{3}\left(3-\sqrt{2}\right)$$
$$=3\sqrt{3}-\sqrt{3}\sqrt{2}$$
$$=3\sqrt{3}-\sqrt{6}$$

99.
$$\left(\sqrt{2}+4\right)\left(\sqrt{5}-1\right)=\sqrt{2}\sqrt{5}-1\sqrt{2}+4\sqrt{5}-4\cdot 1$$
$$=\sqrt{10}-\sqrt{2}+4\sqrt{5}-4$$

100.
$$\left(\sqrt{x}+3\right)^2=\left(\sqrt{x}\right)^2+2\left(\sqrt{x}\right)(3)+(3)^2$$
$$=x+6\sqrt{x}+9$$

101. $\frac{\sqrt{120}}{\sqrt{5}}=\sqrt{\frac{120}{5}}=\sqrt{24}=\sqrt{4\cdot 6}=2\sqrt{6}$

102. $\frac{\sqrt{60x^9}}{\sqrt{15x^4}}=\sqrt{\frac{60x^9}{15x^7}}=\sqrt{4x^2}=2x$

103. $\sqrt{\frac{2}{7}}=\frac{\sqrt{2}}{\sqrt{7}}=\frac{\sqrt{2}}{\sqrt{7}}\cdot\frac{\sqrt{7}}{\sqrt{7}}=\frac{\sqrt{14}}{7}$

104. $\frac{3}{\sqrt{2x}}=\frac{3}{\sqrt{2x}}\cdot\frac{\sqrt{2x}}{\sqrt{2x}}=\frac{3\sqrt{2x}}{2x}$

105. $\frac{3}{\sqrt{x}-6}=\frac{3}{\sqrt{x}-6}\cdot\frac{\sqrt{x}+6}{\sqrt{x}+6}$
$=\frac{3\left(\sqrt{x}+6\right)}{x-36}$
$=\frac{3\sqrt{x}+18}{x-36}$

106. $\frac{\sqrt{7}-5}{\sqrt{5}+3}=\frac{\sqrt{7}-5}{\sqrt{5}+3}\cdot\frac{\sqrt{5}-3}{\sqrt{5}-3}$
$=\frac{\sqrt{7}\sqrt{5}-3\sqrt{7}-5\sqrt{5}+5\cdot 3}{5-9}$
$=\frac{\sqrt{35}-3\sqrt{7}-5\sqrt{5}+15}{-4}$

107. $\sqrt{4x}=2$
$\left(\sqrt{4x}\right)^2=2^2$
$4x=4$
$x=1$

108. $\sqrt{x-4}=3$
$\left(\sqrt{x-4}\right)^2=3^2$
$x-4=9$
$x=13$

109. $\sqrt{4x+8}+6=x$
$\sqrt{4x+8}=x-6$
$\left(\sqrt{4x+8}\right)^2=(x-6)^2$
$4x+8=x^2-12x+36$
$0=x^2-16x+28$
$0=(x-14)(x-2)$
$x-14=0$ or $x-2=0$
$x=14$ $\quad x=2$
x = 2 does not check, so the solution is x = 14.

110. $\sqrt{x-8}=\sqrt{x}-2$
$\left(\sqrt{x-8}\right)^2=\left(\sqrt{x}-2\right)^2$
$x-8=x-4\sqrt{x}+4$
$-12=-4\sqrt{x}$
$3=\sqrt{x}$
$3^2=\left(\sqrt{x}\right)^2$
$9=x$

111. $a^2+b^2=c^2$
$3^2+7^2=c^2$
$9+49=c^2$
$58=c^2$
$\sqrt{58}=\sqrt{c^2}$
$c=\sqrt{58}\approx 7.62$

112. The diagonal is the hypotenuse of a right triangle with the sides of the rectangle as its legs.
$a^2+b^2=c^2$
$2^2+b^2=6^2$
$4+b^2=36$
$b^2=32$
$\sqrt{b^2}=\sqrt{32}$
$b=4\sqrt{2}\approx 5.66$
The length of the rectangle is exactly $4\sqrt{2}$ inches or approximately 5.66 inches.

Chapter 8 Test

1. $\sqrt{16}=4$, because $4^2=16$ and 4 is positive.

2. $\sqrt[3]{125}=5$, because $5^3=125$.

3. $\sqrt[4]{81}=3$, because $3^4=81$.

4. $\sqrt{\frac{9}{16}}=\frac{3}{4}$, because $\left(\frac{3}{4}\right)^2=\frac{9}{16}$.

5. $\sqrt[4]{-81}$ is not a real number since the index 4 is even and the radicand −81 is negative.

6. $\sqrt{x^{10}}=x^5$, because $(x^5)^2=x^{10}$.

7. $\sqrt{54}=\sqrt{9\cdot 6}=\sqrt{9}\sqrt{6}=3\sqrt{6}$

8. $\sqrt{92}=\sqrt{4\cdot 23}=\sqrt{4}\sqrt{23}=2\sqrt{23}$

9. $\sqrt{y^7}=\sqrt{y^6\cdot y}=\sqrt{y^6}\sqrt{y}=y^3\sqrt{y}$

10. $\sqrt{24x^8}=\sqrt{4x^8\cdot 6}=\sqrt{4x^8}\sqrt{6}=2x^4\sqrt{6}$

11. $\sqrt[3]{27}=3$

12. $\sqrt[3]{16}=\sqrt[3]{8\cdot 2}=\sqrt[3]{8}\sqrt[3]{2}=2\sqrt[3]{2}$

13. $\sqrt{\frac{5}{16}}=\frac{\sqrt{5}}{\sqrt{16}}=\frac{\sqrt{5}}{4}$

14. $\sqrt{\frac{y^3}{25}}=\frac{\sqrt{y^3}}{\sqrt{25}}=\frac{\sqrt{y^2\cdot y}}{5}=\frac{\sqrt{y^2}\sqrt{y}}{5}=\frac{y\sqrt{y}}{5}$

15. $\sqrt{13}+\sqrt{13}-4\sqrt{13}=(1+1-4)\sqrt{13}=-2\sqrt{13}$

16. $\sqrt{18}-\sqrt{75}+7\sqrt{3}-\sqrt{8}$

$$\begin{aligned}&=\sqrt{9\cdot 2}-\sqrt{25\cdot 3}+7\sqrt{3}-\sqrt{4\cdot 2}\\&=\sqrt{9}\sqrt{2}-\sqrt{25}\sqrt{3}+7\sqrt{3}-\sqrt{4}\sqrt{2}\\&=3\sqrt{2}-5\sqrt{3}+7\sqrt{3}-2\sqrt{2}\\&=(3-2)\sqrt{2}+(-5+7)\sqrt{3}\\&=\sqrt{2}+2\sqrt{3}\end{aligned}$$

17.
$$\begin{aligned}\sqrt{\frac{3}{4}}+\sqrt{\frac{3}{25}}&=\frac{\sqrt{3}}{\sqrt{4}}+\frac{\sqrt{3}}{\sqrt{25}}\\&=\frac{\sqrt{3}}{2}+\frac{\sqrt{3}}{5}\\&=\frac{5\sqrt{3}}{10}+\frac{2\sqrt{3}}{10}\\&=\frac{5\sqrt{3}+2\sqrt{3}}{10}\\&=\frac{(5+2)\sqrt{3}}{10}\\&=\frac{7\sqrt{3}}{10}\end{aligned}$$

18.
$$\begin{aligned}\sqrt{7}\cdot\sqrt{14}&=\sqrt{7\cdot 14}\\&=\sqrt{98}\\&=\sqrt{49\cdot 2}\\&=\sqrt{49}\sqrt{2}\\&=7\sqrt{2}\end{aligned}$$

19.
$$\begin{aligned}\sqrt{2}\left(\sqrt{6}-\sqrt{5}\right)&=\sqrt{2}\sqrt{6}-\sqrt{2}\sqrt{5}\\&=\sqrt{12}-\sqrt{10}\\&=\sqrt{4\cdot 3}-\sqrt{10}\\&=2\sqrt{3}-\sqrt{10}\end{aligned}$$

20.
$$\begin{aligned}\left(\sqrt{x}+2\right)\left(\sqrt{x}-3\right)&=\sqrt{x}\cdot\sqrt{x}-3\sqrt{x}+2\sqrt{x}-2\cdot 3\\&=x-3\sqrt{x}+2\sqrt{x}-6\\&=x-\sqrt{x}-6\end{aligned}$$

21. $\frac{\sqrt{50}}{\sqrt{10}}=\sqrt{\frac{50}{10}}=\sqrt{5}$

22.
$$\begin{aligned}\frac{\sqrt{40x^4}}{\sqrt{2x}}&=\sqrt{\frac{40x^4}{2x}}\\&=\sqrt{20x^3}\\&=\sqrt{4x^2\cdot 5x}\\&=\sqrt{4x^2}\sqrt{5x}\\&=2x\sqrt{5x}\end{aligned}$$

23. $\sqrt{\frac{2}{3}}=\frac{\sqrt{2}}{\sqrt{3}}=\frac{\sqrt{2}}{\sqrt{3}}\cdot\frac{\sqrt{3}}{\sqrt{3}}=\frac{\sqrt{6}}{3}$

24. $\frac{8}{\sqrt{5y}}=\frac{8}{\sqrt{5y}}\cdot\frac{\sqrt{5y}}{\sqrt{5y}}=\frac{8\sqrt{5y}}{5y}$

25.
$$\begin{aligned}\frac{8}{\sqrt{6}+2}&=\frac{8}{\sqrt{6}+2}\cdot\frac{\sqrt{6}-2}{\sqrt{6}-2}\\&=\frac{8\left(\sqrt{6}-2\right)}{\left(\sqrt{6}\right)^2-2^2}\\&=\frac{8\left(\sqrt{6}-2\right)}{6-4}\\&=\frac{8\left(\sqrt{6}-2\right)}{2}\\&=4\left(\sqrt{6}-2\right)\\&=4\sqrt{6}-8\end{aligned}$$

26. $\frac{1}{3-\sqrt{x}}=\frac{1}{3-\sqrt{x}}\cdot\frac{3+\sqrt{x}}{3+\sqrt{x}}=\frac{1\left(3+\sqrt{x}\right)}{3^2-\left(\sqrt{x}\right)^2}=\frac{3+\sqrt{x}}{9-x}$

27.
$$\begin{aligned}\sqrt{x}+8&=11\\\sqrt{x}&=3\\\left(\sqrt{x}\right)^2&=3^2\\x&=9\end{aligned}$$

28.
$$\begin{aligned}\sqrt{3x-6}&=\sqrt{x+4}\\\left(\sqrt{3x-6}\right)^2&=\left(\sqrt{x+4}\right)^2\\3x-6&=x+4\\2x-6&=4\\2x&=10\\x&=5\end{aligned}$$

29. $\sqrt{2x-2}=x-5$

$$\left(\sqrt{2x-2}\right)^2=(x-5)^2$$
$$2x-2=x^2-10x+25$$
$$0=x^2-12x+27$$
$$0=(x-3)(x-9)$$
$$x-3=0 \text{ or } x-9=0$$
$$x=3 \qquad x=9$$

$x = 3$ does not check, so the solution is $x = 9$.

30. $a^2+b^2=c^2$

$$8^2+b^2=12^2$$
$$64+b^2=144$$
$$b^2=80$$
$$\sqrt{b^2}=\sqrt{80}$$
$$b=4\sqrt{5}$$

The length is $4\sqrt{5}$ inches.

31. $r=\sqrt{\frac{A}{\pi}}$

$$r=\sqrt{\frac{15}{\pi}}$$
$$r\approx 2.19$$

The radius is 2.19 meters.

Cumulative Review Chapters 1–8

1. $-2(-14)=2\cdot 14=28$

2. $9(-5.2)=-9\cdot 5.2=-46.8$

3. $-\frac{2}{3}\cdot\frac{4}{7}=-\frac{2\cdot 4}{3\cdot 7}=-\frac{8}{21}$

4. $-3\frac{3}{8}\cdot 5\frac{1}{3}=-\frac{27}{8}\cdot\frac{16}{3}$

$$=-\frac{27\cdot 16}{8\cdot 3}$$
$$=-9\cdot 2$$
$$=-18$$

5. $4(2x-3)+7=3x+5$

$$8x-12+7=3x+5$$
$$8x-5=3x+5$$
$$5x-5=5$$
$$5x=10$$
$$x=2$$

6. $6y-11+4+2y=8+15y-8y$

$$8y-7=8+7y$$
$$y-7=8$$
$$y=15$$

7. a. The sector for business represents 17% of the circle.

b. These sectors comprise a total of 17% + 4% or 21% of the circle.

c. 17% of 253 = 0.17(253) = 43.01
About 43 of the 253 American travelers would be traveling solely for business.

8. a. $\frac{4(-3)-(-6)}{-8+4}=\frac{-12+6}{-4}=\frac{-6}{-4}=\frac{3}{2}$

b. $\frac{3+(-3)(-2)^3}{-1-(-4)}=\frac{3+(-3)(-8)}{-1+4}$

$$=\frac{3+24}{3}$$
$$=\frac{27}{3}$$
$$=9$$

9. a. $1.02\times 10^5=102,000$

b. $7.358\times 10^{-3}=0.007358$

c. $8.4\times 10^7=84,000,000$

d. $3.007\times 10^{-5}=0.00003007$

10. a. $7,200,000=7.2\times 10^6$

b. $0.000308=3.08\times 10^{-4}$

11. $(3x+2)(2x-5)$

$$=(3x)(2x)-(3x)(5)+2(2x)-2(5)$$
$$=6x^2-15x+4x-10$$
$$=6x^2-11x-10$$

12. $(7x+1)^2=(7x)^2+2(7x)(1)+(1)^2$

$$=49x^2+14x+1$$

13. $xy+2x+3y+6=x(y+2)+3(y+2)$

$$=(y+2)(x+3)$$

14. $xy^2+5x-y^2-5 = x(y^2+5)-(y^2+5)$
$= (y^2+5)(x-1)$

15. $3x^2+11x+6=(3x+2)(x+3)$

16. $3x^2+15x+18=3(x^2+5x+6)=3(x+2)(x+3)$

17. a. The expression is undefined for $x = 3$, since the denominator, $x - 3$, is 0 when $x = 3$.

b. The expression is undefined for $x = 2$ and $x = 1$, since the denominator, $x^2-3x+2=(x-2)(x-1)$, is 0 when $x = 2$ or $x = 1$.

c. There are no values for which the expression is undefined, since the denominator, 3, is never 0.

18. $\frac{2x^2+7x+3}{x^2-9}=\frac{(2x+1)(x+3)}{(x+3)(x-3)}=\frac{2x+1}{x-3}$

19. $\frac{x^2+4x+4}{x^2+2x}=\frac{(x+2)(x+2)}{x(x+2)}=\frac{x+2}{x}$

20.
$$\begin{aligned}\frac{12x^2y^3}{5}\div\frac{3y^3}{x} &= \frac{12x^2y^3}{5}\cdot\frac{x}{3y^3}\\ &= \frac{4x^2}{5}\cdot\frac{x}{1}\\ &= \frac{4x^3}{5}\end{aligned}$$

21. a. $\frac{a}{4}-\frac{2a}{8}=\frac{a}{4}-\frac{a}{4}=0$

b. $\frac{3}{10x^2}+\frac{7}{25x}=\frac{15}{50x^2}+\frac{14x}{50x^2}=\frac{15+14x}{50x^2}$

22. $y = mx + b$
$y = -2x + 4$

23.
$$\begin{aligned}\frac{4x}{x^2+x-30}+\frac{2}{x-5} &= \frac{1}{x+6}\\ (x-5)(x+6)\left(\frac{4x}{(x-5)(x+6)}+\frac{2}{x-5}\right) &= (x-5)(x+6)\left(\frac{1}{x+6}\right)\\ 4x+2(x+6) &= x-5\\ 4x+2x+12 &= x-5\\ 6x+12 &= x-5\\ 5x+12 &= -5\\ 5x &= -17\\ x &= -\frac{17}{5}\end{aligned}$$

24. $4a^2+3a-2a^2+7a-5=(4-2)a^2+(3+7)a-5$
$=2a^2+10a-5$

25.

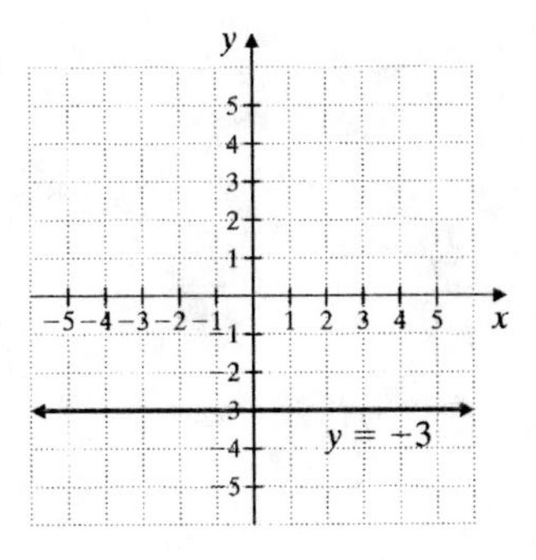

26. For $x = 0$:
$2x+y=6$
$2(0)+y=6$
$y=6$

For $y=-2$:
$2x+y=6$
$2x-2=6$
$2x=8$
$x=4$

For $x = 3$:
$2x+y=6$
$2(3)+y=6$
$6+y=6$
$y=0$

x	y
0	6
4	−2
3	0

27. $y=mx+b$
$y=\frac{1}{4}x-3$

28. The slope of a line perpendicular to $y = 2x + 4$ has slope $-\frac{1}{2}$.

$y-y_1=m(x-x_1)$
$y-5=-\frac{1}{2}(x-1)$
$y=-\frac{1}{2}x+\frac{1}{2}+5$
$y=-\frac{1}{2}x+\frac{11}{2}$

29. $\begin{cases}3x+4y=13\\5x-9y=6\end{cases}$

$15x+20y=65$
$-15x+27y=-18$
$47y=47$
$y=1$

$3x+4y=13$
$3x+4(1)=13$
$3x=9$
$x=3$
The solution is (3, 1).

30. $\begin{cases}\frac{x}{2}+y=\frac{5}{6}\\2x-y=\frac{5}{6}\end{cases}$

$\frac{x}{2}+y=\frac{5}{6}$
$2x-y=\frac{5}{6}$
$\frac{5}{2}x=\frac{10}{6}$
$x=\frac{2}{3}$

$$\frac{x}{2}+y=\frac{5}{6}$$
$$\frac{1}{2}\left(\frac{2}{3}\right)+y=\frac{5}{6}$$
$$\frac{1}{3}+y=\frac{5}{6}$$
$$y=\frac{5}{6}-\frac{1}{3}$$
$$y=\frac{1}{2}$$

The solution is $\left(\frac{2}{3}, \frac{1}{2}\right)$.

31. Let x be Alfredo's walking speed. Then $x + 1$ is Louisa's walking speed.

$$2x+2(x+1)=15$$
$$2x+2x+2=15$$
$$4x+2=15$$
$$4x=13$$
$$x=\frac{13}{4} \text{ or } 3.25$$

$x + 1 = 3.25 + 1 = 4.25$

Alfredo's walking speed is 3.25 miles per hour, and Louisa's is 4.25 miles per hour.

32. Let x be the speed of the slower streetcar. Then $x + 15$ is the speed of the faster streetcar. Twelve minutes is $\frac{1}{5}$ hour.

$$\frac{1}{5}x+\frac{1}{5}(x+15)=11$$
$$\frac{1}{5}x+\frac{1}{5}x+3=11$$
$$\frac{2}{5}x+3=11$$
$$\frac{2}{5}x=8$$
$$x=20$$

$x + 15 = 20 + 15 = 35$

The streetcars are traveling at 20 miles per hour and 35 miles per hour.

33. $\sqrt[3]{1}=1$, because $1^3=1$.

34. $\sqrt{121}=11$, because $11^2=121$ and 11 is positive.

35. $\sqrt[3]{-27}=-3$, because $(-3)^3=-27$.

36. $\sqrt{\frac{1}{4}}=\frac{1}{2}$, because $\left(\frac{1}{2}\right)^2=\frac{1}{4}$.

37. $\sqrt[3]{\frac{1}{125}}=\frac{1}{5}$, because $\left(\frac{1}{5}\right)^3=\frac{1}{125}$.

38. $\sqrt{\frac{25}{144}}=\frac{5}{12}$, because $\left(\frac{5}{12}\right)^2=\frac{25}{144}$.

39. $\sqrt{54}=\sqrt{9\cdot 6}=\sqrt{9}\sqrt{6}=3\sqrt{6}$

40. $\sqrt{63}=\sqrt{9\cdot 7}=\sqrt{9}\sqrt{7}=3\sqrt{7}$

41. $\sqrt{200}=\sqrt{100\cdot 2}=\sqrt{100}\sqrt{2}=10\sqrt{2}$

42. $\sqrt{500}=\sqrt{100\cdot 5}=\sqrt{100}\sqrt{5}=10\sqrt{5}$

43.
$$\begin{aligned}7\sqrt{12}-2\sqrt{75}&=7\sqrt{4\cdot 3}-2\sqrt{25\cdot 3}\\&=7\sqrt{4}\sqrt{3}-2\sqrt{25}\sqrt{3}\\&=7\cdot 2\sqrt{3}-2\cdot 5\sqrt{3}\\&=14\sqrt{3}-10\sqrt{3}\\&=(14-10)\sqrt{3}\\&=4\sqrt{3}\end{aligned}$$

44. $\left(\sqrt{x}+5\right)\left(\sqrt{x}-5\right)=\left(\sqrt{x}\right)^2-5^2=x-25$

45.
$$\begin{aligned}&2\sqrt{x^2}-\sqrt{25x^5}+\sqrt{x^5}\\&=2x-\sqrt{25x^4\cdot x}+\sqrt{x^4\cdot x}\\&=2x-\sqrt{25x^4}\sqrt{x}+\sqrt{x^4}\sqrt{x}\\&=2x-5x^2\sqrt{x}+x^2\sqrt{x}\\&=2x+(-5x^2+x^2)\sqrt{x}\\&=2x-4x^2\sqrt{x}\end{aligned}$$

46.
$$\begin{aligned}\left(\sqrt{6}+2\right)^2&=\left(\sqrt{6}\right)^2+2\left(\sqrt{6}\right)(2)+(2)^2\\&=6+4\sqrt{6}+4\\&=10+4\sqrt{6}\end{aligned}$$

47. $\frac{2}{\sqrt{7}}=\frac{2}{\sqrt{7}}\cdot\frac{\sqrt{7}}{\sqrt{7}}=\frac{2\sqrt{7}}{7}$

48. $\frac{x+3}{\frac{1}{x}+\frac{1}{3}}=\frac{x+3}{\frac{3}{3x}+\frac{x}{3x}}=\frac{x+3}{\frac{3+x}{3x}}=(x+3)\cdot\frac{3x}{3+x}=3x$

49. $\sqrt{x} = \sqrt{5x-2}$

$$\left(\sqrt{x}\right)^2 = \left(\sqrt{5x-2}\right)^2$$
$$x = 5x-2$$
$$-4x = -2$$
$$x = \frac{-2}{-4}$$
$$x = \frac{1}{2}$$

The solution is $\frac{1}{2}$.

50. $\sqrt{x+4} = \sqrt{3x-1}$

$$\left(\sqrt{x+4}\right)^2 = \left(\sqrt{3x-1}\right)^2$$
$$x+4 = 3x-1$$
$$-2x+4 = -1$$
$$-2x = -5$$
$$x = \frac{-5}{-2}$$
$$x = \frac{5}{2}$$

The solution is $\frac{5}{2}$.

Chapter 9

Section 9.1 Practice Exercises

1. $x^2 - 25 = 0$
$(x+5)(x-5) = 0$
$x+5=0$ or $x-5=0$
$x=-5$ $\quad x=5$
The solutions are −5 and 5.

2. $2x^2 - 3x = 9$
$2x^2 - 3x - 9 = 0$
$(2x+3)(x-3) = 0$
$2x+3=0$ or $x-3=0$
$2x=-3$ $\quad x=3$
$x = -\frac{3}{2}$
The solutions are $-\frac{3}{2}$ and 3.

3. $x^2 - 16 = 0$
$x^2 = 16$
$x = \sqrt{16}$ or $x = -\sqrt{16}$
$x = 4$ $\quad x = -4$
The solutions are 4 and −4.

4. $3x^2 = 11$
$x^2 = \frac{11}{3}$
$x = \sqrt{\frac{11}{3}}$ or $x = -\sqrt{\frac{11}{3}}$
$x = \frac{\sqrt{11}\cdot\sqrt{3}}{\sqrt{3}\cdot\sqrt{3}}$ $\quad x = -\frac{\sqrt{11}\cdot\sqrt{3}}{\sqrt{3}\cdot\sqrt{3}}$
$x = \frac{\sqrt{33}}{3}$ $\quad x = -\frac{\sqrt{33}}{3}$
The solutions are $\frac{\sqrt{33}}{3}$ and $-\frac{\sqrt{33}}{3}$.

5. $(x-4)^2 = 49$
$x-4 = \sqrt{49}$ or $x-4 = -\sqrt{49}$
$x-4 = 7$ $\quad x-4=-7$
$x = 11$ $\quad x = -3$
The solutions are 11 and −3.

6. $(x-5)^2 = 18$
$x-5 = \sqrt{18}$ or $x-5 = -\sqrt{18}$
$x-5 = 3\sqrt{2}$ $\quad x-5 = -3\sqrt{2}$
$x = 5+3\sqrt{2}$ $\quad x = 5-3\sqrt{2}$
$x = 5 \pm 3\sqrt{2}$
The solutions are $5 \pm 3\sqrt{2}$.

7. $(x+3)^2 = -5$
This equation has no real solution because the square root of −5 is not a real number.

8. $(4x+1)^2 = 15$
$4x+1 = \sqrt{15}$ or $4x+1 = -\sqrt{15}$
$4x = -1+\sqrt{15}$ $\quad 4x = -1-\sqrt{15}$
$x = \frac{-1+\sqrt{15}}{4}$ $\quad x = \frac{-1-\sqrt{15}}{4}$
$x = \frac{-1\pm\sqrt{15}}{4}$
The solutions are $\frac{-1\pm\sqrt{15}}{4}$.

9. $h = 16t^2$
$650 = 16t^2$
$40.625 = t^2$
$6.4 = t$ or $-6.4 = t$
−6.4 is rejected because time cannot be negative.
It takes the object 6.4 seconds to fall 650 feet.

Vocabulary, Readiness & Video Check 9.1

1. To solve, *a* becomes the radicand and the square root of a negative number is not a real number.

2. A quadratic equation must be in the form of a variable (or polynomial) squared equal to some nonnegative number in order for the property to be used. For Example 6, we use $h = 16t^2$, so this equation can easily be placed in this form. The negative value is rejected because it will not be used in the context of the application.

Exercise Set 9.1

1. $k^2 - 49 = 0$
$(k+7)(k-7) = 0$
$k+7=0$ or $k-7=0$
$k=-7$ $\quad$ $k=7$
The solutions are $k = -7$ and $k = 7$.

3. $m^2 + 2m = 15$
$m^2 + 2m - 15 = 0$
$(m+5)(m-3) = 0$
$m+5=0$ or $m-3=0$
$m=-5$ $\quad$ $m=3$
The solutions are $m = -5$ and $m = 3$.

5. $2x^2 - 32 = 0$
$2(x^2 - 16) = 0$
$2(x+4)(x-4) = 0$
$x+4=0$ or $x-4=0$
$x=-4$ $\quad$ $x=4$
The solutions are $x = -4$ and $x = 4$.

7. $4a^2 - 36 = 0$
$4(a^2 - 9) = 0$
$4(a-3)(a+3) = 0$
$a-3=0$ or $a+3=0$
$a=3$ $\quad$ $a=-3$
The solutions are $a = 3$ and $a = -3$.

9. $x^2 + 7x = -10$
$x^2 + 7x + 10 = 0$
$(x+2)(x+5) = 0$
$x+2=0$ or $x+5=0$
$x=-2$ $\quad$ $x=-5$
The solutions are $x = -2$ and $x = -5$.

11. $x^2 = 64$
$x = \sqrt{64}$ or $x = -\sqrt{64}$
$x = 8$ $\quad$ $x = -8$
The solutions are $x = \pm 8$.

13. $x^2 = 21$
$x = \sqrt{21}$ or $x = -\sqrt{21}$
The solutions are $x = \pm\sqrt{21}$.

15. $x^2 = \frac{1}{25}$
$x = \sqrt{\frac{1}{25}}$ or $x = -\sqrt{\frac{1}{25}}$
$x = \frac{1}{5}$ $\quad$ $x = -\frac{1}{5}$
The solutions are $x = \pm\frac{1}{5}$.

17. $x^2 = -4$ has no real solution because the square root of -4 is not a real number.

19. $3x^2 = 13$
$x^2 = \frac{13}{3}$
$x = \sqrt{\frac{13}{3}}$ or $x = -\sqrt{\frac{13}{3}}$
$x = \frac{\sqrt{13}}{\sqrt{3}} \cdot \frac{\sqrt{3}}{\sqrt{3}}$ $\quad$ $x = -\frac{\sqrt{13}}{\sqrt{3}} \cdot \frac{\sqrt{3}}{\sqrt{3}}$
$x = \frac{\sqrt{39}}{3}$ $\quad$ $x = -\frac{\sqrt{39}}{3}$
The solutions are $x = \pm\frac{\sqrt{39}}{3}$.

21. $7x^2 = 4$
$x^2 = \frac{4}{7}$
$x = \sqrt{\frac{4}{7}}$ or $x = -\sqrt{\frac{4}{7}}$
$x = \frac{\sqrt{4}}{\sqrt{7}} \cdot \frac{\sqrt{7}}{\sqrt{7}}$ $\quad$ $x = -\frac{\sqrt{4}}{\sqrt{7}} \cdot \frac{\sqrt{7}}{\sqrt{7}}$
$x = \frac{2\sqrt{7}}{7}$ $\quad$ $x = -\frac{2\sqrt{7}}{7}$
The solutions are $x = \pm\frac{2\sqrt{7}}{7}$.

23. $2x^2 - 10 = 0$
$2x^2 = 10$
$x^2 = 5$
$x = \sqrt{5}$ or $x = -\sqrt{5}$
The solutions are $x = \pm\sqrt{5}$.

25. $(x-5)^2 = 49$

$x-5 = \sqrt{49}$ or $x-5 = -\sqrt{49}$
$x-5 = 7$ $\quad x-5 = -7$
$x = 12$ $\quad x = -2$
The solutions are $x = -2$ and $x = 12$.

27. $(x+2)^2 = 7$

$x+2 = \sqrt{7}$ or $x+2 = -\sqrt{7}$
$x = -2+\sqrt{7}$ $\quad x = -2-\sqrt{7}$
The solutions are $x = -2 \pm \sqrt{7}$.

29. $\left(m-\frac{1}{2}\right)^2 = \frac{1}{4}$

$m-\frac{1}{2} = \sqrt{\frac{1}{4}}$ or $m-\frac{1}{2} = -\sqrt{\frac{1}{4}}$
$m-\frac{1}{2} = \frac{1}{2}$ $\quad m-\frac{1}{2} = -\frac{1}{2}$
$m = 1$ $\quad m = 0$
The solutions are $m = 0$ and $m = 1$.

31. $(p+2)^2 = 10$

$p+2 = \sqrt{10}$ or $p+2 = -\sqrt{10}$
$p = -2+\sqrt{10}$ $\quad p = -2-\sqrt{10}$
The solutions are $p = -2 \pm \sqrt{10}$.

33. $(3y+2)^2 = 100$

$3y+2 = \sqrt{100}$ or $3y+2 = -\sqrt{100}$
$3y+2 = 10$ $\quad 3y+2 = -10$
$3y = 8$ $\quad 3y = -12$
$y = \frac{8}{3}$ $\quad y = -4$

The solutions are $y = -4$ and $y = \frac{8}{3}$.

35. $(z-4)^2 = -9$ has no real solution because the square root of -9 is not a real number.

37. $(2x-11)^2 = 50$

$2x-11 = \sqrt{50}$ or $2x-11 = -\sqrt{50}$
$2x-11 = 5\sqrt{2}$ $\quad 2x-11 = -5\sqrt{2}$
$2x = 11+5\sqrt{2}$ $\quad 2x = 11-5\sqrt{2}$
$x = \frac{11+5\sqrt{2}}{2}$ $\quad x = \frac{11-5\sqrt{2}}{2}$

The solutions are $x = \frac{11 \pm 5\sqrt{2}}{2}$.

39. $(3x-7)^2 = 32$

$3x-7 = \sqrt{32}$ or $3x-7 = -\sqrt{32}$
$3x-7 = 4\sqrt{2}$ $\quad 3x-7 = -4\sqrt{2}$
$3x = 7+4\sqrt{2}$ $\quad 3x = 7-4\sqrt{2}$
$x = \frac{7+4\sqrt{2}}{3}$ $\quad x = \frac{7-4\sqrt{2}}{3}$

The solutions are $x = \frac{7 \pm 4\sqrt{2}}{3}$.

41. $x^2 - 29 = 0$

$x^2 = 29$
$x = \sqrt{29}$ or $x = -\sqrt{29}$
The solutions are $x = \pm\sqrt{29}$.

43. $(x+6)^2 = 24$

$x+6 = \sqrt{24}$ or $x+6 = -\sqrt{24}$
$x+6 = 2\sqrt{6}$ $\quad x+6 = -2\sqrt{6}$
$x = -6+2\sqrt{6}$ $\quad x = -6-2\sqrt{6}$
The solutions are $x = -6 \pm 2\sqrt{6}$.

45. $\frac{1}{2}n^2 = 5$

$n^2 = 10$
$n = \sqrt{10}$ or $n = -\sqrt{10}$
The solutions are $n = \pm\sqrt{10}$.

47. $(4x-1)^2 = 5$

$4x-1 = \sqrt{5}$ or $4x-1 = -\sqrt{5}$
$4x = 1+\sqrt{5}$ $\quad 4x = 1-\sqrt{5}$
$x = \frac{1+\sqrt{5}}{4}$ $\quad x = \frac{1-\sqrt{5}}{4}$

The solutions are $x = \frac{1 \pm \sqrt{5}}{4}$.

49. $3z^2 = 36$

$z^2 = 12$
$z = \sqrt{12}$ or $z = -\sqrt{12}$
$z = 2\sqrt{3}$ $\quad z = -2\sqrt{3}$
The solutions are $z = \pm 2\sqrt{3}$.

51. $(8-3x)^2-45=0$

$(8-3x)^2=45$

$8-3x=\sqrt{45}$ or $8-3x=-\sqrt{45}$

$8-3x=3\sqrt{5}$ $\quad$ $8-3x=-3\sqrt{5}$

$-3x=-8+3\sqrt{5}$ $\quad$ $-3x=-8-3\sqrt{5}$

$x=\frac{-8+3\sqrt{5}}{-3}$ $\quad$ $x=\frac{-8-3\sqrt{5}}{-3}$

The solutions are $x=\frac{-8\pm3\sqrt{5}}{-3}$.

53. $h=16t^2$

$87.6=16t^2$

$\frac{87.6}{16}=t^2$

$5.475=t^2$

$\sqrt{5.475}=t$ or $-\sqrt{5.475}=t$

$2.3\approx t$ $\quad$ $-2.3\approx t$

Since the time of a dive is not a negative number, reject the solution –2.3. The dive lasted approximately 2.3 seconds.

55. $h=16t^2$

$4000=16t^2$

$250=t^2$

$\sqrt{250}=t$ or $-\sqrt{250}=t$

$15.8\approx t$ $\quad$ $-15.8\approx t$

Since the time of a fall is not a negative number, reject the solution –15.8. It would take approximately 15.8 seconds for the object to reach the bottom of the canyon.

57. $A=\pi r^2$

$36\pi=\pi r^2$

$36=r^2$

$\sqrt{36}=r$ or $-\sqrt{36}=r$

$6=r$ $\quad$ $-6=r$

Since the radius does not have a negative length, reject the solution –6. The radius is 6 inches.

59. $A=s^2$

$20=s^2$

$\sqrt{20}=s$ or $-\sqrt{20}=s$

$2\sqrt{5}=s$ $\quad$ $-2\sqrt{5}=s$

Since the length of a side is not a negative number, reject the solution $-2\sqrt{5}$. The sides have length $2\sqrt{5}\approx4.47$ inches.

61. $A=s^2$

$3039=s^2$

$\sqrt{3039}=s$ or $-\sqrt{3039}=s$

$55.13\approx s$ $\quad$ $-55.13\approx s$

Since the length of a side is not a negative number, reject the solution –55.13. The sides have length 55.13 feet.

63. $x^2+6x+9=(x)^2+2(x)(3)+(3)^2=(x+3)^2$

65. $x^2-4x+4=x^2-2(x)(2)+(2)^2=(x-2)^2$

67. answers may vary

69. $x^2+4x+4=16$

$(x+2)^2=16$

$x+2=\sqrt{16}$ or $x+2=-\sqrt{16}$

$x+2=4$ $\quad$ $x+2=-4$

$x=2$ $\quad$ $x=-6$

The solutions are $x = 2$ and $x = -6$.

71. $x^2=1.78$

$x=\sqrt{1.78}$ or $x=-\sqrt{1.78}$

$x\approx\pm1.33$

The solutions are $x=\pm1.33$.

73. $y=-84(x-12)^2+609,652$

$608,900=-84(x-12)^2+609,652$

$-752=-84(x-12)^2$

$\frac{752}{84}=(x-12)^2$

$\sqrt{\frac{752}{84}}=x-12$ or $-\sqrt{\frac{752}{84}}=x-12$

$12+\sqrt{\frac{752}{84}}=x$ $\quad$ $12-\sqrt{\frac{752}{84}}=x$

$15\approx x$ $\quad$ $9\approx x$

If this trend continues, the first year in which there are 608,900 U.S. highway bridges is 2015 (2006 + 9).

Section 9.2 Practice Exercises

1.
$$x^2+8x+1=0$$
$$x^2+8x=-1$$
$$x^2+8x+16=-1+16$$
$$(x+4)^2=15$$
$$x+4=\sqrt{15} \quad \text{or} \quad x+4=-\sqrt{15}$$
$$x=-4+\sqrt{15} \quad\quad x=-4-\sqrt{15}$$
The solutions are $-4\pm\sqrt{15}$.

2.
$$x^2-14x=-32$$
$$x^2-14x+49=-32+49$$
$$(x-7)^2=17$$
$$x-7=\sqrt{17} \quad \text{or} \quad x-7=-\sqrt{17}$$
$$x=7+\sqrt{17} \quad\quad x=7-\sqrt{17}$$
The solutions are $7\pm\sqrt{17}$.

3.
$$4x^2-16x-9=0$$
$$x^2-4x-\frac{9}{4}=0$$
$$x^2-4x=\frac{9}{4}$$
$$x^2-4x+4=\frac{9}{4}+4$$
$$(x-2)^2=\frac{25}{4}$$
$$x-2=\sqrt{\frac{25}{4}} \quad \text{or} \quad x-2=-\sqrt{\frac{25}{4}}$$
$$x-2=\frac{5}{2} \quad\quad x-2=-\frac{5}{2}$$
$$x=2+\frac{5}{2} \quad\quad x=2-\frac{5}{2}$$
$$x=\frac{9}{2} \quad\quad x=-\frac{1}{2}$$
The solutions are $\frac{9}{2}$ and $-\frac{1}{2}$.

4.
$$2x^2+10x=-13$$
$$x^2+5x=-\frac{13}{2}$$
$$x^2+5x+\frac{25}{4}=-\frac{13}{2}+\frac{25}{4}$$
$$\left(x+\frac{5}{2}\right)^2=-\frac{1}{4}$$
There is no real solution to this equation since the square root of a negative number is not a real number.

5.
$$2x^2=-6x+5$$
$$x^2=-3x+\frac{5}{2}$$
$$x^2+3x=\frac{5}{2}$$
$$x^2+3x+\frac{9}{4}=\frac{5}{2}+\frac{9}{4}$$
$$\left(x+\frac{3}{2}\right)^2=\frac{19}{4}$$
$$x+\frac{3}{2}=\sqrt{\frac{19}{4}} \quad \text{or} \quad x+\frac{3}{2}=-\sqrt{\frac{19}{4}}$$
$$x+\frac{3}{2}=\frac{\sqrt{19}}{2} \quad\quad x+\frac{3}{2}=-\frac{\sqrt{19}}{2}$$
$$x=-\frac{3}{2}+\frac{\sqrt{19}}{2} \quad\quad x=-\frac{3}{2}-\frac{\sqrt{19}}{2}$$
The solutions are $\frac{-3\pm\sqrt{19}}{2}$.

Vocabulary, Readiness & Video Check 9.2

1. By the zero-factor property, if the product of two numbers is zero, then at least one of these two numbers must be zero.

2. If a is a positive number, and if $x^2=a$, then $x=\pm\sqrt{a}$.

3. An equation that can be written in the form $ax^2+bx+c=0$ where a, b, and c are real numbers and a is not zero is called a quadratic equation.

4. The process of solving a quadratic equation by writing it in the form $(x+a)^2=c$ is called completing the square.

5. To complete the square on x^2+6x, add 9.

6. To complete the square on x^2+bx, add $\left(\frac{b}{2}\right)^2$.

7. $\left(\frac{8}{2}\right)^2=4^2=16;\ p^2+8p+16$

8. $\left(\frac{6}{2}\right)^2=3^2=9;\ p^2+6p+9$

9. $\left(\frac{20}{2}\right)^2 = 10^2 = 100;\ x^2 + 20x + \underline{100}$

10. $\left(\frac{18}{2}\right)^2 = 9^2 = 81;\ x^2 + 18x + \underline{81}$

11. $\left(\frac{14}{2}\right)^2 = 7^2 = 49;\ y^2 + 14y + \underline{49}$

12. $\left(\frac{2}{2}\right)^2 = 1^2 = 1;\ y^2 + 2y + \underline{1}$

13. When working with equations, whatever is added to one side must also be added to the other side to keep equality.

14. The coefficient of y^2 is 2. The method of completing the square works only when the coefficient of the squared variable is 1, so we must first divide through by 2.

Exercise Set 9.2

1.
$$x^2 + 8x = -12$$
$$x^2 + 8x + \left(\frac{8}{2}\right)^2 = -12 + \left(\frac{8}{2}\right)^2$$
$$x^2 + 8x + 4^2 = -12 + 4^2$$
$$x^2 + 8x + 16 = -12 + 16$$
$$(x+4)^2 = 4$$

$x + 4 = \sqrt{4}$ or $x + 4 = -\sqrt{4}$

$x = -4 + 2$ $\quad$ $x = -4 - 2$

$x = -2$ $\quad$ $x = -6$

The solutions are $x = -6$ and $x = -2$.

3.
$$x^2 + 2x - 7 = 0$$
$$x^2 + 2x = 7$$
$$x^2 + 2x + \left(\frac{2}{2}\right)^2 = 7 + \left(\frac{2}{2}\right)^2$$
$$x^2 + 2x + 1 = 7 + 1$$
$$(x+1)^2 = 8$$

$x + 1 = \sqrt{8}$ or $x + 1 = -\sqrt{8}$

$x + 1 = 2\sqrt{2}$ $\quad$ $x + 1 = -2\sqrt{2}$

$x = -1 + 2\sqrt{2}$ $\quad$ $x = -1 - 2\sqrt{2}$

The solutions are $x = -1 \pm 2\sqrt{2}$.

5.
$$x^2 - 6x = 0$$
$$x^2 - 6x + \left(\frac{-6}{2}\right)^2 = 0 + \left(\frac{-6}{2}\right)^2$$
$$x^2 - 6x + (-3)^2 = 0 + (-3)^2$$
$$x^2 - 6x + 9 = 9$$
$$(x-3)^2 = 9$$

$x - 3 = \sqrt{9}$ or $x - 3 = -\sqrt{9}$

$x = 3 + 3$ $\quad$ $x = 3 - 3$

$x = 6$ $\quad$ $x = 0$

The solutions are $x = 0$ and $x = 6$.

7.
$$y^2 + 5y + 4 = 0$$
$$y^2 + 5y = -4$$
$$y^2 + 5y + \left(\frac{5}{2}\right)^2 = -4 + \left(\frac{5}{2}\right)^2$$
$$\left(y + \frac{5}{2}\right)^2 = \frac{9}{4}$$

$y + \frac{5}{2} = \sqrt{\frac{9}{4}}$ or $y + \frac{5}{2} = -\sqrt{\frac{9}{4}}$

$y = -\frac{5}{2} + \frac{3}{2}$ $\quad$ $y = -\frac{5}{2} - \frac{3}{2}$

$y = -\frac{2}{2} = -1$ $\quad$ $y = -\frac{8}{2} = -4$

The solutions are $y = -1$ and $y = -4$.

9.
$$x^2 - 2x - 1 = 0$$
$$x^2 - 2x = 1$$
$$x^2 - 2x + \left(\frac{-2}{2}\right)^2 = 1 + \left(\frac{-2}{2}\right)^2$$
$$x^2 - 2x + (-1)^2 = 1 + (-1)^2$$
$$x^2 - 2x + 1 = 1 + 1$$
$$(x-1)^2 = 2$$

$x - 1 = \sqrt{2}$ or $x - 1 = -\sqrt{2}$

$x = 1 + \sqrt{2}$ $\quad$ $x = 1 - \sqrt{2}$

The solutions are $x = 1 \pm \sqrt{2}$.

11.
$$z^2 + 5z = 7$$
$$z^2 + 5z + \left(\frac{5}{2}\right)^2 = 7 + \left(\frac{5}{2}\right)^2$$
$$z^2 + 5z + \frac{25}{4} = 7 + \frac{25}{4}$$
$$\left(z + \frac{5}{2}\right)^2 = \frac{53}{4}$$

$z+\frac{5}{2}=\sqrt{\frac{53}{4}}$ or $z+\frac{5}{2}=-\sqrt{\frac{53}{4}}$

$z=-\frac{5}{2}+\frac{\sqrt{53}}{2}$ $z=-\frac{5}{2}-\frac{\sqrt{53}}{2}$

The solutions are $z=\frac{-5\pm\sqrt{53}}{2}$.

13.
$$3x^2-6x=24$$
$$x^2-2x=8$$
$$x^2-2x+\left(\frac{-2}{2}\right)^2=8+\left(\frac{-2}{2}\right)^2$$
$$x^2-2x+(-1)^2=8+(-1)^2$$
$$x^2-2x+1=8+1$$
$$(x-1)^2=9$$

$x-1=\sqrt{9}$ or $x-1=-\sqrt{9}$

$x=1+3$ $x=1-3$

$x=4$ $x=-2$

The solutions are $x=-2$ and $x=4$.

15.
$$5x^2+10x+6=0$$
$$x^2+2x+\frac{6}{5}=0$$
$$x^2+2x=-\frac{6}{5}$$
$$x^2+2x+\left(\frac{2}{2}\right)^2=-\frac{6}{5}+\left(\frac{2}{2}\right)^2$$
$$(x+1)^2=-\frac{1}{5}$$

This has no real solution because $\sqrt{-\frac{1}{5}}$ is not a real number.

17.
$$2x^2=6x+5$$
$$2x^2-6x=5$$
$$x^2-3x=\frac{5}{2}$$
$$x^2-3x+\left(\frac{-3}{2}\right)^2=\frac{5}{2}+\left(\frac{-3}{2}\right)^2$$
$$x^2-3x+\frac{9}{4}=\frac{5}{2}+\frac{9}{4}$$
$$\left(x-\frac{3}{2}\right)^2=\frac{10}{4}+\frac{9}{4}$$
$$\left(x-\frac{3}{2}\right)^2=\frac{19}{4}$$

$x-\frac{3}{2}=\sqrt{\frac{19}{4}}$ or $x-\frac{3}{2}=-\sqrt{\frac{19}{4}}$

$x=\frac{3}{2}+\frac{\sqrt{19}}{2}$ $x=\frac{3}{2}-\frac{\sqrt{19}}{2}$

$x=\frac{3+\sqrt{19}}{2}$ $x=\frac{3-\sqrt{19}}{2}$

The solutions are $x=\frac{3\pm\sqrt{19}}{2}$.

19.
$$2y^2+8y+5=0$$
$$y^2+4y+\frac{5}{2}=0$$
$$y^2+4y=-\frac{5}{2}$$
$$y^2+4y+\left(\frac{4}{2}\right)^2=-\frac{5}{2}+\left(\frac{4}{2}\right)^2$$
$$(y+2)^2=\frac{3}{2}$$

$y+2=\sqrt{\frac{3}{2}}$ or $y+2=-\sqrt{\frac{3}{2}}$

$y=-2+\frac{\sqrt{6}}{2}$ $y=-2-\frac{\sqrt{6}}{2}$

The solutions are $y=-2\pm\frac{\sqrt{6}}{2}$.

21.
$$x^2+6x-25=0$$
$$x^2+6x=25$$
$$x^2+6x+\left(\frac{6}{2}\right)^2=25+\left(\frac{6}{2}\right)^2$$
$$x^2+6x+3^2=25+3^2$$
$$x^2+6x+9=25+9$$
$$(x+3)^2=34$$

$x+3=\sqrt{34}$ or $x+3=-\sqrt{34}$

$x=-3+\sqrt{34}$ $x=-3-\sqrt{34}$

The solutions are $x=-3\pm\sqrt{34}$.

23.
$$x^2-3x-3=0$$
$$x^2-3x=3$$
$$x^2-3x+\left(-\frac{3}{2}\right)^2=3+\left(-\frac{3}{2}\right)^2$$
$$\left(x-\frac{3}{2}\right)^2=\frac{21}{4}$$

$$x-\frac{3}{2}=\sqrt{\frac{21}{4}} \quad \text{or} \quad x-\frac{3}{2}=-\sqrt{\frac{21}{4}}$$
$$x=\frac{3}{2}+\frac{\sqrt{21}}{2} \qquad x=\frac{3}{2}-\frac{\sqrt{21}}{2}$$

The solutions are $x=\frac{3\pm\sqrt{21}}{2}$.

25.
$$2y^2-3y+1=0$$
$$2y^2-3y=-1$$
$$y^2-\frac{3}{2}y=-\frac{1}{2}$$
$$y^2-\frac{3}{2}y+\left(\frac{-\frac{3}{2}}{2}\right)^2=-\frac{1}{2}+\left(\frac{-\frac{3}{2}}{2}\right)^2$$
$$y^2-\frac{3}{2}y+\left(-\frac{3}{4}\right)^2=-\frac{1}{2}+\left(-\frac{3}{4}\right)^2$$
$$y^2-\frac{3}{2}y+\frac{9}{16}=-\frac{1}{2}+\frac{9}{16}$$
$$\left(y-\frac{3}{4}\right)^2=\frac{1}{16}$$

$$y-\frac{3}{4}=\sqrt{\frac{1}{16}} \quad \text{or} \quad y-\frac{3}{4}=-\sqrt{\frac{1}{16}}$$
$$y=\frac{3}{4}+\frac{1}{4} \qquad y=\frac{3}{4}-\frac{1}{4}$$
$$y=1 \qquad y=\frac{1}{2}$$

The solutions are $y=\frac{1}{2}$ and $y = 1$.

27.
$$x(x+3)=18$$
$$x^2+3x=18$$
$$x^2+3x+\left(\frac{3}{2}\right)^2=18+\left(\frac{3}{2}\right)^2$$
$$\left(x+\frac{3}{2}\right)^2=\frac{81}{4}$$
$$x+\frac{3}{2}=\sqrt{\frac{81}{4}} \quad \text{or} \quad x+\frac{3}{2}=-\sqrt{\frac{81}{4}}$$
$$x=-\frac{3}{2}+\frac{9}{2} \qquad x=-\frac{3}{2}-\frac{9}{2}$$
$$x=\frac{6}{2}=3 \qquad x=-\frac{12}{2}=-6$$

The solutions are $x = 3$ and $x = -6$.

29.
$$3z^2+6z+4=0$$
$$3z^2+6z=-4$$
$$z^2+2z=-\frac{4}{3}$$
$$z^2+2z+\left(\frac{2}{2}\right)^2=-\frac{4}{3}+\left(\frac{2}{2}\right)^2$$
$$z^2+2z+1^2=-\frac{4}{3}+1^2$$
$$z^2+2z+1=-\frac{4}{3}+1$$
$$(z+1)^2=-\frac{1}{3}$$

The equation has no real solution, since the square root of $-\frac{1}{3}$ is not a real number.

31.
$$4x^2+16x=48$$
$$x^2+4x=12$$
$$x^2+4x+\left(\frac{4}{2}\right)^2=12+\left(\frac{4}{2}\right)^2$$
$$(x+2)^2=16$$
$$x+2=\sqrt{16} \quad \text{and} \quad x+2=-\sqrt{16}$$
$$x=-2+4 \qquad x=-2-4$$
$$x=2 \qquad x=-6$$

The solutions are $x = 2$ and $x = -6$.

33. $\frac{3}{4}-\sqrt{\frac{25}{16}}=\frac{3}{4}-\frac{\sqrt{25}}{\sqrt{16}}$
$$=\frac{3}{4}-\frac{5}{4}$$
$$=\frac{3-5}{4}$$
$$=\frac{-2}{4}$$
$$=-\frac{1}{2}$$

35. $\frac{1}{2}+\sqrt{\frac{9}{4}}=\frac{1}{2}+\frac{3}{2}=\frac{4}{2}=2$

37. $\frac{6+4\sqrt{5}}{2}=\frac{6}{2}+\frac{4\sqrt{5}}{2}=3+2\sqrt{5}$

39. $\frac{3-9\sqrt{2}}{6}=\frac{3}{6}-\frac{9\sqrt{2}}{6}=\frac{1}{2}-\frac{3\sqrt{2}}{2}=\frac{1-3\sqrt{2}}{2}$

41. answers may vary

43. a.
$$x^2+6x+9=11$$
$$(x+3)^2=11$$
$$x+3=\sqrt{11} \quad \text{or} \quad x+3=-\sqrt{11}$$
$$x=-3+\sqrt{11} \qquad x=-3-\sqrt{11}$$
The solutions are $x=-3\pm\sqrt{11}$.

b. answers may vary

45. $x^2+kx+\left(\frac{k}{2}\right)^2$ is a perfect square trinomial. If $x^2+kx+16$ is a perfect square trinomial, then
$$\left(\frac{k}{2}\right)^2=16$$
$$\frac{k}{2}=\sqrt{16} \quad \text{or} \quad \frac{k}{2}=-\sqrt{16}$$
$$k=2\cdot 4 \qquad k=2(-4)$$
$$k=8 \qquad k=-8$$
$x^2+kx+16$ is a perfect square trinomial when $k=8$ or $k=-8$.

47.
$$y=3.5x^2+7x+74$$
$$354=3.5x^2+7x+74$$
$$280=3.5x^2+7x$$
$$80=x^2+2x$$
$$\left(\frac{2}{2}\right)^2+80=x^2+2x+\left(\frac{2}{2}\right)^2$$
$$81=(x+1)^2$$
$$\sqrt{81}=x+1 \quad \text{or} \quad -\sqrt{81}=x+1$$
$$9=x+1 \qquad -9=x+1$$
$$8=x \qquad -10=x$$
Since the time will not be a negative number in this context, reject the solution −10. The solution is $x=8$, so the model predicts that Abiomed's revenues from product sales will be $354 million in 2017 (2009 + 8).

49. The solutions are $x=-6$ and $x=-2$.

51. The solutions are $x\approx-0.68$ and $x\approx3.68$

Section 9.3 Practice Exercises

1. $2x^2-x-5=0$

$a=2$, $b=-1$, $c=-5$
$$x=\frac{-b\pm\sqrt{b^2-4ac}}{2a}$$
$$x=\frac{-(-1)\pm\sqrt{(-1)^2-4(2)(-5)}}{2(2)}$$
$$=\frac{1\pm\sqrt{1+40}}{4}$$
$$=\frac{1\pm\sqrt{41}}{4}$$
The solutions are $\frac{1\pm\sqrt{41}}{4}$.

2.
$$3x^2+8x=3$$
$$3x^2+8x-3=0$$
$a=3$, $b=8$, $c=-3$
$$x=\frac{-b\pm\sqrt{b^2-4ac}}{2a}$$
$$x=\frac{-8\pm\sqrt{8^2-4(3)(-3)}}{2(3)}$$
$$=\frac{-8\pm\sqrt{64+36}}{6}$$
$$=\frac{-8\pm\sqrt{100}}{6}$$
$$=\frac{-8\pm10}{6}$$
$$x=\frac{-8+10}{6}=\frac{1}{3} \quad \text{or} \quad \frac{-8-10}{6}=-3$$
The solutions are $\frac{1}{3}$ and −3.

3.
$$5x^2=2$$
$$5x^2-2=0$$
$a=5$, $b=0$, $c=-2$
$$x=\frac{-b\pm\sqrt{b^2-4ac}}{2a}$$
$$x=\frac{-0\pm\sqrt{0^2-4(5)(-2)}}{2(5)}$$
$$=\pm\frac{\sqrt{40}}{10}$$
$$=\pm\frac{2\sqrt{10}}{10}$$
$$=\pm\frac{\sqrt{10}}{5}$$
The solutions are $\pm\frac{\sqrt{10}}{5}$.

4. $x^2 = -2x - 3$

$x^2 + 2x + 3 = 0$

$a = 1, b = 2, c = 3$

$$x = \frac{-b \pm \sqrt{b^2 - 4ac}}{2a}$$

$$x = \frac{-2 \pm \sqrt{2^2 - 4(1)(3)}}{2(1)} = \frac{-2 \pm \sqrt{-8}}{2}$$

There is no real number solution because $\sqrt{-8}$ is not a real number.

5. $\frac{1}{3}x^2 - x = 1$

$\frac{1}{3}x^2 - x - 1 = 0$

$a = \frac{1}{3}, b = -1, c = -1$

$$x = \frac{-b \pm \sqrt{b^2 - 4ac}}{2a}$$

$$x = \frac{-(-1) \pm \sqrt{(-1)^2 - 4\left(\frac{1}{3}\right)(-1)}}{2\left(\frac{1}{3}\right)}$$

$$= \frac{1 \pm \sqrt{\frac{7}{3}}}{\frac{2}{3}}$$

$$= \frac{1 \pm \frac{\sqrt{21}}{3}}{\frac{2}{3}}$$

$$= \frac{3 \pm \sqrt{21}}{2}$$

The solutions are $\frac{3 \pm \sqrt{21}}{2}$.

6. $\frac{1 + \sqrt{41}}{4} \approx 1.9$

$1 - \frac{\sqrt{41}}{4} \approx -1.4$

Vocabulary, Readiness & Video Check 9.3

1. The quadratic formula is $x = \underline{\frac{-b \pm \sqrt{b^2 - 4ac}}{2a}}$.

2. $5x^2 - 7x + 1 = 0$; $a = \underline{5}$, $b = \underline{-7}$, $c = \underline{1}$

3. $x^2 + 3x - 7 = 0$; $a = \underline{1}$, $b = \underline{3}$, $c = \underline{-7}$

4. $x^2 - 6 = 0$; $a = \underline{1}$, $b = \underline{0}$, $c = \underline{-6}$

5. $x^2 + x - 1 = 0$; $a = \underline{1}$, $b = \underline{1}$, $c = \underline{-1}$

6. $9x^2 - 4 = 0$; $a = \underline{9}$, $b = \underline{0}$, $c = \underline{-4}$

7. $$\frac{-1 \pm \sqrt{1^2 - 4(1)(-2)}}{2(1)} = \frac{-1 \pm \sqrt{1+8}}{2} = \frac{-1 \pm \sqrt{9}}{2} = \frac{-1 \pm 3}{2}$$

$\frac{-1+3}{2} = \frac{2}{2} = 1$

$\frac{-1-3}{2} = \frac{-4}{2} = -2$

8. $$\frac{-(-5) \pm \sqrt{(-5)^2 - 4(2)(3)}}{2(2)} = \frac{5 \pm \sqrt{25-24}}{4} = \frac{5 \pm \sqrt{1}}{4} = \frac{5 \pm 1}{4}$$

$\frac{5+1}{4} = \frac{6}{4} = \frac{3}{2}$

$\frac{5-1}{4} = \frac{4}{4} = 1$

9. $$\frac{-5 \pm \sqrt{5^2 - 4(1)(2)}}{2(1)} = \frac{-5 \pm \sqrt{25-8}}{2} = \frac{-5 \pm \sqrt{17}}{2}$$

10. $$\frac{-7 \pm \sqrt{7^2 - 4(2)(1)}}{2(2)} = \frac{-7 \pm \sqrt{49-8}}{4} = \frac{-7 \pm \sqrt{41}}{4}$$

11. a. Yes, in order to make sure you have correct values for a, b, and c.

b. No; it simplifies calculations, but you would still get a correct answer using fraction values in the formula.

12. The exact solution values are keyed into a calculator, and the decimal approximations are then rounded to the requested place value.

Exercise Set 9.3

1. $x^2 - 3x + 2 = 0$

$a = 1, b = -3, c = 2$

$$x = \frac{-b \pm \sqrt{b^2 - 4ac}}{2a}$$

$$x = \frac{-(-3) \pm \sqrt{(-3)^2 - 4(1)(2)}}{2(1)}$$

$$= \frac{3 \pm \sqrt{9-8}}{2}$$

$$= \frac{3 \pm \sqrt{1}}{2}$$

$$= \frac{3 \pm 1}{2}$$

$$x = \frac{3+1}{2} = 2 \text{ or } x = \frac{3-1}{2} = 1$$

The solutions are $x = 2$ and $x = 1$.

3. $3k^2 + 7k + 1 = 0$

$a = 3, b = 7, c = 1$

$$k = \frac{-b \pm \sqrt{b^2 - 4ac}}{2a}$$

$$k = \frac{-7 \pm \sqrt{7^2 - 4(3)(1)}}{2(3)}$$

$$= \frac{-7 \pm \sqrt{49 - 12}}{6}$$

$$= \frac{-7 \pm \sqrt{37}}{6}$$

The solutions are $k = \frac{-7 \pm \sqrt{37}}{6}$.

5. $4x^2 - 3 = 0$

$4x^2 + 0x - 3 = 0$

$a = 4, b = 0, c = -3$

$$x = \frac{-b \pm \sqrt{b^2 - 4ac}}{2a}$$

$$x = \frac{-0 \pm \sqrt{0^2 - 4(4)(-3)}}{2(4)}$$

$$= \frac{0 \pm \sqrt{0 + 48}}{8}$$

$$= \frac{\pm\sqrt{48}}{8}$$

$$= \pm\frac{4\sqrt{3}}{8}$$

$$= \pm\frac{\sqrt{3}}{2}$$

The solutions are $x = \pm\frac{\sqrt{3}}{2}$.

7. $5z^2 - 4z + 3 = 0$

$a = 5, b = -4, c = 3$

$$z = \frac{-b \pm \sqrt{b^2 - 4ac}}{2a}$$

$$z = \frac{-(-4) \pm \sqrt{(-4)^2 - 4(5)(3)}}{2(5)}$$

$$= \frac{4 \pm \sqrt{16 - 60}}{10}$$

$$= \frac{4 \pm \sqrt{-44}}{10}$$

The equation has no solution, since the square root of -44 is not a real number.

9. $y^2 = 7y + 30$

$y^2 - 7y - 30 = 0$

$a = 1, b = -7, c = -30$

$$y = \frac{-b \pm \sqrt{b^2 - 4ac}}{2a}$$

$$y = \frac{-(-7) \pm \sqrt{(-7)^2 - 4(1)(-30)}}{2(1)}$$

$$= \frac{7 \pm \sqrt{49 + 120}}{2}$$

$$= \frac{7 \pm \sqrt{169}}{2}$$

$$= \frac{7 \pm 13}{2}$$

$$y = \frac{7+13}{2} = 10 \text{ or } y = \frac{7-13}{2} = -3$$

The solutions are $y = 10$ and $y = -3$.

11. $2x^2 = 10$

$2x^2 - 10 = 0$

$a = 2, b = 0, c = -10$

$$x = \frac{-b \pm \sqrt{b^2 - 4ac}}{2a}$$

$$x = \frac{-0 \pm \sqrt{0^2 - 4(2)(-10)}}{2(2)} = \frac{\pm\sqrt{80}}{4} = \frac{\pm 4\sqrt{5}}{4} = \pm\sqrt{5}$$

The solutions are $x = \pm\sqrt{5}$.

13. $m^2 - 12 = m$

$m^2 - m - 12 = 0$

$a = 1, b = -1, c = -12$

$$m = \frac{-b \pm \sqrt{b^2 - 4ac}}{2a}$$

$$m = \frac{-(-1) \pm \sqrt{(-1)^2 - 4(1)(-12)}}{2(1)} = \frac{1 \pm \sqrt{1+48}}{2} = \frac{1 \pm \sqrt{49}}{2} = \frac{1 \pm 7}{2}$$

$$m = \frac{1+7}{2} = 4 \text{ or } m = \frac{1-7}{2} = -3$$

The solutions are $m = 4$ and $m = -3$.

15. $3 - x^2 = 4x$

$0 = x^2 + 4x - 3$

$a = 1, b = 4, c = -3$

$$x = \frac{-b \pm \sqrt{b^2 - 4ac}}{2a}$$

$$x = \frac{-4 \pm \sqrt{4^2 - 4(1)(-3)}}{2(1)} = \frac{-4 \pm \sqrt{16+12}}{2} = \frac{-4 \pm \sqrt{28}}{2} = \frac{-4 \pm 2\sqrt{7}}{2} = -2 \pm \sqrt{7}$$

The solutions are $x = -2 \pm \sqrt{7}$.

17. $6x^2 + 9x = 2$

$6x^2 + 9x - 2 = 0$

$a = 6, b = 9, c = -2$

$$x = \frac{-b \pm \sqrt{b^2 - 4ac}}{2a}$$

$$x = \frac{-9 \pm \sqrt{9^2 - 4(6)(-2)}}{2(6)} = \frac{-9 \pm \sqrt{81+48}}{12} = \frac{-9 \pm \sqrt{129}}{12}$$

The solutions are $x = \frac{-9 \pm \sqrt{129}}{12}$.

19. $7p^2 + 2 = 8p$

$7p^2 - 8p + 2 = 0$

$a = 7, b = -8, c = 2$

$$p = \frac{-b \pm \sqrt{b^2 - 4ac}}{2a}$$

$$p = \frac{-(-8) \pm \sqrt{(-8)^2 - 4(7)(2)}}{2(7)} = \frac{8 \pm \sqrt{64-56}}{14} = \frac{8 \pm \sqrt{8}}{14} = \frac{8 \pm 2\sqrt{2}}{14} = \frac{4 \pm \sqrt{2}}{7}$$

The solutions are $p = \frac{4 \pm \sqrt{2}}{7}$.

21. $x^2 - 6x + 2 = 0$
$a = 1,\ b = -6,\ c = 2$
$$x = \frac{-b \pm \sqrt{b^2 - 4ac}}{2a}$$
$$x = \frac{-(-6) \pm \sqrt{(-6)^2 - 4(1)(2)}}{2(1)}$$
$$= \frac{6 \pm \sqrt{36 - 8}}{2}$$
$$= \frac{6 \pm \sqrt{28}}{2}$$
$$= \frac{6 \pm 2\sqrt{7}}{2}$$
$$= \frac{2(3 \pm \sqrt{7})}{2}$$
$$= 3 \pm \sqrt{7}$$
The solutions are $x = 3 \pm \sqrt{7}$.

23. $2x^2 - 6x + 3 = 0$
$a = 2,\ b = -6,\ c = 3$
$$x = \frac{-b \pm \sqrt{b^2 - 4ac}}{2a}$$
$$x = \frac{-(-6) \pm \sqrt{(-6)^2 - 4(2)(3)}}{2(2)}$$
$$= \frac{6 \pm \sqrt{36 - 24}}{4}$$
$$= \frac{6 \pm \sqrt{12}}{4}$$
$$= \frac{6 \pm 2\sqrt{3}}{4}$$
$$= \frac{3 \pm \sqrt{3}}{2}$$
The solutions are $x = \frac{3 \pm \sqrt{3}}{2}$.

25. $3x^2 = 1 - 2x$
$3x^2 + 2x - 1 = 0$
$a = 3,\ b = 2,\ c = -1$
$$x = \frac{-b \pm \sqrt{b^2 - 4ac}}{2a}$$
$$x = \frac{-2 \pm \sqrt{2^2 - 4(3)(-1)}}{2(3)}$$
$$= \frac{-2 \pm \sqrt{4 + 12}}{6}$$
$$= \frac{-2 \pm \sqrt{16}}{6}$$
$$= \frac{-2 \pm 4}{6}$$
$$x = \frac{-2 + 4}{6} = \frac{1}{3} \text{ or } x = \frac{-2 - 4}{6} = -1$$
The solutions are $x = \frac{1}{3}$ and $x = -1$.

27. $4y^2 = 6y + 1$
$4y^2 - 6y - 1 = 0$
$a = 4,\ b = -6,\ c = -1$
$$y = \frac{-b \pm \sqrt{b^2 - 4ac}}{2a}$$
$$y = \frac{-(-6) \pm \sqrt{(-6)^2 - 4(4)(-1)}}{2(4)}$$
$$= \frac{6 \pm \sqrt{36 + 16}}{8}$$
$$= \frac{6 \pm \sqrt{52}}{8}$$
$$= \frac{6 \pm 2\sqrt{13}}{8}$$
$$= \frac{3 \pm \sqrt{13}}{4}$$
The solutions are $y = \frac{3 \pm \sqrt{13}}{4}$.

29. $x^2 + x + 2 = 0$
$a = 1,\ b = 1,\ c = 2$
$$x = \frac{-b \pm \sqrt{b^2 - 4ac}}{2a}$$
$$x = \frac{-1 \pm \sqrt{1^2 - 4(1)(2)}}{2(1)} = \frac{-1 \pm \sqrt{1 - 8}}{2} = \frac{-1 \pm \sqrt{-7}}{2}$$
The equation has no solution, since the square root of -7 is not a real number.

31. $20y^2 = 3 - 11y$

$20y^2 + 11y - 3 = 0$

$a = 20,\ b = 11,\ c = -3$

$$y = \frac{-b \pm \sqrt{b^2 - 4ac}}{2a}$$

$$y = \frac{-11 \pm \sqrt{11^2 - 4(20)(-3)}}{2(20)}$$

$$= \frac{-11 \pm \sqrt{121 + 240}}{40}$$

$$= \frac{-11 \pm \sqrt{361}}{40}$$

$$= \frac{-11 \pm 19}{40}$$

$$y = \frac{-11 + 19}{40} = \frac{1}{5} \text{ or } y = \frac{-11 - 19}{40} = -\frac{3}{4}$$

The solutions are $y = \frac{1}{5}$ and $y = -\frac{3}{4}$.

33. $x^2 - 5x - 2 = 0$

$a = 1,\ b = -5,\ c = -2$

$$x = \frac{-b \pm \sqrt{b^2 - 4ac}}{2a}$$

$$x = \frac{-(-5) \pm \sqrt{(-5)^2 - 4(1)(-2)}}{2(1)}$$

$$= \frac{5 \pm \sqrt{25 + 8}}{2}$$

$$= \frac{5 \pm \sqrt{33}}{2}$$

The solutions are $x = \frac{5 \pm \sqrt{33}}{2}$.

35. $3x^2 - x - 14 = 0$

$a = 3,\ b = -1,\ c = -14$

$$x = \frac{-b \pm \sqrt{b^2 - 4ac}}{2a}$$

$$x = \frac{-(-1) \pm \sqrt{(-1)^2 - 4(3)(-14)}}{2(3)}$$

$$= \frac{1 \pm \sqrt{1 + 168}}{6}$$

$$= \frac{1 \pm \sqrt{169}}{6}$$

$$= \frac{1 \pm 13}{6}$$

$$x = \frac{1 + 13}{6} = \frac{14}{6} = \frac{7}{3} \text{ or } x = \frac{1 - 13}{6} = \frac{-12}{6} = -2$$

The solutions are $x = \frac{7}{3}$ and $x = -2$.

37. $\frac{m^2}{2} = m + \frac{1}{2}$

$\frac{m^2}{2} - m - \frac{1}{2} = 0$

$m^2 - 2m - 1 = 0$

$a = 1,\ b = -2,\ c = -1$

$$m = \frac{-b \pm \sqrt{b^2 - 4ac}}{2a}$$

$$m = \frac{-(-2) \pm \sqrt{(-2)^2 - 4(1)(-1)}}{2(1)}$$

$$= \frac{2 \pm \sqrt{4 + 4}}{2}$$

$$= \frac{2 \pm \sqrt{8}}{2}$$

$$= \frac{2 \pm 2\sqrt{2}}{2}$$

$$= \frac{2\left(1 \pm \sqrt{2}\right)}{2}$$

$$= 1 \pm \sqrt{2}$$

The solutions are $m = 1 \pm \sqrt{2}$.

39. $3p^2 - \frac{2}{3}p + 1 = 0$

$9p^2 - 2p + 3 = 0$

$a = 9,\ b = -2,\ c = 3$

$$p = \frac{-b \pm \sqrt{b^2 - 4ac}}{2a}$$

$$p = \frac{-(-2) \pm \sqrt{(-2)^2 - 4(9)(3)}}{2(9)}$$

$$= \frac{2 \pm \sqrt{4 - 108}}{18}$$

$$= \frac{2 \pm \sqrt{-104}}{18}$$

The equation has no solution, since the square root of -104 is not a real number.

41. $4p^2 + \frac{3}{2} = -5p$

$4p^2 + 5p + \frac{3}{2} = 0$

$8p^2 + 10p + 3 = 0$

$a = 8,\ b = 10,\ c = 3$

$$p = \frac{-b \pm \sqrt{b^2 - 4ac}}{2a}$$

$$p = \frac{-10 \pm \sqrt{10^2 - 4(8)(3)}}{2(8)} = \frac{-10 \pm \sqrt{100 - 96}}{16} = \frac{-10 \pm \sqrt{4}}{16} = \frac{-10 \pm 2}{16}$$

$$p = \frac{-10+2}{16} = -\frac{1}{2} \text{ or } p = \frac{-10-2}{16} = -\frac{3}{4}$$

The solutions are $p = -\frac{1}{2}$ and $p = -\frac{3}{4}$.

43. $5x^2 = \frac{7}{2}x + 1$

$5x^2 - \frac{7}{2}x - 1 = 0$

$10x^2 - 7x - 2 = 0$

$a = 10,\ b = -7,\ c = -2$

$$x = \frac{-b \pm \sqrt{b^2 - 4ac}}{2a}$$

$$x = \frac{-(-7) \pm \sqrt{(-7)^2 - 4(10)(-2)}}{2(10)} = \frac{7 \pm \sqrt{49 + 80}}{20} = \frac{7 \pm \sqrt{129}}{20}$$

The solutions are $x = \frac{7 \pm \sqrt{129}}{20}$.

45. $x^2 - \frac{11}{2}x - \frac{1}{2} = 0$

$2x^2 - 11x - 1 = 0$

$a = 2,\ b = -11,\ c = -1$

$$x = \frac{-b \pm \sqrt{b^2 - 4ac}}{2a}$$

$$x = \frac{-(-11) \pm \sqrt{(-11)^2 - 4(2)(-1)}}{2(2)} = \frac{11 \pm \sqrt{121 + 8}}{4} = \frac{11 \pm \sqrt{129}}{4}$$

The solutions are $x = \frac{11 \pm \sqrt{129}}{4}$.

47. $5z^2 - 2z = \frac{1}{5}$

$5z^2 - 2z - \frac{1}{5} = 0$

$25z^2 - 10z - 1 = 0$

$a = 25,\ b = -10,\ c = -1$

$$z = \frac{-b \pm \sqrt{b^2 - 4ac}}{2a}$$

$$z = \frac{-(-10) \pm \sqrt{(-10)^2 - 4(25)(-1)}}{2(25)} = \frac{10 \pm \sqrt{100 + 100}}{50} = \frac{10 \pm \sqrt{200}}{50} = \frac{10 \pm 10\sqrt{2}}{50} = \frac{1 \pm \sqrt{2}}{5}$$

The solutions are $z = \frac{1 \pm \sqrt{2}}{5}$.

49. $3x^2 = 21$

$3x^2 - 21 = 0$

$3x^2 + 0x - 21 = 0$

$a = 3,\ b = 0,\ c = -21$

$$x = \frac{-b \pm \sqrt{b^2 - 4ac}}{2a}$$

$$x = \frac{-0 \pm \sqrt{0^2 - 4(3)(-21)}}{2(3)}$$
$$= \frac{\pm\sqrt{0+252}}{6}$$
$$= \frac{\pm\sqrt{252}}{6}$$
$$= \frac{\pm 6\sqrt{7}}{6}$$
$$= \pm\sqrt{7}$$

The solutions are $x = \pm\sqrt{7}$ or $x \approx -2.6$ and $x \approx 2.6$.

51. $x^2 + 6x + 1 = 0$

$a = 1,\ b = 6,\ c = 1$

$$x = \frac{-b \pm \sqrt{b^2 - 4ac}}{2a}$$
$$x = \frac{-6 \pm \sqrt{6^2 - 4(1)(1)}}{2(1)}$$
$$= \frac{-6 \pm \sqrt{36-4}}{2}$$
$$= \frac{-6 \pm \sqrt{32}}{2}$$
$$= \frac{-6 \pm 4\sqrt{2}}{2}$$
$$= -3 \pm 2\sqrt{2}$$

The solutions are $x = -3 \pm 2\sqrt{2}$ or $x \approx -5.8$ and $x \approx -0.2$.

53. $x^2 = 9x + 4$

$x^2 - 9x - 4 = 0$

$a = 1,\ b = -9,\ c = -4$

$$x = \frac{-b \pm \sqrt{b^2 - 4ac}}{2a}$$
$$x = \frac{-(-9) \pm \sqrt{(-9)^2 - 4(1)(-4)}}{2(1)}$$
$$= \frac{9 \pm \sqrt{81+16}}{2}$$
$$= \frac{9 \pm \sqrt{97}}{2}$$

The solutions are $x = \dfrac{9 \pm \sqrt{97}}{2}$ or $x \approx 9.4$ and $x \approx -0.4$.

55. $3x^2 - 2x - 2 = 0$

$a = 3,\ b = -2,\ c = -2$

$$x = \frac{-b \pm \sqrt{b^2 - 4ac}}{2a}$$
$$x = \frac{-(-2) \pm \sqrt{(-2)^2 - 4(3)(-2)}}{2(3)}$$
$$= \frac{2 \pm \sqrt{4+24}}{6}$$
$$= \frac{2 \pm \sqrt{28}}{6}$$
$$= \frac{2 \pm 2\sqrt{7}}{6}$$
$$= \frac{1 \pm \sqrt{7}}{3}$$

The solutions are $x = \dfrac{1 \pm \sqrt{7}}{3}$ or $x \approx 1.2$ and $x \approx -0.5$.

57. $y = -3$ is a horizontal line with y-intercept $(0, -3)$.

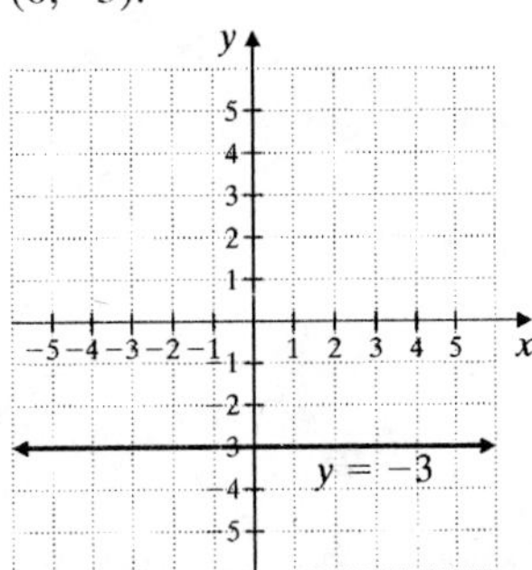

59. $y = 3x - 2$ is a line with slope of 3 and y-intercept $(0, -2)$.

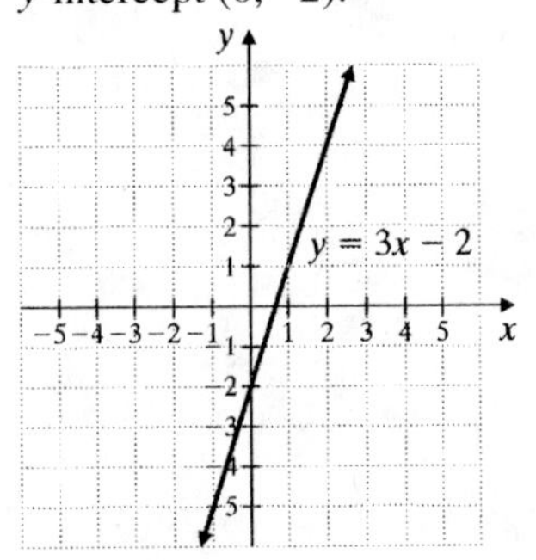

61. $5x^2 + 2 = x$

$5x^2 - x + 2 = 0$

$b = -1$, which is choice c.

63. $7y^2 = 3y$

$0 = -7y^2 + 3y + 0$

$a = -7$, which is choice b.

65. Let x be the width of the chocolate bar. Then the length is $3x - 0.6$.

Area = (length)(width)

$34.65 = (3x - 0.6)(x)$

$34.65 = 3x^2 - 0.6x$

$0 = 3x^2 - 0.6x - 34.65$

$a = 3$, $b = -0.6$, $c = -34.65$

$$x = \frac{-b \pm \sqrt{b^2 - 4ac}}{2a}$$

$$x = \frac{-(-0.6) \pm \sqrt{(-0.6)^2 - 4(3)(-34.65)}}{2(3)}$$

$$= \frac{0.6 \pm \sqrt{0.36 + 415.8}}{6}$$

$$= \frac{0.6 \pm \sqrt{416.16}}{6}$$

$$= \frac{0.6 \pm 20.4}{6}$$

$$x = \frac{0.6 + 20.4}{6} = \frac{21}{6} = 3.5$$

or

$$x = \frac{0.6 - 20.4}{6} = \frac{-19.8}{6} = -3.3$$

Since the width is not a negative number, discard the solution -3.3.

$3x - 0.6 = 3(3.5) - 0.6 = 10.5 - 0.6 = 9.9$

The width was 3.5 feet and the length was 9.9 feet.

67. $x^2 + 3\sqrt{2}x - 5 = 0$

$a = 1$, $b = 3\sqrt{2}$, $c = -5$

$$x = \frac{-b \pm \sqrt{b^2 - 4ac}}{2a}$$

$$x = \frac{-3\sqrt{2} \pm \sqrt{\left(3\sqrt{2}\right)^2 - 4(1)(-5)}}{2(1)}$$

$$= \frac{-3\sqrt{2} \pm \sqrt{18 + 20}}{2}$$

$$= \frac{-3\sqrt{2} \pm \sqrt{38}}{2}$$

The solutions are $x = \dfrac{-3\sqrt{2} \pm \sqrt{38}}{2}$.

69. answers may vary

71. $7.3^2 + 5.4z - 1.1 = 0$

$a = 7.3$, $b = 5.4$, $c = -1.1$

$$z = \frac{-b \pm \sqrt{b^2 - 4ac}}{2a}$$

$$z = \frac{-5.4 \pm \sqrt{(5.4)^2 - 4(7.3)(-1.1)}}{2(7.3)}$$

$$= \frac{-5.4 \pm \sqrt{29.16 + 32.12}}{14.6}$$

$$= \frac{-5.4 \pm \sqrt{61.28}}{14.6}$$

$$z = \frac{-5.4 + \sqrt{61.28}}{14.6} \approx 0.2 \text{ or}$$

$$z = \frac{-5.4 - \sqrt{61.28}}{14.6} \approx -0.9$$

The solutions are $z \approx 0.2$ and $z \approx -0.9$.

73. $h = -16t^2 + 120t + 80$

$30 = -16t^2 + 120t + 80$

$0 = -16t^2 + 120t + 50$

$a = -16$, $b = 120$, $c = 50$

$$t = \frac{-b \pm \sqrt{b^2 - 4ac}}{2a}$$

$$t = \frac{-120 \pm \sqrt{(120)^2 - 4(-16)(50)}}{2(-16)}$$

$$= \frac{-120 \pm \sqrt{14,400 + 3200}}{-32}$$

$$= \frac{-120 \pm \sqrt{17,600}}{-32}$$

$$t = \frac{-120 + \sqrt{17,600}}{-32} \approx -0.4 \text{ or}$$

$$t = \frac{-120 - \sqrt{17,600}}{-32} \approx 7.9$$

Since the time of the flight is not a negative number, discard the solution -0.4. The rocket will reach the height of 30 feet approximately 7.9 seconds after it is launched.

75.
$$y = 450x^2 + 330x + 64,820$$
$$96,000 = 450x^2 + 330x + 64,820$$
$$0 = 450x^2 + 330x - 31,180$$
$$0 = 45x^2 + 33x - 3118$$
$a = 45$, $b = 33$, $c = -3118$
$$x = \frac{-b \pm \sqrt{b^2 - 4ac}}{2a}$$
$$x = \frac{-33 \pm \sqrt{33^2 - 4(45)(-3118)}}{2(45)}$$
$$= \frac{-33 \pm \sqrt{1089 + 561,240}}{90}$$
$$= \frac{-33 \pm \sqrt{562,329}}{90}$$
$$x = \frac{-33 + \sqrt{562,329}}{90} \approx 8 \text{ or}$$
$$x = \frac{-33 - \sqrt{562,329}}{90} \approx -9$$
Since the time is not negative in this context, discard the solution −9. The solution is $x \approx 8$, so the model predicts that Target's total revenue will be $96,000 million in 2016 (2008 + 8).

Integrated Review Practice Exercises

1. $y^2 - 4y - 6 = 0$

Use the quadratic formula with $a = 1$, $b = -4$, and $c = -6$.
$$y = \frac{-b \pm \sqrt{b^2 - 4ac}}{2a}$$
$$y = \frac{-(-4) \pm \sqrt{(-4)^2 - 4(1)(-6)}}{2(1)}$$
$$= \frac{4 \pm \sqrt{16 + 24}}{2}$$
$$= \frac{4 \pm \sqrt{40}}{2}$$
$$= \frac{4 \pm 2\sqrt{10}}{2}$$
$$= 2 \pm \sqrt{10}$$
The solutions are $y = 2 \pm \sqrt{10}$.

2. $(2x+5)^2 = 45$

Use the square root property.
$$(2x+5)^2 = 45$$
$$2x + 5 = \pm\sqrt{45}$$
$$2x + 5 = \pm 3\sqrt{5}$$
$$2x = -5 \pm 3\sqrt{5}$$
$$x = \frac{-5 \pm 3\sqrt{5}}{2}$$
The solutions are $x = \dfrac{-5 \pm 3\sqrt{5}}{2}$.

3.
$$x^2 - \frac{5}{2}x = -\frac{3}{2}$$
$$x^2 - \frac{5}{2}x + \frac{3}{2} = 0$$
$$2x^2 - 5x + 3 = 0$$
$$(2x-3)(x-1) = 0$$
$$2x - 3 = 0 \quad \text{or} \quad x - 1 = 0$$
$$2x = 3 \qquad x = 1$$
$$x = \frac{3}{2} \qquad x = 1$$
The solutions are $x = \dfrac{3}{2}$ and $x = 1$.

Integrated Review

1. $5x^2 - 11x + 2 = 0$
$$(5x-1)(x-2) = 0$$
$$5x - 1 = 0 \quad \text{or} \quad x - 2 = 0$$
$$5x = 1 \qquad x = 2$$
$$x = \frac{1}{5}$$
The solutions are $x = \dfrac{1}{5}$ and $x = 2$.

2. $5x^2 + 13x - 6 = 0$
$$(5x-2)(x+3) = 0$$
$$5x - 2 = 0 \quad \text{or} \quad x + 3 = 0$$
$$5x = 2 \qquad x = -3$$
$$x = \frac{2}{5}$$
The solutions are $x = \dfrac{2}{5}$ and $x = -3$.

3. $x^2 - 1 = 2x$

$x^2 - 2x - 1 = 0$

$a = 1, b = -2, c = -1$

$$x = \frac{-b \pm \sqrt{b^2 - 4ac}}{2a}$$

$$x = \frac{-(-2) \pm \sqrt{(-2)^2 - 4(1)(-1)}}{2(1)}$$

$$= \frac{2 \pm \sqrt{4+4}}{2}$$

$$= \frac{2 \pm \sqrt{8}}{2}$$

$$= \frac{2 \pm 2\sqrt{2}}{2}$$

$$= 1 \pm \sqrt{2}$$

The solutions are $x = 1 \pm \sqrt{2}$.

4. $x^2 + 7 = 6x$

$x^2 - 6x + 7 = 0$

$a = 1, b = -6, c = 7$

$$x = \frac{-b \pm \sqrt{b^2 - 4ac}}{2a}$$

$$x = \frac{-(-6) \pm \sqrt{(-6)^2 - 4(1)(7)}}{2(1)}$$

$$= \frac{6 \pm \sqrt{36-28}}{2}$$

$$= \frac{6 \pm \sqrt{8}}{2}$$

$$= \frac{6 \pm 2\sqrt{2}}{2}$$

$$= 3 \pm \sqrt{2}$$

The solutions are $x = 3 \pm \sqrt{2}$.

5. $a^2 = 20$

$a = \pm\sqrt{20}$

$a = \pm 2\sqrt{5}$

The solutions are $a = \pm 2\sqrt{5}$.

6. $a^2 = 72$

$a = \pm\sqrt{72}$

$a = \pm 6\sqrt{2}$

The solutions are $a = \pm 6\sqrt{2}$.

7. $x^2 - x + 4 = 0$

$a = 1, b = -1, c = 4$

$$x = \frac{-b \pm \sqrt{b^2 - 4ac}}{2a}$$

$$x = \frac{-(-1) \pm \sqrt{(-1)^2 - 4(1)(4)}}{2(1)}$$

$$= \frac{1 \pm \sqrt{1-16}}{2}$$

$$= \frac{1 \pm \sqrt{-15}}{2}$$

The equation has no solution, since the square root of -15 is not a real number.

8. $x^2 - 2x + 7 = 0$

$a = 1, b = -2, c = 7$

$$x = \frac{-b \pm \sqrt{b^2 - 4ac}}{2a}$$

$$x = \frac{-(-2) \pm \sqrt{(-2)^2 - 4(1)(7)}}{2(1)}$$

$$= \frac{2 \pm \sqrt{4-28}}{2}$$

$$= \frac{2 \pm \sqrt{-24}}{2}$$

The equation has no solution, since the square root of -24 is not a real number.

9. $3x^2 - 12x + 12 = 0$

$3(x^2 - 4x + 4) = 0$

$3(x-2)(x-2) = 0$

$x - 2 = 0$ or $x - 2 = 0$

$x = 2$ $\quad$ $x = 2$

The solution is $x = 2$.

10. $5x^2 - 30x + 45 = 0$

$5(x^2 - 6x + 9) = 0$

$5(x-3)(x-3) = 0$

$x - 3 = 0$ or $x - 3 = 0$

$x = 3$ $\quad$ $x = 3$

The solution is $x = 3$.

11. $9 - 6p + p^2 = 0$

$(3-p)(3-p) = 0$

$3 - p = 0$ or $3 - p = 0$

$3 = p$ $\quad$ $3 = p$

The solution is $p = 3$.

12. $49-28p+4p^2=0$
$(7-2p)(7-2p)=0$
$7-2p=0$ or $7-2p=0$
$7=2p$ $\quad 7=2p$
$\frac{7}{2}=p$ $\quad \frac{7}{2}=p$
The solution is $p=\frac{7}{2}$.

13. $4y^2-16=0$
$4(y^2-4)=0$
$4(y-2)(y+2)=0$
$y-2=0$ or $y+2=0$
$y=2$ $\quad y=-2$
The solutions are $y=\pm 2$.

14. $3y^2-27=0$
$3(y^2-9)=0$
$3(y-3)(y+3)=0$
$y-3=0$ or $y+3=0$
$y=3$ $\quad y=-3$
The solutions are $y=\pm 3$.

15. $x^2-3x+2=0$
$(x-2)(x-1)=0$
$x-2=0$ or $x-1=0$
$x=2$ $\quad x=1$
The solutions are $x=1$ and $x=2$.

16. $x^2+7x+12=0$
$(x+3)(x+4)=0$
$x+3=0$ or $x+4=0$
$x=-3$ $\quad x=-4$
The solutions are $x=-3$ and $x=-4$.

17. $(2z+5)^2=25$
$2z+5=\sqrt{25}$ or $2z+5=-\sqrt{25}$
$2z+5=5$ $\quad 2z+5=-5$
$2z=0$ $\quad 2z=-10$
$z=0$ $\quad z=-5$
The solutions are $z=0$ and $z=-5$.

18. $(3z-4)^2=16$
$3z-4=\sqrt{16}$ or $3z-4=-\sqrt{16}$
$3z-4=4$ $\quad 3z-4=-4$
$3z=8$ $\quad 3z=0$
$z=\frac{8}{3}$ $\quad z=0$
The solutions are $z=\frac{8}{3}$ and $z=0$.

19. $30x=25x^2+2$
$0=25x^2-30x+2$
$a=25,\ b=-30,\ c=2$
$$x=\frac{-b\pm\sqrt{b^2-4ac}}{2a}$$
$$x=\frac{-(-30)\pm\sqrt{(-30)^2-4(25)(2)}}{2(25)}$$
$$=\frac{30\pm\sqrt{900-200}}{50}$$
$$=\frac{30\pm\sqrt{700}}{50}$$
$$=\frac{30\pm 10\sqrt{7}}{50}$$
$$=\frac{3\pm\sqrt{7}}{5}$$
The solutions are $x=\frac{3\pm\sqrt{7}}{5}$.

20. $12x=4x^2+4$
$0=4x^2-12x+4$
$0=4(x^2-3x+1)$
$a=1,\ b=-3,\ c=1$
$$x=\frac{-b\pm\sqrt{b^2-4ac}}{2a}$$
$$x=\frac{-(-3)\pm\sqrt{(-3)^2-4(1)(1)}}{2(1)}$$
$$=\frac{3\pm\sqrt{9-4}}{2}$$
$$=\frac{3\pm\sqrt{5}}{2}$$
The solutions are $x=\frac{3\pm\sqrt{5}}{2}$.

21. $\frac{2}{3}m^2 - \frac{1}{3}m - 1 = 0$

$2m^2 - m - 3 = 0$

$(2m-3)(m+1) = 0$

$2m - 3 = 0$ or $m + 1 = 0$

$2m = 3$　　$m = -1$

$m = \frac{3}{2}$

The solutions are $m = \frac{3}{2}$ and $m = -1$.

22. $\frac{5}{8}m^2 + m - \frac{1}{2} = 0$

$5m^2 + 8m - 4 = 0$

$(5m-2)(m+2) = 0$

$5m - 2 = 0$ or $m + 2 = 0$

$5m = 2$　　$m = -2$

$m = \frac{2}{5}$

The solutions are $m = \frac{2}{5}$ and $m = -2$.

23. $x^2 - \frac{1}{2}x - \frac{1}{5} = 0$

$10x^2 - 5x - 2 = 0$

$a = 10, b = -5, c = -2$

$$x = \frac{-b \pm \sqrt{b^2 - 4ac}}{2a}$$

$$x = \frac{-(-5) \pm \sqrt{(-5)^2 - 4(10)(-2)}}{2(10)}$$

$$= \frac{5 \pm \sqrt{25 + 80}}{20}$$

$$= \frac{5 \pm \sqrt{105}}{20}$$

The solutions are $x = \frac{5 \pm \sqrt{105}}{20}$.

24. $x^2 + \frac{1}{2}x - \frac{1}{8} = 0$

$8x^2 + 4x - 1 = 0$

$a = 8, b = 4, c = -1$

$$x = \frac{-b \pm \sqrt{b^2 - 4ac}}{2a}$$

$$x = \frac{-4 \pm \sqrt{4^2 - 4(8)(-1)}}{2(8)}$$

$$= \frac{-4 \pm \sqrt{16 + 32}}{16}$$

$$= \frac{-4 \pm \sqrt{48}}{16}$$

$$= \frac{-4 \pm 4\sqrt{3}}{16}$$

$$= \frac{-1 \pm \sqrt{3}}{4}$$

The solutions are $\frac{-1 \pm \sqrt{3}}{4}$.

25. $4x^2 - 27x + 35 = 0$

$(4x-7)(x-5) = 0$

$4x - 7 = 0$ or $x - 5 = 0$

$4x = 7$　　$x = 5$

$x = \frac{7}{4}$

The solutions are $x = \frac{7}{4}$ and $x = 5$.

26. $9x^2 - 16x + 7 = 0$

$(9x-7)(x-1) = 0$

$9x - 7 = 0$ or $x - 1 = 0$

$9x = 7$　　$x = 1$

$x = \frac{7}{9}$

The solutions are $x = \frac{7}{9}$ and $x = 1$.

27. $(7 - 5x)^2 = 18$

$7 - 5x = \sqrt{18}$

$-5x = -7 + 3\sqrt{2}$

$$x = \frac{-7 + 3\sqrt{2}}{-5} = \frac{7 - 3\sqrt{2}}{5}$$

or

$7 - 5x = -\sqrt{18}$

$-5x = -7 - 3\sqrt{2}$

$$x = \frac{-7 - 3\sqrt{2}}{-5} = \frac{7 + 3\sqrt{2}}{5}$$

The solutions are $x = \frac{-7 \pm 3\sqrt{2}}{-5}$ or $\frac{7 \pm 3\sqrt{2}}{5}$.

28. $(5-4x)^2 = 75$

$$5-4x = \sqrt{75}$$
$$-4x = -5+5\sqrt{3}$$
$$x = \frac{-5+5\sqrt{3}}{-4} = \frac{5-5\sqrt{3}}{4}$$

or

$$5-4x = -\sqrt{75}$$
$$-4x = -5-5\sqrt{3}$$
$$x = \frac{-5-5\sqrt{3}}{-4} = \frac{5+5\sqrt{3}}{4}$$

The solutions are $x = \frac{-5\pm 5\sqrt{3}}{-4}$ or $\frac{5\pm 5\sqrt{3}}{4}$.

29. $3z^2 - 7z = 12$

$$3z^2 - 7z - 12 = 0$$
$$a = 3,\ b = -7,\ c = -12$$
$$z = \frac{-b\pm\sqrt{b^2-4ac}}{2a}$$
$$z = \frac{-(-7)\pm\sqrt{(-7)^2-4(3)(-12)}}{2(3)}$$
$$= \frac{7\pm\sqrt{49+144}}{6}$$
$$= \frac{7\pm\sqrt{193}}{6}$$

The solutions are $z = \frac{7\pm\sqrt{193}}{6}$.

30. $6z^2 + 7z = 6$

$$6z^2 + 7z - 6 = 0$$
$$a = 6,\ b = 7,\ c = -6$$
$$z = \frac{-b\pm\sqrt{b^2-4ac}}{2a}$$
$$z = \frac{-7\pm\sqrt{(7)^2-4(6)(-6)}}{2(6)}$$
$$= \frac{-7\pm\sqrt{49+144}}{12}$$
$$= \frac{-7\pm\sqrt{193}}{12}$$

The solutions are $z = \frac{-7\pm\sqrt{193}}{12}$.

31. $x = x^2 - 110$

$$0 = x^2 - x - 110$$
$$0 = (x-11)(x+10)$$
$$x-11=0 \quad\text{or}\quad x+10=0$$
$$x=11 \qquad\qquad x=-10$$

The solutions are $x = 11$ and $x = -10$.

32. $x = 56 - x^2$

$$0 = 56 - x - x^2$$
$$0 = (8+x)(7-x)$$
$$8+x=0 \quad\text{or}\quad 7-x=0$$
$$x=-8 \qquad\qquad x=7$$

The solutions are $x = 7$ and $x = -8$.

33. $\frac{3}{4}x^2 - \frac{5}{2}x - 2 = 0$

$$3x^2 - 10x - 8 = 0$$
$$(3x+2)(x-4) = 0$$
$$3x+2=0 \quad\text{or}\quad x-4=0$$
$$3x=-2 \qquad\qquad x=4$$
$$x=-\frac{2}{3}$$

The solutions are $x = -\frac{2}{3}$ and $x = 4$.

34. $x^2 - \frac{6}{5}x - \frac{8}{5} = 0$

$$5x^2 - 6x - 8 = 0$$
$$(5x+4)(x-2) = 0$$
$$5x+4=0 \quad\text{or}\quad x-2=0$$
$$5x=-4 \qquad\qquad x=2$$
$$x=-\frac{4}{5}$$

The solutions are $x = -\frac{4}{5}$ and $x = 2$.

35. $x^2 - 0.6x + 0.05 = 0$

$$(x-0.5)(x-0.1) = 0$$
$$x-0.5=0 \quad\text{or}\quad x-0.1=0$$
$$x=0.5 \qquad\qquad x=0.1$$

The solutions are $x = 0.5$ and $x = 0.1$.

36. $x^2 - 0.1x - 0.06 = 0$

$$(x-0.3)(x+0.2) = 0$$
$$x-0.3=0 \quad\text{or}\quad x+0.2=0$$
$$x=0.3 \qquad\qquad x=-0.2$$

The solutions are $x = 0.3$ and $x = -0.2$.

37. $10x^2 - 11x + 2 = 0$

$a = 10,\ b = -11,\ c = 2$

$$x = \frac{-b \pm \sqrt{b^2 - 4ac}}{2a}$$

$$x = \frac{-(-11) \pm \sqrt{(-11)^2 - 4(10)(2)}}{2(10)}$$

$$= \frac{11 \pm \sqrt{121 - 80}}{20}$$

$$= \frac{11 \pm \sqrt{41}}{20}$$

The solutions are $x = \frac{11 \pm \sqrt{41}}{20}$.

38. $20x^2 - 11x + 1 = 0$

$a = 20,\ b = -11,\ c = 1$

$$x = \frac{-b \pm \sqrt{b^2 - 4ac}}{2a}$$

$$x = \frac{-(-11) \pm \sqrt{(-11)^2 - 4(20)(1)}}{2(20)}$$

$$= \frac{11 \pm \sqrt{121 - 80}}{40}$$

$$= \frac{11 \pm \sqrt{41}}{40}$$

The solutions are $x = \frac{11 \pm \sqrt{41}}{40}$.

39. $\frac{1}{2}z^2 - 2z + \frac{3}{4} = 0$

$2z^2 - 8z + 3 = 0$

$a = 2,\ b = -8,\ c = 3$

$$z = \frac{-b \pm \sqrt{b^2 - 4ac}}{2a}$$

$$z = \frac{-(-8) \pm \sqrt{(-8)^2 - 4(2)(3)}}{2(2)}$$

$$= \frac{8 \pm \sqrt{64 - 24}}{4}$$

$$= \frac{8 \pm \sqrt{40}}{4}$$

$$= \frac{8 \pm 2\sqrt{10}}{4}$$

$$= \frac{4 \pm \sqrt{10}}{2}$$

The solutions are $z = \frac{4 \pm \sqrt{10}}{2}$.

40. $\frac{1}{5}z^2 - \frac{1}{2}z - 2 = 0$

$2z^2 - 5z - 20 = 0$

$a = 2,\ b = -5,\ c = -20$

$$z = \frac{-b \pm \sqrt{b^2 - 4ac}}{2a}$$

$$z = \frac{-(-5) \pm \sqrt{(-5)^2 - 4(2)(-20)}}{2(2)}$$

$$= \frac{5 \pm \sqrt{25 + 160}}{4}$$

$$= \frac{5 \pm \sqrt{185}}{4}$$

The solutions are $z = \frac{5 \pm \sqrt{185}}{4}$.

41. answers may vary

Section 9.4 Practice Exercises

1. $y = -3x^2$

x	y
−2	−12
−1	−3
0	0
1	−3
2	−12

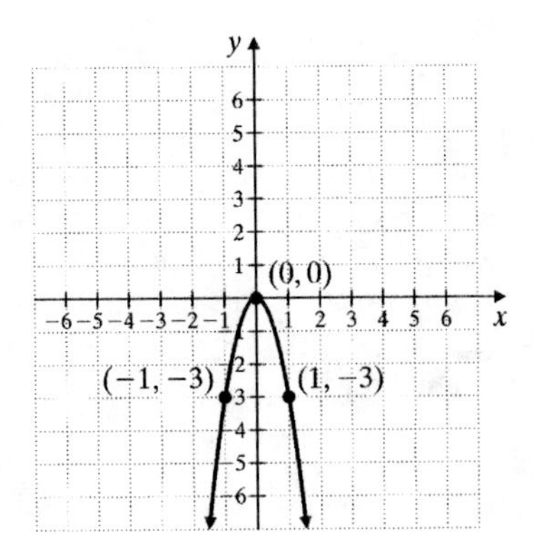

2. $y = x^2 - 9$

Let $x = 0$.

$y = 0^2 - 9 = -9$

The y-intercept is (0, −9).
Let $y = 0$.

$0 = x^2 - 9$

$0 = (x-3)(x+3)$

$x - 3 = 0$ or $x + 3 = 0$

$x = 3$ $\quad x = -3$

The x-intercepts are (3, 0) and (−3, 0).

x	y
−3	0
−2	−5
−1	−8
0	−9
1	−8
2	−5
3	0

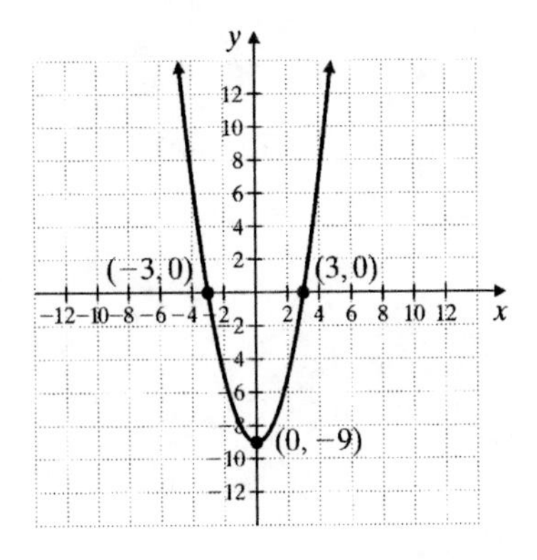

3. $y = x^2 - 2x - 3$

$a = 1, b = -2, c = -3$

$$x = \frac{-b}{2a} = \frac{-(-2)}{2(1)} = \frac{2}{2} = 1$$

$y = 1^2 - 2(1) - 3 = 1 - 2 - 3 = -4$

The vertex is (1, −4).
Let $x = 0$.

$y = 0^2 - 2(0) - 3 = -3$

The y-intercept is (0, −3).
Let $y = 0$.

$0 = x^2 - 2x - 3$

$0 = (x-3)(x+1)$

$x - 3 = 0$ or $x + 1 = 0$

$x = 3$ $\quad x = -1$

The x-intercepts are (3, 0) and (−1, 0).

x	y
−1	0
0	−3
1	−4
2	−3
3	0

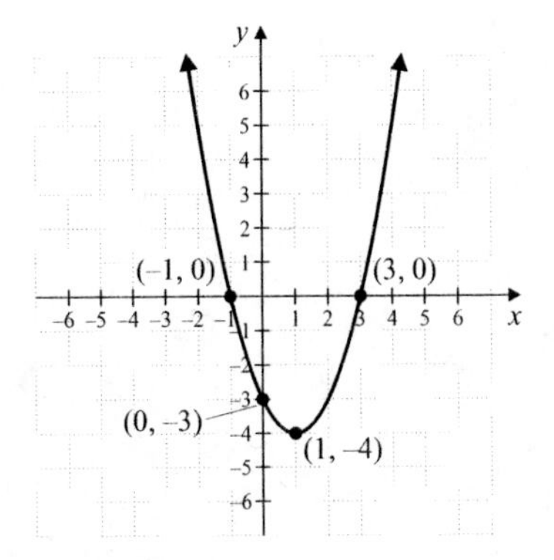

4. $y = x^2 - 4x + 1$

$a = 1, b = -4, c = 1$

$$x = \frac{-b}{2a} = \frac{-(-4)}{2(1)} = \frac{4}{2} = 2$$

$y = 2^2 - 4(2) + 1 = 4 - 8 + 1 = -3$

vertex = (2, −3)

$$x = \frac{-b \pm \sqrt{b^2 - 4ac}}{2a}$$

$$x = \frac{-(-4) \pm \sqrt{(-4)^2 - 4(1)(1)}}{2(1)}$$
$$= \frac{4 \pm \sqrt{16-4}}{2}$$
$$= \frac{4 \pm \sqrt{12}}{2}$$
$$= \frac{4 \pm 2\sqrt{3}}{2}$$
$$= 2 \pm \sqrt{3}$$
$$= 3.7 \text{ or } 0.3$$

x	y
2	−3
$2+\sqrt{3} \approx 3.7$	0
$2-\sqrt{3} \approx 0.3$	0
0	1
4	1

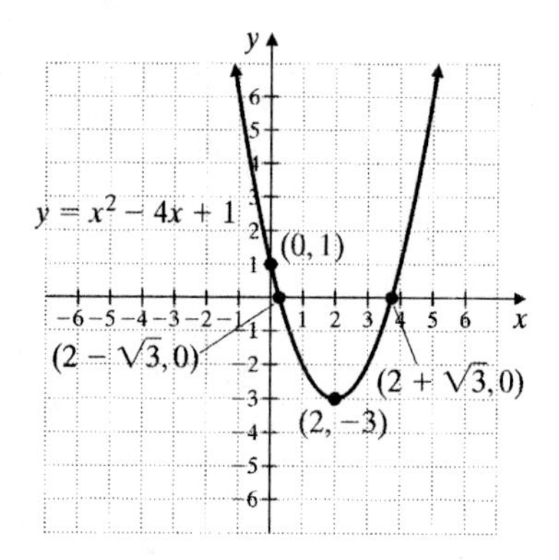

Calculator Explorations

1. $x^2 - 7x - 3 = 0$ $x \approx -0.41, 7.41$
2. $2x^2 - 11x - 1 = 0$ $x \approx -0.09, 5.59$
3. $-1.7x^2 + 5.6x - 3.7 = 0$ $x \approx 0.91, 2.38$
4. $-5.8x^2 + 2.3x - 3.9 = 0$ No real solutions
5. $5.8x^2 - 2.6x - 1.9 = 0$ $x \approx -0.39, 0.84$
6. $7.5x^2 - 3.7x - 1.1 = 0$ $x \approx -0.21, 0.70$

Vocabulary, Readiness & Video Check 9.4

1. If a parabola opens upward, the lowest point is called the vertex; if a parabola opens downward, the highest point is called the vertex. If a graph can be folded along a line such that the two sides coincide or form mirror images of each other, we say that the graph is symmetric about that line and that line is the line of symmetry.

2. The vertex; it is a very useful point to plot since it is the highest or lowest point on your graph.

3. For example, if the vertex is in quadrant III or IV and the parabola opens downward, then there won't be any x-intercepts and there's no need to let $y = 0$ and solve the equation for x.

Exercise Set 9.4

1.

x	$y = 2x^2$
−2	$2(-2)^2 = 2(4) = 8$
−1	$2(-1)^2 = 2(1) = 2$
0	$2(0)^2 = 2(0) = 0$
1	$2(1)^2 = 2(1) = 2$
2	$2(2)^2 = 2(4) = 8$

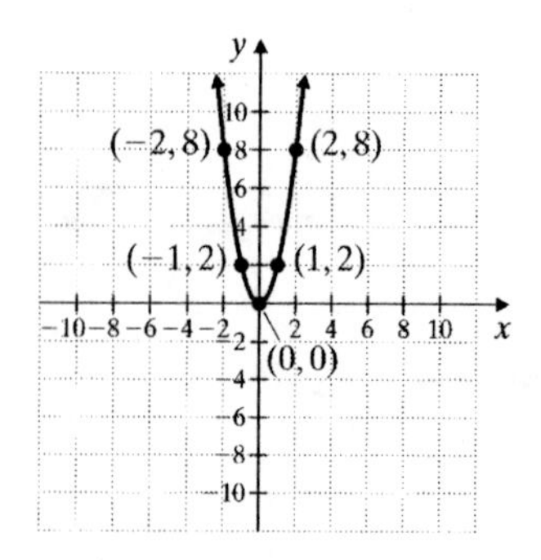

3.

x	$y=-x^2$
−2	$-(-2)^2=-(4)=-4$
−1	$-(-1)^2=-(1)=-1$
0	$-(0)^2=-(0)=0$
1	$-(1)^2=-(1)=-1$
2	$-(2)^2=-(4)=-4$

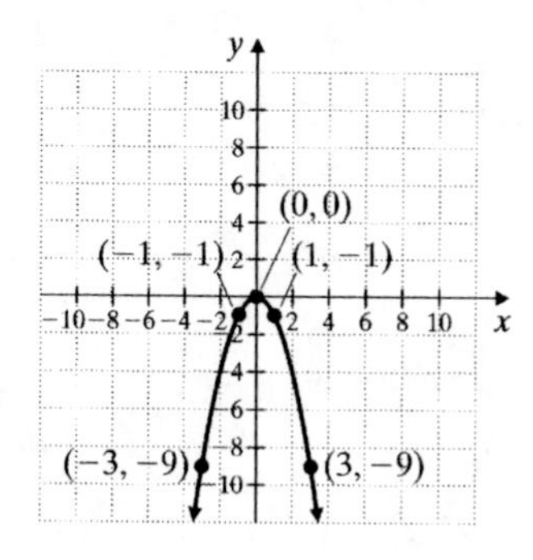

5. $y=x^2-1$

$y=0$: $0=x^2-1$

$1=x^2$

$\pm\sqrt{1}=x$

$\pm 1=x$

x-intercepts: $(-1, 0)$, $(1, 0)$

$x=0$: $y=0^2-1=-1$

y-intercept: $(0, -1)$

$y=x^2+0x-1$

$a=1$, $b=0$, $c=1$

$$\frac{-b}{2a}=\frac{-0}{2(1)}=\frac{0}{2}=0$$

$x=0$: $y=0^2-1=-1$

The vertex is $(0, -1)$.

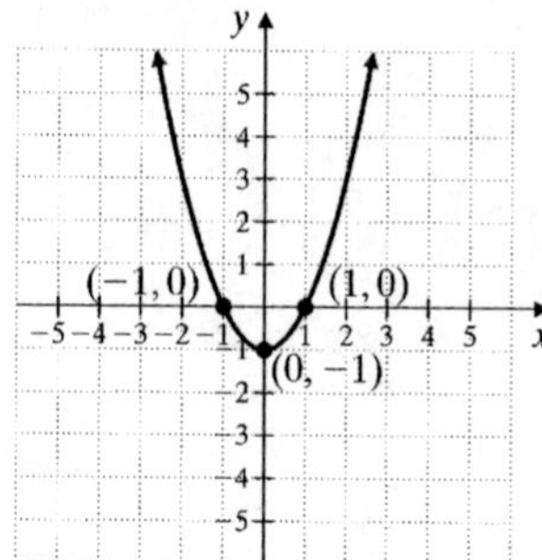

7. $y=x^2+4$

$y=0$: $0=x^2+4$

$-4=x^2$

$\pm\sqrt{-4}=x$

x-intercepts: none

$x=0$: $y=0^2+4=4$

y-intercept: $(0, 4)$

$y=x^2+0x+4$

$a=1$, $b=0$, $c=4$

$$\frac{-b}{2a}=\frac{-0}{2(1)}=\frac{0}{2}=0$$

$x=0$: $y=0^2+4=4$

The vertex is $(0, 4)$.

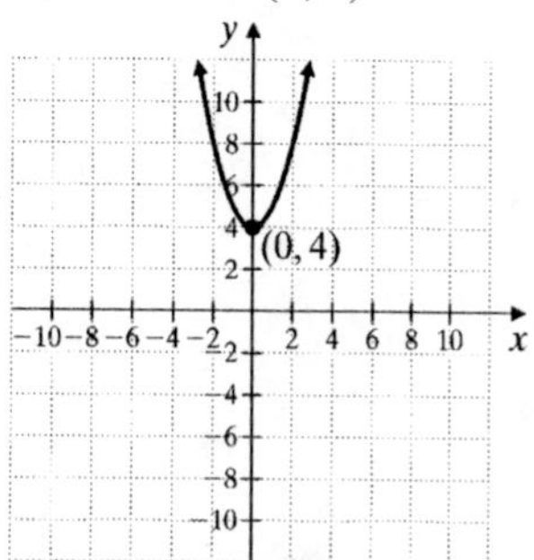

9. $y=-x^2+4x-4$

$y=0$: $0=-x^2+4x-4$

$0=x^2-4x+4$

$0=(x-2)^2$

$0=x-2$

$2=x$

The x-intercept is $(2, 0)$.

$x=0$: $y=-0^2+4(0)-4=-4$

The y-intercept is $(0, -4)$.

$y=-x^2+4x-4$

$a=-1$, $b=4$, $c=-4$

$$\frac{-b}{2a}=\frac{-4}{2(-1)}=\frac{-4}{-2}=2$$

$x=2$: $y=-(2)^2+4(2)-4=-4+8-4=0$

The vertex is $(2, 0)$.

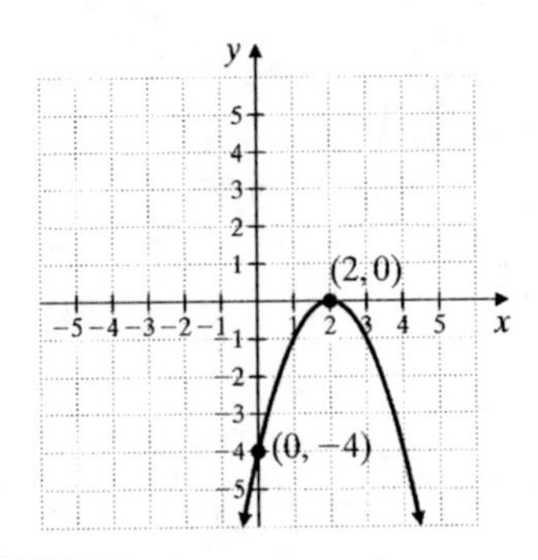

11. $y = x^2 + 5x + 4$

$y = 0$: $0 = x^2 + 5x + 4$

$0 = (x+4)(x+1)$

$0 = x + 4$ or $0 = x + 1$

$-4 = x$ $\quad -1 = x$

x-intercepts: $(-4, 0)$ and $(-1, 0)$

$x = 0$: $y = 0^2 + 5(0) + 4 = 4$

y-intercept: $(0, 4)$

$y = x^2 + 5x + 4$

$a = 1, b = 5, c = 4$

$$\frac{-b}{2a} = \frac{-5}{2(1)} = \frac{-5}{2}$$

$$x = \frac{-5}{2}: \; y = \left(-\frac{5}{2}\right)^2 + 5\left(-\frac{5}{2}\right) + 4$$

$$= \frac{25}{4} - \frac{25}{2} + 4$$

$$= \frac{25}{4} - \frac{50}{4} + \frac{16}{4}$$

$$= -\frac{9}{4}$$

The vertex is $\left(-\frac{5}{2}, -\frac{9}{4}\right)$ or $\left(-2\frac{1}{2}, -2\frac{1}{4}\right)$.

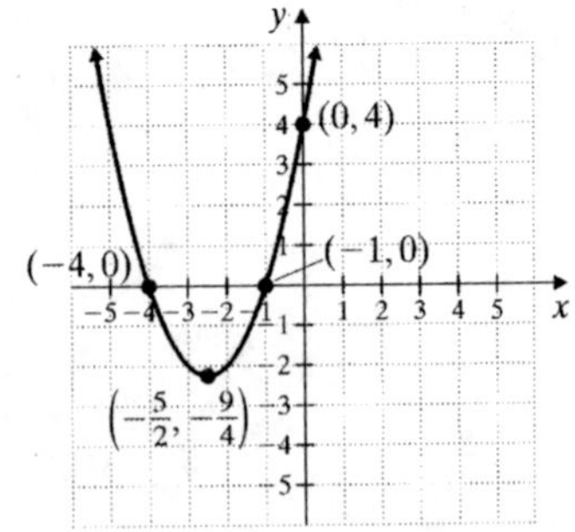

13. $y = x^2 - 4x + 5$

$y = 0$: $0 = x^2 - 4x + 5$

$a = 1, b = -4, c = 5$

$$y = \frac{-b \pm \sqrt{b^2 - 4ac}}{2a}$$

$$y = \frac{-(-4) \pm \sqrt{(-4)^2 - 4(1)(5)}}{2(1)}$$

$$= \frac{4 \pm \sqrt{16 - 20}}{2}$$

$$= \frac{4 \pm \sqrt{-4}}{2}$$

There are no x-intercepts, since $\sqrt{-4}$ is not a real number.

$x = 0$: $y = 0^2 - 4(0) + 5 = 5$

y-intercept: $(0, 5)$

$y = x^2 - 4x + 5$

$a = 1, b = -4, c = 5$

$$\frac{-b}{2a} = \frac{-(-4)}{2(1)} = \frac{4}{2} = 2$$

$x = 2$: $y = 2^2 - 4(2) + 5 = 4 - 8 + 5 = 1$

vertex: $(2, 1)$

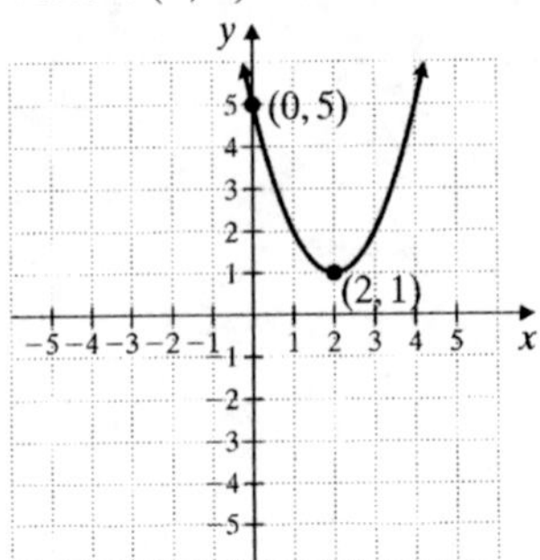

15. $y = 2 - x^2$

$y = 0$: $0 = 2 - x^2$

$x^2 = 2$

$x = \pm\sqrt{2}$

x-intercepts: $(\sqrt{2}, 0)$ and $(-\sqrt{2}, 0)$

$x = 0$: $y = 2 - 0^2 = 2$

y-intercept: $(0, 2)$

$y = 2 - x^2$

$a = -1, b = 0, c = 2$

$$\frac{-b}{2a} = \frac{-0}{2(-1)} = \frac{0}{-2} = 0$$

$x = 0$: $y = 2 - 0^2 = 2$

vertex: $(0, 2)$

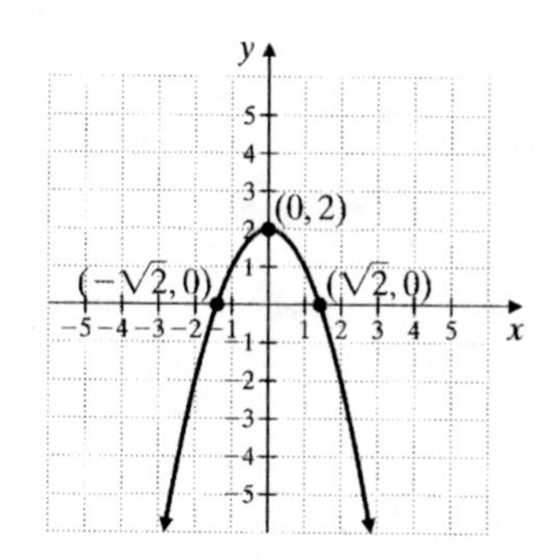

17. $y = \frac{1}{3}x^2$

$y = 0$: $0 = \frac{1}{3}x^2$
$0 = x^2$
$0 = x$

x-intercept: (0, 0)
The y-intercept is also (0, 0).

$y = \frac{1}{3}x^2 + 0x + 0$

$a = \frac{1}{3}$, $b = 0$, $c = 0$

$\frac{-b}{2a} = \frac{-0}{2\left(\frac{1}{3}\right)} = \frac{0}{\frac{2}{3}} = 0$

$x = 0$: $y = \frac{1}{3}(0)^2 = \frac{1}{3}(0) = 0$

vertex: (0, 0)
For additional points, use $x = \pm 3$.

$x = -3$: $y = \frac{1}{3}(-3)^2 = \frac{1}{3}(9) = 3$

$x = 3$: $y = \frac{1}{3}(3)^2 = \frac{1}{3}(9) = 3$

(−3, 3) and (3, 3) are also on the graph.

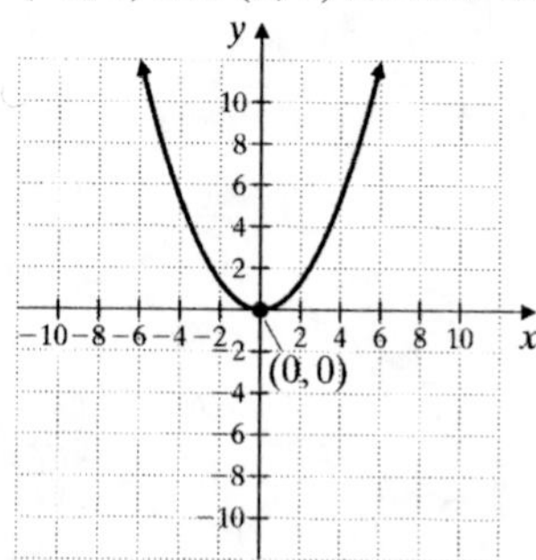

19. $y = x^2 + 6x$

$y = 0$: $0 = x^2 + 6x$
$0 = x(x + 6)$
$x = 0$ or $x + 6 = 0$
$x = -6$

x-intercepts: (0, 0) and (−6, 0)

$x = 0$: $y = 0^2 + 6(0) = 0$

y-intercept: (0, 0)

$y = x^2 + 6x + 0$

$a = 1$, $b = 6$, $c = 0$

$\frac{-b}{2a} = \frac{-6}{2(1)} = \frac{-6}{2} = -3$

$x = -3$: $y = (-3)^2 + 6(-3) = 9 - 18 = -9$

vertex: (−3, −9)

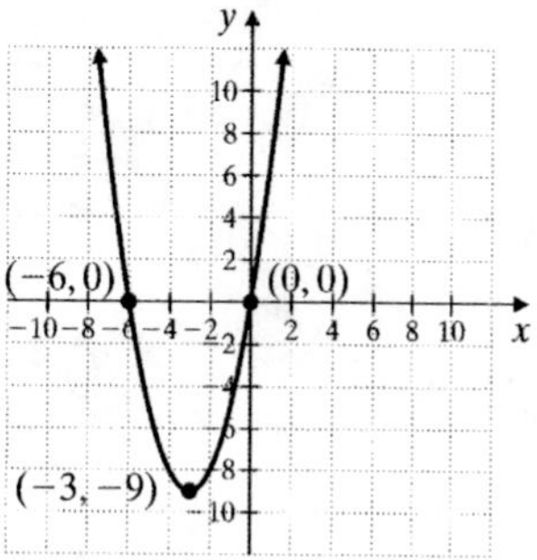

21. $y = x^2 + 2x - 8$

$y = 0$: $0 = x^2 + 2x - 8$
$0 = (x + 4)(x - 2)$
$0 = x + 4$ or $0 = x - 2$
$-4 = x$ $\quad 2 = x$

x-intercepts: (−4, 0), (2, 0)

$x = 0$: $y = 0^2 + 2(0) - 8 = -8$

y-intercept: (0, −8)

$y = x^2 + 2x - 8$

$a = 1$, $b = 2$, $c = -8$

$\frac{-b}{2a} = \frac{-2}{2(1)} = \frac{-2}{2} = -1$

$x = -1$: $y = (-1)^2 + 2(-1) - 8 = 1 - 2 - 8 = -9$

vertex: (−1, −9)

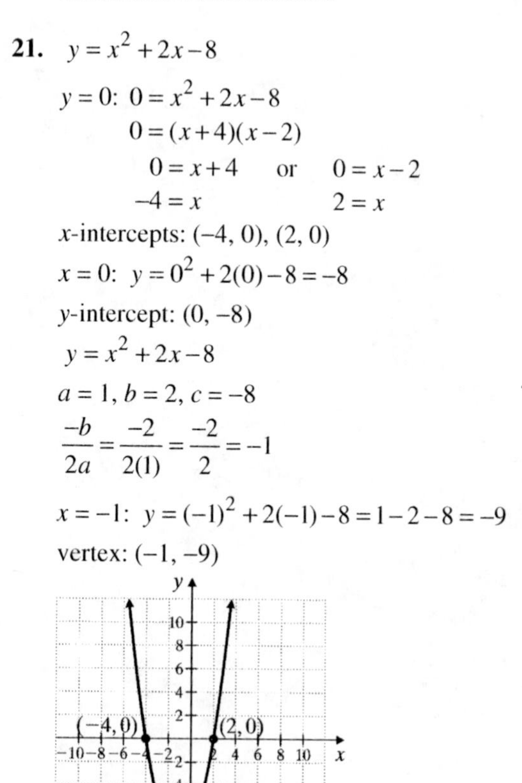

23. $y=-\frac{1}{2}x^2$

$y=0$: $0=-\frac{1}{2}x^2$

$0=x^2$

$0=x$

x-intercept: (0, 0)

The y-intercept is also (0, 0).

$y=-\frac{1}{2}x^2+0x+0$

$a=-\frac{1}{2},\ b=0,\ c=0$

$\frac{-b}{2a}=\frac{-0}{2\left(-\frac{1}{2}\right)}=\frac{0}{-1}=0$

$x=0$: $y=-\frac{1}{2}(0)^2=-\frac{1}{2}(0)=0$

vertex: (0, 0)

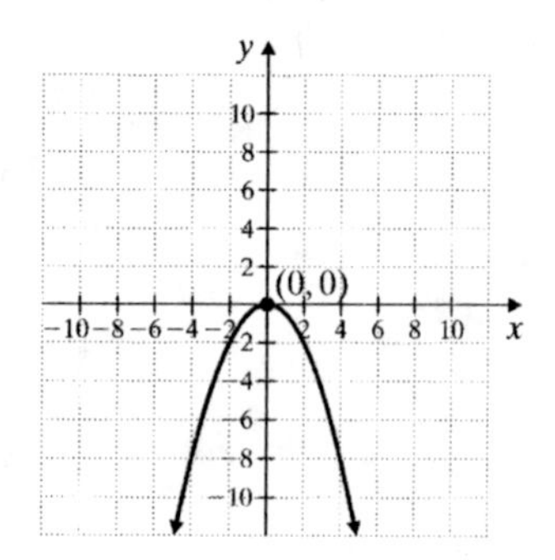

25. $y=2x^2-11x+5$

$y=0$: $0=2x^2-11x+5$

$0=(2x-1)(x-5)$

$0=2x-1$ or $0=x-5$

$1=2x$ $\quad 5=x$

$\frac{1}{2}=x$

x-intercepts: $\left(\frac{1}{2}, 0\right)$, (5, 0)

$x=0$: $y=2(0)^2-11(0)+5=5$

y-intercept: (0, 5)

$y=2x^2-11x+5$

$a=2,\ b=-11,\ c=5$

$\frac{-b}{2a}=\frac{-(-11)}{2(2)}=\frac{11}{4}$

$x=\frac{11}{4}$: $y=2\left(\frac{11}{4}\right)^2-11\left(\frac{11}{4}\right)+5$

$=2\left(\frac{121}{16}\right)-\frac{121}{4}+5$

$=\frac{121}{8}-\frac{121}{4}+5$

$=\frac{121}{8}-\frac{242}{8}+\frac{40}{8}$

$=-\frac{81}{8}$

vertex: $\left(\frac{11}{4}, -\frac{81}{8}\right)$

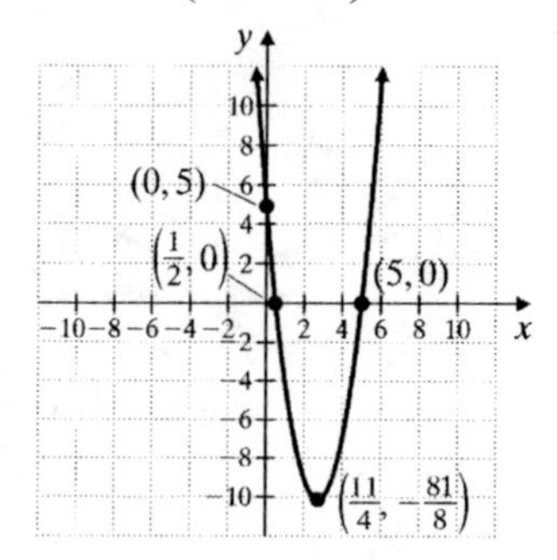

27. $y=-x^2+4x-3$

$y=0$: $0=-x^2+4x-3$

$0=x^2-4x+3$

$0=(x-3)(x-1)$

$x-3=0$ or $x-1=0$

$x=3$ $\quad x=1$

x-intercepts: (3, 0) and (1, 0)

$x=0$: $y=-0^2+4(0)-3=-3$

y-intercept: (0, −3)

$y=-x^2+4x-3$

$a=-1,\ b=4,\ c=-3$

$\frac{-b}{2a}=\frac{-4}{2(-1)}=\frac{-4}{-2}=2$

$x=2$: $y=-(2)^2+4(2)-3=-4+8-3=1$

vertex: (2, 1)

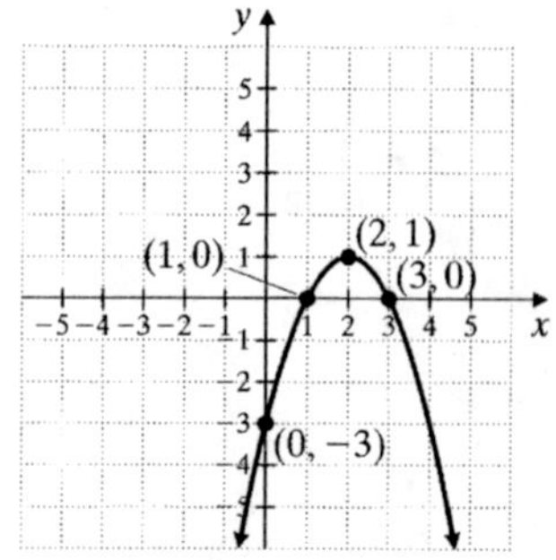

29. $\dfrac{\frac{1}{7}}{\frac{2}{5}} = \dfrac{1}{7} \div \dfrac{2}{5} = \dfrac{1}{7} \cdot \dfrac{5}{2} = \dfrac{5}{14}$

31. $\dfrac{\frac{1}{x}}{\frac{2}{x^2}} = \dfrac{1}{x} \div \dfrac{2}{x^2} = \dfrac{1}{x} \cdot \dfrac{x^2}{2} = \dfrac{x}{2}$

33. $\dfrac{2x}{1-\frac{1}{x}} = \dfrac{x(2x)}{x\left(1-\frac{1}{x}\right)} = \dfrac{2x^2}{x-1}$

35. $\dfrac{\frac{a-b}{2b}}{\frac{b-a}{8b^2}} = \dfrac{a-b}{2b} \div \dfrac{b-a}{8b^2} = \dfrac{a-b}{2b} \cdot \dfrac{8b^2}{b-a} = -4b$

37. $y = x^2 + 2x - 2$

$y = 0$: $0 = x^2 + 2x - 2$

$a = 1, b = 2, c = -2$

$$x = \frac{-b \pm \sqrt{b^2 - 4ac}}{2a}$$

$$x = \frac{-2 \pm \sqrt{2^2 - 4(1)(-2)}}{2(1)}$$

$$= \frac{-2 \pm \sqrt{4+8}}{2}$$

$$= \frac{-2 \pm \sqrt{12}}{2}$$

$$= \frac{-2 \pm 2\sqrt{3}}{2}$$

$$= -1 \pm \sqrt{3}$$

x-intercepts: $\left(-1-\sqrt{3}, 0\right) \approx (-2.7, 0)$,

$\left(-1+\sqrt{3}, 0\right) \approx (0.7, 0)$

$x = 0$: $y = 0^2 + 2(0) - 2 = -2$

y-intercept: $(0, -2)$

$\dfrac{-b}{2a} = \dfrac{-2}{2(1)} = \dfrac{-2}{2} = -1$

$x = -1$: $y = (-1)^2 + 2(-1) - 2 = 1 - 2 - 2 = -3$

vertex: $(-1, 3)$

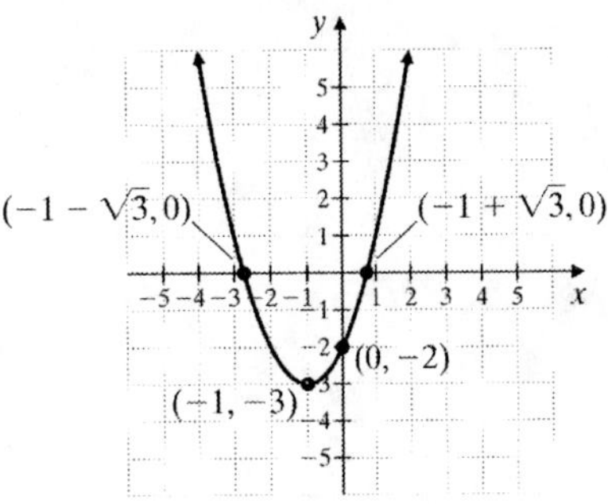

39. $y = x^2 - 3x + 1$

$y = 0$: $0 = x^2 - 3x + 1$

$a = 1, b = -3, c = 1$

$$x = \frac{-b \pm \sqrt{b^2 - 4ac}}{2a}$$

$$x = \frac{-(-3) \pm \sqrt{(-3)^2 - 4(1)(1)}}{2(1)}$$

$$= \frac{3 \pm \sqrt{9-4}}{2}$$

$$= \frac{3 \pm \sqrt{5}}{2}$$

x-intercepts: $\left(\dfrac{3-\sqrt{5}}{2}, 0\right) \approx (0.4, 0)$,

$\left(\dfrac{3+\sqrt{5}}{2}, 0\right) \approx (2.6, 0)$

$x = 0$: $y = 0^2 - 3(0) + 1 = 1$

y-intercept: $(0, 1)$

$\dfrac{-b}{2a} = \dfrac{-(-3)}{2(1)} = \dfrac{3}{2}$

$x = \dfrac{3}{2}$: $y = \left(\dfrac{3}{2}\right)^2 - 3\left(\dfrac{3}{2}\right) + 1$

$$= \frac{9}{4} - \frac{9}{2} + 1$$

$$= \frac{9}{4} - \frac{18}{4} + \frac{4}{4}$$

$$= -\frac{5}{4}$$

vertex: $\left(\dfrac{3}{2}, -\dfrac{5}{4}\right)$

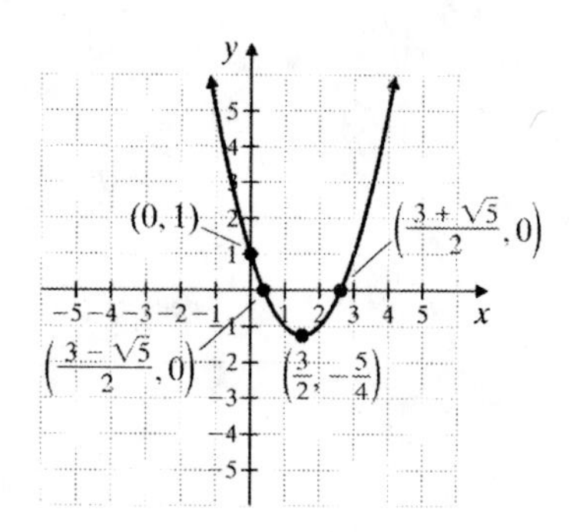

41. a. The maximum height appears to be about 256 feet.

b. The fireball appears to reach its maximum height when $t = 4$ seconds.

c. The fireball appears to return to the ground when $t = 8$ seconds.

43. With $a > 0$, the parabola opens upward. An upward-opening parabola that crosses the x-axis twice is graph **A**.

45. With $a < 0$, the parabola opens downward. A downward-opening parabola that does not touch or cross the x-axis (no x-intercept) is graph **D**.

47. With $a > 0$, the parabola opens upward. An upward-opening parabola that crosses the x-axis once is graph **F**.

Chapter 9 Vocabulary Check

1. If $x^2 = a$, then $x = \sqrt{a}$ or $x = -\sqrt{a}$. This property is called the square root property.

2. The graph of $y = x^2$ is called a parabola.

3. The formula $\frac{-b}{2a}$, where $y = ax^2 + bx + c$, is called the vertex formula.

4. The process of solving a quadratic equation by writing it in the form $(x+a)^2 = c$ is called completing the square.

5. The formula $x = \frac{-b \pm \sqrt{b^2 - 4ac}}{2a}$ is called the quadratic formula.

6. The lowest point on a parabola that opens upward is called the vertex.

7. The zero-factor property states that if the product of two numbers is zero, then at least one of the two numbers is zero.

Chapter 9 Review

1.
$$x^2 - 121 = 0$$
$$x^2 = 121$$
$$x = \pm\sqrt{121}$$
$$x = \pm 11$$
The solutions are $x = \pm 11$.

2.
$$y^2 - 100 = 0$$
$$y^2 = 100$$
$$y = \pm\sqrt{100}$$
$$y = \pm 10$$
The solutions are $y = \pm 10$.

3.
$$3m^2 - 5m = 2$$
$$3m^2 - 5m - 2 = 0$$
$$(3m+1)(m-2) = 0$$
$$3m+1 = 0 \quad \text{or} \quad m-2 = 0$$
$$3m = -1 \qquad m = 2$$
$$m = -\frac{1}{3}$$
The solutions are $m = -\frac{1}{3}$ and $m = 2$.

4.
$$7m^2 + 2m = 5$$
$$7m^2 + 2m - 5 = 0$$
$$(7m-5)(m+1) = 0$$
$$7m-5 = 0 \quad \text{or} \quad m+1 = 0$$
$$7m = 5 \qquad m = -1$$
$$m = \frac{5}{7}$$
The solutions are $m = \frac{5}{7}$ and $m = -1$.

5.
$$x^2 = 36$$
$$x = \pm\sqrt{36}$$
$$x = \pm 6$$
The solutions are $x = \pm 6$.

6.
$$x^2 = 81$$
$$x = \pm\sqrt{81}$$
$$x = \pm 9$$
The solutions are $x = \pm 9$.

7. $k^2 = 50$

$$k = \pm\sqrt{50}$$
$$k = \pm 5\sqrt{2}$$

The solutions are $k = \pm 5\sqrt{2}$.

8. $k^2 = 45$

$$k = \pm\sqrt{45}$$
$$k = \pm 3\sqrt{5}$$

The solutions are $k = \pm 3\sqrt{5}$.

9. $(x-11)^2 = 49$

$$x - 11 = \pm\sqrt{49}$$
$$x - 11 = \pm 7$$
$$x = 11 + 7 \quad \text{or} \quad x = 11 - 7$$
$$x = 18 \qquad x = 4$$

The solutions are $x = 18$ and $x = 4$.

10. $(x+3)^2 = 100$

$$x + 3 = \pm\sqrt{100}$$
$$x + 3 = \pm 10$$
$$x = -3 + 10 \quad \text{or} \quad x = -3 - 10$$
$$x = 7 \qquad x = -13$$

The solutions are $x = 7$ and $x = -13$.

11. $(4p+5)^2 = 41$

$$4p + 5 = \pm\sqrt{41}$$
$$4p = -5 \pm \sqrt{41}$$
$$p = \frac{-5 \pm \sqrt{41}}{4}$$

The solutions are $p = \frac{-5 \pm \sqrt{41}}{4}$.

12. $(3p+7)^2 = 37$

$$3p + 7 = \pm\sqrt{37}$$
$$3p = -7 \pm \sqrt{37}$$
$$p = \frac{-7 \pm \sqrt{37}}{3}$$

The solutions are $p = \frac{-7 \pm \sqrt{37}}{3}$.

13. $h = 16t^2$

$$100 = 16t^2$$
$$6.25 = t^2$$
$$\pm 2.5 = t$$

Reject -2.5 because time is not negative. It will take Kara 2.5 seconds to hit the water.

14. $5 \text{ miles} \cdot \frac{5280 \text{ ft}}{1 \text{ mile}} = 26,400 \text{ feet}$

$$y = 16t^2$$
$$26,400 = 16t^2$$
$$1650 = t^2$$
$$\pm 40.6 = t$$

Reject -40.6 because time is not negative.
A 5-mile freefall will take 40.6 seconds.

15. $x^2 - 9x = -8$

$$x^2 - 9x + \left(-\frac{9}{2}\right)^2 = -8 + \left(-\frac{9}{2}\right)^2$$
$$\left(x - \frac{9}{2}\right)^2 = -8 + \frac{81}{4}$$
$$\left(x - \frac{9}{2}\right)^2 = \frac{49}{4}$$
$$x - \frac{9}{2} = \pm\sqrt{\frac{49}{4}}$$
$$x - \frac{9}{2} = \pm\frac{7}{2}$$
$$x = \frac{9}{2} + \frac{7}{2} = \frac{16}{2} = 8 \quad \text{or} \quad x = \frac{9}{2} - \frac{7}{2} = \frac{2}{2} = 1$$

The solutions are $x = 8$ and $x = 1$.

16. $x^2 + 8x = 20$

$$x^2 + 8x + \left(\frac{8}{2}\right)^2 = 20 + \left(\frac{8}{2}\right)^2$$
$$(x+4)^2 = 36$$
$$x + 4 = \pm\sqrt{36}$$
$$x + 4 = \pm 6$$
$$x = -4 + 6 = 2 \quad \text{or} \quad x = -4 - 6 = -10$$

The solutions are $x = 7$ and $x = -10$.

17. $x^2 + 4x = 1$

$$x^2 + 4x + \left(\frac{4}{2}\right)^2 = 1 + \left(\frac{4}{2}\right)^2$$
$$(x+2)^2 = 5$$
$$x + 2 = \pm\sqrt{5}$$
$$x = -2 \pm \sqrt{5}$$

The solutions are $x = -2 \pm \sqrt{5}$.

18.
$$x^2-8x=3$$
$$x^2-8x+\left(-\frac{8}{2}\right)^2=3+\left(-\frac{8}{2}\right)^2$$
$$(x-4)^2=19$$
$$x-4=\pm\sqrt{19}$$
$$x=4\pm\sqrt{19}$$
The solutions are $x=4\pm\sqrt{19}$.

19.
$$x^2-6x+7=0$$
$$x^2-6x=-7$$
$$x^2-6x+\left(-\frac{6}{2}\right)^2=-7+\left(-\frac{6}{2}\right)^2$$
$$(x-3)^2=2$$
$$x-3=\pm\sqrt{2}$$
$$x=3\pm\sqrt{2}$$
The solutions are $3\pm\sqrt{2}$.

20.
$$x^2+6x+7=0$$
$$x^2+6x=-7$$
$$x^2+6x+\left(\frac{6}{2}\right)^2=-7+\left(\frac{6}{2}\right)^2$$
$$(x+3)^2=2$$
$$x+3=\pm\sqrt{2}$$
$$x=-3\pm\sqrt{2}$$
The solutions are $-3\pm\sqrt{2}$.

21.
$$2y^2+y-1=0$$
$$2y^2+y=1$$
$$y^2+\frac{1}{2}y=\frac{1}{2}$$
$$y^2+\frac{1}{2}y+\left(\frac{1}{4}\right)^2=\frac{1}{2}+\left(\frac{1}{4}\right)^2$$
$$\left(y+\frac{1}{4}\right)^2=\frac{9}{16}$$
$$y+\frac{1}{4}=\pm\sqrt{\frac{9}{16}}$$
$$y+\frac{1}{4}=\pm\frac{3}{4}$$
$$y=-\frac{1}{4}+\frac{3}{4}=\frac{2}{4}=\frac{1}{2} \text{ or}$$
$$y=-\frac{1}{4}-\frac{3}{4}=-\frac{4}{4}=-1$$
The solutions are $y=\frac{1}{2}$ and $y=-1$.

22.
$$4y^2+3y-1=0$$
$$4y^2+3y=1$$
$$y^2+\frac{3}{4}y=\frac{1}{4}$$
$$y^2+\frac{3}{4}y+\left(\frac{3}{8}\right)^2=\frac{1}{4}+\left(\frac{3}{8}\right)^2$$
$$\left(y+\frac{3}{8}\right)^2=\frac{25}{64}$$
$$y+\frac{3}{8}=\pm\sqrt{\frac{25}{64}}$$
$$y+\frac{3}{8}=\pm\frac{5}{8}$$
$$y=-\frac{3}{8}+\frac{5}{8}=\frac{2}{8}=\frac{1}{4} \text{ or } y=-\frac{3}{8}-\frac{5}{8}=-\frac{8}{8}=-1$$
The solutions are $y=\frac{1}{4}$ and $y=-1$.

23. $9x^2+30x+25=0$
$a=9$, $b=30$, $c=25$
$$x=\frac{-b\pm\sqrt{b^2-4ac}}{2a}$$
$$x=\frac{-30\pm\sqrt{30^2-4(9)(25)}}{2(9)}$$
$$=\frac{-30\pm\sqrt{900-900}}{18}$$
$$=\frac{-30\pm\sqrt{0}}{18}$$
$$=\frac{-30}{18}$$
$$=-\frac{5}{3}$$
The solution is $x=-\frac{5}{3}$.

24. $16x^2-72x+81=0$
$a=16$, $b=-72$, $c=81$
$$x=\frac{-b\pm\sqrt{b^2-4ac}}{2a}$$

$$x = \frac{-(-72) \pm \sqrt{(-72)^2 - 4(16)(81)}}{2(16)}$$
$$= \frac{72 \pm \sqrt{5184 - 5184}}{32}$$
$$= \frac{72 \pm \sqrt{0}}{32}$$
$$= \frac{72}{32}$$
$$= \frac{9}{4}$$

The solution is $x = \frac{9}{4}$.

25. $7x^2 = 35$
$7x^2 - 35 = 0$
$a = 7, b = 0, c = -35$

$$x = \frac{-b \pm \sqrt{b^2 - 4ac}}{2a}$$
$$x = \frac{-0 \pm \sqrt{0^2 - 4(7)(-35)}}{2(7)}$$
$$= \frac{\pm\sqrt{980}}{14}$$
$$= \frac{\pm 14\sqrt{5}}{14}$$
$$= \pm\sqrt{5}$$

The solutions are $x = \pm\sqrt{5}$.

26. $11x^2 = 33$
$11x^2 - 33 = 0$
$a = 11, b = 0, c = -33$

$$x = \frac{-b \pm \sqrt{b^2 - 4ac}}{2a}$$
$$x = \frac{-0 \pm \sqrt{0^2 - 4(11)(-33)}}{2(11)}$$
$$= \frac{\pm\sqrt{1452}}{22}$$
$$= \frac{\pm 22\sqrt{3}}{22}$$
$$= \pm\sqrt{3}$$

The solutions are $x = \pm\sqrt{3}$.

27. $x^2 - 10x + 7 = 0$
$a = 1, b = -10, c = 7$

$$x = \frac{-b \pm \sqrt{b^2 - 4ac}}{2a}$$
$$x = \frac{-(-10) \pm \sqrt{(-10)^2 - 4(1)(7)}}{2(1)}$$
$$= \frac{10 \pm \sqrt{100 - 28}}{2}$$
$$= \frac{10 \pm \sqrt{72}}{2}$$
$$= \frac{10 \pm 6\sqrt{2}}{2}$$
$$= 5 \pm 3\sqrt{2}$$

The solutions are $x = 5 \pm 3\sqrt{2}$.

28. $x^2 + 4x - 7 = 0$
$a = 1, b = 4, c = -7$

$$x = \frac{-b \pm \sqrt{b^2 - 4ac}}{2a}$$
$$x = \frac{-4 \pm \sqrt{4^2 - 4(1)(-7)}}{2(1)}$$
$$= \frac{-4 \pm \sqrt{16 + 28}}{2}$$
$$= \frac{-4 \pm \sqrt{44}}{2}$$
$$= \frac{-4 \pm 2\sqrt{11}}{2}$$
$$= -2 \pm \sqrt{11}$$

The solutions are $x = -2 \pm \sqrt{11}$.

29. $3x^2 + x - 1 = 0$
$a = 3, b = 1, c = -1$

$$x = \frac{-b \pm \sqrt{b^2 - 4ac}}{2a}$$
$$x = \frac{-1 \pm \sqrt{(1)^2 - 4(3)(-1)}}{2(3)}$$
$$= \frac{-1 \pm \sqrt{1 + 12}}{6}$$
$$= \frac{-1 \pm \sqrt{13}}{6}$$

The solutions are $x = \frac{-1 \pm \sqrt{13}}{6}$.

30. $x^2+3x-1=0$
$a=1, b=3, c=-1$
$$x=\frac{-b\pm\sqrt{b^2-4ac}}{2a}$$
$$x=\frac{-3\pm\sqrt{3^2-4(1)(-1)}}{2(1)}=\frac{-3\pm\sqrt{9+4}}{2}=\frac{-3\pm\sqrt{13}}{2}$$
The solutions are $x=\frac{-3\pm\sqrt{13}}{2}$.

31. $2x^2+x+5=0$
$a=2, b=1, c=5$
$$x=\frac{-b\pm\sqrt{b^2-4ac}}{2a}$$
$$x=\frac{-1\pm\sqrt{1^2-4(2)(5)}}{2(2)}=\frac{-1\pm\sqrt{1-40}}{4}=\frac{-1\pm\sqrt{-39}}{4}$$
The equation has no solution, since the square root of -39 is not a real number.

32. $7x^2-3x+1=0$
$a=7, b=-3, c=1$
$$x=\frac{-b\pm\sqrt{b^2-4ac}}{2a}$$
$$x=\frac{-(-3)\pm\sqrt{(-3)^2-4(7)(1)}}{2(7)}=\frac{3\pm\sqrt{9-28}}{14}=\frac{3\pm\sqrt{-19}}{14}$$
The equation has no solution since the square root of -19 is not a real number.

33. $x=\frac{-1+\sqrt{13}}{6}\approx 0.4$ or $x=\frac{-1-\sqrt{13}}{6}\approx -0.8$

34. $x=\frac{-3+\sqrt{13}}{2}\approx 0.3$ or $x=\frac{-3-\sqrt{13}}{2}\approx -3.3$

35.
$$y=-66x^2+368x+3432$$
$$3264=-66x^2+368x+3432$$
$$0=-66x^2+368x+168$$
$$0=33x^2-184x-84$$
$a=33, b=-184, c=-84$
$$x=\frac{-b\pm\sqrt{b^2-4ac}}{2a}$$
$$x=\frac{-(-184)\pm\sqrt{(-184)^2-4(33)(-84)}}{2(33)}=\frac{184\pm\sqrt{33,856+11,088}}{66}=\frac{184\pm\sqrt{44,944}}{66}=\frac{184\pm 212}{66}$$
$$x=\frac{184+212}{66}=\frac{396}{66}=6 \text{ or}$$
$$x=\frac{184-212}{66}=\frac{-28}{66}=-\frac{14}{33}$$
Since time will not be a negative number in this context, discard the solution $-\frac{14}{33}$. The solution is $x=6$, so the model predicts that the number of visitors to Yosemite National Park will be 3264 thousand in 2014 (2008 + 6).

36.
$$y=538x^2+19,421x+54,762$$
$$302,772=538x^2+19,421x+54,762$$
$$0=538x^2+19,421x-248,010$$
$a=538, b=19,421, c=-248,010$
$$x=\frac{-b\pm\sqrt{b^2-4ac}}{2a}$$
$$x=\frac{-19,421\pm\sqrt{19,421^2-4(538)(-248,010)}}{2(538)}=\frac{-19,421\pm\sqrt{377,175,241+533,717,520}}{1076}=\frac{-19,421\pm\sqrt{910,892,761}}{1076}=\frac{-19,421\pm 30,181}{1076}$$
$$x=\frac{-19,421+30,181}{1076}=\frac{10,760}{1076}=10 \text{ or}$$
$$x=\frac{-19,421-30,181}{1076}=\frac{-49,602}{1076}=-\frac{24,801}{538}$$

Since time will not be a negative number in this context, discard the solution $-\frac{24,801}{538}$. The solution is $x = 10$, so the model predicts that the amount of electricity generated by wind power will be 302,772 thousand megawatt hours in 2018 (2008 + 10).

37. $y = 5x^2$

$y = 0$: $0 = 5x^2$
$0 = x^2$
$0 = x$

x-intercept: (0, 0)
The y-intercept is also (0, 0).

x	$y = 5x^2$
−2	$5(-2)^2 = 5(4) = 20$
−1	$5(-1)^2 = 5(1) = 5$
0	$5(0)^2 = 5(0) = 0$
1	$5(1)^2 = 5(1) = 5$
2	$5(2)^2 = 5(4) = 20$

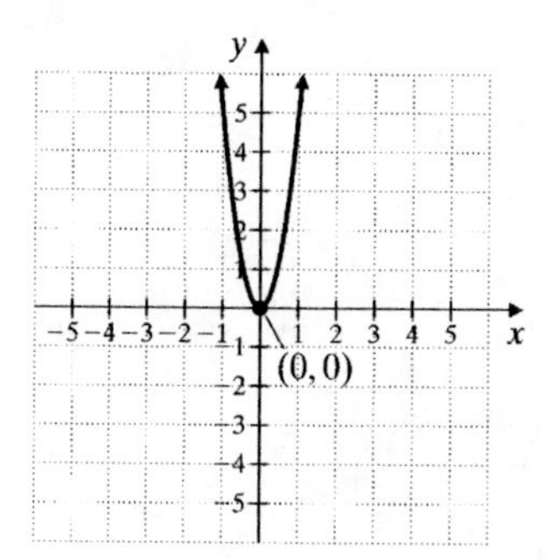

38. $y = -\frac{1}{2}x^2$

$y = 0$: $0 = -\frac{1}{2}x^2$
$0 = x^2$
$0 = x$

x-intercept: (0, 0)
The y-intercept is also (0, 0).

x	$y = -\frac{1}{2}x^2$
−2	$-\frac{1}{2}(-2)^2 = -\frac{1}{2}(4) = -2$
−1	$-\frac{1}{2}(-1)^2 = -\frac{1}{2}(1) = -\frac{1}{2}$
0	$-\frac{1}{2}(0)^2 = -\frac{1}{2}(0) = 0$
1	$-\frac{1}{2}(1)^2 = -\frac{1}{2}(1) = -\frac{1}{2}$
2	$-\frac{1}{2}(2)^2 = -\frac{1}{2}(4) = -2$

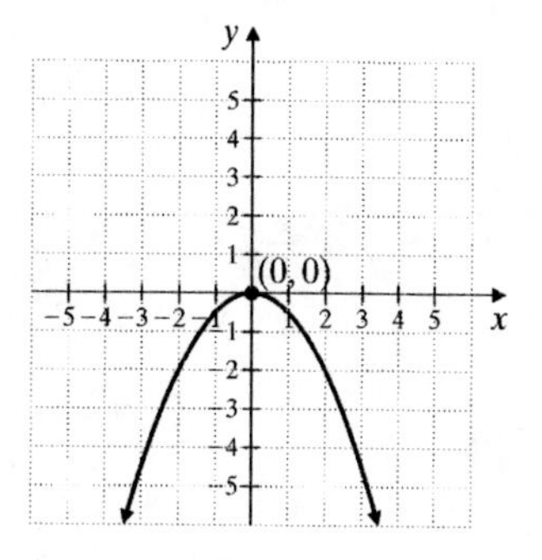

39. $y = x^2 - 25$

$y = 0$: $0 = x^2 - 25$
$25 = x^2$
$\pm\sqrt{25} = x$
$\pm 5 = x$

x-intercepts: (5, 0) and (−5, 0)

$x = 0$: $y = 0^2 - 25 = -25$

y-intercept: (0, −25)

$y = x^2 + 0x - 25$

$a = 1$, $b = 0$, $c = -25$

$\frac{-b}{2a} = \frac{-0}{2(1)} = \frac{0}{2} = 0$

$x = 0$: $y = 0^2 - 25 = -25$

vertex: (0, −25)

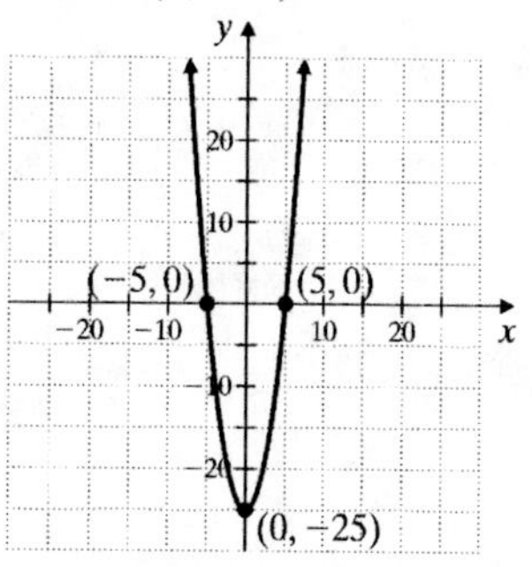

40. $y = x^2 - 36$

$y = 0$: $0 = x^2 - 36$

$36 = x^2$

$\pm\sqrt{36} = x$

$\pm 6 = x$

x-intercepts: (6, 0) and (−6, 0)

$x = 0$: $y = 0^2 - 36 = -36$

y-intercept: (0, −36)

$y = x^2 + 0x - 36$

$a = 1, b = 0, c = -36$

$$\frac{-b}{2a} = \frac{-0}{2(1)} = \frac{0}{2} = 0$$

$x = 0$: $y = 0^2 - 36 = -36$

vertex: (0, −36)

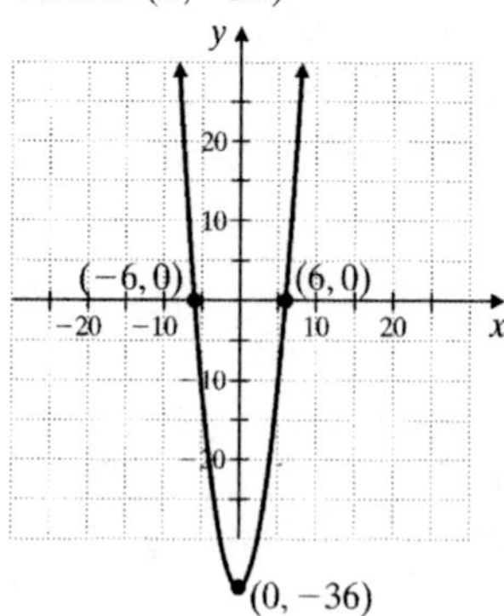

41. $y = x^2 + 3$

$y = 0$: $0 = x^2 + 3$

$-3 = x^2$

$\pm\sqrt{-3} = x$

x-intercepts: none

$x = 0$: $y = 0^2 + 3 = 3$

y-intercept: (0, 3)

$y = x^2 + 0x + 3$

$a = 1, b = 0, c = 3$

$$\frac{-b}{2a} = \frac{-0}{2(1)} = \frac{0}{2} = 0$$

$x = 0$: $y = 0^2 + 3 = 3$

vertex: (0, 3)

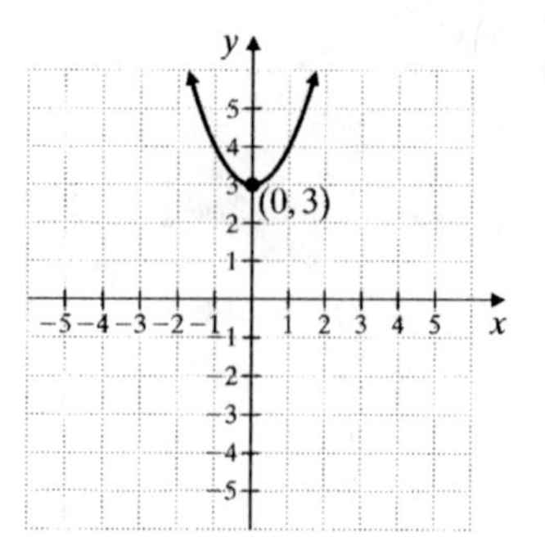

42. $y = x^2 + 8$

$y = 0$: $0 = x^2 + 8$

$-8 = x^2$

$\pm\sqrt{-8} = x$

x-intercepts: none

$x = 0$: $y = 0^2 + 8 = 8$

y-intercept: (0, 8)

$y = x^2 + 0x + 8$

$a = 1, b = 0, c = 8$

$$\frac{-b}{2a} = \frac{0}{2(1)} = \frac{0}{2} = 0$$

$x = 0$: $y = 0^2 + 8 = 8$

vertex: (0, 8)

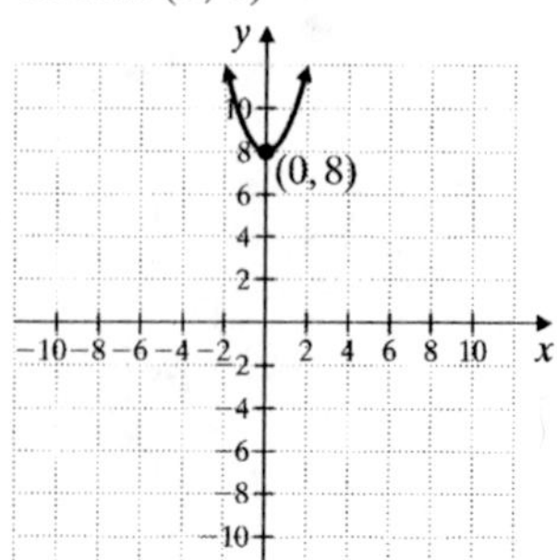

43. $y = -4x^2 + 8$

$y = 0$: $0 = -4x^2 + 8$

$0 = -4(x^2 - 2)$

$0 = x^2 - 2$

$2 = x^2$

$\pm\sqrt{2} = x$

x-intercepts: $\left(\sqrt{2}, 0\right)$ and $\left(-\sqrt{2}, 0\right)$

$x = 0$: $y = -4(0)^2 + 8 = 8$

y-intercept: (0, 8)

$y = -4x^2 + 0x + 8$

$a = -4, b = 0, c = 8$

$$\frac{-b}{2a} = \frac{-0}{2(-4)} = \frac{0}{-8} = 0$$

$x = 0$: $y = -4(0)^2 + 8 = 8$

vertex: (0, 8)

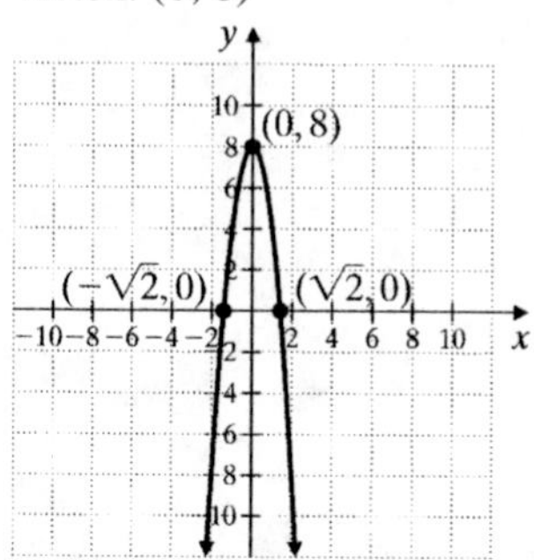

44. $y = -3x^2 + 9$

$y = 0$: $0 = -3x^2 + 9$

$0 = -3(x^2 - 3)$

$0 = x^2 - 3$

$3 = x^2$

$\pm\sqrt{3} = x$

x-intercepts: $\left(\sqrt{3}, 0\right)$ and $\left(-\sqrt{3}, 0\right)$

$x = 0$: $y = -3(0)^2 + 9 = 9$

y-intercept: (0, 9)

$y = -3x^2 + 0x + 9$

$a = -3$, $b = 0$, $c = 8$

$$\frac{-b}{2a} = \frac{-0}{2(-3)} = \frac{0}{-6} = 0$$

$x = 0$: $y = -3(0)^2 + 9 = 9$

vertex: (0, 9)

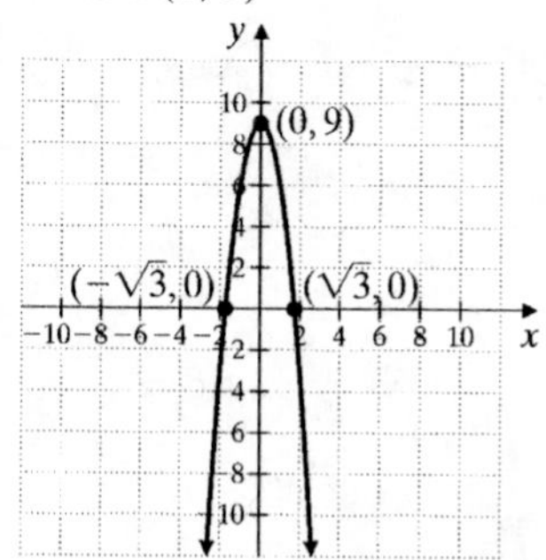

45. $y = x^2 + 3x - 10$

$y = 0$: $0 = x^2 + 3x - 10$

$0 = (x + 5)(x - 2)$

$0 = x + 5$ or $0 = x - 2$

$-5 = x$ $\quad 2 = x$

x-intercepts: (−5, 0) and (2, 0)

$x = 0$: $y = 0^2 + 3(0) - 10 = -10$

y-intercept: (0, −10)

$y = x^2 + 3x - 10$

$a = 1$, $b = 3$, $c = -10$

$$\frac{-b}{2a} = \frac{-3}{2(1)} = -\frac{3}{2}$$

$$x = -\frac{3}{2}:\ y = \left(-\frac{3}{2}\right)^2 + 3\left(-\frac{3}{2}\right) - 10$$

$$= \frac{9}{4} - \frac{9}{2} - 10$$

$$= \frac{9}{4} - \frac{18}{4} - \frac{40}{4}$$

$$= -\frac{49}{4}$$

vertex: $\left(-\frac{3}{2}, -\frac{49}{4}\right)$

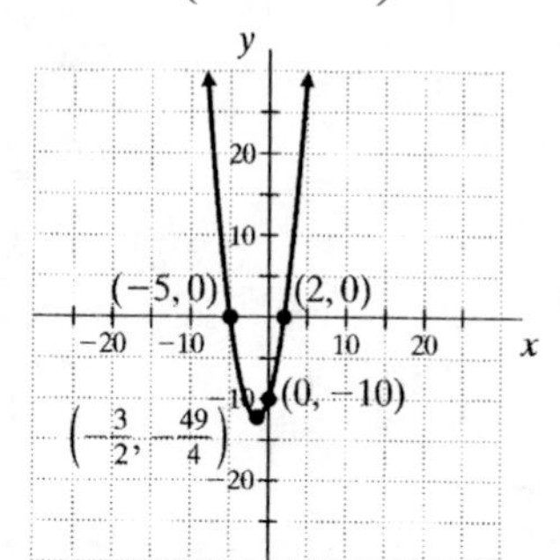

46. $y = x^2 + 3x - 4$

$y = 0$: $0 = x^2 + 3x - 4$

$0 = (x + 4)(x - 1)$

$0 = x + 4$ or $0 = x - 1$

$-4 = x$ $\quad 1 = x$

x-intercepts: (−4, 0) and (1, 0)

$x = 0$: $y = 0^2 + 3(0) - 4 = -4$

y-intercept: (0, −4)

$y = x^2 + 3x - 4$

$a = 1$, $b = 3$, $c = -4$

$$\frac{-b}{2a} = \frac{-3}{2(1)} = \frac{-3}{2}$$

$x = -\frac{3}{2}$: $y = \left(-\frac{3}{2}\right)^2 + 3\left(-\frac{3}{2}\right) - 4$

$= \frac{9}{4} - \frac{9}{2} - 4$

$= \frac{9}{4} - \frac{18}{4} - \frac{16}{4}$

$= -\frac{25}{4}$

vertex: $\left(-\frac{3}{2}, -\frac{25}{4}\right)$

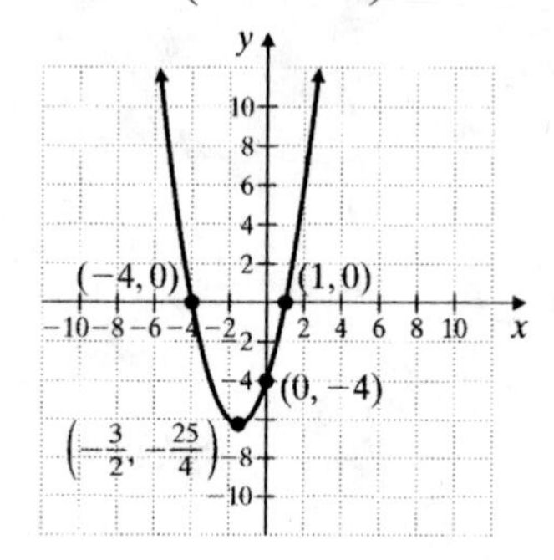

47. $y = -x^2 - 5x - 6$

$y = 0$: $0 = -x^2 - 5x - 6$

$0 = x^2 + 5x + 6$

$0 = (x+3)(x+2)$

$0 = x + 3$ or $0 = x + 2$

$-3 = x$ $\quad -2 = x$

x-intercepts: (–3, 0) and (–2, 0)

$x = 0$: $y = -0^2 - 5(0) - 6 = -6$

y-intercept: (0, –6)

$y = -x^2 - 5x - 6$

$a = -1, b = -5, c = -6$

$\frac{-b}{2a} = \frac{-(-5)}{2(-1)} = \frac{5}{-2} = -\frac{5}{2}$

$x = -\frac{5}{2}$: $y = -\left(-\frac{5}{2}\right)^2 - 5\left(-\frac{5}{2}\right) - 6$

$y = -\frac{25}{4} + \frac{25}{2} - 6$

$y = -\frac{25}{4} + \frac{50}{4} - \frac{24}{4}$

$y = \frac{1}{4}$

vertex: $\left(-\frac{5}{2}, \frac{1}{4}\right)$

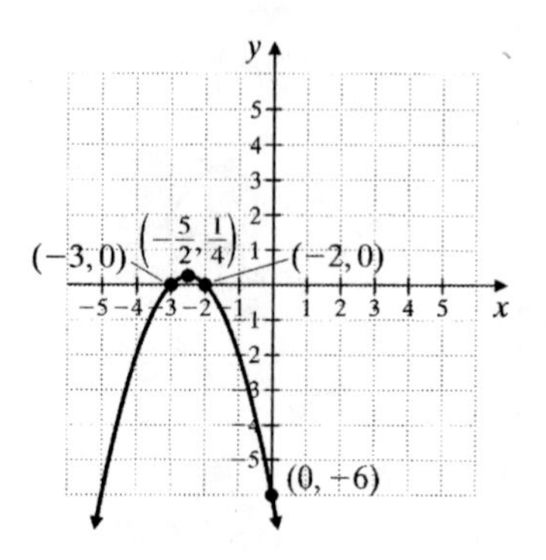

48. $y = 3x^2 - x - 2$

$y = 0$: $0 = 3x^2 - x - 2$

$0 = (3x+2)(x-1)$

$0 = 3x + 2$ or $0 = x - 1$

$-2 = 3x$ $\quad 1 = x$

$-\frac{2}{3} = x$

x-intercepts: $\left(-\frac{2}{3}, 0\right)$ and (1, 0)

$x = 0$: $y = 3(0)^2 - 0 - 2 = -2$

y-intercept: (0, –2)

$y = 3x^2 - x - 2$

$a = 3, b = -1, c = -2$

$\frac{-b}{2a} = \frac{-(-1)}{2(3)} = \frac{1}{6}$

$x = \frac{1}{6}$: $y = 3\left(\frac{1}{6}\right)^2 - \frac{1}{6} - 2$

$y = \frac{3}{36} - \frac{1}{6} - 2$

$y = \frac{3}{36} - \frac{6}{36} - \frac{72}{36}$

$y = -\frac{75}{36} = -\frac{25}{12}$

vertex: $\left(\frac{1}{6}, -\frac{25}{12}\right)$

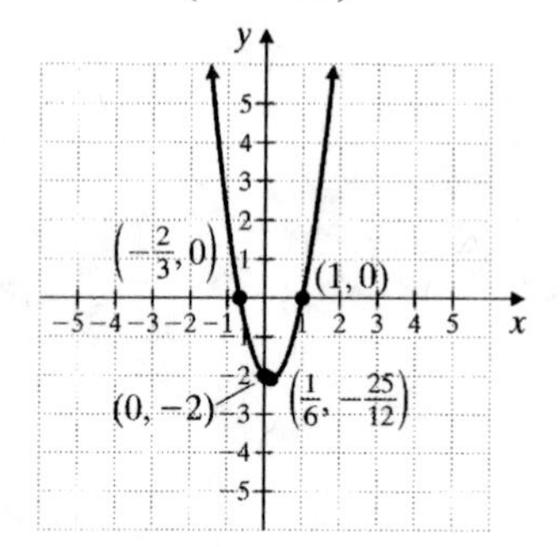

49. $y = 2x^2 - 11x - 6$

$y = 0$: $0 = 2x^2 - 11x - 6$
$0 = (2x+1)(x-6)$
$0 = 2x+1$ or $0 = x-6$
$-1 = 2x$ $\quad 6 = x$
$-\frac{1}{2} = x$

x-intercepts: $\left(-\frac{1}{2}, 0\right)$ and $(6, 0)$

$x = 0$: $y = 2(0)^2 - 11(0) - 6 = -6$

y-intercept: $(0, -6)$

$y = 2x^2 - 11x - 6$

$a = 2, b = -11, c = -6$

$\frac{-b}{2a} = \frac{-(-11)}{2(2)} = \frac{11}{4}$

$x = \frac{11}{4}$: $y = 2\left(\frac{11}{4}\right)^2 - 11\left(\frac{11}{4}\right) - 6$

$y = \frac{242}{16} - \frac{121}{4} - 6$

$y = \frac{242}{16} - \frac{484}{16} - \frac{96}{16}$

$y = -\frac{338}{16} = -\frac{169}{8}$

vertex: $\left(\frac{11}{4}, -\frac{169}{8}\right)$

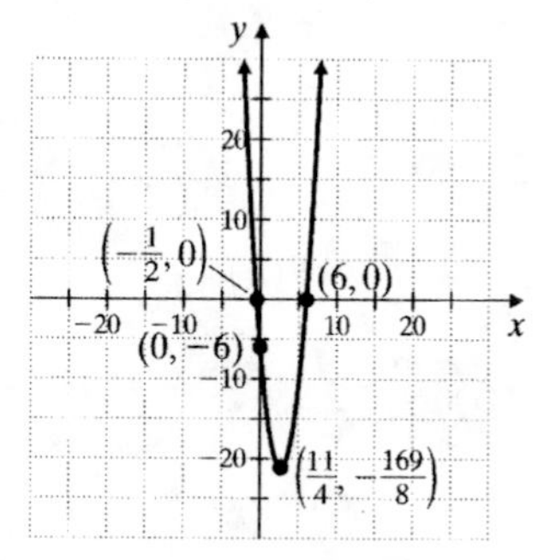

50. $y = -x^2 + 4x + 8$

$y = 0$: $0 = -x^2 + 4x + 8$

$a = -1, b = 4, c = 8$

$x = \frac{-b \pm \sqrt{b^2 - 4ac}}{2a}$

$x = \frac{-4 \pm \sqrt{4^2 - 4(-1)(8)}}{2(-1)}$

$= \frac{-4 \pm \sqrt{16+32}}{2}$

$= \frac{-4 \pm \sqrt{48}}{-2}$

$= \frac{-4 \pm 4\sqrt{3}}{-2}$

$= 2 \pm 2\sqrt{3}$

x-intercepts: $\left(2 + 2\sqrt{3}, 0\right)$ and $\left(2 - 2\sqrt{3}, 0\right)$

$x = 0$: $y = -0^2 + 4(0) + 8 = 8$

y-intercept: $(0, 8)$

$y = -x^2 + 4x + 8$

$a = -1, b = 4, c = 8$

$\frac{-b}{2a} = \frac{-4}{2(-1)} = \frac{-4}{-2} = 2$

$x = 2$: $y = -(2)^2 + 4(2) + 8 = -4 + 8 + 8 = 12$

vertex: $(2, 12)$

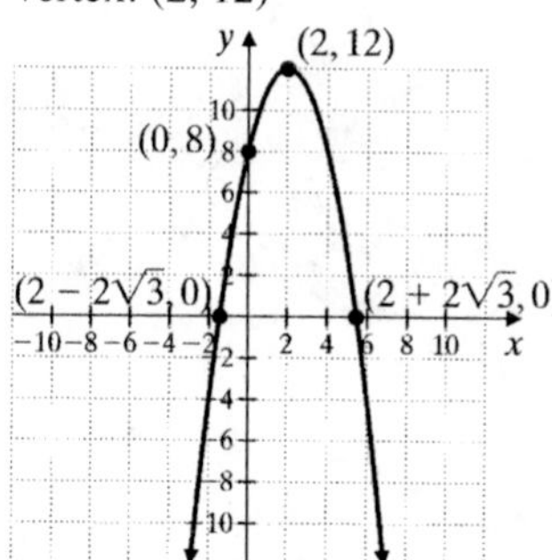

51. $y = 2x^2$ matches with graph A.

52. $y = -x^2$ matches with graph D.

53. $y = x^2 + 4x + 4$ matches with graph B.

54. $y = x^2 + 5x + 4$ matches with graph C.

55. The graph crosses the x-axis once so there is one real solution.

56. The graph crosses the x-axis twice so there are two real solutions.

57. The graph doesn't cross the x-axis so there are no real solutions.

58. The graph crosses the x-axis twice so there are two real solutions.

59. $x^2 = 49$

$x = \pm\sqrt{49}$

$x = \pm 7$

The solutions are $x = \pm 7$.

60. $y^2 = 75$

$y = \pm\sqrt{75}$

$y = \pm 5\sqrt{3}$

The solutions are $y = \pm 5\sqrt{3}$.

61. $(x-7)^2 = 64$

$x - 7 = \pm\sqrt{64}$

$x - 7 = \pm 8$

$x - 7 = 8$ or $x - 7 = -8$

$x = 15$ $x = -1$

The solutions are $x = 15$ and $x = -1$.

62.

$$x^2 + 4x = 6$$

$$x^2 + 4x + \left(\frac{4}{2}\right)^2 = 6 + \left(\frac{4}{2}\right)^2$$

$$(x+2)^2 = 10$$

$$x + 2 = \pm\sqrt{10}$$

$$x = -2 \pm \sqrt{10}$$

The solutions are $x = -2 \pm \sqrt{10}$.

63.

$$3x^2 + x = 2$$

$$x^2 + \frac{1}{3}x = \frac{2}{3}$$

$$x^2 + \frac{1}{3}x + \left(\frac{1}{6}\right)^2 = \frac{2}{3} + \left(\frac{1}{6}\right)^2$$

$$\left(x + \frac{1}{6}\right)^2 = \frac{25}{36}$$

$$x + \frac{1}{6} = \pm\sqrt{\frac{25}{36}}$$

$$x + \frac{1}{6} = \pm\frac{5}{6}$$

$$x = -\frac{1}{6} \pm \frac{5}{6}$$

$$x = -\frac{1}{6} + \frac{5}{6} = \frac{4}{6} = \frac{2}{3} \quad \text{or} \quad x = -\frac{1}{6} - \frac{5}{6} = -\frac{6}{6} = -1$$

The solutions are $x = \frac{2}{3}$ and $x = -1$.

64.

$$4x^2 - x - 2 = 0$$

$$x^2 - \frac{1}{4}x - \frac{1}{2} = 0$$

$$x^2 - \frac{1}{4}x = \frac{1}{2}$$

$$x^2 - \frac{1}{4}x + \left(-\frac{1}{8}\right)^2 = \frac{1}{2} + \left(-\frac{1}{8}\right)^2$$

$$\left(x - \frac{1}{8}\right)^2 = \frac{33}{64}$$

$$x - \frac{1}{8} = \pm\sqrt{\frac{33}{64}}$$

$$x - \frac{1}{8} = \pm\frac{\sqrt{33}}{8}$$

$$x = \frac{1}{8} \pm \frac{\sqrt{33}}{8}$$

$$x = \frac{1 \pm \sqrt{33}}{8}$$

The solutions are $x = \frac{1 \pm \sqrt{33}}{8}$.

65. $4x^2 - 3x - 2 = 0$

$a = 4$, $b = -3$, $c = -2$

$$x = \frac{-b \pm \sqrt{b^2 - 4ac}}{2a}$$

$$x = \frac{-(-3) \pm \sqrt{(-3)^2 - 4(4)(-2)}}{2(4)}$$

$$= \frac{3 \pm \sqrt{9 + 32}}{8}$$

$$= \frac{3 \pm \sqrt{41}}{8}$$

The solutions are $x = \frac{3 \pm \sqrt{41}}{8}$.

66. $5x^2 + x - 2 = 0$

$a = 5$, $b = 1$, $c = -2$

$$x = \frac{-b \pm \sqrt{b^2 - 4ac}}{2a}$$

$$x = \frac{-1 \pm \sqrt{1^2 - 4(5)(-2)}}{2(5)}$$
$$= \frac{-1 \pm \sqrt{1+40}}{10}$$
$$= \frac{-1 \pm \sqrt{41}}{10}$$

The solutions are $x = \frac{-1 \pm \sqrt{41}}{10}$.

67. $4x^2 + 12x + 9 = 0$
$a = 4, b = 12, c = 9$

$$x = \frac{-b \pm \sqrt{b^2 - 4ac}}{2a}$$
$$x = \frac{-12 \pm \sqrt{12^2 - 4(4)(9)}}{2(4)}$$
$$= \frac{-12 \pm \sqrt{144 - 144}}{8}$$
$$= \frac{-12 \pm \sqrt{0}}{8}$$
$$= -\frac{12}{8}$$
$$= -\frac{3}{2}$$

The solution is $x = -\frac{3}{2}$.

68. $2x^2 + x + 4 = 0$
$a = 2, b = 1, c = 4$

$$x = \frac{-b \pm \sqrt{b^2 - 4ac}}{2a}$$
$$x = \frac{-1 \pm \sqrt{1^2 - 4(2)(4)}}{2(2)}$$
$$= \frac{-1 \pm \sqrt{1-32}}{4}$$
$$= \frac{-1 \pm \sqrt{-31}}{4}$$

The equation has no solution, since the square root of −31 is not a real number.

69. $y = 4 - x^2$

$y = 0$: $0 = 4 - x^2$
$x^2 = 4$
$x = \pm\sqrt{4}$
$x = \pm 2$

x-intercepts: (2, 0) and (−2, 0)
$x = 0$: $y = 4 - 0^2 = 4$
y-intercept: (0, 4)
$y = 4 + 0x - x^2$
$a = -1, b = 0, c = 4$

$$\frac{-b}{2a} = \frac{-0}{2(-1)} = \frac{0}{-2} = 0$$

$x = 0$: $y = 4 - 0^2 = 4$
vertex: (0, 4)

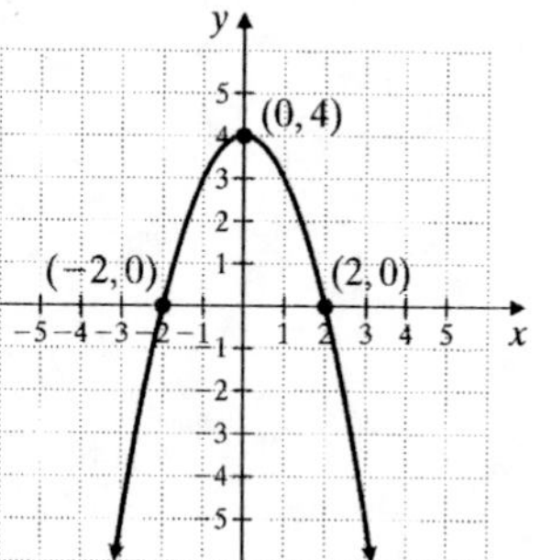

70. $y = x^2 + 4$

$y = 0$: $0 = x^2 + 4$
$-4 = x^2$
$\pm\sqrt{-4} = x$
x-intercepts: none
$x = 0$: $y = 0^2 + 4 = 4$
y-intercept: (0, 4)
$y = x^2 + 0x + 4$
$a = 1, b = 0, c = 4$

$$\frac{-b}{2a} = \frac{-0}{2(1)} = \frac{0}{2} = 0$$

$x = 0$: $y = 0^2 + 4 = 4$
vertex: (0, 4)

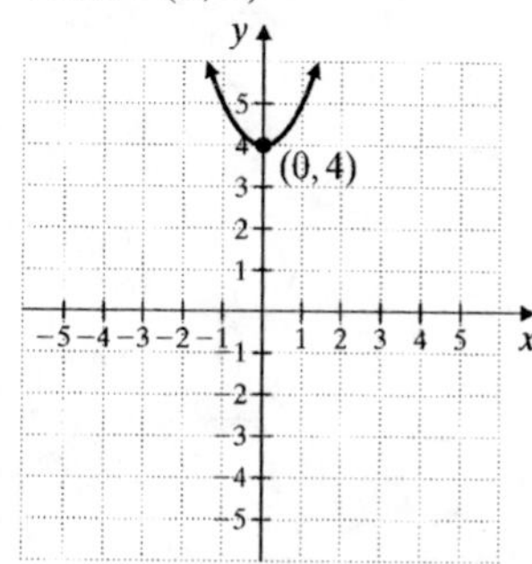

71. $y = x^2 + 6x + 8$

$y = 0$: $0 = x^2 + 6x + 8$

$0 = (x+4)(x+2)$

$x + 4 = 0$ or $x + 2 = 0$

$x = -4$ $\quad$ $x = -2$

x-intercepts: $(-4, 0)$ and $(-2, 0)$

$x = 0$: $y = 0^2 + 6(0) + 8 = 8$

y-intercept: $(0, 8)$

$y = x^2 + 6x + 8$

$a = 1, b = 6, c = 8$

$$\frac{-b}{2a} = \frac{-6}{2(1)} = \frac{-6}{2} = -3$$

$x = -3$: $y = (-3)^2 + 6(-3) + 8 = 9 - 18 + 8 = -1$

vertex: $(-3, -1)$

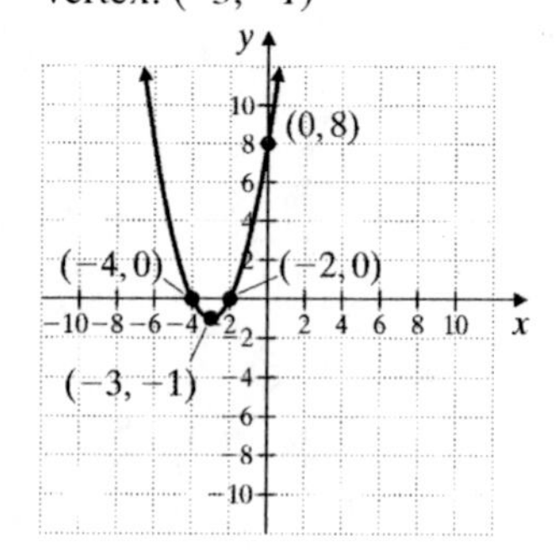

72. $y = x^2 - 2x - 4$

$y = 0$: $0 = x^2 - 2x - 4$

$a = 1, b = -2, c = -4$

$$x = \frac{-b \pm \sqrt{b^2 - 4ac}}{2a}$$

$$x = \frac{-(-2) \pm \sqrt{(-2)^2 - 4(1)(-4)}}{2(1)}$$

$$= \frac{2 \pm \sqrt{4 + 16}}{2}$$

$$= \frac{2 \pm \sqrt{20}}{2}$$

$$= \frac{2 \pm 2\sqrt{5}}{2}$$

$$= 1 \pm \sqrt{5}$$

x-intercepts: $\left(1 + \sqrt{5}, 0\right)$ and $\left(1 - \sqrt{5}, 0\right)$

$x = 0$: $y = 0^2 - 2(0) - 4 = -4$

x-intercept: $(0, -4)$

$y = x^2 - 2x - 4$

$a = 1, b = -2, c = -4$

$$\frac{-b}{2a} = \frac{-(-2)}{2(1)} = \frac{2}{2} = 1$$

$x = 1$: $y = 1^2 - 2(1) - 4 = 1 - 2 - 4 = -5$

vertex: $(1, -5)$

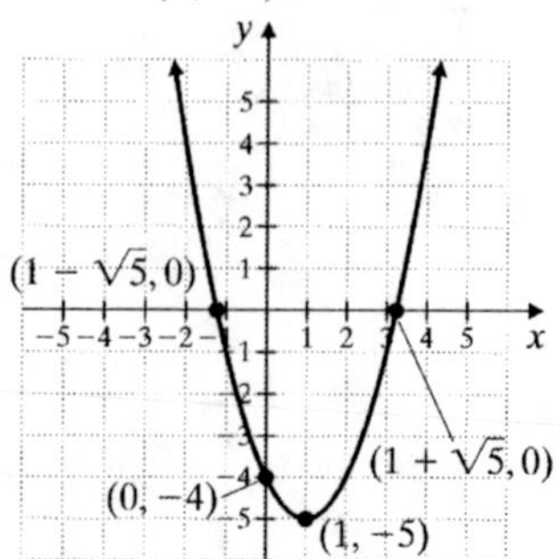

Chapter 9 Test

1. $x^2 - 400 = 0$

$(x + 20)(x - 20) = 0$

$x + 20 = 0$ or $x - 20 = 0$

$x = -20$ $\quad$ $x = 20$

The solutions are $x = \pm 20$.

2. $2x^2 - 11x = 21$

$2x^2 - 11x - 21 = 0$

$(2x + 3)(x - 7) = 0$

$2x + 3 = 0$ or $x - 7 = 0$

$2x = -3$ $\quad$ $x = 7$

$x = -\frac{3}{2}$

The solutions are $x = -\frac{3}{2}$ and $x = 7$.

3. $5k^2 = 80$

$k^2 = 16$

$k = \pm\sqrt{16}$

$k = \pm 4$

The solutions are $k = \pm 4$.

4. $(3m - 5)^2 = 8$

$3m - 5 = \pm\sqrt{8}$

$3m - 5 = \pm 2\sqrt{2}$

$3m = 5 \pm 2\sqrt{2}$

$$m = \frac{5 \pm 2\sqrt{2}}{3}$$

The solutions are $m = \frac{5 \pm 2\sqrt{2}}{3}$.

5.
$$x^2-26x+160=0$$
$$x^2-26x=-160$$
$$x^2-26x+\left(\frac{-26}{2}\right)^2=-160+\left(\frac{-26}{2}\right)^2$$
$$x^2-26x+(-13)^2=-160+(-13)^2$$
$$x^2-26x+169=-160+169$$
$$(x-13)^2=9$$
$$x-13=\sqrt{9} \text{ or } x-13=-\sqrt{9}$$
$$x-13=3 \qquad x-13=-3$$
$$x=16 \qquad x=10$$
The solutions are $x = 10$ and $x = 16$.

6.
$$3x^2+12x-4=0$$
$$x^2+4x-\frac{4}{3}=0$$
$$x^2+4x=\frac{4}{3}$$
$$x^2+4x+\left(\frac{4}{2}\right)^2=\frac{4}{3}+\left(\frac{4}{2}\right)^2$$
$$x^2+4x+(2)^2=\frac{4}{3}+2^2$$
$$(x+2)^2=\frac{16}{3}$$
$$x+2=\pm\sqrt{\frac{16}{3}}$$
$$x+2=\pm\sqrt{\frac{16}{3}}\cdot\sqrt{\frac{3}{3}}$$
$$x+2=\pm\frac{4\sqrt{3}}{3}$$
$$x=-2\pm\frac{4\sqrt{3}}{3}$$
The solutions are $x=-2\pm\frac{4\sqrt{3}}{3}$.

7. $x^2-3x-10=0$
$a = 1, b = -3, c = -10$
$$x=\frac{-b\pm\sqrt{b^2-4ac}}{2a}$$
$$x=\frac{-(-3)\pm\sqrt{(-3)^2-4(1)(-10)}}{2(1)}$$
$$=\frac{3\pm\sqrt{9+40}}{2}$$
$$=\frac{3\pm\sqrt{49}}{2}$$
$$=\frac{3\pm 7}{2}$$
$$x=\frac{3+7}{2}=\frac{10}{2}=5 \text{ or } x=\frac{3-7}{2}=\frac{-4}{2}=-2$$
The solutions are $x = 5$ and $x = -2$.

8. $p^2-\frac{5}{3}p-\frac{1}{3}=0$
$a=1,\ b=-\frac{5}{3},\ c=-\frac{1}{3}$
$$p=\frac{-b\pm\sqrt{b^2-4ac}}{2a}$$
$$p=\frac{-\left(-\frac{5}{3}\right)\pm\sqrt{\left(-\frac{5}{3}\right)^2-4(1)\left(-\frac{1}{3}\right)}}{2(1)}$$
$$=\frac{\frac{5}{3}\pm\sqrt{\frac{25}{9}+\frac{4}{3}}}{2}$$
$$=\frac{\frac{5}{3}\pm\sqrt{\frac{25}{9}+\frac{12}{9}}}{2}$$
$$=\frac{\frac{5}{3}\pm\sqrt{\frac{37}{9}}}{2}$$
$$=\frac{\frac{5}{3}\pm\frac{\sqrt{37}}{3}}{2}$$
$$=\frac{5\pm\sqrt{37}}{6}$$
The solutions are $p=\frac{5\pm\sqrt{37}}{6}$.

9.
$$(3x-5)(x+2)=-6$$
$$3x^2+6x-5x-10=-6$$
$$3x^2+x-10=-6$$
$$3x^2+x-4=0$$
$a = 3, b = 1, c = -4$
$$x=\frac{-b\pm\sqrt{b^2-4ac}}{2a}$$

$$x=\frac{-1\pm\sqrt{1^2-4(3)(-4)}}{2(3)}$$
$$=\frac{-1\pm\sqrt{1+48}}{6}$$
$$=\frac{-1\pm\sqrt{49}}{6}$$
$$=\frac{-1\pm 7}{6}$$
$$x=\frac{-1+7}{6}=1 \text{ or } x=\frac{-1-7}{6}=-\frac{4}{3}$$

The solutions are $x = 1$ and $x=-\frac{4}{3}$.

10. $(3x-1)^2=16$
$$3x-1=\pm\sqrt{16}$$
$$3x-1=\pm 4$$
$$3x=1\pm 4$$
$$x=\frac{1\pm 4}{3}$$
$$x=\frac{1+4}{3}=\frac{5}{3} \quad \text{or} \quad x=\frac{1-4}{3}=\frac{-3}{3}=-1$$

The solutions are $x=\frac{5}{3}$ and $x=-1$.

11. $3x^2-7x-2=0$

$a = 3$, $b = -7$, $c = -2$
$$x=\frac{-b\pm\sqrt{b^2-4ac}}{2a}$$
$$x=\frac{-(-7)\pm\sqrt{(-7)^2-4(3)(-2)}}{2(3)}$$
$$=\frac{7\pm\sqrt{49+24}}{6}$$
$$=\frac{7\pm\sqrt{73}}{6}$$

The solutions are $x=\frac{7\pm\sqrt{73}}{6}$.

12. $x^2-4x-5=0$
$$(x-5)(x+1)=0$$
$$x-5=0 \quad \text{or} \quad x+1=0$$
$$x=5 \qquad x=-1$$

The solutions are $x = 5$ and $x = -1$.

13. $3x^2-7x+2=0$

$a = 3$, $b = -7$, $c = 2$
$$x=\frac{-b\pm\sqrt{b^2-4ac}}{2a}$$
$$x=\frac{-(-7)\pm\sqrt{(-7)^2-4(3)(2)}}{2(3)}$$
$$=\frac{7\pm\sqrt{49-24}}{6}$$
$$=\frac{7\pm\sqrt{25}}{6}$$
$$=\frac{7\pm 5}{6}$$
$$x=\frac{7+5}{6}=2 \text{ or } x=\frac{7-5}{6}=\frac{1}{3}$$

The solutions are $x=\frac{1}{3}$ and $x = 2$.

14. $2x^2-6x+1=0$

$a = 2$, $b = -6$, $c = 1$
$$x=\frac{-b\pm\sqrt{b^2-4ac}}{2a}$$
$$x=\frac{-(-6)\pm\sqrt{(-6)^2-4(2)(1)}}{2(2)}$$
$$=\frac{6\pm\sqrt{36-8}}{4}$$
$$=\frac{6\pm\sqrt{28}}{4}$$
$$=\frac{6\pm 2\sqrt{7}}{4}$$
$$=\frac{3\pm\sqrt{7}}{2}$$

The solutions are $x=\frac{3\pm\sqrt{7}}{2}$.

15. Let x = the length of the base.

$4x$ = height
$$A=\frac{1}{2}bh$$
$$18=\frac{1}{2}(x)(4x)$$
$$36=4x^2$$
$$9=x^2$$
$$\pm\sqrt{9}=x$$
$$\pm 3=x$$

$4x = 4(3) = 12$

The base is 3 feet. The height is 12 feet.

16. $y=-5x^2$

$y=0$: $0=-5x^2$
$0=x^2$
$0=x$

x-intercept: (0, 0)
The y-intercept is also (0, 0).

$y=-5x^2+0x+0$

$a=-5$, $b=0$, $c=0$

$$\frac{-b}{2a}=\frac{-0}{2(-5)}=\frac{0}{-10}=0$$

$x=0$: $y=-5(0)^2=0$

vertex: (0, 0)

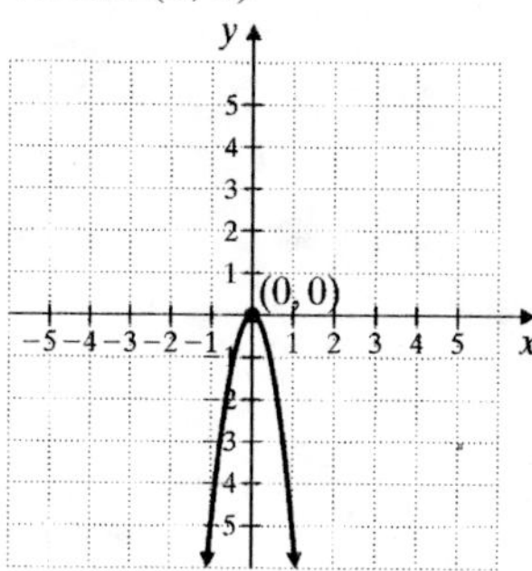

17. $y=x^2-4$

$y=0$: $0=x^2-4$
$0=(x+2)(x-2)$
$0=x+2$ or $0=x-2$
$-2=x$ $\quad 2=x$

x-intercepts: (−2, 0), (2, 0)

$x=0$: $y=0^2-4=-4$

y-intercept: (0, −4)

$y=x^2+0x-4$

$a=1$, $b=0$, $c=-4$

$$\frac{-b}{2a}=\frac{-0}{2(1)}=\frac{0}{2}=0$$

$x=0$: $y=0^2-4=-4$

vertex: (0, −4)

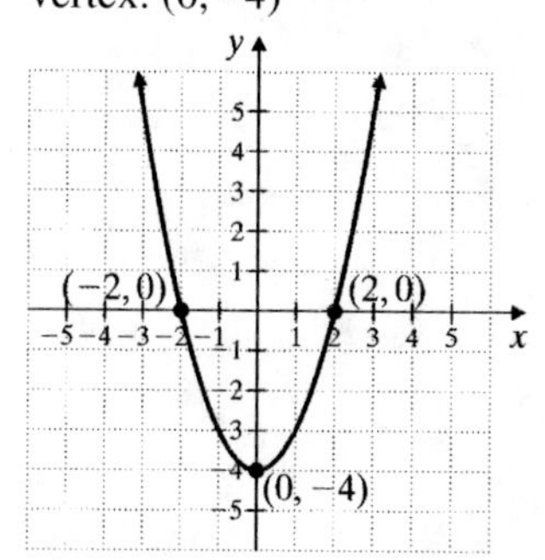

18. $y=x^2-7x+10$

$y=0$: $0=x^2-7x+10$
$0=(x-5)(x-2)$
$x-5=0$ or $x-2=0$
$x=5$ $\quad x=2$

x-intercepts: (5, 0) and (2, 0)

$x=0$: $y=0^2-7(0)+10=10$

y-intercept: (0, 10)

$y=x^2-7x+10$

$a=1$, $b=-7$, $c=10$

$$\frac{-b}{2a}=\frac{-(-7)}{2(1)}=\frac{7}{2}$$

$$x=\frac{7}{2}:\ y=\left(\frac{7}{2}\right)^2-7\left(\frac{7}{2}\right)+10=\frac{49}{4}-\frac{49}{2}+10=\frac{49}{4}-\frac{98}{4}+\frac{40}{4}=-\frac{9}{4}$$

vertex: $\left(\frac{7}{2}, -\frac{9}{4}\right)$

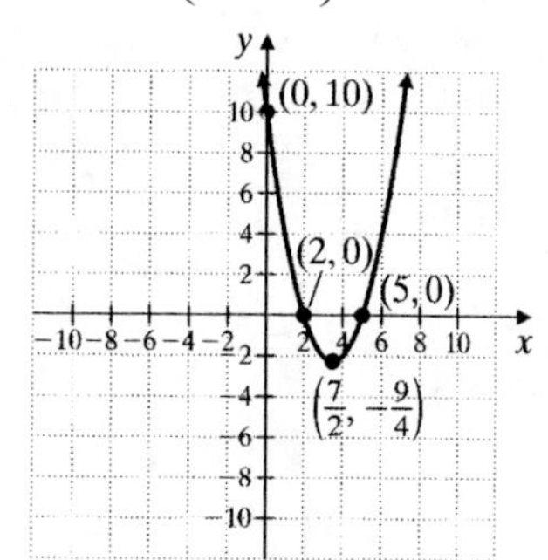

19. $y=2x^2+4x-1$

$y=0$: $0=2x^2+4x-1$

$a=2$, $b=4$, $c=-1$

$$x=\frac{-b\pm\sqrt{b^2-4ac}}{2a}$$

$$x = \frac{-4 \pm \sqrt{4^2 - 4(2)(-1)}}{2(2)}$$
$$= \frac{-4 \pm \sqrt{16+8}}{4}$$
$$= \frac{-4 \pm \sqrt{24}}{4}$$
$$= \frac{-4 \pm 2\sqrt{6}}{4}$$
$$= \frac{-2 \pm \sqrt{6}}{2}$$

x-intercepts: $\left(\frac{-2-\sqrt{6}}{2}, 0\right)$ and $\left(\frac{-2+\sqrt{6}}{2}, 0\right)$

$x = 0$: $y = 2(0)^2 + 4(0) - 1 = -1$

y-intercept: $(0, -1)$

$y = 2x^2 + 4x - 1$

$a = 2, b = 4, c = -1$

$\frac{-b}{2a} = \frac{-4}{2(2)} = \frac{-4}{4} = -1$

$x = -1$: $y = 2(-1)^2 + 4(-1) - 1 = 2 - 4 - 1 = -3$

vertex: $(-1, -3)$

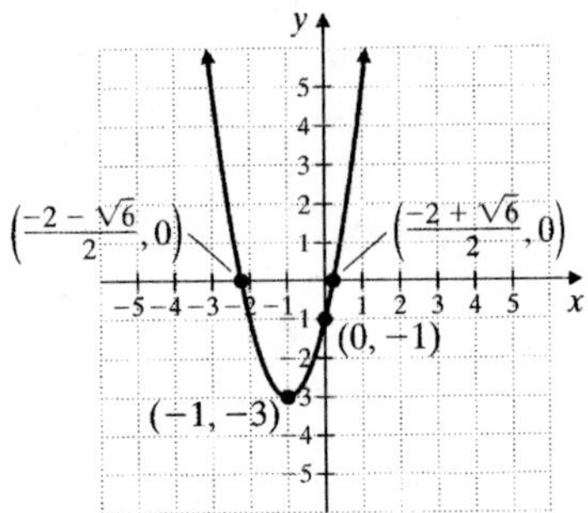

20. $d = \frac{n^2 - 3n}{2}$

$9 = \frac{n^2 - 3n}{2}$

$18 = n^2 - 3n$

$0 = n^2 - 3n - 18$

$0 = (n-6)(n+3)$

$n - 6 = 0$ or $n + 3 = 0$

$n = 6$ $\quad$ $n = -3$

A 6-sided polygon has 9 diagonals.

21. $h = 16t^2$

$120.75 = 16t^2$

$\frac{120.75}{16} = t^2$

$7.546875 = t^2$

$\sqrt{7.546875} = t$ or $-\sqrt{7.546875} = t$

$2.7 \approx t$ $\quad$ $-2.7 \approx t$

Since the time of the dive is not a negative number, discard the solution $t \approx -2.7$. The dive took approximately 2.7 seconds.

Cumulative Review Chapters 1–9

1. $y + 0.6 = -1.0$

$y = -1.6$

2. $8x - 14 = 6x - 20$

$2x = -6$

$x = -3$

3. $8(2 - t) = -5t$

$16 - 8t = -5t$

$16 = 3t$

$\frac{16}{3} = t$

4. $2(x + 7) = 5(2x - 3)$

$2x + 14 = 10x - 15$

$29 = 8x$

$\frac{29}{8} = x$

5. Let x be the first number and y be the second number. Then $x + y = 37$ and $x - y = 21$.

$$\begin{cases} x + y = 37 \\ x - y = 21 \end{cases}$$

Add the equations.

$$\begin{array}{r} x + y = 37 \\ x - y = 21 \\ \hline 2x \quad = 58 \end{array}$$

$\frac{2x}{2} = \frac{58}{2}$

$x = 29$

$x + y = 37$

$29 + y = 37$

$y = 8$

The numbers are 29 and 8.

6. Let n be the first integer. Then $n + 1$ and $n + 2$ are the next two consecutive integers.
$$\begin{aligned} n+(n+1)+(n+2) &= 438 \\ 3n+3 &= 438 \\ 3n &= 435 \\ n &= 145 \end{aligned}$$
$n + 1 = 146$
$n + 2 = 147$
The three consecutive integers are 145, 146, and 147.

7. $3^0 = 1$

8. $\left(\frac{-6x}{y^3}\right)^3 = \frac{(-6x)^3}{(y^3)^3} = \frac{(-6)^3x^3}{y^{3\cdot3}} = -\frac{216x^3}{y^9}$

9. $(5x^3y^2)^0 = 1$

10. $\frac{a^2b^7}{(2b^2)^5} = \frac{a^2b^7}{32b^{10}} = \frac{a^2}{32b^3}$

11. $-4^0 = -1$

12. $\frac{(3y)^2}{y^2} = \frac{9y^2}{y^2} = 9$

13.
$$\begin{aligned} (3y+2)^2 &= (3y)^2 + 2(2)(3y) + 2^2 \\ &= 9y^2 + 12y + 4 \end{aligned}$$

14. $(x^2+5)(y-1) = x^2y - x^2 + 5y - 5$

15. $\frac{x^2+7x+12}{x+3} = x+4$

$$\begin{array}{r} x+4 \\ x+3\overline{\big)\,x^2+7x+12} \\ \underline{x^2+3x} \\ 4x+12 \\ \underline{4x+12} \\ 0 \end{array}$$

16.
$$\begin{aligned} 2+8.1a+a-6 &= 8.1a+a+2-6 \\ &= (8.1+1)+(2-6) \\ &= 9.1a-4 \end{aligned}$$

17. $r^2 - r - 42 = (r-7)(r+6)$

18. a. $\frac{x-y}{7-x} = \frac{-4-7}{7-(-4)} = \frac{-11}{11} = -1$

b. $x^2 + 2y = (-4)^2 + 2(7) = 16 + 14 = 30$

19. $10x^2 - 13xy - 3y^2 = (2x-3y)(5x+y)$

20.
$$\begin{aligned} \frac{1}{x+2} + \frac{7}{x-1} &= \frac{1(x-1)}{(x+2)(x-1)} + \frac{7(x+2)}{(x-1)(x+2)} \\ &= \frac{x-1+7x+14}{(x+2)(x-1)} \\ &= \frac{8x+13}{(x+2)(x-1)} \end{aligned}$$

21.
$$\begin{aligned} 8x^2 - 14x + 5 &= 8x^2 - 4x - 10x + 5 \\ &= 4x(2x-1) - 5(2x-1) \\ &= (2x-1)(4x-5) \end{aligned}$$

22.
$$\begin{aligned} \frac{x^2+7x}{5x} \cdot \frac{10x+25}{x^2-49} &= \frac{x(x+7)}{5x} \cdot \frac{5(2x+5)}{(x-7)(x+7)} \\ &= \frac{2x+5}{x-7} \end{aligned}$$

23. a. $4x^3 - 49x = x(4x^2 - 49) = x(2x-7)(2x+7)$

b.
$$\begin{aligned} 162x^4 - 2 &= 2(81x^4 - 1) \\ &= 2(9x^2-1)(9x^2+1) \\ &= 2(9x^2+1)(3x-1)(3x+1) \end{aligned}$$

24.
$$\begin{aligned} \frac{2x+7}{3} &= \frac{x-6}{2} \\ 6\left(\frac{2x+7}{3}\right) &= 6\left(\frac{x-6}{2}\right) \\ 2(2x+7) &= 3(x-6) \\ 4x+14 &= 3x-18 \\ x &= -32 \end{aligned}$$

25.
$$\begin{aligned} (5x-1)(2x^2+15x+18) &= 0 \\ (5x-1)(2x+3)(x+6) &= 0 \end{aligned}$$
$$\begin{array}{lllll} 5x-1=0 & \text{or} & 2x+3=0 & \text{or} & x+6=0 \\ 5x=1 & & 2x=-3 & & x=-6 \\ x=\frac{1}{5} & & x=-\frac{3}{2} & & \end{array}$$

The solutions are $x = \frac{1}{5}$, $x = -\frac{3}{2}$, and $x = -6$.

26. a. $4x - 3 + 7 - 5x = -x + 4$

b. $-6y + 3y - 8 + 8y = 5y - 8$

c. $7+10.1a-a-11 = 10.1a-a+7-11$
$= 9.1a-4$

d. $2x^2-2x = 2x^2-2x$

27. $\dfrac{x^2+8x+7}{x^2-4x-5} = \dfrac{(x+7)(x+1)}{(x-5)(x+1)} = \dfrac{x+7}{x-5}$

28.
$$2x^2+5x=7$$
$$2x^2+5x-7=0$$
$$(2x+7)(x-1)=0$$
$$2x+7=0 \quad \text{or} \quad x-1=0$$
$$2x=-7 \qquad x=1$$
$$x=-\frac{7}{2}$$
The solutions are $x=-\dfrac{7}{2}$ and $x=1$.

29.
$$\frac{n}{6}-\frac{5}{3}=\frac{n}{2}$$
$$6\left(\frac{n}{6}-\frac{5}{3}\right)=6\left(\frac{n}{2}\right)$$
$$n-10=3n$$
$$-10=2n$$
$$-5=n$$
The number is –5.

30.
$$d=\sqrt{(x_2-x_1)^2+(y_2-y_1)^2}$$
$$d=\sqrt{(2-(-7))^2+(5-4)^2}$$
$$d=\sqrt{(9)^2+(1)^2}$$
$$d=\sqrt{81+1}$$
$$d=\sqrt{82}$$
The distance is $\sqrt{82}$ units.

31.

	x	$y=3x$
a.	-1	-3
b.	0	0
c.	-3	-9

32. a. x-intercept: (4, 0)
y-intercept: (0, 1)

b. x-intercepts: (–2, 0), (0, 0), (3, 0)
y-intercept: (0, 0)

33. a.
$$y=-\frac{1}{5}x+1$$
$$2x+10y=3$$
$$10y=-2x+3$$
$$y=\frac{-2}{10}x+\frac{3}{10}$$
$$y=-\frac{1}{5}x+\frac{3}{10}$$
These two equations have the same slope, therefore they are parallel.

b. $x+y=3 \Rightarrow y=-x+3$
$-x+y=4 \Rightarrow y=x+4$
These two equations have slopes whose product is –1, therefore they are perpendicular.

c. $3x+y=5 \Rightarrow y=-3x+5$
$2x+3y=6 \Rightarrow 3y=-2x+6$
$\Rightarrow y=-\dfrac{2}{3}x+2$
These two equations have different slopes (and their product is $\neq -1$), so they are neither parallel nor perpendicular.

34. $y=3x+7$
$x+3y=-15 \Rightarrow 3y=-x-15$
$$y=-\frac{x}{3}-5$$
These two equations have slopes whose product is –1; therefore these lines are perpendicular.

35. a. {(–1, 1), (2, 3), (7, 3), (8, 6)}
This is a function because each x-value is assigned to only one y-value.

b. {(0, –2), (1, 5), (0, 3), (7, 7)}
This is not a function because the x-value 0 is paired with two different y-values.

36. a. $\sqrt{80}+\sqrt{20}=4\sqrt{5}+2\sqrt{5}=6\sqrt{5}$

b. $2\sqrt{98}-2\sqrt{18}=2\cdot 7\sqrt{2}-2\cdot 3\sqrt{2}$
$=14\sqrt{2}-6\sqrt{2}$
$=8\sqrt{2}$

c. $\sqrt{32}+\sqrt{121}-\sqrt{12}=4\sqrt{2}+11-2\sqrt{3}$

37. $\begin{cases} 2x+y=10 \\ x=y+2 \end{cases}$

Substitute $y + 2$ for x in the first equation.

$$\begin{aligned} 2(y+2)+y&=10 \\ 2y+4+y&=10 \\ 3y&=6 \\ y&=2 \end{aligned}$$

Solve for x.

$x = y + 2 = 2 + 2 = 4$

The solution for this system is (4, 2).

38. $\begin{cases} 5x+y=3 \\ y=-5x \end{cases}$

Substitute $-5x$ for y is the first equation. Solve for x.

$$\begin{aligned} 5x-5x&=3 \\ 0&=3 \quad \text{False} \end{aligned}$$

The system has no solution.

39. $\begin{cases} 2x-y=7 \\ 8x-4y=1 \end{cases}$

Solve the first equation for y.

$$\begin{aligned} 2x-y&=7 \\ 2x-7&=y \end{aligned}$$

Substitute $2x - 7$ for y in the second equation and solve for x.

$$\begin{aligned} 8x-4(2x-7)&=1 \\ 8x-8x+28&=1 \\ 28&=1 \quad \text{False} \end{aligned}$$

The system has no solution.

40. $\begin{cases} -2x+y=7 \\ 6x-3y=-21 \Rightarrow -2x+y=7 \end{cases}$

The two equations are the same. Therefore, there is an infinite number of solutions.

41. $\sqrt{36}=6$ because $6^2=36$.

42. $\sqrt{\frac{4}{25}}=\frac{2}{5}$ because $\left(\frac{2}{5}\right)^2=\frac{4}{25}$.

43. $\sqrt{\frac{9}{100}}=\frac{3}{10}$ because $\left(\frac{3}{10}\right)^2=\frac{9}{100}$.

44. $\sqrt{\frac{16}{121}}=\frac{4}{11}$ because $\left(\frac{4}{11}\right)^2=\frac{16}{121}$.

45.
$$\begin{aligned} \frac{2}{1+\sqrt{3}}&=\frac{2}{1+\sqrt{3}}\cdot\frac{1-\sqrt{3}}{1-\sqrt{3}} \\ &=\frac{2-2\sqrt{3}}{1-3} \\ &=\frac{2-2\sqrt{3}}{-2} \\ &=-1+\sqrt{3} \end{aligned}$$

46. $\frac{5}{\sqrt{8}}=\frac{5}{2\sqrt{2}}\cdot\frac{\sqrt{2}}{\sqrt{2}}=\frac{5\sqrt{2}}{4}$

47.
$$\begin{aligned} (x-3)^2&=16 \\ x-3&=\pm\sqrt{16} \\ x-3&=\pm 4 \\ x&=3\pm 4 \end{aligned}$$

$x = 3 + 4 = 7$ or $x = 3 - 4 = -1$

The solutions are $x = 7$ and $x = -1$.

48.
$$\begin{aligned} 3(x-4)^2&=9 \\ (x-4)^2&=3 \\ x-4&=\pm\sqrt{3} \\ x&=4\pm\sqrt{3} \end{aligned}$$

The solutions are $x=4\pm\sqrt{3}$.

49.
$$\begin{aligned} \frac{1}{2}x^2-x&=2 \\ \frac{1}{2}x^2-x-2&=0 \end{aligned}$$

$a=\frac{1}{2},\ b=-1,\ c=-2$

$$\begin{aligned} x&=\frac{-b\pm\sqrt{b^2-4ac}}{2a} \\ x&=\frac{-(-1)\pm\sqrt{(-1)^2-4\left(\frac{1}{2}\right)(-2)}}{2\left(\frac{1}{2}\right)} \\ &=\frac{1\pm\sqrt{1+4}}{1} \\ &=1\pm\sqrt{5} \end{aligned}$$

The solutions are $x=1\pm\sqrt{5}$.

50. $x^2 + 4x = 8$

$$x^2 + 4x - 8 = 0$$

$a = 1$, $b = 4$, $c = -8$

$$x = \frac{-b \pm \sqrt{b^2 - 4ac}}{2a}$$

$$x = \frac{-4 \pm \sqrt{4^2 - 4(1)(-8)}}{2(1)}$$

$$= \frac{-4 \pm \sqrt{16 + 32}}{2}$$

$$= \frac{-4 \pm \sqrt{48}}{2}$$

$$= \frac{-4 \pm 4\sqrt{3}}{2}$$

$$= -2 \pm 2\sqrt{3}$$

The solutions are $x = -2 \pm 2\sqrt{3}$.

Appendices

Appendix A Exercise Set

1. $a^3+27=a^3+3^3$
$=(a+3)[a^2-(a)(3)+3^2]$
$=(a+3)(a^2-3a+9)$

3. $8a^3+1=(2a)^3+1^3$
$=(2a+1)[(2a)^2-(2a)(1)+1^2]$
$=(2a+1)(4a^2-2a+1)$

5. $5k^3+40=5(k^3+8)$
$=5(k^3+2^3)$
$=5(k+2)[k^2-(k)(2)+2^2]$
$=5(k+2)(k^2-2k+4)$

7. $x^3y^3-64=(xy)^3-4^3$
$=(xy-4)[(xy)^2+(xy)(4)+4^2]$
$=(xy-4)(x^2y^2+4xy+16)$

9. $x^3+125=x^3+5^3$
$=(x+5)[x^2-(x)(5)+5^2]$
$=(x+5)(x^2-5x+25)$

11. $24x^4-81xy^3$
$=3x(8x^3-27y^3)$
$=3x[(2x)^3-(3y)^3]$
$=3x(2x-3y)[(2x)^2+(2x)(3y)+(3y)^2]$
$=3x(2x-3y)(4x^2+6xy+9y^2)$

13. $27-t^3=3^3-t^3$
$=(3-t)[3^2+(3)(t)+t^2]$
$=(3-t)(9+3t+t^2)$

15. $8r^3-64=8(r^3-8)$
$=8(r^3-2^3)$
$=8(r-2)[r^2+(r)(2)+2^2]$
$=8(r-2)(r^2+2r+4)$

17. $t^3-343=t^3-7^3$
$=(t-7)[t^2+(t)(7)+7^2]$
$=(t-7)(t^2+7t+49)$

19. $s^3-64t^3=s^3-(4t)^3$
$=(s-4t)[s^2+(s)(4t)+(4t)^2]$
$=(s-4t)(s^2+4st+16t^2)$

Appendix B Exercise Set

1. mean $=\dfrac{21+28+16+42+38}{5}=\dfrac{145}{5}=29$
16, 21, <u>28</u>, 38, 42
median = 28
no mode

3. mean $=\dfrac{7.6+8.2+8.2+9.6+5.7+9.1}{6}$
$=\dfrac{48.4}{6}$
≈ 8.1
5.7, 7.6, <u>8.2, 8.2</u>, 9.1, 9.6
median $=\dfrac{8.2+8.2}{2}=8.2$
mode: 8.2

5. mean
$=\dfrac{0.2+0.3+0.5+0.6+0.6+0.9+0.2+0.7+1.1}{9}$
$=\dfrac{5.1}{9}$
≈ 0.6
0.2, 0.2, 0.3, 0.5, <u>0.6</u>, 0.6, 0.7, 0.9, 1.1
median = 0.6
mode: 0.2 and 0.6

7. mean
$=\dfrac{231+543+601+293+588+109+334+268}{8}$
$=\dfrac{2967}{8}$
≈ 370.9
109, 231, 268, <u>293, 334</u>, 543, 588, 601
median $=\dfrac{293+334}{2}=313.5$
no mode

9. $\frac{1454+1250+1136+1127+1107}{5}=\frac{6074}{5}$
$=1214.8$
The mean height of the five tallest buildings is 1214.8 feet.

11. 1002, 1023, 1046, <u>1107, 1127</u>, 1136, 1250, 1454
$\text{median}=\frac{1107+1127}{2}=1117$
The median height of the eight tallest buildings is 1117 feet.

13. $\frac{7.8+6.9+7.5+4.7+6.9+7.0}{6}=\frac{40.8}{6}=6.8$
The mean time was 6.8 seconds.

15. The mode time is 6.9 seconds.

17. 74, 77, <u>85, 86</u>, 91, 95
$\text{median}=\frac{85+86}{2}=85.5$
The median test score was 85.5.

19. $\frac{78+80+66+\cdots+72}{15}=\frac{1095}{15}=73$
The mean pulse rate is 73.

21. The values 70 and 71 both occur twice while other values only occur once, so the mode of the pulse rates is 70 and 71.

23. There were 9 rates lower than the mean of 73.

25. Since the mode is 21, the value 21 must occur at least twice in the set of numbers. Since there are an odd number of numbers in the set, the median is the middle number. That is, the median of 20 is one of the numbers in the set.
The missing numbers are 21, 21, and 20.

Appendix C Exercise Set

1. The set of negative integers from −10 to −5 in roster form is {−9, −8, −7, −6}.

3. The set of the days of the week starting with the letter *T* in roster form is {Tuesday, Thursday}.

5. The set of whole numbers in roster form is {0, 1, 2, 3, 4, ...}.

7. There are no integers between 1 and 2, so the answer is { } or ∅.

9. Since 3 is a listed element of the set {1, 3, 5, 7, 9}, the statement $3 \in \{1, 3, 5, 7, 9\}$ is true.

11. $\{3\} \subseteq \{1, 3, 5, 7, 9\}$ since every element of the left-hand set is also an element of the right-hand set.

13. $\{a, e, i, o, u\} \subseteq \{a, e, i, o, u\}$ since every element of the left-hand set is also an element of the right-hand set.

15. {May} ⊆ the set of the days of the week is false since the element in the left-hand set is not an element of the right-hand set. Note that the days of the week are Sunday, Monday, Tuesday, Wednesday, Thursday, Friday, and Saturday.

17. Since 9 is not an even number, it is not an element of the set $\{x|x$ is an even number$\}$. Thus the statement $9 \notin \{x|x$ is an even number$\}$ is true.

19. $\{a\} \not\subseteq$ the set of vowels is false since the element in the left-hand set is a vowel.

21. $A \cup B = \{1, 2, 3, 4, 5, 6\} \cup \{2, 4, 6\}$
$= \{1, 2, 3, 4, 5, 6\}$

23. $A \cap B = \{1, 2, 3, 4, 5, 6\} \cap \{2, 4, 6\}$
$= \{2, 4, 6\}$

25. $C \cup D = \{1, 3, 5\} \cup \{7\} = \{1, 3, 5, 7\}$

27. $B \cap D = \{2, 4, 6\} \cap \{7\} = \varnothing$

29. Since every element of $B = \{2, 4, 6\}$ is also an element of $A = \{1, 2, 3, 4, 5, 6\}$, the statement $B \subseteq A$ is true.

31. Since the element of $D = \{7\}$ is not an element of $C = \{1, 3, 5\}$, the statement $D \subseteq C$ is false.

33. Since 2 is an element of $A = \{1, 2, 3, 4, 5, 6\}$, but not of $C = \{1, 3, 5\}$, the statement $A \subseteq C$ is false.

35. Since not every element of $B = \{2, 4, 6\}$ is also an element of $C = \{1, 3, 5\}$, the statement $B \not\subseteq C$ is true.

37. Since the empty set is a subset of every set, the statement $\varnothing \subseteq A$ is true.

39. $A \cup D = \{1, 2, 3, 4, 5, 6\} \cup \{7\}$
$= \{1, 2, 3, 4, 5, 6, 7\}$
The statement is true.

41. Since the union of a set with the empty set is the original set, the statement that $\{a, b, c\} \cup \{\ \}$ is $\{a, b, c\}$ is true.

Appendix D Exercise Set

1. $90° - 19° = 71°$
The complement of a 19° angle is a 71° angle.

3. $90° - 70.8° = 19.2°$
The complement of a 70.8° angle is a 19.2° angle.

5. $90° - 11\frac{1}{4}° = 78\frac{3}{4}°$
The complement of an $11\frac{1}{4}°$ angle is a $78\frac{3}{4}°$ angle.

7. $180° - 150° = 30°$
The supplement of a 150° angle is a 30° angle.

9. $180° - 30.2° = 149.8°$
The supplement of a 30.2° angle is a 149.8° angle.

11. $180° - 79\frac{1}{2}° = 100\frac{1}{2}°$
The supplement of a $79\frac{1}{2}°$ angle is a $100\frac{1}{2}°$ angle.

13. $\angle 1$ and the angle marked 110° are vertical angles, so $m\angle 1 = 110°$.
$\angle 2$ and the angle marked 110° are supplementary angles, so
$m\angle 2 = 180° - 110° = 70°$.
$\angle 3$ and the angle marked 110° are supplementary angles, so
$m\angle 3 = 180° - 110° = 70°$.
$\angle 4$ and $\angle 3$ are alternate interior angles, so
$m\angle 4 = m\angle 3 = 70°$.
$\angle 5$ and the angle marked 110° are alternate interior angles, so $m\angle 5 = 110°$.
$\angle 6$ and $\angle 5$ are supplementary angles, so
$m\angle 6 = 180° - m\angle 5 = 180° - 110° = 70°$.
$\angle 7$ and the angle marked 110° are corresponding angles, so $m\angle 7 = 110°$.

15. $180° - 11° - 79° = 90°$
The third angle measures 90°.

17. $180° - 25° - 65° = 90°$
The third angle measures 90°.

19. $180° - 30° - 60° = 90°$
The third angle measures 90°.

21. Since the triangle is a right triangle, one angle measures 90°.
$180° - 45° - 90° = 45°$
The other two angles of the triangle measure 45° and 90°.

23. Since the triangle is a right triangle, one angle measures 90°.
$180° - 17° - 90° = 73°$
The other two angles of the triangle measure 90° and 73°.

25. Since the triangle is a right triangle, one angle measures 90°.
$180° - 39\frac{3}{4}° - 90° = 50\frac{1}{4}°$
The other two angles of the triangle measure $50\frac{1}{4}°$ and 90°.

27. $\frac{12}{4} = \frac{18}{x}$
$12x = 72$
$x = \frac{72}{12}$
$x = 6$

29. $\frac{6}{9} = \frac{3}{x}$
$6x = 27$
$x = \frac{27}{6}$
$x = 4.5$

31. $a^2 + b^2 = c^2$

$6^2 + 8^2 = c^2$

$36 + 64 = c^2$

$100 = c^2$

Since c represents a length, we assume that c is positive. Since $c^2 = 100$, $c = 10$. The hypotenuse of the right triangle has length 10.

33. $a^2 + b^2 = c^2$

$a^2 + 5^2 = 13^2$

$a^2 + 25 = 169$

$a^2 = 144$

Since a represents a length, we assume that a is positive. Since $a^2 = 144$, $a = 12$. The other leg of the right triangle has length 12.

Practice Final Exam

1. $-3^4 = -(3 \cdot 3 \cdot 3 \cdot 3) = -81$

2. $4^{-3} = \frac{1}{4^3} = \frac{1}{64}$

3.
$$\begin{aligned}6[5+2(3-8)-3] &= 6[5+2(3+(-8))-3]\\ &= 6[5+2(-5)-3]\\ &= 6[5+(-10)+(-3)]\\ &= 6[-5+(-3)]\\ &= 6[-8]\\ &= -48\end{aligned}$$

4.
$$\begin{array}{r} 5x^3 + x^2 + 5x - 2 \\ -\ (8x^3 - 4x^2 + x - 7) \\ \hline \end{array} \qquad \begin{array}{r} 5x^3 + x^2 + 5x - 2 \\ -8x^3 + 4x^2 - x + 7 \\ \hline -3x^3 + 5x^2 + 4x + 5 \end{array}$$

5.
$$\begin{aligned}(4x-2)^2 &= (4x)^2 - 2(4x)(2) + 2^2\\ &= 16x^2 - 16x + 4\end{aligned}$$

6.
$$\begin{aligned}&(3x+7)(x^2+5x+2)\\ &= 3x(x^2+5x+2)+7(x^2+5x+2)\\ &= 3x(x^2)+3x(5x)+3x(2)+7(x^2)+7(5x)+7(2)\\ &= 3x^3+15x^2+6x+7x^2+35x+14\\ &= 3x^3+22x^2+41x+14\end{aligned}$$

7.
$$\begin{aligned}6t^2-t-5 &= 6t^2+5t-6t-5\\ &= t(6t+5)-1(6t+5)\\ &= (6t+5)(t-1)\end{aligned}$$

8.
$$\begin{aligned}3x^3-21x^2+30x &= 3x(x^2-7x+10)\\ &= 3x(x-5)(x-2)\end{aligned}$$

9.
$$\begin{aligned}180-5x^2 &= 5(36-x^2)\\ &= 5(6^2-x^2)\\ &= 5(6-x)(6+x)\end{aligned}$$

10.
$$\begin{aligned}3a^2+3ab-7a-7b &= 3a(a+b)-7(a+b)\\ &= (a+b)(3a-7)\end{aligned}$$

11.
$$\begin{aligned}x-x^5 &= x(1-x^4)\\ &= x[1^2-(x^2)^2]\\ &= x(1+x^2)(1-x^2)\\ &= x(1+x^2)(1+x)(1-x)\end{aligned}$$

12.
$$\begin{aligned}\left(\frac{4x^2y^3}{x^3y^{-4}}\right)^2 &= \frac{4^2x^{2\cdot2}y^{3\cdot2}}{x^{3\cdot2}y^{-4\cdot2}}\\ &= \frac{16x^4y^6}{x^6y^{-8}}\\ &= 16x^{4-6}y^{6-(-8)}\\ &= 16x^{-2}y^{14}\\ &= \frac{16y^{14}}{x^2}\end{aligned}$$

13.
$$\frac{5-\frac{1}{y^2}}{\frac{1}{y}+\frac{2}{y^2}} = \frac{y^2\left(5-\frac{1}{y^2}\right)}{y^2\left(\frac{1}{y}+\frac{2}{y^2}\right)} = \frac{5y^2-1}{y+2}$$

14.
$$\begin{aligned}&\frac{x^2-13x+42}{x^2+10x+21} \div \frac{x^2-4}{x^2+x-6}\\ &= \frac{x^2-13x+42}{x^2+10x+21} \cdot \frac{x^2+x-6}{x^2-4}\\ &= \frac{(x-6)(x-7)}{(x+3)(x+7)} \cdot \frac{(x+3)(x-2)}{(x+2)(x-2)}\\ &= \frac{(x-6)(x-7)}{(x+7)(x+2)}\end{aligned}$$

15.
$$\begin{aligned}&\frac{5a}{a^2-a-6} - \frac{2}{a-3}\\ &= \frac{5a}{(a-3)(a+2)} - \frac{2}{a-3}\\ &= \frac{5a}{(a-3)(a+2)} - \frac{2(a+2)}{(a-3)(a+2)}\\ &= \frac{5a-2(a+2)}{(a-3)(a+2)}\\ &= \frac{5a-2a-4}{(a-3)(a+2)}\\ &= \frac{3a-4}{(a-3)(a+2)}\end{aligned}$$

16.
$$\begin{aligned} 4(n-5) &= -(4-2n) \\ 4n-20 &= -4+2n \\ 4n-20-2n &= -4+2n-2n \\ 2n-20 &= -4 \\ 2n-20+20 &= -4+20 \\ 2n &= 16 \\ \frac{2n}{2} &= \frac{16}{2} \\ n &= 8 \end{aligned}$$

17.
$$\begin{aligned} x(x+6) &= 7 \\ x^2+6x &= 7 \\ x^2+6x-7 &= 0 \\ (x+7)(x-1) &= 0 \end{aligned}$$
$x+7=0$ or $x-1=0$
$x=-7$ $\quad x=1$
The solutions are −7 and 1.

18.
$$\begin{aligned} 3x-5 &\ge 7x+3 \\ 3x-5-3x &\ge 7x+3-3x \\ -5 &\ge 4x+3 \\ -5-3 &\ge 4x+3-3 \\ -8 &\ge 4x \\ \frac{-8}{4} &\ge \frac{4x}{4} \\ -2 &\ge x \end{aligned}$$
$\{x|x \le -2\}$

19. $2x^2-6x+1=0$
$a = 2, b = -6, c = 1$
$$\begin{aligned} x &= \frac{-b \pm \sqrt{b^2-4ac}}{2a} \\ x &= \frac{-(-6) \pm \sqrt{(-6)^2-4(2)(1)}}{2(2)} \\ &= \frac{6 \pm \sqrt{36-8}}{4} \\ &= \frac{6 \pm \sqrt{28}}{4} \\ &= \frac{6 \pm 2\sqrt{7}}{4} \\ &= \frac{3 \pm \sqrt{7}}{2} \end{aligned}$$
The solutions are $x = \frac{3 \pm \sqrt{7}}{2}$.

20. The LCD is $3 \cdot 5 \cdot y = 15y$.
$$\begin{aligned} \frac{4}{y} - \frac{5}{3} &= -\frac{1}{5} \\ 15y\left(\frac{4}{y} - \frac{5}{3}\right) &= 15y\left(-\frac{1}{5}\right) \\ 60-25y &= -3y \\ 60 &= 22y \\ \frac{60}{22} &= y \\ \frac{30}{11} &= y \end{aligned}$$
The solution is $\frac{30}{11}$.

21.
$$\begin{aligned} \frac{5}{y+1} &= \frac{4}{y+2} \\ 5(y+2) &= 4(y+1) \\ 5y+10 &= 4y+4 \\ y &= -6 \end{aligned}$$
The solution is −6.

22. The LCD is $2(a - 3)$.
$$\begin{aligned} \frac{a}{a-3} &= \frac{3}{a-3} - \frac{3}{2} \\ 2(a-3)\left(\frac{a}{a-3}\right) &= 2(a-3)\left(\frac{3}{a-3} - \frac{3}{2}\right) \\ 2a &= 6-3(a-3) \\ 2a &= 6-3a+9 \\ 5a &= 15 \\ a &= 3 \end{aligned}$$
Since $a = 3$ causes the denominator $a - 3$ to be 0, the equation has no solution.

23.
$$\begin{aligned} \sqrt{2x-2} &= x-5 \\ \left(\sqrt{2x-2}\right)^2 &= (x-5)^2 \\ 2x-2 &= x^2-10x+25 \\ 0 &= x^2-12x+27 \\ 0 &= (x-3)(x-9) \end{aligned}$$
$x-3=0$ or $x-9=0$
$x=3$ $\quad x=9$
$x = 3$ does not check, so the solution is $x = 9$.

24. $5x - 7y = 10$
$y = 0$: $5x-7(0)=10$
$$\begin{aligned} 5x-0 &= 10 \\ 5x &= 10 \\ x &= 2 \end{aligned}$$
x-intercept: (2, 0)

$x = 0$: $5(0) - 7y = 10$
$0 - 7y = 10$
$-7y = 10$
$y = -\frac{10}{7}$

y-intercept: $\left(0, -\frac{10}{7}\right)$

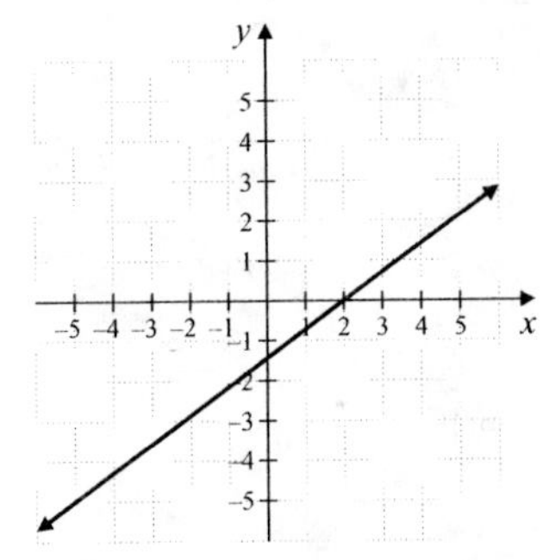

25. The graph of $y = -1$ is a horizontal line with a y-intercept of $(0, -1)$.

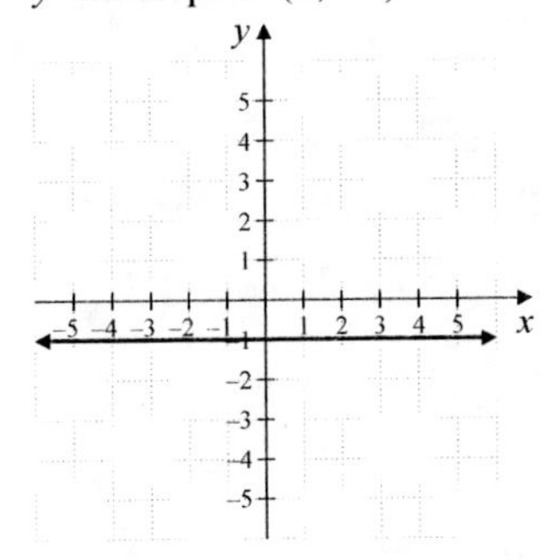

26. Graph the boundary line, $y = -4x$, with a solid line.

Test $(1, 1)$: $y \geq -4x$
$1 \geq -4(1)$
$1 \geq -4$ True

Shade the half-plane containing $(1, 1)$.

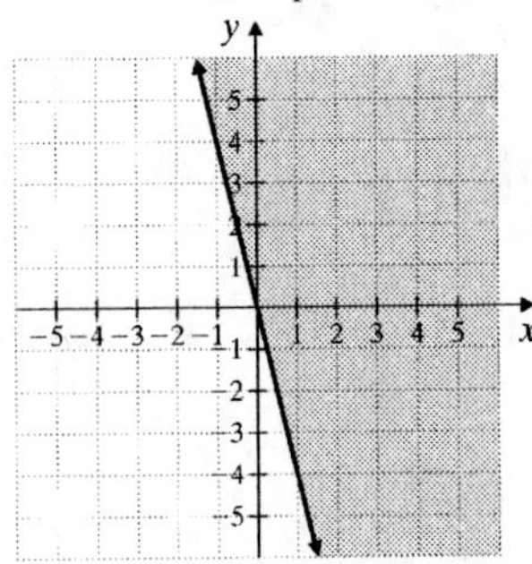

27. $m = \frac{y_2 - y_1}{x_2 - x_1} = \frac{2-(-5)}{-1-6} = \frac{7}{-7} = -1$

28. $y = mx + b$
$-3x + y = 5$
$y = 3x + 5$
$m = 3$

29. $m = \frac{y_2 - y_1}{x_2 - x_1} = \frac{3-(-5)}{1-2} = \frac{3+5}{-1} = -8$

$y - y_1 = m(x - x_1)$
$y - (-5) = -8(x - 2)$
$y + 5 = -8x + 16$
$y = -8x + 11$
$8x + y = 11$

30. $m = \frac{1}{8}$; $b = 12$

$y = \frac{1}{8}x + 12$
$8y = x + 8(12)$
$8y = x + 96$
$x - 8y = -96$

31. $\begin{cases} 3x - 2y = -14 \\ y = x + 5 \end{cases}$

Substitute $x + 5$ for y in the first equation and solve for x.

$3x - 2y = -14$
$3x - 2(x + 5) = -14$
$3x - 2x - 10 = -14$
$x - 10 = -14$
$x = -4$

$y = x + 5 = -4 + 5 = 1$

The solution of the system is $(-4, 1)$.

32. $\begin{cases} 4x - 6y = 7 \\ -2x + 3y = 0 \end{cases}$

Multiply the second equation by 2.

$\begin{cases} 4x - 6y = 7 \\ 2(-2x + 3y) = 2(0) \end{cases} \rightarrow \begin{cases} 4x - 6y = 7 \\ -4x + 6y = 0 \end{cases}$

Add the equations to eliminate x.

$$\begin{array}{r} 4x - 6y = 7 \\ -4x + 6y = 0 \\ \hline 0 = 7 \end{array}$$

Since the statement $0 = 7$ is false, the system has no solution.

33. $f(x) = x^3 - x$

a. $f(-1) = (-1)^3 - (-1) = -1 + 1 = 0$

b. $f(0) = 0^3 - 0 = 0 - 0 = 0$

c. $f(4) = 4^3 - 4 = 64 - 4 = 60$

34. No vertical line will intersect the graph more than once, so the graph is the graph of a function.
From the graph, the domain is all real numbers and the range is $\{2\}$.

35. $\sqrt{16} = 4$, because $4^2 = 16$ and 4 is positive.

36. $\sqrt[3]{125} = 5$, because $5^3 = 125$.

37. $\sqrt{\frac{9}{16}} = \frac{3}{4}$, because $\left(\frac{3}{4}\right)^2 = \frac{9}{16}$.

38. $\sqrt{54} = \sqrt{9 \cdot 6} = \sqrt{9}\sqrt{6} = 3\sqrt{6}$

39. $\sqrt{24x^8} = \sqrt{4x^8 \cdot 6} = \sqrt{4x^8}\sqrt{6} = 2x^4\sqrt{6}$

40.
$$\begin{aligned}
&\sqrt{18} - \sqrt{75} + 7\sqrt{3} - \sqrt{8} \\
&= \sqrt{9 \cdot 2} - \sqrt{25 \cdot 3} + 7\sqrt{3} - \sqrt{4 \cdot 2} \\
&= \sqrt{9}\sqrt{2} - \sqrt{25}\sqrt{3} + 7\sqrt{3} - \sqrt{4}\sqrt{2} \\
&= 3\sqrt{2} - 5\sqrt{3} + 7\sqrt{3} - 2\sqrt{2} \\
&= (3-2)\sqrt{2} + (-5+7)\sqrt{3} \\
&= \sqrt{2} + 2\sqrt{3}
\end{aligned}$$

41.
$$\begin{aligned}
\frac{\sqrt{40x^4}}{\sqrt{2x}} &= \sqrt{\frac{40x^4}{2x}} \\
&= \sqrt{20x^3} \\
&= \sqrt{4x^2 \cdot 5x} \\
&= \sqrt{4x^2}\sqrt{5x} \\
&= 2x\sqrt{5x}
\end{aligned}$$

42.
$$\begin{aligned}
\sqrt{2}\left(\sqrt{6} - \sqrt{5}\right) &= \sqrt{2}\sqrt{6} - \sqrt{2}\sqrt{5} \\
&= \sqrt{12} - \sqrt{10} \\
&= \sqrt{4 \cdot 3} - \sqrt{10} \\
&= 2\sqrt{3} - \sqrt{10}
\end{aligned}$$

43. $\frac{8}{\sqrt{5y}} = \frac{8}{\sqrt{5y}} \cdot \frac{\sqrt{5y}}{\sqrt{5y}} = \frac{8\sqrt{5y}}{5y}$

44.
$$\begin{aligned}
\frac{8}{\sqrt{6}+2} &= \frac{8}{\sqrt{6}+2} \cdot \frac{\sqrt{6}-2}{\sqrt{6}-2} \\
&= \frac{8\left(\sqrt{6}-2\right)}{\left(\sqrt{6}\right)^2 - 2^2} \\
&= \frac{8\left(\sqrt{6}-2\right)}{6-4} \\
&= \frac{8\left(\sqrt{6}-2\right)}{2} \\
&= 4\left(\sqrt{6}-2\right) \\
&= 4\sqrt{6} - 8
\end{aligned}$$

45. Let n be the number.
$$\begin{aligned}
n + 5\left(\frac{1}{n}\right) &= 6 \\
n + \frac{5}{n} &= 6 \\
n\left(n + \frac{5}{n}\right) &= n(6) \\
n^2 + 5 &= 6n \\
n^2 - 6n + 5 &= 0 \\
(n-5)(n-1) &= 0
\end{aligned}$$
$n - 5 = 0$ or $n - 1 = 0$
$n = 5$ $\quad n = 1$
The number is 1 or 5.

46. Let x = area code 1, then $2x$ = area code 2.
$$\begin{aligned}
x + 2x &= 1203 \\
3x &= 1203 \\
\frac{3x}{3} &= \frac{1203}{3} \\
x &= 401
\end{aligned}$$
$2x = 2(401) = 802$
The area codes are 401 and 802.

47. Let x be speed of one hiker and y be the speed of the other hiker.
$$\begin{cases} 4x + 4y = 36 \\ y = 2x \end{cases}$$
Substitute $2x$ for y in the first equation; then solve for x.

$$
\begin{aligned}
4x+4y&=36\\
4x+4(2x)&=36\\
4x+8x&=36\\
12x&=36\\
x&=3
\end{aligned}
$$

Now solve for y.

$y = 2x = 2(3) = 6$

One hiker hikes at 3 mph and the other hikes at 6 mph.

48. Let x = number of cc of 12% solution and y = number of cc of 16% solution.

$$
\begin{cases} x+80=y \\ 0.12x+0.22(80)=0.16y \end{cases}
$$

Substitute $x + 80$ for y in the second equation; then solve for x.

$$
\begin{aligned}
0.12x+17.6&=0.16y\\
0.12x+17.6&=0.16(x+80)\\
0.12x+17.6&=0.16x+12.8\\
4.8&=0.04x\\
120&=x
\end{aligned}
$$

120 cc of the 12% saline is needed to make the 16% solution.